ASTRONOMY

A JOURNEY INTO SCIENCE

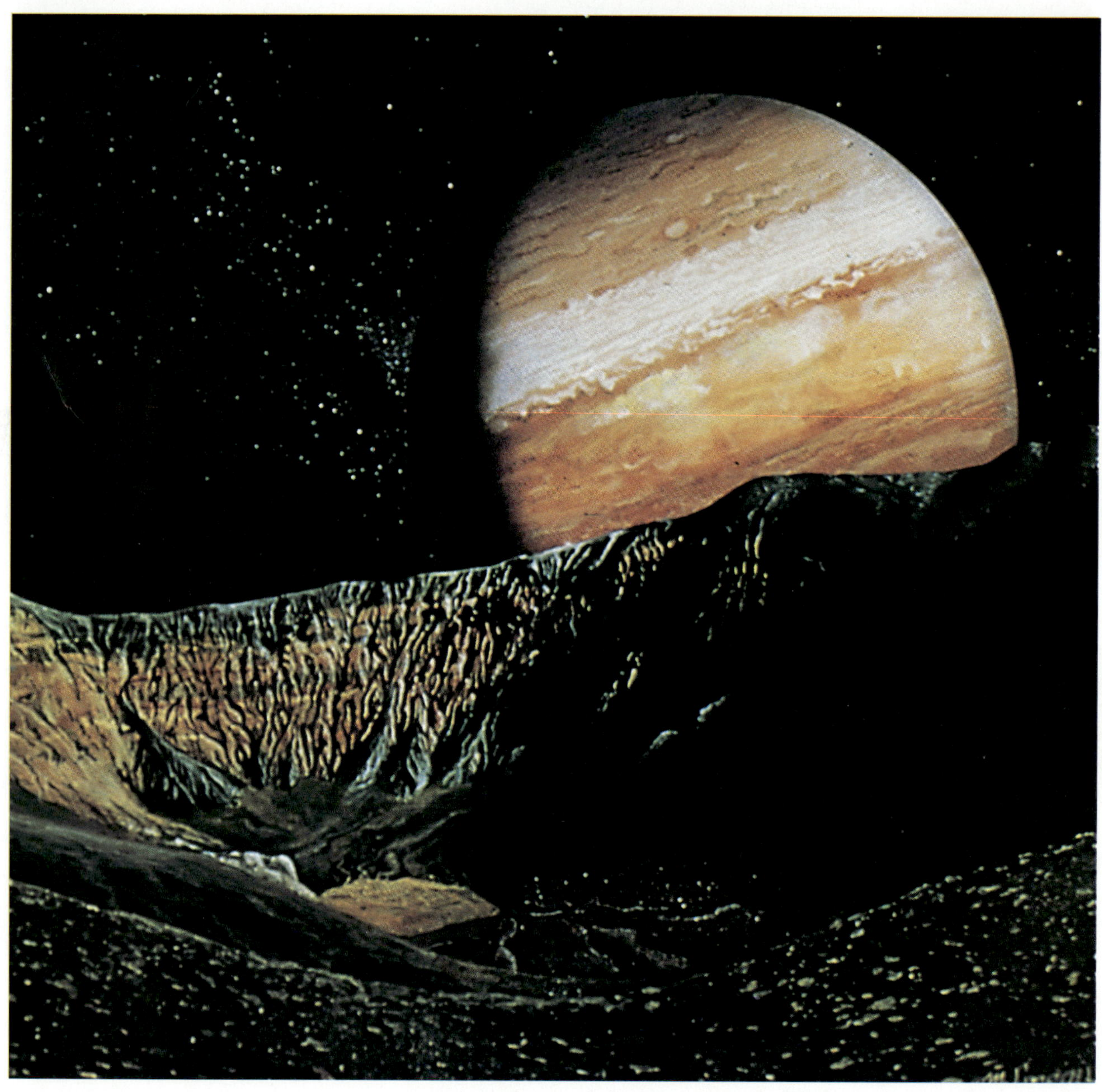

Artist's conception of Jupiter seen from Io, one of its moons.

CLOSE UPS

Extraterrestrial Life Close Ups

ASTRONOMY

A JOURNEY INTO SCIENCE

Karl F. Kuhn
Eastern Kentucky University

West Publishing Company
St. Paul ■ New York ■ San Francisco ■ Los Angeles

Production Credits

Copyediting: Catalyst Communication Arts
Design/Art Direction: The Book Company/Wendy Calmenson
Technical Illustration: John Foster, Victor Royer
Cartoons: Dennis McCarthy
Composition: Parkwood Composition
Cover Design: The Book Company/Wendy Calmenson
Cover Image: "Comet Passing Earth." Painting by Michael W. Carroll.

By WEST PUBLISHING COMPANY
50 W. Kellogg Boulevard
P.O. Box 64526
St. Paul, MN 55164-1003

Printed in the United States of America
96 95 94 93 92 91 90 89 8 7 6 5 4 3 2 1

Library of Congress Cataloging-in-Publication Data

Kuhn, Karl F.
Astronomy: a journey into science.
Includes index.
1. Astronomy. I. Title.
QB43.2.K835 1989 520 88-28013
ISBN 0-314-47009-3

Photo Credits

Front Matter
Page ii Painting by Michael W. Carroll; **Page iv** NASA; **Page xvi** National Optical Astronomy Observatories (NOAO).

Prologue
Chapter Opening © 1980, Royal Observatory, Edinburgh; **Fig. P–1** © Jerry Schad/Science Source, Photo Researchers, Inc.; **Fig. P–2** Jim Baumgardt, Burlingame, CA; **Figure P–3** NOAO; **Fig. P–5a** Tersch Enterprises; **Fig. P–5b** Tersch Enterprises; **Fig. P–7** Dennis Milon/Photo by George East; **Fig. P–8** Dennis Milon and Dewey Burkes; **Fig. P–9** Robin Scagell/Science Photo Library, Photo Researchers, Inc.; **Fig. P–10** *TASS* from Sovfoto; **Fig. P–12** National Aeronautics and Space Administration (NASA); **Fig. P–13** Dan Gordon; **Fig. P–14a** NOAO; **Fig. P–14b** NOAO; **Fig. P–14c** NOAO; **Fig. P–14d** NOAO; **Fig. P–16** Author photo; **Fig. P–17** Author photo; **Fig. P–18** Author photo; **Fig. P–19b** © Paolo Koch/Photo Researchers, Inc.; **Fig. TP–1** Ronald Sheridan/The Ancient Art & Architecture Collection; **Fig. TP-2** Ronald Sheridan/The Ancient Art & Architecture Collection.

Chapter One
Chapter Opening Courtesy of the Adler Planetarium. Hand-colored engraving by Johann Bayer (1572–1625); **Fig. 1–1a** Jim Baumgardt; **Fig. 1–1b** Dennis Milon/Photo by Allan E. Morton;

(continued after index)

Dedication

To three generations of women in my life:
Sharon,
Karyn and Kim,
and Mandy.

CONTENTS

PROLOGUE: INTRODUCTION TO THE UNIVERSE

Section I. THE SYSTEM OF THE HEAVENS

CHAPTER 1: AN EARTH-CENTERED UNIVERSE 22

CHAPTER 2: A SUN-CENTERED SYSTEM 58

CHAPTER 3: THE TRIUMPH OF THE HELIOCENTRIC SYSTEM 88

CHAPTER 4: MOTION, INERTIA, AND FORCE 120

CHAPTER 5: GRAVITATION AND SPACEFLIGHT 146

Section II: PLANETARY ASTRONOMY

CHAPTER 8: THE TERRESTRIAL PLANETS 258

CHAPTER 9: THE JOVIAN PLANETS 304

Section III. STARS—HOW WE KNOW WHAT WE KNOW

CHAPTER 15: THE LIVES OF STARS AND THE SOLAR SYSTEM 546

CHAPTER 16: THE DEATH OF STARS 580

Section IV: GALAXIES AND COSMOLOGY

APPENDIX A: RELATIVITY A-1

PREFACE

THE OVERWHELMING MAJORITY OF students enrolled in introductory astronomy courses are not majoring in astronomy nor in any other field of science. Astronomy is an ideal introduction to science for non-science majors, and *Astronomy: A Journey into Science* was written with these students in mind. For most beginning astronomy students, the class will be one of very few science courses in their college careers. For some, it will be their only physical science course, their only contact with physical science during four years of college.

In an astronomy course, therefore, students should come to understand not only the unique insight on the universe that astronomy provides, but they should learn what science—in general—is, what scientific hypotheses and theories are, how they are tested and developed, and what motivates scientists. Astronomy is an ideal introduction to science because astronomy incorporates *all* of the sciences.

As a result of the goal of introducing science to students, *Astronomy: A Journey into Science* is different from other texts in a number of ways.

DISTINCTIVE FEATURES

More Background Material

Most elementary astronomy texts assume that students understand the science that underlies astronomical phenomena. This, in most cases, is an unwarranted assumption, and *Journey* includes background material omitted by other texts. One of many possible examples is the discussion of heat transfer in everyday circumstances before describing energy transport within the Sun. Such background material is not only necessary to appreciate astronomy, but it is basic scientific knowledge that few students will have encountered in other courses, and it illustrates the close relationship that exists between the various sciences.

Today's student is visual-minded, and much attention has been given to the photo and illustration program. The use of color throughout—not just on separate color plates—not only makes the text more attractive, but full-color illustrations help to clarify difficult concepts. With few exceptions, the art is totally original. Clarity, color, and quality have been stressed.

Complete Color Art Program

The logic of science is emphasized. Wherever possible, we examine the observations that support present theory. For example, there is an explanation of how small asteroids are measured and a careful discussion of the evidence for the nature of pulsars.

Emphasis on Methods by which Science Advances

The heliocentric/geocentric question is just the first of many examples of conflicting scientific theories that are used to show how scientists evaluate hypotheses and develop theories. (Other examples: solar system formation, star formation, the cause for galactic spiral arms, and cosmology). Astronomy is seen as an ongoing, growing activity.

Frequent Examples of Conflicting Theories

Students who will not become astronomers should spend their time studying the logic of science rather than the technical terminology of astronomy. In deciding whether to introduce each new term I have asked two questions: (1) Is the word necessary for a clear discussion of the topic, and (2) Are the students likely to encounter the word in their reading in later life? If the answer to both questions is "No," I have not defined the term in the body of the text. Realizing, however, that some instructors may disagree on the application of my criteria, I have defined some of these terms in the margin so that they may be included or omitted as the particular course requires.

Less Jargon

When new terminology is introduced, it is defined in the margin. This provides students with a quick, easy reference (and instructors with a quick outline of terms used.) In addition, an alphabetic glossary is provided at the back of the book.

Margin Glossary

Some margin notes give additional facts about the material under discussion; some point the student to previous (or future) references in the text; and some provide study hints for the student.

Margin Notes

THREE CATEGORIES OF THEMATIC BOXES

In an ideal world, all science courses probably would have laboratories. Because our world is real, optional activity boxes are included. It is my experience that

Activities

even if they are not assigned, the interested student will take advantage of them and better appreciate the active nature of science.

Close Ups

Many interesting aspects of astronomy are external to the main thrust of the text. Sometimes the material is too mathematical, and sometimes—although it is interesting—it is simply not of major importance. Chapters typically have two or more Close Ups that discuss complementary material. (A few examples: Chaos Theory, The Small Angle Formula, Astrology, Fission and Fusion Power on Earth.)

Historical Notes

Science is a people activity; students should read about the scientists behind the observations and theories. Historical Notes present personal accounts of selected astronomers. A list of Historical Notes—as well as the other two categories of thematic boxes—is on the front end-pages.

More Chapter-Ending Questions

This text contains many more questions at the end of each chapter than is common. There are typically 15 to 20 Recall Questions, 10 Questions to Ponder, and 2 to 5 Calculations. Instructors can announce to the students which ones they consider important in their courses, or they can assign some questions one term and others the next. Answers to all even-numbered Recall Questions and Calculations are found in Appendix M.

Separate Close Ups on Extraterrestrial Life

There are 7 Close Ups concerning extraterrestrial life. (Their titles are listed separately from other Close Ups on the front end-pages.) ET Life Close Ups are dispersed throughout the text, each appearing near material to which it relates. For example, the Close Up regarding the search for extraterrestrial intelligence is located near the discussion of radio telescopes. If an instructor chooses, he/she may cover this material as a unit, treating it as another chapter. Or it may be omitted entirely.

More Stellar Photos and Maps

When specific celestial objects are mentioned, there is often an accompanying photo (and sometimes a chart) to show where the object is located in the sky. This helps to make astronomy real to students.

OTHER FEATURES

Optional Mathematics

More mathematical examples are included in this text than in most others. After each example is a parallel problem for the student, a "Try One Yourself." I believe that the best way to understand a principle is by using it, and mathematical abilities of students—after years of declining—are rising once again. Realizing that some instructors will want to omit them, the examples are set apart from the rest of the text so that some or all may readily be skipped without loss of continuity. So that a student may check his/her understanding immediately upon completing a Try One Yourself, solutions to all of these are included in Appendix M.

SI Units

Standard SI units are used throughout rather than the cgs units normally used by astronomers. The difference is slight, but SI units are what the student will encounter in everyday life. Since most students are more comfortable with British units, these are occasionally used, especially in the beginning of the text, and many quantities are stated in both SI and British units.

Ready Reference of Astronomical Values

The back end-pages contain often-referenced constants and quantities. A more complete list is found in the numerous appendixes.

Pull-Out Star Charts

Perforated star charts are included at the end of the book to aid students in enjoying the night sky.

TEXT ORGANIZATION

The text employs the solar system-to-stars-to-galaxies approach. The advantages for the intended audience are many: Students move from the more familiar to the less familiar; the historical development of science can be emphasized; and the heliocentric/geocentric conflict serves as an excellent introduction to the development of scientific theories.

Optional Appendixes

There are two appendixes that provide an option to cover their topics in greater detail. Appendix A is a conceptual discussion of Einstein's theories of relativity and has the format of a regular chapter. Although relativity is discussed briefly along with black holes and again with cosmology, instructors may wish to use the more complete treatment of this appendix before beginning these topics. Likewise, Appendix L discusses geometrical optics and might be included along with the coverage of telescopes.

A TYPICAL CHAPTER

Each chapter has the following features:

Opening Vignette

On the opening page is a short selection that is related to the subject of the chapter and that may be of interest to students.

Chapter Outline

An outline (similar to the Table of Contents) gives a preview of what is ahead.

Theme Boxes

Three types of theme boxes—Activities, Close Ups, and Historical Notes—are dispersed through the chapter near the relevant text material.

Mathematical Examples and Exercises

These are separated from the normal text so that they may easily be skipped if the instructor wishes. (Solutions to Try One Yourself Exercises are in Appendix M.)

Recall Questions Numerous questions that ask the student to recall important facts, relationships, and theories help greatly in the learning process.

Questions to Ponder These questions fall into three categories: (1) questions that are more complex than the Recall Questions and ask students to relate a concept to others either in that chapter or in earlier chapters, (2) questions that ask students to evaluate their own positions on a subject that has no right/wrong answer, and (3) research questions requiring students to consult other books.

Calculations A few quantitative questions give students a chance to practice calculations similar to those shown in the chapter or to apply quantitative ideas from previous chapters to material at hand.

ANCILLARIES

Instructor's Manual For each chapter, the Instructor's Manual contains (1) teaching tips, (2) a large test bank, (3) a list of prerequisites, and (4) notes about particularly important or difficult concepts.

***Great Ideas* Guide** Adopters of the text may obtain a free copy of *Great Ideas for Teaching Astronomy* (West Publishing Company, 1989).

Computerized Test Bank The test bank contains 25 to 30 multiple choice items for each chapter as well as section exams. The program allows the instructor to add his/her own items and to delete or edit those provided. Questions can be printed in an order chosen by the instructor or in random order.

Study Guide For each chapter, the guide provides key terms, a self test, a mathematical exercise, answers and a list of additional readings.

Transparency Masters Masters of important line art are provided for making transparencies.

Slides Full color slides of some of the more colorful illustrations are provided for classroom use.

ACKNOWLEDGMENTS

First, I wish to thank my students who, with their questions, have helped me discover which concepts are most difficult and require more care in explanation. (In many cases, it was a student who taught me how to explain those difficult concepts.) My colleagues at Eastern Kentucky University have been of great help, particularly Charles Teague and Bruce MacLaren, both of whom provided valuable insights. Without the sabbatical leave granted by the university, the book would have been greatly delayed.

The folks at West Publishing Company give a new meaning to the word *teamwork*. First, Jerry Westby, the indefatigable editor, kept me on my toes with his questions, comments, suggestions, and sense of humor.

Pam McClanahan, production editor extraordinaire, not only attended to the details of production, but she provided numerous, very helpful suggestions. She was there to cheerfully reassure me when snags occurred.

Tom Hilt was a tremendous help in locating photos from various sources. Thanks to Liz Lee for overseeing the development of the ancillaries and to Dr. Fred Thomas of Sinclair Community College for preparing the Study Guide. Three very talented artists—Victor Royer, John Foster, and Dennis McCarthy—worked with Wendy Calmenson to prepare the original artwork which adds so much to the appearance and ease-of-understanding of the book.

Karyn West spent countless hours on her typewriter and telephone tracking sources of photos. Without her work, there might be embarrassing blank spots in the book.

The following reviewers read (and in some cases re-read) the manuscript in its various stages. I thank them for their many suggestions, corrections, and encouragement.

Bill T. Adams
Baylor University

Parviz Ansari
Seton Hall University

Bruce L. Bonneau
Foothill College

Donald J. Bord
University of Michigan, Dearborn

Arvind Borde
Syracuse University

John Broderick
Virginia Polytechnic Institute and State University

George Crawford
Southern Methodist University

John J. Dykla
Loyola University, Chicago

Duane K. Fowler
Northern Michigan University

Vasken Hagopian
Florida State University

Richard Herr
University of Delaware

Richard Hills
Weber State University

Darrel Hoff
University of Northern Iowa

Hollis R. Johnson
Indiana University

Burton F. Jones
University of California, Santa Cruz

Terry Jay Jones
University of Minnesota

John Kenny
Bradley University

John Kielkopf
University of Louisville

Steve Lattanzio
Orange Coast College

Paul Lee
Louisiana State University

Robert J. Manning
Davidson College

Robert C. Mitchell
Central Washington University

Leonard Muldawer
Temple University

Robert Mutel
University of Iowa

John Pappademos
University of Illinois at Chicago

David A. Pierce
El Camino College

James Pierce
Mankato State University

Hans S. Plendl
Florida State University

John Schopp
San Diego State University

Paul P. Sipiera
Wm. Rainey Harper Community College

Michael L. Sitko
University of Cincinnati

Mark H. Slovak
University of Wisconsin, Madison

Dale Smith
Bowling Green State University

Fred Thomas
Sinclair Community College

Charles Tolbert
University of Virginia

David W. Wingert
Georgia State University

Louis Winkler
Pennsylvania State University

Lois J. Zimring
Michigan State University

An author seldom hears from users of his/her text. I invite adopters of *Astronomy: A Journey into Science* to write me with comments and suggestions (and complaints). In particular, if you know of better illustrations or better examples of everyday phenomena that help students understand an astronomical concept, I would appreciate hearing your idea. It is you who know best what should be changed in future editions. My address:

Dept. of Physics & Astronomy
Eastern Kentucky University
Richmond, KY 40475

When I began this project, I didn't realize the demands that it would place on my family. I thank them for their support and I apologize for not being as available to them as I would like.

Karl F. Kuhn

PROLOGUE

The Horsehead Nebula is an area of dust obscuring the bright nebula behind it.

What are stars made of? How did the solar system form? Is there life elsewhere in the universe? What is it like on Venus? What provides the Sun's power? How long has the universe existed? How will it all end?

This is an exciting time in science, especially in astronomy. Since before the dawn of civilization, people have wondered about the heavens, but only recently have we had the ability to search for answers to many of the fundamental questions. Today's tools and techniques have solved problems that were considered impossible just a hundred years ago. During the lifetimes of today's astronomers, entirely new astronomical objects have been discovered, and we have found many more questions that beg for solution. We live in an exciting time.

Introduction to the Universe

ASTRONOMY IS THE OLDEST of the sciences, and because it is still changing rapidly, it is also among the newest. Its long history and its recent advances make it an excellent example of how scientists gather data, how those data are developed into theories, how various theories compete against one another, and how and why some theories are retained and others abandoned. Data and theories are the working materials of all of the sciences, and throughout the study of astronomy one finds many examples of the interplay between data and theory.

Science is a very human endeavor, and the science produced by a particular culture is in part a reflection of the characteristics of that culture. Science cannot be divorced from people. In astronomy it is easy to see the human aspects of science.

Astronomy is a particularly interesting science because it studies BIG questions. Many of these questions are big not only in the sense of the size of their subject—planets, stars, and galaxies—but they are also big in the sense of importance. Their answers tell us something about ourselves, both as individuals and as a species. Throughout this text we will see that our mental outlook changes as we better appreciate where humanity fits into this magnificent universe. We will see that in size our species and our planet are dwarfed to near nothingness in comparison to what is out there. In a sense, astronomy teaches humility.

On the other hand, we will marvel that humans have been able to learn so much about the universe in which we live, and that such tiny beings are able to comprehend the huge, complex universe. From this we may decide that size is not important after all, at least not when compared to the intelligence and spirit of our species. Maybe we need not be quite so humble.

Before embarking on our astronomical journey through the universe, let us take a quick look at what is ahead.

THE VIEW FROM EARTH

In our modern world, few of us are able to sit outside on a clear night, watch the sky, and think. Fortunately, our forefathers had the time and inclination to do this. Figure P–1 shows a portion of the sky as it might be seen on a clear night. To ancient humans such a sight was common, but it could never have been commonplace. Before electric lights and television, watching the sky must have been a very popular activity.

When the night sky is observed over a few hours, we see that the stars appear to move, rising higher in the sky in the east and getting lower in the west. Figure P–2 is a time exposure of the eastern sky as seen from the central U.S. Each of the lines you see in the sky is caused by a star that moved while the photo was being taken.

If we point a camera at the northern sky and take a time exposure photo, we obtain one like that in Figure P–3. Notice that the stars form concentric circles. The bright star near the center of the circles—the short bright line—is the star we call Polaris, or the North Star. It is almost in the center of the circular motion of the stars, and thus it changes its position only very slightly

Figure P–1. *Although it is difficult to show with a camera what the eye sees, this view of the sky is approximately what you might see with the naked eye on a very dark night.*

Figure P–2. *This is a time exposure of the constellation Orion rising in the east.*

HISTORICAL NOTE

Stonehenge

ON A PLAIN ABOUT 70 miles west of London, England, stands Stonehenge, the remains of an ancient monument (see Figure TP–1). Stonehenge is ringed with a ditch about 5 feet deep, and inside this is a circle of white spots on the ground. These are called Aubrey holes, after John Aubrey, who discovered them in 1666. They are filled with chalk, but also contain some rock and cremated human bones. The Aubrey holes once formed a circle of 56 holes, 285 feet in diameter. In the central portion of the figure you can see what is left of a circle of upright stones, some with connecting stones across their tops. This ring is about 14 feet high, and once was a complete circle of 30 stones with a complete cap around its top. Inside this are the largest stones, the remains of a horseshoe-shaped configuration of 5 arches, each about 17 feet tall.

Radioactive dating shows that Stonehenge was built in three stages starting between 2500 and 3000 B.C. and continuing to 2075 B.C., about the same time that the oldest pyramids were being built in Egypt. The largest of the stones weigh as much as 50 tons. When we consider that no such stones exist in the area, and that they must have been transported at least 20 miles to the site (over water and land), we realize that the building of Stonehenge was truly a remarkable achievement.

Of more importance to us, however, is the purpose for which Stonehenge was built. It was long considered to have been primarily of religious significance to the people who built it. Since the late 1700s, however, we have known that if one stands at the center of Stonehenge early in the morning of the first day of summer, the Sun will be seen rising over the Heelstone

Figure TP–1. *Stonehenge. The large circle of stones is about 25 yards across.*

(seen in Figure TP–2). In the 1960s, British astronomer Gerald Hawkins gave the subject of ***archaeoastronomy*** a boost when he used a computer to look for other astronomical alignments. (Archaeoastronomy, a combination of archaeology and astronomy, studies the astronomical knowledge of ancient peoples.) Hawkins found many alignments corresponding to special rising and setting places for the Sun and Moon. For example, two of the rocks in the outer circle are aligned with the most southerly rising position of the Moon, and two others align with sunset on the first day of winter.

Hawkins proposed that Stonehenge was an astronomical monument, used to keep track of the time of year. Such timekeeping was important to agricultural communities, for it was critical to know the correct times for planting and harvesting.

There are some possible objections to Hawkins's hypothesis. One, why would people use such massive stones for this purpose? A few smaller stones could have marked the alignments rather than giant arches and circles. The answer might be that ancient astronomy was closely connected to religion. The Sun and Moon held important religious meanings to many early peoples, and perhaps the builders of Stonehenge used it also as a tribute to their gods.

Figure TP–2. *The Heelstone.*

Secondly, one might ask if the alignments found by Hawkins could not be simply coincidental. This is unlikely, for there are just too many alignments to be accounted for by coincidence.

Hawkins also had an explanation for there being 56 Aubrey holes. The number seemed to have no significance until Hawkins pointed out that there is a pattern of eclipses of the Moon that repeats in a 56-year cycle. He proposed that this cycle was known by the people who built Stonehenge and that holes were used as an eclipse predictor. Hawkins's ideas have not been completely accepted, especially by archaeologists, and his explanation for the Aubrey holes is regarded with particular skepticism.

Stonehenge is unique only in that it is the largest and most sophisticated of Stone Age monuments that seem to have astronomical significance. Hundreds of such monuments are found over the British Isles, and there are many more on the European continent. In North America, there are about three dozen Indian *medicine wheels*. These are circular patterns of stones (and sometimes wood), many of which have been found to show alignments with particular positions of solar and lunar rising and setting and with similar positions for the brightest stars. In Central and South America, alignments of buildings of the Mayan and Aztec cultures indicate that these people also placed major importance on rising and setting positions of heavenly objects.

Figure P–3. *Notice that the stars in this time exposure of the northern sky seem to move around one point. Polaris, the North Star, caused the bright streak near the center.*

during the night. This made it very special to travelers before the invention of modern direction-finding techniques. Other stars appear to move around the sky, so they cannot easily be used to determine direction, but that single star, Polaris, stays at (almost) the same place: directly north.

When the moon is in the sky, it rises and sets following a path similar to those of the stars seen near it. Likewise, during the day, the Sun takes a parallel journey across the sky.

Today we know the reason for the heavenly objects' daily westward motion across the sky: the Earth rotates toward the east, carrying us with it (Figure P–4). Since the Earth is so large, its motion is not at all obvious and humans did not learn of this motion until about 400 years ago. The struggle involved in learning this, and the conflict between the two leading theories explaining the heavenly motions, forms the subject matter of the first section of this text.

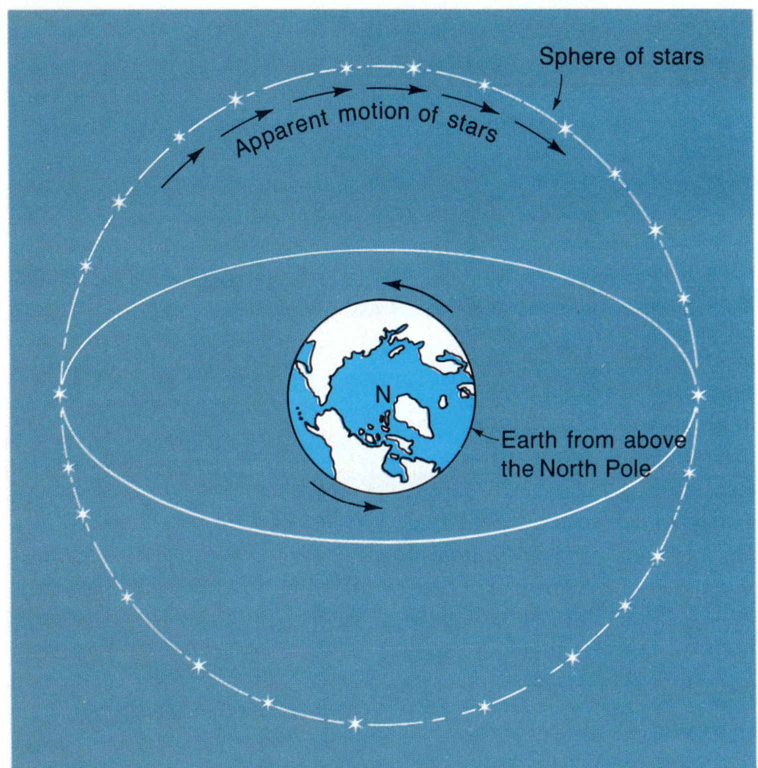

Figure P–4. *The Earth's rotation toward the east makes it appear to us riding on the Earth that the sky is rotating to the west. Hence stars appear to rise in the east and move westward during the night.*

THE SOLAR SYSTEM

Solar system: The Sun and all objects revolving around it, as well as gas and dust between those objects.

When we watch the sky night after night, we find that it is not quite as permanent as it first appears. Although the stars remain in the same patterns year after year, other objects—although they look like stars to the naked eye—move slowly among the stars so that they change their positions noticeably over a few weeks (Figure P–5). These objects are the planets, and their observed motion among the stars is caused by the fact that both the Earth and the planets revolve around the Sun. As we ride on the moving Earth and observe the moving planets, we see them as tiny spots of light gradually moving against the background of the distant stars (Figure P–6).

The wandering of the planets among the stars presented a problem that was explained differently by the two theories mentioned above. We will see in Chapters 1 through 4 how each theory explained planetary motions, and we will see that Isaac Newton, in the 1600s, developed theories of gravitation and motion that form the basis for today's common model of the solar system.

The exploration of the Earth's neighborhood by space probes has opened new avenues for astronomers. In Chapter 5 we will apply Newton's laws to these probes and explain the nature of weightlessness.

In Section II, we study the individual objects that make up the solar system, starting with the Moon (Figure P–7). We then move on to consider the remainder of the solar system: the other planets, the asteroids, and the comets.

Figure P–5. *These photos of the same area of the sky were taken 2 weeks apart. Notice that during the 2 weeks, the planet Mars has moved relative to the background stars.*

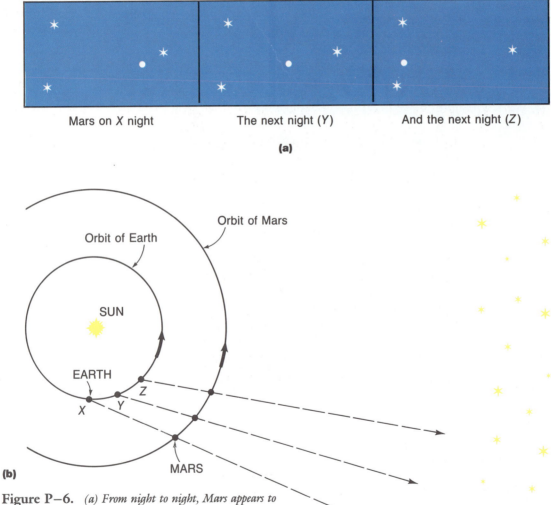

Mars on *X* night

The next night (*Y*)

And the next night (*Z*)

(a)

Orbit of Mars

Orbit of Earth

SUN

EARTH

Z

Y

X

MARS

(b)

Figure P–6. *(a) From night to night, Mars appears to shift its position relative to the stars. (b) The explanation lies in the fact that both Earth and Mars are moving around the Sun.*

Figure P–7. *The Moon, referred to in numerous romantic songs, is the second most prominent object in our sky.*

Figure P–8. *A small telescope reveals Jupiter as a disk with some faint surface features.*

THE STARS

Although the stars and the planets look nearly the same to the naked eye, the light from a star is produced by an energy source within it, while a planet's brightness is the result of reflected sunlight. The Sun, easily the brightest object in our sky, is simply a star, not basically different from the thousand or so stars you see when you look up at night. If we try to imagine how far away we would have to move the Sun to make it appear as dim as a typical star, we begin to appreciate how far away the stars are.

Using even a small telescope, the planets can be seen as disks (Figure P–8), but through even the largest telescope, the stars still appear as tiny specks of light. Figure P–9 is a photo of a small area of the sky. Although some stars appear larger than others in the photo, this is caused by their greater brightness and not by greater size.

In studying Section III, you may be surprised at how much we have been able to learn about objects as distant as the stars. We are able to measure a star's composition, its size, and its mass. In addition, we are fairly confident of our theory about stars' life cycles. We know how stars form, we know the sources of their energy during their lifetimes of billions of years, and we know how they end their lives: some in tremendous explosions called ***supernovae***. Many questions remain, of course; including details of how the solar system formed

Supernova: (plural, supernovae): The explosive destruction of a star.

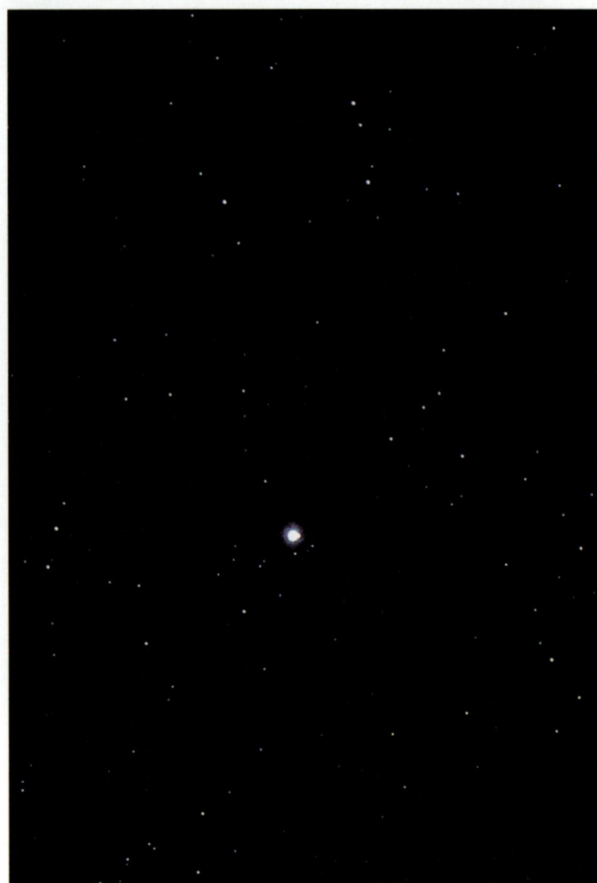

Figure P–9. *This photo of the stars indicates that some stars are brighter than others, but the stars' size is not apparent through a telescope.*

Figure P–10. *This telescope in the Soviet Union is the largest in the world. Note the size of the person.*

along with the Sun, the whether such planet systems are common around other stars in the universe.

To understand how we have been able to learn about the stars we must examine the instruments that astronomers use. These include not only the large optical telescopes (Figure P–10), but even larger telescopes that are designed to receive radio waves from space (Figure P–11) and various types of satellite telescopes designed to detect radiation from the heavens that does not reach Earth's surface (Figure P–12). We study the nature of radiation from the stars in Chapter 11 and telescopes in Chapter 13.

GALAXIES

The Milky Way Galaxy (or "the Galaxy"): The group of a few hundred billion stars of which our Sun is one.

Our Sun is one of perhaps 200 billion stars that make up the ***Milky Way Galaxy***. These stars are arranged in our galaxy in a spiral pattern such as that shown in Figure P–13. Astronomers learned of the nature of the galaxy in the

Figure P–11. *The radio telescope at left, which had a dish 300 feet in diameter, collapsed in November 1988. The telescopes are located at Green Bank, West Virginia.*

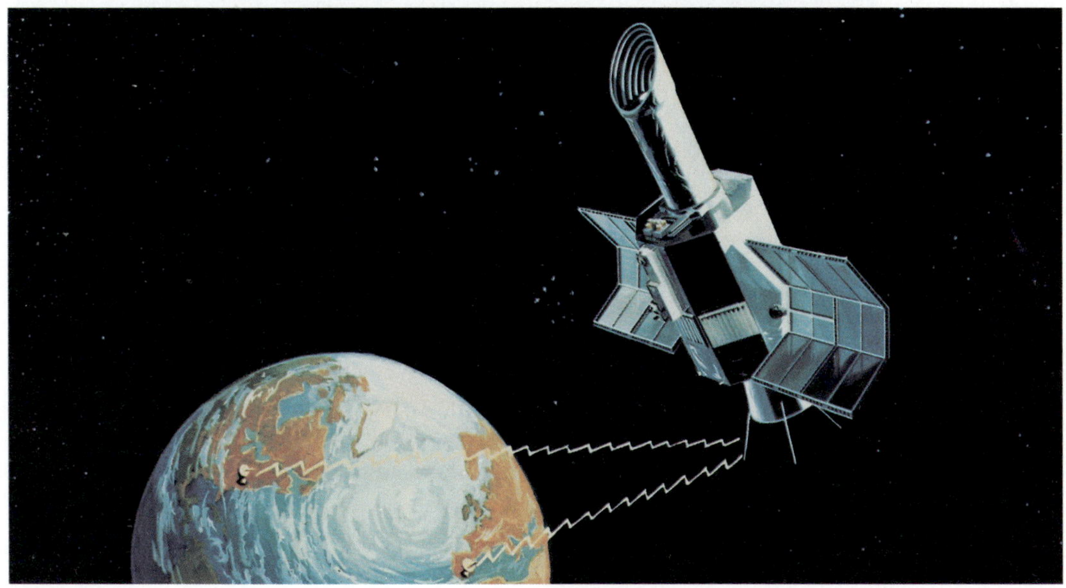

Figure P–12. *Ultraviolet light from celestial objects is detected by the International Ultraviolet Explorer, launched in 1978. The large flat panels on the satellite are solar cells that provide power.*

Figure P–13. *This is a spiral galaxy (named NGC 5194) similar to our galaxy. The nature of the Milky Way Galaxy was not understood until this century.*

first quarter of this century, and they then quickly discovered that our galaxy is not alone; that there are many such "island universes" far beyond our own. Figure P–14 shows photos of a number of these galaxies, illustrating the great differences in structure that exist among them.

In Section IV, we study the galaxies. We will see that although our knowledge about these distant objects is less extensive than what we know about our neighboring planets or even the stars, today's astronomical instruments are bringing us more and more information about them and theories of their nature and development are beginning to fall into place.

Finally, at the end of Chapter 18, we consider the biggest subject of all in astronomy: the nature of the universe as a whole, where it came from, and where it is going. Such a question may seem unanswerable by scientific means, but the tools and understanding provided by Einstein's theories of relativity are allowing us to study such questions.

Figure P–14. *Each of these photos is of a galaxy consisting of thousands to hundreds of billions of stars. There are patterns of similarities in their structure, but also great differences.*

THE SCALE OF THE UNIVERSE

In our study of astronomy, we will move progressively outward from the Earth. This follows the historical development of the subject, proceeding from the solar system to the stars, to the galaxies, and finally to the universe considered as a whole. It is difficult for the inhabitants of tiny Earth to appreciate the change of scale that is necessary when we move from one of these subjects to the next (see Figure P–15). To try to get an idea of the sizes involved in astronomy, let us make an imaginary scale model of the universe.

Suppose we start our model with a standard Earth globe, as found in many classrooms (Figure P–16). Most of these are about 1 foot in diameter, so we will let the Earth's 8000-mile diameter be represented by 1 foot. On this scale, the Moon is a sphere 3 inches in diameter (about like a baseball), located about 30 feet away (Figure P–17). This is easy to imagine. In trying to picture the Sun on this scale, we run into trouble, however, for the Sun's diameter is more than 100 times the Earth's, and the Sun would be a 100-foot sphere located 2 miles away. To picture the Sun on this scale, think of a 10-story building 2 miles from the globe. This imaginary model is already very large, and we have not yet even begun to consider the distant planets, much less other stars. This scale simply will not work.

Let's change the scale, letting the model for the Sun be a large grapefruit. Now the Earth is a BB, or the head of a shirtpin, about 35 feet away (Figure P–18). The Moon is a speck about a foot from Earth and is invisible in the figure. Jupiter, the largest planet, is a chewing gum ball or child's marble about 150 feet away—half the length of a football field. Pluto, the farthest planet, is a small pinhead about a fourth of a mile away. Thus the entire solar system requires an area about a half mile in diameter.

Where is the nearest star? About 1500 miles away! If our grapefruit Sun is located in Chicago, the nearest star might be in Arizona's Grand Canyon (Figure P–19). This represents the average distance between stars in our galaxy, so we must imagine grapefruits spread randomly about, with approximately 1500 miles between adjacent ones. The diameter of the Galaxy on this scale

CLOSE UP

Powers of Ten

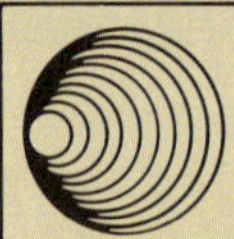

NUMBERS IN SCIENCE—AND particularly in astronomy—are sometimes extremely small or extremely large. For example, the diameter of a typical atom is about 0.0000000002 meter, and the diameter of the Galaxy is about 1,000,000,000,000,000,000,000 meters. Both are rather awkward numbers to use. One way to avoid such clumsy numbers is to use different units of measure. Thus, in astronomy, we define the *light year* as the distance light travels in a year, about 10,000,000,000,000,000 meters. So the Galaxy has a diameter of about 100,000 light years. Sometimes, however, it is inconvenient to switch units to avoid dealing with large and small numbers, and so scientists use *powers of ten* notation, also called *scientific* notation, or *exponential* notation. (The exponent is simply the power to which a number is raised.)

Scientific notation is simple because when the number 10 is raised to a positive power, the exponent is simply the amount of zeroes in the number. For example:

$$10^1 = 10$$
$$10^2 = 100$$
$$10^3 = 1000$$

Written in meters, the diameter of the Galaxy contains 21 zeroes, so it is written as 10^{21} meters.

If the number to be expressed in this notation is not a simple power of 10, it is written as shown below:

$$2{,}100 = 2.1 \times 1000 = 2.1 \times 10^3$$
$$305{,}000 = 3.05 \times 100{,}000 = 3.05 \times 10^5$$

Notice that to change a number from scientific notation to regular notation, one simply moves the decimal place the number of digits indicated by the exponent, filling in zeroes if necessary.

Rather than explain negative exponents, we will just give some examples and let you see the pattern:

$$10^0 = 1$$
$$10^{-1} = 0.1$$
$$10^{-2} = 0.01$$
$$10^{-3} = 0.001$$
$$2 \times 10^{-8} = 0.00000002$$
$$4.67 \times 10^{-5} = 0.0000467$$

In this case, the same rule is followed, moving the decimal point the number of places indicated by the exponent, but this time moving it to the left and supplying any needed zeroes.

In the Close Up "Significant Figures" in Chapter 3, we will see another advantage of using powers of ten notation.

would be about 30,000 miles. Again, we find that our scale is too large. The 30,000-mile galaxy of grapefruits is beyond our imagining. By developing such scale models, we begin to appreciate the distances involved in the real universe.

We will return a number of times to constructing imaginary scale models of parts of the universe. If you try each time to imagine the distances involved, the repetition will be far from boring, but will instead be mind-expanding.

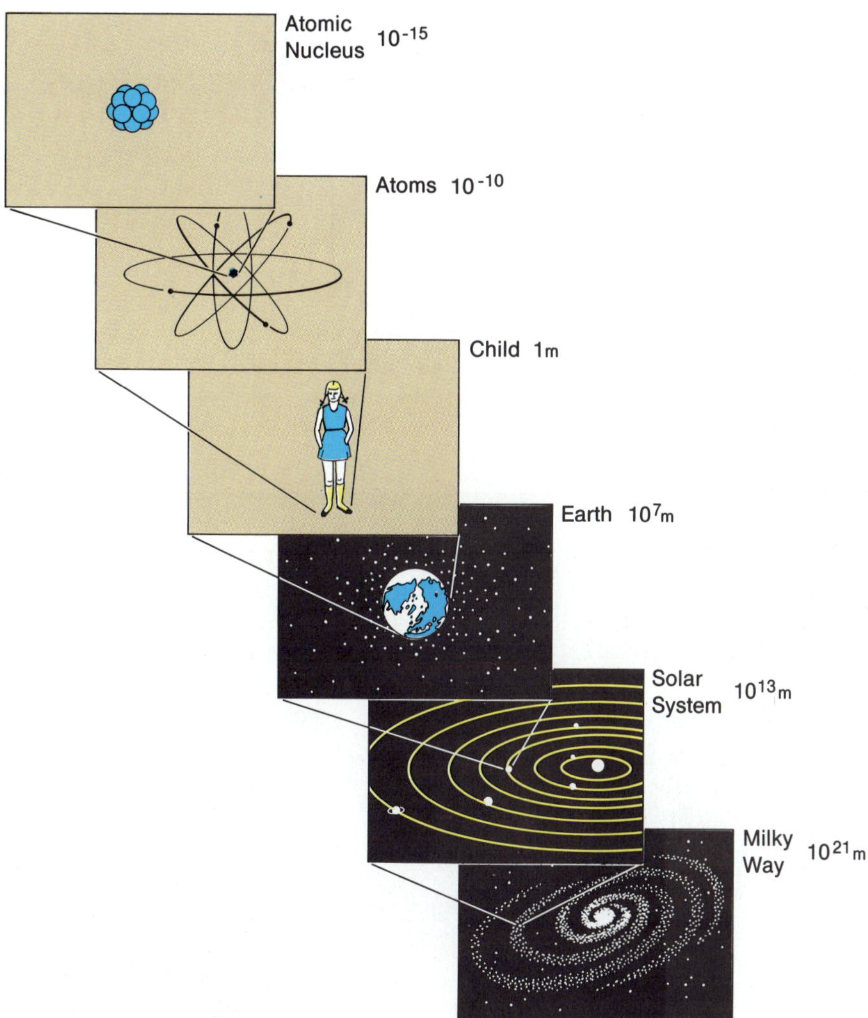

Figure P–15. *The use of powers-of-ten notation makes it much easier to describe the sizes of the objects illustrated, but be careful that the notation does not hide the tremendous differences that exist.*

Figure P–16. *We will let this globe represent the real Earth, and calculate the sizes and distances of other astronomical objects.*

Figure P–17. *The baseball is the correct size for the Moon if the globe represents the Earth. The baseball must be placed 30 feet from the globe for the scale to be correct.*

Figure P–18. *If the Sun is a grapefruit, the Earth is the head of a shirtpin 35 feet away.*

Figure P–19. *If the grapefruit Sun is located in Chicago, the nearest star is at the distance of Arizona's Grand Canyon. Can you find the grapefruit in the photo of the Grand Canyon?*

ASTRONOMY, SCIENCE, AND THE UNITY OF THE UNIVERSE

Astronomy is a natural science. This means that it is a study based on observations and experiments of the natural world, just as are the other natural sciences: chemistry, geology, physics, biology, meteorology, etc. Although the various natural sciences differ somewhat in method, the major difference between them is their subject matter. The subject of astronomy is primarily the universe beyond the Earth. The Earth, however, is an astronomical object—a planet—and therefore is included in the study also.

Astronomy shows us that the universe in which we live is a unit, not just a collection of separate objects. Each of the natural sciences studies a different aspect of the universe, but because the unity exists, there are very many areas of overlap between the various sciences. Astronomy is particularly close to physics, with the two fields becoming indistinguishable at times. Likewise, astronomy and geology have many areas in common, as we attempt to understand the formation, evolution, and surface features of the planets. Chemistry and biology become areas of the astronomer's concern as he or she studies the compositions of astronomical objects and the possibility of extraterrestrial life. And meteorology aids the astronomer in studying weather patterns on other planets. Conversely, what we learn about the weather on those planets is helpful in understanding weather on Earth.

Sometimes the word "astrophysics" is used synonymously with "astronomy."

Perhaps the most striking example of the unity of the universe will be provided in Chapter 16 of this book, where we will see that the very atoms that make up our human bodies were once inside stars, and that if stars did not undergo supernova explosions, many of the chemical elements that make up our bodies would not even exist.

CONCLUSION

The subject matter of astronomy ranges in size from the unimaginably small to the unimaginably large. Of primary interest to the astronomer are objects ranging from somewhat less than human size (asteroids and meteroids) to the very largest objects (galaxies), and even the universe as a whole. In order to understand these objects, however, astronomers must know about the smallest of objects—the atoms and the subatomic particles of which they are made.

Astronomy does not stand alone however. Because nature itself is not subdivided into categories, astronomy cannot be divorced from other fields of science, and astronomers are involved with all of them, from anthropology to zoology.

RECALL QUESTIONS

1. What is unique about the star Polaris?
2. Describe a time exposure photo of the northern sky.
3. What causes the nightly apparent motion of the stars across the sky?
4. Describe the motion of the Moon in the sky during the night.
5. Besides a difference in "twinkling" (to be explained later in the book), how can a planet be distinguished from a star using only the naked eye?
6. Who developed the theory of gravitation, and in what century did this occur?
7. How do the planets and stars differ as to the cause of their brightness in the night sky?
8. Name some attributes of stars which astronomers are able to measure.
9. How do stars and planets differ when they are seen in a telescope?
10. Name two instruments, besides visible-light telescopes, that are used to receive information from the stars.
11. What is a galaxy?
12. About how many stars make up the Milky Way Galaxy?
13. Describe a scale model that includes the Earth, Moon, Sun, and nearest star.

QUESTIONS TO PONDER

1. Discuss the differences between the various natural sciences, using insights you may have gained in studying another science.
2. There is no bright pole star in the southern hemisphere. Describe the appearance of a time exposure photo of the sky above the south pole.
3. How would you estimate the duration of a time exposure of the stars in the northern sky?
4. If you live in the U.S. or Canada, consult "Staying Up Nights" in the June 1987 edition of *Astronomy* magazine and report on astronomy clubs in your state, province, or part of the country.

CALCULATIONS

1. Use data on planetary, stellar, and galactic distances from the appendices to calculate the dimensions of a model of the universe where the Sun is pinhead-size (about 3 millimeters, or 1/8 inch). Include the size of the Earth and the Galaxy, and the distances between Earth and Sun, between the Sun and a nearby star, and between our galaxy and the Andromeda Galaxy.
2. Write the following numbers in scientific notation: 256,000, 345.6, 0.0056, 0.987.
3. Write the following numbers in standard decimal format: 6.42×10^5, 6.88×10^{-4}, 5.55×10^2, 4.76×10^{-1}.
4. The mass of the Earth is 5.977×10^{24} kilograms. Write this number in everyday decimal form.

CHAPTER 1

This hand-colored engraving of *Ursa Major* appeared in *Uranometria* by Johann Bayer in 1661.

The name Ursa Major *is the Latin translation of the Greek name of the constellation of the Great Bear. This group of stars has been associated with a bear since before the times of Homer and it is referred to in the* Odyssey. *The Greek story of the origin of the constellation held that the bear had once been a nymph who attracted the attention of Zeus, the father of the other gods. This caused Hera (the wife and sister of Zeus) to be jealous and to change the nymph into a bear. Finally, to protect the bear from hunters, Zeus grabbed it by the tail and flung it into the sky. This is why Ursa Major has a longer tail than Earthly bears do.*

A number of American Indian tribes also saw the stars as a giant bear. One legend held that hunters had chased the big bear onto a mountain from which it jumped into the sky. The bowl of our dipper is now the bear and the handle is made up of the hunters who followed.

The Iroquois, like the Greeks, saw the dipper's handle as the tail of the bear, explaining that Earthly bears once had long tails like the great one in the sky, but that their tails had frozen off when they dipped them through holes in the ice to catch fish.

Other Indian tribes had far different myths concerning the stars of Ursa Major. Some saw the four stars of the bowl of the Big Dipper as a stretcher holding a wounded warrior. The four stars we identify as the handle of the dipper were the medicine man and his wife, the wounded man's wife, and his dog.

An Earth-Centered Universe

IN THE LAST DECADE of the twentieth century, we have good explanations for what we see in the sky and we understand the Earth's relationship to other astronomical objects. The search for this understanding began long ago, long before telescopes were invented. We will begin by examining what can be learned about the heavens without the use of modern instruments. Then we will consider a theory of the universe that is nearly 2000 years old.

Why do we begin an astronomy text by looking back into history instead of plunging into today's astronomy? Furthermore, why start with an outmoded theory? There is good reason. In order to understand how astronomy—and science in general—works, we must look at its progression through the ages. We must see what led up to today's ideas.

Our starting place in the study of astronomy will be the same as that of people of long ago. We will consider the heavens as seen by the naked eye and will examine the observed motions of stars and planets. In the first five chapters we will study two major theories that explain these motions, examining them in two different ways: First, we will study them to answer the question of where the Earth fits into the scheme of things; second, we will examine the criteria for a good scientific theory and see how well these theories match the criteria. Finally, we will see how and why one of the major theories won out over the other, and how the understanding of nature provided by the successful theory eventually allowed humans to travel beyond the Earth and to leave footprints on the Moon. The journey to the Moon actually began centuries ago.

(a) (b)

Figure 1–1. *A time exposure shows the motion of stars as they rise in the east (a) and set in the west (b). You can make photos like these with any camera capable of taking time exposures.*

Figure 1–2. *Notice that the stars in this time exposure of the southern sky seem to move around one point: the south celestial pole.*

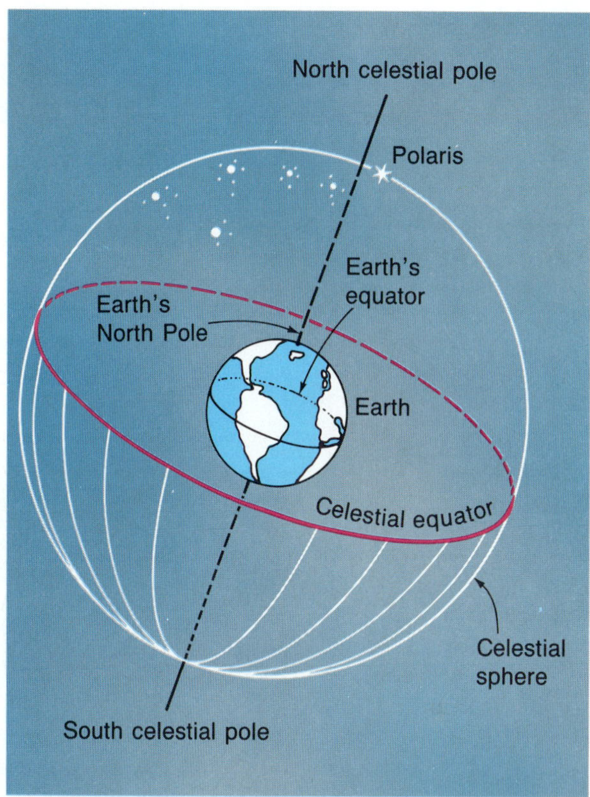

Figure 1–3. *Because of the daily motion of objects in the sky, it appears that those objects are on a sphere surrounding the earth. (The celestial sphere is really much larger compared to Earth than it appears in this diagram, of course.)*

THE CELESTIAL SPHERE

As we watch the sky during the night, we see stars rising in the east and setting in the west (Figure 1–1). Stars above the poles of the Earth move in concentric circles, centered on a spot in the sky above each pole (Figure 1–2).

Another observation that can be made by even a casual observer is that the stars stay in the same patterns night after night and year after year. The Big Dipper seems to retain its shape through the ages as it moves around and around the North Star. It is easy to see why the ancients concluded that the stars act as if they are on a huge sphere that surrounds the Earth and rotates around it. Figure 1–3 illustrates this ***celestial sphere.*** Its axis of rotation passes through the sphere at the center of the circles of Figure 1–2. That photo was taken from the Southern Hemisphere, and the point at the center of the circles, the ***south celestial pole,*** is exactly above the south pole of the Earth. Above the Earth's north pole we find a corresponding point on the celestial sphere called the ***north celestial pole.*** The motions of the stars make it appear that the Earth is sitting

Celestial sphere: The sphere of heavenly objects that seems to center on the observer.

Celestial pole: The point on the celestial sphere directly above a pole of the Earth.

CLOSE UP

Light Pollution

Figure T1–1. *A composite photo of the United States at night indicates how much light we let escape into space. If you live in the United States, see how closely you can identify your area of the country.*

MOST OF US TODAY are much more ignorant of the night sky than were the people of ancient times. In many cases it is simply too difficult to see the sky. From inside our cities we can see only the Moon and the brightest of stars and planets, if any at all. Even a few hundred years ago these problems did not exist. Cities had neither tall buildings that block so much of our sky, nor did they have the bright lights that obscure the stars. These lights are a major factor in limiting modern people's enjoyment of the night skies.

Figure T1–1 is a composite photo of the United States at night taken from a satellite. It illustrates just how much light escapes from Earth into space. It is not this escaped light that prohibits us from seeing the stars, however; but it is the portion of this light that is scattered by the air and smog back toward our eyes. This effect is termed *light pollution* by astronomers. It lights up our nighttime sky and prohibits us from seeing the stars from urban areas in much the same way that the Sun does during the day. Light pollution changes the naturally black night into milky gray, against which the fainter stars have insufficient contrast to be seen. The light pollution above a city is especially noticeable when you are driving at night five to ten miles from the city. From a distance you can see the horizon lit up. Light pollution not only represents a waste of energy caused by a lack of reflectors on street lights (to reflect the light downward where it is useful), but it keeps us from appreciating one of the most beautiful sights of nature.

Some communities now have ordinances that prohibit commercial lighting systems from shining light into the sky. Figure T1–2 shows a street light with a shield that saves energy and reduces light pollution.

When viewed far from the smog and light pollution of our urban areas, the sky is truly an awesome sight. You are encouraged to take any opportunity you have to get in the country, as far from city lights as you can, lie on your back, and absorb the heavens.

Figure T1–2. *Street lights with shields like this please both the astronomer and the person interested in saving energy.*

still in the center of the celestial sphere as it rotates around us. To picture the motion of the celestial sphere, you might think of it as spinning on rods that extend straight out from the Earth's north and south poles. These rods would connect to the celestial sphere at the north and south celestial poles.

Constellations

It is natural, when we look at the sky, to look for order—for a pattern. This desire for order is felt by all people, and it is basic to science. Indeed, we can see (or imagine) patterns in the stars. The ancients saw similar patterns and identified them by associating them with beings in their particular mythology. Figure 1–4 is a drawing of the hunter Orion in Greek mythology. He is fighting off Taurus, the bull at upper right. If you look at the evening sky of December, January, February, or March, you will see the stars of Orion. It might be easy for you to imagine the three closely spaced stars to be the belt of the hunter. And if you have a good dark sky, you can see the stars that form his sword's sheath hanging from the belt, as well as the stars forming his upraised arm.

Constellation: An area of the sky containing a pattern of stars named for a particular object, animal, or person.

The mythological creatures of most other ***constellations*** are much more difficult to imagine. Ursa Major, the Big Bear, is a constellation of the northern sky. It probably wouldn't remind you of a bear, but you probably recognize the pattern of stars in the bear's rear and tail (Figure 1–5). This is the pattern that today we call the Big Dipper.

The word constellation *comes from Latin, meaning "stars together."*

About half of today's officially designated 88 constellations are those identified by the ancient Greeks, and the names we know them by are Latin translations of the original names. Today we realize that the constellations have no real identity. They are simply accidental patterns of stars, much the same as patterns you might have seen in the clouds when you were a young dreamer watching them passing overhead. The stars don't change their patterns as quickly as the clouds, however; so perhaps it was natural for ancient peoples to attribute real meaning to them. Besides, the stars were in the "heavens," and their association with the gods seemed natural.

Figure 1–4. *Orion the Hunter was pictured as warding off Taurus the bull.*

Figure 1–5. *The Big Dipper is part of the constellation of Ursa Major (the Big Bear).*

Why do we say that the patterns are accidental? To begin with, the various stars are located at different distances from Earth. This means that if the Earth were in a greatly different position, we would see different patterns. For example, the stars of the Big Dipper are not all close to one another; one of them, named *Alkaid,* is much, much farther away than the others (Figure 1–6).

In spite of their artificial nature, constellations are used by astronomers today to identify parts of the sky. We might say, for example, that Halley's Comet was in the constellation Aquarius on Christmas Day, 1985. The ancient Greeks had no constellations in areas of the sky which did not have bright stars, nor in areas they could not see from their location in the northern hemisphere, so others have had to be added to their list. Today's 88 constellations fill the entire celestial sphere. In addition, the constellations of the Greeks had poorly defined boundaries, so astronomers have had to establish definite ones. Figure 1–7 shows the constellation Cygnus.

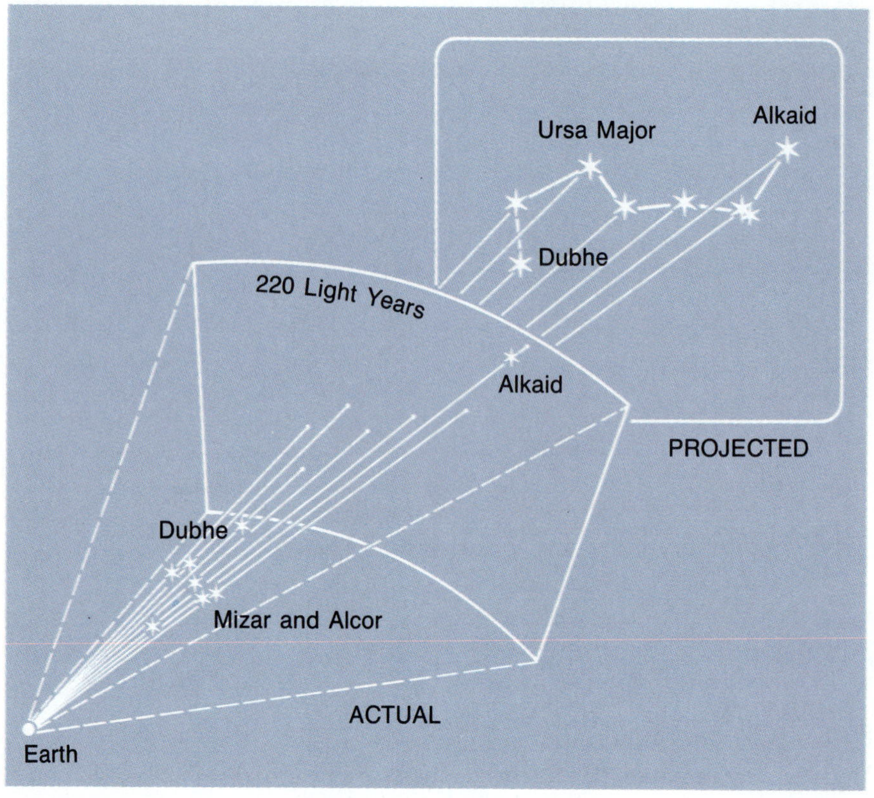

Figure 1–6. *The stars of the Big Dipper are at the relative distances from us as indicated, the farthest being 220 light years away. Because we cannot perceive their distances, they appear as shown on the screen at upper right.*

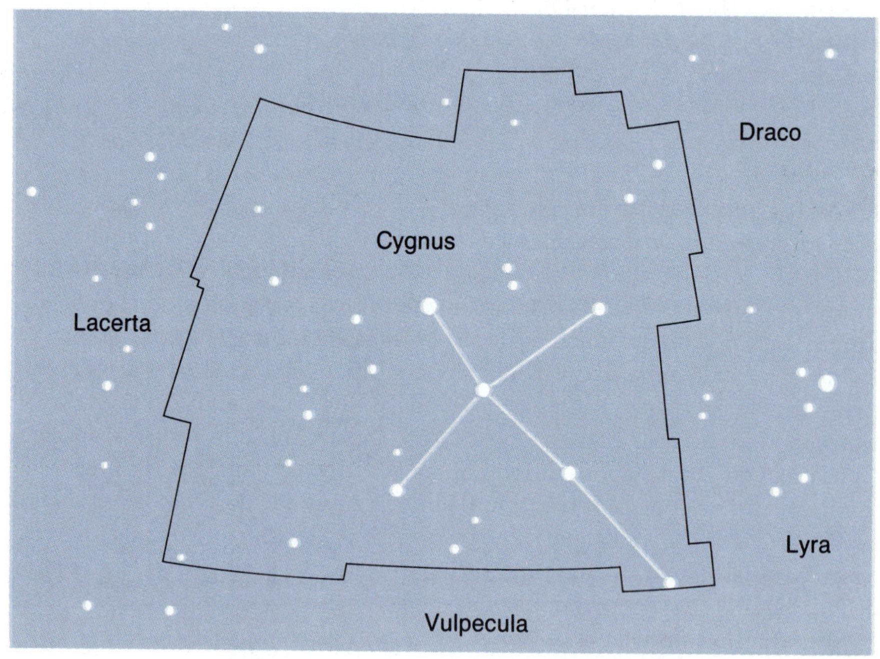

Figure 1–7. *The constellation Cygnus was seen as a swan, but we often call it the Northern Cross, as outlined here. The official boundaries of Cygnus are shown as dashed lines. (You can find Cygnus in the evening sky in late summer and early fall.)*

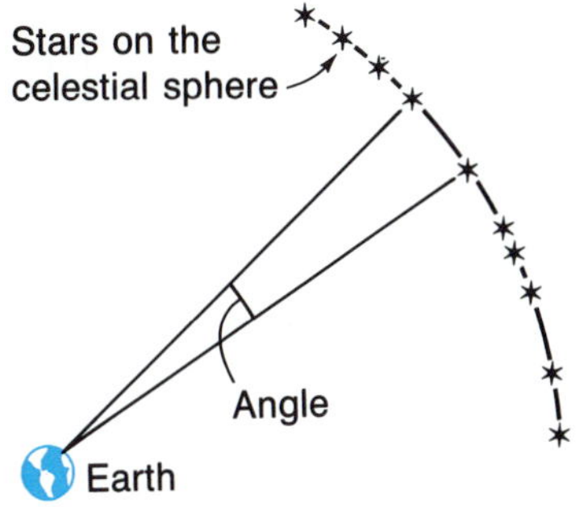

Figure 1–8. *The two stars, when viewed from Earth, have an angular separation as shown.*

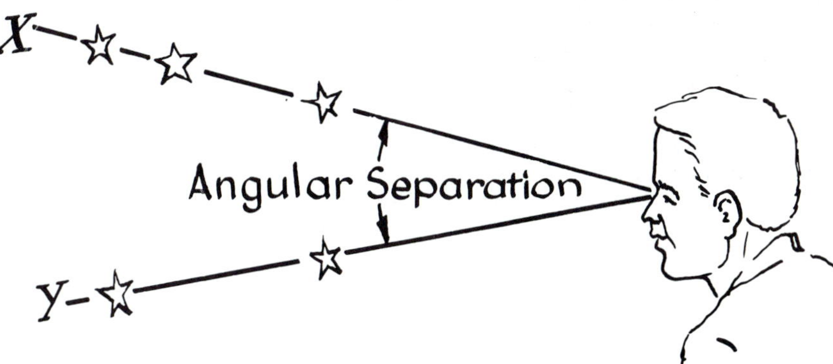

Figure 1–9. *The angular separation of stars says nothing about their distances apart. All of the stars that lie along line* X *have the same angular separation from the stars that lie along line* Y.

Measuring the Positions of Celestial Objects

When people speak of objects in the sky, they sometimes talk about how far apart they are. What they are probably referring to is their ***angular separation,*** or the angle between the objects as seen from here on Earth. Figure 1–8 illustrates this angle. We say, for example, that the stars forming Orion's shoulders have an angular separation of about ten degrees. As Figure 1–9 indicates, this says nothing about the actual distance between the stars. It tells us their angular separation, but not their distance apart.

Angular separation: Measured from the observer, the angle between lines toward two objects.

It is often necessary in astronomy to discuss angles much smaller than one degree. For this purpose, each degree is divided into 60 ***minutes of arc*** (or 60 ***arcminutes,*** or 60′), and each minute is divided into 60 ***seconds of arc*** (or 60 ***arcseconds,*** or 60″). Notice that although these units are very similar in definition to units of time, they are not units of time, but are units of angle. As an example of the use of these smaller units, a good human eye can detect that two stars appearing close together are indeed two stars if they are separated by about one arcminute or more. We will see in Chapter 10 that the use of a telescope allows us to detect such double stars if they are separated by as little as one arcsecond—3600 times smaller than one degree.

Minute of arc: One-sixtieth of a degree of arc.

Second of arc: One-sixtieth of a minute of arc.

EXAMPLE Mizar and Alcor are two stars in the Big Dipper that can be distinguished by the naked eye (see Figure 1–5). They are separated by 12 minutes of arc. Mizar, the brighter of the two, reveals itself in a telescope to be two stars, separated by 14 arcseconds. Express each of these angles in degrees.

Solution Since 60 arcminutes is one degree:

$$\frac{1^\circ}{60'} = 1$$

We multiply the 12 arcminutes by this ratio, thus converting its value to degrees:

$$12' \times \frac{1^\circ}{60'} = 0.20^\circ$$

We do the second part of the problem in like manner, but another factor is needed to convert arcseconds to arcminutes:

$$14'' \times \frac{1'}{60''} \times \frac{1^\circ}{60'} = 0.0039^\circ$$

The answer to the Try One Yourself is in Appendix M.

Try One Yourself. The top star in the head of Orion (Figure 1–4) is a telescopic double star with a separation of 4.5 seconds of arc. Express this angle in degrees.

There is an easy way to estimate angles in the sky. Make a fist and hold it at arm's length. The angle you see between the opposite sides of your fist is about 10 degrees. (See Figure 1–10). For estimating smaller angles, the angle made by the end of your little finger held at arm's length is about one degree. You can use these rules to estimate the angular separations of stars.

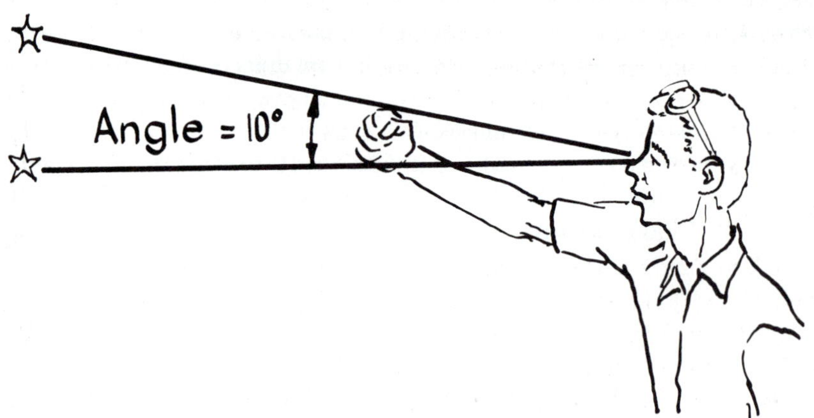

Figure 1–10. *Your fist held at arm's length yields an angle of about 10 degrees.*

THE SUN'S MOTION: HOW LONG IS A YEAR?

Like the stars, the Sun and Moon also seem to move around the Earth as the hours pass, rising in the east and setting in the west. Watching the Sun's motion over a few days might lead us to conclude that it is moving at the same rate as the stars, but if we carefully observe what stars are in the sky immediately after sunset, we will see that they change as the weeks and months go by. It is as though the Sun were out there among the stars, but not staying in the same place on the sphere. Instead, it seems to move constantly eastward among the stars. To explain: If we have a map of the stars that surround the Earth, we can locate the position of the Sun on this map by observing the stars which appear just after the Sun sets and again just before the Sun rises. The Sun is between those two groups of stars. If we do this again two weeks later, we would find that the Sun has moved and is now farther toward the east on our map. So it appears that the Sun does not participate fully in the motion of the celestial sphere. It does appear to move around the Earth, but not quite as fast as the sphere of stars.

How long does the Sun take to get all the way around the sphere of stars? You could determine the approximate time simply by drawing the pattern of stars you see in the western sky after sunset tonight, and then waiting until you see that same pattern again. If you do that, you would probably determine the time to be somewhere between 350 days and 380 days. If you did this over a number of cycles, perhaps you could determine that the time is about 365 days. And, being an alert observer, you would undoubtedly notice that this cycle coincides with the cycle of the seasons here on Earth. Finally, you—or perhaps your descendants to whom you hand down your data—would decide that the time for the Sun to cycle through the stars exactly fits the cycle of the seasons, and that 365 days is the length of the year. (The time for the Sun to get back to the same place among the stars is not a whole number of days. It is about 365¼ days. Every four years we add one day to our year to make up for this difference. This is our leap year.)

See the Historical Note for more details on leap years.

Early in history it was noticed that certain star patterns appear in the sky at the same time every year. Even before the development of the calendar, the arrival of these patterns was used as an indication of a coming change of seasons. For example, the arrival of the constellation Leo in the evening sky meant that spring was coming, and this warned people that even though the weather may not yet indicate it, warm days were on the way.

The Ecliptic

As the Sun moves among the stars, it traces out the same path year after year. Figure 1–11(a) is a map of the stars in a band above the Earth's equator out to 30° on either side of the equator. The second part of the figure (b) shows the relationship of these stars to the Earth. In fact, the line drawn straight across the map is directly above the Earth's equator and is called the ***celestial equator.*** The other line on the map shows the Sun's path among the stars and shows that it is sometimes north of the equator and sometimes south. The path that

Celestial equator: A line on the celestial sphere directly above the Earth's equator.

HISTORICAL NOTE

Leap Year and the Calendar

OUR PRESENT CALENDAR COMES primarily from the Roman calendar. That calendar started its year in March; the Latin words for the numbers seven through ten are "septem," "octo," "novem," and "decem." By the time of Christ, January and February had been added, giving us the twelve months we have now. The months of the calendar were based on the Moon's period of revolution around the Earth and alternated in length between 29 and 30 days in order to match the average of 29½ days between full Moons. Twelve of these lunar months totaled only 354 days, which was defined as the length of the standard year. To make up for the fact that this calendar quickly got out of synchronization with the seasons, an entire month was inserted about every three years—a sort of "leap month."

No single authority controlled the calendar, and by the time of Julius Caesar (100–44 B.C.), the date assigned to a specific day was vastly different between different communities. In some countries, nearby communities did not even agree on what year it was. Caesar reformed the calendar in 46 B.C., making the months alternate between 31 and 30 days, except for February, which had 29. Thus the *Julian Calendar* had 365 days. And (almost) like ours, it added one day at the end of February every four years to make it correspond more closely to the time it takes the Sun to return to the vernal equinox. This latter time determines the seasons and is called the *tropical year*.

After Julius Caesar's death, what had originally been the fifth month (Quintilis) was named after him and is called July today. Then later, the following month was named after Augustus Caesar, the emperor who followed Julius. At the time, July had 31 days and August only 30. This would seem to slight Augustus, and so a day was added to August and subtracted from February. The months following August were changed so that the alternation between 30 and 31 continued again, starting with August.

The tropical year is 365.242199 days long. This means that the seasons on Earth repeat after that period of time, and the difference between the tropical year and the average 365.25 days of the Julian calendar caused the calendar to gradually get out of synchronization with the seasons. By the year 1582, the vernal equinox* occurred on March 11 rather than March 25, as it had when Julius Caesar instituted the calendar. It was time for more reform. Pope Gregory XIII declared that ten days would be dropped from the month of October, so that October 15 followed October 4 that year. That restored the vernal equinox to March 21, which corresponded to what the church wanted for establishing the date of Easter each year. To keep the calendar from having to be adjusted this way again, Pope Gregory instituted a new leap year rule: every year whose number was divisible by four would be a leap year unless that year was an even century year which was not divisible by 400 (such as 1800 or 1900). Thus 1700, 1800, and 1900 were not leap years, but the year 2000 will be a leap year.

The Roman Catholic countries went along with the *Gregorian calendar,* but most other countries chose to stick with the old calendar. England changed in 1752, at which time it was necessary to omit 11 days. Russia did not adopt the Gregorian calendar until this century.

There has been one more change since Pope Gregory's time: The years 4000, 8000, and 12000, and so on will not be leap years, as they would have according to the original Gregorian calendar. The present calendar is accurate enough that it will not have to be revised for 20,000 years.

---

*The vernal equinox, defined later in this chapter, determines the moment when spring begins.

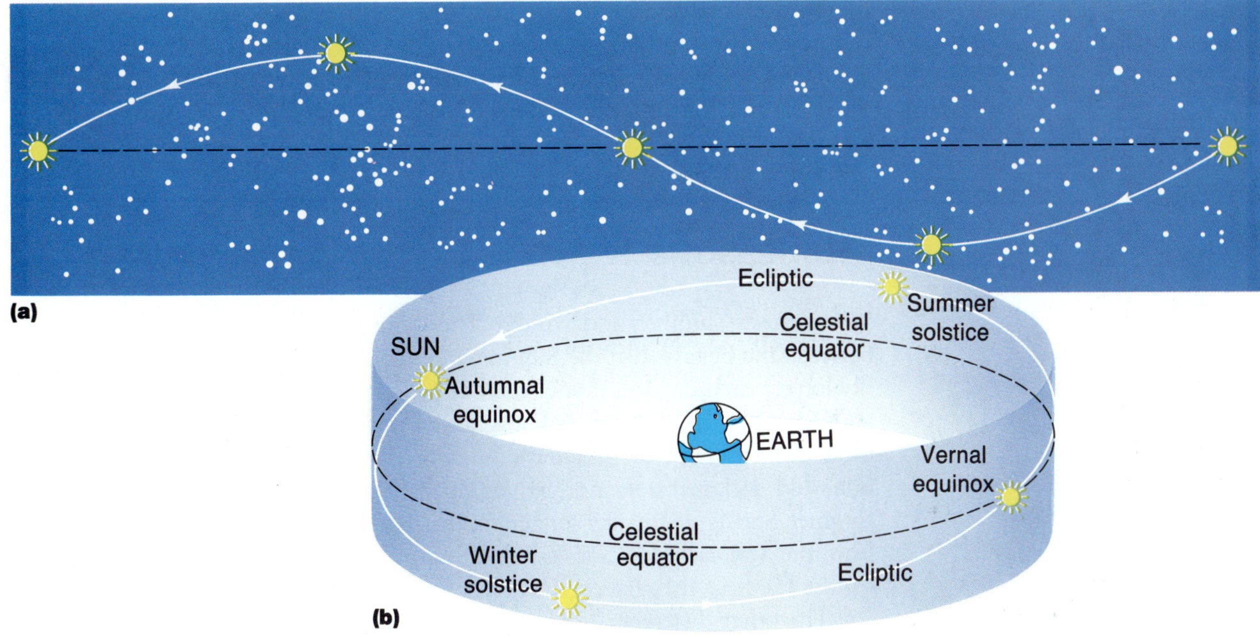

Figure 1–11. *(a) is a map of the stars within 30° of the equator. Picture this map wrapped around the Earth as shown in (b).*

the Sun takes among the stars is called the ***ecliptic,*** and it is not the same as the celestial equator. (The name comes from the fact that it is only when the Moon is on or very near this line that an eclipse can occur. This will be discussed further in Chapter 6.)

Ecliptic: The apparent path of the Sun on the celestial sphere.

The constellations through which the Sun passes as it moves along the ecliptic are called constellations of the ***zodiac.*** Figure 1–12 shows the 12 major constellations of the zodiac. (Actually, the Sun spends almost three weeks in a 13th constellation, Ophiuchus, which is a much longer time than it spends in Scorpius.)

Zodiac: The band that lies 9° either side of the ecliptic on the celestial sphere.

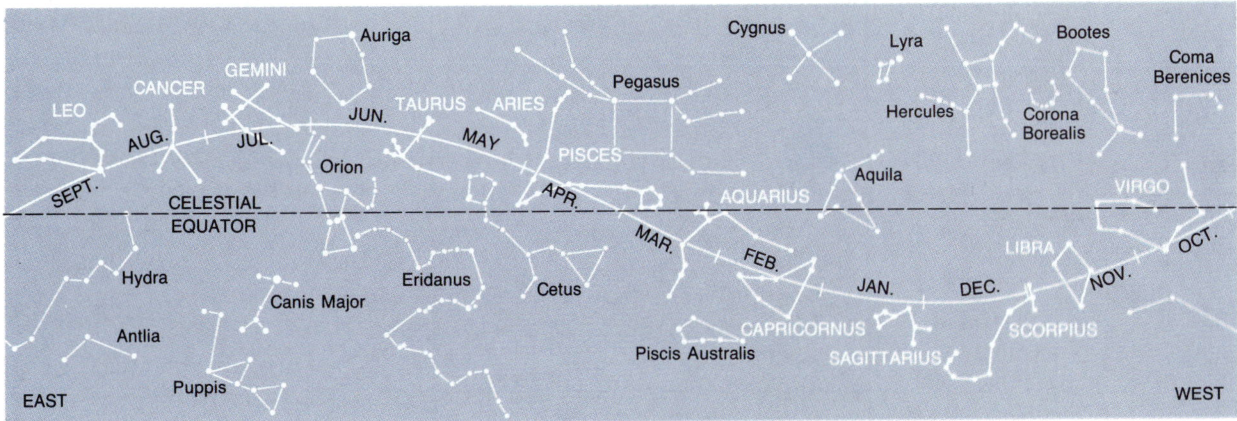

Figure 1–12. *The constellations of the zodiac are highlighted and outlined along the ecliptic. Don't confuse the constellations of the zodiac with the astrological signs of the zodiac.*

Notice the months shown along the ecliptic. These show when the Sun is at these locations among the stars. On March 21 the Sun is in the constellation Pisces, and is crossing the equator on its way north. The changing position of the Sun on the celestial sphere is what causes the seasons, as we will see below.

THE SEASONS

What causes summer and winter? There are two easily observed differences between the behavior of the Sun in winter and summer. First, the Sun is in the sky longer each day in summer than in winter. In December you may leave for an 8:00 A.M. class in near-darkness, and it may be dark again for your evening meal. Contrast this to June days, when it is more difficult to arise before the Sun, and darkness does not arrive until late in the evening. The fact that the summer Sun is in the sky longer means that more energy pours down on us from the Sun each day, and less time is available at night for our surroundings to lose the heat they have gained.

The other difference in the Sun's observed behavior is just as important in causing our seasonal differences. Figure 1–13 contains two drawings showing the location of the Sun at midday in the Kansas sky in late June and in late December. Notice how much higher the Sun is in June. This means that the solar rays are hitting the ground at a more direct angle in June than in December. Figure 1–14 is a drawing that shows the path of the Sun across the sky in June and again in December.

If you live in the southern hemisphere, you know that what was explained above is backwards for your part of the Earth. In the southern hemisphere, the Sun gets higher in the sky in December than in June, and the seasons are

Figure 1–13. *In the northern hemisphere, shown here, the Sun is much higher in the sky in June than in December. In the southern hemisphere, the reverse is true.*

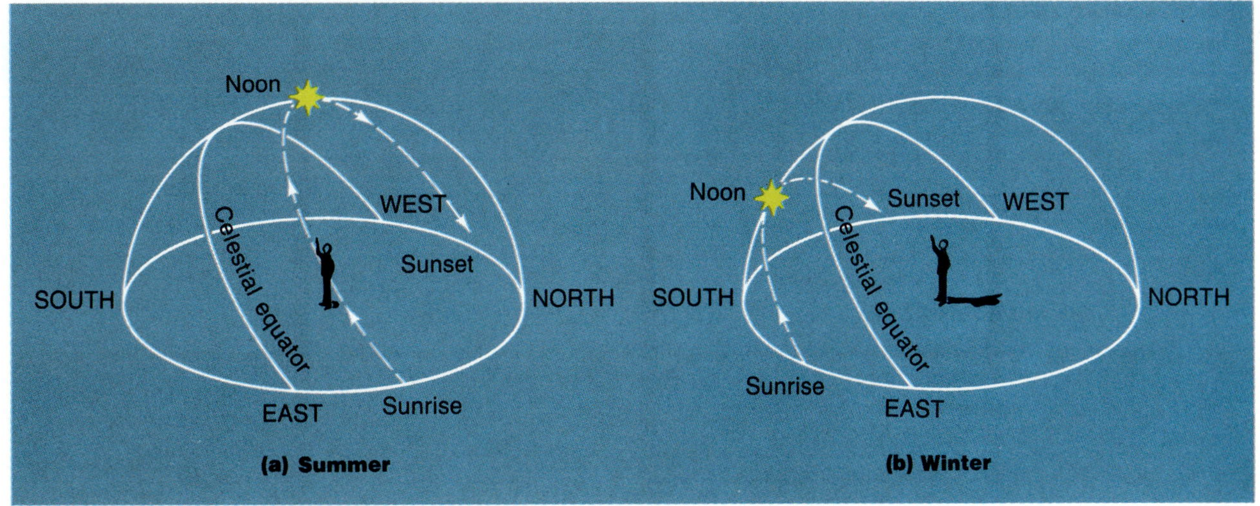

Figure 1–14. *The Sun's apparent path across the sky (in the northern hemisphere) in June (a) and December (b).*

reversed from those in the northern hemisphere. While people are enjoying summer in Canada, Australians are feeling the chill of winter.

To see how the path of the Sun along the ecliptic relates to the seasons, refer to the star map of Figure 1–12 and notice that the Sun is at its northernmost position on about June 21. If you picture the celestial sphere circling the Earth in late June, you can see that it will carry the Sun to a much higher position in the sky of North America. The Sun reaches its southernmost position on about December 22, and on that date it reaches the least altitude in the sky of the northern hemisphere (Figure 1–15).

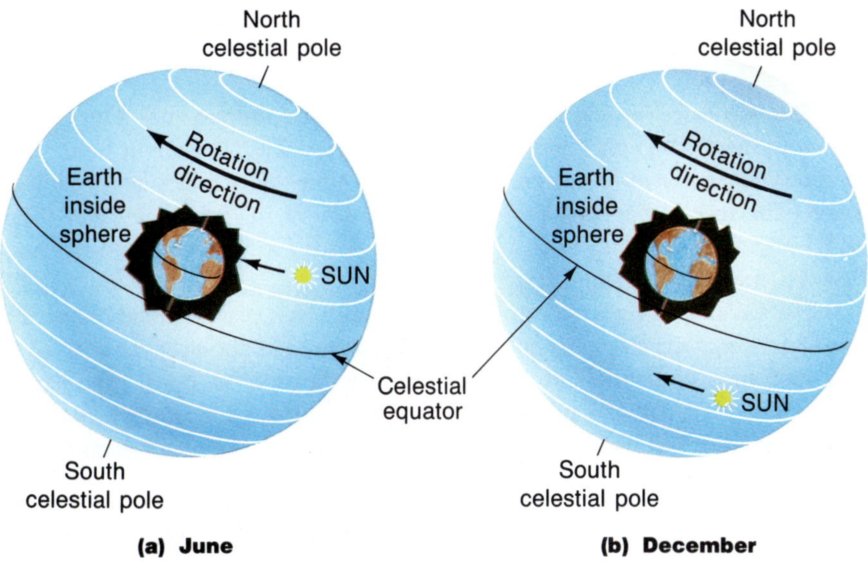

Figure 1–15. *The height the Sun reaches in the sky is explained by its position on the celestial sphere. In (a), the Sun is shown in its June position, and in (b) it is in its December position on the celestial sphere.*

ACTIVITY

Observing the Moon's Phases

THE MOON AND ITS phases are a very easy-to-observe astronomical phenomenon. Actually observing the changing phases of the Moon will make them much more real to you than simply reading about them.

This exercise will take a number of nights. Find a calendar that lists Moon phases, or ask your instructor for this information, so that you can begin your observations three or four nights after the new Moon. Observations should start at sunset or shortly thereafter. At least four observations should be made, continuing for at least three more clear nights during the two weeks following your first night.

It is important that the observations be made from exactly the same place and at exactly the same time of night. Stand at the same place, not just in the same parking lot, for example. You might even go so far as to mark your location with chalk.

1. On each night of your observations, after you have arrived at your observing location, use a full sheet of paper to make a sketch of the position of the Moon relative to buildings, trees, etc. on the horizon. This sketch should show how things look to you, and should not be a map. Label buildings, such as "Student Union." Also include prominent stars you see near the Moon. Draw the Moon in the shape it appears, and with the correct apparent size relative to objects on the horizon. Finally, write the date and time on your paper.

2. On one of the nights, repeat the observation after waiting about an hour. Use the same sketch and show the new position of the Moon.

3. When you have finished your four (or more) observations, make a general statement about how the Moon changed position and phase from one night to another. Look for a pattern in the Moon's behavior and explain that pattern based on the Moon's motion around the Earth.

4. You might try continuing your observation for an entire phase cycle (about a month), and note any problems that arise or any modifications you have to make to your observing scheme.

Summer and winter solstice: The points on the celestial sphere where the Sun reaches its northernmost and southernmost positions, respectively.

Vernal and autumnal equinoxes: The points on the celestial sphere where the Sun crosses the celestial equator while moving north and south, respectively.

The two dates discussed in the last paragraph are unique ones. During the spring the Sun gets higher and higher in the midday sky. Then about June 21—at the ***summer solstice***—it stops climbing, and after that it starts getting lower again. The reverse happens about December 22—at the ***winter solstice.*** (The word *solstice* is a conjunction of the Latin words "sol" meaning "Sun" and "sistere" meaning "to stand still.")

The star map reveals two more unique events each year: Around March 21 the Sun crosses the celestial equator moving north, and about September 23 it crosses it moving south. On these dates every location on Earth experiences equal periods of day and night. The events are called the ***vernal equinox*** (or ***spring equinox***) and the ***autumnal equinox*** respectively. (*Equinox* means "equal night.")

Why did we estimate each of the four dates above rather than state them exactly? Because they may vary a day or so from year to year. The fact that there are not exactly 365 days per year causes our calendars to get out of synchronization with astronomical events until leap year allows us to catch up. In addition, your "today" may be the previous day or the next day somewhere else on the Earth.

THE MOON AND ITS PHASES

The Moon orbits the Earth in such a way that its same face points toward Earth at all times. "The Man in the Moon" always faces us. At first thought, you might be tempted to say that the Moon does not rotate, but this is not so. Consider Figure 1–16. The Moon is shown with a spot on its surface. In (a), you see that if the Moon did not rotate, this spot would always face the same way in space. In (b), that spot continues to face the Earth as the Moon goes around its orbit. But this means that the Moon must rotate once for every revolution around the Earth. The fact that the rotation period and revolution period of the Moon are exactly equal seems remarkable and cannot simply be attributed to mere coincidence. In fact, this is another phenomenon which can be explained by the law of universal gravitation, as will be described in Chapters 5 and 6.

Rotation: The spinning of an object about an axis passing through it.

Revolution: The orbiting of one object around another.

Phases (of the Moon): The changing appearance of the Moon during its cycle, caused by the relative positions of Earth, Moon, and Sun.

Figure 1–17 is a number of photos showing the ***phases*** of the Moon at various times during a period of about a month. The cause for the Moon's phases is fairly straightforward and has been known since antiquity. To explain the Moon's phases, we need only consider three objects: the Earth, the Sun, and the Moon.

The words "month" and "Moon" have the same root, as you might guess by their similarity.

The Moon circles the Earth, completing one orbit in slightly less than a month. This causes its position in the sky relative to the Sun to change with

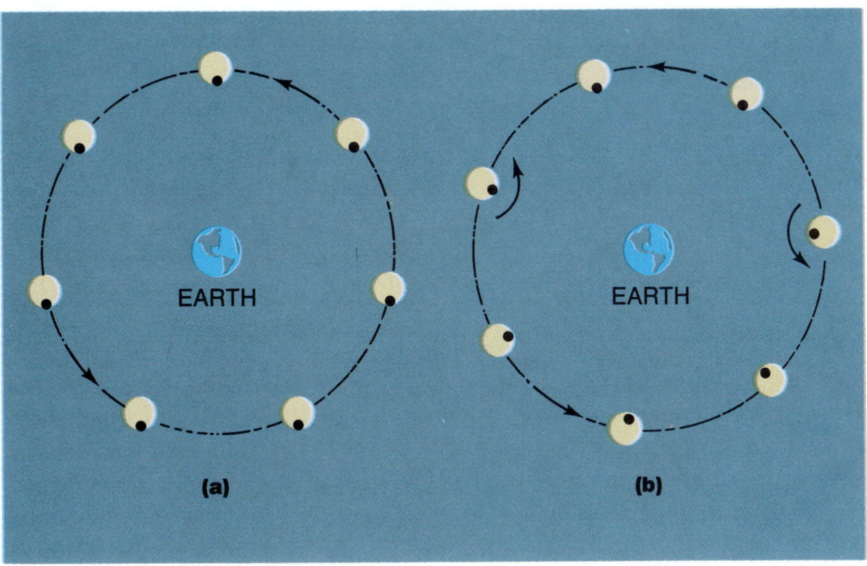

Figure 1–16. *The Moon rotates as it revolves. If it didn't rotate, a point on its surface would always point in the same direction in space as in (a). Instead, the Moon always keeps the same face pointed toward Earth (b).*

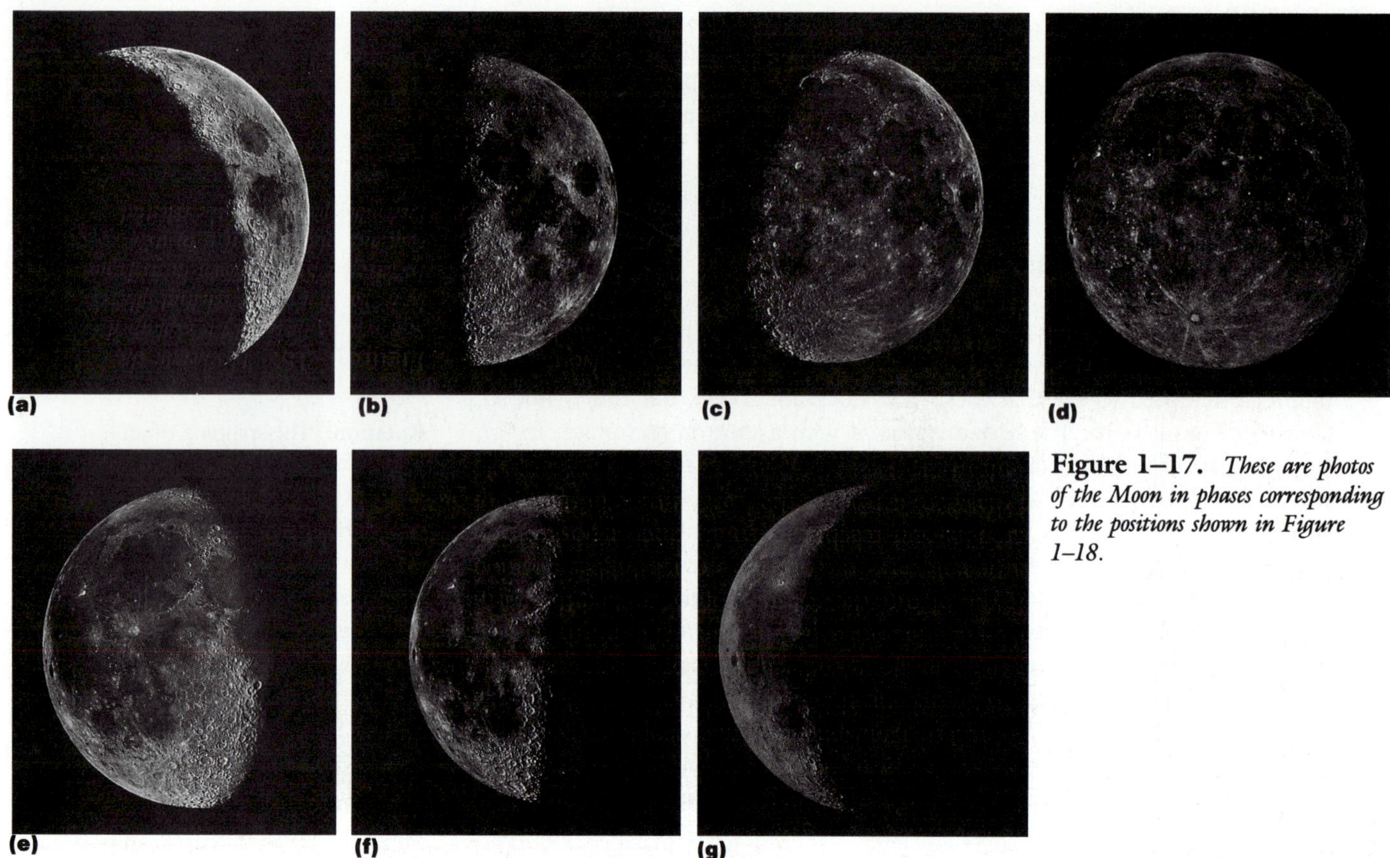

Figure 1–17. *These are photos of the Moon in phases corresponding to the positions shown in Figure 1–18.*

Terms related to the phases of the Moon are defined based on Table 1–1 and Figure 1–18.

time. Figure 1–18 shows various positions of the Moon in its orbit around the Earth. The Sun is out of the picture, far to the left. The drawing of the Earth, as well as each of the Moon drawings, is dark on the side away from the Sun, because sunlight does not reach that side of the Earth and Moon. Now try to imagine what the Moon in position A looks like to someone on Earth. Most of the side of the Moon that faces the Earth is dark. Only a small portion is lit by the Sun. Figure 1–17(a) shows how the Moon appears from Earth when it is in position A. We call such a Moon a ***waxing crescent*** Moon.

When the Moon is at position B on its monthly trip around the Earth, we call it a ***first quarter*** Moon, for if we start the cycle when the Moon is at position H, it is now one quarter of the way around.

Position C will appear from Earth like the photo of Figure 1–17(c) and is called a ***waxing gibbous*** Moon. "Waxing" is an old German term which means "growing." Between points H and D, the Moon is waxing; the lit portion growing from night to night from a thin crescent near H toward the ***full Moon*** (when the Moon is at position D). When we see the Moon in a gibbous phase, most of its sunlit side is facing the Earth.

Observe the photo of the Moon when it is in position E. It is again in a gibbous phase, but the gibbous phase here is called ***waning gibbous*** because

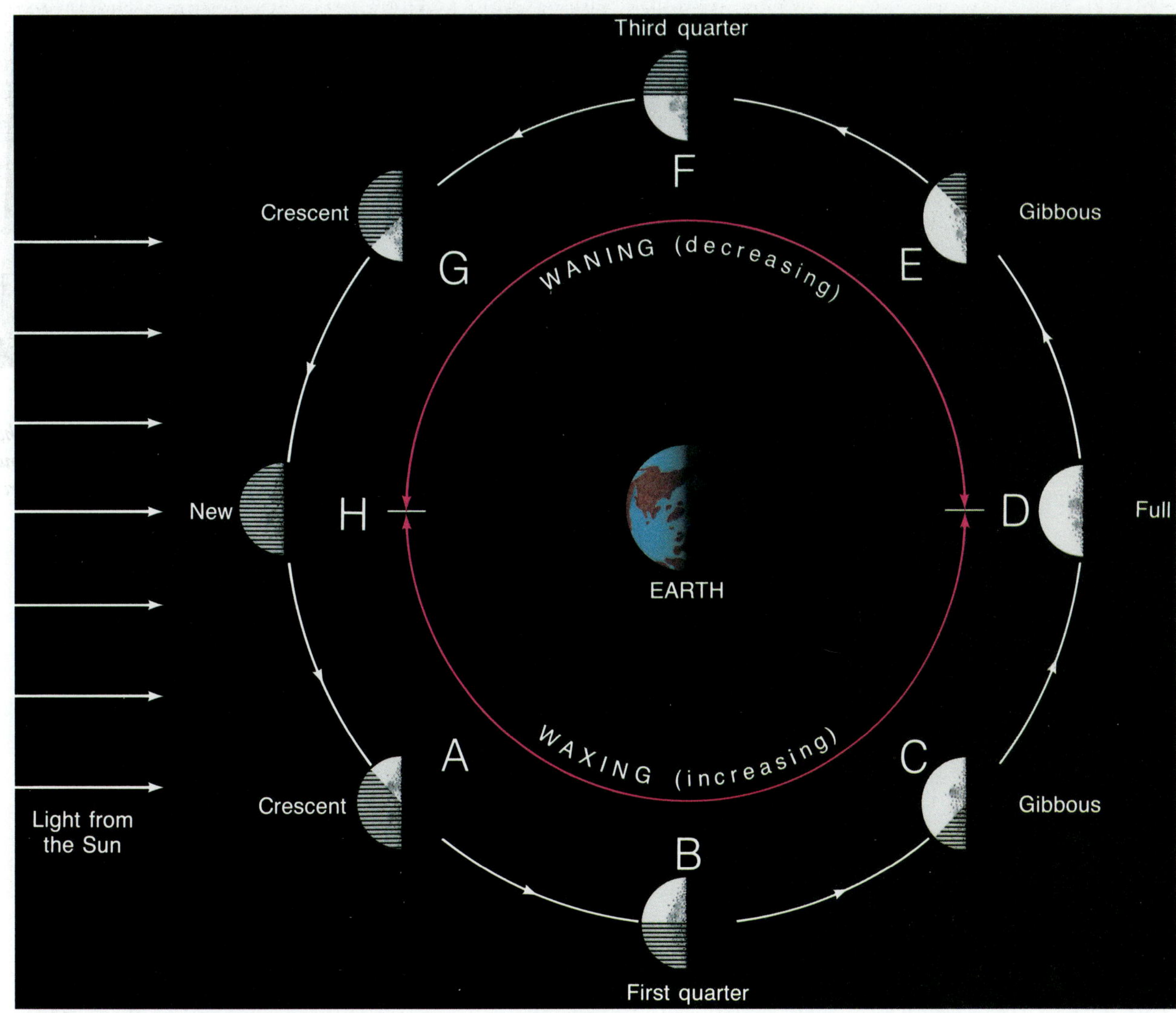

Figure 1–18. *The Moon in various phases. No light reaches the half of the Moon shown as black. The shaded portion, though lit, cannot be seen from Earth.*

from night to night the lit portion that we observe is decreasing (waning) in size.

At position F the Moon is again in a quarter phase, the ***third quarter,*** or ***last quarter.*** Then around position G we have the ***waning crescent*** phase, and finally the Moon is back to where we start the cycle, at position H. Because we (arbitrarily) begin the cycle here, we call this a ***new Moon.*** In this position, the Moon is not visible in our sky because no sunlight strikes the side facing Earth. Only at this phase can an eclipse of the Sun occur. (In Chapter 6 we will see why such an eclipse does not occur every time there is a new Moon.)

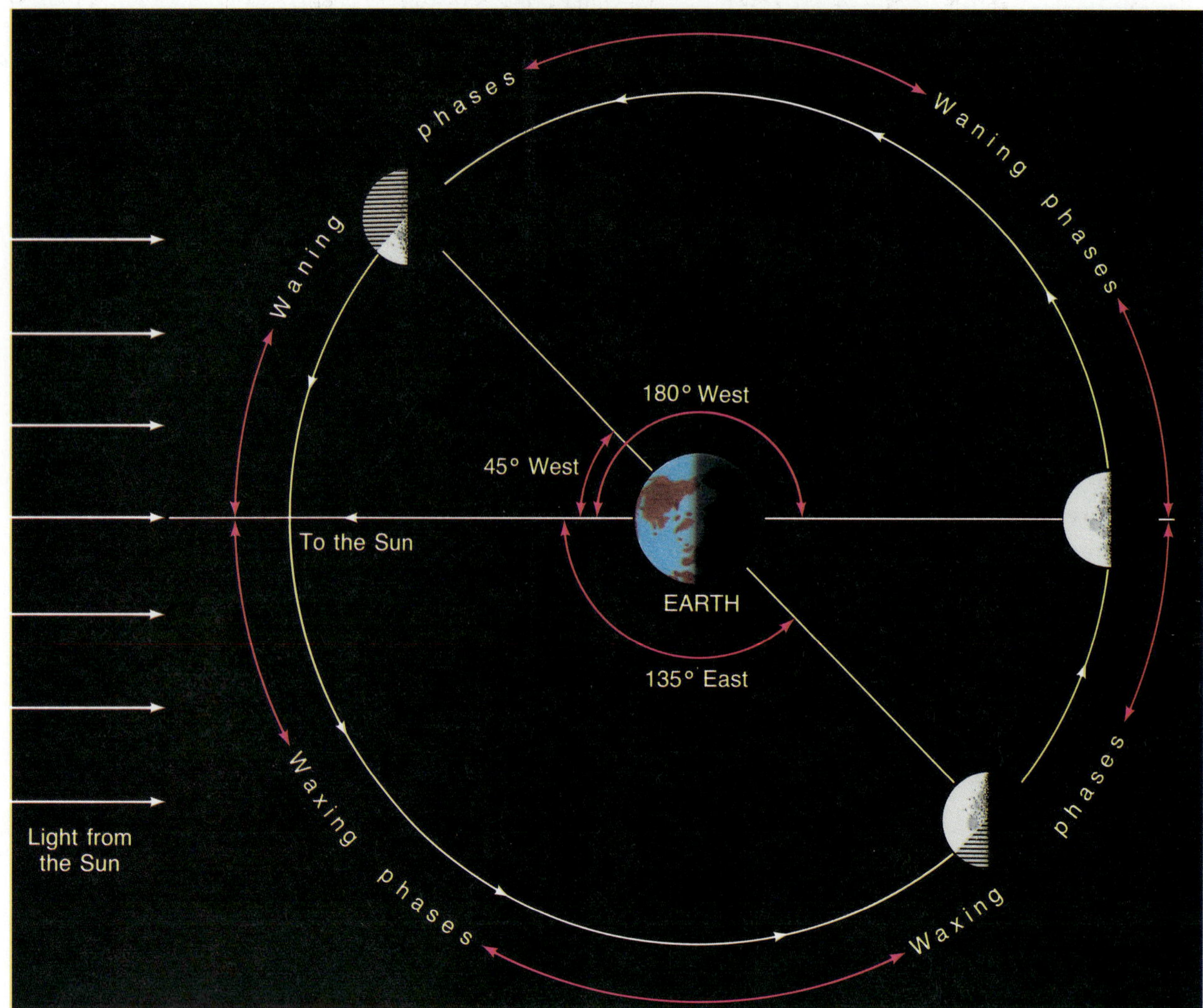

Figure 1–19. *The elongation of the Moon is shown for three positions. Elongations are stated as an angle either east or west of the Sun. We are viewing the system from above the north, so that east is counterclockwise.*

Elongation: The angle of the Moon (or a planet) from the Sun in the sky.

Table 1–1 Terms Relating to Moon Phases

Phase	Elongation (in degrees)
Waxing	0–180° East
Waning	0–180° West
Crescent	0–90° East or West
Gibbous	90°–180° East or West
New	0
First Quarter	90° East
Full	180°
Third Quarter	90° West

Elongation is the angle shown in Figure 1–19.

ACTIVITY

Do-It-Yourself Phases

THIS IS AN IMPORTANT activity and is worth the trouble if you want to understand Moon phases. We will simulate the Sun/Earth/Moon system as it is shown in Figure 1–18. For the Moon you will need something like an orange, a grapefruit, or a softball. The softball would be best because it is rounder than the other choices. For the Sun, you can use a bright light across an otherwise dark room. (You could go outside and use the real Sun, but you must be careful not to look directly at it. This method would work best when the Sun is low in the sky.) Your head will be the Earth, and one of your eyes will be you.

Hold the ball out at arm's length so that it is nearly between you and your Sun. Now observe it as you move it around to the left until it is at 90 degrees to the Sun, the first quarter position, as is shown in Figure T1–3. Did you see its growing crescent as you were moving it? Continue to move it around your head and observe it as it changes phase. (When it gets directly behind you from the Sun, you will eclipse it.)

Now fix the Moon at the first quarter position by laying it down on something in that position. Turn your head (and body) slowly around and around toward the left to simulate the Earth rotating. When the Sun is directly in front of you, it is noontime. As you lose sight of the Sun, it is sunset, about 6 PM. It is midnight when the back of your head is toward the Sun. When the Moon is at first quarter and it is sunset on Earth, where do you see the Moon?

Put the moon in various other positions and observe it as you rotate your head to simulate different times of day. Answer the following questions:

1. If the Moon is at third quarter, at about what time will it rise? At about what time will it set?

2. At about what time will a full Moon appear highest overhead?

3. If you see the Moon in the sky in midafternoon, about what phase will it be?

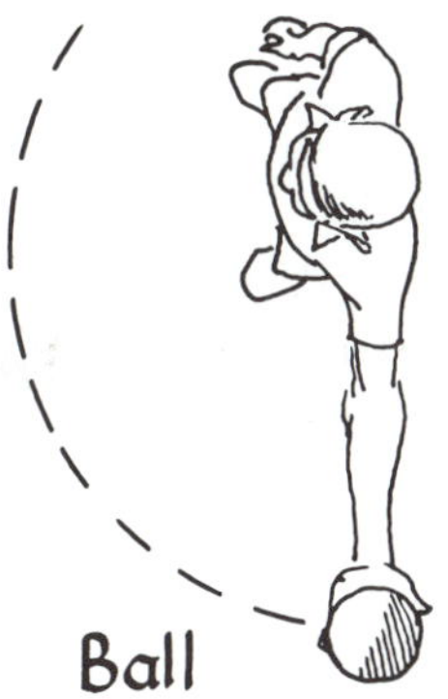

Figure T1–3. *The person is holding a ball that represents the Moon at first quarter. When doing this, you should put the light bulb farther away than indicated here, perhaps 10 to 15 feet away.*

SCIENTIFIC MODELS—A GEOCENTRIC MODEL

Scientific model: A theory that accounts for a set of observations in nature.

Thus far in this chapter we have explained the celestial motions we see by describing ***scientific models.*** The idea of the stars residing on a giant celestial sphere is a model that explains the observation of the daily motion of the stars. Likewise, the changing position of the Sun in the sky explains the changing of the seasons and the changing position of the Moon relative to the Sun explains the phases of the Moon.

A scientific model is not necessarily a physical model, in the sense of a model car or a model airplane. It may even be impossible to construct an actual physical object to represent a scientific model, because a scientific model is basically a mental picture that attempts to use analogy to explain a set of observations in nature. Thus we are able to say that the stars appear as if they are on a sphere rotating around the Earth. Later in the book we will encounter some scientific models that cannot be represented well by a physical construction.

It is easy to combine our explanation for the seasons with our model of the celestial sphere. Imagine a track—perhaps like a railroad track—along the ecliptic. As the celestial sphere rotates around the Earth, the Sun stays on this track (Figure 1–20), following the sphere's general motion, but gradually slipping back eastward along the track so that as the months pass, it changes its position among the stars. It moves along the track with such speed that in about 365 days it is back to where it started.

The Sun in our model moves from north to south of the equator and back again. This corresponds to the motion of the real Sun. We are not saying that there is actually—in real life—a track on which the Sun rides, but our model gives us this picture to help us feel comfortable with the motion of the Sun. That is one of the functions of a scientific model: to allow us to feel comfortable with observations. Our model does not actually explain the Sun's motion in the sense of telling us why it occurs, but it provides a mechanism that allows us to say, "OK, that makes sense now."

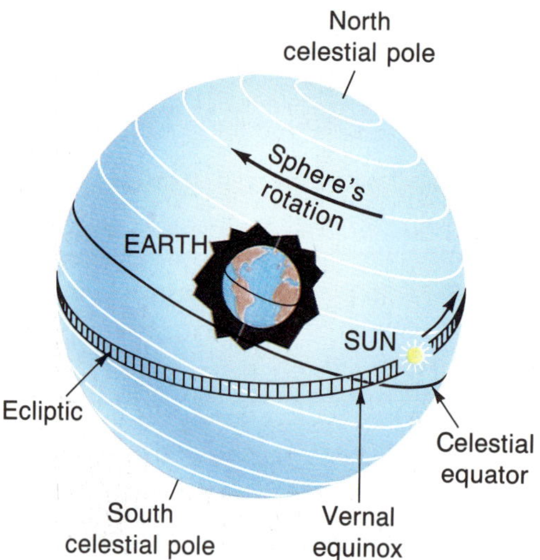

Figure 1–20. *The Sun acts as if it follows a track around the celestial sphere. The sphere rotates toward the west as the Sun gradually moves along it toward the east.*

The model we have constructed is a ***geocentric model;*** an Earth-centered model. As we saw in the Prologue, this is not today's model of the universe. In order to see how scientific models are developed, changed, and replaced, we will describe the geocentric model developed by the Greeks some 2000 years ago, and then we will examine reasons why the model has been replaced (but not entirely abandoned).

Geocentric model: A model of the universe with the Earth at its center.

ARISTOTLE

To trace the evolution of our present model of the system of the heavens, we begin with the advanced Greek culture that lasted from about 600 B.C. until a few hundred years after Christ. Before looking directly at the Greek model, however, we must begin with a discussion of the world as seen by Aristotle, a Greek thinker who lived from about 384 to 322 B.C., but who still influences today's thinking.

Aristotle's system of the world made a great distinction between earthly things and the things of the heavens. He observed that matter on Earth seems fundamentally different in behavior from matter in the sky. For example, objects on Earth have an undeniable tendency to fall to Earth. They fall *down*. In fact, we might define the word "down" as the direction things fall. Things in the sky don't do that; they seem instead to move in circles around the Earth.

Aristotle spoke of the "natural" behavior of objects. He said that it is natural for earthly objects to move downward, that they seem to *naturally* seek the downward direction. On the other hand, it seems reasonable that heavenly objects move in circles. They never move downward. Something in the very nature of the objects seems to make them behave differently.

(a)

(b)

Figure 1–21. *(a) Aristotle (ca. 384–322 B.C.). (b) Ptolemy (ca. 100–200 A.D.).*

Aristotle saw another basic difference between the two types of objects: While earthly objects always seem to come to a stop, heavenly objects just keep going. Thus the natural motion of the two classes of objects appears to be different. This idea from Aristotle was a basic part of the world view of the Greeks, and it can be seen in their celestial models. They believed that there were two different sets of rules: one for earthly objects and one for celestial objects.

THE GREEK CELESTIAL MODEL

One of the first detailed celestial models was developed by Eudoxus (ca. 408–355 B.C.), a Greek mathematician who predated Aristotle by nearly a century. Similar to the model we developed, the model of Eudoxus placed the stars on a sphere, but instead of having the Sun move on a track on that sphere, he placed the Sun on another sphere that rotates around the Earth inside the sphere of the stars, as shown in Figure 1–22. Since the Sun is sometimes north of the Earth's equator and sometimes south, the figure shows this sphere turning on a different axis from the one on which the stellar (celestial) sphere rotates.

There was good reason for the Greeks to use a sphere to carry the Sun, and it is linked to Aristotle's distinction between earthly and celestial natures. Today, we have grown up with the idea that objects in our sky are part of the same universe in which we live. They may be vastly different from our Earth, yet they are still explainable in earthly terms. This is a fairly new way in which to perceive the universe. (In Chapter 4 we will discuss how this new view came about.) Today we still sometimes speak like Aristotle when we refer to the sky as "the heavens," as if it were not part of our universe.

A second thing that we must appreciate about the Greek culture of 2000 years ago is its admiration of geometry. This is traceable back to the time of Pythagorus, who died around 500 B.C. He is known today principally for his

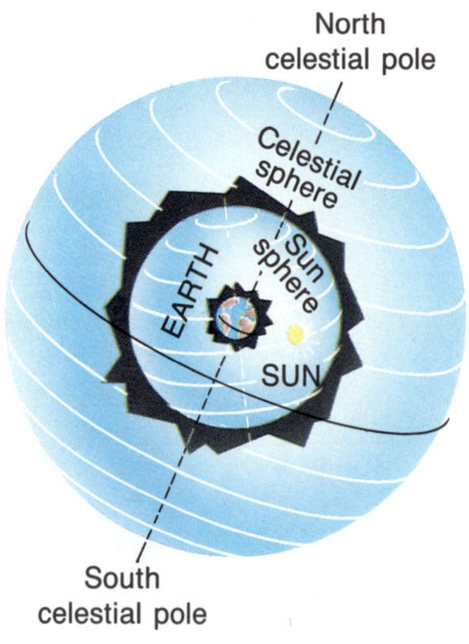

Figure 1–22. *In order to account for the Sun's path around the Earth (and among the stars), Eudoxus located the Sun on a sphere that moves around the stationary Earth inside the celestial sphere of stars. Notice that the axis of the Sun's sphere is tilted with respect to the axis of the celestial sphere.*

study of geometry. Pythagorus argued that the Earth must be a sphere because of the shape of the shadow it casts on the Moon during a lunar eclipse. He then carried the idea of spheres much farther and developed a model of the Earth, Sun, Moon, and planets using many spheres rotating around one another. To the Greek thinkers, geometry was considered the queen of mathematics. Geometric explanations were sought for all natural phenomena.

Pythagorus, of course, was the discoverer of the Pythagorean theorem.

The Greek model we will discuss most thoroughly is that of Claudius Ptolemy, who lived around 150 A.D. In his book (which we call the *Almagest)* he presented a comprehensive model that lasted for more than 1300 years after his death. Today we know this geocentric model as the **Ptolemaic model.** Ptolemy abandoned the spheres of earlier Greek models, for he saw no need to imagine actual physical objects carrying the Sun, Moon, and planets. The stars, however, are still spoken of as being on the celestial sphere.

Ptolemaic model: The theory of the heavens devised by Claudius Ptolemy.

In accordance with the thinking of Aristotle, people of Ptolemy's time thought that things in the sky (the "heavens") must be perfect. It seemed reasonable that the heavens would feature the circle, a symmetrical shape with no beginning and no end. Indeed, the stars seem to move in circles around the Earth. It seemed natural that they lie in a spherical arrangement, and that the sphere moves around us at a constant speed.

Notice that we still find symbolism in the circle: we use the never-ending circle for our wedding rings.

The Moon fit the Ptolemaic scheme perfectly. All that is necessary is that it moves around the Earth as the closest heavenly body. In fact, the Moon fit the overall Greek philosophical approach. Being the closest to the imperfect Earth, the Moon might be expected to have imperfections on it. Indeed, it has dark and light areas. It is somewhere between the imperfect Earth and the perfect heavens.

THE PLANETS

Thus far we have barely mentioned the other major class of objects in the sky: the planets. Without a telescope we can see five planets: Mercury, Venus, Mars, Saturn, and Jupiter. The Greeks, of course, knew of these planets. In fact, the word "planet" comes from a Greek word meaning "wanderer." And wander they do. Like the Sun and Moon, the planets move around among the stars on the celestial sphere. The Sun and Moon always move eastward among the stars, but the planets sometimes stop their eastward motion and move westward for a while. They lack the simple, uniform motion of the Sun and Moon.

The Greeks listed the Sun and Moon as planets. That made seven planets, and this is the origin of our week having seven days. In fact, some of the days are named after planets.

Figure 1–23 is a photo of the constellation Cancer on the night of October 23, 1977. Mars is the brightest of the "stars" in that photo. Find it about two-thirds of the way from the left and a little above center. Figure 1–24 is a similar photo taken about 2½ weeks later, on the night of November 10. Notice that Mars has moved down and to the left, and is near the small cluster of stars (which is called the Beehive, incidentally). These photos show the sky as it appears to someone in the northern hemisphere: East is toward the left, south at the bottom. So Mars has moved eastward, and a little toward the south. A photo taken a few weeks later would show that it is still moving eastward and slightly southward. Things would be fine if Mars continued on in this uniform

Figure 1–23. *On October 23, 1977, Mars appeared as a bright star in the constellation Cancer. It is a little above center, about two-thirds of the way from the left.*

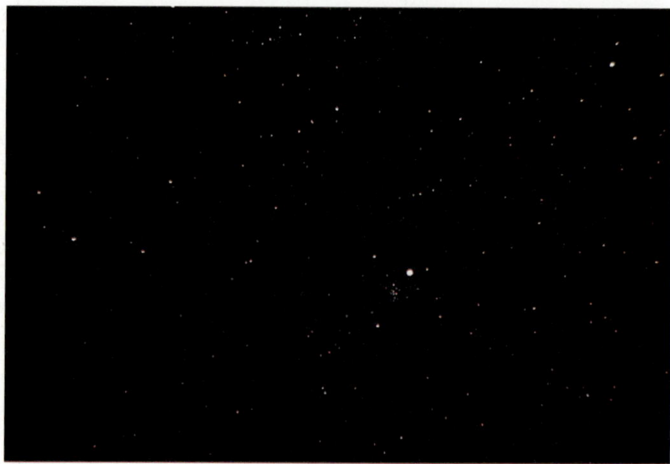

Figure 1–24. *Find Mars in this photo taken on November 10, 1977. Notice that it has moved down and to the left.*

Figure 1–25. *This drawing shows the motion of Mars through Cancer during its retrograde motion of winter 1977–78.*

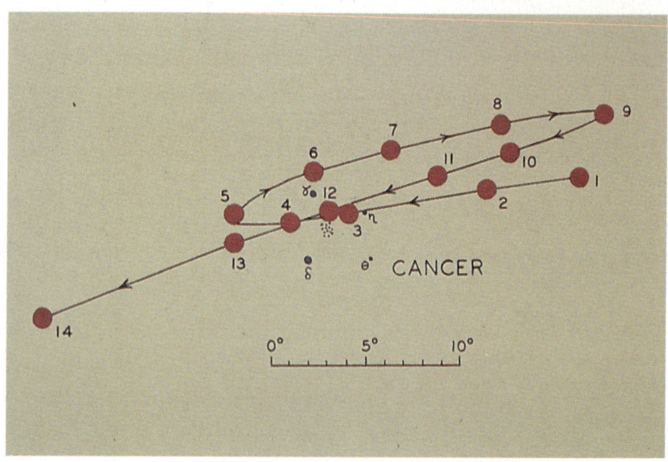

way. Uniform motions fit the idea of simple, circular motions in the heavens. (The model can account for the southward motion of Mars if the plane of Mars's motion is placed at an angle with the celestial equator. Mars moves northward again later in the year. This is how the model explains the Sun's motion.)

Figure 1–25 is a drawing that shows the progress of Mars through the fall and winter sky of 1977–78. Positions 2 and 3 correspond to the previous two photos. Notice that Mars continued its eastward motion until it got to position 5 (which happened on December 18, 1977), and then it started moving *backwards,* heading west! It moved backwards until March 5, position 9 on the figure, and then it resumed its eastward motion again. It was finally at position 14 on May 29. And—not shown in the drawing—it continued in this direction for almost two years, after which it again did one of its backwards loops. We call this backwards motion ***retrograde motion.*** Retrograde motion is characteristic of all the planets, including those discovered in modern times.

Figure 1–26. *Venus and Jupiter are seen here next to the crescent moon. Mercury can be seen at the center of the sunset's glow. In situations like this, Venus is bright enough that it has frequently been the cause of UFO reports.*

Although the planets move among the stars in a seemingly irregular manner, there are limits to where they move, for they never get more than a few degrees from the ecliptic. Mercury and Venus have an additional limitation on their motion. These two never appear very far from the position of the Sun in the sky. We only see them either in the western sky shortly after the Sun has set or in the eastern sky shortly before sunrise (Figure 1–26). Mercury appears so close to the Sun that it is hard to find, even if we know where to look. This is because even when it is at its maximum elongation, we have to look for it in the semibright sky during dawn or twilight. You will never see Mercury or Venus high in the sky at night.

Retrograde motion: The east-to-west motion of a planet against the background of stars.

Any model of the planets must explain not only their observed retrograde motion, but why they stay near the ecliptic, and also the peculiar behavior of Mercury and Venus. How did the Greeks account for the planets' unique motions? Were the heavens perhaps imperfect, with the planets not moving uniformly in a smooth circle?

Figure 1–27. *Mars seems to move in a circle on a second rod that rotates on the longer one.*

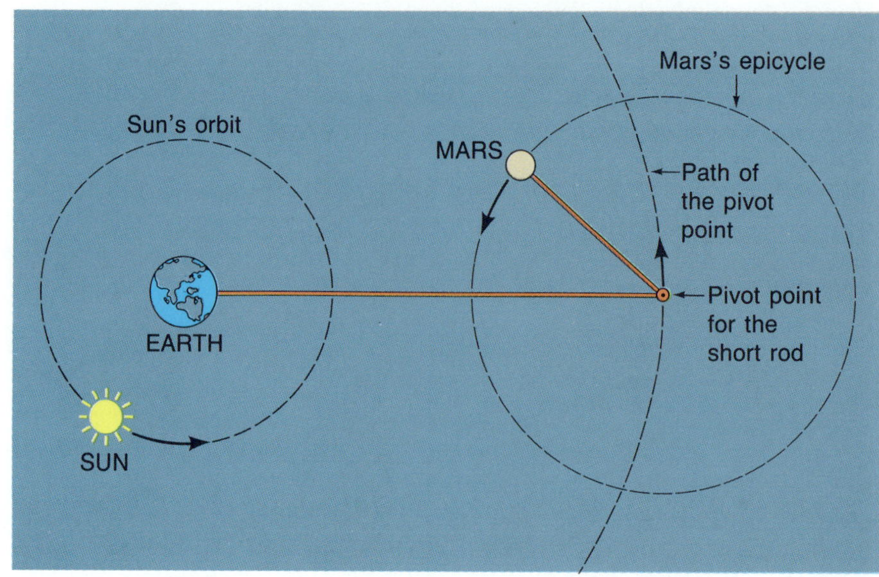

Figure 1–28. *Mars's motion on its epicycle results in a looping path, which—when seen from Earth—appears as retrograde motion.*

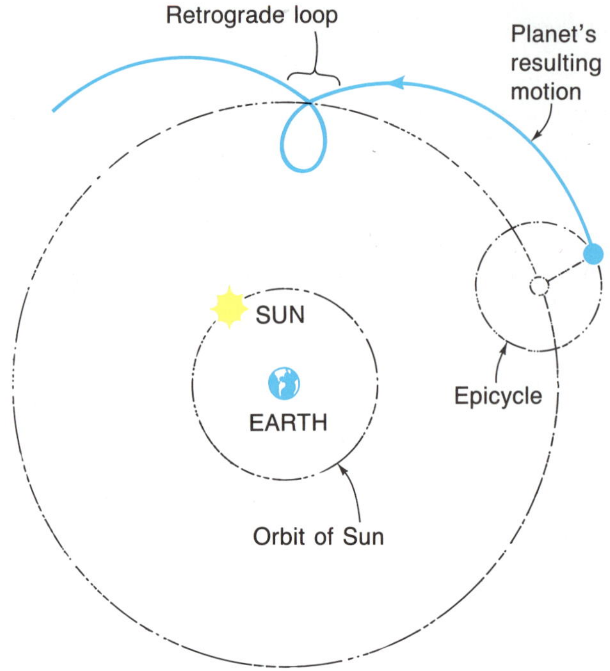

Epicycles

Epicycle: The circular orbit of a planet in the Ptolemaic model, the center of which revolves around the Earth in another circle.

Ptolemy was indeed able to use circles to make a model which fit the planets' motions. (Though this idea was not original with him, he did elaborate on it and he is generally given credit for it.) It simply took more than one circle for each planet. Figure 1–27 shows a rod extending out from Earth with a second rod attached to its end. The second rod can swing around its connecting point

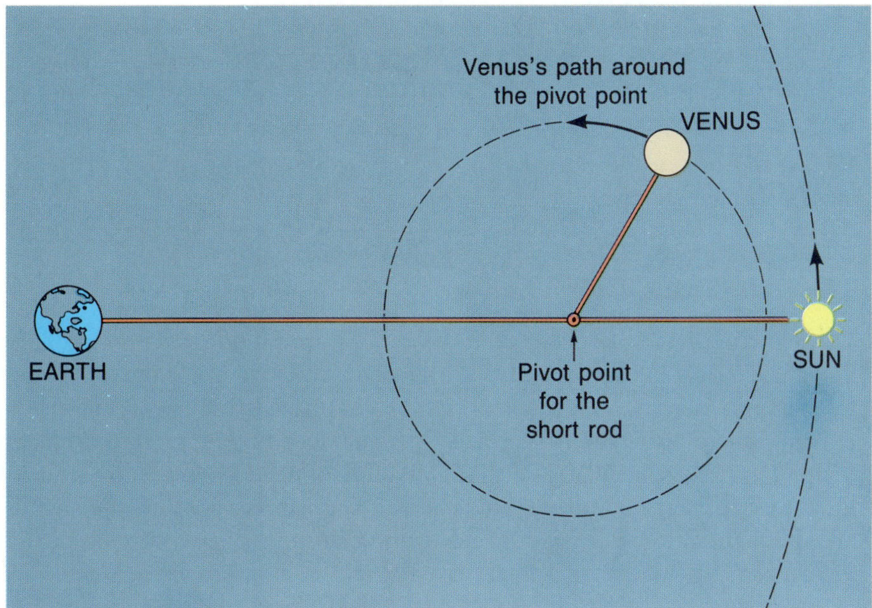

Figure 1–29. *The center of Venus's (and Mercury's) epicycle stays between the Earth and the Sun. This accounts for the fact that Venus (and Mercury) are never seen in the sky far from the Sun.*

on the longer rod. At the other end of this second rod, we place Mars. Now we rotate the big rod uniformly around the Earth. At the same time, we rotate the small rod around its pivot point on the large one. By a correct choice of rotation speeds, an observed motion for Mars such as shown in Figure 1–28 can be produced.

When a person on Earth views the motion of Mars among the stars, the person sees the planet moving eastward most of the time. This corresponds to the long rod turning around counterclockwise—to the left—in the drawing. Occasionally, however, the planet retrogrades. This occurs when the planet on the end of the little rod passes closest to Earth, so that it appears to back up for a short time.

The Ptolemaic model did not contain the rods referred to above, of course. Again, the model did not use actual physical objects guiding the planets. A spot moving around the Earth served as the center of motion for the smaller circle of the planet's motion. This smaller circle is called an ***epicycle.*** In this way, Ptolemy was able to preserve the idea of perfect heavenly circles and perfect, uniform, heavenly speeds.

How, then, does the model explain why the planets never move far from the ecliptic? The answer is that the plane of their circular motion lies very close to the plane of the Sun's motion. The greater the angle between these two planes, the farther the planet will move from the ecliptic as it moves among the stars.

The different behavior of Mercury and Venus was explained by having the centers of their epicycles remain along a line between Earth and the Sun (Figure 1–29). While the center of other planets' epicycles could be anywhere on their circle around the Earth without regard to where the Sun is, these two are special cases.

CRITERIA FOR SCIENTIFIC MODELS

The Greek model had many features which we look for today in a scientific model. We will discuss the Ptolemaic model in light of three criteria that are applied to scientific models today. Even though these are the standards of today's science and not of the thinking of Ptolemy's times, it is instructive to apply them to Ptolemy's model.

The Data-fitting Criterion

Actually, to make the model fit the data more accurately, Ptolemy had to either make the planets vary their speeds along their paths or to move the Earth off-center among the planets' paths. He chose the latter method. After this and a few similar adjustments, the model fit pretty well.

The first criterion: A model must fit the data. It must fit what is observed. The Ptolemaic model fits pretty well, for it gives an explanation for the motions of the heavenly objects. The stars' motion was explained by the rotation of the celestial sphere on which they reside. The Sun and the Moon were theorized to move in circles around the Earth. Explanation of the planets' motion required the use of epicycles, and it required special treatment for Mercury and Venus, but when these features were included, the planets' motions were accounted for. So the Ptolemaic model passes the first test. It fits the available data.

The Prediction Criterion

The second criterion that today's scientists place on a model in judging whether it is a good one is as follows: The model must make predictions that allow it to be tested, and the model must be of such a nature that it would be possible to disprove it—that is, to show that it needs to be modified to fit new data, or perhaps that it needs to be discarded entirely. There are many testable predictions built into the Ptolemaic model. First, the model made predictions for the locations of the planets at times in the future. One could use the model to predict that Jupiter would be at a particular place in the sky at a particular time the next year. Another example: the Ptolemaic model holds that the Earth is stationary. It would predict that as knowledge advances and new methods are found to measure the motion of the Earth—either rotational motion or motion through space—no motion would be found. If further experimentation did not confirm this, the model would have to be adjusted drastically or abandoned.

"Prediction" here does not necessarily refer to the future. It means that the theory itself indicates an observation which will either support it or disprove at least part of it. Sometimes the observation has already been made, and needs only to be checked against the theory. For example, one might check the Ptolemaic model's prediction of where Jupiter was ten years ago.

The Aesthetic Criterion

The simplicity spoken of here is not accepted as a necessary attribute of beauty in art, so we are speaking of a different type of beauty.

The third and final feature of a good model is that it should be aesthetically pleasing. This concept is difficult to define. It generally means that the model should be simple. Neat. Beautiful. As an example, we mentioned that Pythagorus was the one who is credited with the idea of placing the heavenly objects on a number of spheres rotating within one another. Later followers of Pythagorus wrote that the motion of these spheres resulted in "music of the spheres" that could be heard by those attuned to it. Such a claim would be quite unconventional—to say the least—in today's science. Scientists today are considered suspect if they hear things no one else hears. We might even lock them up. However, the fact that people of that time spoke of such things does illustrate the necessity of a good scientific model having a pleasing quality or a beauty. Today's idea of beauty in a scientific model is more along the lines of looking for symmetry and simplicity. A model should be as simple as possible; containing the fewest arbitrary assumptions.

Let's test Ptolemy's model against this criterion. In his model, the Sun and Moon travel around the Earth in different paths than the stars, but they obey basically the same rules, moving from east to west around the Earth, and they all use circles. This idea that all objects obey the same rules is what would be described by scientists as a form of symmetry.

As pointed out in a previous margin note, Ptolemy's actual model was even more complicated; even less aesthetically pleasing.

To explain the motions of the planets, epicycles were required. The model still uses the circles so dear to Greek thinking, but somehow one wonders if there couldn't be a better method than epicycles to represent what is observed. In addition, the fact that different rules are required for Mercury and Venus makes the Ptolemaic model less pleasing to the senses. The need for different rules for certain planets means that the model lacks an aspect of simplicity. How many new special rules will be needed? Each one makes the model less appealing in the scientific sense. So we see that the Ptolemaic model begins to look less attractive as a good scientific model.

Remember, though, that the primary rule for a good model is that it fits the observations. An aesthetically pleasing model that did not come close to the real world would be an almost useless model. The Ptolemaic model *did* fit the data, and so we must judge it to be an acceptable model. It just lacks that certain neatness we would like.

We will return a number of times to a discussion of the features of a good scientific model.

ANOTHER MODEL

About 400 years before Ptolemy, the Greek philosopher Aristarchus had proposed a moving-Earth solution to explain the motions of the heavens. This model explained that the reason that the sky seems to move westward was that the spherical Earth is spinning eastward. Why was this model not given much consideration? Ptolemy was aware of this model, so we'll let him answer the question:

> *Now some people, although they have nothing to oppose to these [Ptolemy's] arguments, agree on something, as they think, more plausible. And it seems to them there is nothing against their supposing, for instance, the heavens immobile and the Earth as turning on the same axis [as the stars] from west to east very nearly one revolution per day. . . . But it has escaped their notice that, indeed, as far as the appearances of the stars are concerned, nothing would perhaps keep things from being in accordance with this simpler conjecture, but that in the light of what happens around us in the air such a notion would seem altogether absurd.*

Ptolemy was arguing that if the Earth turned on an axis, it would be moving through the air around it, and therefore there should be a tremendous wind observed in the opposite direction. Notice, however, that Ptolemy refers to that model as being a "simpler conjecture." He saw that it was more aesthetically pleasing, but argued that it was not a good model because it presented obvious contradictions with the observation that the air stays where it is, along with everything loose on Earth. Ptolemy was unable to see that the Earth might be carrying the air and everything on it along as it rotates.

MODEL, THEORY, AND HYPOTHESIS

The three terms in the title of this section are words used not only by scientists, but by the general public. They mean a different thing in science from what they normally do in everyday language, however. We have been calling the work of Ptolemy a model, and have discussed how this does not mean a physical model, but rather a developed set of ideas used to describe some aspect of nature. Instead of calling Ptolemy's creation a model, we could have called it a theory. The two words mean about the same thing in scientific usage, and we could have spoken of the "Ptolemaic theory."

Notice, then, that in science the word *theory* is used differently than in everyday language. In science a scheme is not usually called a theory until its ideas are shown to fit observed data successfully. In nonscientific use, the word *theory* is often used to refer to ideas that are much more fanciful and less secure. I might say, for example, that I have a theory about what caused yesterday's car wreck. In science we wouldn't call my car wreck idea a theory. We would call it an ***hypothesis.***

Hypothesis: A tentative theory.

An hypothesis is a guess—perhaps a very intelligent guess—at a theory or part of a theory. Generally a theory starts as an hypothesis. The idea of epicycles to explain planetary motion had been used by Hipparchus, who lived about 300 years before Ptolemy. It was then an hypothesis. The hypothesis was developed and included with others to form Ptolemy's final model. According to today's scientific use of the words *theory* and *model,* Ptolemy's plan would not even have been given those names until it was shown to be able to explain the heavenly motions.

The distinction between hypothesis and theory is not totally clear cut in science, though. Sometimes the word theory *is used very loosely.*

You might hear someone refer to Einstein's theory of relativity or to the theory of biological evolution and say, "Well, it is *only* a theory," meaning that you mustn't put too much faith in it. It is correct that these two concepts are only theories, but keep in mind what a scientist means in calling something a theory. Later in this text we will see that Einstein's theory is well-founded and has been experimentally verified many times. Although we will not discuss the theory of evolution here, virtually all biologists form the same conclusion about it.

CLOSE UP

Astrology

ASTROLOGY IS THE BELIEF that the relative positions of the Sun, planets, and Moon affect the destiny of humans. It is used by astrologers to indicate the nature of any epoch of history and to determine which of the celestial bodies is dominating at the time. Most often it is used to determine the nature of personalities at the time of birth or of circumstances in the future, and it is traditionally used for timing or resolving problems in the areas of marriage, business, medicine, or politics. Astrology apparently had its origins more than 3000 years ago with the Chaldeans near the Persian Gulf. Early astrology held that the heavens affected only nations and national leaders, but the Greeks expanded the idea to include every person. In particular, the Greeks contributed the idea that the positions of the planets, Moon, and Sun at the moment of a person's birth have an effect on the life of that person. To discover what the effect is on an individual, a horoscope is cast.

Casting a horoscope involves determining the positions of the heavenly objects at the moment of birth of the person. To do so, the zodiac is divided into 12 equal-sized *signs*. Although these signs have the same names as the 12 principal constellations of the zodiac (omitting Ophiuchus), they lack correspondence with the constellations in two ways. First, the constellations differ from one another in size, while all signs are the same size. The constellation Cancer, for example, is much smaller than Virgo or Pisces, but their signs are the same size. Second, the signs are at different locations in the sky from the constellations. At one time there was fairly good correspondence in position, but because of *precession* (to be discussed in Chapter 6), the signs and constellations are now "out of whack" by about one position. Thus the sign of Aries is located mostly in the constellation of Pisces. This means that a person born on March 22 is said by astrologers to be an Aries, while the Sun actually appeared in the constellation Pisces at the person's birth. (Some astrologers, called sidereal astrologers, use the constellations rather than the signs.)

The sky above us is divided by astrologers into 12 *houses,* starting with the eastern horizon and proceeding around the sky. The houses do not rotate with the stars, but remain in the same positions relative to Earth. Thus the first house stays above the eastern horizon. The houses are arbitrarily associated with various areas of our lives, such as children, health, and enemies. Likewise, the planets are associated with various good or evil influences. Mars, for example, is associated with war.

Finally, a significance is placed on the angular spacings (called *aspects)* of the planets, Moon, and Sun in the sky.

When the above information is determined for the moment of a person's birth, we have what is called the *natal chart* for the person. This chart could be made by anyone who knows the sky and has tables of solar, lunar, and planetary positions. It might be determined, for example, that the planet Mars was just above the eastern horizon at the time of your birth. So what? This is where interpretation by the astrologer comes in. After such interpretation, the astrologer is supposed to be able to tell you about your personality and life pattern.

It may appear from what I have written thus far that astronomers accept astrology as part of their science. Not so! Having read this section, you are hereby required to read the Close Up in the next chapter regarding the application of scientific criteria to astrology.

CONCLUSION

There is a tendency to think that Greek science was bad science because it is not today's science. This is simply not true. It is true that the Ptolemaic model is more primitive than today's. The Greeks did not have the accurate observations and extensive data we have today. Their model, however, did fit the data they had. In fact, it fits the casual observations of most people today. Can you think of any direct evidence obtainable without a large telescope that contradicts Ptolemy's model? What observations can you personally make, without relying on reference materials, that will show the 2000-year-old Ptolemaic model to be a poor model?

RECALL QUESTIONS

1. In what century did Ptolemy live?
2. As you watch a planet during the night, in what direction does it appear to move across the sky?
3. What is the elongation of the Moon at first quarter? When it is full?
4. What is a constellation?
5. Name the planets that are never seen far from the Sun in the sky.
6. How do stars near Polaris appear to move as we watch them through the night?
7. List three criteria for a good scientific model.
8. Define *retrograde motion* of a planet.
9. What is the ecliptic?
10. Name eight different phases of the Moon in the order in which they occur.
11. Which of the criteria of a scientific model is the most important one?
12. What observation convinced Pythagorus that the Earth was spherical?
13. What group supposedly heard the "music of the spheres?"
14. Name the constellations of the zodiac.
15. In what direction across the background of stars do the planets normally move?
16. What is an epicycle?
17. Define *geocentric.*
18. What is the origin of the word *planet?*
19. What are the approximate dates of the summer and winter solstices and what is their significance?
20. In what part of the sky must you look to find the planets Mercury and Venus?

QUESTIONS TO PONDER

1. In your personal case, why are you probably less familiar with the night sky than your great grandparents?
2. Explain how you can determine the length of the year by astronomical observation.
3. Explain why it is necessary to have a leap year every four years.
4. What is meant by saying that one criterion for a good scientific model is its aesthetic quality?
5. How did Ptolemy's model account for the Sun moving back and forth between the northern and southern hemispheres?
6. How did the Ptolemaic model account for retrograde motion?
7. Explain why the Greeks used circles and spheres to account for heavenly motions.
8. How did the Ptolemaic model explain the different behavior of Mercury and Venus?

9. The reason we have 360 degrees in a circle is related to an ancient calendar that had 360 days in its year. Check references and report on this.
10. Discuss the relative importance of the three criteria for a good scientific model.
11. In what direction do we see the Moon move across the sky during the night? In what direction does it move relative to the stars? Explain the discrepancy.
12. Describe one or two observations that would disprove the Ptolemaic model.
13. Does the Sun ever appear in the constellation Orion? (See Figure 1–12.) If so, at what time of year?
14. At about what time does the full Moon rise? Set?
15. Why can we only see one side of the Moon? Is the other side always dark?
16. If you see the Moon high overhead shortly after sunset, about what phase is it in?
17. Will the Moon appear crescent or gibbous when it is at position X in Figure 1–30? At position Y? At about what time will the Moon rise if it is at position Y?
18. The star in Orion's left knee is much farther away than the other bright stars in the constellation. (See Figure 1–4.) If you look at Orion in the sky, and then move far into space in the direction to your left, how will that star shift relative to the others?

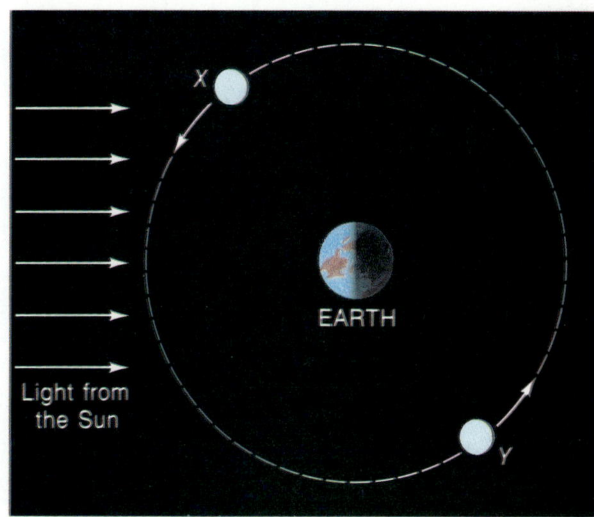

Figure 1–30. *(Question 17) Is the Moon crescent or gibbous at point* X? *At* Y? *What time does it rise when at* Y?

19. Which days of the week are named for which planets? You may have to consult another book for this, because although you might guess (correctly) that Sunday was named for the Sun, some of the corresponding names do not sound alike in English.

CALCULATIONS

1. Express an angle of 15 arcseconds in degrees.
2. Four degrees of angle is how many arcminutes?
3. Suppose one star is 3 arcminutes from another. What is this angle in degrees?

CHAPTER 2

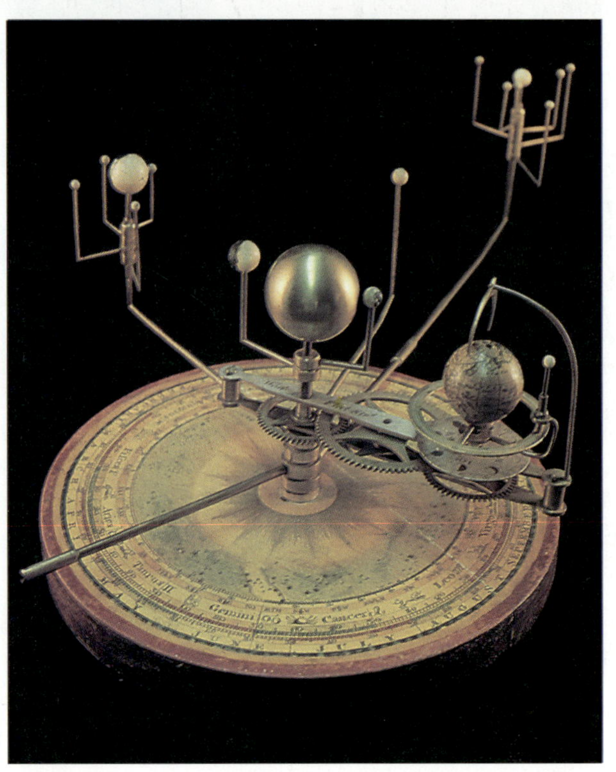

This is an early 19th century device which simulates celestial motions.

At the time of Mikolaj Koppernigk (1447–1543), Latin was the universal language in much the same way that English is today. It was the language that scholars learned so that they could communicate with people from other countries. It was also common for people to "Latinize" their names. Thus Mikolaj Koppernigk became Nicolaus Copernicus. (Some people went farther than this. One of Copernicus' teachers had been named Wodka, which means vodka; *he Latinized his name to Abstemius, which means* abstainer.*)*

Copernicus entered Bologna University in Italy in 1496. It was operated differently from today's universities in that the students were completely in charge. Students elected a Rector who served a two-year term governing the university. At one time it was traditional for the students to celebrate the election of a new Rector by tearing off his clothes and selling the pieces. In order to teach, professors had to swear to obey the Rector. A professor could be fined for being late to class, for omitting anything from his lecture, or for lecturing too long.

(This material is from *The World of Copernicus,* by Angus Armitage, The New American Library, 1947.)

A Sun-Centered System

IN THE LAST CHAPTER we described the development of Ptolemy's geocentric system of the world. Notice that although we call this model the Ptolemaic Model, we must not think of it as being entirely due to Ptolemy. He worked in Alexandria, Egypt, and had access to the great Greek library there. His model was constructed by molding the thought and tradition of the past—much of it the work of Aristotle and Hipparchus—and adding his own insights to arrive at the final model. This is typical of scientific development. When we look with hindsight to the great advances in science, we see that they occur when the time is ripe for them; when previous thought in the area seems ready to be brought together with new insight to take a great step forward. It seems that all that is needed is the right person to take this step; the mind that sees a little more clearly than others, the mind that is ready to take the leap of imagination into the future.

Ptolemy's model, developed in about A.D. 125, served as the accepted model for more than 1300 years. In this chapter we will look at the introduction of a new model and examine the arguments for and against each of the two competing models. In chapter 3, we will see how the new model finally won out because of adjustments that helped it pass the primary test of a scientific model: the model's ability to fit the observations more accurately.

SCIENCE AND ASTRONOMY FROM A.D. 200 To 1200

You might wonder why no major advances in our understanding of the astronomical world were made from the time of Greek civilization until the 1500s. The reason lies in the social, religious, and political changes taking place during those times, and in the interests of the cultures that came into possession of the documents of Greek culture.

Our name for Ptolemy's great work, Almagest, *is a combination Greek-Arabic word, and means "the greatest."*

By Ptolemy's time, the center of Greek learning had shifted to Egypt, particularly around Alexandria. Gradually the Greek culture declined as it was replaced by that of the conquering Romans, who were a much more practical people—less philosophically minded and little interested in astronomy. Centuries after Ptolemy (in A.D. 640) Alexandria was captured by the Muslims, who seized the Greek documents found in the libraries there. The Muslims took them back to the Middle East, where many were translated into Arabic. The Muslim astronomers made careful observations of the heavens, adding to the store of knowledge about the motions of the planets but making no advances in Ptolemy's theoretical model.

In western Europe during this time, the works of the Greeks were unknown or at least forgotten. Around the year 1200, however, writings from the Greek culture were reintroduced to Europeans and many were translated into western languages. It was common for *Almagest* to be found in libraries of Europe during the thirteenth century.

THE MARRIAGE OF ARISTOTLE AND CHRISTIANITY

It was during the thirteenth century that Saint Thomas Aquinas, one of the greatest theologians and philosophers of the Christian church, incorporated the works of Aristotle and Ptolemy into Christian thinking. Aquinas insisted that there must be no conflict between faith and reason, and he blended the natural philosophy of Aristotle with Christian revelation. For the Aristotelian and Ptolemaic ideas we have discussed, the blend was an easy one. The idea of an Earth-centered world fit comfortably with literal biblical interpretation, for it placed humans at the center of God's creation: the ultimate expression of the divine will.

Figure 2–1. *Thomas Aquinas (1224?–1275).*

As we discussed in the last chapter, the idea of a central, unmoving Earth was natural for early humans. Through the work of Aquinas, this easily accepted idea was shown to fit perfectly with Christian beliefs. So Aristotle's science—and with it the Ptolemaic model—became even more entrenched in our western culture. It was no longer just a natural, normal way of thinking about the world, but was part of Christian thinking; part of religious dogma.

Why are we studying material that seems to be more nearly church history than science? Because science does not exist separately from culture and, in particular, from religious beliefs. Science and scientists are part of the society in which they live, and it is necessary to know some of the flavor of an age to appreciate the work of the thinkers of that age—including those thinkers who today might be labelled as scientists. So it is important to emphasize that after the thirteenth century, the teachings of Aristotle and the Ptolemaic model were an ingrained part of western thinking.

A very important aspect of the times was the great reliance on authority, particularly on authorities of the past. Today most of us have a much greater tendency to rely on our own thoughts, observations, experiences, and feelings than did people of the times we are discussing. Thus Aquinas was using the authority of the Bible, the authority of earlier churchmen, the authority of his superiors in the church, and the authority of Aristotle. Arguments were often settled by reference to authorities rather than by personal experience or independent experimentation.

Into this world came a man who was to cause a revolution in the way people think.

NICOLAUS COPERNICUS

At the time when Columbus was journeying to America, a brilliant Polish student was studying astronomy, mathematics, medicine, economics, law, and theology. As was common for scholars of that time, Nicolaus Copernicus did not limit himself to studies in a particular discipline like we do today (perhaps necessarily). Later in life his primary responsibility was that of a churchman, Canon of the Cathedral of Frauenberg. He is known today, however, for initiating the revolution that finally resulted in replacing the geocentric system of Ptolemy with a ***heliocentric***—sun-centered—system.

Heliocentric: Centered on the Sun.

For nearly 40 years, Copernicus worked on his heliocentric system. Even after all this work, he was slow to publish it. It was published with the title, *On the Revolutions of the Heavenly Spheres* but it is often called *De Revolutionibus,* a shortened form of the original Latin title. The preface was written by Andreas Osiander, a mathematician and theologian who consulted with Copernicus. In that preface, Osiander speaks in Copernicus's name and attempts to explain the delay in publication as follows:

Figure 2–2. *Nicolaus Copernicus (1473–1543).*

> *. . . thinking therefore within myself that to ascribe movement to the Earth must indeed seem an absurd performance on my part to those who know that many centuries have consented to the establishment of the contrary judgment, namely that the Earth is placed immovably as the central point in the middle of the Universe, I hesitated long whether, on the one hand, I should give to the light these my commentaries written to prove the Earth's motion . . . .*

and again:

> *The thought of the scorn which I had to fear on account of the novelty and incongruity of my theory, well-nigh induced me to abandon my project. These misgivings and actual protests have been overcome by my friends.*

A system with a moving Earth was not an entirely new idea. We pointed out at the end of the last chapter that in the third century B.C., Aristarchus had proposed just such a system. In his book, Copernicus refers often to early thinkers who had proposed the same ideas. He seems to have done so in order to justify his ideas by the authority of these thinkers. That he felt the need for such justification may seem curious in a man who proposed such a revolutionary idea, but it simply points out that each of us is a creature of the age in which we live. Although it was a time of revolutions, revolution takes time, and it never occurs without pain.

There apparently were two primary reasons that Copernicus decided to develop and publish his model. First, as the centuries had passed, it was found that Ptolemy's predicted positions for celestial objects were not in agreement with the carefully observed positions. For example, Ptolemy's model could be used to predict the position of Saturn on some night in the future. The prediction of its position on a night a few years later was accurate enough, but when the prediction was made for a few centuries in the future, the predicted position may have differed from the carefully observed position by as much as 2 degrees.

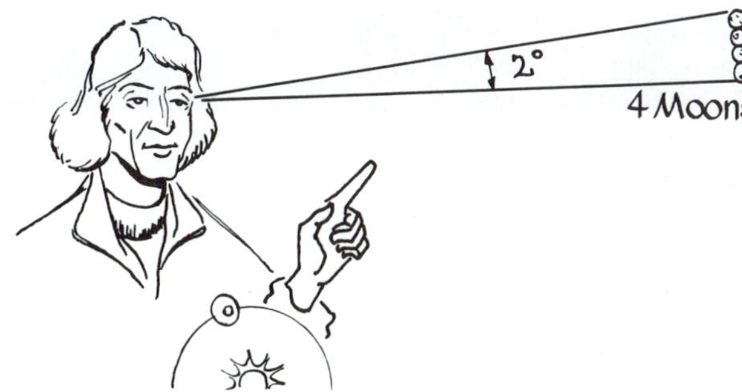

Figure 2–3. *When the difference between the predicted position of a planet and the observed position is 2 degrees, four moons could fit between the two positions.*

This difference corresponds to about four times the diameter of the Moon (Figure 2–3). Ptolemy's model had been corrected a number of times through the ages to bring it up to date so that it would be fairly accurate for a while. These corrections were not refinements in the way it worked, but were "resettings" of each planet's position that were made in order to fit the latest data. A good model would not require such updating. Copernicus recognized this, and he sought a more accurate model.

A second reason Copernicus sought another model was that he did not feel that the Ptolemaic model was aesthetically pleasing enough. We will return to a discussion of the criteria for a good model after we have described the model developed by Copernicus.

THE COPERNICAN SYSTEM

Copernicus's system is heliocentric, with the Sun at its center. The Earth assumes the role of just another one of the planets, all of them revolving around the Sun. The Earth becomes the third planet from the Sun. Figure 2–4 shows the order of the planets that were known in the sixteenth century.

If we consider the view of Figure 2–4 to be that of someone far out in space above the northern hemisphere of the Earth, the planets all move around in a counterclockwise direction as shown by the arrows in the figure. Another feature of Copernicus's model is that the planets closer to the Sun move faster than the planets farther out. In the figure, the lengths of the arrows represent the speeds of the planets, with the longer arrows representing greater speeds.

Let us check the Copernican model to see how well it accounts for the motions of the heavenly bodies. Then in the next chapter we will see how this model finally gained acceptance.

The Rotation of the Celestial Sphere

Ptolemy had explained the daily motions of the heavenly objects as being due to the rotation of the celestial sphere from east to west around a stationary Earth. But the same apparent motion results if the sphere stands still and the Earth rotates from west to east under it. As recognized by Ptolemy, a rotating Earth under a stationary sky produces the same result as the opposite model

ACTIVITY

The Rotating Earth

On a clear dark night, find a good observing location and draw a quick sketch of some stars high over your head. Look for a pattern so that you can remember these stars later. If you can, mark your map so that it shows which stars are toward the north, east, south and west.

Take about two hours off, and then go back to your observing location and find the stars you drew. How has their position changed? You probably know what this answer should be from having read the text. Your real job, however, is to stand under the stars and imagine the cause of their motion as being due to the rotation of the Earth. Try to "feel" the Earth turning under the stars. Which is easiest to imagine, the stars on a sphere rotating around the Earth, or the Earth spinning under the stars?

Finally, spend a half hour watching either a sunrise or a sunset. Or better yet, watch the Moon rise or set, particularly when it is full or nearly full. Now picture this phenomenon as explained by the heliocentric system. Try to think of the Earth turning instead of the Sun or Moon moving. Imagine yourself on a little ball that is turning so that there is a change in where the Sun's light hits the ball.

Perhaps such thinking required that you stretch your imagination a little. Does the activity help you to appreciate why the heliocentric system of Copernicus was not quickly accepted?

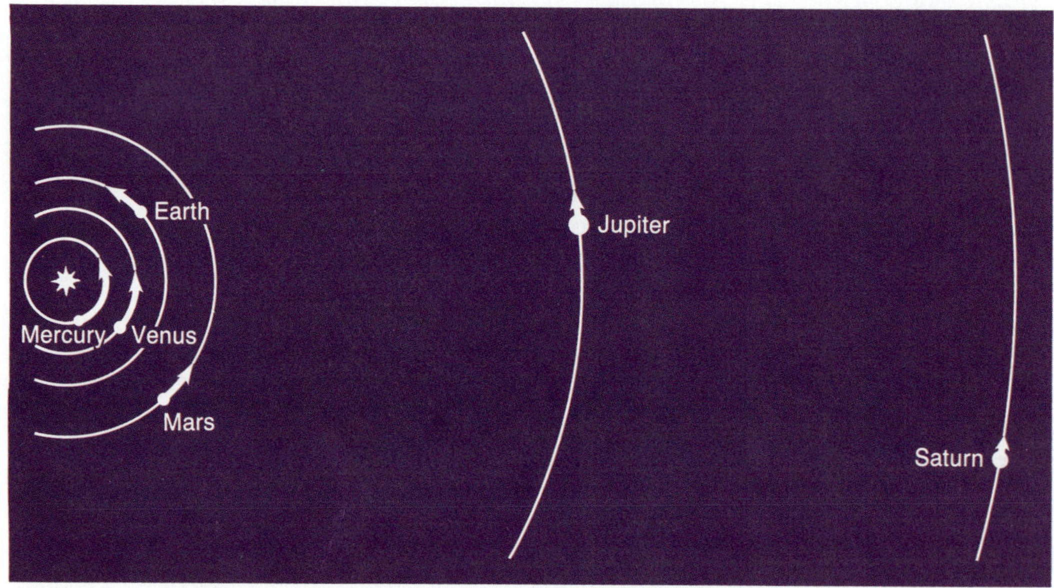

Figure 2–4. *The naked-eye planets, shown in order from the Sun, with their relative orbital speeds indicated by the length of the arrows.*

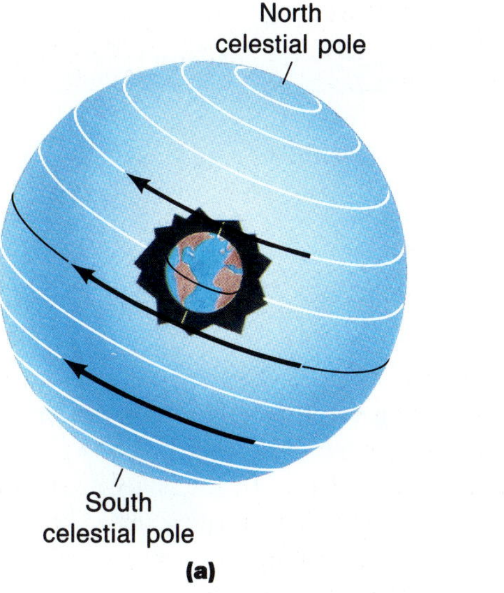

Figure 2–5 *The celestial sphere rotating toward the west around a stationary Earth (a) produces the same observed motion of stars as the Earth rotating under a stationary celestial sphere (b).*

(see Figure 2–5). What about the great wind that should be produced by a moving Earth? To solve this problem, Copernicus stated as an assumption that the air around the Earth simply follows the Earth around. How high above the Earth does this air extend? Copernicus does not answer the question, but notice that his theory requires that the air does not extend all the way to the stars, or even to the Moon.

If the air extended to the Moon, it would tend to drag the Moon along with the Earth.

The Motion of the Sun Among the Stars

Figure 2–6 shows the Earth in a circular orbit around the Sun. Its path does not show as a circle in the drawing because we are seeing it at an angle. If our point of view in the drawing were straight above the system, the Earth's orbit would appear as a circle and the band of stars would likewise appear as a circle. Notice in the drawing that when viewed from Earth, the Sun can be thought of as located among certain stars. As the Earth continues around the Sun, the position of the Sun among those stars appears to change. The Sun seems to be moving across the background of the stars. This, of course, is what is observed. Ptolemy explained it by having the Sun revolve around the Earth independent of the celestial sphere. Copernicus explained it by having the Earth move around a stationary Sun.

It is interesting to note that Copernicus retained many of the Greek ideas. For example, he pictured the Earth as lying on a sphere that rotates around the Sun, for the sphere was considered necessary to hold the Earth in its circular path around the Sun. Today, the idea of an invisible heavenly sphere seems odd, and its use by Copernicus is another indication of the difficulty he had in breaking with the past. In our drawings, we will show the motion of the Earth (and other objects) without including the spheres carrying those objects.

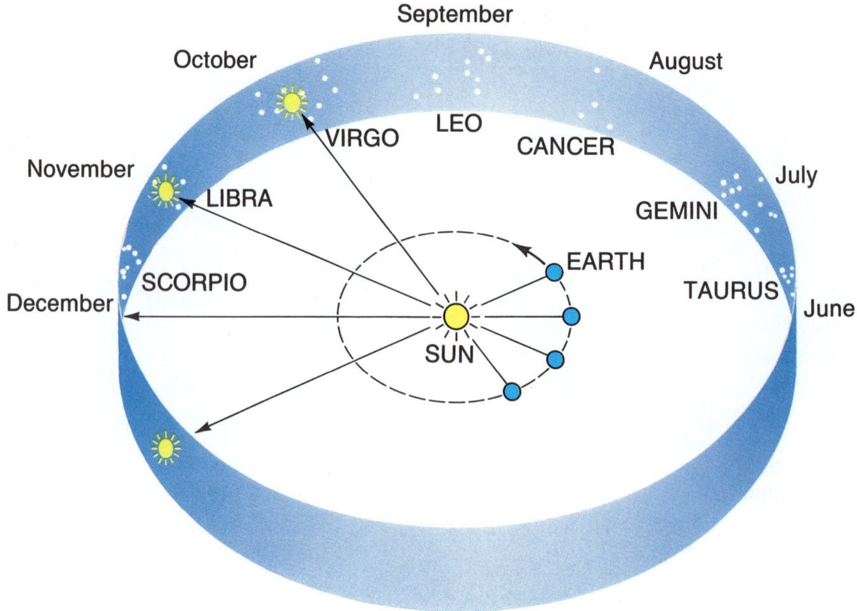

Figure 2–6. *As the Earth moves around the Sun during the year, the Sun appears to move among the background stars.*

As pointed out in the last chapter, the Sun appears to move from among the stars south of the equator to those north of the equator and back again. To fit this observation, Copernicus's model had the plane of the Earth's equator tilted with respect to that of the plane of its orbit around the Sun (Figure 2–7.)

Look carefully at Figure 2–7 to see that the tilt of the Earth's axis would cause the ecliptic to be sometimes above and sometimes below the celestial equator.

MOTIONS OF THE PLANETS

As stated in the last chapter, someone watching the planets from Earth would see that these wanderers spend most of their time moving from west to east against the background of the stars. How does Copernicus' model handle this motion? Figure 2–8 shows three positions of the Earth in its orbit, along with the corresponding positions of Mars. The lines from Earth to Mars illustrate the direction in which we on Earth see Mars. As time goes by, Mars appears to move among the background stars on the figure. Now picture a globe in the position of the Earth, and you will see that this motion of the planet is from west to east among the background stars. The model fits.

A greater problem presented by the planets is their retrograde motion. On a schedule that is regular, but different for each planet, the planets stop their west-to-east motion and move for a time in the opposite direction. It was this retrograde motion that forced Ptolemy to introduced his epicycles, and it is here that we see a major difference between the two models. Copernicus explained retrograde motion in an entirely different way.

Figure 2–9 shows the Earth and Mars as they continue their motion beyond that shown in the previous figure. Look at what happens when Earth passes Mars. Mars no longer seems to be moving among the stars in the same direction it was before. It seems to have reversed its direction. This means that while the

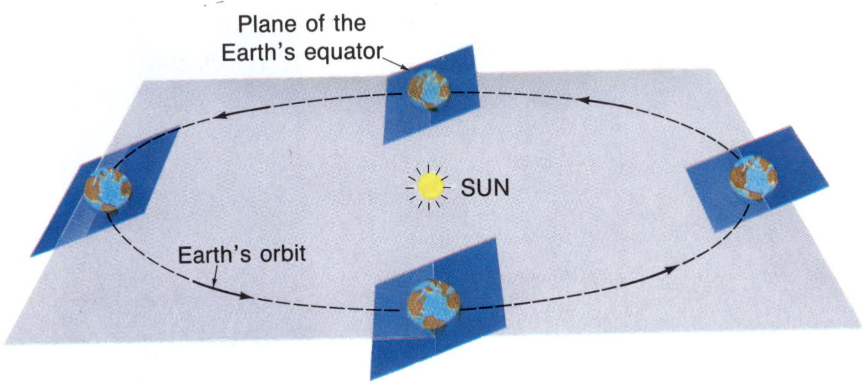

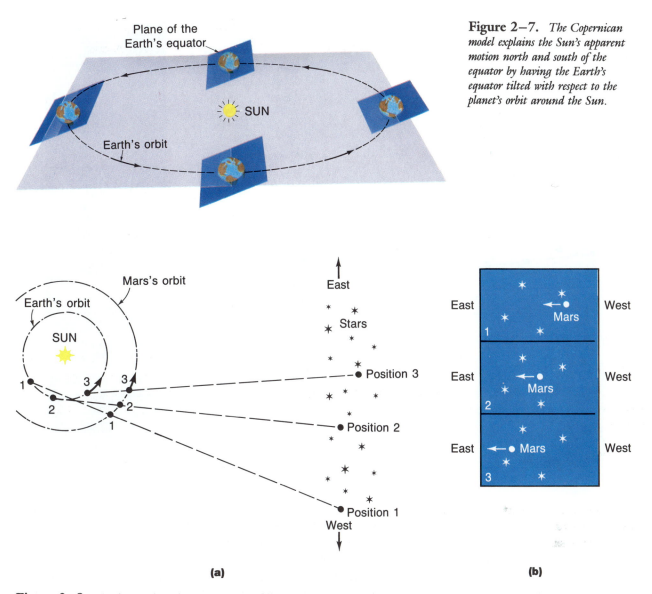

Figure 2–7. *The Copernican model explains the Sun's apparent motion north and south of the equator by having the Earth's equator tilted with respect to the planet's orbit around the Sun.*

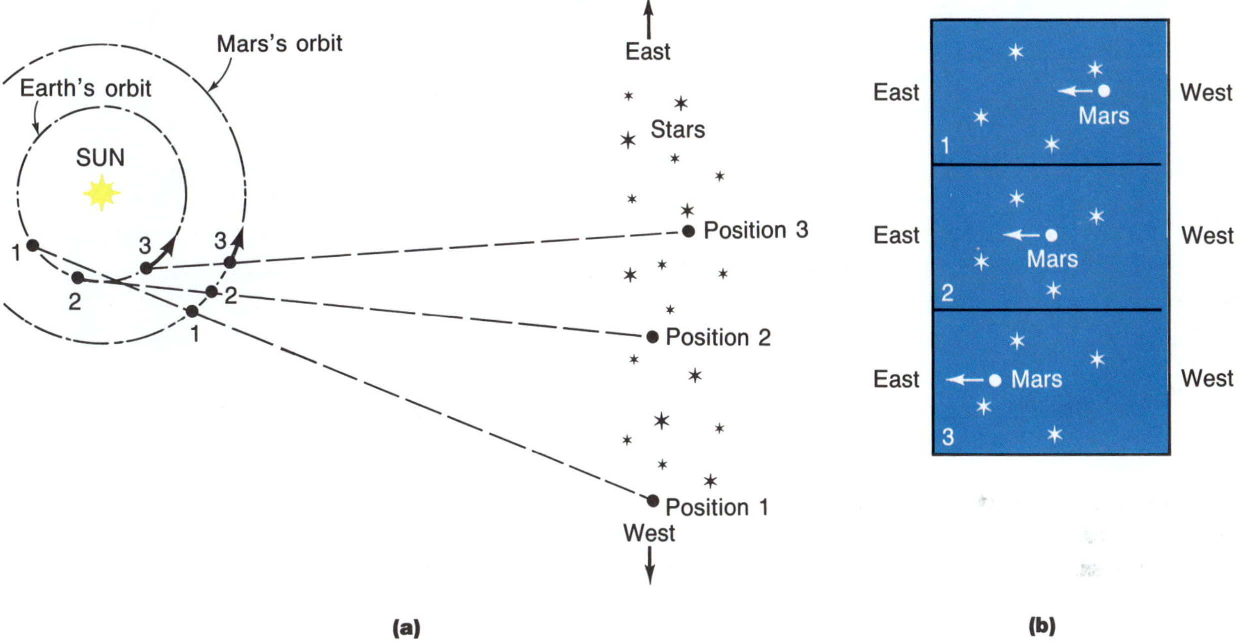

Figure 2–8. *As the Earth and Mars move, earthlings see Mars move among the stars. During the time of the three positions shown, Mars is moving up among the stars in (a) (eastward on the celestial sphere). This causes our view of it to change as shown in (b).*

Earth is passing Mars, Mars appears to move from east to west. Retrograde motion!

To picture how this works, think of riding along a freeway in the fast lane and quickly passing a slow-moving car. Now suppose you watch that car against a distant background; perhaps some trees across a field or even some distant mountains (Figure 2–10). As you are approaching the car, it will seem to move forward along the background. Then as you pass the car, you will see it appear to sweep backwards against the background. In order to see this, you must be sure to concentrate on the background against which you see the car. That is

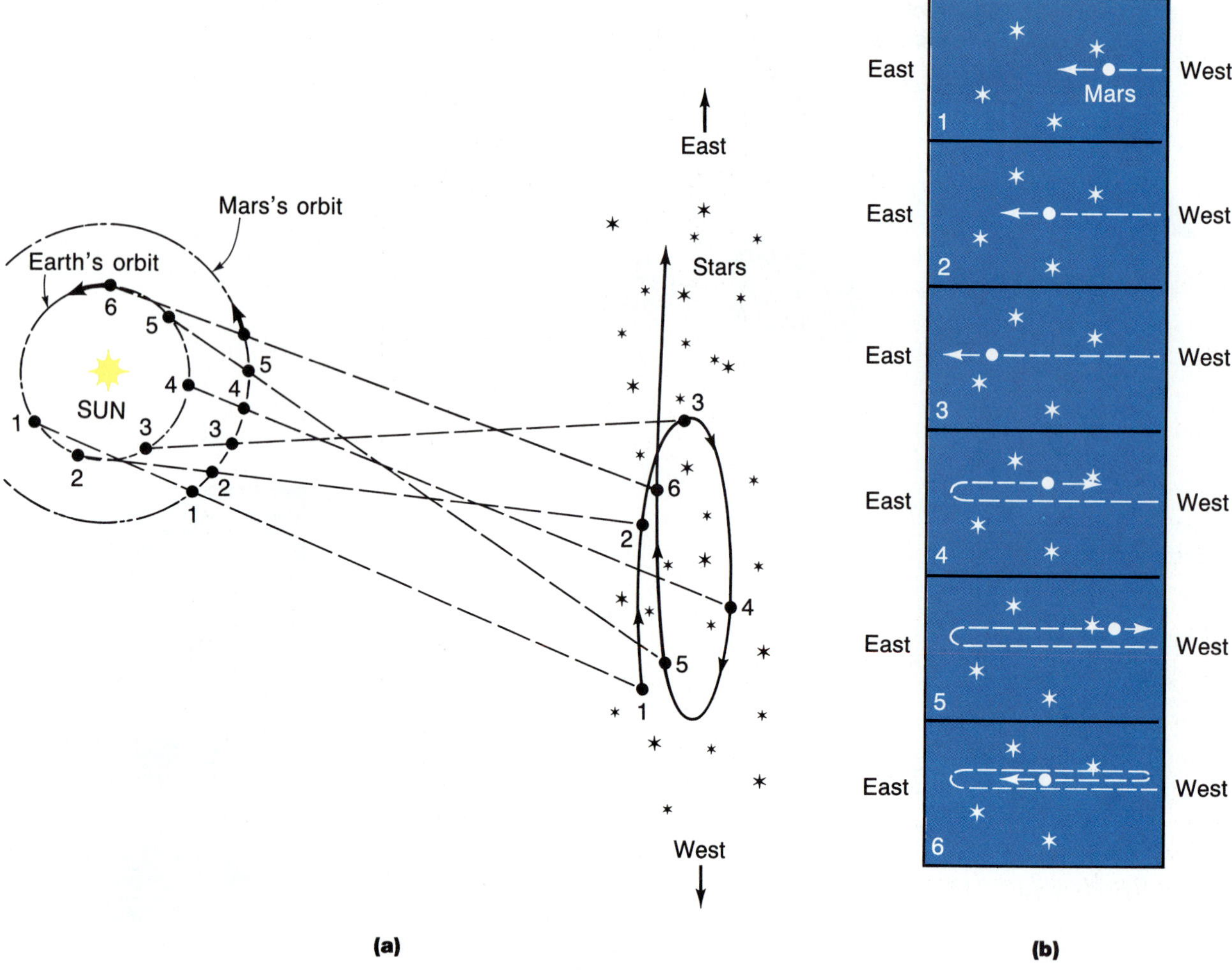

Figure 2–9. *(a) Continuing the motions begun in the last figure, as Earth passes Mars, Mars appears to move backwards among the stars. This is the heliocentric explanation for retrograde motion. (b) The view from Earth of Mars's motion among the stars.*

what we are doing as we observe retrograde motion of a planet: We watch as it appears to move among faraway stars.

The Moon seems to move around the Earth about once each month. Copernicus accounted for this in his model by having the Moon on a sphere that revolves around the Earth in that time. For Copernicus, everything else circles the Sun but the Moon revolves around the Earth.

ACCURACY OF THE COPERNICAN MODEL

The heliocentric model as we have described it is able to account qualitatively for the observed motions of the stars, the Moon, and the planets. If we are going to compare it to the Ptolemaic model, however, we must ask how ac-

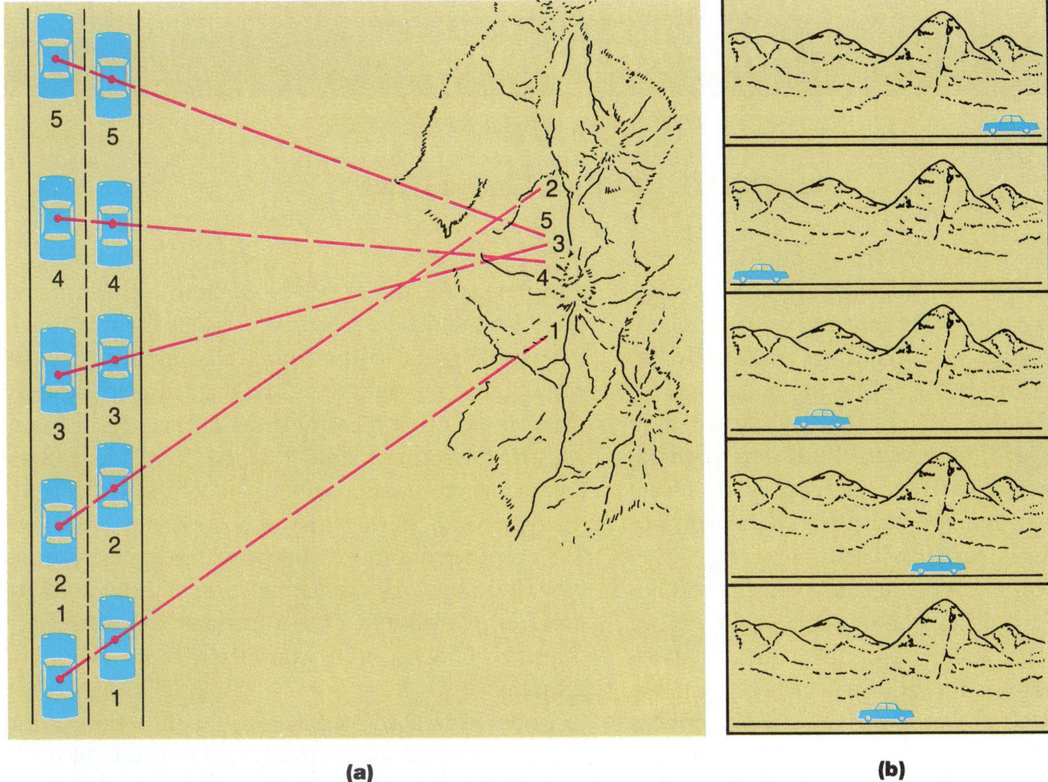

Figure 2–10. *As the car in the left lane speeds up and passes the other car, its driver sees that car appear to move backwards against the distant mountains. Part (a) illustrates the lines of sight, and (b) shows what is seen from the car in the left lane.*

curately it fits the data. To be a good model, it must account for the most accurate positions recorded for the planets. Fitting a model to the observations is particularly difficult in the case of planetary retrograde motions. Do the positions of the planets according to the Copernican model correspond to accurate observations of the planets' positions? Well, not quite. There are differences between predicted positions and observed positions.

Copernicus sought to increase the accuracy of his model by inserting small epicycles for the celestial objects to move on. This was necessary because Copernicus assumed—like Ptolemy—that the planets move at a constant speed, and the observed motions do not correspond exactly to planets moving at constant speed. The epicycles of Copernicus were smaller than those of Ptolemy, and the motion of a planet around them was much slower, so that the planets moved in an oblong path, similar to the exaggerated case shown in Figure 2–11.

Why did he not abandon the idea of circles altogether? The reason lies partly in the fact that the circle had such a long tradition. Beyond this, however, Copernicus felt that there was good reason for using the circle. He argued that the motions in the heavens are repetitive, and "a circle alone can thus restore the place of a body as it was."

HISTORICAL NOTE

Copernicus and His Times

MIKOLAJ KOPPERNIGK (OR NICOLAUS Copernicus by his Latinized name) was born in Torun, Poland, and was educated in Poland and Italy. During most of his life he worked for the church, serving as Canon of the Cathedral of Frauenberg, Poland. Copernicus lived during an exciting—and disturbing—time, as revealed by listing some of his contemporaries: Henry VIII, Martin Luther*, Michelangelo, and Raphael.

As we pointed out in the text, Copernicus was liberally educated in a number of subjects. What, we might ask, was the source of his great interest in astronomy? It was not the field in which he earned a living. The answer is that at the time of Copernicus astronomy was not far divorced from a churchman's everyday life. There were two reasons for this: First, it was important to the church that its feast days be celebrated at the right time, and this required a correct calendar—a matter for astronomers. (The dates of Easter and the feasts that follow it depend even today upon accurate determination of the time of the vernal equinox.) Second, at the time of Copernicus, astrology was considered a more important study than astronomy, but astrology cannot function without accurate data from astronomy. Thus astronomy was a necessary stepsister of what were considered more important subjects.

Copernicus lived long before the invention of the telescope. Measurements of the heavens had to be made with the naked eye, and yet accurate measurements were necessary. One important measurement was the

*Although Luther was himself a revolutionary, he called Copernicus "the fool who would overturn the whole science of astronomy."

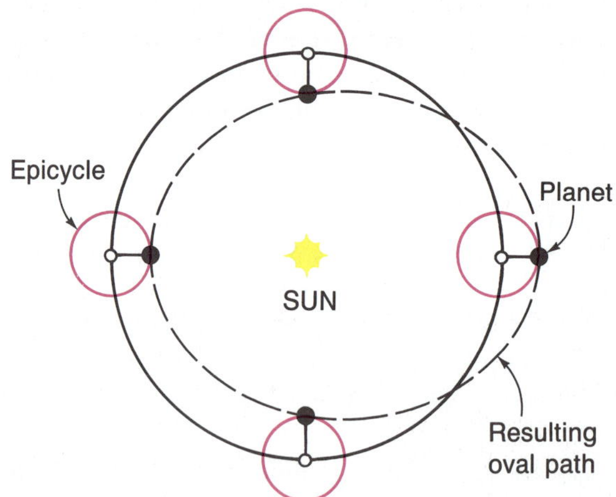

Figure 2–11. *Copernicus added epicycles (whose size is greatly exaggerated here) in order to make the planets move in noncircular paths so that their motion fit the observations. The dashed line shows the path taken by the planet.*

exact time that a celestial object crossed the meridian. (The *meridian* is an imaginary line drawn on the sky from south to north passing directly over the observer's head.) Copernicus made this measurement by having his house constructed with a narrow slot in one of its walls. By placing himself at the correct location and looking through the slot, he could determine accurately when a given object was crossing (or *transiting*) the meridian.

It had long been known that planets change their brightnesses as they shift positions among the stars. The Ptolemaic system accounted for this with its epicycles, for they would cause a planet to change its distance from the Sun and from Earth. Copernicus noticed, however, that Mars seemed to change in brightness even more than would be predicted by that model. The epicycles of Ptolemy were simply not large enough to explain the great changes in Mars's brightness. Thus Ptolemy's model explained the basic phenomenon, but it did not explain it very precisely when accurate data was used. It was this observation that first prompted Copernicus to reconsider the heliocentric system of Aristarchus, which had lain hidden in obscurity for 2000 years.

Copernicus studied the ancient writings and became convinced that a Sun-centered system would not only give better data for use by the church and by astrologers, but that it would be a more aesthetically pleasing system. He linked astronomy very closely with his religious faith, arguing that placing the Sun at the center of everything is completely logical, because the Sun is the source of light and life. The Creator, he said, would naturally place it at the center.

So Copernicus was unable to abandon the epicycles of Ptolemy. And even with his epicycles, Copernicus's model has an error that typically comes to about two degrees over a few centuries of planetary motion. This is about the same error as in Ptolemy's model.

COMPARING THE TWO MODELS: THE FIRST CRITERION

The most important criterion for a good scientific model is the first one: The model must fit the observations. We have shown that both models are able to account quantitatively for all the heavenly motions observed, but that in making a prediction of a planetary position a few centuries into the future (or into the past), each system produces an error of about two degrees of arc. So how do we choose between the two models? One way is to search for more data that might distinguish between them. If we could find an observation that could be explained by one model and not the other, we would have evidence upon which to build a case.

Mercury and Venus

Venus is often called the "morning star" or "evening star" because it is so bright and obvious in the morning or evening sky. It easily outshines surrounding stars.

Recall that Mercury and Venus never get very far from the Sun in the sky. In order to explain this observation, the Ptolemaic theory treated them as special cases by requiring that the center of their epicycles remain along a line between the Earth and the Sun (Figure 2–12). The Copernican system explains the observation in a different manner. As seen in Figure 2–13, neither planet can get very far from the Sun as seen from Earth simply because they orbit the Sun

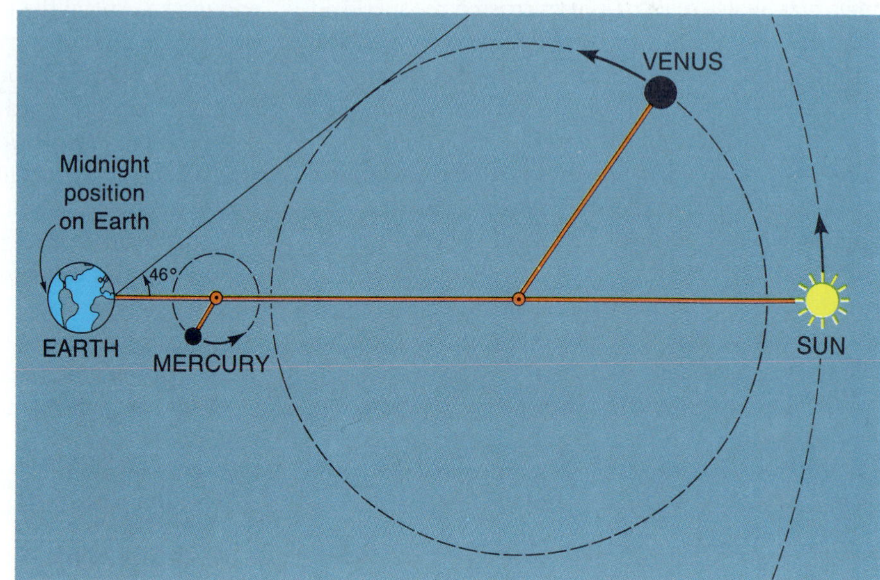

Figure 2–12. *In the Ptolemaic model, the centers of Mercury's and Venus's epicycles stay between the Earth and the Sun. This accounts for the fact that the planets never are seen far from the Sun. Venus reaches a maximum of 46 degrees elongation.*

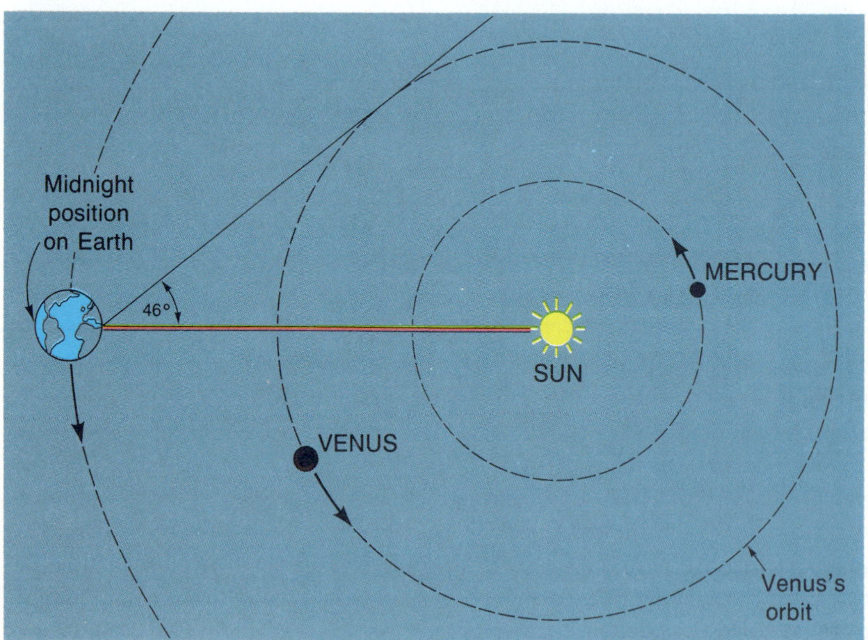

Figure 2–13. *This figure illustrates the heliocentric model's explanation for the fact that Mercury and Venus never get far from the Sun and can never be seen in the sky late at night.*

at a distance less than the Earth's distance. Notice that no matter where Venus is in its orbit, it can never appear farther than 46 degrees away from the Sun as viewed from Earth.

Now refer to the figure and locate the position on Earth at which the time would be midnight. Notice that there would be no way for a person to see Mercury or Venus at midnight. Thus both models are able to explain why the observed motion of Mercury and Venus differ from those of the other planets.

The Aberration of Starlight

As an example of evidence that favors one model over the other, we will present some information that was not available to people of the sixteenth century, both because they did not have instruments sensitive enough to detect it and because they did not have sufficient understanding of the nature and the speed of light. We will begin by considering an analogy.

Suppose it is raining on a calm day, so that the rain is falling straight down. If you stand in the rain and hold a tube vertically, as shown in Figure 2–14(a), raindrops will fall straight through the tube. If, however, you want raindrops to fall through the tube while you are running, you will have to hold the tube at a forward angle, as shown in Figure 2–14(b). This idea is the basis for a test of whether or not the Earth is moving or standing still.

Figure 2–15 is a view of the Earth in orbit around the Sun. As before, we are looking at the circular orbit at an angle so that it appears noncircular. Consider light striking the Earth from a star far above the plane of the Earth's orbit. Because of the Earth's motion, light from the star will seem to act just as the raindrops in our analogy. The figure shows how a telescope would have to be oriented in order for light from the star to pass down the telescope's tube.

Figure 2–14. *(a) The stationary man holds the tube vertically so that raindrops fall straight through it. (b) The moving woman must hold the tube at an angle if raindrops are to fall through it, because she is running forward.*

Figure 2–15. *The aberration of starlight. Because of the Earth's motion in its orbit, a telescope has to be aimed in a very slightly different direction on opposite sides of the Earth's orbit.*

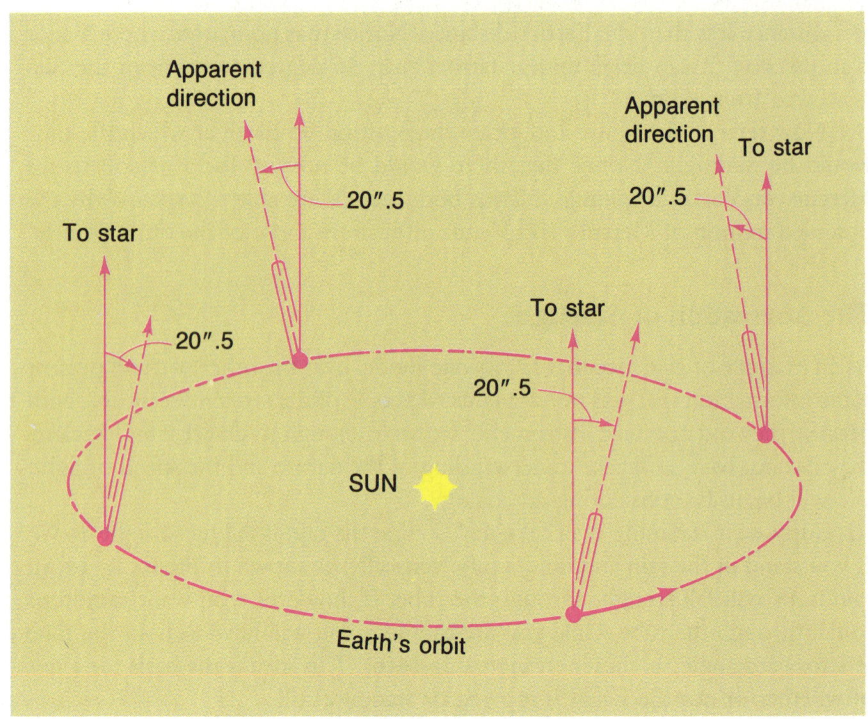

Aberration of starlight: The apparent shift in the direction of a star due to the Earth's orbital motion.

Notice that the telescope must be pointed in a different direction depending upon the motion of the Earth. This phenomenon is called the ***aberration of starlight***.

The Ptolemaic model holds that the Earth stands still, so there would be no aberration of starlight. The Copernican model, however, would predict that aberration does exist. In practice, the phenomenon is very difficult to detect, because the speed of light is so much greater than the speed of the Earth's motion around the Sun. This results in there being very little difference in the direction the telescope must be pointed at different parts of the Earth's orbit. (Actually the telescope must be changed by an angle of only 41 arcseconds when on opposite sides of the Sun.)

In the analogy, the faster you run with the tube, the greater its angle will have to be in order for raindrops to fall through it.

The aberration of starlight was discovered in 1729, and the phenomenon definitely favors the Copernican system, but since it could not be observed in the 1500s, it did not help people of that time to decide between the two models. Before considering the second criterion for a good model, we will look at one more observation we can use today to select the better model.

Parallax

Hold your thumb in front of your face while you close one eye and look at the wall across from you (Figure 2–16). Now, without moving your thumb, view the wall with the other eye. Wink with one eye and then the other. You will

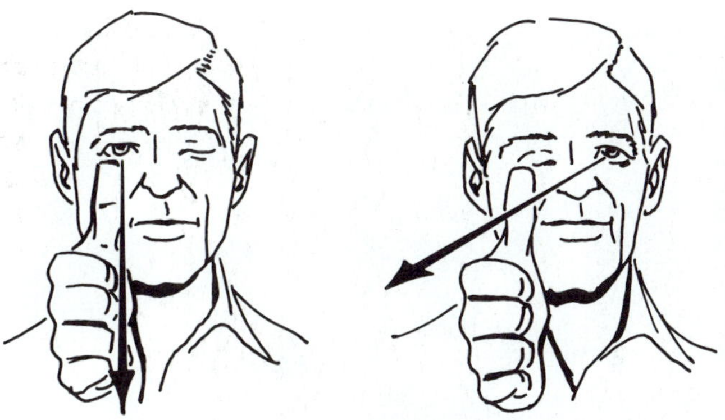

Figure 2–16. *To observe parallax, hold one thumb in front of you and look at it first with one eye and then with the other.*

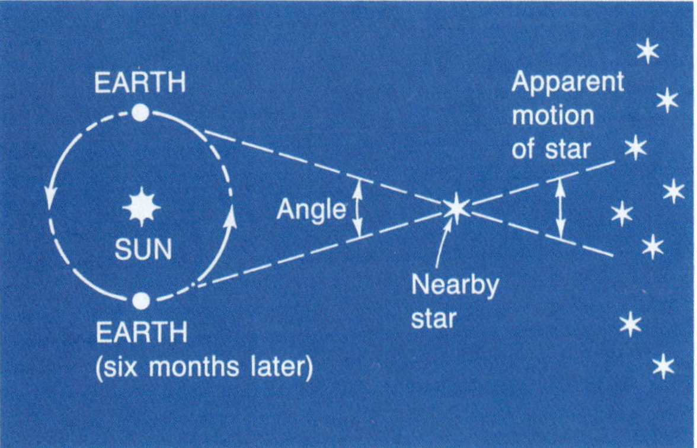

Figure 2–17. *As the Earth goes around the Sun, we see parallax of a nearby star as it shifts its position against background stars. The amount of parallax is very much exaggerated here. For the nearest star, the angle shown is only about 1.5 arcseconds (0.0004 degree).*

see that your thumb seems to move from one spot on the wall to the other. The explanation, of course, is that you are looking at it from a different location each time you change eyes. In general, if our observing location changes, nearby objects will appear to move with respect to distant objects. This observation is given the general name ***parallax***.

Parallax: The apparent shifting of nearby objects with respect to distant ones as the position of the observer changes.

Both the Copernican model and the Ptolemaic model held that the stars are on a sphere a great distance from the Earth. Neither model included the idea that some stars may be much farther from Earth than others. Parallax provided such evidence, as we will see.

Figure 2–17 shows the Earth in June and again six months later. Notice that, just as your finger shifted its apparent position when you alternately winked your eyes, if stars differ in their distances from Earth, a nearby star would be expected to shift its position relative to very distant stars. Such ***stellar parallax*** was first observed in 1838. All of the stars are so far away, however, that the greatest annual shift observed for any star is only 1.5 arcseconds, and this is why stellar parallax was not observed earlier.

Stellar parallax: The apparent annual shifting of nearby stars with respect to background stars. Later this term will be used to refer to the angle of shift.

As we have said, parallax was not observed until it was well established that Ptolemy's model was unsatisfactory, but it serves as a good example of an observation that was contrary to that model. The Ptolemaic model held that the Earth does not move, but the observation of stellar parallax showed that it does. Although it would have surprised Copernicus to learn that the stars are at such different distances, this new idea could have easily been incorporated into his model and in this case his model would have predicted stellar parallax.

The first star observed to show parallax was 61 Cygni, shown in Figure 2–18.

In the 1500s there was no definitive observation that could be made to decide which of the two competing models fit the data more accurately. For that reason, using the evidence available in the 1500s, we still must call our comparison of the models a draw, based on the criterion of data-fitting.

The aberration of starlight was discovered in 1729 when an astronomer, James Bradley, was looking for stellar parallax. He was unable to observe parallax.

Figure 2–18. *Stellar parallax was first observed in 1838. The star named 61 Cygni was seen to change its position annually.*

(a)

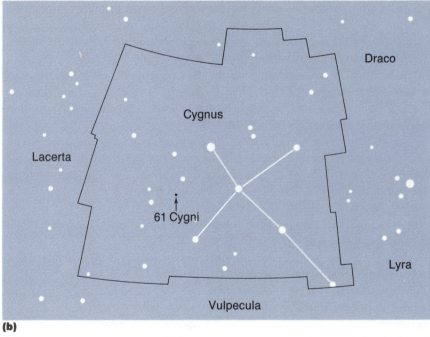

Draco
Cygnus
Lacerta
61 Cygni
Lyra
Vulpecula

(b)

THE SECOND CRITERION

The second criterion of a good scientific model is that it must make verifiable predictions that would allow the model to be disproved. In the last section we saw that each model makes a prediction concerning the aberration of starlight.

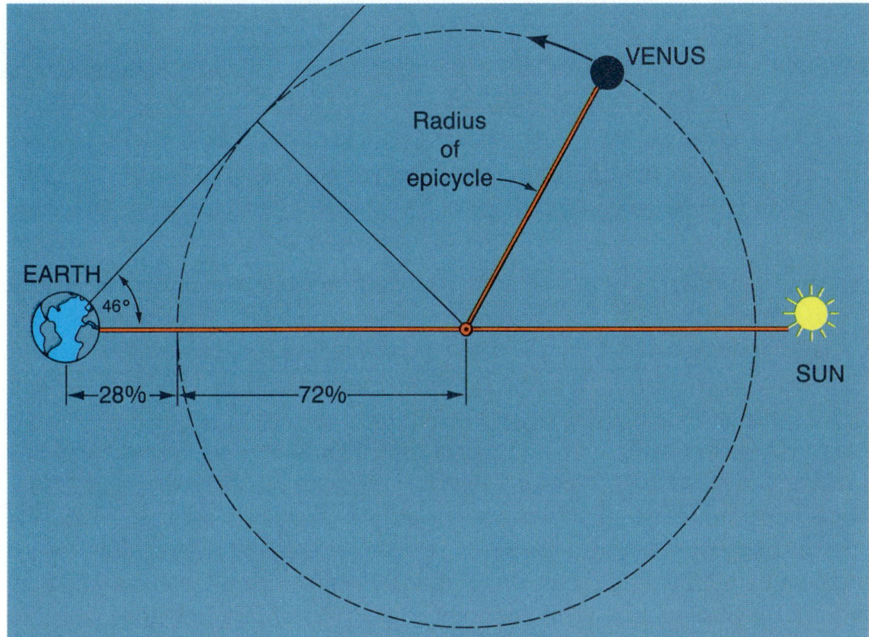

Figure 2–19. *Using the Ptolemaic system, one can determine that the radius of Venus's epicycle is 72 percent of the distance from Earth to the center of the epicycle.*

Indeed, both models are of the type that make verifiable predictions, so both pass this test. But wouldn't all models make this type of prediction? Here is an example of a model that would not be considered a good scientific model because it does not make verifiable predictions. Suppose someone presents a model that says that the planets appear to move the way they do because of magical spirits that move them around at will. Using this model, whatever we later find out about a planet can be explained by saying that the magical spirits are doing it. There would be no way to prove this theory wrong. Whatever observation is made, one could attribute it to the whim of a magical spirit. This model cannot be accepted as a good scientific model because it does not contain within itself predictions that allow it to be disproved.

As shown by the example of aberration of starlight, predictions are inherent in models. They need not be stated by the model's designer.

This seems to be an odd criterion. It says that a theory must contain the potential seeds of its own destruction. If a theory's prediction turns out to be correct, this serves as further evidence that the theory is good, but it does not "prove" the theory. On the other hand, a theory—or at least some aspects of a theory—can always be proven wrong by further evidence. We can disprove a theory, but we can never absolutely prove one.

DISTANCES TO PLANETS

Recall that Venus (and Mercury) are never seen far from the Sun in our sky. In fact, Venus gets a maximum of 46 degrees from the Sun. One can use this fact along with the Ptolemaic system to calculate for each planet the relative size of its epicycle compared to its distance from Earth. For example, it could be calculated that the radius of Venus's epicycle would be 72 percent of the

ACTIVITY

The Radius of Mars's Orbit

LET US DETERMINE THE radius of Mars's orbit from data available to Copernicus. To do so you will make a scale drawing and will assume that the Earth and Mars move in circular orbits. Start by marking a location for the Sun on one side of a piece of paper. As in Figure T2–1(a), draw an arc of a circle about halfway across the paper to represent the orbit of the Earth. Then draw a line from the Sun horizontally across the paper. Suppose that when the Earth crosses your horizontal line, Mars is seen to be halfway around the sky from the Sun, such that the Sun and Mars are on opposite sides of the Earth. (Mars is then said to be in *opposition*.) Mars and the Earth are both moving in the direction shown in Figure T2–1(a). Now suppose that Mars is observed night after night until it is at *quadrature*. Quadrature is defined as occurring when Mars is 90 degrees from the Sun in the sky, as shown in Figure T2–1(c). This is observed to occur 106 days after opposition. This observation, along with the sidereal period of Mars, is all that we need to determine the radius of its orbit. (The *sidereal period* of a planet is the time required for the planet to return to the same position among the stars as viewed from the Sun.) The sidereal period of Mars is 1.88 years.

To draw your figure to scale, you must first calculate the angle through which the Earth moves in 106 days. This is obtained by dividing 106 days by 365 days and multiplying by 360 degrees. Draw a line from the Sun at the angle you have calculated (angle X in part (b) of the figure), and determine the position of Earth (point E) when Mars is at quadrature.

Since when the Earth reaches point E, Mars is at quadrature, you should start at E and draw a line toward the right 90 degrees from the line to the Sun. This line points to the planet Mars. You don't yet know where Mars is located on this line. You can determine this by calculating what angle Mars moves through in 106 days. Use its sidereal period of 1.88 years to calculate this angle. (1.88 years is 686 days, so Mars moves 106/686 of a complete circle in 106 days.) Starting from the Sun, draw a line at the angle you have calculated (angle Y in Figure T2–1(d)). Since Mars lies along this line, it must be located where the two lines cross.

Measure the distance from the Sun to Mars, divide this by the Earth's distance to the Sun and compare your result to that in Table 2–1. Are you close?

distance from the center of its epicycle to Earth (see Figure 2–19). No relationship could be calculated for distances from Earth to the various planets, however. That is, this model does not allow us to calculate how much farther Jupiter is from Earth than Mars is.

Copernicus's heliocentric system, however, does allow such a calculation. We will show how the calculation can be done for Venus or Mercury. A different method must be used for planets whose orbits lie beyond Earth's, and an example for Mars is presented in the accompanying Activity.

With pencil, paper, ruler, and protractor, one can repeat the calculations Copernicus did to determine the relative distances of planets from the Sun. We will assume that the paths of the planets are circles. This assumption will prove accurate enough for our purposes.

We'll determine Mercury's distance from the Sun compared to Earth's.

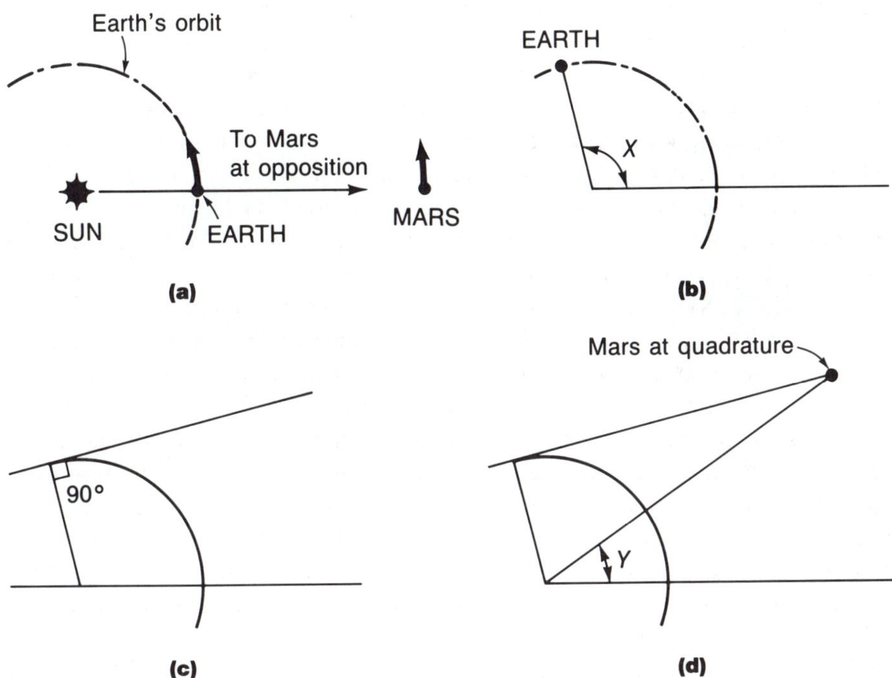

Figure T2–1. *Steps involved in determining the radius of an outer planet's orbit. This figure is not drawn to scale, as yours must be.*

There is only one observation we need in order to do this: the maximum angular separation of Mercury and the Sun. That angle is 23 degrees. To do the calculation, we make a scale drawing, using data for Mercury. We begin, in Figure 2–20(a), by making a dot to represent the Earth and another to represent the Sun and connecting them with a line. In part (b), we start at the Earth's dot and draw a line at an angle of 23 degrees from the line to the Sun (instead of the 46 degree angle of Figure 2–13.) This line represents the greatest angular separation of Mercury and the Sun. All that is left to do is to draw a circle centered on the Sun and tangent to that line (Figure 2–20(c)). Now we measure the radius of that circle and divide it by the Earth-Sun distance on the drawing, obtaining a value very close to 0.4.

The value we have obtained is the radius of Mercury's orbit expressed as a fraction of the distance from Earth to Sun. The distance from the Earth to the

Figure 2–20. *Steps in determining the radius of Mercury's orbit in terms of astronomical units.*

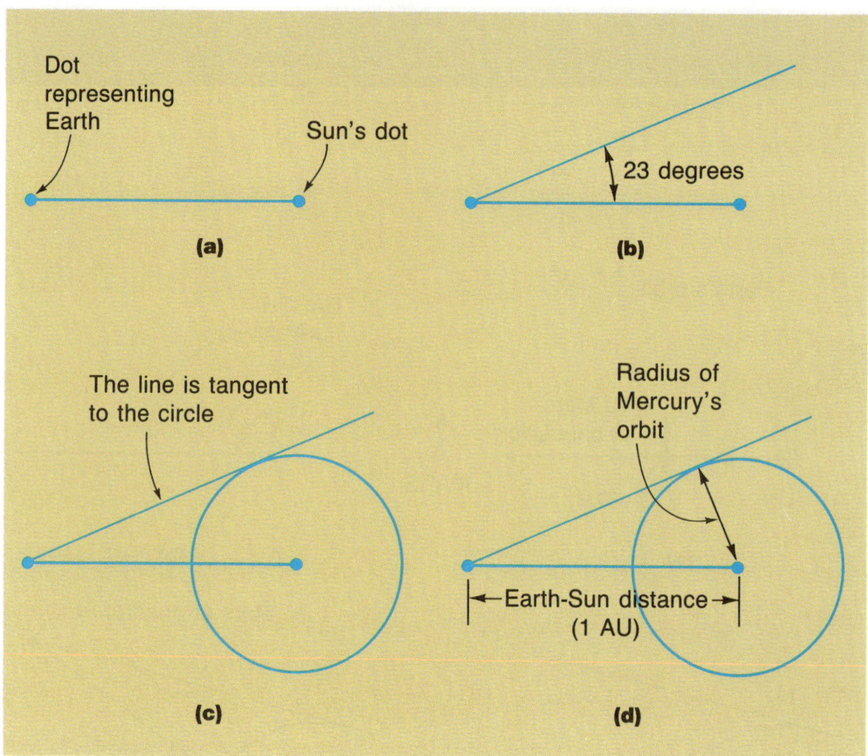

Astronomical unit (AU): A unit of distance equal to the average distance between the Earth and the Sun.

We could have defined "the Mercurian Unit (MU)." This would have made Mercury 1.0 MU from the Sun and Earth 2.6 MU from the Sun. We haven't done anything fishy in defining the AU. Its definition just makes the discussion simpler.

Sun is defined as one ***astronomical unit*** (abbreviated ***AU***). At the time of Copernicus, this distance was not known in other distance units (miles, for example), so all that could be said is that Mercury is 0.4 astronomical unit from the Sun. Today we know Earth's distance, and we can multiply it by 0.4 to obtain Mercury's distance. We still retain the AU as a unit of measure in the solar system, however, for it is a very convenient unit. Table 2–1 lists the distances to the planets that were known at the time of Copernicus, showing the values he obtained for the radii of their orbits as well as today's values.

How do we know the actual distances today? One way to determine this is to bounce radar beams from the planets and measure the time it takes for the beam to return. Knowing the speed of the radar signals, we can calculate the distance to the planet.

Notice that although both models allow a calculation of distance, the Copernican model permits a more system-wide calculation. Using the Ptolemaic model, one could calculate the ratio of the radius of a planet's epicycle to the planet's distance from Earth. The system said nothing, however, about a comparison of the various planets' distances from Earth, except that it did list them in order of distance.

The Copernican model, on the other hand, allows us to calculate the radius of each planet's orbit relative to that of other planets. This feature of the Copernican model makes it more appealing to us as a scientific model and it leads us to the final criterion: aesthetics.

Table 2.1 Comparison of Copernicus's Values of Planetary Distances to Today's Values

Sun-to-Planet Distances		
Planet	*Copernicus's Values*	*Today's Values*
Mercury	0.38 AU	0.387 AU
Venus	0.72	0.723
Earth	1.00	1.000
Mars	1.52	1.524
Jupiter	5.2	5.203
Saturn	9.2	9.539

THE THIRD CRITERION: AESTHETICS

Copernicus worked for years to develop his heliocentric model. He, and other thinkers of his time and the years following his death, preferred it to the Ptolemaic model in spite of the fact that it had no particular advantage as far as the criteria we have discussed thus far. One of the reasons he preferred it was that he thought it was somewhat more accurate (which really is not the case). But his—and others—real reason for preferring this model is due to our third criterion for a good model: neatness, simplicity, beauty, and aesthetic quality.

By the very nature of aesthetics, there can be disagreement over which of two things is more aesthetically pleasing. Let us, however, look at the two models in question and try to make a judgment.

We have already seen that both models are able to explain why Mercury and Venus are never seen high in the night sky. The Ptolemaic model, however, required a special rule to bring this about. In the Copernican model, this was a natural consequence of the fact that Mercury's and Venus's orbits are inside the Earth's orbit. They are not treated in any special manner.

A person on Mars would never see Mercury, Venus, or Earth in his or her night sky.

Ptolemy uses epicycles to explain retrograde motion. Each planet must be assigned an epicycle of a certain size, a speed for its motion around the epicycle, and another speed for the center of the epicycle moving around the Earth. There was no pattern in what speeds were assigned. And, as we have seen, Mercury and Venus were special cases. Copernicus, on the other hand, saw a pattern in planetary speeds. According to his model, the farther from the Sun a planet is, the slower it moves in its orbit. This rule of motion was all that was needed to account for retrograde motion. It was because of this simplicity that Copernicus and others felt that it was a better model for explaining the heavens.

In our discussion of the two systems, we have almost neglected the fact that Copernicus also used epicycles in his model. They were much smaller epicycles than Ptolemy's, but he did find them necessary in order to make his

CLOSE UP

The Criteria Applied to Astrology

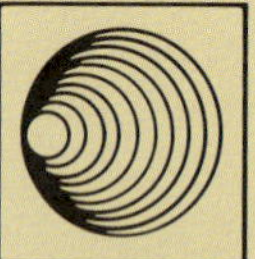

Having discussed astrology in a Close Up in Chapter 1, we will now discuss astrology as a scientific theory, applying the criteria for a good theory. The first criterion: Does astrology fit the observations? It claims to be able to account (at least in part) for our personalities and our characteristics. Can it? This question has been tested many times. Such tests have often been criticized as being unfair to astrology, however, in that astrologers were not involved in designing them.

Shawn Carlson, while a graduate student in physics at the University of California, set out to involve astrologers in such a test. He worked with the National Council for Geocosmic Research, which conducts astrological research and has the respect of astrologers worldwide. Carlson's experiment involved determining the natal charts for 116 persons who also had their personality traits measured by the California Personality Inventory (CPI). Twenty-eight astrologers agreed to see if they could match their interpretation of the natal charts with the personalities as measured by the CPI.

To do the experiment, each of the 28 participating astrologers was given the natal chart of a person along with three CPI results, one of which was for the person whose natal chart was being examined. The astrologer was asked to choose the CPI results that matched his or her interpretation of the natal chart. Each astrologer repeated the test with data from a number of people.

The scientific hypothesis was that the astrologers would be able to choose the correct CPI result in only about one-third of the trials (since there were three choices). Allowing for statistical fluctuation, it was agreed that if the astrologers were correct in more than 51 of the 116 cases, the scientific hypothesis would be shown as probably wrong. Astrologers predicted that they would be able to choose the right CPI result in at least half the cases, about 58 out of 116. Allowance was also made here for statistical variation by saying that if they did not choose at least 45 of the

model as accurate as he could. They were not needed to explain retrograde motion.

Recall from a margin note in the last chapter that Ptolemy's model actually contained further complications, such as the Earth being slightly off-center of the planets' motions. This contributed to Copernicus's conviction that his own system was simpler.

This leaves us with a dilemma: the basic idea of the heliocentric system would have to be judged as neater. But when one considers it in its full detail, it loses some of its beauty. So what is the final verdict? Probably in favor of the Copernican model, though the decision seems less than overwhelming.

Osiander, the writer of the preface of *De Revolutionibus,* however, felt that there was no contest; the Copernican model won in the aesthetics contest. In the preface he gives an analogy to the way that the developers of the Ptolemaic model had to make different assumptions to account for the motions of different planets:

> *With them it is as though an artist were to gather the hands, feet, head and other members for his images from divers models, each part excellently drawn, but not related to a*

116 correctly, the astrological hypothesis would be shown as probably wrong.

Of the 116 choices first made by the astrologers, 40 were correct. This was hardly more than the one-third that would represent complete chance, as predicted by the scientists. Carlson's study also included other tests, and we recommend that you read his report in the December 5, 1985 issue of *Nature*. In essence, the experiment shows that astrology does not fit the data. It fails our first criterion.

The second criterion for a good scientific theory is that it makes verifiable predictions which could possibly prove the theory wrong. There is confusion as to whether astrology does this or not. In Carlson's study, astrologers claimed that if enough cases were considered, they could make verifiable predictions. Astrologers would say that it is not fair to test the accuracy of a horoscope of an individual person, because a horoscope only predicts tendencies, and people do not necessarily follow their tendencies. In other words, you might be very different from what is predicted by your horoscope and still not contradict astrological predictions. If this is the case, there is no way that astrology can be proved incorrect. It therefore fails the second criterion.

The third criterion is one of aesthetics. Astrology doesn't pass this test, for it is not a single, unified theory at all, but is a group of arbitrary rules as to the effect of heavenly objects on people. One could hardly call it aesthetically pleasing.

Since astrology is a faith system it cannot be disproved to those who believe. People who believe in it do not care whether scientists can explain the belief or not.

Scientists do not accept astrology as a valid theory. Some scientists strongly object to their being astrological columns in newspapers, but others claim that astrology does no harm if people use it only as a pastime and do not let it actually determine what they do with their lives.

single body, and since they in no way match each other, the result would be monster rather than man.

Some people at the times of Copernicus saw in his theory a unity which the Ptolemaic theory lacked. One reason they felt so strongly the need for this unity was because of a belief in an orderly Creator. Thus the feeling that an aesthetically pleasing model is preferable to a less pleasing model was at least partly a religious feeling. Scientists today would express their desire for simple models in different terms, but the desire is still present. One of the reasons for the search for a simple model in science is that our experience since the time of Copernicus tells us that when there are two competing models, the simpler one will be more likely to fit new data—data from observations not yet made. That this pattern will continue with new models is a faith different from that of Copernicans. It is a faith in the unity and beauty of nature, and a faith based on past experience.

OPPOSITION TO THE COPERNICAN MODEL

As a scientific model, the Copernican system had an aesthetic advantage over the Ptolemaic system, but it was not readily accepted. The reason is that it seemed to violate common sense. It seemed impossible to people of Copernicus's time that the Earth could possibly be moving because it feels so stationary. Today we have grown up accustomed to the idea of planet Earth hanging in space. We have seen pictures of it taken from space (Figure 2–21), and so the idea of a moving Earth does not seem to contradict our common sense.

This serves as an example of the danger of considering something as nebulous as common sense when evaluating a scientific model. Ignoring common sense is indeed a difficult thing to do, for what we call common sense is simply the sum total of our lifetime learning and prejudice, much of which we have not even tried to evaluate for ourselves and which is very difficult for us to evaluate. We are indeed part of our culture. And if our culture says that the Earth is immovable, it is very unlikely that we will seriously dream that things could be otherwise.

Another reason for people of the sixteenth century to reject the Copernican model is that it was contrary to Aristotle's physics, with its view of natural motions. This had been an established, accepted idea for 1500 years. The Earth was the center of everything. Things fell down to the Earth. Why would they fall to the Earth if it were not the center? Copernicus was in serious disagreement with these accepted ideas.

Finally, we must remember that Thomas Aquinas had successfully incorporated Aristotle's physics into his theological view of the world. Aristotle's ideas were not only part of what we would today call science, but of Christian theology and philosophy besides. The Earth was considered not only to be the physical center of the world, but the philosophical center, the heart of the world, the place where God's all-important human beings reside. If the Earth is simply one of the planets, where does that leave humankind? Are we simply riders on one of five planets, rather than the rulers of creation?

The history of the human dream of being at the center of everything probably dates back to the first primitive tribes. Undoubtedly, they thought that the land on which they lived was at the center of everything that existed. Some very primitive tribes, just discovered in this century, referred to the borders of their known territory as the edge of the world. Other examples come from geographical names—the ancient Chinese symbol for the word *China* was formed from two symbols meaning "middle country;" and *Mediterranean* means "middle of the Earth."

On a more personal level, there is a sense in which each of us thinks of himself or herself as the center of everything. We speak of a conceited person as being "self-centered." One might argue that part of achieving maturity is the recognition that we are not at the center. Humankind has had to—and continues to—mature in much the same manner.

The fact that Copernicus moved the Earth from the center of everything was the primary reason for opposition to his model. Taking the Earth from the center of everything takes humans away from the center. This not only contradicts a possible biblical interpretation, but it seems to damage the self-image of our species.

Figure 2–21. *Photos of Earth were first made only about 20 years ago. Taken by Apollo astronauts, they changed mankind's attitude about its planet.*

It should be easy to see, then, why people did not quickly accept the model of the world with which we have grown up. In fact, the Copernican model had to undergo a major change—and parallel models had to be developed—before the Copernican system would be seen as a model so scientifically superior that it necessarily had to replace the old one.

CONCLUSION

To emphasize the importance of Copernicus's work, it is instructive to consider the word "revolution." Today there are two primary meanings of that word. One is the definition used in this chapter: the orbiting of one object around some point. The other everyday use of the word is in reference to an upheaval, as a social or political revolution. Before Copernicus's book was published, the word had only the first meaning. It was not used to describe upheavals. Recall the title of Copernicus's book: *On the Revolutions of the Heavenly Spheres,* or *De Revolutionibus*. This single book started such an upheaval in people's thinking that the word "revolution" took on a second meaning; the meaning which is perhaps more common today.

Copernicus began the revolution, but it was up to others to carry it forward to its completion. That is the story of the next few chapters.

RECALL QUESTIONS

1. Identify or define: heliocentric, geocentric, astronomical unit, *De Revolutionibus.*
2. Who was Thomas Aquinas, and what part did he play in determining which model was accepted?
3. Who wrote *De Revolutionibus*?
4. According to Osiander, why did Copernicus hesitate to publish his findings?
5. How did the Copernican model explain the daily motion of the heavens?
6. How did the Copernican model explain retrograde motion?
7. Why was Copernicus forced to use epicycles in his model?
8. Name two discoveries made in the eighteenth and nineteenth centuries that the Copernican model fits (or can be made to fit), but the Ptolemaic model does not.
9. What rule did the Copernican model have concerning the speed of one planet compared to another?
10. Define parallax and describe a method of demonstrating it.
11. About how great was the error between the predictions of Copernicus's model and the observed positions of planets?
12. Define *aberration of starlight*. In what century was it discovered?
13. How did each model explain why Mercury and Venus are never seen high in the night sky?

QUESTIONS TO PONDER

1. Was stellar parallax observed at the time of Copernicus? Which model, the heliocentric or the geocentric, was in agreement with this finding?
2. Explain why the Copernican model is generally considered more aesthetically pleasing than the Ptolemaic model.
3. Discuss the apparent need of humans to be at center stage, and the effect of this need on the choice between the two competing models discussed in this chapter.
4. Discuss the reason that today's scientists seek aesthetically pleasing models.
5. Explain why Copernicus was dissatisfied with Ptolemy's model.
6. Explain what is meant when we say that the Sun is "in" a certain constellation, and how the Copernican model explains the Sun's change of position among the stars.
7. Describe some features of Copernicus's model that can be taken as an indication of his reluctance to break completely with the old mode.
8. Compare the two models as to their accuracy in fitting the data available in the sixteenth century.
9. Devise a theory that explains retrograde motion, but which would not be considered a good scientific theory because it does not produce verifiable predictions.
10. When you read of how each of the models explained retrograde motion, you may have found it easier to understand how the Ptolemaic model explained it than how the Copernican model explained it. If so, how can Copernicus's explanation for retrograde motion be called the simpler one?

11. Ptolemy observed parallax of the Moon against the distant stars. Explain how this can be observed. (Hint: It is not due to the motion of the Moon or of the Earth around the Sun.)

12. Why is the Copernican explanation for the unusual motions of Venus and Mercury considered better than the Ptolemaic explanation?

CALCULATIONS

1. For those who have studied trigonometry: Ptolemy calculated the radius of Mercury's epicycle to be 0.39 times the distance from the center of the epicycle to Earth. Use this fact to calculate the maximum elongation angle between Mercury and the Sun as seen from Earth. (Hint: See Figure 2–19.)
2. Calculate the distance to Jupiter in astronomical units. Data: Jupiter's sidereal period is 11.86 years, and quadrature occurs 87.5 days after opposition.
3. Suppose that intelligent beings live on the fourth planet from their star in another planetary system. In terms of the radius of their planet's orbit, calculate the radius of the third planet's orbit, if that planet is never seen at an angle greater than 54 degrees from the star.

CHAPTER 3

Tycho Brahe (1546–1601) was the most accurate observer of planetary positions prior to the invention of the telescope. Perhaps this is because he threw himself into his work so vigorously. As evidence that he took his work seriously, consider that when he was 20 years old he got into an argument with a fellow student concerning some point in mathematics and he ended up fighting a duel over the disagreement. In the duel, a slash of the sword resulted in part of his nose being cut off. He devised a false nose out of gold, silver, and wax and he wore it for the rest of his life.

Tycho's dedication to his work resulted in him becoming Europe's foremost astronomer by age 25. The same vigor continued throughout his life and it apparently extended even to his partying; he died of a burst bladder that resulted from drinking too much at a banquet.

The Triumph of the Heliocentric System

COPERNICUS DIED IN 1543, just as *De Revolutionibus* was being released. As we saw, his model was considered by some people to be better than the old Ptolemaic model, but his model was no better in satisfying the first requirement of a good theory: It did not fit the data any more accurately. So in the mid-sixteenth century, we still did not have a good model for explaining the motions in the heavens. In one sense, all Copernicus had done was to confuse things more.

In this chapter you will study a revised version of the Copernican model. You will see that although this revised model was not perfect, it was certainly a major improvement.

TYCHO BRAHE

Three years after Copernicus died, in 1546, Tycho Brahe (pronounced "bra" or "bra-uh") was born in Denmark. When he was a teenager studying law, he developed an interest in astronomy and in studying that subject he learned that both the Ptolemaic and Copernican models were based on recorded planetary data that were inaccurate. That is, he found discrepancies between different tables of observed planetary positions. He was convinced that before the question could be settled as to which model was better, or before a completely new and better model could be devised, there was a need for more accurate observations of the positions of planets as they move across the background of stars.

People of Tycho's time are often known by their first names. For example, do you know the last name of Michelangelo?

In 1572 Tycho observed what we know today as a supernova ("nova" is from the Latin word for "new"), a bright object in the sky that had not been seen before. It was bright enough to be seen even during the day and its lack of motion across the stars made it appear to be on the celestial sphere rather than being another planet. Tycho observed it for 16 months before it faded from his view. Finding a new object in the heavens was an astounding observation, for it cut at the heart of people's ideas about the nature of the heavens. The ancient Greek idea was that the heavens were perfect and unchangeable, and this notion went hand-in-hand with the Ptolemaic model. Although Tycho's observation was not directly relevant to the solar system debate, it was important for shaping a world view concerning the permanence of the stars. The Copernican model was more comfortable with change in the heavens because it did not draw such a great distinction between the heavens and the Earth. From the time of his observation, Tycho Brahe was hooked on astronomy. Although he developed his own model of the heavens (a cross between Copernicus's and Ptolemy's), we remember him today primarily for his contribution of accurate measurements of planetary positions through the years.

Tycho achieved some fame because of his discovery of the new star, and to keep him from leaving Denmark, King Frederick II built an observatory for Tycho on the island of Hveen. And what an observatory! It contained four observing buildings, a library, shops, and living quarters for students and the staff that maintained the facility. At this time, the telescope had not yet been invented, so all of Tycho's observations were with the naked eye. To get the most accurate observations, he built the largest observing instruments yet con-

Figure 3–1. *This is a mural of Tycho's quadrant, designed to measure the elevation angle of an object in the heavens. The quadrant itself consists of the slot at upper left and the large quarter-circle. Tycho is shown as the observer at the edge of the mural at right center. In addition, he had himself painted on the quadrant (the large figure on the wall near the center) in order to remind his assistants that he was watching. Tycho's quadrant had a radius of over 2 meters.*

structed. Figure 3–1 is a drawing that shows some of his equipment. The large quarter-circle in the foreground is drawn to the scale of the people in front. The person shown partially on the drawing at right center is looking past a pointer on the degree markings and through the slot in the wall at upper left. Because of the large size of the degree marks, Tycho was able to measure angles to an accuracy of better than 0.1 degree, far more accurate than any measurements up to his time and close to the limit the human eye can observe.

In addition to making such accurate measurements, Brahe was careful to determine just how great was the accuracy of each measurement. For example, he would state not only the position of a particular planet, but would also give a value indicating the amount of uncertainty in his measurement. When such data is later compared to a particular model to test the model for accuracy, it is thus possible to know how close the predictions of the model should be expected to come to the measurements reported. If a particular measurement made by Tycho was stated as being accurate to 0.1 degree of angle, then all that could be expected of a model would be that its predictions come within 0.1 degree of that measurement. On the other hand, if the model did not make the prediction within 0.1 degree of the measurement, it would have to be considered as not accurate enough. The inclusion of a statement concerning the accuracy of measurements is now a common practice in science, but it wasn't in Tycho's time.

One tenth of a degree is the angle between opposite sides of a quarter when viewed from a distance of 50 feet.

HISTORICAL NOTE

Tycho Brahe

TYGE BRAHE (TYCHO IS his Latinized name) was born in Denmark on December 14, 1546. His parents had little interest in learning, however, and it is fortunate that Tycho was raised by his uncle, George Brahe, an educated man. On August 21, 1560 the young Tycho saw an eclipse of the Sun. Such an event was regarded at that time as being primarily of astrological significance rather than having any scientific importance. Tycho, however, was greatly impressed that science had been able to predict its occurrence. He resolved to learn how such predictions could be made and he took up the study of astronomy. Then in 1563 a *conjunction* of Jupiter and Saturn occurred. (A conjunction occurs when two planets are seen to pass one another in the sky.) Old astronomical tables were wrong by as much as a month in predicting this event, however, and Copernicus's tables were several days off. Tycho saw the need to improve the accuracy of such tables and he devoted his life to making accurate observations of planetary positions.

In 1572 Tycho observed the first new star that had been reported in 1600 years. This quote shows his excitement at this discovery:

> In the evening, after sunset, when, according to my habit, I was contemplating the stars in a clear sky, I noticed that a new and unusual star, surpassing all the other stars in brilliancy, was shining almost directly above my head; and since I had, almost from boyhood, known all the stars of the heavens perfectly (there is no great difficulty in attaining that knowledge), it was quite evident to me that there had never before been any star in that place in the sky, even the smallest, to say nothing of a star so conspicuously bright as this. I was so astonished at this sight that I was not ashamed to doubt the trustworthiness of my own eyes. But when I observed that others, too, on having the place pointed out to them, could see that there really was a star there, I had no further doubts. A miracle indeed, either the greatest of all that have occurred in the whole range of nature since the beginning of the world, or one certainly that is to be classed with those attested by the Holy Oracles.

Although Brahe is known primarily for his accurate observations, he proposed a scientific model that he thought combined the best features of the Copernican and Ptolemaic systems. In Brahe's model, the Sun and Moon revolved around the Earth, but the other planets

Night after night for 20 years, Tycho recorded data concerning the positions of planets, with particular emphasis on Mars. As I mentioned above, Tycho Brahe is known especially for his accurate observations. The other thing he is most known for today was something he did the year before he died: In 1600 he hired Johannes Kepler as an assistant.

JOHANNES KEPLER

Johannes Kepler was born in what is now Germany, and he concentrated on the study of theology. During this study, he learned of the Copernican system and became an advocate for it.

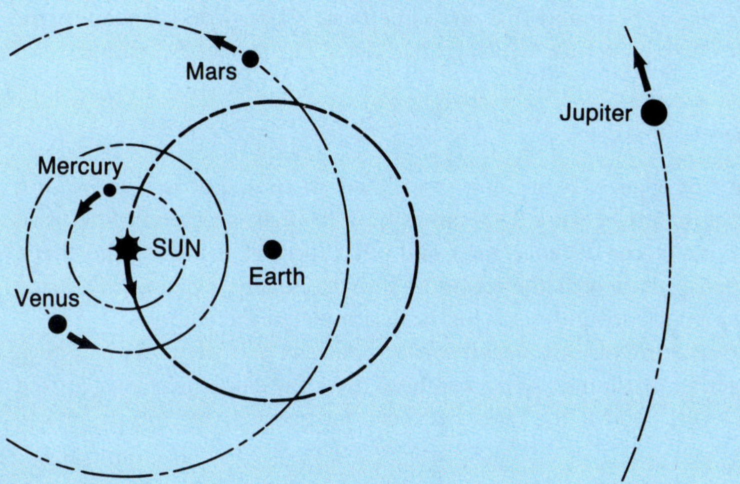

Figure T3–1. *Brahe's model of the solar system was a combination of Ptolemy's and Copernicus's. The Earth is stationary, with the Sun revolving around it. The other planets revolve around the Sun.*

revolved around the Sun. Figure T3–1 illustrates such a system. Notice that it is exactly like Copernicus's model except that in Brahe's system the stationary object is the Earth instead of the Sun. This retained the idea of the Earth being the center of the universe, and it solved many problems that a moving Earth seemed to pose, such as why there is no east-to-west wind due to the spinning Earth and why objects fall straight down rather than falling to the west of where they were released. Tycho's model, however, never got much consideration among thinkers of the time, and it died with him in 1601. (You might read the selection on the opening page of this chapter to see how he died.)

In 1600 Kepler accepted a position as assistant to Tycho Brahe, and Tycho assigned him the job of working on models of planetary motion. After Tycho's death, Kepler took over most of Tycho's records and continued the search he began under Tycho. After four years of work, during which time he tried 70 different combinations of circles and epicycles, he finally was able to devise a combination for Mars that would predict its position—when compared to Tycho's observations—to within 0.13 degrees.

Kepler's accuracy of 0.13 degrees (8 arcminutes) is about one-fourth the diameter of the Moon. Recall that prior to this, a typical error between predicted positions and observed positions was about two degrees of angle. Kepler, however, was not satisfied. The error of 0.13 degrees still exceeded the likely error in Tycho's measurements (about 0.1 degree). Kepler knew enough about Tycho

$$0.13° \times \frac{60'}{1°} = 8'$$

Figure 3–2. *Johannes Kepler (1571–1630)*

Ellipse: A geometrical shape every point of which is the same total distance from two fixed points (the foci).

Eccentricity of an ellipse: The result obtained by dividing the distance between the foci by the longest distance across an ellipse (the *major axis*).

Brahe's methods to know that the data were accurate enough that an error of 8 arcminutes was too much, and he sought a model that would fit the data to the limits of its precision. This is a tribute both to Kepler's persistence and to his belief in the accuracy of Tycho's careful measurements.

Finally, Kepler decided to abandon the circle as the basic motion of the planets and to try other shapes. He tried various ovals, with the speed of the planet changing in different ways as it went around the oval. After another nine years of work, he found a shape that fit satisfactorily with the observed path of Mars. What's more, he found that the same basic shape worked not just for Mars, but for every planet for which he had data. The shape? An ellipse.

The Ellipse

An ***ellipse,*** at first glance, is nothing more than an oval. But it is more; not every oval is an ellipse. Figure 3–3 shows how to draw an ellipse. Stick two tacks into a board some distance apart and put a loop of string around them. Then use a pencil to stretch the string as shown. Finally, keeping the string taut, move the pencil around until you have completed the ellipse.

Ellipses can be of various ***eccentricities*** (flatnesses). If the tacks had been put closer together or a longer string had been used, the ellipse may have looked like Figure 3–4(a). And if the tacks had been farther apart or the string had

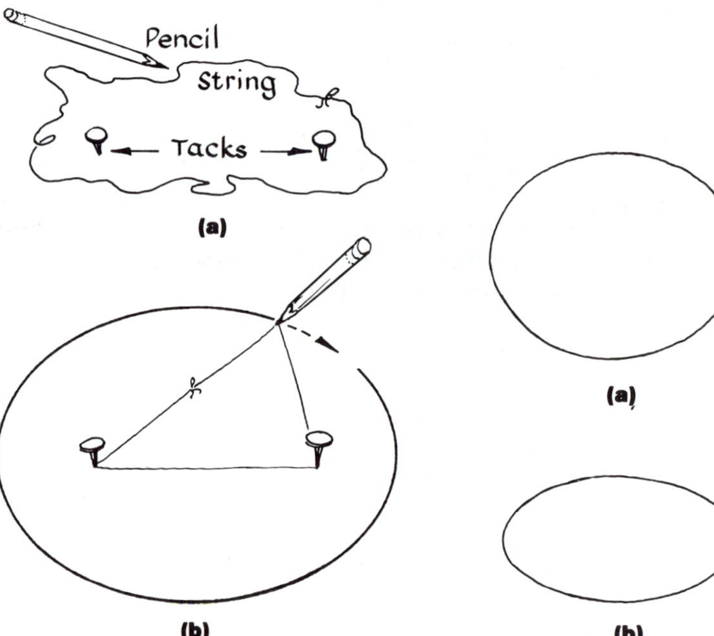

Figure 3–3. *To draw an ellipse (a) start with two tacks, a pencil, and a string. (b) Stretch the string taut with the pencil and swing it around.*

Figure 3–4. *(a) An ellipse with a small eccentricity. (b) A more eccentric ellipse.*

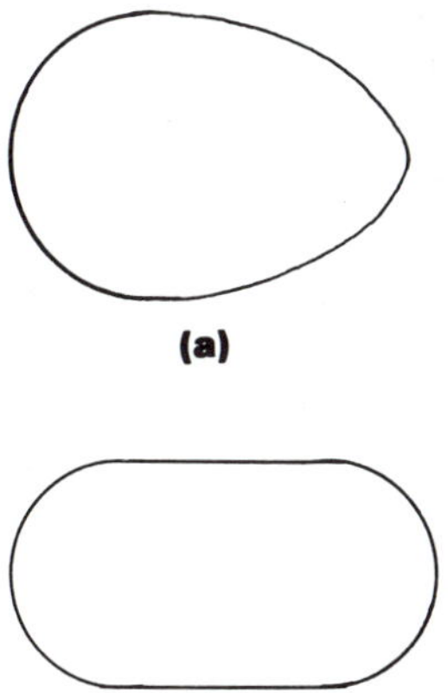

Figure 3–5. *An ellipse is not egg shaped (a), nor is it shaped like a racetrack (b).*

Figure 3–6. *The edge of a round plate appears as an ellipse when viewed at an angle.*

been shorter, the ellipse may have looked like Figure 3–4(b). Although different ellipses have different eccentricities, it is important to see that there are definite rules for the shape of an ellipse. It is not egg-shaped (Figure 3–5a), or two half-circles connected by straight lines (Figure 3–5b).

An example of an elliptical shape you see every day is that of a circle seen at an angle. Figure 3–6 shows a round plate seen in perspective. The shape made by its edge—the shape the artist drew to correctly represent the plate—is an ellipse. The eccentricity of such an apparent ellipse is changed by changing your angle of viewing the plate. Viewing it from above, you see a circle but by viewing it at an angle, you see its apparent shape become an eccentric ellipse.

A circle has zero eccentricity.

Finally, one more point about the ellipse: There is a name for the points where we placed our tacks to make the drawing. Each point is called a ***focus*** of the ellipse (plural: "foci," so an ellipse has two foci).

Focus of an ellipse: One of the two fixed points that define an ellipse. (See the definition of *ellipse*.)

KEPLER'S FIRST TWO LAWS OF PLANETARY MOTION

Kepler published his first results in 1609 in his book *The New Astronomy*. Today we summarize his findings for planetary motion into three laws, the first of which is:

> **Kepler's First Law: Each planet's path around the Sun is an ellipse, with the Sun at one focus of the ellipse.**

Figure 3–7 shows an elliptical path of a planet around the Sun. We have exaggerated the eccentricity of the ellipse, however, to make it more obvious. The elliptical paths of most planets are not very eccentric; they are nearly circles.

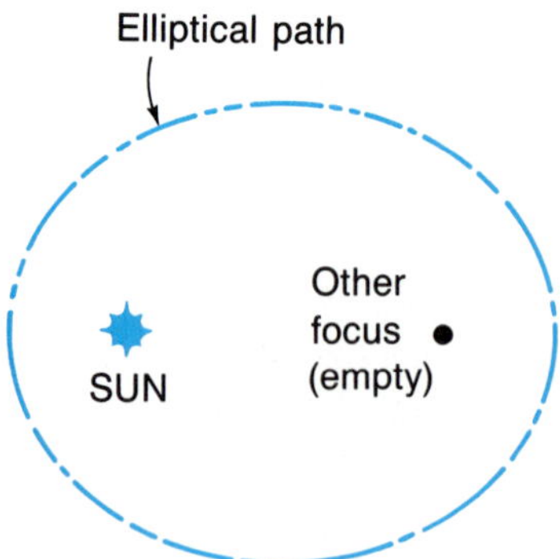

Figure 3–7. *Kepler's first law states that a planet moves in an elliptic orbit with the Sun at one focus of the ellipse. The orbits of actual planets are much more circular than the ellipse shown here.*

The second law tells us about the speed of a planet as it moves around its ellipse. Kepler found that a planet moves faster when it is closer to the Sun and slower when farther away. He was able to make a more definite statement than this, however; one which would allow a calculation of the speed.

> **Kepler's Second Law: A planet moves along its elliptical path with a speed that changes in such a way that a line from the planet to the Sun sweeps out equal areas in equal time.**

The second law takes some explanation. Suppose we consider that the Earth moves in an elliptical path such as shown in Figure 3–8. Further suppose that point *A* in the figure represents the position of the Earth on January 1, and point *B* is its position on February 1. The yellow-shaded area represents the area "swept out" during those 31 days. Then suppose that on July 1 the planet is at point *C*. Kepler's second law tells us that in the next 31 days, the line from the Earth to the Sun must sweep out the same area as it did in January. Thus the Earth must move more slowly at its greater distance from the Sun, so that it gets only to point *D* on August 1. In fact, during any 31 days, the area swept out must be the same.

The time that is applied in Kepler's second law need not be 31 days, of course. The point of the law is that no matter what time interval is selected, if we compare the area swept out by the imaginary line during that amount of time at different places on the elliptical journey of the planet, we will get the same area everywhere. Again, in Figure 3–8 we have greatly exaggerated the eccentricity of the Earth's orbit. In actuality, the elliptical path of the Earth has a very small eccentricity, so much so that it is almost a circle. Copernicus's circle very nearly fits for the path of the Earth.

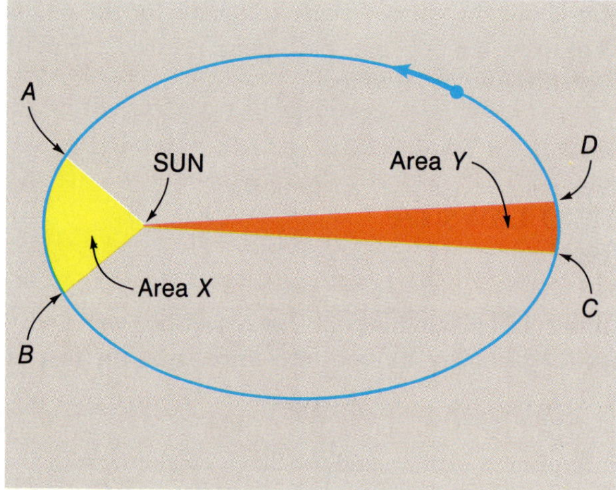

Figure 3–8. *Kepler's second law: A line from the Sun to a planet sweeps out equal areas in equal times. In the drawing, area* X = *area* Y.

KEPLER'S THIRD LAW

The first two laws proposed by Kepler tell us about the paths of each individual planet. Although they propose the same basic shape for the path of each planet (the ellipse), they do not tell us anything about how the speed of each planet along its path compares to that of another. The third law does. Before stating it, however, we need to define what is meant by the period of an object's motion. The ***period*** is simply the amount of time taken by an object—here, a planet—to complete one cycle around its orbit and return to the same position.

Period (of a regularly repeating cycle): The time that elapses between successive repetitions.

Kepler's Third Law: The ratio of the square of a planet's orbital period to the cube of its average distance from the Sun is the same for each planet.

The third law is most easily understood when it is expressed in symbols, as follows:

Let P = planet's orbital period,
d = average distance of the planet from the Sun,
and C = a constant whose value depends on the physical units used for P and d.

$$\text{Then } \frac{P^2}{d^3} = C \text{ .}$$

The distance in Kepler's Third Law is actually the semimajor axis, *or half the length of the major axis of the orbit. This is very nearly the same as the planet's average distance from the Sun.*

Kepler's third law states that P^2/d^3 has the same value for every planet. We'll do a sample calculation to show how the expression is used.

EXAMPLE

Let us use the fact that the period of Mars's orbit is 1.88 years to calculate the average distance of Mars from the Sun. In using Kepler's third law, any units of measure can be used for the period and the distance. We will choose units that make the calculation simple, expressing time in units of the year so that P for the Earth is one year. We could state d in miles, but this would lead to very large numbers, and so we will use the astronomical unit. Recall that one astronomical unit is defined as the average distance from the Earth to the Sun. Now let's apply the calculation to the Earth:

Kepler, of course, did not know the Earth-Sun distance. Today we know that an astronomical unit is about 93 million miles.

$$\frac{P^2}{d^3} = \frac{(1 \text{ year})^2}{(1 \text{ AU})^3} = 1 \text{ year}^2/\text{AU}^3$$

Notice the advantage of choosing the units we did: The ratio is equal to 1 for the Earth. That is, the numerical value for the square of the period is equal to that of the cube of the distance. $P^2 = d^3$ (in years and astronomical units).

Let's apply this to Mars. The third law says that the ratio of P^2/d^3 is also equal to 1 for Mars when the above units are used. We will substitute the value of Mars's period and solve for its distance from the Sun.

$$\frac{P^2}{d^3} = 1 \text{ year}^2/\text{AU}^3$$

$$\frac{(1.88 \text{ year})^2}{d^3} = 1 \text{ year}^2/\text{AU}^3$$

$$d^3 = 3.53 \text{ AU}^3$$

$$d = 1.52 \text{ AU}$$

This agrees with the measured value of the average distance of Mars from the Sun.

Try One Yourself. Calculate the radius of Venus's orbit using the fact that its period of revolution around the Sun is 0.615 year. (Venus's orbit is very close to circular, so that its average distance from the Sun is essentially the radius of its orbit.)

In doing the above "Try One Yourself," you may get a slightly different answer than that given in Table 3–1. The explanation for this lies in the idea of significant figures in measurement and is discussed in a Close Up. Table 3–1 shows modern data for each of the planets known to Kepler. In each case, Kepler's third law tells us that the distance cubed should be equal to the period squared. Compare the last two columns.

Kepler had fairly accurate data for the orbital periods of the planets. The actual distances to planets were unknown to him, but remember that the Copernican system allows us to calculate the *relative* distances to the planets. For example, it was known that Mars is 1.52 times farther from the Sun than the Earth is. This, however, is all that is needed to do calculations with Kepler's third law, because 1.52 is the radius of Mars's orbit in astronomical units. So Kepler didn't need to know the actual distances. To the limits of the accuracy of the data he had obtained from Tycho Brahe, Kepler's third law fit.

KEPLER'S MODEL AND THE SCIENTIFIC CRITERIA

Should we think of Kepler's work as a new model or as a refinement of Copernicus's heliocentric model? It certainly is a heliocentric model but it so drastically changes the model of Copernicus that perhaps we should refer to it as a new model, Kepler's. Let's subject it to the three tests we have discussed for a good scientific model.

Table 3–1 Testing Kepler's Third Law

Planet	Distance from Sun	Period of Revolution	Distance Cubed	Period Squared
Mercury	0.387 AU	0.241 yr	0.0580 AU^3	0.0581 yr^2
Venus	0.723	0.615	0.378	0.378
Earth	1.000	1.000	1.000	1.000
Mars	1.524	1.881	3.540	3.538
Jupiter	5.203	11.86	140.9	140.7
Saturn	9.539	29.46	868.0	867.9

The first criterion is that the model should fit the observations. This was the great success of Kepler's model. It fit the data well. In fact, Kepler's model fit Tycho's accurate data, a test more stringent than had ever been applied to a model before. Copernicus had not even attempted to obtain results more accurate than 10 arcminutes but Kepler had rejected his first model because it was in error by only 8 arcminutes.

The second criterion is that the model should make predictions that can be checked, and that might possibly prove the model wrong. Kepler's laws certainly do this. The third law predicts, for example, not only that new, more accurate data will still fit it but that if a new planet were found it also would follow the period-vs.-distance rule.

We turn to the third criterion, the aesthetic one. Is Kepler's model simpler than those of Ptolemy and Copernicus? Recall that Copernicus's model was simple only so long as we did not require a high degree of accuracy. When accuracy was demanded, Copernicus was forced to use epicycles just as Ptolemy had. Kepler's model has no messy epicycles; no secondary paths for the planets. Every planet, without exception, moves in an elliptical path. One rule, and one rule only, can be applied to tell us how the speed of a planet changes as it goes around the Sun. And his third law even gives us a simple rule relating the periods of the planets and their distances from the Sun.

Philosophy, Theology, and Kepler's System

Although Kepler's model fit the criteria for a good scientific model, it deviated greatly from generally accepted ideas. First, Kepler abandoned the circle. Copernicus had begun a revolution, but he had retained the circle of the ancient Greeks, feeling that it was a necessary part of the nature of the heavens. And Copernicus had followed some older Greek models with the idea that giant spheres carry the planets around. If a planet is associated with the surface of a rotating sphere, it necessarily has a circular motion. Kepler abandoned these spheres along with the circles.

CLOSE UP

Significant Figures

WHEN I, AS A scientist, write that I have measured the diameter of a round table to be 72 centimeters, I am telling you more than just the value of the diameter. I am also telling you something about the accuracy of my measurement method. You would know from my stated result that the measurement was made to the nearest centimeter. Thus I would be claiming that the table is between 71.5 and 72.5 cm in diameter. But how do you know that I did not mean that the desk was *exactly* 72 cm across? For two reasons: First, no measurement can be exact. Every measurement has a built-in error; the amount of error will depend upon the measuring device and the method of measurement. Second, if I had meant that the desk was measured accurately to 1/10 of a centimeter, I would have written the result as 72.0 cm. This would have meant that I claimed the desk's width was between 71.95 cm and 72.05 cm. Notice, then, that a value of 72.0 has a slightly different meaning in science than it does in mathematics. Mathematically, 72 and 72.0 mean exactly the same thing; but in science, we know that the latter value is expressed to a greater degree of precision.

This method of stating measurements has definite implications upon how the numbers are handled. Suppose, for example, that you wish to calculate the circumference of a table that has a diameter of 72 cm. Recall that the circumference of a circle can be calculated by multiplying its diameter by π, which is about 3.1415927. When I use my calculator to do this calculation, I obtain a value of 226.19467 cm. According to my rules, if I write the number this way, I am claiming that the circumference of the table is between 226.194665 cm and 226.194675 cm. Could it be possible that I have such a precise result when I started by simply measuring the diameter to the nearest centimeter? Of course not. How, then, should I express my answer? There is a rule for doing this. To discuss the rule, it is necessary to first define significant figures.

We say that the number of *significant figures* in a number is equal to the number of digits in the number, not counting those which exist only because they are needed to locate the decimal place. According to my statement, the number 72 has two significant figures. If it had been written as 72.0 we would say that it has three significant figures. Thus the number of significant figures in a number tells us something about how precisely the number was measured.

The number 0.0000457 has three significant figures. The zeroes exist only to show where the decimal place is. What about a number like 74000? The num-

When Kepler abandoned the sphere and the circular motion, his model was contradicting some philosophical notions about the nature of the heavens—the perfect heavens. The circle seemed to be the logical motion for planets (and for all of the heavens) because of its perfection. Kepler's ideas contradicted this logic.

What reason would there be for planets to move along elliptic paths? Although the ellipse had been known since Greek times, it would seem a strange choice by the Creator as a path for the planets.

ber of significant figures here is ambiguous. If someone tells me that his car has 74000 miles on it, I would assume that he means only two significant figures; that the car probably has more than 73500 miles and less than 74500 miles on it. But I can't be sure. Maybe it happens to have 74000 miles, within a half mile. In that case the number was meant to have 5 significant figures. This problem of ambiguity is solved in science by using powers of ten notation, discussed in a Close Up in the Prologue. Using this notation, such a number would be expressed as 7.4×10^4 to indicate two significant figures and as 7.4000×10^4 to indicate 5 significant figures. I don't suggest that you write mileage numbers like this, but it does solve a problem when precision is called for.

The rule for multiplying two numbers while accounting for significant figures states that the product of two numbers should contain as many significant figures as the multiplier with the least number of significant figures, or perhaps one more. (The rule for dividing is similar.) Applying this to the value for the circumference of the table with a diameter of 72 cm, we see that we should write it simply as 230 cm, or perhaps 226 cm.

Finally, as another example of the use of significant figures, we will calculate the area of the table. This is π times the radius squared. Calculate the radius by halving the diameter, square that value, and multiply by π. A calculator yields 4071.5041. According to the rule, I should write the result as 4100 square cm, or better yet, 4.1×10^3 square cm. (We rounded the second digit up because of the "7" in the next place.) If we choose to keep an extra significant digit, we would write it as 4070 square cm or 4.07×10^3 square centimeters.

The effect of the rules for significant figures is seen in Table 3–1. The numbers in the last two columns are not exactly equal. Consider the case of Mars. Remember that when its distance from the Sun is given as 1.524 AU, this could be anywhere from 1.5235 to 1.5245. This means that the cube of the distance could be anywhere from 3.5361 to 3.5431. We cannot expect the numbers to match exactly. Measurements are not exact—and the way we state numbers reflects this.

Then consider the law of equal areas. Why would the planets change speeds while moving around their ellipses? One must agree that Kepler's second law is an odd one. It is as if there is a string running from a planet to the Sun, and someone keeps track of the area this string sweeps across. This was an unusual idea; an idea with no relation to anything else in nature, or in philosophy, or theology, or in music, or in any other area of study.

Kepler's third law is almost as strange. He found a relationship between each planet's period of revolution and its distance from the Sun, but the law seems almost like playing magical numbers. Why would nature use such a peculiar rule, with squares of times and cubes of distances? Because of these differences from accepted ideas, we might anticipate difficulty in Kepler's model being accepted.

THE VALUE OF KEPLER'S WORK AND ITS ACCEPTANCE

Kepler himself evaluated the importance of his findings as follows:

6000 years was the accepted age of the universe.

> *What sixteen years ago I urged as a thing to be sought, that for which I joined Tycho Brahe . . . at last I have brought to light and recognize its truth beyond my fondest expectations. . . . The die is cast, the book is written, to be read either now or by posterity, I care not which. It may well wait a century for a reader, as God has waited six thousand years for an observer.*

We might wonder whether Kepler really did not care when someone read his book. There is no evidence that he shunned fame, and the last part of the quote indicates his realization of the importance of his work: In a roundabout way he is making a comparison of his creation with THE creation. Try to imagine the excitement that he must have felt when he realized that he had used observations that had lain hidden since the beginning of time and that he had devised a model that actually agreed with what was seen in nature. Finally, after all those centuries, this model—his model—worked!

Kepler's model had the disadvantage that it did not seem to fit in with any nonastronomical field of knowledge, but it fulfilled its own purpose. We might assume that although his model would not be accepted immediately by the general public, the thinkers of Kepler's time would see its value. This was not the case at all. To see why his model was not accepted, we must consider the person of Kepler and the nature of his writings.

The biographer referred to is Arthur Koestler, whose book, The Watershed, *is a biography of Kepler. Recommended.*

Kepler's methods were far different from what would be called science today. A biographer has compared his method of work to sleepwalking: no system, no order, just hit or miss. In our discussion of Kepler's laws, we organized them much more logically than Kepler himself did. Kepler was a mystic and he believed in numerology. Along with the three laws we have discussed, his books are filled with theorems that today seem ridiculous. For example, he associated with each planet a certain musical tune. Figure 3–9 shows these. This is reminiscent of the "music of the spheres" heard by the Pythagoreans of Greek times.

We normally call these natural satellites "moons," but strictly speaking, "Moon" is the name of the Earth's satellite. Calling Jupiter's satellites moons is like calling all tissues Kleenexes® and all photocopiers Xeroxes®. We do it, nonetheless.

When Kepler learned of the discovery of moons circling Jupiter, he made predictions as to the number of moons around other planets. He based his prediction on geometric harmony. He argued that since the Earth had one moon and Jupiter four, Mars must have two and Saturn eight. He thought that nature would follow the numerical pattern he saw: one for Earth, two for Mars, four for Jupiter, and eight for Saturn. This kind of playing with numbers is considered nonsense today, and it was considered the same by many intelligent

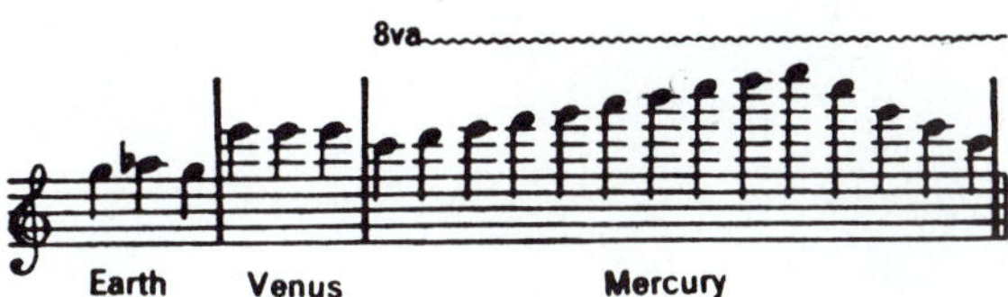

Figure 3–9. *Kepler associated certain musical notes with each planet.*

people of Kepler's day also. We will see in Chapter 7 a similar law that shows a pattern in the various planets' distances from the Sun.

Because of Kepler's strange methods, the laws that today bear his name and that are studied by every astronomy student were so hidden in his books that they received little attention. There was one person in particular who would have profited from Kepler's findings: Galileo.

GALILEO GALILEI AND THE TELESCOPE

Figure 3–10. *Galileo Galilei (1564–1642)*

Discoveries and Opinions of Galileo, *by Stillman Drake, is a recommended source of information about Galileo.*

Galileo Galilei was born in Italy in 1564, three days before Michelangelo died. He lived at the same time as Kepler, and the two men wrote to each other and exchanged ideas. Galileo's contribution to our understanding of the solar system consisted both of observations and theory. In the remainder of this chapter we will discuss his observations, and in the next chapter we will show how he—and later Isaac Newton—made sense of Kepler's model.

The first use of lenses to form a telescope apparently occurred in the Netherlands in 1608, but the instrument was not used to study the sky until Galileo heard about it and, in a surprisingly short time, built his own in 1609. Let's quote Galileo about his first use of the telescope (which had a magnification of three).

> *Forsaking terrestrial observations, I turned to celestial ones, and first I saw the Moon from as near at hand as if I were scarcely two terrestrial radii away. After that I observed often with wondering delight both the planets and the fixed stars.*

Anyone who has been able to use a telescope under a dark clear sky can appreciate Galileo's "wondering delight," especially in view of the fact that he was the first person to see the wonders of which he writes. Here are a few of the things he saw in the heavens:

Mountains and valleys on the Moon (Figure 3–11)

More stars than can be observed with the naked eye

Four moons of Jupiter

Figure 3–11. *Even a small telescope shows features such as mountains and craters on the Moon. This photo, though, was made with the large Lick Observatory telescope.*

Sunspots—dark areas on the Sun that move across its surface

The complete cycle of phases of the planet Venus (similar to the Moon's phases)

Each of these five observations had a bearing on the choice between the geocentric and heliocentric models. We will discuss each in turn.

Mountains and Valleys on the Moon

Galileo must have been astonished at what he saw when he turned his telescope to the Moon. If you have never seen the Moon through a telescope, borrow one—or use binoculars—to look at it. You will find that you can see definite features: mountain ranges, valleys, craters. This was much more astonishing to Galileo than it is to us when we first hear of it, because he had never considered that the Moon had Earth-like features.

Neither the heliocentric nor the geocentric world model made a direct prediction regarding features on the Moon, and so in that sense Galileo's observation cannot be used to make a decisive choice between the two. When we consider the basic philosophy behind each model, however, the Moon's features do make a statement. The Ptolemaic model was based on the idea of the perfection of the heavens and upon there being a basic difference between earthly things and heavenly things. Although the Moon had been considered somewhat on the boundary line between the two, the existence of mountains on the Moon, mountains somewhat similar to earthly mountains, did not fit well with that philosophy. Copernicus broke somewhat with this heaven/Earth distinction but he still retained the perfect circles and spheres in the heavens. Kepler, however, used the "imperfect" ellipse and his theory did not rely so much on the idea of perfection in the heavens. His theory would be more comfortable with Galileo's Moon mountains and although the discovery of Earth-like irregularities on the Moon is not overwhelming evidence to help us choose between the models, it does seem to favor the Keplerian world view over that of Ptolemy.

The Great Multitude of Stars

Galileo's telescope revealed that the sky contains many more stars than had previously been imagined. The use of this observation to help select the better model yields results much like Galileo's Moon observation. Recall that Thomas Aquinas had incorporated the Ptolemaic model into Christian theology. Part of this idea was the centrality of mankind, not only in position, but in importance. Passages in the first book of the Bible can be interpreted to indicate that the stars were put in the sky for the exclusive purpose "to shed light on the Earth" (Gen. 1:17). Why, then, do stars exist that are so dim that they cannot even be seen by the unaided eye? What is their purpose?

The existence of these stars seemed to undermine the Ptolemaic model, and with it a literal interpretation of the Bible. For this reason, there were many people in Galileo's time who simply refused to look through a telescope at the stars. Galileo reports that when these people were offered a look through his telescope they refused, saying that whatever was being seen was caused by the instrument itself. While it is true that Galileo's telescopes had imperfections and sometimes showed ghost images, when used for viewing things on Earth they were perfectly reliable. Why would they suddenly deceive people when pointed to the heavens? This argument did not sway those who refused to look. They argued that these "extra" stars simply could not exist in the sky and that therefore there was no reason to look for them. A remark made by Galileo soon after the death of one of those who had refused to look was typical of him: "[He] did not choose to see my celestial trifles while he was on Earth. Perhaps he will do so now that he has gone to heaven."

Perhaps the reason these people refused to use the telescope was that they were afraid that they would see something that might cause them to reexamine what they believed. There probably have been times in each of our lives when we refused to consider new ideas because we were comfortable with our own and did not want them challenged, but avoiding all new ideas will result in living in a sealed box.

In Galileo's time, the newly invented lenses were also used by magicians and they were therefore associated with unreliable observations. This was a factor in reasonable people refusing to look through Galileo's telescope.

Satellites of Jupiter

In January 1610, Galileo turned his attention to Jupiter. Beside that planet, he saw three very faint stars. They were all in line, two east of the planet and one west. As he continued to observe them night after night, he noted that there were four stars, rather than three and it became clear to him that they were not

ACTIVITY

Features on the Moon

OBTAIN A PAIR OF binoculars and use them to observe the Moon. Good quality binoculars will be equal to or better than the telescopes Galileo used. It is best to observe the Moon within two or three days of its first quarter, when you can see the Moon high in the sky around sunset.

The line separating the dark part of the Moon from the illuminated portion is called the *terminator.* Why can you see more detail along the terminator than at other places?

The major features you will find are craters and *maria* (the plural of *mare,* which is a lunar plain formed by molten lava), but you may also be able to find some mountain ranges, *rays* (streaks of light material thrown out of craters), and *rilles* (winding valleys that resemble Earth's canyons). Can you find any craters that have peaks in their centers?

Use the binoculars to observe the Moon over a few nights. Because of the changing position of the terminator, you'll see an ever-changing landscape.

Figure T3–2 is a simple map of the Moon. Before using it while viewing the Moon, compare it to the photo of the Moon in Figure 3–11 in order to familiarize yourself with translating from the map to an actual view.

The Moon in the figure is inverted, for that is how it appears in an astronomical telescope. If you are using binoculars, you should turn the drawing upside-down to help you determine the names of features you find on the Moon. We will see in Chapter 13 why the inversion occurs in a telescope.

stars at all, but were natural satellites—moons—of Jupiter. Figure 3–12 is a reproduction from his notebook of their positions relative to Jupiter on successive nights. He concluded that they were objects that revolved around the planet just as the planets themselves revolve around the Sun in the heliocentric model. The fact that they never appear north or south of the planet just indicated to Galileo that their orbital plane is aligned with the Earth. Figure 3–13 shows

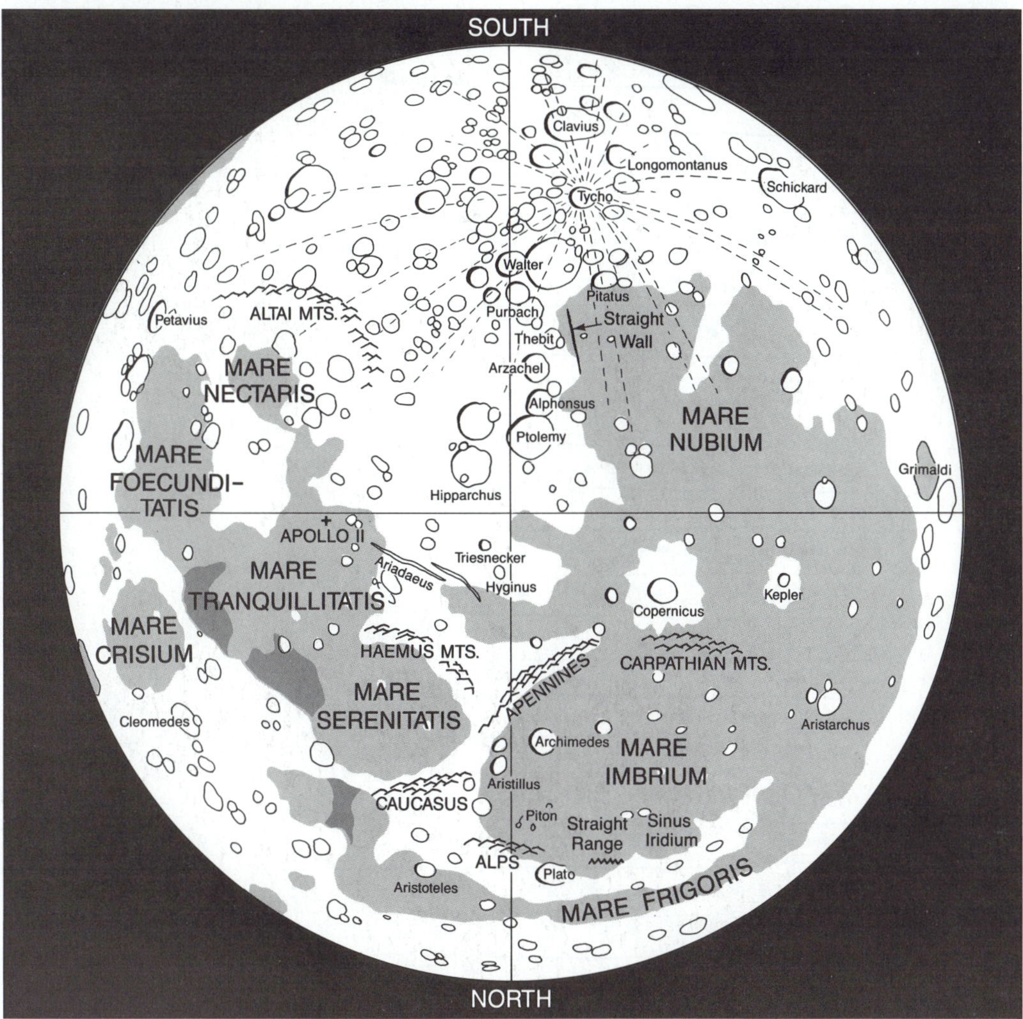

Figure T3–2. *A map of the Moon showing a telescopic view, inverted from what is seen with the naked eye.*

that someone viewing Jupiter's satellites from Earth would see them moving back and forth from one side of the planet to the other.

The Ptolemaic model held that the Earth is the center of everything and that everything revolves around the Earth. Galileo's observation of these satellites indicated otherwise. In the heliocentric system, everything revolves around the Sun except for our Moon, which circles the Earth. It further holds that the

Figure 3–12. *Galileo's original drawings of his first observations of the motions of Jupiter's moons. The writing down the left is the date of the observation, the circles are Jupiter, and the dots (or stars) are Jupiter's moons.*

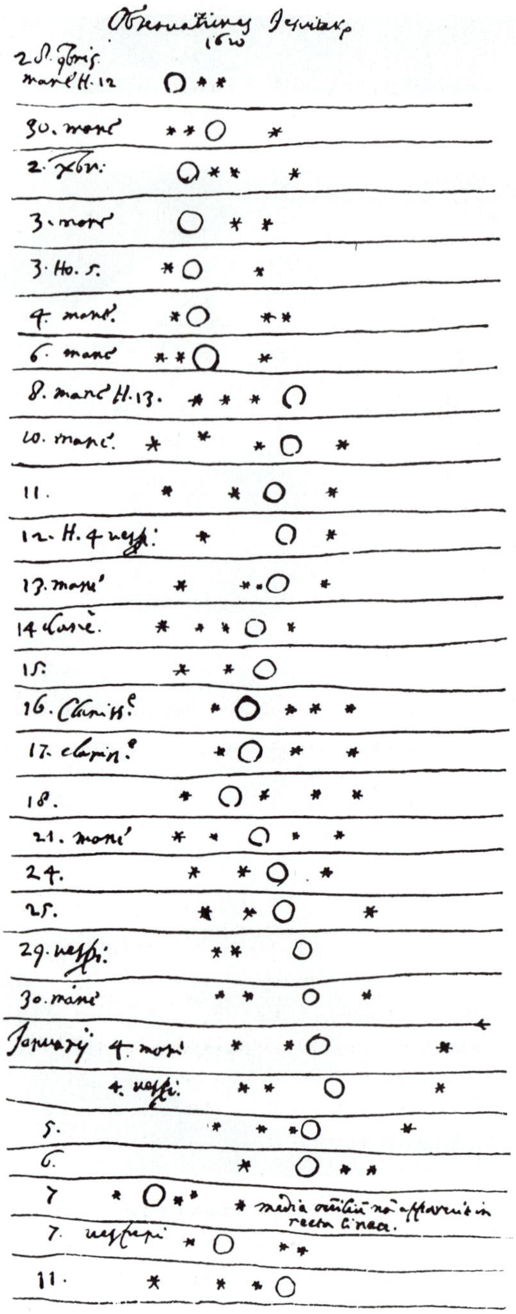

Earth is just one of the planets. We see, therefore, that the heliocentric model is much more comfortable with this Galilean observation of a "solar system" in miniature.

In addition, the fact that Jupiter is able to move through space without leaving its satellites behind conflicts with the Aristotelian/Ptolemaic view, which held that if the Earth moved through space it would leave the Moon behind.

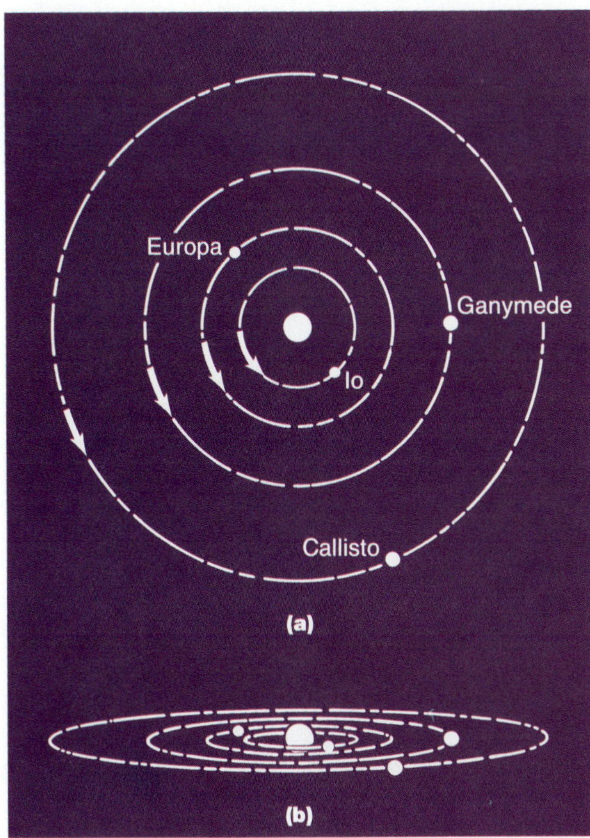

Figure 3–13. *(a) Jupiter's Galilean satellites as seen face on. (b) The same satellites and orbits seen almost edge on. From Earth, this is how this system appears.*

Sunspots

Galileo observed the Sun by using his telescope to project an image of the Sun on a surface (Figure 3–14a). (Viewing the Sun directly through a telescope will destroy your eyesight—immediately!) He saw that there are dark spots on the Sun; that the Sun is not a perfect, uniformly lit source of light. Further, over a period of time these sunspots move across the surface of the Sun, indicating that the Sun is rotating.

This observation by Galileo was not so much an argument for the heliocentric system as it was another bit of data that created questions about the Ptolemaic system. What of the perfect heavens? Was the Sun also "tainted?"

A rotating Sun also made a rotating Earth more plausible.

The Phases of Venus and the Moon

The observation that was most convincing in choosing between the contending models was Galileo's observation of a complete set of phases for the planet Venus. Figure 3–15 shows a number of photos of Venus at different times. Notice how it exhibits phases similar to the Moon. To the naked eye, Venus appears as a dot of light but Galileo's telescope revealed its phases. As we will see, this observation was very important evidence for the heliocentric theory.

Figure 3–14. *(a) A telescope can be used to project an image of the Sun onto a piece of paper. (b) A photo of the Sun showing sunspots.*

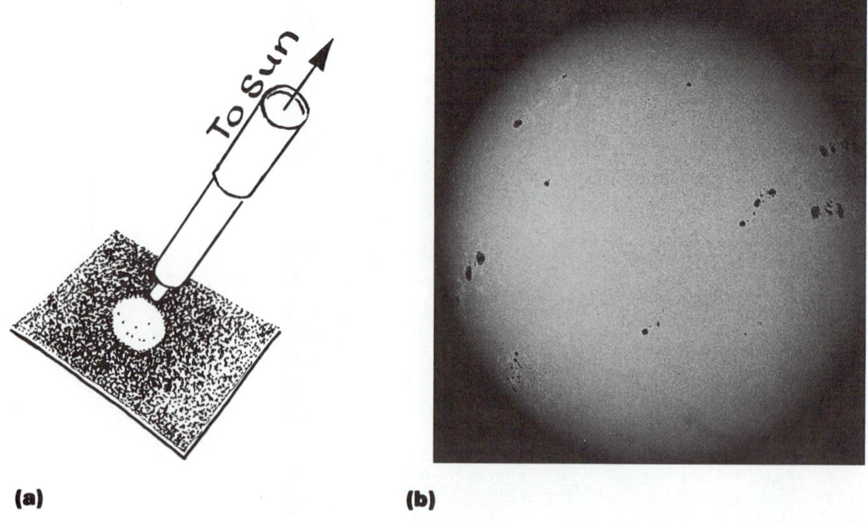

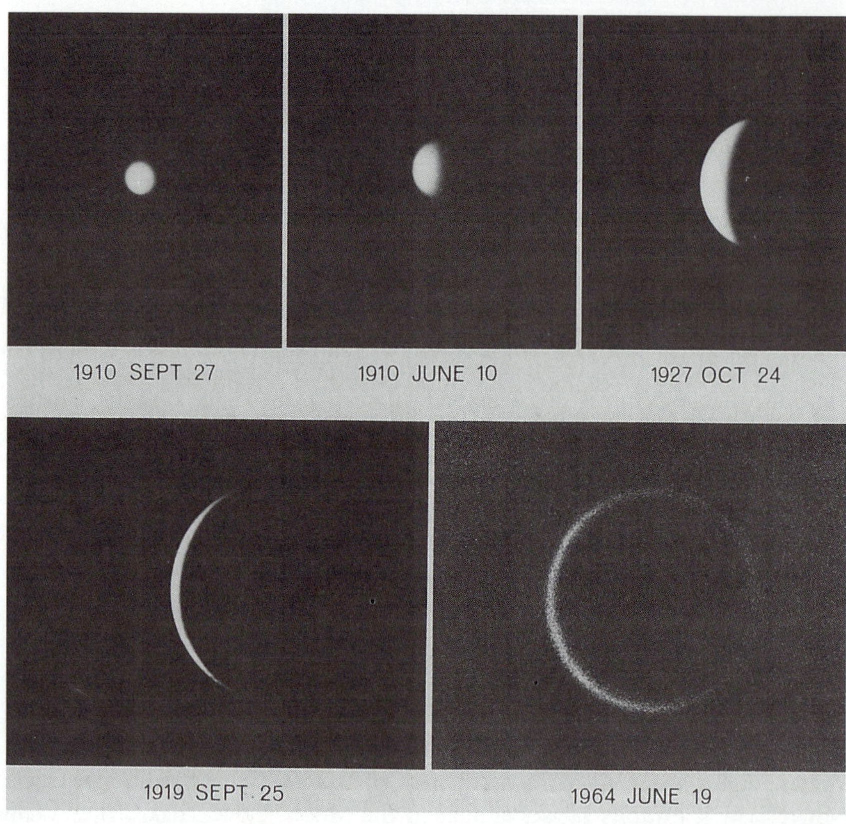

Figure 3–15. *When viewed through a telescope, Venus shows phases. Notice how its apparent size changes as it goes through its phases.*

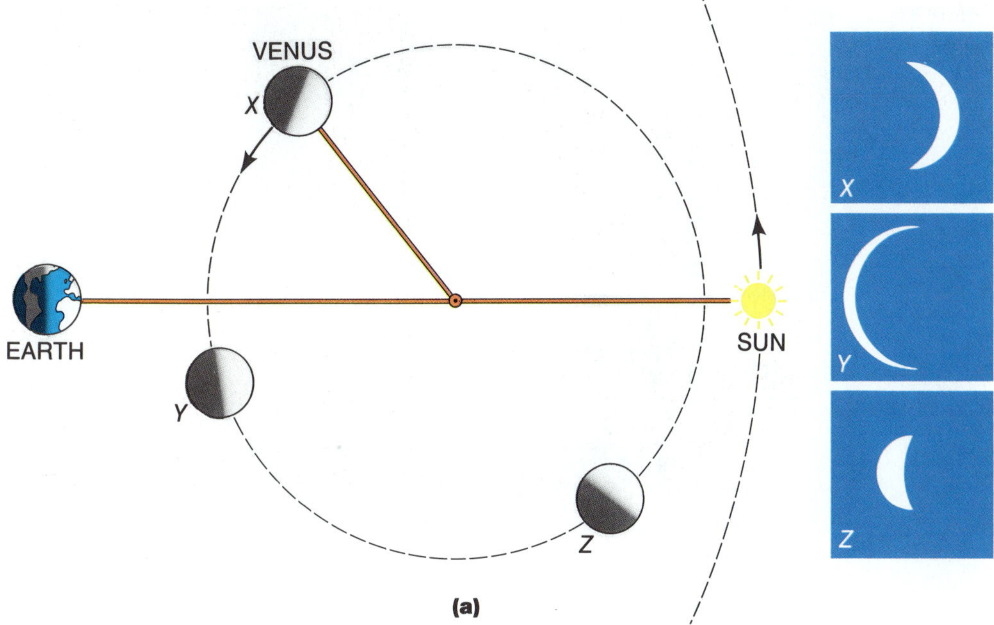

Figure 3–16. *(a) Venus's motion according to Ptolemy. (b) This is how Venus would appear from Earth when it is at each of the three points shown in part (a).*

Phase Predictions Made by the Two Models Recall that repeated observation of Venus leads us to conclude that Venus never gets far from the Sun in the sky. It always appears either in the east shortly before sunrise or in the west shortly after sunset and it is never seen high overhead at night. The Ptolemaic model explained this by saying that the center of its epicycle always remains on a line that joins the Earth and the Sun. Figure 3–16 shows this.

What phases of Venus would be seen according to the Ptolemaic model? Notice that the Sun-lit side of Venus never faces the Earth. At times we should be able to see part of its illuminated surface, but never much of it. Venus should never get beyond a crescent phase. Galileo, however, saw Venus in a gibbous phase; an observation unexplainable by the Ptolemaic model. Let's look at the heliocentric model and see if it can account for the observation.

Figure 3–17 is a diagram of the orbit of Venus according to the heliocentric model. Before reading further, predict in what phase it will appear from Earth when it is at each of the three positions shown in the figure.

At point *X* in Figure 3–17, Venus would be seen in a crescent phase by an Earth-bound viewer. As it moves farther around and reaches point *Y*, it should appear in quarter phase. Then when it reaches point *Z*, it should appear in a gibbous shape, similar to the gibbous Moon.

So the heliocentric system of Copernicus (and Kepler) is able to explain Galileo's observation of gibbous phases of Venus. Here, finally, we have an observation that gives us data for choosing one model over the other. A model must be able to explain the observations. Ptolemy's model cannot explain the gibbous phases that can be observed for Venus. The heliocentric model can.

Figure 3–17. *Venus at various places in its orbit. Predict what phase is seen from Earth in each case.*

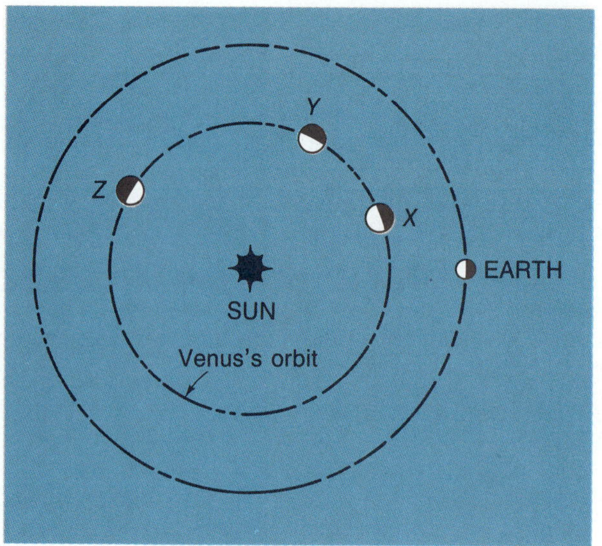

OTHER TELESCOPIC OBSERVATIONS BY GALILEO

When a telescope is used to view the stars, they still seem to be the tiny specks of light we see with the naked eye. They appear brighter but the telescope does not indicate that a star has a definite size. Galileo noted that when he looked at a planet, however, he could see that it appeared larger and not just brighter. The planets appeared as tiny disks. This observation had no bearing on choosing between the two celestial models but indicated another difference between planets and stars.

When Galileo observed Saturn, he noted that it seemed to have two bulges on opposite sides. His telescopes were not good enough for him to see that these were rings around the planet, but he was able to tell that this planet was not simply a circular disk. He thought that the shape he observed was caused by two satellites circling Saturn. (Figure 3–18 is a much better view of Saturn than Galileo obtained with his homemade telescopes.)

KEPLER AND GALILEO

Galileo was a believer in the Copernican system long before he ever looked through a telescope. He was aware of its failings, but he still believed that it was a better system than the geocentric system, because he saw that it was better aesthetically.

Galileo wrote to Kepler: "You are the first and almost the only person who . . . has . . . given entire credit to my statements."

It is interesting that although Galileo corresponded with Kepler and received Kepler's writings that contained what we today refer to as his three laws, he never adopted them. As we explained earlier, Kepler's scientific gems were hidden by a maze of mysticism and numerology. Even Galileo, who had both the interest and the ability to understand Kepler's laws, did not realize the value of his work. In Galileo's studies of Jupiter's satellites, he had observed that the

Figure 3–18. *This is how Saturn appears in a small telescope.*

ones farther from Jupiter revolve more slowly than the closer ones. If he had applied Kepler's third law to this situation, he would have found it in agreement with his observation.

GALILEO'S PUBLICATIONS AND THE CONSEQUENCES

Galileo's work with his telescope confirmed his belief in the Copernican system. In the beginning, however, he was cautious in proclaiming that he was a Copernican. In his book *The Starry Messenger* (1610), he described his observations of Earth's Moon and of Jupiter's satellites but he did not use his findings to support the heliocentric system. The book sold as fast as it could be printed and it resulted in great fame for Galileo. On the opening page of Chapter 4 is a copy of an English translation of the first page.

When Galileo wrote *The Letters on Sunspots* to report on finding the imperfections on the Sun, he declared his belief in the Copernican system. This announcement by Galileo began an uproar which would cause Galileo, and society in general, much grief. One factor contributing to the opposition engendered by the letters was that Galileo wrote them in Italian, his native language, while most scholarly writings of the time were written in the "international" language of Latin. Latin was the accepted language for scholarly papers because it was understood by well-educated people in all countries. The common literate person, however, was less likely to be familiar enough with Latin to read texts written in that language. Galileo choose to write in the language of the people of his country because he was convinced that people other than scholars could understand his arguments and his evidence that the Copernican system was correct.

HISTORICAL NOTE

Johannes Kepler

JOHANNES KEPLER WAS BORN in a small town in southern Germany on December 21, 1571 (or December 27, depending upon which calendar you use—see "Leap Year and the Calendar" in Chapter 1). His father was poor and unreliable, his mother had a violent temper, and Johannes was a very sickly child. When he was four, he contracted smallpox and nearly died. The disease left him with poor eyesight, a condition that prohibited him from enjoying the astronomical sights reported to him by Galileo. So rather than becoming an observer, he made his contributions in theoretical astronomy by applying his great mathematical abilities and his insights to the results of others' observations. The most accurate information was the nontelescopic data from Tycho Brahe, for whom Kepler worked during his last year of Brahe's life. (Tycho hired Kepler in hopes that the mathematician would be able to verify Tycho's Earth-centered model, but Kepler apparently took the position with the idea of gaining access to the accurate data he needed to test his own ideas.)

Today we know Kepler for the three laws to which we give his name, but these laws seem to have been developed almost accidentally as he searched for other, deeper patterns in the heavens. For example, his first great endeavor was to try to find out why the planets were spaced as they were. When he was about 24 years old and was working as a professor of mathematics, an idea occurred to him while he was lecturing. He was discussing the size of the largest circle which could be drawn inside a triangle and the smallest which could be drawn outside the triangle when he wondered if this could possibly be the basis for the spacings of the planets. After working with various shapes, he decided that it was not two-dimensional figures that were significant, but solids. It was known that only five symmetric solids were possible (the tetrahedron, cube, octahedron, dodecahedron, and icosahedron). That number is exactly what would be needed to fill the spaces between six planets. Figure T3–3 shows a construction that illustrates the plan Kepler finally worked out. Kepler might have saved a lot of time if all of today's planets had been known then, for he would have had too many planets for the five solids. (Kepler might have spent the rest of his life searching for the three missing symmetric solids.)

Ptolemy and Copernicus had sought to work out

Primarily because of this controversy, the Roman Catholic Church in 1616 declared that the Copernican doctrine was "false and absurd" and issued a proclamation prohibiting Galileo from holding or defending it. Why did the Church make such a strong statement? We must recall the times, and the fact that the Ptolemaic theory and Aristotle's physics were a definite part of both religious doctrine and the general culture.

It was not only the Catholic Church that voiced opposition to the Copernican theory. Martin Luther, who was a contemporary of Copernicus, had called Copernicus "the fool who would overturn the whole science of astronomy." John Calvin asked, "Who will venture to place the authority of Copernicus above that of the Holy Spirit?" To appreciate why religious leaders were so a model for the heavenly motions, but Kepler went further. He also asked *why* the planets would have such motions. It was Kepler—not Isaac Newton, whom we will discuss in the next chapter—who first hypothesized that there was a force that kept the planets near the Sun: a force we today call gravity. In addition to this force, however, Kepler felt that there must be a force sweeping the planets around the Sun. Just like the gravitational force, this force should get weaker with distance. That is why, he argued, the more distant planets move around the Sun more slowly. It was his consideration of this sweeping force that led him to his third law. Although no such sweeping force exists, it helped Kepler to find a rule which later led Newton to discover the forces that *do* exist.

Kepler remained poor and unappreciated throughout his life. His difficulties were many: His first marriage was unhappy, his wife and one of his children died of smallpox, his mother was convicted of witchcraft, and he was forced to cast horoscopes—which he personally ridiculed—to support himself and his family. His consolation was that he knew that his scientific findings were important and would be recognized after he died.

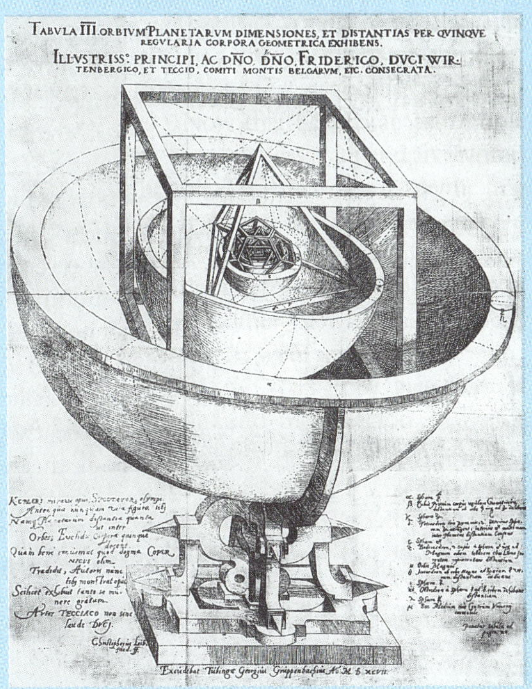

Figure T3–3. *Kepler worked out a plan using the five regular three-dimensional figures to fix the spheres of the five known planets at their correct distances.*

concerned about this issue, we must realize that they considered the salvation of the individual of paramount importance; more important than answering the question of what is the best world theory. They feared that the idea of a nongeocentric world might confuse people as to the supremacy of humankind in God's plan and therefore threaten the salvation of those people.

The Catholic Church became particularly involved in Galileo's case for at least three reasons. First, Galileo sought to be a faithful Catholic all his life. Second, the Catholic Church was a hierarchical and authoritarian church and it had much more authority in Galileo's time than it has today. Third, the Catholic Church was still recoiling from the Reformation and there was a tendency to overreact to anything that could be seen as heresy.

It was during Galileo's time, in 1600, that Giordano Bruno was burned at the stake for heresy.

In 1623 a friend of Galileo was chosen as Pope taking the name Urban VIII. Probably because of this friendship, Galileo was able to persuade the censors to allow him to publish a book that would lay out the two world models in an unbiased fashion. Galileo wrote *Dialogue Concerning Two Chief World Systems* in 1632. In this book there are three characters who discuss the models: Salviati, who favors the Copernican system; Sagredo, who is neutral but ends up being easily persuaded to the Copernican viewpoint; and Simplicio, who defends the Ptolemaic system. Just how unbiased the book was is indicated by the name of the third character. You do not have to know Italian to appreciate the implications of naming a character "Simplicio."

Some of Simplicio's arguments against the Copernican model were those of Pope Urban VIII (often using his very own words).

In the preface of *Two Chief World Systems,* Galileo states that the models should be considered only mathematical fantasy but a reader soon realizes that Galileo considered them more than that. Church authorities considered the book an insult and Galileo was called before the Roman Inquisition, the church's committee charged with examining suspected heresies and heretics. Galileo was charged with violating the prohibition against discussing the issue (though there is controversy even today as to whether such a prohibition was ever made). After a five-month trial, he was found guilty. He was forced to sign a statement that he did not believe in the Copernican theory and he was sentenced to confinement in his villa for the rest of his life. At this time he was nearly 70 years old and in failing health.

Even in the twilight of his life, Galileo made important contributions, writing a book that we will discuss in the next chapter.

LESSONS FROM GALILEO'S CASE

Albert Einstein once gave the following advice: "Do not pride yourself on the few great men who, over the centuries, have been born on the Earth through no merit of yours. Reflect, rather, on how you treated them at the time, and how you have followed their teachings."

Much has been written about the case of Galileo, and we do not have nearly enough space to discuss it fully here. What is important to realize, however, is that the case is not simple. The Copernican model was not the only model that was capable of describing the world and Ptolemy's model did indeed have some advantages over Copernicus's model. Remember that the work of Kepler was not well known and that Galileo's findings were new and controversial. In addition, Galileo was not a diplomatic fellow. He wrote in the language of the people, so that theoretical discussions which in the past had taken place freely among scholars were now exposed for all to see.

On the other hand, church leaders saw these new ideas as a threat both to their own authority and to the authority of the Holy Scripture. By silencing Galileo, they sought to quell these threatening ideas they regarded as false. And perhaps some of their motives were not this pure. There is evidence that part of Galileo's troubles stemmed from a grudge against him by some church authorities.

The problem is not a simple one, but it is hoped that both the Church and science have learned from it.

Issues like the one presented by the Copernican model are not a thing of the past. In this century, the theory of evolution has brought up the same kinds of questions and controversy.

Galileo is reported to have said that "the Bible was written to show us how to go to heaven, and not how the heavens go." He regarded the Scriptures and nature as two representations of the mind of the Creator and was convinced that there can be no contradiction between the two when each is read correctly. To quote him:

Galileo was quoting Cardinal Baronius, the head of the Vatican Library, on the purpose of the Bible.

> *I do not feel obliged to believe that that same God who has endowed us with sense, reason, and intellect has intended to forgo their use and by some other means to give us knowledge which we can attain by them. He would not require us to deny sense and reason in physical matters which are set before our eyes and minds by direct experience or necessary demonstrations.*

If you are a religious person, you must decide for yourself how you read Scripture and how you read nature. As an educated person, you must work out your beliefs for yourself. In so doing, you should attempt to unify what you know in the various fields of knowledge rather than to compartmentalize your knowledge. Saying, "I hold this belief in religion and this contrary belief in science," is not a very comfortable position.

CONCLUSION

We have discussed two primary models: the geocentric model of Ptolemy and the heliocentric model as presented by Copernicus and revised by Kepler. We concluded that Kepler's model fit the data better and appeared to be the better theory. It suffered one flaw, however: Kepler's ellipses and his rules concerning the speeds of planets did not "make sense," in that they did not correspond to anything else in nature.

Finally, we saw that Galileo's observations lent further support to the Copernican/Keplerian theory. In particular, his observations of the phases of Venus should have made it hard to defend the Ptolemaic theory. Nevertheless, there were numerous defenders of that system and although few of them were scientists, they were important people. The world was not ready to accept the new theory. People still clung to their belief in an Earth/heaven system, rather than in a *solar* system.

RECALL QUESTIONS

1. What made Tycho Brahe decide that more accurate data were needed?
2. What observation first brought Tycho fame?
3. What was the major contribution made by Tycho to our knowledge of the world?
4. Explain how to draw an ellipse, including how to draw one with more or less eccentricity.
5. State and explain Kepler's second law, the law of equal areas.
6. What is an astronomical unit?
7. Which criteria for a good model did Kepler's model satisfy better than previous ones had?
8. List five telescopic observations made by Galileo.

9. Most of Galileo's observations argued *against* the Ptolemaic system rather than *for* a heliocentric system. Which was the exception?
10. What difference appears between a planet and a star when they are observed in a telescope?
11. How did the language chosen for Galileo's books have a bearing on the Church's opposition to him?
12. Did Galileo use Kepler's model? Why or why not?
13. List the following men in the order in which they lived: Copernicus, Aristotle, Ptolemy, Galileo, Brahe, Kepler. (Hint: you may have a "tie" between two of them.)
14. List the planets known in Kepler's time in order of their speeds in orbit. List them in order of their distance from the Sun.
15. Copernicus had stated that planets farther from the Sun move slower than nearer ones. What did Kepler's third law add to this statement?

QUESTIONS TO PONDER

1. Why did it take so long for Kepler to abandon the circle as the governing shape for planetary motions?
2. Is a circle an ellipse? That is, would a planet moving in a circular path be violating Kepler's laws?
3. Discuss the importance of Kepler's work in our understanding of the solar system, and the reasons his findings were not immediately accepted.
4. Kepler was a believer in a heliocentric system and he wrote books supporting it. He was not prosecuted by the Catholic Church because he was Lutheran, but his views did not meet much other opposition. Why?
5. In our examples using Kepler's law, we used astronomical units and years in the calculations. Would we have obtained the same results if we had used miles and hours? Why or why not?
6. No celestial model we have discussed makes predictions concerning the smoothness of the Moon or the number of stars in the sky. Why did these two observations by Galileo have a bearing on a choice of models?
7. Figure 3–15 shows that Venus appears to be larger when it is in certain phases. Why does this occur?
8. Suppose you measure the diameter of a round table to be 52 centimeters, and you wish to calculate its circumference. Do you get different results if you use π to be 3.14 rather than 3.1415927? Explain. (See the Close Up.)
9. How would you judge Brahe's model (described in an Historical Note) on the criterion of aesthetics?
10. Figure 3–19 shows Venus's orbit inside the Earth's. At position *X*, would Venus appear in a crescent or a gibbous phase? At *Y*? At *Z*?

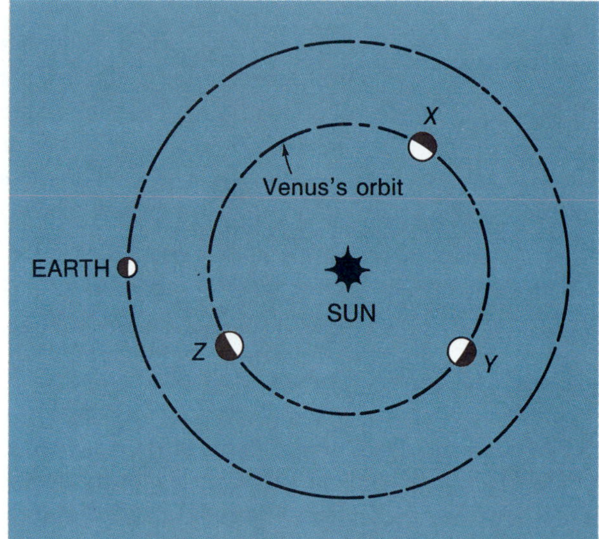

Figure 3–19. *Is Venus in a crescent or a gibbous phase at* X, Y, *and* Z? *(Questions to Ponder #10)*

11. As a planet follows an elliptical path around the Sun, does it move faster when it is closest to the Sun or farthest from the Sun?
12. Is Jupiter ever seen from Earth in a crescent phase? Explain.
13. The Roman Catholic Church reinvestigated the Galileo case a few years ago. Report on the results of this study.
14. The Earth is closest to the Sun in January. Why does this not cause us to experience the hottest weather in January?

CALCULATIONS

1. Suppose that a new planet were discovered at a distance of 7.0 astronomical units from the Sun. What would be its period of revolution around the Sun?
2. Use Kepler's third law to calculate the average distance of Mercury from the Sun, given that Mercury's period of revolution is 0.241 year.
3. The planet Uranus was discovered in 1781. The semimajor axis of its orbit is 19.2 AU. Use this to calculate its period of revolution around the Sun.
4. If the diameter of a circle is measured to be 4.6 inches, calculate its circumference and state it to the correct number of significant figures. (See the Close Up.)

CHAPTER 4

A translation of the first page of a book written by Galileo Galilei:
THE STARRY MESSENGER
Revealing great, unusual, and remarkable spectacles, opening these to the consideration of every man, and especially of philosophers and astronomers;

AS OBSERVED BY GALILEO GALILEI Gentleman of Florence, Mathematics Professor at the University of Padua,

WITH THE AID OF A SPYGLASS recently invented by him, in the surface of the Moon, in innumerable Fixed Stars, in Nebulae, and above all in FOUR PLANETS swiftly revolving about Jupiter at different distances and periods, and known to no one before the Author perceived them and named them

THE MEDICEAN STARS

Venice 1610

Motion, Inertia, and Force

HOW DO WE DESCRIBE motion? What makes things move? What keeps them moving? What kind of motion do falling objects have? These are basic questions we must answer if we are to understand the Earth and its relationship to the planets and stars. To study these questions, we will take a quick look at the Aristotelian solution to the problem and then examine the major contribution made by Galileo. Finally, we will study the work of Sir Isaac Newton, the genius who wove the various threads together into a single cloth of understanding.

ARISTOTLE

There was a theory of motion that prevailed for 2000 years before our modern theory was developed and that old idea is still part of our thinking today. It began with the ancient Greek civilization we discussed briefly in the previous chapters and it was developed by Aristotle in his writings around 350 B.C. Aristotle sought to pull together all of the fields of learning—fields we think of as separate—such as philosophy, natural science, poetry, and architecture.

In accordance with the society in which he lived, Aristotle emphasized order and the natural place of things. He wrote that all matter on Earth is made up of four elements: air, earth (dirt and rock), water, and fire. Things in the heavens are made up of another element, called ***quintessence*** ("fifth element") or ***aether***. The four Earthly elements each have a natural place. The natural place of the element earth is as low as possible. Next in order upward is water, then air, and then fire. This explains why dirt and rocks sink in water: They are seeking their natural place. In fact, Aristotle argued that a heavier object seeks its natural place—downward—with more vigor than does a lighter one, and that the heavier one will therefore fall faster. All elements except fire fall through the air, seeking their natural place (which is below the air). Fire, on the other hand, rises above the air toward its natural place: up. And the fifth element? It is free to do anything.

Quintessence (or aether): In ancient philosophy, the essence of which the heavens are made, higher in value than the four elements of earth, air, fire, and water.

Although these ideas comprise a theory relating to the physical world, they are also in harmony with ideas of society in Aristotle's time. Just as each of the five elements was said to have its natural place, each group in society had its natural place and it was expected to stay there.

The idea of natural place will sound familiar to anyone who has studied slavery and its social consequences.

The natural thing for a moving object to do, according to Aristotle, is to stop. To get an object to move in any direction other than toward its natural place requires that we force it to move. Thus a stone falls because it is seeking its natural place. If we throw a stone, however, we force it to move other than towards its natural place (Figure 4–1).

Motions in the Heavens

In the first chapter we discussed the Greek model of the heavens. The heavens were not expected to conform to the rules described above because the heavens were said to be made of quintessence, an element that had its own natural rules. In the heavens, the natural motion was considered to be circular instead of downward (or upward, in the case of fire-like materials). As discussed earlier,

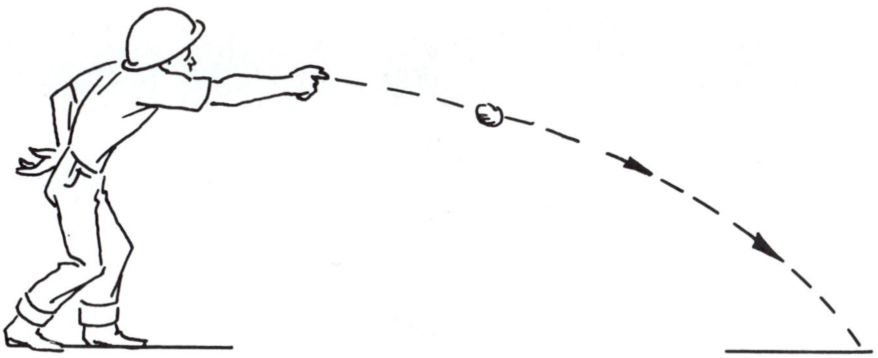

Figure 4–1. *A thrown stone follows a curved path.*

this conformed to the philosophical and theological views of the perfection of the heavens. Aristotle explained that the heavens have different rules than the Earth because they are made of a different natural element.

The Aristotelian model of nature seemed to account for normal observations. In addition, it harmonized with philosophical ideas about the distinction between the heavens and the Earth. Aristotle's theories were so generally accepted for so long (20 centuries!) that by the sixteenth century Aristotle was referred to simply as "The Philosopher." He was quoted as the authority on any number of subjects, from medicine to law to natural science.

Recall that the basic world model of the Greek culture had been challenged by Copernicus and those who followed him, and Galileo's publication of *Dialogue Concerning Two Chief World Systems* in 1632 finally presented compelling arguments (and evidence) that the Greek model of the heavens was not satisfactory. The sixteenth and seventeenth centuries were times of change. Even the theories of "The Philosopher" were questioned in this revolutionary age. The questioning began with Galileo.

GALILEO ON FALLING OBJECTS

After writing *The Two Chief World Systems,* Galileo was sentenced by the Inquisition to house arrest. Although old and nearly blind, he did not quit. He began to concentrate on matters other than astronomy, and his work resulted in a major book, *Discourses and Mathematical Demonstrations Concerning Two New Sciences Pertaining to Mechanics and Local Motion*. It is usually called simply *Two New Sciences.*

Aristotle had argued that if a heavy object and a light object are dropped at the same time, the heavier one will reach the ground before the light one. Galileo stated that experimentation with dropping two stones, one weighing ten times the other, showed that they fall side by side, at least as far as can be observed. Figure 4–2 is a multiple-exposure photo of two objects dropped at the same time. It shows them falling together.

One might argue that the heavier object in the figure seems to be only 10 to 20 times heavier than the lighter one and that if they were more different

HISTORICAL NOTE

Galileo Galilei

GALILEO WAS BORN IN Pisa, Italy, on February 15, 1564, the same year that Shakespeare was born (and that Michelangelo died). He was the eldest of seven children born to Vincenzio Galilei, who was an accomplished musician and the author of works on musical theory. The family had been a very influential and wealthy one during the previous century, but Vincenzio was not a man of wealth. At age 12 Galileo went away to school, studying the usual course of Greek, Latin, and logic. When he was 17, he entered medical school but lost interest in it and took up mathematics. He was forced to leave school after four years and before graduating, however, because of a lack of money.

His creativity began to show when he was in medical school. While there he invented a pendulum device for measuring pulse rates. After leaving school, he continued to study mathematics, applying it to the physics of motion and of liquids. In 1589, at the age of 25, he was appointed professor of mathematics at Pisa, and he spent the next two decades as a college professor.

In his early years, Galileo had little interest in astronomy but medical students whom he taught were required to learn some astronomy in order to use it in medical astrology. He therefore became quite familiar with the Ptolemaic model. In 1597, however, he obtained a book published by Kepler concerning the Copernican theory. He apparently preferred the Copernican theory from the time he learned of it but he kept these ideas secret (except from Kepler and a few others) to avoid controversy. Not until his publication of *Letters on Sunspots* in 1613, well after many of his telescopic discoveries had been made, did he openly espouse Copernicus' ideas.

Galileo was of average height, stout, and had reddish hair. His first biographer, who had been one of his pupils, reported that he was quick to anger but just as quickly calmed. His wit is apparent in his writings and he sometimes used it very sarcastically against those with whom he disagreed. He never married, perhaps because his father died before Galileo was 30 and he was then burdened with financial obligations from his family, particularly from a brother, Michelangelo, who was a musician. He did have a mistress, however, and the two had three children.

Galileo's interests were not confined to science. He once considered painting as a career and a painter of the time, Lodovico Gigoli, credited Galileo with teaching him to use perspective in his painting.

As discussed in the text, Galileo was a very controversial figure and therefore had many enemies. On the other hand, many people appreciated his genius, including the philosopher Thomas Hobbes and the writer John Milton (Figure T4–1), both of whom visited

in weight they would surely not fall together; that a grain of sand would not fall as rapidly as a concrete block. Galileo's reply to arguments like this was that although you might be able to see a small difference in the rate of fall of two objects that are very different in weight, Aristotle had argued that if one object is 1000 times heavier than another, it should fall 1000 times as fast. This is obviously not true. Aristotle's idea, then, must be wrong.

But what about Galileo's statement that the objects fall side by side? Galileo

Figure T4–1. *The English poet Milton visited Galileo in 1638. Galileo is shown pointing something out to Milton, who is using the telescope.*

him while he was serving his sentence of confinement during the last years of his life.

Galileo became blind in 1638 but even then he continued to work, by dictating to others. He died on January 8, 1642 (the same year that Isaac Newton was born) and he was buried in his parish church. A century later his remains were moved into the church and a monument was constructed.

admitted that we might be able to tell some difference in the rate of fall of objects that differ greatly in weight, and he had an answer for why this happens. It happens for the same reason that a piece of paper falls much more slowly than a concrete block: air resistance is involved.

Galileo proposed that it is the air that prohibits a piece of paper from falling faster; that the air has much more effect on the paper than it does on the block. Likewise, any difference between the falling times of the grain of sand and the

Figure 4–2. *This multiple exposure photo of two steel balls falling shows that a large and a small one fall side by side.*

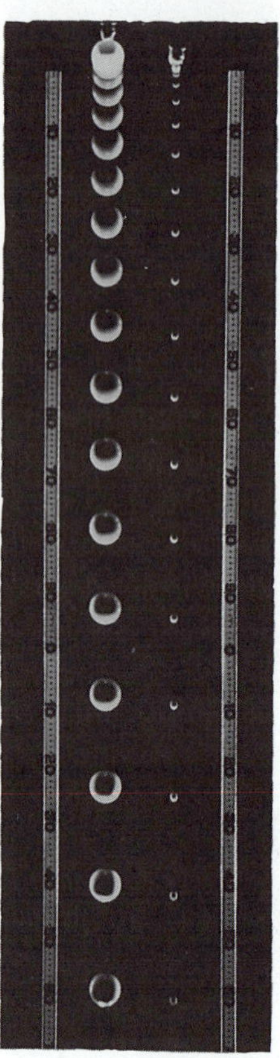

concrete block are due to the effects of air friction. If we could drop these objects in a perfect vacuum where there is no air to affect them, they would fall exactly together, Galileo argued.

Notice here that Aristotle's theory predicted something that worked in some cases (the paper and the block) but not in others. Galileo's rule started with a different premise: Under ideal circumstances all objects would fall together. He then gave a reason that some objects don't act this way. The problem was that Galileo was unable to test his idea because the vacuum pump had not yet been invented.

Before going any farther with Galileo's ideas, we must stop and carefully define a word we will be using often: speed.

Speed and Velocity

Until the time of Johannes Kepler and Galileo, descriptions of nature had relied primarily on geometry, a practice traceable to the Greek thinkers of old. Recall from the last chapter that Kepler used an equation to describe the relationship between the planets' periods of revolution and their distances from the Sun. Galileo also used equations in his description of nature. He defined concepts

ACTIVITY

Estimating Speeds

To give you a real feeling for the calculation of speed, it is worthwhile to estimate some speeds in your everyday environment. As people walk by, make an estimate of how far (in feet) they walk in some particular time and calculate the speed in ft/sec. Do an estimation of someone's speed using metric units. Finally, use the fact that 1 m/sec is a little more than 2 miles/hr to express this speed in miles per hour.

Make similar estimates of at least two other motions and convert them to miles per hour.

carefully, using definitions that permitted quantities to be measured and related mathematically to one another. In the physical sciences today, such definitions are a necessity. One such mathematically defined term is speed. The ***average speed*** of an object is defined to be the result of dividing the distance an object travels by the time taken to go that distance. For example, if you drive 120 miles, taking three hours to complete the trip, your average speed is calculated by

Average speed: The distance traveled by an object divided by the time elapsed during the travel.

$$\frac{120 \text{ miles}}{3 \text{ hours}} = 40 \text{ miles per hour, or } 40 \text{ mi/hr.}$$

Notice that the units in which speed is expressed indicate the division: miles divided by (per) hours. We will not use the more common abbreviation "mph" for "miles per hour" because it obscures the fact that the word "per" indicates a division.

For speeds other than those of cars, trains, and planes, it is convenient to use different units for distance and time. We might use feet for distance and seconds for time, yielding speed units of feet/second. In the metric system, we will use meters and seconds, giving speed units of meters/second, or m/s. The length of a hallway in your house might be 5 meters (about 16 feet, since a meter is just a little more than 3 feet), and if you roll a ball down the hall so that it gets to your little sister in 2 seconds, the average speed of the ball is (5 meters)/(2 seconds), or 2.5 m/s (Figure 4–3). This is a little greater than 5 miles/hour.

In science we normally use SI (from the French "Systeme International d'Unites"), which are based on meters, kilograms, and seconds. We will use them often in this book.

If you want to change speeds given in m/s to mi/hr, which you may be more comfortable with, you can double the m/s speed and get an approximately correct mi/hr speed.

In everyday language, speed and velocity mean the same thing, but in science a distinction is made between them. If we wish to state an object's ***velocity,*** we must state not only its speed, but also the direction in which it is moving. Thus, if we say that a car is moving at 45 mi/hr *northward,* we are stating the car's velocity. We will see later that taking such care with the definition is useful.

Velocity: Speed, along with a statement of the direction in which the object is moving.

Figure 4–3. *The ball takes 2 seconds to travel the 5 meters. This is an average speed of 2.5 meters per second.*

Back to Falling Objects

It is a common experience that objects continue to speed up as they fall. You know in trying to catch something dropped from above that the farther it falls the faster it is going. This thought-experiment does not indicate how long objects continue to speed up, but we see that they certainly do so over the first few feet.

Galileo wanted to know just how a falling object gains speed. Is there any pattern to it? For example, does it gain the same amount of speed for each foot it falls, so that if it gains 8 ft/s while falling the first foot, will it gain another 8 ft/s while falling the second foot? Or perhaps it gains the same amount of speed during each second it falls. In this case, if it gained 32 ft/s during the first second of fall, it would gain another 32 ft/s during the next second, and so on. The problem was that things fall so fast that Galileo was unable to measure the times and distances involved. He solved the speed problem in a unique way. Figure 4–4 shows a ball rolling down an incline. In this case, the ball also gains speed, but slowly enough so that we can make time and distance measurements. Galileo made such measurements and found out that the ball speeds up in the second manner: It gains the same amount of speed each second.

The mechanical clock had not yet been invented. In fact, it was none other than Galileo who invented it—shortly before he died.

Next Galileo used a slightly steeper incline. In this case, the ball increased its speed more rapidly but still it followed the same pattern; changing its speed by the same amount each second. At even steeper inclines, measurement became more difficult but as near as he could measure, the same pattern held. Galileo argued that if the incline were straight up and down, the same thing would happen. That is, the ball would increase its speed uniformly as time passes.

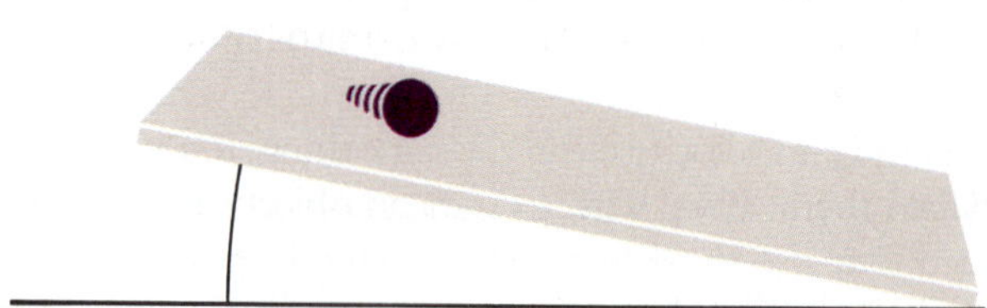

Figure 4–4. *Galileo used an inclined plane to investigate acceleration.*

CLOSE UP

Simplicity and Nature

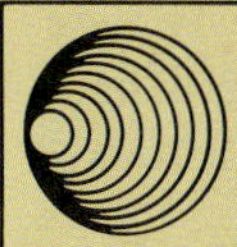

WHY DID GALILEO EVEN dare to hope that there would be a pattern in the motion of falling objects? Why did he think that a falling object should gain speed with any uniformity in the first place? This goes back to what we have discussed earlier. Science is based on a faith that nature is basically simple and understandable. It has been argued that the origin of this faith stems from the belief in a benevolent Creator; one who created a world with simple, understandable patterns. This was the belief of the ancient Greek culture and it is continued in Christianity.

Galileo stated that in studying nature we are learning something about the mind of the Creator. He used this argument to justify his studies to church authorities. Few modern scientists tie their work so closely to their theology—if, indeed, Galileo did—but all have a faith in the basic simplicity of nature. Recall that one of the criteria for a good theory is that it be aesthetically pleasing, which translates into it having a neat, understandable nature.

Acceleration

To better describe the motion of a falling object, we must define acceleration in a more precise way than the word is used in our everyday language. It was Galileo who gave us the definition of acceleration we use today.

The ***acceleration*** of an object is defined as the result obtained by dividing the change in the object's velocity by the time used to make the change. Since velocity includes both speed and direction, this means that if an object changes either its speed or its direction (or both), it will be accelerating. For now, however, we will ignore the change of direction and concentrate only on changes of speed.

Acceleration: The result obtained by dividing the change in velocity of an object by the time required to make that change.

If the definition of acceleration is new to you, you should reread this section until you are comfortable with it.

EXAMPLE

Suppose you are moving along in your car at 25 mi/hr. You realize you are late for dinner (and besides that, the guy behind you is blowing his horn). You push on the gas to speed up and four seconds later you are moving at 55 mi/hr. What is your average acceleration?

Solution

$$\text{acceleration} = \frac{55 \text{ mi/hr} - 25 \text{ mi/hr}}{5 \text{ seconds}}$$
$$= \frac{30 \text{ mi/hr}}{5 \text{ seconds}}$$
$$= 6 \text{ mi/hr per second}$$

Try One Yourself. Suppose that a person is walking at 3 mi/hr and begins to run, changing her speed to 7 mi/hr in 1.5 seconds. What is her acceleration?

Figure 4–5. *A speedometer changing from 25 mi/hr to 55 mi/hr in 5 seconds represents an average acceleration of 6 mi/hr/s, since:*

$$\frac{55 \text{ mi/hr} - 25 \text{ mi/hr}}{5 \text{ s}} = \frac{30 \text{ mi/hr}}{5 \text{ s}} = 6 \text{ mi/hr/s}$$

Look at the result of the example above. It tells us that, on the average, during each second you were accelerating your car changed its speed by 6 mi/hr. Since during each second you increased the speed by 6 mi/hr, it took 5 seconds to change from 25 mi/hr to 55 mi/hr (see Figure 4–5).

We have stated the units of acceleration as "mi/hr per s." Recall, however, that the symbol / means "divided by" or "per." Thus we could have written "miles per hour per second" or even "(mi/hr)/s." In the metric system, speed is usually expressed in m/s. Acceleration is then m/s per second, or (m/s)/s.

The last unit of acceleration we will mention is the common one used in the British system of measurement. Its unit of speed is ft/s, and its acceleration unit is (ft/s)/s. If an object has an acceleration of 20 (ft/s)/s, this means that during each second, its speed is increasing by 20 ft/s.

Rather than writing (m/s)/s and (ft/s)/s, these units of acceleration are usually written m/s^2 and ft/s^2, which mean the same thing. Thus in the last example, the object has an acceleration of 20 ft/s^2.

Falling Objects Again

Galileo discovered that balls rolling down an incline increase their speed the same amount during each second of their fall. This means that they have a constant acceleration as they roll. On a steeper incline, the acceleration would be greater but it would still be uniform for that case (Figure 4–6).

Galileo concluded that a freely falling object would also accelerate uniformly. He was unable to measure what that acceleration would be because a falling

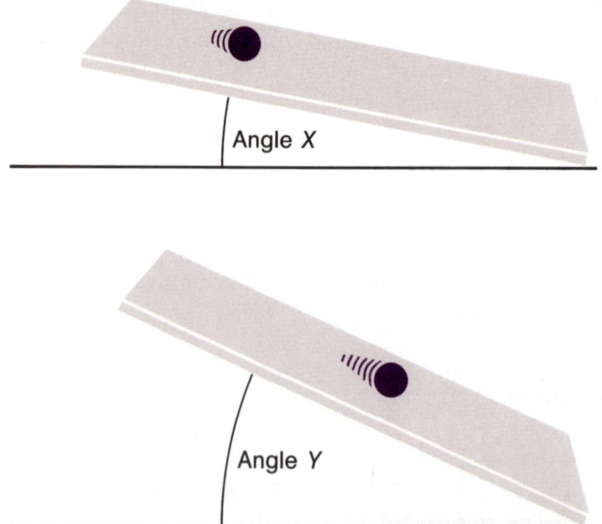

Figure 4–6. *The acceleration of the ball rolling down the upper plane is less than the acceleration of the ball on the lower.*

object very quickly reaches a speed greater than he was able to measure. Today we easily measure such speeds and many elementary science laboratories include an exercise where students measure the acceleration of a falling object. The result? *A freely falling object has an acceleration of about 9.8 m/s²,* (about 32 ft/s²). This is a much greater acceleration than we have used in examples thus far. It means that only one second after starting its fall, an object has reached a speed of 9.8 m/s, or 32 ft/s. This is about 22 miles per hour!

Free fall: The condition of an object when there is no force exerted on it except for gravitational force.

EXAMPLE

How much speed does an object have three seconds after it is dropped? Find the answer in both the metric system and the British system of units.

Solution

Since a falling object accelerates at 9.8 m/s², it gains 9.8 m/s each second. We can calculate its speed after three seconds, therefore, as follows:

$$\begin{aligned}\text{Speed after 3 seconds} &= \text{acceleration} \times 3 \text{ seconds} \\ &= 9.8 \text{ m/s}^2 \times 3 \text{ s} \\ &= 29.4 \text{ m/s}\end{aligned}$$

To calculate the speed in ft/s, we could convert 29.4 meters to feet, but instead, we will repeat the above calculation using the acceleration in ft/s²:

$$\begin{aligned}\text{Speed after 3 seconds} &= 32 \text{ ft/s}^2 \times 3 \text{ s} \\ &= 96 \text{ ft/s}\end{aligned}$$

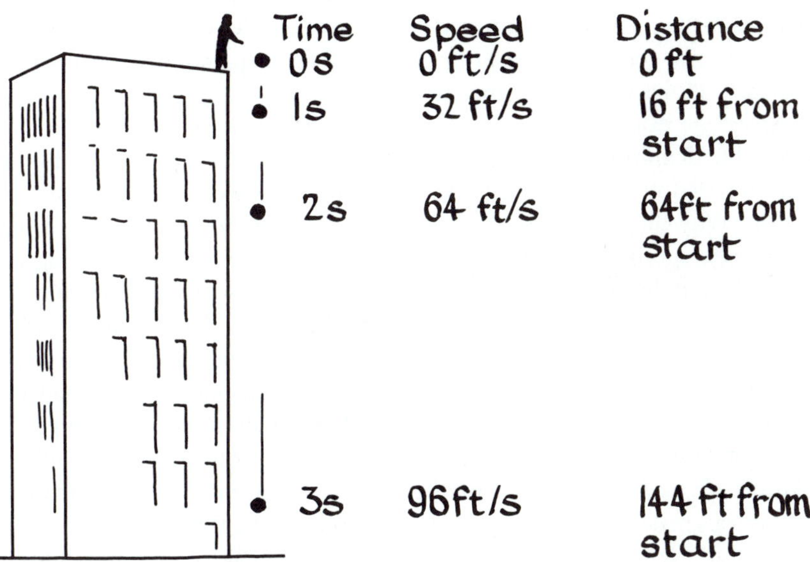

Figure 4–7. *Under frictionless conditions, a ball would fall as shown. In practice, air friction prohibits it from gaining speed this quickly.*

Try One Yourself. If an object falls freely, what will be its speed 5 seconds after being released?

Acceleration of gravity: The acceleration of an object due only to gravitational force. Usually the term refers to an object near Earth's surface, where the acceleration is 9.8 m/s^2, or 32 ft/s^2.

Does this work for all objects? Yes, as long as they are falling freely; that is, as long as there is nothing aiding or restricting their fall. Air is the most likely culprit here. As an object increases its speed, the air resistance on it increases. You can experience this effect by holding your hand outside the window of a moving car. The faster the car moves, the greater the force of the air on your hand. For some falling objects, such as a piece of paper, air resistance is an important factor at low speeds. Other objects, such as a spoon or a bowling ball, are not greatly affected by the air until they reach great speeds. Until such speeds are reached, these objects fall with an acceleration of 9.8 m/s^2. We call this acceleration "the acceleration of gravity on Earth." (On the surface of the Moon, the acceleration of gravity is less; about 1.6 m/s^2.)

DEFINITIONS

There are two things to note about the way Galileo defined acceleration. First, the term is defined in a manner that tells us how to calculate its value. The definition says: "To calculate acceleration, divide the change in velocity by the time required to make that change." This manner of defining terms is common in physical science. We make careful definitions that are mathematical in nature. In your study of astronomy, therefore, you must be sure to study definitions very carefully. Most of the terms used in astronomy are also part of nonscientific language, but in nonscientific use they are usually neither defined nor used as carefully and they may have different meanings there as well. In addition, in nonscientific use terms are usually not defined in a quantitative manner—a manner that allows them to be measured.

The second point is that Galileo could have defined the word differently. One possible definition could have been the change in velocity divided by the distance traveled while the velocity was changing. This would be a meaningful definition. Why did he choose the other one? Because he had concluded that falling objects speed up by the same amount ***each second*** and by choosing the definition he did, he was able to say that falling objects "accelerate uniformly"—that their acceleration stays constant. If he had defined acceleration in the other manner suggested, falling objects would not have a constant acceleration at all. Their acceleration (by that definition of the term) would change as they fall and a description of their manner of falling would be much more complicated. In general, our choice of definitions determines how simple our statements will be.

At first glance, it may seem unfair to define a term in a particular manner just so we can make a simple statement. We do it all the time in science but it is also done in everyday language. The purpose of language is to communicate, and we define our words in whatever way will contribute to good communication. When the tube on which we watch the "soaps" and the football games was invented, there was no word for it. A word was invented, and a definition given to it which helps us to tell a friend how we wasted the weekend.

NATURAL MOTION ACCORDING TO GALILEO

Aristotle held that the natural state of Earthly objects is to be at rest as close as possible to their natural position. We saw earlier that this idea can be used to explain many of the motions we see, because everything in our experience does indeed eventually come to rest. In *Two New Sciences* Galileo proposed a more fruitful way of looking at things. In direct contradiction of Aristotle, he stated that the natural way for an object to behave is to continue whatever motion it has. He said that the only reason a moving object stops is that something causes it to stop. If there were nothing causing an object to stop moving it would continue moving forever at the same velocity, according to Galileo.

Compare Galileo's idea to your experience. It is common for objects to stop moving rather than to remain in motion. If Galileo is right, something must be stopping the objects. In some cases the "stopper" is obvious! When you drop your wallet, the floor stops it. When you pitch a baseball, the catcher's glove stops it (unless the bat gets to it first). These are the easy cases. If you slide a book across the floor, it also stops, but the stopping agent is not quite so obvious. In this case, friction stops it. Friction is the name we give to the force that results from the rubbing of the book against the floor. It is a force that always opposes motion.

There is even friction between a rolling ball and the floor. Friction here is far less obvious then in the other cases, though, and there is so little friction that we may have to roll a ball quite a distance before it slows significantly, as indicated in Figure 4–8.

Friction is present in every case we can imagine here on Earth. We can reduce it by lubrication (you might put oil on your sliding book) but we can never eliminate it entirely. Notice, therefore, that Galileo was telling us about an idealized case. He was saying that without friction or anything else to stop an object, it would keep going. Of what use is such an idealization? Friction is part of our everyday experience and we cannot neglect it, so why state such an idealized law? The answer is that an ideal case is *simpler* than the real case, and if we can understand how things work in an idealized situation, we can then add the complications and understand the real case.

Figure 4–8. *The ball rolls farther across the floor than the book does, but the ball eventually stops, too, because of friction.*

THE LAW OF INERTIA

Inertia: The property of an object whereby it tends to maintain whatever velocity it has.

The "net" force is the total force, taking into account all forces and their directions. Equal forces in opposite directions cancel one another.

There is a name we give to the natural tendency of objects to keep moving: We call it inertia. ***Inertia*** is the tendency of an object to maintain its velocity. With this term in mind, we will state Galileo's rule formally and call it the law of inertia:

> **The Law of Inertia: Unless an object is acted on by a net, outside force, it will continue in a straight line at a constant speed.**

or,

> **Unless an object is acted on by a net, outside force, it will continue at the same velocity.**

(Thus we see that the term *velocity* simplified our statement slightly.) We must point out here that the speed referred to includes the speed of zero. Thus if an object is at rest (at zero speed), it will continue to stay at rest unless there is a force that causes it to move.

Examples of the Law of Inertia

The reason that we have used the terms *net* and *outside* in the above statement of the law of inertia can be seen by considering some examples. First, consider what happens if two thugs are trying to steal a safe by pushing on it as shown in Figure 4–9. If they push equally hard, the safe will not accelerate in either direction because the force exerted by one person is cancelled by the other. In order to change the velocity of the safe, there must be a net force on it.

Next, suppose that you are inside a closed truck. If you push against the front wall of the truck, can you make it move? Figure 4–10 shows a number of people trying to push the truck forward in this way. They can never make it move in this manner because their feet are pushing backwards on the floor of the truck as their hands push forward. No matter what their strength, they

Figure 4–9. *Two not-so-bright crooks at work. Their forces cancel one another, so there is no net force on the safe and it will therefore not accelerate.*

Figure 4–10. *No matter how hard the people push on the front wall of the truck, they cannot make the vehicle move because they exert no force from outside on the truck. Although they push forward on the wall, their feet push backwards on the floor.*

can't make the truck move because they are not an *outside* force. An object can be accelerated only by forces exerted by a second, outside object. "Inside" forces always cancel one another (or push the object apart). Notice that in the case of the people pushing from inside the truck, the force exerted by their feet cancels that exerted by their hands.

You surely have had the experience of riding in a car when the driver unexpectedly slammed on the brakes. What happened? You felt like you were thrown forward (Figure 4–11). What could push you forward? In fact, nothing pushed you forward. You simply continued forward when the car slowed down. Your body has inertia; it tends to maintain its velocity. The car slowed because of a force on it: the force of friction that resulted when the brakes were applied. Your body continued forward until some outside force could act on it. If you were wearing a seat belt, it supplied the force. If not, perhaps the windshield had to do the job of slowing you down.

On the other hand, why do you feel pressed back into the seat when a car speeds up quickly from rest? The answer is that your body tends to continue at rest; it has inertia and it therefore stays at rest until something exerts a force on it to change its motion (or lack of motion, in this case). The seat back exerts the force to get it going.

Figure 4–11. *When a car stops suddenly, a rider feels thrown forward. There must be a force exerted on the rider if she is to stop with the car.*

Figure 4–12. *It is the inertia of a hammer that results in the nail being driven.*

As a final example, let us consider what it is about a hammer that makes it helpful in driving a nail into a wall (Figure 4–12). It is not simply the force provided by your arm, for it is difficult to drive a nail with a child's toy hammer. It is the inertia of the hammer's head that drives the nail. Although we have not yet discussed weight, it is important to see that it is not the weight of the hammer that drives the nail. Rather it is the inertia of the hammer that is the important factor. As we will discuss in the next chapter, weight is a force pulling downward on the hammer. If we had been driving the nail into the floor, we might argue that the weight of the hammer would have been of value to us. But when driving a nail into a wall, the hammer's weight cannot help, for we do not need a downward force. We need a sideways force. What happens is that your arm gets the hammer moving, and once the hammer is moving it tends to keep moving: it has inertia. It will continue at the same velocity until something changes its velocity. That something is the nail. The hammer, in turn, exerts a force on the nail to drive it into the wall.

In the next chapter we will apply ideas of inertia to cases where the *direction* of motion of an object is changed rather than simply the speed of the object.

GALILEO YIELDS TO ISAAC NEWTON

Galileo was a leader in breaking the bonds of Aristotelian thought. He stirred up controversy not only by advocating the Copernican system but by contradicting Aristotle in describing the manner in which objects move and fall. He and Johannes Kepler also gave a new direction to the methods of stating natural laws. Both of them stated their laws in a manner that allowed measurement and testing by the methods of mathematics.

In addition, Galileo referred constantly to experiments that would check his hypotheses. This was a new procedure in the study of nature. Prior to the times of Galileo, the primary method of discussing nature was to refer to the authorities, primarily to Aristotle. Such a reliance on observation and experimentation rather than authority is a cornerstone of science today. Its beginning is usually credited to Galileo.

The same year that Galileo died, Isaac Newton (Figure 4–13) was born. Newton was the genius who took Galileo's findings and tied them together into one grand theory of motion. Galileo had predicted this, writing that he had "opened up to this vast and most excellent science, of which my work is merely the beginning, ways and means by which other minds more acute than mine will explore its remote corners."

Today, most of Newton's work has been superseded by modern quantum mechanics and the theories of Albert Einstein. Both scientists and nonscientists continue to use Newton's theories for two reasons, however. First, Newton's ideas are so much a part of our culture and our thinking that for a long time to come we will continue to think of nature in Newtonian terms. Second, in our everyday world, Newton's theories work as well as the newer theories. It is only when one considers the very small world of the atom or the very largest and/or very fastest aspects of the universe, that quantum mechanics and Einstein's relativity are necessary.

Figure 4–13. *Sir Isaac Newton (1642–1727) worked in many fields. In this painting he is shown experimenting with a prism to investigate the nature of light.*

For these reasons, even though many scientists understand the new theories, they find that the ideas of Newton are easier to use and are accurate enough for almost all everyday thinking. Newtonian physics is at the heart of most of the mechanical aspects of today's world and of the astronomy we will study. It has so changed the world that it will not soon be replaced in our thinking.

NEWTON'S FIRST LAW OF MOTION

The first of Isaac Newton's accomplishments that are of interest in our study of astronomy are his laws of motion. Newton was able to describe the motion of objects in terms of three laws—known as "Newton's Laws of Motion"—that are the basic laws relating force and motion. In science, the word *force* means about the same thing as in everyday language; you can think of a force as a push or a pull. We have used the term in stating Galileo's Law of Inertia. And in fact, as his first law of motion, Newton simply restated Galileo's law. So, in the interest of completeness, we will rewrite that law under its new title:

> **Newton's First Law (The Law of Inertia): Unless an object is acted upon by a net, outside force, the object will maintain a constant velocity.**

HISTORICAL NOTE

Isaac Newton

TO GIVE AN IDEA of the importance of Isaac Newton (1642–1727) to the history of thought, we quote the English poet, Alexander Pope, who lived at the same time as Newton. He wrote:

"Nature and nature's laws lay hid in night.
God said, Let Newton be! and all was light."

Newton was born prematurely on Christmas day, 1642, in a small village in England. He was so small at birth that his mother said that he would have fit into a quart pot. His father, a fairly prosperous farmer, had died before Isaac was born and his mother decided that he was too frail to become a farmer. He was not a particularly good student, however; "wasting" much of his time tinkering with mechanical things, including model windmills, sundials, and kites that carried lanterns and scared the local people at night. When her second husband died, Isaac's mother called him home from school to help on the farm, where he spent most of his teenage years.

At the age of 19 he was admitted to Cambridge University. There he studied mathematics and natural philosophy (known as science today). In 1665, after Newton had received his degree and was serving as a junior faculty member, England was swept by the bubonic plague. This incurable disease killed more than 10 percent of the population of London in only three months, and those who could afford to do so escaped it by moving away from population centers. Newton returned to his mother's home at Woolsthorpe. There he worked feverishly on science for the next two years. These must have been two of the most productive years in the history of science, for during this time Newton made discoveries in light and optics, in force and motion, in gravitation and planetary motion. He also devised a theory of color. To solve a problem in gravitation, he invented calculus. During this time he outlined what would become his major book, *Philosophiae Naturalis Principia Mathematica,* usually called *The Principia.* In later years, Newton wrote, "All this was in the two plague years of 1665 and 1666, for in those days I was in the prime of my age for invention, and

NEWTON'S SECOND LAW

Newton's First Law indicates that a force is needed to change the velocity of an object—to accelerate it. Newton's second law goes beyond this, telling us more about what a force is and how we can measure a force. It tells us how much force is necessary to produce a certain acceleration of an object.

Consider the brick shown in Figure 4–14 and imagine that the wheels allow the brick to move without friction. (Notice that we are idealizing things again. We want to consider a case with no friction whatsoever.) The brick is at rest in part (a). It has inertia, however, so it will stay at rest even though there is no friction. In (b), you give the brick a push. While you are pushing, the brick accelerates. What determines how great its acceleration is? The force you exert. If you exert a greater force, the brick's acceleration is greater (part c). A tiny minded Mathematics and Philosophy more than at any time since."

Newton's manner of attacking a problem was simply to concentrate his mind on it with such intensity that he finally solved it. His ability to concentrate must have been tremendous. This resulted in him appearing to be what we might call absentminded. It was not uncommon for him to work all day, forgetting to eat. The story is told that once when riding his horse, he got off the horse to unlatch a gate, led the horse through the gate, relatched the gate, and then led his horse home, forgetting to get back on. He was in deep concentration about some problem and walking probably gave him more time to think than did riding, anyway.

As a young man, Newton was not interested in publishing his work. He seemed to want to discover the mysteries of nature simply for his own curiosity, and he often had to be persuaded by friends to publish his findings. Besides, he did not like having to defend his views against criticism and he wanted to avoid getting involved in arguments over who was the first to make certain discoveries. Nevertheless he *did* get involved quite fiercely in such disputes, including one with Gottfried Leibniz over which of them first developed the calculus. This dispute lasted long after both had died.

It is interesting to note that Newton had a very practical side as well as being a theoretician. He served as Warden of the Mint and while in this office, he began the practice of making coins with small notches around their edges. (Check the quarters in your pocket.) This was done to discourage people from illegally shaving off and retaining the valuable metal of the coins.

In many cases, a person is not recognized as a genius until after his or her death. Such was not the case for Newton. He received honors and was given positions of authority. He was elected president of the Royal Society, an organization of scientists, in 1703 and every year thereafter until his death in 1727. In 1705 he was knighted; the first scientist to receive this honor. When he died, Sir Isaac Newton was buried in Westminster Abbey after a state funeral.

force will produce little acceleration. This idea gives us the first part of Newton's Second Law: The acceleration of an object is proportional to the force on it. "Proportional to" actually goes beyond what we deduced from our thought experiment. It tells us that twice the force will cause twice the acceleration.

Refer to Figure 4–15. Here we show one hand pushing on a frictionless brick as before and another hand pushing on a stack of two bricks. If the hands push with equal forces, which will cause the most acceleration? It should be intuitive that the greater acceleration will be produced by the one pushing on the single brick. It is important to see that friction has nothing to do with this. It is tougher to accelerate two bricks than one simply because the two have more inertia. In fact, two bricks have twice as much inertia as one brick. This is an idea we didn't discuss above: Some objects have more inertia than others. So before continuing with the second law, we must introduce the term *mass*.

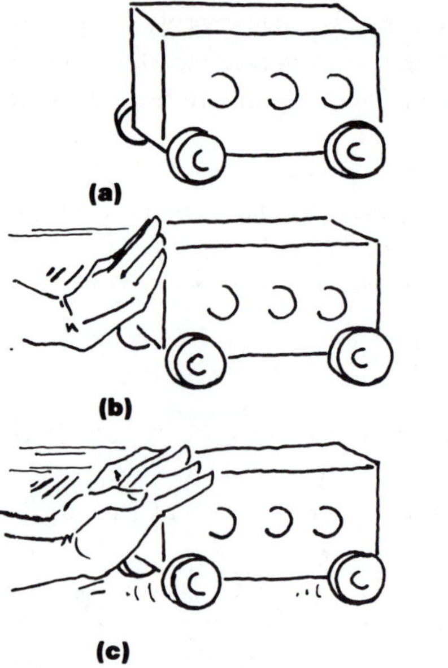

Figure 4–14. *The wheeled brick (a) will accelerate if a force is exerted on it (b). If twice as much force is exerted on it, it will accelerate at twice the rate (c).*

Figure 4–15. *The same amount of force will give twice as much mass only half the acceleration.*

An Important Digression—Mass and Weight

Mass: The quantity of inertia possessed by an object.

Mass is another of those terms, like speed, that is used in everyday language but is given a more specific definition in science. We have already laid the foundation for the scientific definition because an object's *mass* is simply the measure of the amount of inertia that object has. Instead of saying that one object has twice the inertia of another, we normally say that it has twice the mass.

Rotational inertia—for a spinning object—involves more than just mass and will be discussed in a later chapter.

It is important to say what mass is not. Mass is not volume. Suppose that the brick in the last figure is not a true brick, but is instead a piece of styrofoam in disguise. If the person does not know this, however, he will have quite a surprise when he pushes on the brick. He will find that it has much more acceleration than he expected. Why? Because a styrofoam brick has much less mass than a real brick even though they both have the same size—the same volume.

Mass also is *not* weight! Remember that in the example of driving a nail with a hammer, we distinguished between inertia and weight. And in our examples of pushing the frictionless bricks, the weight of the brick was not a factor; weight is simply the force of gravity pulling downward on the bricks. Weight did not oppose the pushing hand. This is a subtle distinction, but a very important one. It is worth one more example.

We will discuss weight more fully later, but you are familiar with the idea that if a person gets far enough away from Earth she can be considered weightless. Imagine yourself in a spacecraft in a weightless environment. A squabble breaks out among the crew, and an astronaut with a violent streak throws an iron skillet at you from behind. When the skillet hits you, will it hurt? Now, probably none of us has experienced this, so we are asking you to use your intuition to give an answer. If your guess is that the skillet would not hurt you, imagine instead that a freight train hits you in this weightless environment. We hope that your intuition tells you that the object has inertia even if it doesn't have weight and since it has inertia, a force is required to stop its motion. (Yes, it would hurt.)

Thus mass is a more fundamental quantity than weight. An object may be weightless, but it is never massless.

The worldwide unit of mass measurement is the kilogram. At the International Bureau of Weights and Measures near Paris is a platinum cylinder which has, by definition, a mass of *exactly* one kilogram (see Figure 4–16). To give you an idea of what one kilogram of mass is, a kilogram weighs about 2.2 pounds at the surface of the Earth. It is not correct to say that a kilogram is *about equal to* 2.2 pounds, because a pound is a unit that expresses weight and a kilogram is a unit that expresses mass. A kilogram weighs about 2.2 pounds on the surface of the Earth but at some other location it might weigh a different amount.

Figure 4–16. *The official standard kilogram is kept at the International Bureau of Weights and Measures near Paris, France. This is a secondary standard kept at the National Institute of Standards and Technology (NIST) near Washington D.C.*

Back to Newton's Second Law

Figure 4–15 showed one hand pushing on one brick and another hand pushing on two. It was stated that we would expect less acceleration from the two bricks. In fact, measuring the accelerations would show that if the forces were equal, the acceleration of the two bricks would be exactly half of that of the single brick and the same force applied to three bricks would result in one-third the acceleration. Thus, acceleration is inversely proportional to the mass being accelerated. If the force is the same, more mass means less acceleration, in exact proportion.

We have thus far discussed the relationships between force and acceleration and between mass and acceleration. We can sum up these relationships in a mathematical statement:

$$\text{ACCELERATION} = \frac{\text{NET FORCE}}{\text{MASS}}$$

This tells us that the greater the net force, the greater the acceleration, but the greater the mass, the less the acceleration. It also makes it apparent that if the net force is zero, there is no acceleration, which agrees exactly with Newton's first law. The expression above is usually written:

$$\text{FORCE} = \text{MASS} \times \text{ACCELERATION}$$

or in symbols:

$$F = ma$$

NEWTON'S THIRD LAW

Newton's third law is simple to state:

> **Newton's Third Law: When object *X* exerts a force on object *Y*, object *Y* exerts an equal and opposite force back on *X*.**

This law seems very innocent but its implications are great. It is sometimes stated as: "For every action there is an equal and opposite reaction." In this form it is then applied to a wide variety of fields—politics, religion, sports. Newton meant it only for physical forces, however, so our statement above makes its meaning much more explicit than the way it is commonly given.

You are probably sitting on a chair. In doing so, you exert a force downward on the chair. The third law tells us that the chair exerts an equal force upward on you. We often call the force you exert the action force and the force of the chair on you the reaction force but this is arbitrary. The chair's force could just as well be called the action force and your force on it the reaction force. The point is that one of these forces cannot exist without the other and neither "comes first."

Notice that the statement of Newton's third law indicates that two objects, (called *X* and *Y* here), are always involved in the application of forces. *Always.* A force cannot be exerted without an object to exert it and an object on which it is exerted. The word "object" here might refer to an individual atom or it might refer to a collection of atoms and the collection might be a gas or a liquid as well as a solid object. For example, when you held your hand out of the car window to feel the force of air resistance, it was the air that exerted a force on your hand. The third law tells us that your hand therefore exerted a force on the air. This was not easy to see, but we know that your hand must have deflected the air as it came by and a force was necessary to deflect the air. Calling the air an object may seem odd, but air is simply a collection of objects—atoms.

THE VALUE OF NEWTON'S THREE LAWS

Aristotle explained motion in a far different way than did Galileo and Newton. Was Aristotle wrong? Science today is learning to avoid calling a theory wrong. Aristotle's rules of motion did explain some things, although they left many things unexplained.

Newton's laws of motion are much more satisfactory. They not only explain motion, but they do so in a measurable, mathematical manner. This is important in science. We wish to be able to measure things and to predict events quantitatively. Newton's laws do this. Can we then say that they are right? Again, we avoid this in science. We say that they are good laws. They fit the data. They make predictions that can be checked. And, as we will see in the next chapter, they fit in with other laws to make an overall theory that is successful in explaining many things in nature. But we must avoid thinking that they are the final answer; that they are complete. In fact, as mentioned earlier, Newton's laws have been replaced today by those of Einstein and by quantum mechanics. The latter ideas, however, are much more difficult to understand and since they disagree with Newton's laws only in what most of us would call unusual circumstances, we can continue to use the laws of Newton in our everyday life.

The idea of using a theory when we know that it is not the most up-to-date may seem unsatisfactory. But you and I do it all the time. We speak of the Sun rising even though we know that this is part of the old geocentric system. The heliocentric system tells us that sunrise occurs when the Earth turns so that our part of the Earth is exposed to its light. The old Earth-centered theory is a handy way to picture the skies, however, and even astronomers still use it, even though there is no doubt whatsoever that the heliocentric system is a much better overall model of the solar system. There are many cases in science where we use whatever model is best for our purpose, and a mental picture of the Sun rising in the east, moving across the sky, and setting in the west is perfectly good for many purposes. Likewise, your children and your children's children and . . . will be studying Newtonian science.

CONCLUSION

In order to understand the motions of astronomical objects, both within our solar system and beyond, we must understand the principles of force and motion. Aristotle's ideas on the subject had very limited applicability and it was not until Galileo formulated the law of inertia that a consistent theory of mechanics began to be developed. Galileo's investigation of falling objects led him to the definition of acceleration that we still use today, and it paved the way for Isaac Newton's three laws of motion. These three laws, along with the law of universal gravitation (to be discussed in Chapter 5) gave humanity a different outlook on its world and provided scientists with a means of understanding the motions of objects, not only here on Earth but in the heavens as well.

RECALL QUESTIONS

1. In what century did Aristotle live?
2. What were the five elements proposed by Aristotle?
3. Why, according to Aristotle, do objects in the sky not behave like those on Earth?
4. Define speed.
5. Name three possible units you can use to express speed.
6. Describe how the speed of a freely falling object changes.
7. Define velocity and distinguish it from speed.
8. Define acceleration.
9. What is the numerical value (with units) of the acceleration of gravity?
10. Define inertia and give an example of its action.
11. How did Galileo determine the nature of the changes in the speed of falling objects without being able to measure them?
12. What aspect(s) of science had their beginning with Kepler and Galileo?
13. State the law of inertia.
14. In what century did Isaac Newton live?

15. What caused Newton to be free for the two years in which he did his most productive work? (Hint: See the Historical Note.)
16. Which is the more fundamental quantity, mass or weight? Describe an observation that confirms your answer.
17. State Newton's three laws and cite an example of each.
18. Give an example of Newton's third law using a hammer hitting a nail.
19. Why do we teach Newton's laws even though they have been replaced by more modern scientific concepts?

QUESTIONS TO PONDER

Figure 4–17. *Can you use Aristotle's theory to explain why a hot air balloon rises?*

1. Use Aristotle's theory to explain the rising of a hot air balloon and the eventual falling of such a balloon after the fire goes out.
2. Why did Galileo even suspect that there would be a pattern to the speed changes of a falling object?
3. When a rock is thrown straight upward, does it have an acceleration at the instant when it is at the top of its path?
4. Suppose that air friction is negligible, and that a movie is taken of a ball as it goes up and comes back down. In viewing the film, is there any way to tell whether it is being run forward or backward? If so, how? If not, why not? Answer the same questions in the case of a thrown styrofoam ball, where significant air friction is present.
5. Use the criteria of a good theory to evaluate Galileo's rule of falling objects. Why do we not call it a "theory?"
6. Give an example (other than those in the text) of the action of inertia.
7. Newton's laws tell us that no force is needed to keep something moving. Why, then, don't we turn off the engine of an airplane once the plane has reached the desired speed?
8. When a bug hits the windshield of a moving car, the bug and car exert equal forces on one another. Why, then, does the bug accelerate so much more than the car?
9. If we wad up a piece of paper, it will fall much faster than an unwadded one. Explain this by analyzing it in terms of forces.

10. Can you think of any cases (other than those given in this chapter) of common use of a theory that is not up-to-date?

11. Since we cannot avoid air friction in everyday life, what is the value of worrying about how objects fall in these conditions?

CALCULATIONS

1. If you drive a car 12 miles in 15 minutes, what is your average speed in miles per hour?
2. Suppose you travel 210 kilometers in 2 hours. What is your average speed for the trip? Is this a realistic speed?
3. If a boat travels at a constant speed of 15 kilometers an hour for 30 minutes, quickly slows to 10 kilometers per hour, and travels at that speed for another 15 minutes, what is its average speed for the whole trip?
4. If an object starts at rest and falls freely for 4 seconds, what speed does it attain?
5. The average speed of the object in the last question is half of its final speed. How far does it fall during the four seconds?
6. How far does an object fall (from rest) during its first second of fall?
7. Suppose a car changes its speed from 12 m/s to 20 m/s, taking 4 seconds to make the change. How great is its acceleration during this time?

CHAPTER 5

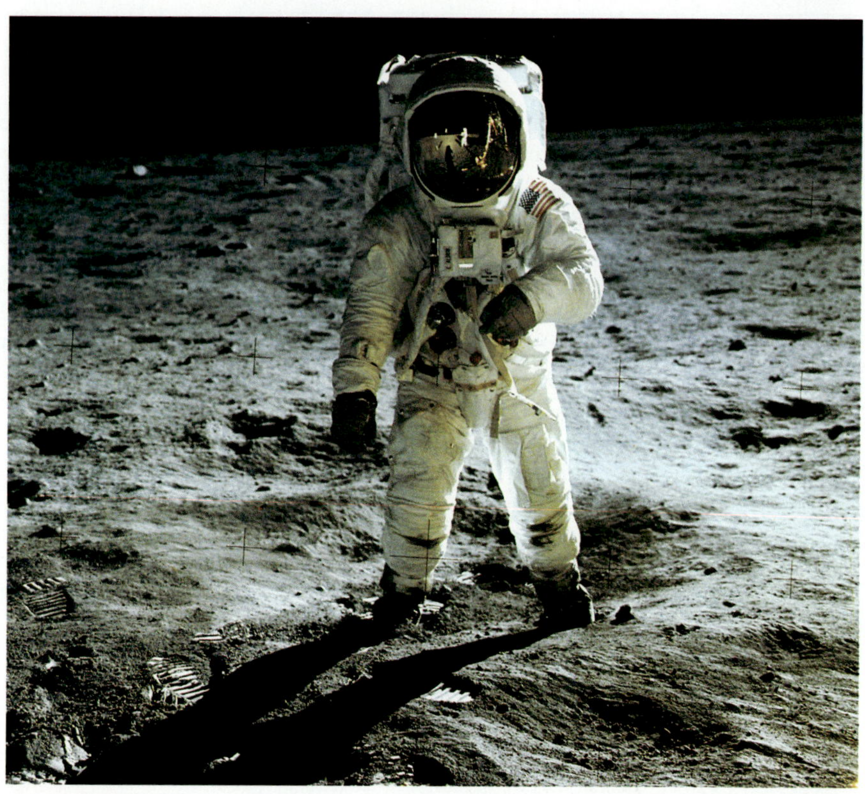

Neil A. Armstrong took this photo of Edwin E. Aldrin Jr. walking on the Moon.

On July 20, 1969, the first humans landed on another astronomical object. From the Sea of Tranquility they reported to mission headquarters in Houston, "Houston, Tranquility Base here. The Eagle has landed." Shortly thereafter, Neil A. Armstrong stepped onto the surface of the Moon saying, "That's one small step for a man, one giant leap for mankind." (Historians might note that although Armstrong meant to say what was quoted above, he actually left out the word "a.")

Armstrong and Edwin (Buzz) Aldrin Jr. stayed on the Moon for 21.6 hours. When they left the lunar module, each wore a 90 kilogram spacesuit. On Earth this weighed 200 pounds, but on the Moon it weighed a mere 33 pounds. During their 2½ hours of walking on the Moon outside the lunar module, they deployed a number of instruments to measure various features of the Moon and they unveiled a plaque that reads "Here Men From Planet Earth First Set Foot Upon the Moon. July 1969 A. D. We Came in Peace For All Mankind." As they left the Moon to rejoin Michael Collins in the command module and return to Earth, the plaque remained behind to express for ages to come a lofty ambition of the human race.

Gravitation and Spaceflight

ONE OF THE GREATEST of Isaac Newton's many accomplishments was his development of the law of universal gravitation. This law was important for a number of reasons. First, it connected Kepler's laws with other aspects of nature. Second, it gave us a tool that is used even today and that is the basis for calculations of the motions of today's spacecraft. But most important, it gave us a different understanding of the nature of our universe.

Before we can study the law of gravitation, we must examine the application of Newton's laws of motion to circular motion. Then we will proceed to the weighty subject of gravity, and finally to space flight, particularly flights to the Moon.

MOTION IN A CIRCLE

Recall that acceleration is defined as a change in *velocity* divided by the time used to make the change and that velocity is defined to include direction as well as speed. Thus far, all of our examples of acceleration have involved changes in speed and none have involved direction changes. Now we will look at the latter.

Recall also that Newton's second law says that an unbalanced force always produces an acceleration. Consider what happens if you whirl a rock on a string. You are exerting a force on the rock toward the center of its circular motion as you pull inward on the string (Figure 5–1). If you are careful, you might be able to make the rock go around you with a constant speed. What about

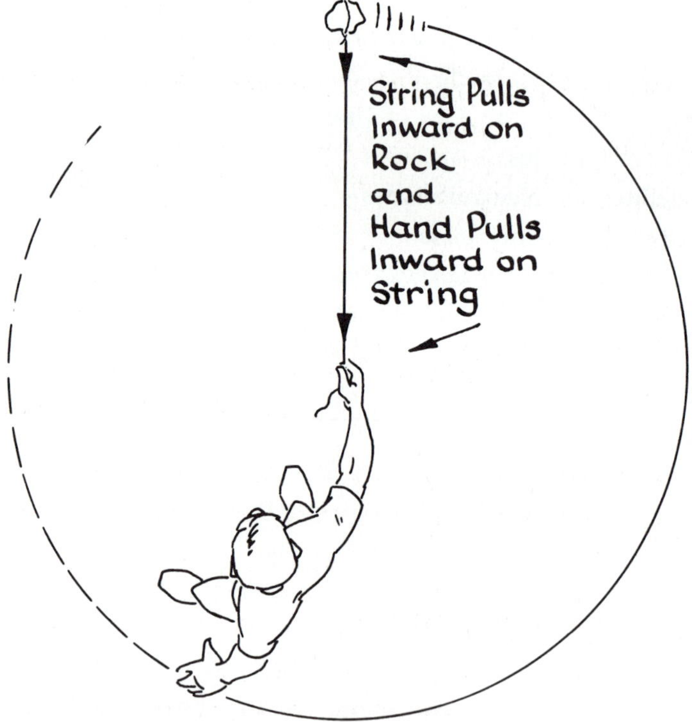

Figure 5–1. *The boy exerts a force inward on the string, which in turn pulls inward on the rock.*

ACTIVITY

Circular Motion

THIS SHOULD BE DONE outside, away from anything breakable. Tie some object such as a rock or a small tool on the end of a fairly long (6 or 8 feet) string or rope. Now whirl the object in a horizontal circle around you. Feel the pull you must exert to keep the object in the circle. Whirl it faster. Do you have to increase the force you exert on the string?

Now let go of the string and carefully observe the path taken by the object. Forget the downward motion (caused by gravity) and concentrate on how the object travels horizontally. Figure 5–2 shows a number of potential paths for the object. Which of these did your object take?

the acceleration which, according to Newton's law, must be produced by the force you are exerting? If the rock moves at a constant speed, the force is not causing a change in speed. Instead, it changes the ***direction*** of the rock's motion.

Motion of an object in a circle at constant speed is an example of acceleration by changing direction. It may seem odd to call this an acceleration, but notice that if acceleration is defined this way, Newton's law holds in all cases. As we discussed before, words are defined in such a way that they allow for efficient statements. When we define acceleration as we have, the force you exert toward the center of the circle when you whirl a rock on a string does indeed cause an acceleration.

Figure 5–2 shows a boy whirling a rock horizontally on a string (as you are encouraged to do in the accompanying Activity box). In 5–2(a), the boy is just releasing the string. The rest of the drawings show a number of potential paths for the rock when the string is released. Which of these is the way you expect the object to travel based on Newton's laws? Which way did the object travel in your activity?

Once the string breaks, there is no horizontal force on the rock. If this is the case, there can be no horizontal acceleration of the rock. Thus, according to Newton's first law, it will continue in a straight line. When the rock was moving in the circle, the only horizontal force on it was exerted by the string. (The rock's weight pulls down, but we were considering only horizontal forces and horizontal motions.) The horizontal acceleration produced was therefore due only to the string.

Centripetal Force and Centripetal Acceleration

A force is necessary to make something move in a curve. We give this force a name, calling it the ***centripetal*** (meaning "center-seeking") ***force.*** This is not

Centripetal force: The force directed toward the center of the curve along which the object is moving.

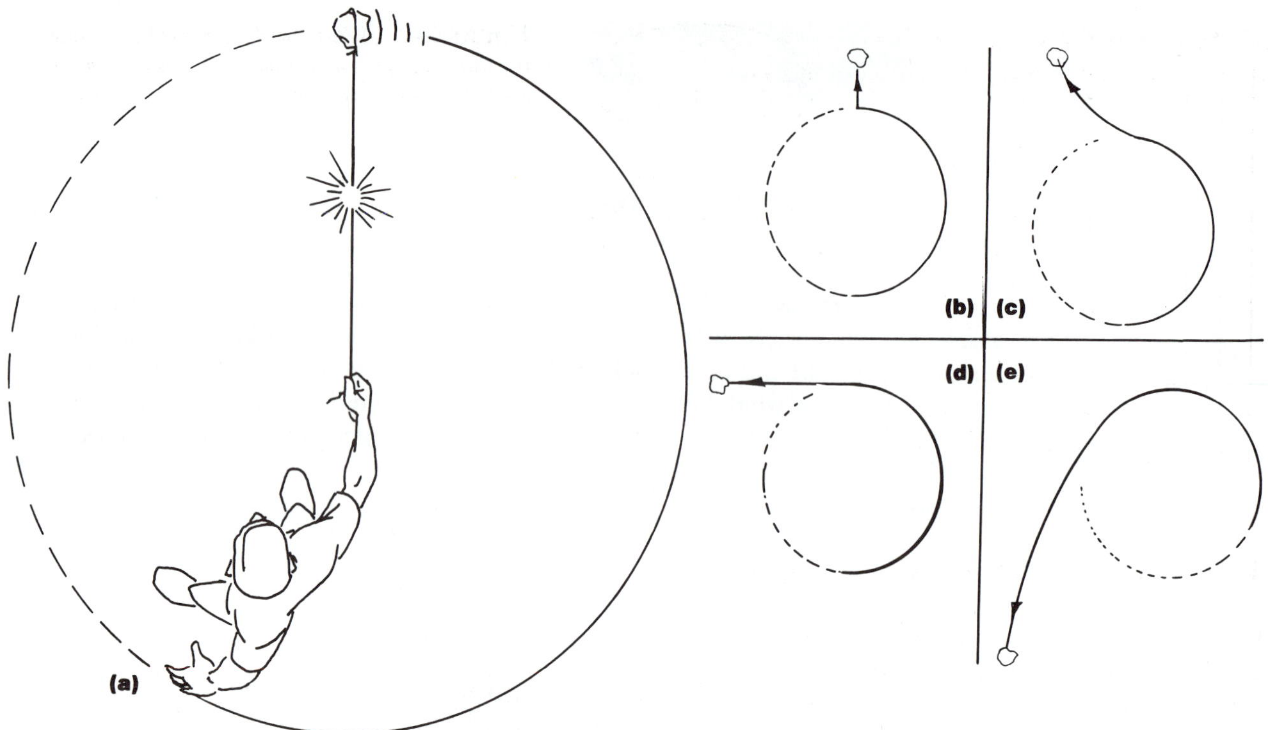

Figure 5–2. *(a) The string breaks as the rock is whirled in a circle. (b–e). Which way does the rock go after the string breaks?*

really a new kind of force, but rather it is the name we apply to a force if that force is causing something to move in a curve. For example, the centripetal force involved when you were whirling the rock was the force exerted by the string on the rock. The centripetal force on a car rounding a curve is the frictional force of the road acting on the tires in the direction toward the center of curvature. The roadway exerts a force on the car, keeping it in the curve. Do not think of centripetal force as another type of force similar to a push, a pull by a string, or a friction force. It might be any of those three or others; it is simply what we call whatever force is responsible for a motion along a curved path.

Centripetal acceleration: The center-directed acceleration of an object moving in a circle.

The acceleration of an object that is moving in a curve is called the ***centripetal acceleration.*** What about the direction of the centripetal force and centripetal acceleration? The direction of the force on the whirling rock is obvious: it is along the string toward the center of the circle, or perpendicular to the curved path. The acceleration must also be in this direction because the force and acceleration in Newton's second law are always in the same direction. Centripetal force and centripetal acceleration are always directed perpendicular to the curved path of the object (Figure 5–3). That is what "centripetal" means.

The centripetal acceleration for any object moving in a circle—not just an object in orbit—is equal to v^2/r *where* v *is the speed of the object and* r *is the radius of the circle.*

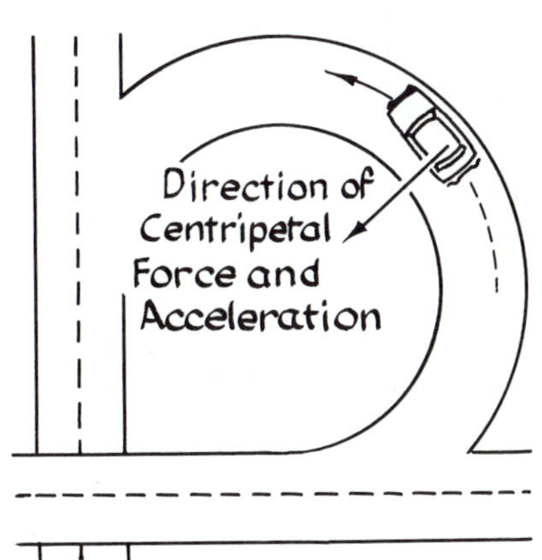

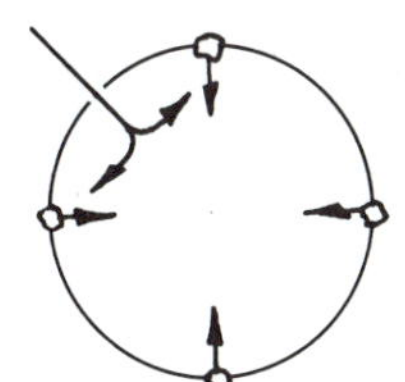

Figure 5–3. *When an object moves in a curve, the centripetal force on the object and its centripetal acceleration are always perpendicular to the curve.*

THE LAW OF UNIVERSAL GRAVITATION

We have seen that Newton's book *The Principia* contained his three laws of motion and we have seen their importance in understanding motion. Their real importance, however, shows up when they are combined with another of Isaac Newton's accomplishments: the law of universal gravitation, or "the law of gravity."

"Plato is my friend, Aristotle is my friend, but my better friend is truth."—Isaac Newton as an undergraduate.

Newton's first law states that an object continues at the same speed in the same direction unless some unbalanced force acts on it. This law is applicable to objects on Earth, but what about objects in the sky? Aristotle held that Earthly laws do not hold in the heavens, but Newton sought to apply his laws of motion there too. It was during his great, productive years of 1665 and 1666 that, in Newton's words, he "began to think of gravity extending to the orb of the Moon." The Moon follows a nearly circular path around the Earth, so if the laws of motion apply to the Moon, there must be a force on it toward the center of its circle—a centripetal force.

Newton hypothesized that there is a force of attraction between the Earth and the Moon and that this serves as the centripetal force. What causes this force? Newton didn't know, but he hypothesized that it is the same force that causes an object here on Earth to fall to the ground. He proposed that there is an attractive force between every two objects in the universe and that the magnitude of this force depends on the masses of the two objects and the distances between their centers.

Let's stop a moment and emphasize what this means. Newton was saying that *every* object in the universe attracts every other one. This includes the book you are reading and the pen in your hand. According to Isaac Newton, they attract one another. Naturally you can't feel the force pulling these two objects toward one another because they have such little mass and the strength of the force depends upon the amount of mass. When one of the attracting objects is the Earth, however, one of the masses is certainly large and we experience the force. This is the force that causes what we call *weight*.

CLOSE UP

Centrifugal Force

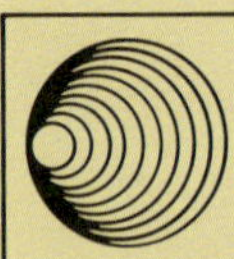

THERE IS A VERY common misconception about circular motion involving the term *centrifugal force.* (Note that this is not *centripetal* force.) Consider a rock being whirled on the end of a string. It is sometimes said that there is a force—called the centrifugal ("center-fleeing") force—pulling outward on the rock. It is explained that this outward force cancels the inward force exerted by the string. This, however, is entirely inconsistent with Newton's laws and with our understanding of the nature of force. If a centrifugal force exists on an object moving in a circle and if this force cancels the centripetal force, the object would no longer move in a circle, for when all forces on an object cancel one another, the object moves (as described by Newton's first law) in a straight line rather than in a circle. When the forces on an object total zero, the object has a constant velocity. In order to move in a circle, an object must have a ***nonzero net force*** on it. This force acts to accelerate the object toward the center of the circle.

There is another problem in arguing that a centrifugal force is exerted on an object moving in a circle: there is nothing to exert this force. In the case of a rock being whirled on a string, the string exerts the inward centripetal force, but there is nothing to pull outward on the rock. A force cannot exist unless it is exerted on one object by another object. An object cannot exert a force on itself. This rule follows from the nature of forces, as described by Newtonian mechanics. The person in Figure T5–1 will never be able to pick himself up by his bootstraps.

Thus there is no centrifugal force pulling outward on a rock (or on any other object moving in a circle). There is, of course, an outward force exerted ***on the string by the rock.*** This is simply the reaction force (by Newton's third law) to the one exerted inward on the rock by the string. In addition, the person whirling the rock feels the string pulling outward on his or her hand—the reaction force to the hand pulling inward on the string. Figure T5–2 indicates these forces.

How can we say that a specific type of force does not exist? Since Newton's laws are part of a theory, couldn't someone propose an alternative theory which includes centrifugal forces? The answer is, "Yes, but this new theory should be a consistent theory." It should not say that in one case a net force of zero results in no acceleration and in another case there is an acceleration. A theory must be self-consistent. It might be possible to devise a consistent theory that includes outward forces on circling objects, but mixing such an idea with Newton's laws of motion will not work.

Centrifugal forces as we defined them above simply are not consistent with Newton's laws. The term centrifugal force is sometimes defined differently, however. If centrifugal force is defined as the reaction force to the centripetal force, then centrifugal forces cer-

In equation form this is

$$F = \frac{Gm_1m_2}{d^2}$$

G *here is a constant number that depends upon what units are being used for* m_1, m_2, *and* d. (m_1 *and* m_2 *are the masses of the objects and* d *is the distance between their centers.)*

A more complete statement of what we now call Newton's law of universal gravitation (or the law of gravity) is as follows:

The Law of Universal Gravitation: Between every two objects there is an attractive force, the magnitude of which is directly proportional to the mass of each object and inversely proportional to the square of the distance between the centers of the objects.

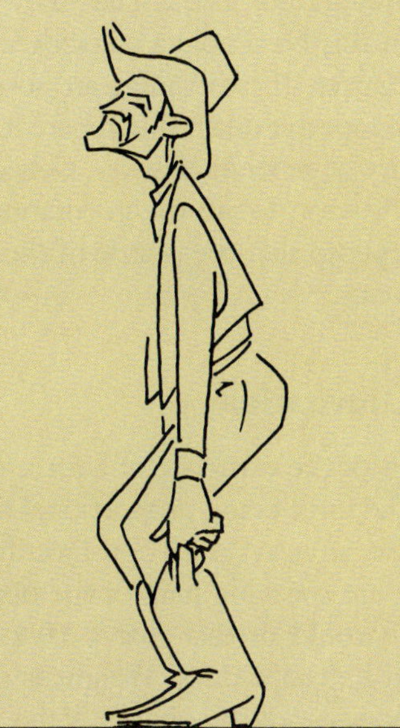

Figure T5–1. *A person cannot pick himself up by his bootstraps no matter how strong he is because a force from outside is needed in order to accomplish this.*

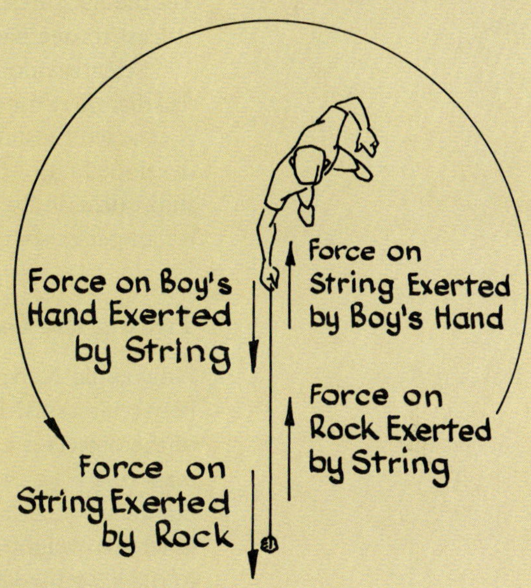

Figure T5–2. *All four forces shown are equal in magnitude but they are exerted on different objects.*

tainly do exist. A rock being whirled on a string exerts a force outward on the whirler's hand (through the string) as his hand pulls inward on the rock. Likewise, as the Earth circles the Sun, a gravitational (centripetal) force is exerted on the Earth towards the Sun, and an equal (centrifugal) force is exerted on the Sun towards the Earth. There is no problem with centrifugal force defined in this manner, but we see here an example of the necessity for carefully defining what we mean by a term. Since there is no general agreement on what is meant by *centrifugal* force, we will avoid the problem by not using the term again in this book.

It is this force, according to Newton, that not only makes objects fall to Earth, but that keeps the Moon in orbit around the Earth and keeps the planets in orbit around the Sun. Newton proposed that this law, along with his laws of motion, could explain the motions of the planets as well as the falling of objects on Earth. If he could explain the planets' motions, this would clear up the mystery of why Kepler's laws worked and bring the motions of the heavenly planets within the realm of our scientific understanding.

Gravity and Mass: How Science Progresses

Some recent experiments provide very tentative evidence for another force—much weaker than gravity—exerted by one object on another. This force, if it exists, depends upon something other than the masses of the objects.

Mass was defined in the previous chapter as the measure of the amount of inertia an object has. In stating the law of gravity, however, Newton proposed that mass is also the quantity that determines the strength of gravitational attraction. Why should the same quantity be the measure of two seemingly different physical properties? It is certainly not obvious that *inertia* (the resistance to a change in motion) should have anything to do with *gravitational attractiveness*. Yet the measures of these two properties are not just similar; they are identical at least to one part in 100 billion. Can this be simply a coincidence?

Scientists do not like such coincidences. If two things are so similar, they feel that there must be a reason. The attempt to explain this apparent coincidence is what led Einstein to develop his theory of general relativity. General relativity, developed more than 200 years after Newton found the gravitational relationship, successfully relates the two concepts so that the puzzle of the coincidence no longer exists.

Testing the Law of Universal Gravitation

How could Newton test his hypothesis of the existence of a force of attraction between masses? First, the part about the force being proportional to the masses of the objects is easy to test; in one case, anyway. A 10 kg object has twice the mass of a 5 kg object. Remember, we are speaking here of the objects' *masses,* not their weights. The law of gravity would predict that a 10 kg object has twice the weight of a 5 kg object, and indeed it does. Weight is proportional to mass, so the law seems to work when one of the objects is the Earth.

To test the force's dependence on distance, we must be able to compare forces on objects at different distances. The law states that the force is inversely proportional to the square of the distance between the objects' centers. For example, if the two objects are the Earth and your book and if you take your

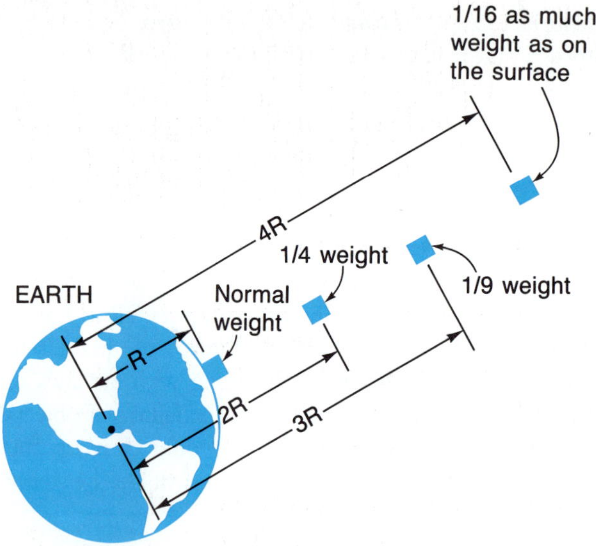

Figure 5–4. *When an object gets farther from Earth, its gravitational force toward Earth decreases as the square of the distance to Earth's center.*

HISTORICAL NOTE

Newton, Hooke, and Halley

IT IS INTERESTING THAT there were other thinkers of Newton's time who hypothesized a law of gravitation, and one of them, Robert Hooke, even suggested an inverse square relationship for distance. Newton went farther than this, however, and checked the idea of an inverse square law by using the Moon calculations we discuss in the text. It happened that in doing the check, Newton made some slight error, so that he found that his hypothesis was only "pretty nearly" right, in his words. Because he was not satisfied with being only pretty nearly right, he put his calculations aside. This was about the year 1666. Eighteen years later, others were still speculating on the nature of a possible gravitational force, and Edmund Halley (of Halley's Comet fame) went to see Newton to ask him what type orbit a planet would follow it if obeyed an inverse-square-distance law. Newton replied that such a law would result in an elliptical path, and that he had done the calculations some time back. Since that time he had lost them! Upon the urging of Halley, he reproduced his work and included it with his other findings in *The Principia*. We might say that while the rest of the world was searching for the law of gravity, Newton had already found it and lost it.

book to a location twice as far from the center of the Earth as it is now, the law says that the book will weigh only one-fourth (which is $1/2^2$) as much. At three times as far from the Earth, it will weigh one-ninth as much. (See Figure 5–4.)

Newton, of course, was unable to change an object's distance to Earth's center significantly. He could have taken an object up on a mountain and measured its weight, but his theory predicted that the weight change in this case would be so small that it would be unmeasurable with the methods he had available. Instead, he used an object already in the sky: the Moon. And rather than comparing forces directly, he compared accelerations produced by the forces.

We have seen that every object on Earth accelerates toward the Earth with the same acceleration, 9.8 m/s^2, or 32 ft/s^2. From his second law, Newton knew that this acceleration would change in the same way that the force does, that the acceleration would be less at greater distances from the Earth, and that it would decrease as the square of the distance. The Moon's path is actually an ellipse, but it is close enough to circular that we can consider it a circle to do a rough test of the law of gravity. Newton knew that the distance from the center of the Earth to the Moon was about 60 times the distance from the center of the Earth to its surface. (In the next chapter we will see how this was known.) According to his law, then, the centripetal acceleration of the Moon should be $(1/60^2)$, or 1/3600 of the acceleration of gravity on Earth. But how could he measure the acceleration of the Moon? He did this by using the fact

Using equations, with m = *mass of the object and* m_E = *mass of Earth:*

$$F = \frac{Gmm_E}{d^2}$$

The object accelerates, however, so F = ma. *Thus:*

$$ma = \frac{Gmm_E}{d^2}$$

and

$$a = \frac{Gm_E}{d^2} .$$

This equation tells us that the gravitational acceleration towards Earth depends only upon the mass of the Earth and the distance from Earth.

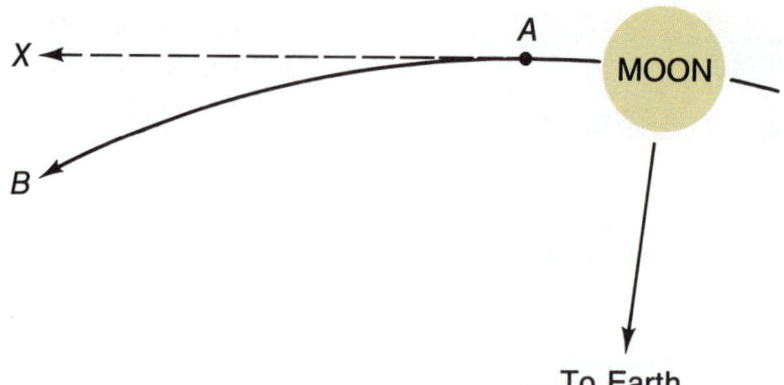

Figure 5–5. *If the gravitational force on the moon suddenly ceased at point* A, *the moon would continue on to point* X *instead of curving to point* B. *The distance from* X *to* B *is how far the moon "falls" toward the earth during the time it moves from* A *to* B.

that the Moon moves in (nearly) a circle and therefore has a centripetal acceleration—it is accelerating toward the Earth.

An object with half the acceleration will gain half the speed and thus get half as far.

If one object has half the acceleration of another, it will fall half as far in the first second of fall. Figure 5–5 shows a very small part of the Moon's orbit. The dashed arrow is the path the Moon would take if there were no force on it—it would go in a straight line from *A* to *X*. Instead, it curves along the path of the solid arrow, going from *A* to *B*. If the amount of time that passes between points *A* and *B* is one second, we can consider that the Moon has "fallen" from *X* to *B* during that second. Knowing that the Moon takes 27.3 days to complete its orbit, we can calculate that the distance is about 0.14 centimeters. Now, Newton predicted that the distance would be 1/3600 of the distance an object falls on Earth during the first second of fall. One can measure (or calculate) that on Earth an object travels a distance of 4.9 meters during its first second of fall. And 1/3600 of 4.9 meters is 0.00136 meters, or 0.136 centimeters. The law works!

The speed after one second is 9.8 m/sec, and so the average speed is 4.9 m/s during the one second. Thus the object travels 4.9 m in the second.

Newton, in fact, did calculations similar to those above to check his hypothesis concerning gravitational force. This was one of the calculations that convinced him his law was valid.

NEWTON'S LAWS AND KEPLER'S LAWS: THE ELLIPSE

The derivation of Kepler's laws from Newton's requires calculus, a branch of mathematics whose discovery will be discussed later.

We have shown how Newton was able to use circular motion to apply his laws to the motion of the Moon. In fact, Newton was able to show that based on his laws of motion and his law of gravity, objects in orbit around the Sun would be compelled to move in elliptical orbits. This, of course, is what Kepler had found to be their paths. We will not show the mathematics involved in Newton's calculations, but it is important to emphasize that if Kepler's first law, the law of ellipses, had not been known beforehand, it could have been derived from the laws formulated by Newton.

Kepler's second law states that as a planet orbits the Sun, a line from it to the Sun sweeps out equal areas in equal times. Again, we will not show the mathematics that prove that this law necessarily follows from the laws of Newton, but we will show, without math, that Newton's laws at least make Kepler's

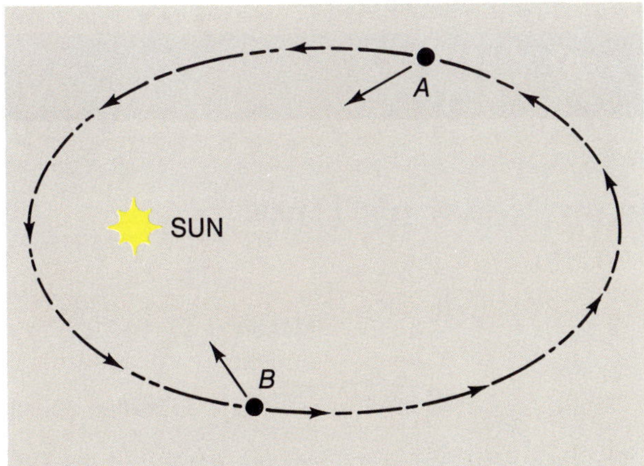

Figure 5–6. *At point* A *the planet is moving closer to the sun and the gravitational force on it causes it to speed up as well as curve from a straight line. At point* B, *the force of gravity slows the planet as well as curving its path.*

second law seem reasonable. Figure 5–6 shows the exaggerated elliptical path of a planet. Consider the planet when it is at position *A,* getting closer to the Sun as it moves along. The arrow pointing from the planet toward the Sun represents the gravitational force exerted on the planet by the Sun. Notice that when the planet is in this position, the direction of the force is not perpendicular to the motion of the planet; it is not simply a centripetal force. If it were a centripetal force it would be exerted perpendicular to the motion of the planet and it would have only one effect: to cause the planet to curve in its path. Instead, it has two effects: it causes the planet to curve, and it also causes the planet to speed up. This is because it pulls *forward* as well as sideways on the planet.

At location *B* in that figure the planet is moving away from the Sun. Here the arrow representing gravitational force shows us that the force pulls both backwards and sideways on the planet, thus slowing it down as well as curving its path. So we see that as the planet moves toward the Sun it speeds up and as it moves away it slows down. This is what is predicted by Kepler's second law.

We have not shown that Newton's laws would actually result in Kepler's law of equal areas, but only that both laws would result in the planet going fastest when it is closest to the Sun. In fact, it can be shown mathematically that application of Newton's laws of motion and his law of gravitation *necessarily* result in the law of equal areas. Here again, if this law of Kepler had not already been known by Newton, he could have deduced it from his laws.

Kepler's Third Law

Finally, we come to the third law of Kepler, the one that relates a planet's period of revolution to its distance from the Sun. It was this law of Kepler that led Newton and others to surmise that any force of gravity must obey an inverse square law. It is shown in the accompanying Close Up that if nature obeys an inverse square law for gravitation, planets must necessarily have the period-distance relationship of Kepler's third law. In addition, Newton added some-

CLOSE UP

Kepler's Third Law from Newton's Laws

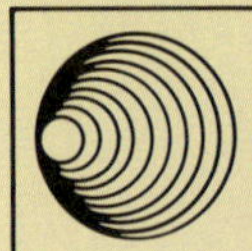

KEPLER'S THIRD LAW CAN be written as follows:

$$\frac{d^3}{P^2} = C$$

where P is the period of a planet's motion around the Sun, d is its average distance from the Sun, and C is a constant. We saw that when units of years and astronomical units are used in the equation, $C = 1$. Newton's laws show us why this expression works and they also allow us to learn more about the value of C, as we will show below.

It was stated in a margin note that the centripetal acceleration (a) of any object moving in a circle is given by

$$a = \frac{v^2}{r} \qquad \begin{aligned} v &= \text{the speed of the object} \\ r &= \text{the radius of the circle} \end{aligned}$$

We also saw in a margin note that the acceleration caused by the gravitational force on an object in orbit is

$$a = \frac{GM}{d^2} \qquad \begin{aligned} G &= \text{constant, depending on the units used} \\ M &= \text{the mass of the central object} \\ d &= \text{the distance between the objects} \end{aligned}$$

(In the expression shown in the margin, "m_E" was used for the mass of the central object that attracts the orbiting object because the Earth was the central object in that case.)

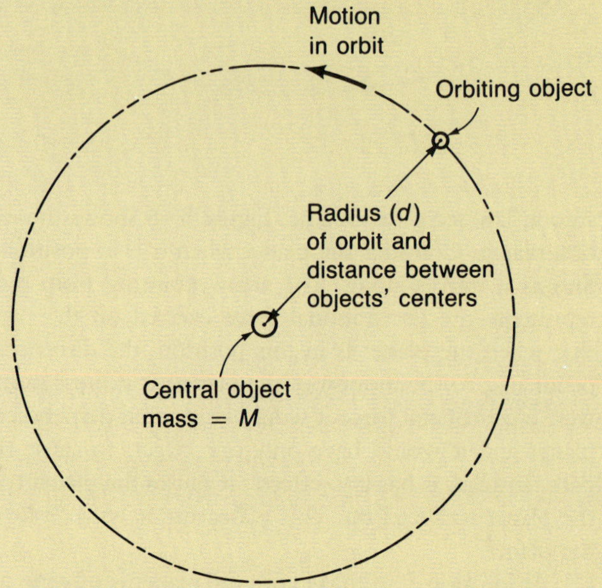

Figure T5–3. *As one object moves in a circular orbit around a central object, we use* d *to represent the radius of the orbit (which is also the distance between the centers of the objects).*

If we consider an object in circular orbit, these two accelerations are the same, for gravitational force supplies the centripetal acceleration. In addition, the radius of the circle is d, the distance between the objects (Figure T5–3). We set the two expressions equal:

$$\frac{v^2}{d} = \frac{GM}{d^2}$$

or

$$v^2 = \frac{GM}{d}$$

Now, the speed of an object in a circle can be expressed in a different way, using the definition of speed. The circumference of a circle is $2\pi r$, or in this case $2\pi d$, since d is the radius of the circle (*not* the diameter). Thus, using P for the time to circle the orbit,

$$v = \frac{2\pi d}{P}$$

v = velocity in a circular orbit
d = radius of the orbit
P = period of the circular motion

Next, we put this in for v in the previous expression and simplify it:

$$\frac{(2\pi d)^2}{P^2} = \frac{GM}{d}$$

$$\frac{4\pi^2 d^3}{P^2} = \text{GM}$$

$$\frac{d^3}{P^2} = \frac{GM}{4\pi^2}$$

Finally, since $G/4\pi^2$ is a constant, we will replace it with the symbol K:

$$\frac{d^3}{P^2} = KM$$

d = radius of orbit
P = period of orbit
K = a constant whose value depends on the units used
M = mass of the central object

Compare this to the first equation in this Close Up. We see that the constant C in the original statement of Kepler's third law depends upon the mass of the object at the center of the orbit. In the case of the solar system, this is the mass of the Sun, but for Jupiter's system of moons the mass of Jupiter would be the mass of importance.

Finally, we must point out that we have oversimplified things slightly. We will see in a later chapter that M is not simply the mass of the central object but the total mass of the central object and the object in orbit. In the cases we are now considering, however, the mass of the orbiting object is so much less than that of the central object that it is unimportant in the calculation. (The mass of any planet is very small compared to the mass of the Sun.)

The importance of the above derivation is that it shows not only that Kepler's third law follows directly from Newton's laws but that Newton's laws tell us something more about the situation than did the original law of Kepler.

It was by means of Newton's statement of Kepler's third law that the masses of the planets were first calculated. We will use the equations above in Chapter 7 to calculate the mass of Saturn.

thing to Kepler's third law, showing that the masses of the objects are important in the relationship, and that the law is expressed more fully as

Actually, M *is the sum of the masses of the two objects, both the Sun and the planet in this case. The difference is unimportant to us now, but will be discussed in the next chapter.*

$$\frac{d^3}{P^2} = KM$$

d = average radius or orbit
P = period or orbit
K = a constant whose value depends upon the units used in the equation
M = mass of object at center of orbit

Science seeks to show that the various phenomena we observe are not independent of one another but are instead manifestations of a relatively few basic principles. We see here the success of Newton's work in relating the laws of Kepler with more basic ideas: those of gravitation and the motion and mass of objects. Once Newton's laws are known, all three of Kepler's laws necessarily follow.

Kepler's laws only described what was observed without stating *why* nature worked as observed in terms of other phenomena. Newton's laws supplied the "why" of Kepler's laws. Notice, however, that Newton did not answer "why" in the ultimate sense, for we can still ask why universal gravitation exists. Science does not answer the ultimate "why," but rather ties together the phenomena we observe. Thus Newton's laws allow us to see that the orbits of planets are related to the pull our bodies feel toward the Earth.

Falling Objects (Once More)

Recall that Aristotle had argued that a heavier object should fall faster because there is a greater force on it, whereas Galileo demonstrated that all objects fall at the same speed. Now we can use the law of gravity and Newton's second law to see why Aristotle's prediction doesn't happen. The law of gravity tells us that the weight of an object is proportional to its mass. That is, if one object has twice the mass of another, it will weigh twice as much. Now apply Newton's second law, $a = F/m$. If we drop the two objects, we see that since the one with twice the weight (force) also has twice the mass, it will accelerate just the same as the lighter object (because F/m is the same). Aristotle's problem is explained!

THE IMPORTANCE OF NEWTON'S LAW OF GRAVITY

Up until the time of Newton, Aristotle's ideas of the separateness of the Earth and the heavens was the accepted world view. We cannot pretend that it was only Isaac Newton who thought differently, for from the time of Galileo the idea of the oneness of nature had begun to take shape. In fact, other thinkers of Newton's time were working on theories of gravitation. It was Newton, however, who put it all together. The law of gravity and his three laws of motion make up the heart of what we call "Newtonian mechanics" (or "classical mechanics," to distinguish it from the mechanics of Einsteinian relativity). It is difficult to overemphasize the effect that Newton's work has had on our thinking.

For one thing, these laws were the first ever that could be shown to hold for both the heavens and the Earth. The idea that there were two natures, one up there and one down here, had been a part of western culture since before Greek times. Now it was broken; now it was possible to look upon the universe as ONE. In fact, the word universe begins with the prefix "uni," meaning "one, single" (*uni*t, *uni*fy, *uni*que, *uni*ted, and so on). Such a concept was foreign before the time of Newton. He showed us that we live in one cosmos, that nature is singular, that things up there might be expected to be like things down here. In a sense, Newton changed our world from a "duoverse" to a *uni*verse. Our modern science of astronomy could not exist without this basic understanding.

In addition, the work of Sir Isaac Newton gave us the idea that nature is explainable; that if we work hard enough at it, we have the ability to understand the many seemingly mysterious things that occur in nature. No longer do we have to fear the darkness, for the objects in the skies are part of our universe.

Change comes slowly. We can't pretend that Newton caused societal beliefs to change overnight, but he led us in a giant step forward. In many U.S. newspapers there is a daily astrology column but few newspapers have a daily article on astronomy. Even today, three hundred years after Newton, there are many who continue to look to the stars and planets for signs.

It has been claimed that in the U.S. there are more astrologers making a living casting horoscopes than there are astronomers doing science.

THE INVENTION OF CALCULUS

Newton hypothesized, in his law of gravity, that every object in the universe attracts every other one. The most obvious example of this attractive force is, of course, the force between objects on Earth and the Earth itself. Newton realized, however, that the Earth is not a single entity but is made up of countless objects. He must have asked what constitutes an object. Is a rock a single object, or is a rock made up of tiny grains of smaller rock? To carry this farther, today we know that these tiny grains are made up of atoms. It seems that his law would have to be applied to individual particles—perhaps atoms—rather than to the Earth as a whole. Figure 5–7 indicates forces pulling an object on Earth toward a number of particles within the Earth. The number of particles, of course, is immense and the drawing only shows a few, but the overall attraction the object feels toward the Earth must result from the sum of these individual attractions.

Newton wondered how all these individual attractions add up. The problem is one of adding an uncountable number of forces, each having its own magnitude and direction. Since the Earth is spherical (or very close to it), it is reasonable to expect that the overall force should be directed toward the center. (In the figure, we show as many forces pulling downward toward the left of center as we show toward the right of center, and the two sets of forces should be equal in magnitude.)

Figure 5–7. *The doll feels a force toward each little particle within the Earth. It is the sum of all these forces that is the total force of gravity toward the Earth.*

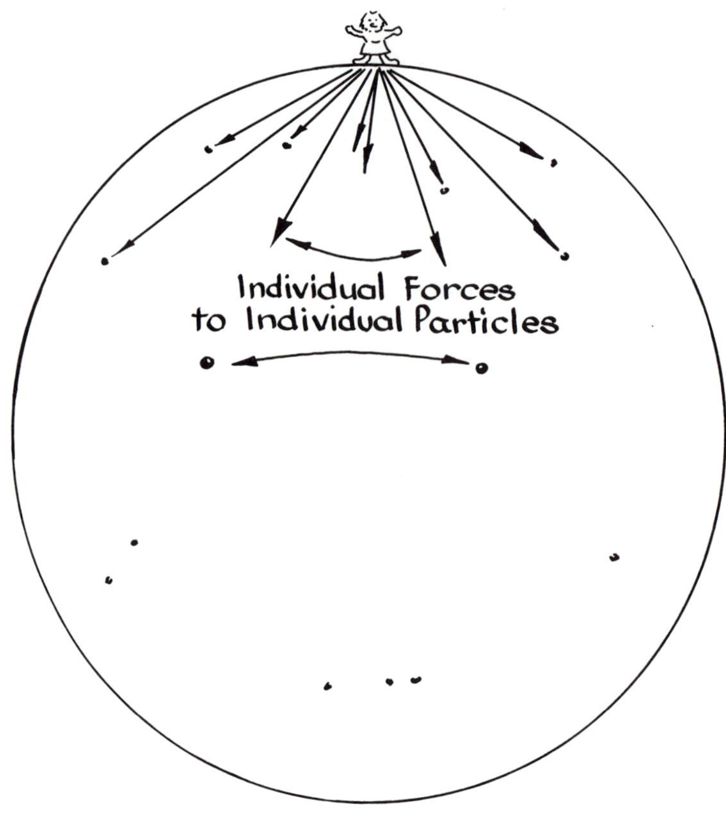

There are many cases of scientific advances being made independently by two different people at about the same time. The invention of calculus is an example of this. Baron von Leibniz was the other inventor. His system was apparently simpler than Newton's and the calculus studied today is closer to the system of Leibniz than to Newton's.

Newton wanted to be able to add all the attractive forces and see if the result agreed with his hypothesis that the entire mass could be considered as being at the center. The necessary mathematics did not exist, however, that would allow him to calculate the overall effect of an unimaginably large number of unimaginably small forces. So he invented a new branch of mathematics—calculus—to solve this problem. As we have indicated, the result of his use of the calculus was that the distance used in the law of gravity, when applied to astronomical objects, is the distance between the *centers* of the objects, so that when the Earth is one of the objects, distances must be taken to the center of the Earth. We use the entire mass of the Earth in determining the force and we act as if that entire mass were at a point at the center of the Earth (see Figure 5–8).

ARTIFICIAL SATELLITES AND NEWTON'S LAWS

The reason for getting above the atmosphere is that he did not want air friction to hinder the object's motion.

Figure 5–9 is a copy of a drawing in Isaac Newton's *Principia*. Newton proposed that we imagine that a mountain could be found high enough that it extended above the Earth's atmosphere. On that mountain is a cannon. If we fire the cannon, the cannonball will follow a path somewhat like the one that strikes the Earth at *D*. Put more gunpowder in the cannon and the cannonball might

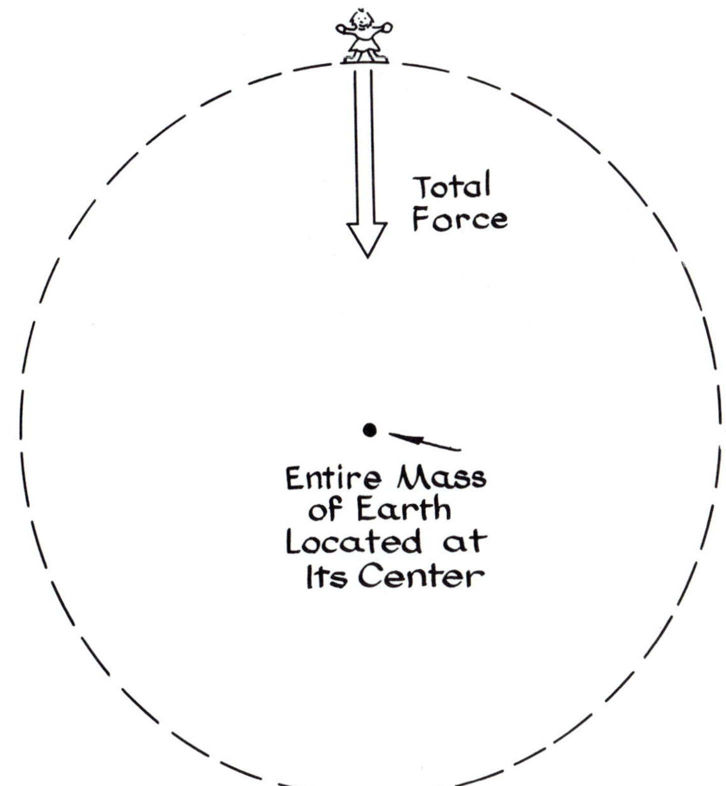

Figure 5–8. *The effect of the forces in Figure 5–7 is a single force toward the center of the Earth, as if the entire mass of the Earth is located there.*

reach point *E* or even point *F*. The cannonball is going part of the way around the Earth. Add even more gunpowder and the cannonball might go as far as point *G*. Finally, we put in even more gunpowder—just the right amount. The cannonball moves in the circular path shown as *V*. The ball will continue going around the Earth (assuming that we move the gun out of the way). The ball will be in orbit. Newton, more than 300 years ago, predicted that such an orbit was possible. He knew, though, that there was no mountain large enough and that such a powerful cannon was impossible (or at least impractical). Today, however, we use another of his principles to propel objects into orbit around the Earth.

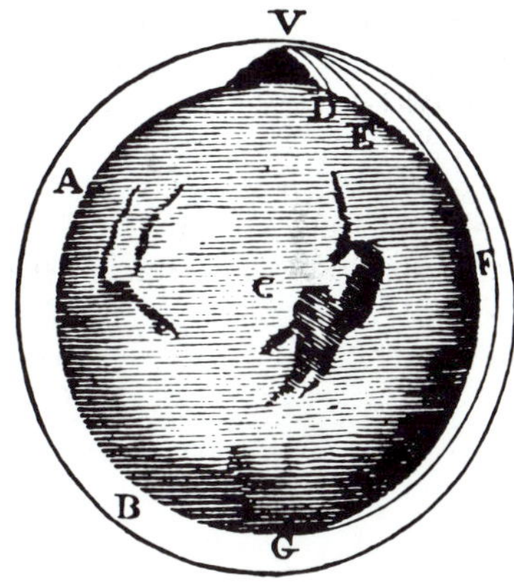

Figure 5–9. *Isaac Newton predicted that if a cannon could be made strong enough and be placed above the atmosphere, it could put a cannonball into orbit.*

Recall what happened the last time you blew up a balloon and then released it. Can you explain what made it take off across the room? The force propelling a released balloon across a room has much in common with the force lifting a spacecraft from Earth or changing its speed or direction in space. Newton's third law states that when object *X* exerts a force on object *Y*, *Y* exerts an equal and opposite force back on *X*. In the case of the balloon, the stretched rubber of the balloon exerts a force on the air to squeeze it out. But if the rubber exerts a force on the air, pushing it out the hole, the air in turn exerts an equal and opposite force on the rubber of the balloon. So the air goes one way and the balloon the other.

In the case of a rocket, it is the burning propellant that is forced out the back of the rocket. And if there is a force pushing the propellant out in one direction, there must be a force pushing the rocket the other direction. Notice that the force does not depend on there being something behind the rocket (or balloon) for the escaping gas to push against. Therefore a rocket works in a vacuum as well as in air. The simple fact that the gas is being pushed backwards forces the rocket (and the balloon) forward.

HUMANS IN SPACE: WEIGHTLESSNESS

In October 1957, the USSR launched the first artificial satellite, named Sputnik (which translates to "travelling companion"). This spurred the United States to quicken its space activities, and in the years since Sputnik, the National Aeronautics and Space Administration (NASA) has launched more than 600 satellites into orbit. The USSR has launched about the same number and other nations now have their own satellites orbiting the Earth and probing space well beyond the Earth.

In later chapters we will discuss space probes sent to examine planets and other objects in our solar system, as well as probes that have been sent beyond the solar system.

The first satellites were, of course, without passengers (as are most of today's). Since the 1960s however, humans have become so common in space that today (except for the launching of NASA's space shuttle) humans in orbit receive little attention in the press.

Figure 5–10 shows the liftoff of a space shuttle. What happens to the astronauts in the spacecraft when the rockets are firing? They are accelerating forward along with the craft, of course, but to accelerate them requires a force and since the acceleration is great, the force is great. The force is provided by the couch on which the astronaut lies. He or she feels pressed back into the seat in much the same way that you feel pushed against a car seat when the car accelerates—just more so.

You have seen photos or televised segments of astronauts floating freely in their spacecraft. Figure 5–11 is one. It may seem that this apparent weightlessness does not square with Newton's law of gravity. Today's space shuttles orbit the Earth at a height above the surface of between about 150 and 200 miles. This means that they are between 4150 and 4200 miles from Earth's center instead of the 4000 miles they were when they were on the surface. This relatively small change in the distance from the Earth's center results in a fairly small decrease in the gravitational attraction of the astronauts toward the Earth—their weight. In the accompanying Close Up we show that a person in orbit 200 miles above Earth's surface has only 10 percent less attraction toward Earth at that height. A 150-pound astronaut still has a gravitational force of 135 pounds on him in orbit; he certainly is not weightless. Then why do the astronauts float around?

Suppose that Newton had two cannons on his mythical mountain and had fired a small cannonball along with the big one. If he had fired them at the same time and at the same speed, they would have followed exactly the same path, moving along side by side. (Recall that a heavy and a light object fall with

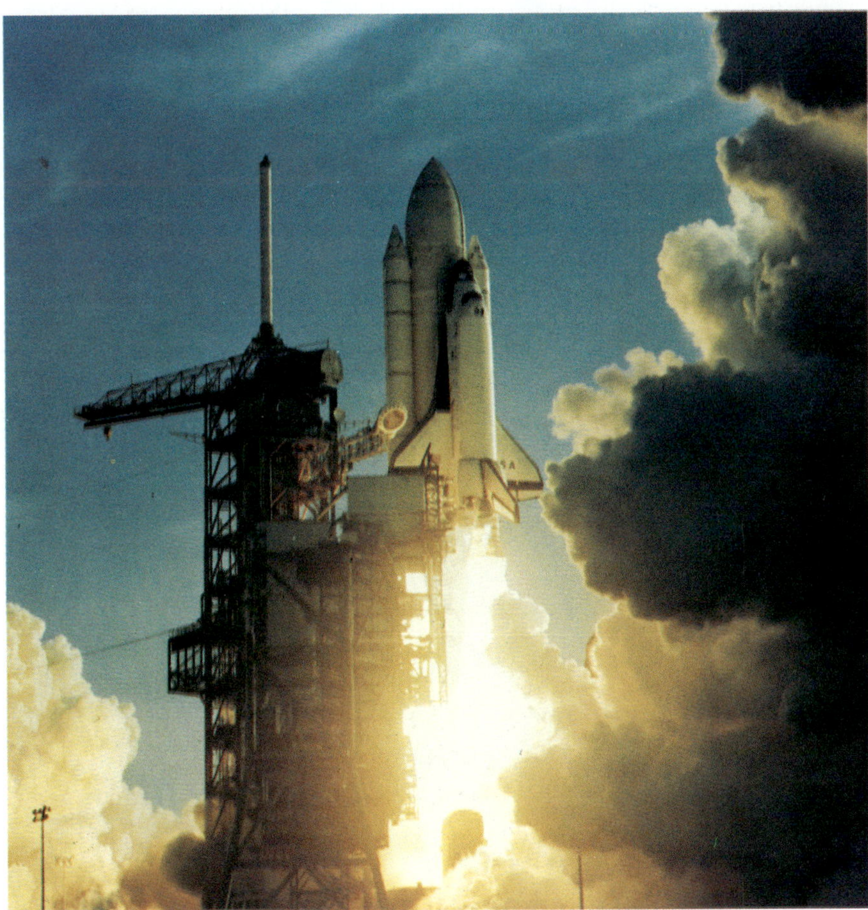

Figure 5–10. *The liftoff of the Space Shuttle Columbia.*

the same acceleration.) Now suppose that the larger cannonball had been hollow and the smaller one had been inside it. Since small and large objects follow the same path, the small one would simply ride along in the larger one without pressing against the sides. If a cannonball could feel, it would *feel* weightless. It would be freely falling along with its container, the larger cannonball. We say that it is in free fall, the condition of an object on which the only force acting is the force of gravity.

A similar effect would be felt by a person in an elevator falling freely after its cable broke. The person would feel weightless, floating about in the elevator.

It may seem odd to refer to an object in orbit as *falling,* for the object gets no closer to the Earth as it falls. The reason the satellite never gets to the surface, of course, is that the Earth's surface is curved by the same amount that the satellite curves in its path. For the satellite to remain in a circular orbit at a given distance from Earth's center, there is a certain definite speed the satellite must have. Newton's laws allow us to calculate that speed.

At this point we must decide upon a precise definition of the word *weight.* At least two definitions are in common use, one being the gravitational force exerted on an object according to Newton's law of gravity, and the other being the force with which an object pushes downward on whatever is supporting it.

Figure 5–11. *Astronaut Gerald P. Carr is in apparent weightlessness in Skylab 4.*

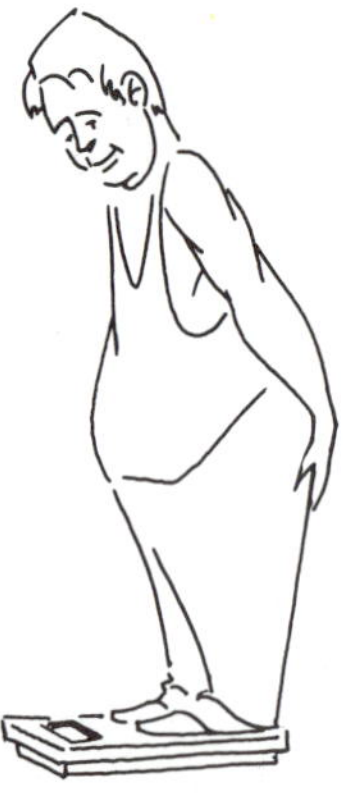

Figure 5–12. *Weight is usually thought of as the force with which we push down on our support (the scale, here), which is equal to the force with which the support pushes up on us. This is not the way we define it in this text.*

As Figure 5–12 indicates, the latter is the everyday working definition of weight. Here on the surface of Earth the two definitions very nearly correspond, but as we have seen, an object in orbit 200 miles above Earth's surface has 90 percent as much gravitational attraction as it does on the surface, even though it has no need for a support at all. According to the first definition, the orbiting object weighs 90 percent of what it did on Earth but according to the second definition, it is weightless. Both definitions are used at various times by scientists, but in this book we will use the first one, defining ***weight*** as the gravitational attraction (according to Newton's law of gravity) exerted on an object by the astronomical body on which it is located. (Reference to an "astronomical body" allows us to speak of the weight of an object located on something other than Earth—the Moon or Mars, for example.)

According to our definition of weight, astronauts in a spacecraft are not truly weightless. There is definitely a force of gravity on them. (If there were no force of gravity on them, they would not remain in orbit but would move away from Earth following a straight line path. Astronauts and spacecraft can't disregard Newton's laws!) Astronauts *feel* weightless because they are in free fall, just as if they jumped off a very tall building.

Thus far we have considered only circular orbits, but the same conditions

CLOSE UP

Calculating Gravitational Force on an Astronaut

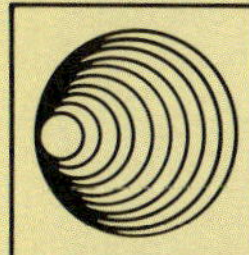

WE WILL USE NEWTON'S law of universal gravitation to calculate how much the gravitational force exerted by the Earth on a person changes when the person moves from Earth's surface to an orbiting height 200 miles above the surface. Since the mass of the person does not change when he or she is raised above the surface and since the mass of the Earth certainly does not change, we need only consider the change in distance. The law of gravity tells us that the force of gravitational attraction depends inversely upon the square of the distance between the objects' centers.

On Earth's surface, we are about 4000 miles from its center. In moving 200 miles up, we increase the distance from center by a factor of 4200/4000. The gravitational force changes by the inverse square of this, or $(4000/4200)^2$. This is about 0.9, and it tells us that an object 4200 miles from Earth's center has a gravitational force on it equal to 9/10 the force on it at the surface. In other words, an astronaut in orbit 200 miles above Earth has a gravitational force on her equal to 90 percent of her weight on Earth. A 130 pound astronaut still has a gravitational force of 117 pounds on her in orbit. This, in fact, is the centripetal force which keeps the person in orbit. Without it, the person (and the spacecraft) would move in a straight-line path and move away from Earth.

Figure T5–4. *Astronaut Buzz Aldrin Jr., the second person on the moon, steps from Eagle's ladder, 21 July, 1969.*

apply in the actual orbits of artificial satellites. And those orbits are none other than ellipses, the same ellipses discovered by Kepler and explained by Newton.

Even on Earth, an object would have a slightly different weight by the two definitions. This is due to the rotation of the Earth.

Weight: The gravitational force between an object and the planet, moon, etc. on—or near—which it is located.

The Beginnings of Extraterrestrial Exploration

Four months after Sputnik was orbited in 1957, the U.S. followed with Explorer I. These orbiting satellites were primarily a triumph of technology rather than of basic science, because the basic science of orbital motion had been worked out by Newton three hundred years before. The first satellites were quick to

Figure 5–13. *A photo of part of New York City taken from space.*

Van Allen Belts: Region around the Earth where charged particles are trapped by the Earth's magnetic field. (Discussed more fully in Chapter 7.)

begin their contributions to science, however, starting with Explorer's discovery that there are areas around the Earth where nuclear particles and electrons are trapped by the magnetic field of the Earth. The discovery of these ***Van Allen Belts*** was the first scientific discovery of the Space Age. Since that first discovery, the space programs of the U.S. and other nations have contributed greatly to our knowledge, particularly in the fields of Earth science and astronomy. Figure 5–13 is a photo of part of New York City taken from an orbiting satellite.

As you recall from Chapter 1, the Moon orbits the Earth in such a way that the same face of the Moon points toward Earth at all times. We cannot see its backside from Earth. In 1959, the Soviet Union launched the first spacecraft to go around the Moon. Although the Moon is the most familiar nighttime object in our sky, we had seen only half of its surface until the USSR sent cameras to its other side.

From 1964 through 1968 NASA spacecraft orbited the Moon and landed a number of robot spacecraft on it. These took some 17,000 pictures of the Moon's surface and analyzed the chemical composition of its surface. In addition, the USSR has used robot spacecraft to retrieve samples of the lunar soil and return them to Earth. The most exciting of the lunar adventures, however, were the Apollo missions that landed astronauts on the Moon.

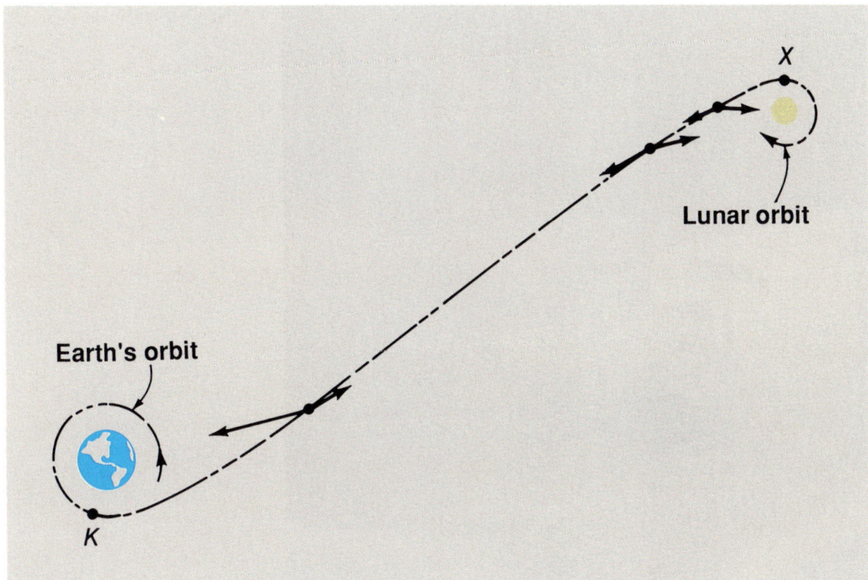

Figure 5–14. *A craft sent to orbit the moon starts out in earth orbit. A rocket is fired at point* K *to send it toward the moon. At point* X, *a rocket is fired to put it in Moon orbit. The arrows along the path show the approximate relative gravitational pull to the earth and moon. Distances are vastly out of scale here.*

TRAVEL TO THE MOON

> *I believe that this nation should commit itself to achieving the goal before this decade is out, of landing a man on the Moon and returning him safely to Earth.*
>
> —President John F. Kennedy, 1961

The motivation for President Kennedy's commitment to place a man on the Moon was not primarily scientific; it involved national and international politics. Nonetheless, science—particularly astronomy—certainly benefitted from that program.

The gravitational aspects of a flight to the Moon are not vastly more complicated that those of Earth orbit. The mission begins with the spacecraft orbiting the Earth. As indicated in Figure 5–14, a rocket then launches the spacecraft out of Earth orbit to begin its trip to the Moon. As it gets farther from the Earth and nearer the Moon, the gravitational pull of the Moon on the spacecraft becomes significant. Recall that the law of gravity tells us that *every* object in the universe attracts every other. This means that the Moon exerts a gravitational force on a satellite even when it is in Earth-orbit (and in fact on us here on the surface of the Earth). This force is very small compared to the gravitational attraction toward the Earth until the craft gets closer to the Moon. Along the path of the Moon-bound craft in Figure 5–14 we have drawn arrows indicating the relative amount of gravitational force exerted by the Earth and Moon.

The Moon's gravitational force on your body is only three parts per million of your weight. That is, the Moon exerts only 3/1,000,000 of the gravitational pull on you that the Earth does.

There is one point where the gravitational pulls toward Earth and Moon exactly cancel one another. Up until this point, the force pulling the craft toward the Earth has been greater than that pulling it toward the Moon and the spacecraft has been slowing down. After passing that point, the pull toward the

Figure 5–15. *Photograph of the Gemini VII spacecraft taken through the hatch window of Gemini VI during rendezvous maneuvers at an altitude of approximately 160 miles in December 1965.*

Moon is greater and the spacecraft speeds up. It is interesting to note that the astronauts feel no different when passing through that point. They simply feel weightless all the time. They are in free fall all the time, coasting right along with the craft, and as it changes speed they change speed right along with it, completely unaware of any change. Finally, when the spacecraft has reached the appropriate point in its journey (at about X in Figure 5–14), a rocket is fired to slow it down so that it remains in orbit around the Moon. Without this firing, the craft would have so much speed that it would swing right past the Moon.

The Apollo craft which was used for the Moon landings was a three-part spacecraft: the command module (where the crew normally lived and worked), the service module (the propulsion and support—air conditioning, etc.—system), and the lunar module that landed on the Moon.

Now the astronauts are in orbit around the Moon. At this point the lunar module (LM—often pronounced "lem") disconnects from the command/service module that remains in orbit, fires its rockets to slow down, and descends to the Moon's surface. The lunar module does not need to be very large and does not need particularly powerful rockets. This is because the force of gravity on the surface of the Moon is only one sixth of that on the Earth, so the fall toward the Moon is not as fast and the liftoff requires much less energy. In addition, the Moon does not have an atmosphere, so the frictional effects of an atmosphere need not be considered like they do for takeoff and landing on Earth.

To leave the Moon, part of the lunar module is left behind when a rocket launches the small remaining craft into lunar orbit to reconnect (to "dock") with the command module. Then rockets fire again to send the craft out of lunar orbit and back toward the Earth.

Notice that very little of the trip to and from the Moon is spent with the rockets firing. They are fired only to begin and end each portion of the trip and to make minor midcourse corrections. On one of the missions to the Moon an astronaut was speaking with his son by radio. His son asked who was driving. The reply was that Isaac Newton was doing most of the driving at the time.

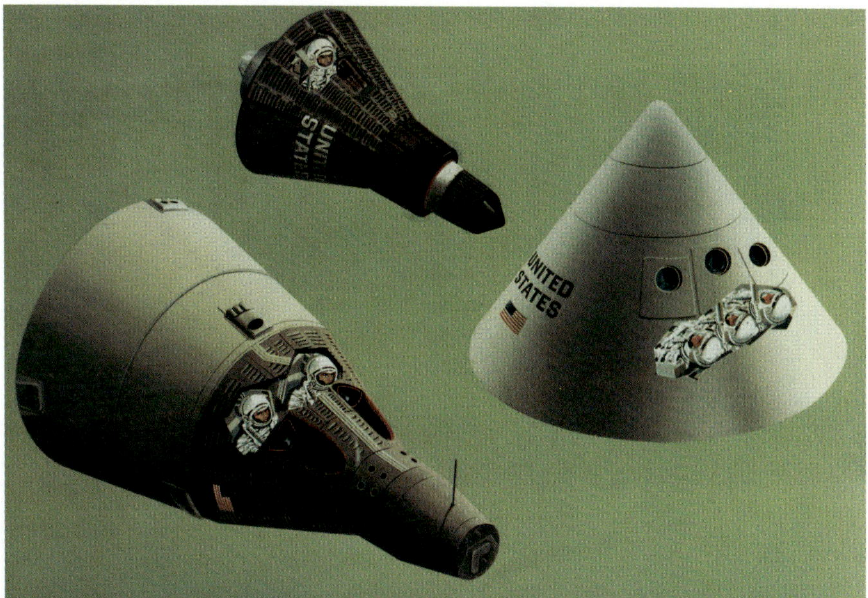

Figure 5–16. *The Mercury, Gemini, and Apollo capsules drawn to scale.*

The Apollo Missions

While the gravitational aspects of landing on the Moon are not particularly complicated, actually doing so is tremendously complicated, especially since we so value the lives of the astronauts. In planning for the Moon landings, NASA had to develop not only physical apparatus such as new computer hardware but new techniques, including some which are now being utilized by business, medicine, and industry. The job of landing people on the Moon and returning them to Earth was the most complicated endeavor ever undertaken. At the time of President Kennedy's commitment, Project Mercury was underway. This was a series of launches of the Mercury capsule, a very small spacecraft in which one person rode more as a passenger than as a pilot. Next came Project Gemini (Figure 5–15), which put pairs of astronauts into Earth-orbit. Finally, Project Apollo was initiated. This had the capability to launch three astronauts into orbit in a fairly comfortable vehicle with its own rocket power, with a lunar landing module, and with enough supplies for an eight- to twelve-day round trip to the Moon. Figure 5–16 shows the Mercury, Gemini, and Apollo craft to scale and Figure 5–17 shows the relative sizes of rockets used to launch various satellites, including the Space Shuttle.

The first flight, originally designated Apollo 1, suffered an 18-month delay because of the tragic fire that cost the lives of three astronauts during preflight tests. Finally, during the Christmas season of 1968, astronauts Frank Borman, James Lovell, and William Anders became the first humans to fly around the Moon. Their Apollo craft orbited the Moon ten times before returning to Earth.

Up until that time, no human had seen the Earth from afar. Recall that in Earth-orbit, astronauts are only about 200 miles above the Earth's surface. From that distance, only a fairly small portion of the Earth's surface is visible at one time. The Apollo missions gave us photos such as that in Figure 5–18, an

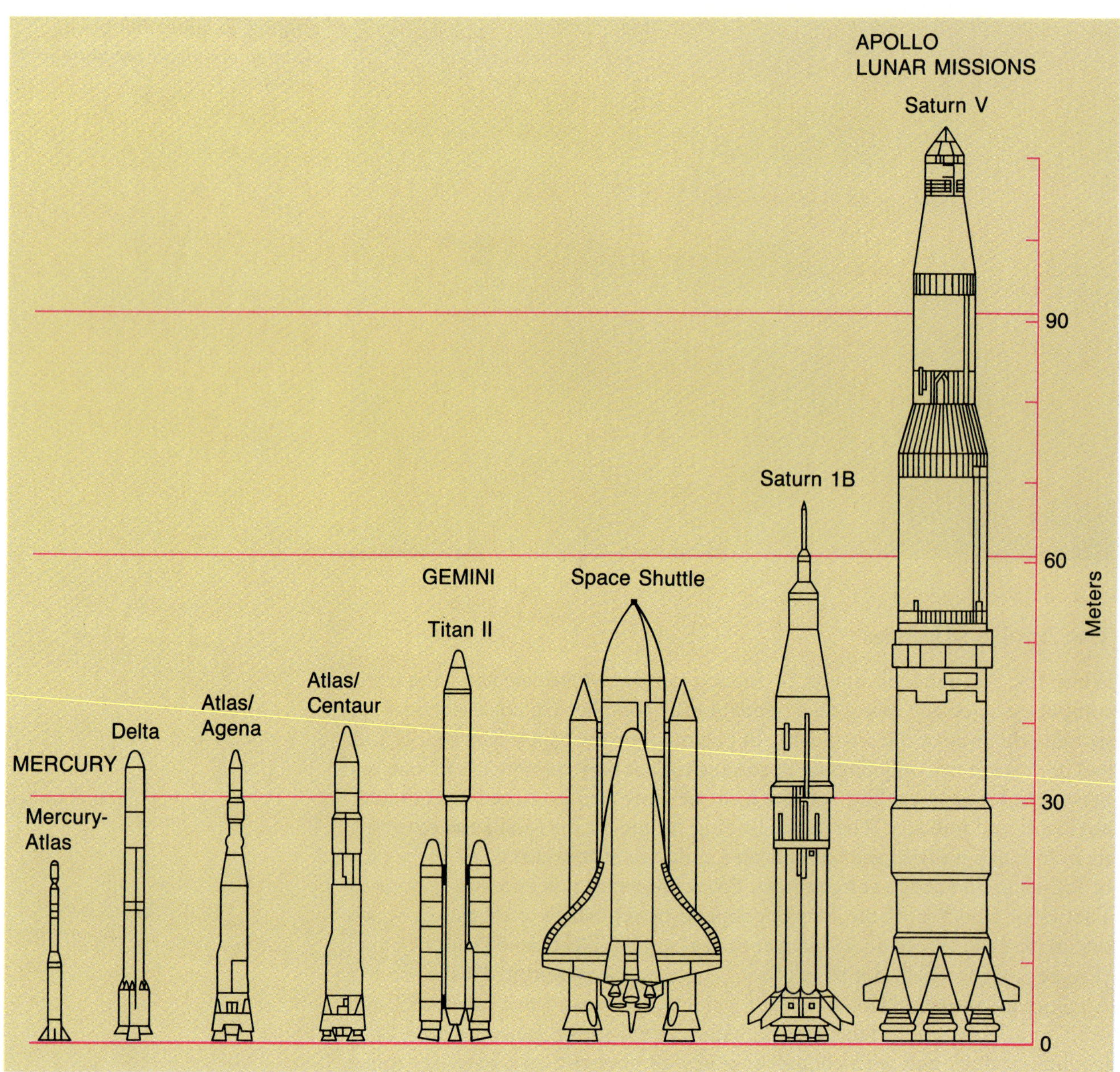

Figure 5–17. *In this scale drawing you can compare the sizes of various rockets to the Space Shuttle. The rockets used to launch Mercury, Gemini, and Apollo are marked (although somewhat different configurations of each particular rocket were used for different missions).*

entirely different view of Earth than humans had ever had. The word *view* here should be taken not only in the photographic sense. Such photos made us realize for the first time the small size of our planet, alone in the vastness of space. They emphasized the fragility of our planet and its beauty. Since the times of Galileo and Newton, we had known that the Earth was but one of a number of planets, that it was not the center of the universe, and that in size it was not particularly important. This abstract knowledge and the actual realization of the

Table 5–1 Manned Apollo Missions

Mission	Astronauts	Date	Primary Goals
7	Schirra Eisele Cunningham	Oct. 11–22 1968	First manned Apollo flight; First live TV from space; (163 orbits).
8	Borman Lovell Anders	Dec. 21–27 1968	10 Lunar orbits—20 hours; First use of Saturn V for manned vehicle; Earth and Moon photos.
9	McDivitt Scott Schweickart	March 3–13 1969	Earth orbit; First manned flight in lunar module; 37 minutes outside spacecraft.
10	Cernan Young Stafford	May 18–26 1969	Lunar module tests in lunar orbit; 61.6 hrs in lunar orbit; First live color TV from space.
11	Armstrong Collins Aldrin	July 16–24 1969	First landing, July 20; 2 hr, 31 min outside module; 20 kg of lunar soil returned.
12	Conrad Gordon Bean	Nov 14–24 1969	Found parts of Surveyor 3; 31 hrs on Moon; 34 kg of lunar soil returned.
13	Lovell Swigert Haise	April 11–17 1970	Mission aborted. One lunar orbit; used life support of LM to survive.
14	Shepard Roosa Mitchell	Jan + 31–9 1971	Outside LM 9 hrs. 25 minutes Used hand cart to gather 42 kg of lunar material.
15	Scott Irwin Worden	July + 26–7 1971	Mountainous area; used Lunar Roving Vehicle; 66 kg of lunar soil returned.
16	Young Mattingly Duke	April 16–27 1972	Studied highland area; 71 hrs on Moon; 95.8 kg of lunar soil returned.
17	Cernan Evans Schmitt	Dec 7–19 1972	First scientist on Moon (Figure 5–19); rode 30.5 km on Moon; 110.4 kg of lunar soil returned.

Figure 5–18. *Photos such as this taken by the Apollo 17 astronauts changed our attitude about the Earth.*

Figure 5–19. *This photo shows scientist-astronaut Harrison Schmitt standing next to a boulder during the Apollo 17 mission. Schmitt was the only scientist to go to the Moon.*

fact are vastly different, however. Certainly an increased appreciation of our place in the universe resulted from photos taken by the astronauts and from their descriptions of the Earth from space. It may be that the most important benefit from the space program is the change in humanity's outlook concerning its home planet.

PLANETARY EXPLORERS

Of the many space achievements by the U.S. and the USSR, we are particularly interested in spacecraft that have been sent to other planets. These include the following: landers, which have descended to the surfaces of Venus and Mars and sent back valuable information about the conditions there; orbiters, by which we have been able to map the surfaces of Venus and Mars; and flybys, which have produced tremendous amounts of information on all the planets (except Neptune and Pluto), numerous moons of other planets, and Comet Halley. (At the time of writing we were still awaiting Voyager's passage by Neptune in August, 1989.)

Figure 5–20 shows the various missions the United States has sent to other planets. Triangles in the figure show the dates of launch. In looking at the figure, one might wonder why it has not been updated since 1978. In fact, the U.S. has not launched a craft to another planet since then. This lack of planetary exploration cannot be blamed on the Space Shuttle Challenger accident of

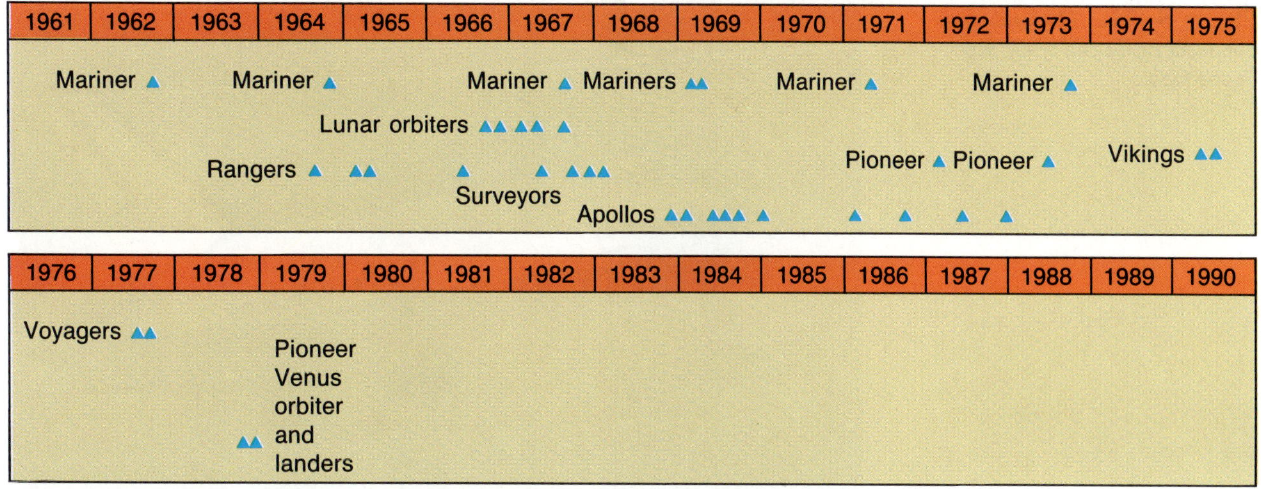

Figure 5–20. *Dates and brief descriptions of U.S. lunar and planetary explorers.*

January 1986, for such missions require many years of preparation prior to their launch. The long hiatus from this endeavor actually began in the early 1970s. Many scientists are very concerned about the lack of effort in this area and are seeking to revitalize exploration of the planets. Astronomers look forward to the launching of Magellan (the radar mapper of Venus), and Galileo (the craft that will orbit Jupiter and jettison a probe into that planet's atmosphere).

THE SPACE SHUTTLE

In 1981 the United States began the first orbital flights of the Space Transportation System, the Space Shuttle. This first reusable spacecraft has a number of parts; a reusable orbiter with wings to enable it to land, a huge fuel tank on which the orbiter rides as it is launched and which is jettisoned and not reused, and two solid rocket boosters that are jettisoned into the ocean during launch and then retrieved. The orbiter (Figure 5–21) weighs 75 tons and has a cargo bay 18.3 meters (60 feet) long and 4.6 meters (15 feet) in diameter. The vehicle is designed to be used to retrieve and repair satellites in orbit, to serve as a base from which to deploy satellites, and to be a platform on which to do research in space. Of special interest in astronomy will be the use of the Shuttle to launch the Space Telescope, the Gamma Ray Observer, and the Galileo and Magellan craft mentioned above. We will discuss these in later chapters.

The Space Shuttle is limited to Earth orbit. It is not designed to carry people into deeper space and therefore will have a difficult time bringing back the excitement of the Apollo Moon landings. At the time of writing the Space Shuttle program is just emerging from the delay caused by the disastrous accident to the Challenger. The scientific community is waiting to see how much the shuttle is used in serving humanity's need to know more about the universe.

Figure 5–21. *The landing of Columbia at the end of the first shuttle flight.*

CONCLUSION

Newtonian mechanics can be used to understand circular motion and when combined with Newton's law of universal gravitation, it provides the principles that explain Kepler's laws. In more modern times we have seen the applicability of Newton's laws to artificial satellites, to the landing of the first humans on another world, and to spacecraft traveling to other planets and beyond. In future chapters we will see how the law of universal gravitation is used in astronomers' search for information about the very distant objects in our universe and how the law allows us to calculate the masses of planets, of distant stars, and even of galaxies.

RECALL QUESTIONS

1. In what direction does a string exert a force on a rock being whirled in a circle?
2. If a rock is being whirled on a string and the string breaks, in what direction does the rock go? Answer in terms of Newton's laws of motion.
3. How does the scientific definition of acceleration differ from the everyday usage?
4. What is the direction of a centripetal force?
5. What determines the magnitude of the gravitational force between two objects?
6. How did Newton use an astronomical object to check his hypothesized law of gravity?
7. Explain how the law of gravity accounts for the fact that planets move fastest when closest to the Sun.
8. What development in mathematics was made by Newton and what produced the need for this?
9. Did Newton's laws conflict with Kepler's laws? Explain.
10. What provides the centripetal force needed to keep the Earth in orbit around the Sun?
11. What was the name and year of launch of the first artificial satellite? What nation launched it?
12. It seems that a tremendous amount of fuel would be burned during the nearly two weeks of rocket firing during a trip to the Moon. Explain the misconception here.
13. Why is landing on the Moon simpler than landing on Earth?
14. On what date was the first Moon landing?

QUESTIONS TO PONDER

1. Use the idea of centripetal force to explain why acceleration was defined to include a change in direction as well as a change in speed. (Hint: Recall Newton's second law.)
2. Name the object providing the centripetal force in each case below:
   a. A rock is whirled on a string.
   b. A planet orbits the Sun.
   c. A person sits on a bench riding on a merry-go-round.
   d. A track star rounds a turn heading home.
   e. An airplane turns to make its final approach.
3. Explain why the work of Newton had implications far beyond science.
4. How does the period of revolution of an artificial satellite depend on the radius of its orbit? (Hint: Artificial satellites must follow the same laws of nature as natural satellites.)
5. If weight is defined as the force of gravity, are astronauts in orbit around the Earth truly weightless? What about astronauts on a trip to the Moon?
6. Consult some other books or journals to find benefits of the Apollo program. What do *you* think was the most important benefit from this program? Do you think that it was worth the money spent?
7. We were not as conscious of sexism in language in 1969 as we are today. Propose better wording for the plaque on the Moon.
8. What keeps a satellite moving as it orbits the Earth?
9. Report on the latest scientific accomplishments of NASA's space program.
10. Write to The Planetary Society, 65 North Catalina Ave., Pasadena, CA, 91106, for information and report on the Society's activities promoting space exploration.

CALCULATIONS

1. On the surface of the Earth we are about 4000 miles from the center. If you weigh 160 pounds on the surface, how much will you weigh 4000 miles above the surface? Twelve thousand miles above the surface?
2. Mars has a radius about half that of Earth's and a mass about 0.1 of Earth's. A person who weighs 110 pounds on Earth would weigh how much on the surface of Mars?

CHAPTER 6

When large-scale space colonies—space cities—are constructed, great amounts of material will be needed, for the cities must be large enough to provide a comfortable environment for their inhabitants. The building material will come from the Moon. The reason for getting raw materials from the Moon rather than from Earth is simply that the Moon's weaker gravitational field makes it much easier to lift an object from the Moon than from Earth.

The lunar soil contains all of the chemicals needed for space construction. For example, the soil of the lunar highlands contains as much as 18 percent aluminum (by weight) and oxygen is plentiful in the soil. The material will be processed on the Moon and then launched toward a "catcher" in space by a "mass driver." The Moon's lack of an atmosphere will allow the mass driver to throw the building supplies into space.

Detailed plans for space colonies have been drawn up by Gerard O'Neill and his colleagues at the Space Studies Institute at Princeton. They have constructed working models of mass drivers and have proven their feasibility. Their calculations show that large space colonies—large enough to support thousands of inhabitants—can be built from lunar material using technology that exists today.

The Earth-Moon System

ASTRONOMERS HAVE LEARNED A lot about the Moon since we first sent space probes to it more than 20 years ago, but you might be more impressed with how much we knew about the Moon before the advent of the space age. There is much to learn about the Moon by simple naked eye and telescopic observation. Even before the invention of the telescope, people had calculated the distance to the Moon and its size. As you study the methods used in determining these values, admire the ingenuity of the people who devised them.

After examining how we know the Moon's size and distance, we will discuss solar and lunar eclipses and will look at the cause and effect of tides. In so doing, we will see that Newton's laws force us to make revisions to the original laws of Kepler. Finally, we will consider some theories concerning the origin of the Moon.

THE SIZE OF THE MOON

When we look up at the Moon from Earth we have no way of judging its actual size. If you ask children how big the Moon looks, one might say that it is about the size of her play ball. Another might claim that it is gumball size. Just how do we judge the size of something we see? For example, suppose you see two objects like those shown in Figure 6–2(a). could you tell me how large they are? One looks larger than the other, but in fact they are the same size and one looks smaller simply because it is farther from the camera. When we view an object at a distance, we estimate its size in a number of ways. One way is by comparing it to objects of known size that are near it. That makes the judgement of the size of the same two balls in Figure 6–2(b) easier.

When we look at the sky, there are no familiar objects near the things we see. In this case, we can estimate the size of the object only if we know how far away it is. By combining knowledge of its angular size and its distance, we make a judgement as to the object's size.

Figure 6–1.

(a)

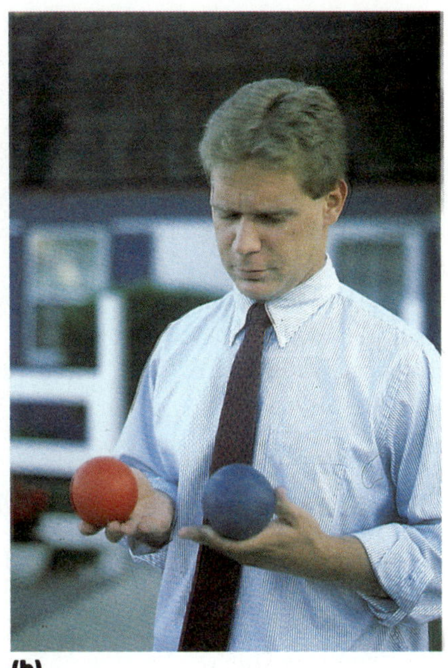

(b)

Figure 6–2. *(a) It is difficult to judge the real size of the two balls here. They are actually the same size. (b) Here we can use the surroundings to judge the size of the balls.*

Angular Size

The ***angular size*** of an object is the angle between two lines which start at the observer and go to opposite sides of the object. Figure 6–3 shows someone looking at the Moon. The angle between the lines to the top and bottom of the Moon is indicated. (Note that angular size is defined very similarly to angular separation in Chapter 1.) The angular size determines how big the image of the Moon is on the retina of your eye. Refer again to Figure 6–2(a). As you look at the two balls, the angular size of the ball on the left is less than that of

Angular size of an object: The angle at the viewer between two lines, each on opposite sides of the object.

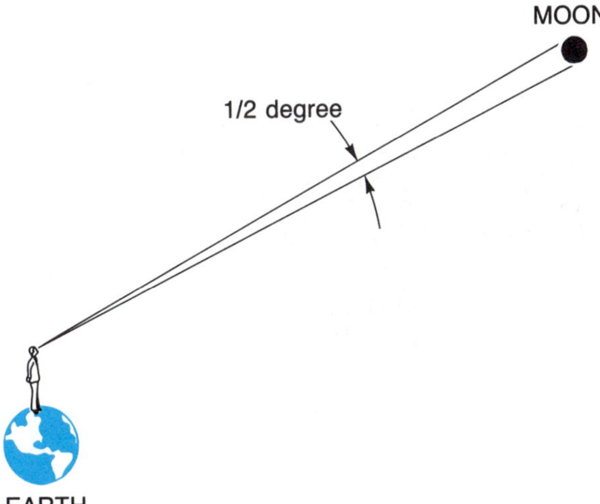

Figure 6–3. *The Moon's angular diameter is about ½ degree.*

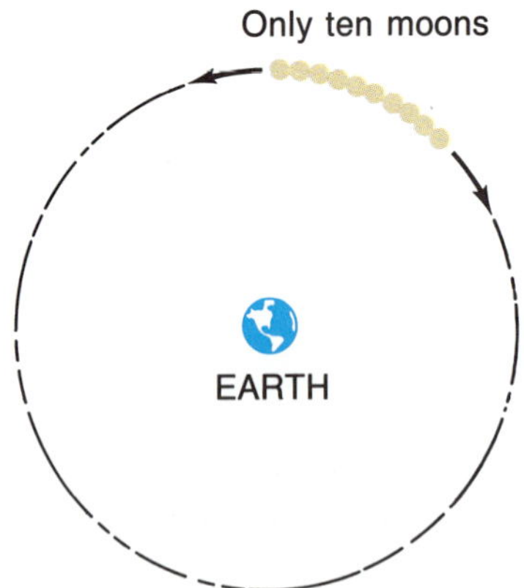

Figure 6–4. *A total of about 720 Moons could be drawn around the Earth if they were drawn at the correct distance. They wouldn't fit as drawn here, because these are much too close to Earth.*

the ball on the right. So, not knowing its distance, you would say that it looks smaller.

The angular size of the Moon seen from Earth is very close to one half degree. Since there are 360 degrees in a complete circle, this means that about 720 Moons could be fitted in a circle around the Earth (Figure 6–4).

The angular size of the Moon might be a nice fact to know, but it doesn't really tell us anything about the Moon itself. The angular size depends not only on the actual size of the Moon, but on how far away the observer is. The actual size of the Moon—its size in miles—would be much more valuable knowledge. In order to determine this, we must know the distance to the Moon. This distance was estimated some 1800 years ago by a man familiar to you from Chapter 1: Ptolemy.

Measuring the Distance to the Moon

Recall our discussion of parallax in Chapter 2. Parallax is the phenomenon we observe when we hold a finger in front of us and look at it first with one eye

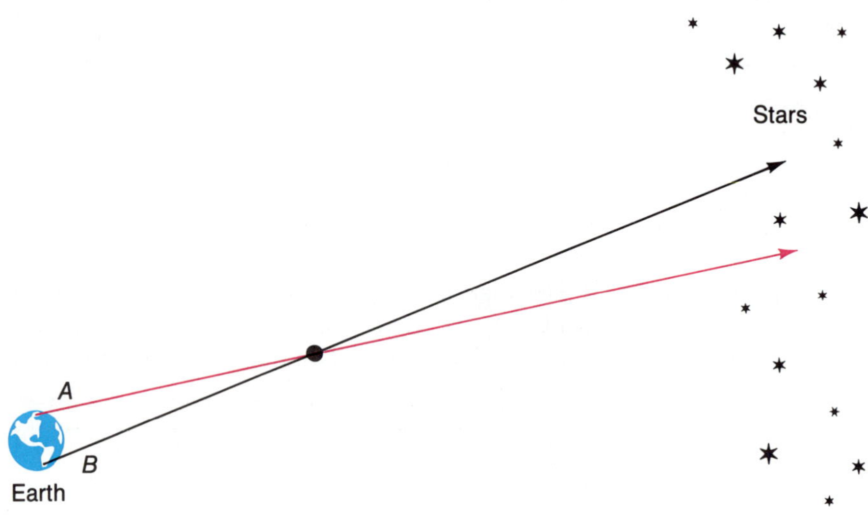

Figure 6–5. *When viewed from two different places on Earth (A and B in the figure above), the Moon seems to be at two different places among the stars. The effect is greatly exaggerated in the drawings.*

and then with the other. The finger's position changes with reference to the background when we do this. The Moon exhibits parallax as seen from different positions on the Earth. For example, if a person in Chicago and another person in Paris, France, happen to be looking at the Moon at the same time, they will observe it in slightly different positions against the background of stars. Figure 6–5 shows this effect but exaggerates it greatly. In reality, the Moon is far enough away that the parallactic shift of the Moon among the stars is very small. But it can be observed, and the observation allows us to measure the distance from the Earth to the Moon.

To show how we can measure this distance, suppose you are vacationing in Kenya and you notice that the Moon happens to be directly overhead. Refer to Figure 6–6 and imagine that you are at position *A*. The Earth turns as time passes, and three hours later you are at position *B*. You measure the angle between a line to the Moon and to the ***zenith*** (the point directly overhead). That is the angle indicated at point *B* in the Figure. Now you have everything you need for determining the distance to the Moon. If you know trigonometry, you can use that branch of math but you don't even need it. To see how it is done, do the Activity "The Distance to the Moon."

Zenith: The point in the sky located directly overhead.

The Moon moves slightly during the three hours and this would have to be taken into account for more accurate results.

In the Activity it is stated that the Moon's distance is about 30 Earth diameters. Notice that the method used there does not yield the actual distance to the Moon, but the distance in terms of the diameter of the Earth. It doesn't allow us to determine how many *miles* it is to the Moon. The measurement and calculation we have described was first made by Ptolemy. His result was 27.3

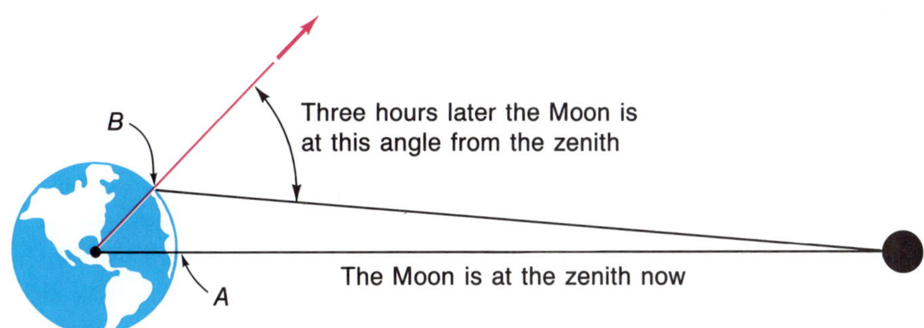

Figure 6–6. *The observer is at the equator. The points* A *and* B *show the observer's motion in three hours. Since the Earth rotates through 45 degrees in this time, if the Moon were infinitely far away the angle indicated would be 45 degrees. It is greater than that, and its value lets us calculate the distance to the Moon.*

ACTIVITY

The Distance to the Moon

LET'S USE SOME IMAGINARY data to calculate the distance to an imaginary Moon. For this activity you will need a protractor to measure angles, a ruler, and a sharp pencil. Suppose that we take data in Kenya as described in the text and that the angle indicated in Figure 6–6 is 50.5 degrees.

To determine the distance to the Moon, you will draw a figure similar to the upper part of Figure 6–6. Start near the left edge of a large sheet of paper by drawing the Earth as a circle one inch in diameter. (Drawing a circle around a quarter will give you very close to the right size.) Next draw a line straight across your paper from the center of the Earth as shown in Figure T6–1(a). According to our data, the Earth rotated for three hours to move us from point *A* to point *B* in Figure 6–6. So it went through one-eighth of a revolution. The angle at the center of the Earth must therefore be 45 degrees. Locate point *B* by measuring an angle of 45 degrees counterclockwise from the horizontal line and draw there a line straight out from the Earth (Figure T6–1(b)). This line points to the zenith at that location on Earth. Measure a 50.5 degree angle clockwise from that zenith line as shown in Figure T6–1(c) and extend the line outward until it meets the original line you drew. Where the lines meet represents the location of the Moon.

To determine the distance to the Moon, measure the distance along the bottom line from the center of the Earth to the Moon. Since you drew the Earth with a one-inch diameter, the length of this line, in inches, tells you the distance to the Moon in Earth diameters.

If you were careful in your drawing, you should have determined that the distance to the imaginary Moon is about eight Earth diameters. Accurate drawings are difficult, however, and if your results are between 7 and 9 Earth diameters you have done a good job. Actually, the parallax observed for the Moon over a three-hour time period is less than indicated by the 50.5 degrees you were given. We did not use the true value in the activity because you would have needed a much larger paper and extreme care in your measurement of angles. In practice, methods of trigonometry yield the best precision. If you had used the correct angle, you would have found that the distance to the Moon is about 30 Earth diameters.

We picked Kenya for our measuring location for a reason: it is on the equator. If a nonequatorial location is chosen the calculation gets tougher because the 45 degree angle we used would not be appropriate. There is another small complication we have ignored. During the three hours between our measurements, the Moon moves. The amount of this motion is know accurately, however, and it can be taken into account. Although we have greatly simplified the procedure, the basic idea is still valid. The method can be used to determine the distance to the Moon relative to the diameter of the Earth.

Earth diameters—very close to today's value of 30.16 for the average distance to the Moon. All Ptolemy needed to know in order to determine the actual distance to the Moon was the size of the Earth. And he did know it. The Earth's size had been measured some 350 years before Ptolemy.

Measuring the Size of the Earth

Eratosthenes (276–195 B.C.) devised a clever way to measure the size of the Earth. It was known that at noon on the first day of summer, the Sun shone

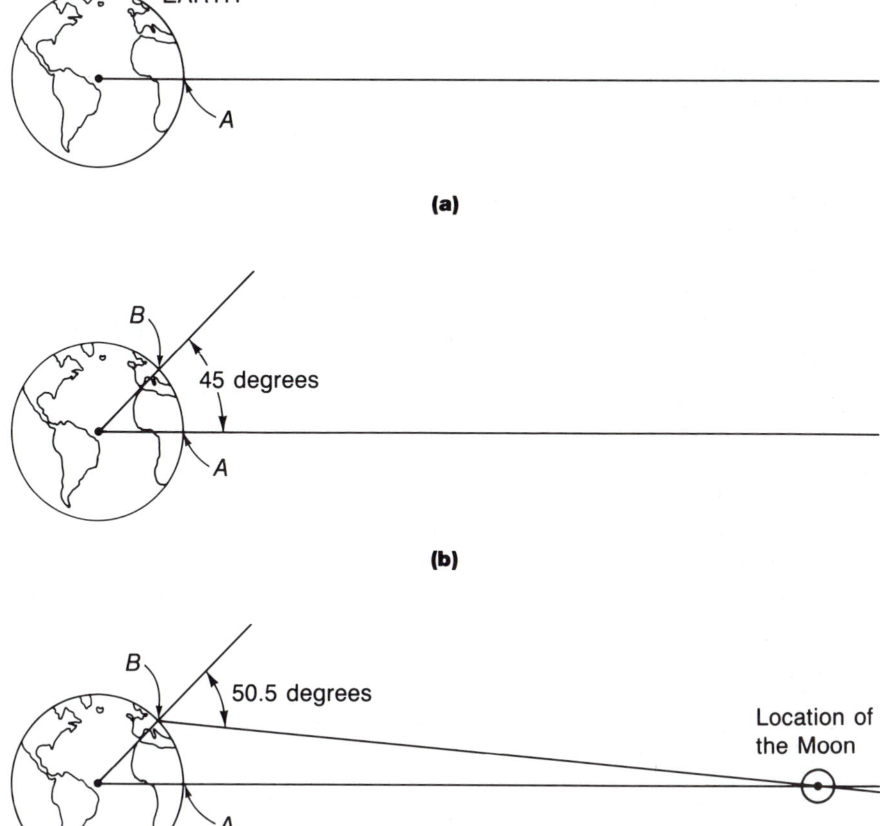

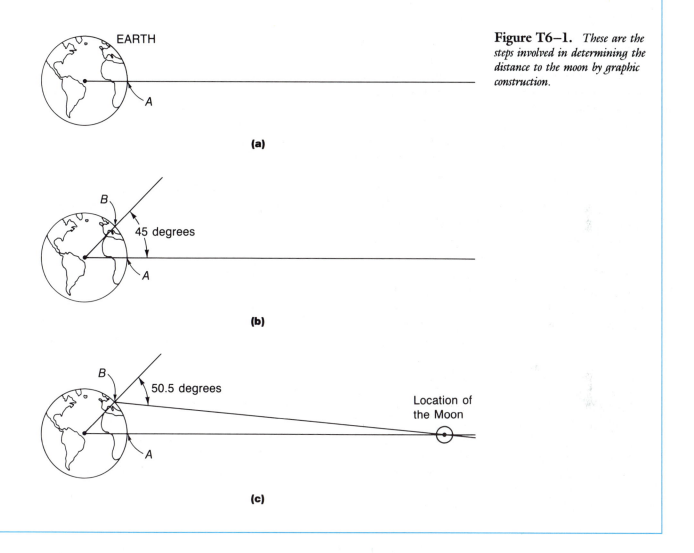

Figure T6–1. *These are the steps involved in determining the distance to the moon by graphic construction.*

straight down a well in the town of Syene. Thus the Sun was directly overhead; on the zenith at that time. Now the town of Alexandria lies about 500 miles north of Syene, and at noon on that day, the Sun was located at an angle of 7 degrees away from the zenith (see Figure 6–7).

Syene was located near the Nile and is now named Aswan.

It was known that the Sun is very far away, so Eratosthenes could consider that sunlight striking both towns was coming from the same direction. (This is different from the case for the Moon, where the difference in direction was used to measure the distance to the Moon.) As indicated in Figure 6–7, this showed that Alexandria was 7 degrees around the Earth from Syene. Since 7

Figure 6–7. *It was observed that when the Sun was overhead at Syene, it was 7 degrees from overhead at Alexandria. This permitted calculation of the circumference of the Earth.*

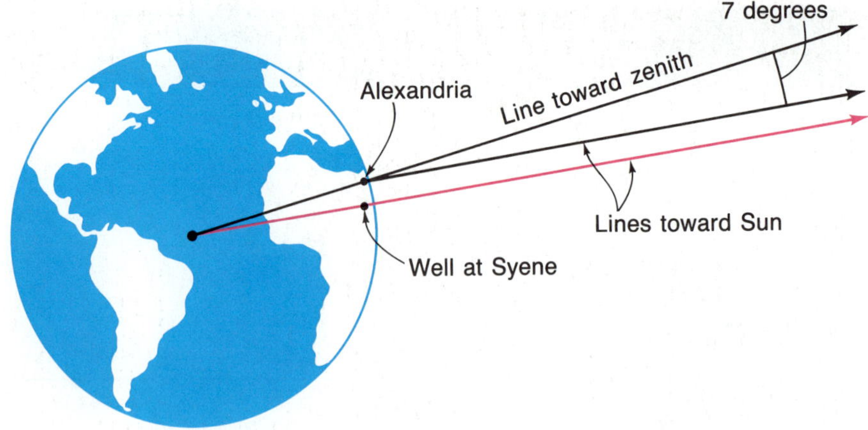

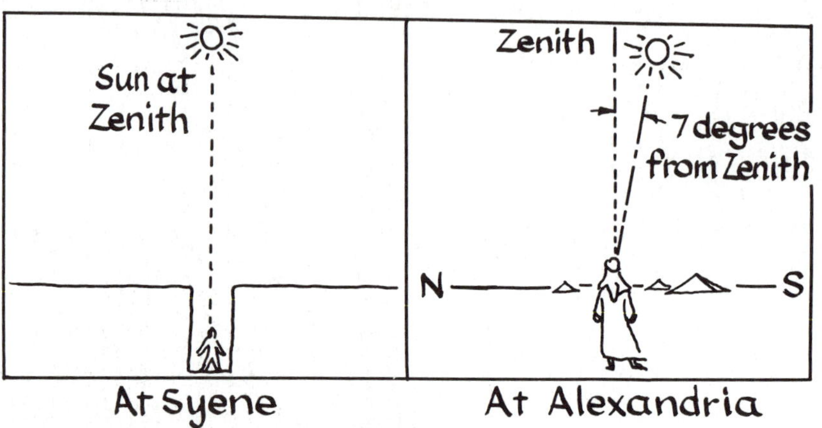

Stadium: An ancient Greek unit of length, perhaps equal to 0.15 to 0.2 kilometers. Various stadia were in use.

Eratosthenes apparently used a stadium equal to about 0.2 kilometers. If so, he measured the Earth's circumference to be about 50,000 kilometers, which is 25 percent too high.

degrees is about 1/50th of the angle around the Earth (7/360), the entire distance around the Earth should be 50 times the distance from Syene to Alexandria. The Greek unit of distance at the time was the ***stadium*** (plural: stadia) and measurements along the surface indicated that Alexander and Syene were 5000 stadia apart. Thus it was calculated that the distance around the Earth was 50 × 5000 stadia, or 250,000 stadia.

It would be nice to be able to know how accurate Eratosthenes's measurement was. The problem is that we don't know the precise length of the stadium he used. Units were not well defined at the time. Besides that, the distance between the towns was determined by measuring how long it took runners to go from one city to the other. Today we don't consider this a particularly good way to measure distances, especially distances as great as 500 miles. The important point is, however, that the method is correct and the Greeks were able to get a realistic value for the circumference of the Earth. Today we know this distance to be about 40,000 kilometers, or 25,000 miles. This means that the diameter is about 13,000 kilometers—8000 miles. This value is worth memorizing. A more accurate value can be found in Appendix F.

The Moon's Actual Distance and Size

The distance to the Moon is 30 Earth diameters, and the diameter of the Earth is 13,000 kilometers. This makes the Moon about 390,000 kilometers, or 240,000 miles, from Earth.

Now we can use the distance to the Moon and the angular size of the Moon (as seen from Earth) to calculate the size of the Moon. To do so we will make use of a simple equation relating the angular size, the distance, and the width of the object. (We will use the term *width* instead of diameter so that our equation will be general for all objects. In the case of the Moon, this will be the diameter.) The equation:

$$\text{Width} = \frac{(\text{angular size in degrees}) \times (\text{distance})}{57.3}$$

This equation is not pulled out of a hat. Its derivation is in a Close Up in Chapter 14. You are encouraged to look ahead.

In symbols:

$$W = \frac{\theta d}{57.3}$$

W = width
θ = angular size in degrees
d = distance away

This equation cannot be used accurately for large angles (more than about 5 degrees) but in most cases in astronomy, angular sizes are much less than this. Before using the equation for the Moon, let's use it in a terrestrial example.

EXAMPLE

Suppose you look at a picture on the wall across a large room. You happen to have a device with you that allows you to measure the angular width of the picture and you determine it to be 1.5 degrees. Suppose further that you know that the picture is 18 feet away from you. How wide is the picture? Substituting into the equation:

$$W = \frac{\theta d}{57.3}$$

$$= \frac{(1.5) \times 18 \text{ feet}}{57.3}$$

$$= 0.47 \text{ feet, or about 6 inches}$$

Try One Yourself. Use the equation to calculate the diameter of the Moon, using the fact that the Moon's angular size is 0.52 degree and its distance is 384,000 kilometers (more accurate values than given above).

Figure 6–8. *The larger ball has a diameter four times the smaller. Would you say that its size is four times larger? Its volume is actually 64 times greater.*

MOON
(ping-pong ball)

EARTH (a grapefruit)

Figure 6–9. *The Moon's diameter is about one-fourth of the Earth's and it is 30 Earth diameters away. This drawing shows the approximate correct scale.*

The small angle formula, when used with the data given in the "Try One Yourself," yields 3480 kilometers for the Moon's diameter (this is 2160 miles). This is very close to one fourth of the diameter of the Earth. Figure 6–8 is a photo of two balls, one four times the diameter of the other. Form a mental image of these two balls. You would probably not say that the large one is four times the size of the small one. The word *size* is an indefinite term, so that one doesn't know whether it refers to diameter, area, or volume. We will not use it in a quantitative sense; that is, when we mean that one object has four times the diameter of another, we will be careful not to say that its *size* is four times greater.

To appreciate the scale of the Moon-Earth system, imagine the Earth to be a large grapefruit (about 5 inches in diameter). On this scale, the Moon would be a ping-pong ball 10 feet away. Between the Earth and Moon there would be nothing but empty space (see Figure 6–9).

REVIEW: TWO MEASURING TECHNIQUES

As we continue our study of astronomy, we will see a number of relationships that allow us to measure features of things in the heavens. We have seen two here. We described how parallax was used to measure the distance to the Moon. This method is often called triangulation because it involves using a triangle to find a distance. We will see in a later chapter that this method is also used to measure distances to nearby stars. Parallax is an important phenomenon in astronomy. Notice that using it involves a relationship among three quantities: the size of the baseline, the angle of parallax, and the distance to the object. Knowing any two of the three, one can calculate the third.

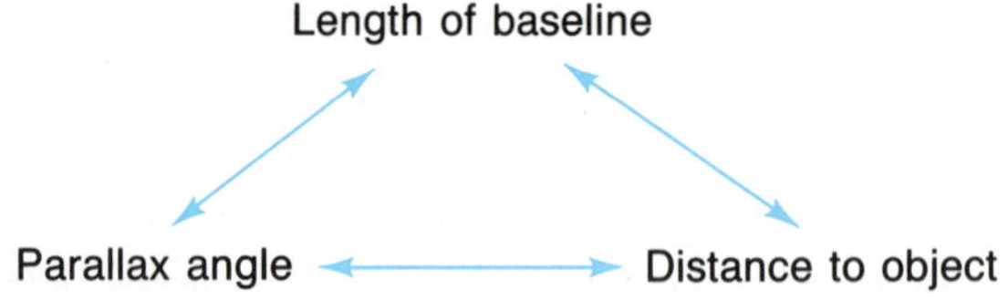

We also saw that there is a relationship among angular size, actual size, and distance. Again, if you know any two of the quantities, you can calculate the third. This is an important relationship, and later we will see its use in measuring the sizes of planets.

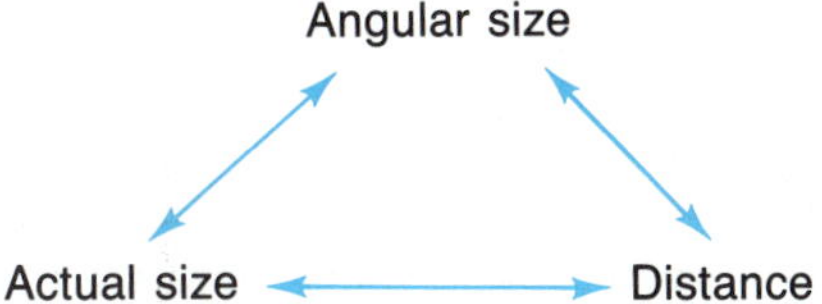

THE MOON'S CHANGING SIZE

Figure 6–10 is a composite photo of the Moon taken at two different times by the same camera. The photo makes it obvious that the apparent size of the Moon changes. But apparent size depends on distance and actual size. The actual size of the Moon doesn't change, of course, so it must be that our distance from the Moon changes. This occurs because the Moon's orbit is an ellipse, just like the orbits of all orbiting objects. And the Moon's orbit is fairly eccentric; that is, it is not very circular. The larger apparent size occurs when the Moon is closest to the Earth, a distance of 356,300 kilometers. The smaller apparent size occurs at the Moon's maximum distance from Earth, 406,600 kilometers.

Perigee: The point in its orbit when an Earth satellite is closest to Earth.

Apogee: The point in the orbit of an Earth satellite where it is farthest from Earth.

Figure 6–10. *These are two photos of the moon put side by side, one taken when the moon is closest to Earth (356,300 km) and the other when it is farthest (406,600 km).*

CLOSE UP

The Moon Illusion

PERHAPS YOU HAVE OBSERVED the full Moon when it was near the horizon and noticed that it seemed unusually large. This is not an effect caused by its distance from Earth. The reason for this extra large appearance of the Moon is not completely understood. We can show, however, that the actual angular size of the rising or setting Moon is no greater than normal. One way to do this would be to use a surveyor's transit to measure the angle between one side of the Moon and the other. There is an easier way to make the comparison, however. Suppose you take a photo of the Moon when it is near the horizon and then take another photo later that same night when the Moon is high in the sky and appears its "normal" size. You could then measure the size of the Moon in each photo. This will give you a comparison of the two angular sizes. The two photos of Figure T6–2 were made in this way. Measure the size of the Moon in the two photos. You will see that they are the same. Thus the effect you see when the Moon is near the horizon is a psychological effect. It seems to be caused by the apparent proximity of the Moon to objects on the horizon. Somehow we subconsciously compare it to trees or buildings in the foreground and conclude that the Moon is extra large.

This psychological effect is an example of the way our senses can be fooled and of the care we must take in forming conclusions regarding objects we see in the heavens.

Figure T6–2. *When the Moon is on the horizon, it appears to be larger, but measurement shows that its angular size is the same. Try it.*

LUNAR ECLIPSES—WHY NOT MONTHLY?

In one of the Chapter 1 activity boxes you studied the phases of the Moon by standing some distance from a lamp and moving a ball (portraying the Moon) around your head. You undoubtedly noticed that at the point depicting the full Moon the ball got into your shadow so that it could not be illuminated by the

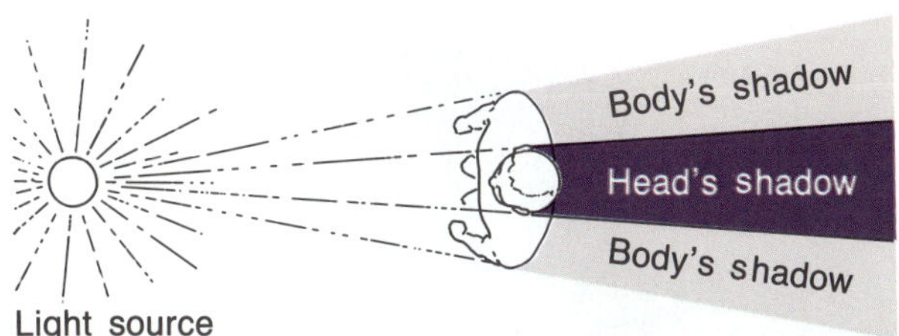

Figure 6–11. *With a small lamp as a source of light, your shadow gets larger and larger behind you.*

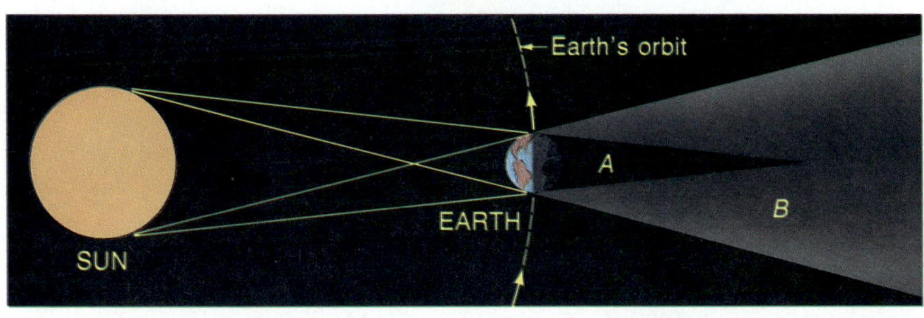

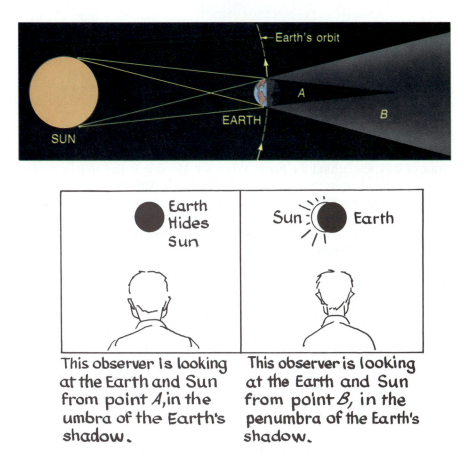

Figure 6–12. *Point* A *is in the umbra of the Earth's shadow. Point* B *is in the penumbra, where light from part of the sun hits it. Distances are not to scale in the drawing.*

light source. This represented an eclipse of the Moon, a ***lunar eclipse***. Perhaps you wondered why it doesn't happen at the time of every full Moon. There are several reasons. First, as we have seen, the Moon and Earth are very small compared to their distance apart: the Moon is 30 Earth diameters away. Thus it is unlikely that they will align so accurately that one eclipses the other. Think of trying to align a grapefruit, a ping-pong ball 12 feet away, and a distant object.

Lunar eclipse: An eclipse in which the Moon passes into the shadow of the Earth.

Another reason that the ball/head simulation is misleading is that the shadow of your head is larger than your actual head (Figure 6–11), while the Earth's shadow is smaller than the Earth. This occurs because the source of light—the Sun—is so much larger than the Earth. Look at Figure 6–12. Although distances

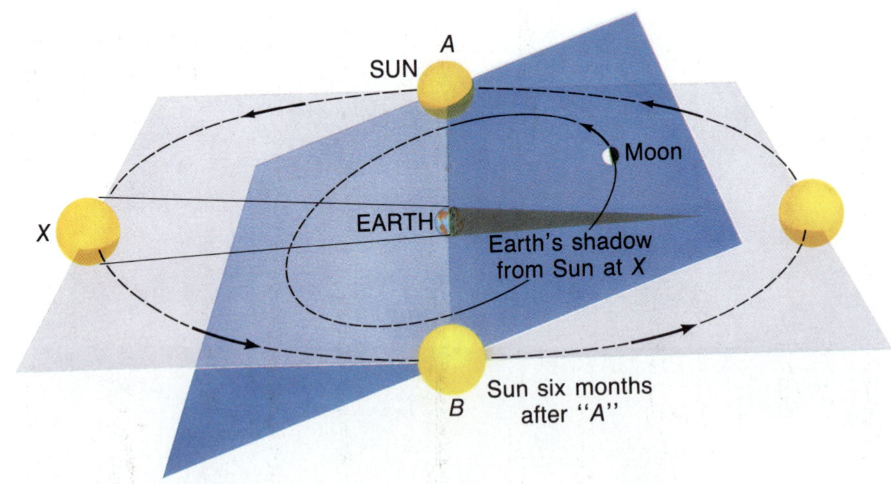

Figure 6–13. *When the Sun is at* X, *the moon's orbit does not pass through the Earth's shadow so there can be no eclipse. When the sun is at* A *or* B, *the Moon can pass through the Earth's shadow and produce a lunar eclipse. Thus potential eclipses occur about six months apart.*

are not nearly to scale, the principle illustrated is correct. Notice that there is a cone of darkness behind the Earth. When the Moon is in this area (at point *A*, for example) it is in the full shadow of the Earth. This full shadow of the Earth—called the ***umbra***—tapers down to a point. At the distance of the Moon, the width of the umbra is only three-fourths of the diameter of the Earth. So the Moon is much less likely to pass through the shadow than was the pseudo-Moon of your simulation.

Umbra: The portion of a shadow that receives no direct light from the light source.

If the Moon is at point *B* of Figure 6–12, on the other hand, it is only in partial shadow, for light from the lower part of the Sun is hitting it. When the Moon is here, in the ***penumbra,*** it will not receive the full light from the Sun and will appear dim to Moon-watchers on Earth. The penumbral shadow increases in size at greater distances from Earth but it is not equally dark across its width. Right next to the umbra it is very dark and it gets brighter and brighter out toward its edge. When the Moon passes through the outer penumbra, we don't even notice the darkening.

Penumbra: The portion of a shadow that receives direct light from only part of the light source.

There is a more important factor in explaining why a lunar eclipse does not occur at each full Moon, however. The plane of revolution of the Moon is tilted as compared to the plane of the Earth's revolution around the Sun. This is easiest to see from a geocentric perspective. Figure 6–13 shows the Earth with the Sun appearing to circle it along path *A-X-B*. Notice that as the Moon circles the Earth, it goes north and south of the plane of the Sun's path. The Earth's shadow—shown for when the Sun is a *X*—is directly behind the Earth from the Sun. Thus in most cases of a full Moon, the Moon will not be in that shadow but will be either north or south of it. The plane of the Moon's orbit changes relatively little as the year progresses and so eclipses can only occur about twice a year, when the Sun is at positions *A* and *B* in Figure 6–13.

Eclipse season: A time of the year during which a solar or lunar eclipse is possible.

These occur when the Moon is near points A *and* B *in Figure 6–13.)*

Types of Lunar Eclipses

The Earth's umbra gets smaller as the distance from Earth increases. At the Moon's average distance, the umbra has a diameter of about 5700 miles. Since

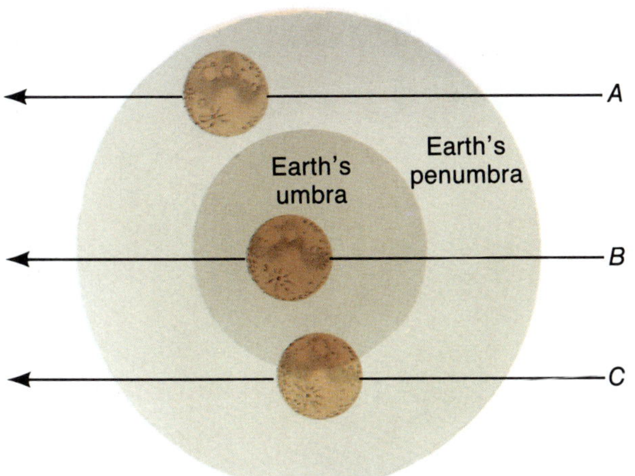

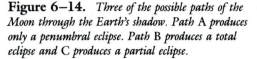

Figure 6–14. *Three of the possible paths of the Moon through the Earth's shadow. Path* A *produces only a penumbral eclipse. Path* B *produces a total eclipse and* C *produces a partial eclipse.*

the diameter of the Moon is less than 2200 miles, the Moon can easily be covered by the umbra. But the Moon might not pass right through the umbra. Figure 6–14 shows three possible paths of the Moon through the shadow of the Earth. If the Moon moves along path *A*, it will only pass through the penumbra. In this case, it will darken slightly as it does so but such an effect will not be obvious from Earth and will only be noticeable if the Moon passes into the darkest part of the penumbra, near the umbra.

Penumbral lunar eclipse: An eclipse of the Moon in which the Moon passes through the Earth's penumbra but not through its umbra.

If the Moon follows path *B*, it will slowly darken as it moves toward the umbra but the eclipse will not become obvious until the Moon contacts the umbra. Figure 6–15(a) is a photo of the Moon when part of it is in the umbra. As the Moon continues moving into the umbra, the shadow slowly moves across the surface, until the Moon appears as in 6–15(b). We are now seeing a ***total lunar eclipse***. Depending upon the Moon's distance from Earth, it may take about an hour from the time of first contact with the umbra until it reaches totality, when the entire Moon is in the Earth's umbral shadow. It can stay in the shadow up to about 1½ hours.

Total lunar eclipse: An eclipse of the Moon in which the Moon is completely in the umbra of the Earth's shadow.

Partial lunar eclipse: An eclipse of the Moon in which only part of the Moon passes through the umbra of the Earth's shadow.

If the Moon follows path *C* of Figure 6–14, it will never be entirely covered by the umbra, and we will see only a ***partial lunar eclipse.*** The dark shadow will creep across the Moon, covering (in the case shown) only the top part of the Moon.

Observing Lunar Eclipses

An eclipse of the Moon, especially a total eclipse, is a beautiful sight. The totally eclipsed Moon is not completely dark, however. Even when the Moon is completely in the umbra, some sunlight strikes the Moon. This light has been bent by the Earth's atmosphere as shown in Figure 6–16. As we will discuss in a later chapter, light that passes through many miles of atmosphere is red in color. (This is why sunsets are red.) Because of this, the sunlight reaching the totally eclipsed Moon is red and the Moon appears a dark red color.

If you have never observed a total eclipse of the Moon, you should make

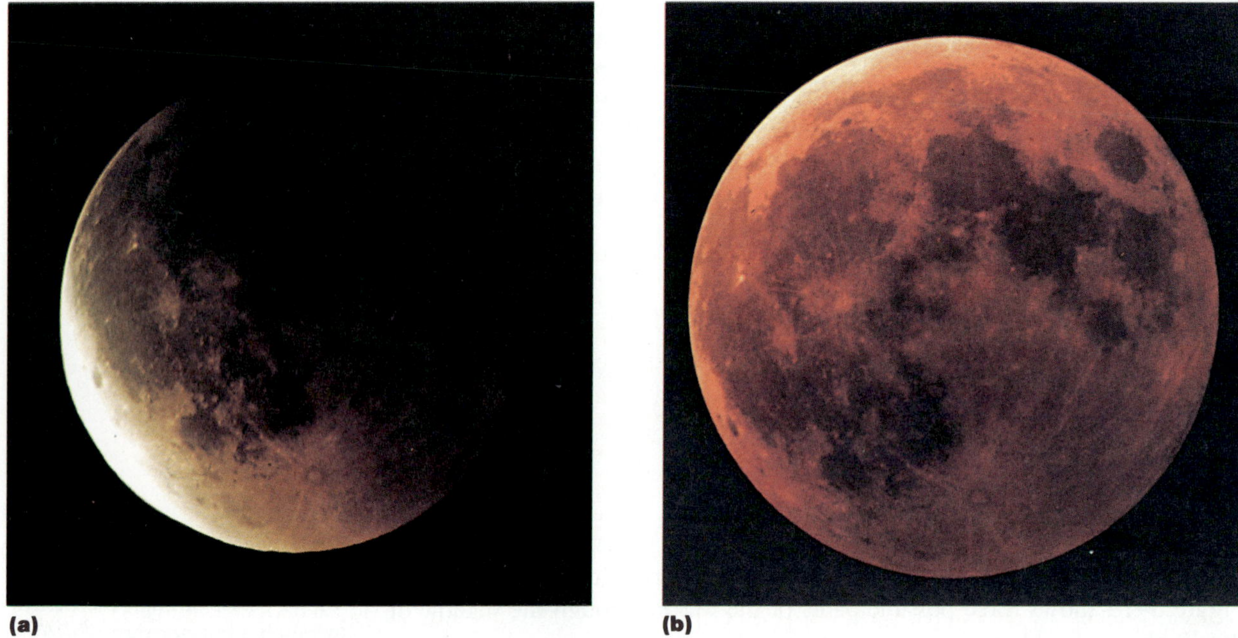

Figure 6–15. *Photo (a) was taken before the Moon was totally eclipsed. Photo (b) shows the totally eclipsed Moon. The red color is due to light that has passed through the Earth's atmosphere before striking and being scattered from the Moon.*

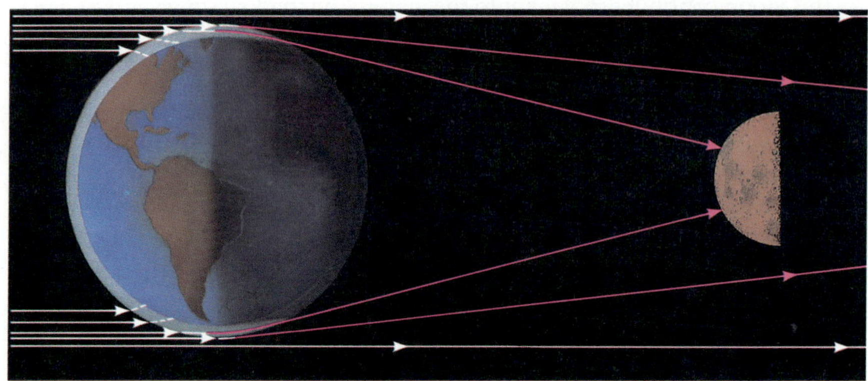

Figure 6–16. *Though the Moon is in total eclipse, some light is refracted toward it by the Earth's atmosphere. Red light refracts more than the other visible colors, so the Moon is seen as having a dark red color during a total lunar eclipse.*

an effort to do so. Table 6–1 shows the dates of coming lunar eclipses. There are two reasons you cannot be sure that you will be able to see any particular lunar eclipse. First, in order to be able to see it, you must be on the dark side of the Earth when the eclipse occurs. So on the average, only half the people on Earth have a chance to see a given lunar eclipse. For this reason, the last column of the table indicates whether each eclipse will be visible from at least part of North America. Second, there is the weather factor. A cloudy night can ruin a long-planned eclipse-viewing party.

Table 6–1 Dates of Coming Lunar Eclipses

Date	Type of Eclipse	Visible from Part of North America?
Aug. 17, 1989	Total	Yes
Feb. 9, 1990	Total	No
Aug. 6, 1990	Partial	No
Dec. 21, 1991	Partial	Yes
June 15, 1992	Partial	Yes
Dec. 9, 1992	Total	Yes
June 4, 1993	Total	No
Nov. 19, 1993	Total	Yes
May 25, 1994	Partial	Yes
April 15, 1995	Partial	No
April 4, 1996	Total	Yes
Sept. 27, 1996	Total	Yes
Mar. 24, 1997	Partial	Yes
Sept. 16, 1997	Total	No
July 28, 1999	Partial	Yes
Jan. 21, 2000	Total	Yes
July 16, 2000	Total	No

While watching a lunar eclipse, try to notice the curve of the Earth's shadow on the Moon. Does it convince you that the Earth is round, or can you think of another reasonable explanation for the curved shadow?

TOTAL ECLIPSE OF THE SUN

We have seen that a lunar eclipse occurs when the shadow of the Earth falls on the full Moon. An eclipse of the Sun—a ***solar eclipse***—occurs when the Moon, in its new phase, gets directly between the Sun and Earth so that its shadow falls on the Earth. There is a major difference in the events, however. The Earth's size is such that its umbral shadow reaches back into space nearly a million miles and at the distance of the Moon, it is easily large enough to cover the entire Moon. The umbral shadow of the Moon, however, reaches only about one fourth that far, or about 377,000 kilometers. But recall that the average distance from the Earth to the Moon is 384,000 kilometers. Thus if the Moon stayed at this distance, its umbra would never reach the Earth. There would never be a dark shadow of the Moon on the Earth.

Solar eclipse (or eclipse of the Sun): An eclipse in which light from the Sun is blocked by the Moon.

377,000 km = 234,000 mi and 384,000 km = 239,000 mi.

The Moon, though, follows an eccentric elliptic orbit, getting as close as 356,300 kilometers to Earth. So it does get close enough that its umbra can reach the Earth. When this occurs, we can experience one of the most spectacular

Figure 6–17. *In (a), if the Moon's umbra strikes the Earth at an angle, the resulting width of the total eclipse is greater. In (b), the Moon's motion causes the path of the total solar eclipse to sweep across the Earth. The eclipse shown moves primarily across the land and would be seen by many people.*

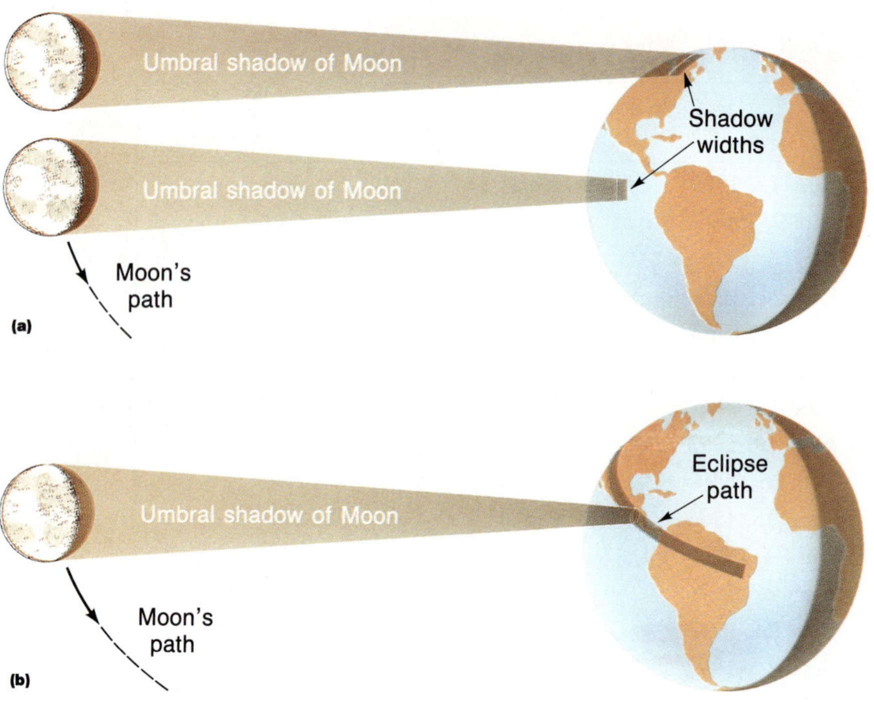

Total solar eclipse: An eclipse in which light from the normally visible portion of the Sun (the photosphere) is completely blocked by the Moon.

270 kilometers is about 170 miles.

of natural phenomena, a ***total solar eclipse***. Figure 6–17(a) shows two cases of the Sun, Moon, and Earth being aligned when the Moon is close enough for its umbra to reach Earth. Even when the Moon is at its closest, the width of its umbral shadow at the Earth's distance is only about 270 kilometers. If the shadow hits the Earth as does the lower shadow in the figure, its width will be 270 kilometers. If the shadow happens to strike the Earth as does the upper one, the shadow on the surface may measure wider than 270 kilometers but it seldom exceeds 400 kilometers (250 miles).

This explains why relatively few people ever experience a total solar eclipse. Only people inside that small area of Earth covered by the umbra see the Sun entirely hidden by the Moon. As the Moon moves along, its shadow swings in an arc across the surface of the Earth. (See Figure 6–17(b).) Thus there is a strip across the Earth's surface in which the total eclipse may be seen. This strip may be many thousands of miles long, but its width can reach a maximum of only about 250 miles. You have to be within this strip at the exact moment of totality, in clear weather, in order to see the total eclipse of the Sun.

Experiencing a Total Solar Eclipse

An eclipse begins when the Moon starts to cover the disk of the Sun in the sky. When this happens, only people prepared to view the eclipse will notice the "bite" being taken out of the Sun. The change in brightness of the sky is not particularly noticeable, and since the Sun is still so bright that our eyes instinc-

(a)

(b)

Figure 6–18. *The photo in (a) shows an uneven circle because of the Moon's rough surface. (The bright spots are called* Baily's beads.*) It is obvious why the effect shown in photo (b) is called the diamond ring effect.*

tively close when we look up at it, we cannot observe its shape directly. By using one of the methods described in the accompanying Close Up for viewing the eclipse, one can see the Sun slowly being covered. Note that you do not actually see the Moon; its dark side is facing Earth. Instead, you see what appears to be a circular disk that is slowly moving across the disk of the Sun.

Finally, just before the Moon entirely covers the Sun, the day begins to darken. In warm weather, a noticeable cooling takes place. It is at this time that some animals are fooled into behaving like they do at sunset. Some birds go to roost. Domestic animals accustomed to being fed at sunset may come to their feeding area. Some flowers begin to close.

Other strange things happen. Because of the sudden cooling of the atmosphere in the dark umbral shadow, a sudden breeze may spring up. Bands of shadows can be seen sweeping across the ground. These are apparently caused by light from the Sun being bent by currents of warm and cool air. This is similar to the waviness you can observe looking across the top of a hot burner on a stove. The light coming across the stove is bent in that case by the hot air currents rising from the burner.

Just before totality, light from the edge of the Sun can come to us only through mountain valleys on the Moon. Figure 6–18(a) shows this effect. Finally, the last portion of the Sun is seen. Figure 6–18(b) makes it obvious why this is called the ***diamond ring effect***.

Diamond ring: The bright ring seen just before and after totality in a solar eclipse.

Finally, a few hours after the start of the eclipse, totality occurs. At totality, the sky is dark enough that planets and the brightest stars can be seen in the

CLOSE UP

Observing a Solar Eclipse

THERE IS A MISCONCEPTION that the Sun emits especially harmful rays of some kind during a solar eclipse. This is an especially geocentric idea, for it would mean that somehow the Sun knows when Earth's Moon is about to block sunlight from the Earth. Naturally, the radiation emitted by the Sun during an eclipse is no different from that emitted at any other time.

Like most misconceptions, however, there is an element of truth in this idea. In fact, the Sun continuously emits radiation which is harmful to our eyes: *infrared* radiation.* If you were to look at the Sun any time, this radiation would harm your eyes but normally you are not able to look at the Sun. If you attempt it, your eye will quickly close because of the intense light. When the Sun is nearly totally eclipsed, however, its light is dim enough that you are able to look at it. So during an eclipse it would be possible for a person to stare directly at the Sun for some time, all the while unknowingly absorbing the harmful infrared rays in his or her eyes.

There are a number of safe methods of observing the Sun during an eclipse. First, we might use a telescope with a solar filter attached. This is a filter that blocks out some 99.99 percent of the Sun's light, allowing just enough through for us to see the Sun.

A more convenient way to use a telescope to observe an eclipse is illustrated in Figure T6–3, which shows the author using a telescope to project the partially eclipsed Sun onto a screen. In this case, a reflecting device was mounted on the telescope to cause the image to appear off to the side. This photo was taken when the Moon had progressed much of the way across the solar disk. It was near noon at the time but the sky had still not darkened noticeably. Even a little of the Sun's light is enough to give us a bright day on Earth.

Figure T6–3. *The telescope has a mirror (called a* star diagonal*) attached to it so that it projects the image of the partially eclipsed Sun onto the screen at the side.*

A third method of safely observing a solar eclipse—one requiring little equipment—is by pinhole projection. To use this method, all you need is a piece of cardboard with a hole in it and a piece of paper to use as a screen. Figure T6–4 illustrates the method. Try holes of different sizes, from one millimeter to one made with a paper punch. (A hole made by a pin will probably be too small.) To see the image better, you might use large pieces of cardboard to shield your screen from reflected light or work in a dark room that has an opening facing the Sun. Block all of the opening except for your pinhole.

During the eclipse shown in Figure T6–3, we noticed that a small mirror would serve as a pinhole of sorts. The mirror was held so that it reflected the Sun

*The Sun also emits ultraviolet radiation that can harm the eyes. The primary danger during solar eclipses is infrared radiation, however.

Figure T6–4. *A pinhole projector can be used to view a solar eclipse but you must shield your screen from scattered light better than is done here.*

(a)

(b)

Figure T6–5. *Part (a) shows a partial solar eclipse being projected on a distant wall. A simple small mirror was used to do this and although it was much larger than an actual pinhole, it was small compared to the distances involved. Part (b) shows multiple images produced by "pinholes" between leaves in a tree. Note their crescent shapes.*

onto the inside wall of a building. Figure T6–5(a) shows the image—about 1 foot tall—on the brick wall. Figure T6–5(b) shows multiple pinhole images that appeared on a sidewalk because of small holes between leaves on a tree overhead.

A fourth method of observing an eclipse is to obtain a safe filter through which you can view the Sun directly. Extreme care must be exercised here, however. If your filter does not block the Sun's infrared radiation, it may damage your eyes. (Smoked glass is definitely not recommended.) To avoid the possibility of injury, it is recommended that you be very sure of your filter or that you use one of the other methods.

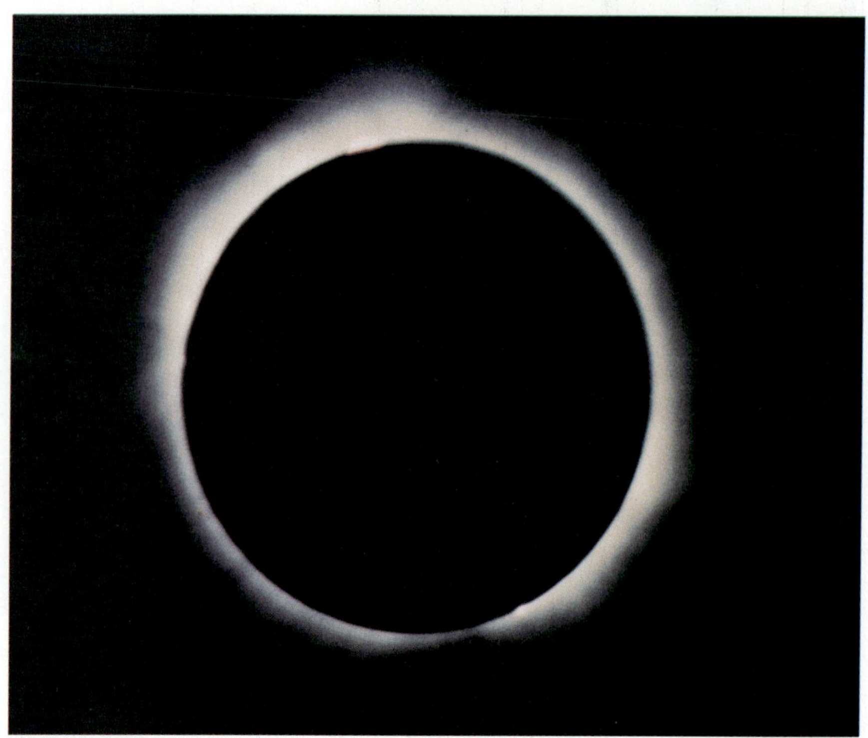

Figure 6–19. *During a total eclipse, the glowing light of the Sun's atmosphere—the corona—is visible.*

Corona: The outer atmosphere of the Sun (to be discussed in Chapter 12).

sky. The appearance of the Sun is shown in Figure 6–19. What you see around the dark disk where the Moon blocks out the Sun is the glowing outer atmosphere of the Sun, called the ***corona***. This is a layer of gas that extends for millions of miles above what normally appears to be the surface of the Sun. The gas glows because of its high temperature but the glow is so much dimmer than the light we receive from the main body of the Sun that it is observed only during an eclipse. The opportunity to observe the corona is one of the scientific values of an eclipse, although today we are able to block out the Sun by artificial means in order to observe its outer layers. We will discuss the Sun further in Chapter 12.

The experience of a total solar eclipse is truly an awesome experience. As Table 6–2 shows, it will be some time before one is visible in the United States. If you get a chance to travel to the path of totality of a solar eclipse, don't pass it up. It is one of nature's grandest spectacles.

A PARTIAL SOLAR ECLIPSE

Partial solar eclipse: An eclipse in which only part of the Sun's disk is covered by the Moon.

Figure 6–20 shows not only the Moon's umbra but also its penumbra. People on Earth who are in the umbra see a total eclipse but anyone within the penumbra will see a ***partial solar eclipse***. The penumbra covers a much greater portion of the Earth's surface, stretching about 3000 kilometers—2000 miles—from the central path of totality, so most of us have the opportunity to see a

Table 6–2 Dates and Locations of Coming Solar Eclipses

Date	Location	Type Eclipse
Jan. 26, 1990	South Atlantic	Annular
July 22, 1990	Finland, Siberia	Total
Jan. 15, 1991	Australia	Annular
July 11, 1991	Hawaii, S. America	Total
Jan. 4, 1992	Pacific Ocean	Annular
June 30, 1992	South America, Africa	Total
May 10, 1994	Pacific, U.S.	Annular
Nov. 3, 1994	South America	Total
April 29, 1995	South America	Annular
Oct. 24, 1995	India, S.E. Asia	Total
March 9, 1997	Siberia	Total
Feb. 26, 1998	South America	Total
Aug. 22, 1998	Indian Ocean	Annular
Feb. 16, 1999	Australia	Annular
Aug. 11, 1999	Europe to India	Total

There are about twice as many total or annular solar eclipses as total lunar eclipses, but you are much less likely to see a solar eclipse because of the narrow path of the shadow.

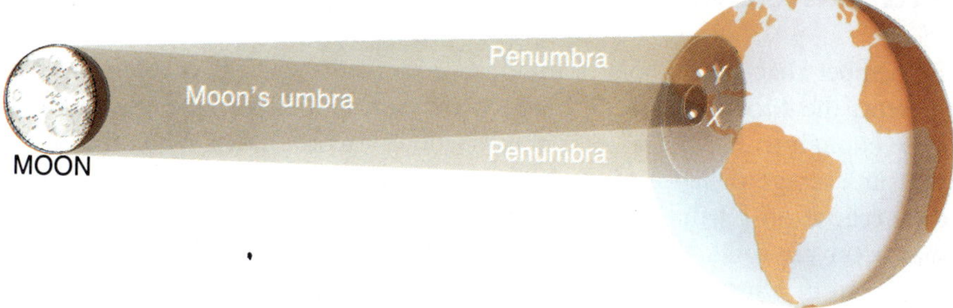

Figure 6–20. *A person at point X sees a total solar eclipse while a person at Y sees a partial eclipse, seeing only the southern part of the Sun blocked by the Moon.*

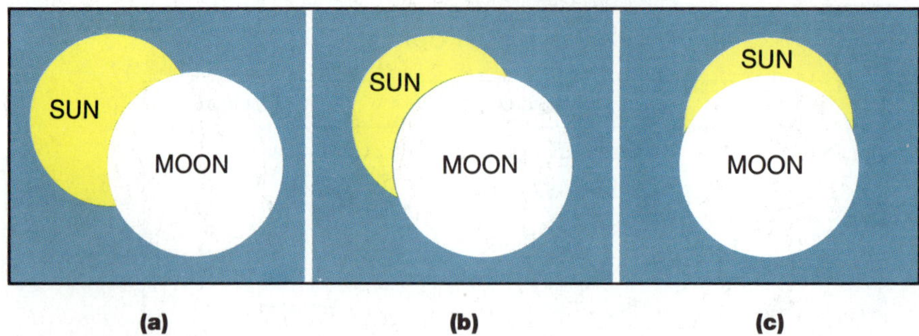

Figure 6–21. *The disk of the Moon gradually moves across that of the Sun until maximum partial eclipse is reached (c).*

number of partial solar eclipses during our lifetimes. Though a partial eclipse is not as impressive as a total eclipse, it is still worth viewing.

In the case of a partial eclipse, the dark disk of the Moon moves across the Sun but its path is not perfectly aligned with the Sun and it does not move across the center of the Sun. Figure 6–21 illustrates what is seen from Earth. The closer you are to the path of totality, the more of the Sun will be blocked out by the Moon.

THE ANNULAR ECLIPSE

Remember that a total solar eclipse can occur only when two conditions are present: the Moon is directly between the Sun and Earth, and the Moon is close enough to Earth that its umbral shadow reaches Earth. When the Moon is at its average distance from the Earth, it is a little too far away to cause a total eclipse on Earth. Therefore somewhat fewer than half the solar eclipses that occur are total. Figure 6–22 illustrates what happens when the Moon is too far from Earth to allow a total eclipse. A person at point *P* on the Earth will still see a partial eclipse where the disk of the Moon crosses that of the Sun. The fact that the Moon's disk would appear slightly smaller would not change the appearance of this partial eclipse greatly from that indicated in Figure 6–21.

Think of what an observer at point *A* on Earth would see in this case. The Moon is so far away that its disk is not large enough to cover the Sun even when it gets directly centered on it. Figure 6–23 is a photo of what is seen at eclipse maximum in this case. The Latin word *annulus* means *ring* and from the figure you can see why such an eclipse is called an ***annular eclipse***. Note the spelling; this is not an annual eclipse. Slightly over half of the solar eclipses are annular. Table 6–2 indicates which of the coming eclipses will be total and which annular. Figure 6–24 shows the paths of total solar eclipses through 2017. Notice that the next total solar eclipse to cross the continental U.S. will occur on August 21, 2017 (and then the next will be on April 8, 2024).

Annular eclipse: An eclipse in which the Moon is too far from Earth for its disk to completely cover that of the Sun, so that the outer edge of the Sun is seen as a ring.

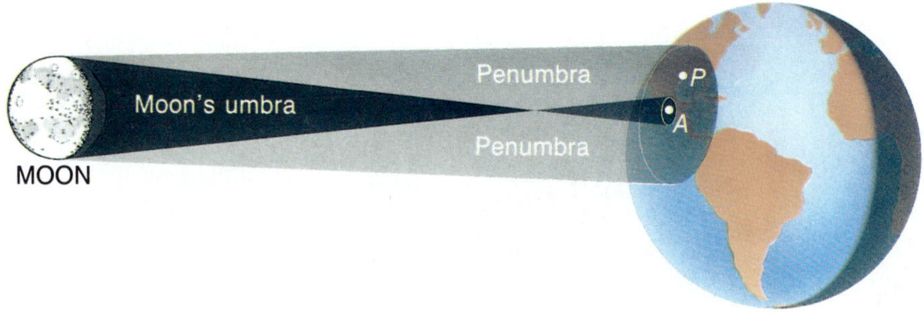

Figure 6–22. *When the Moon is far away during a solar eclipse, the eclipse will be annular. The person at point* A *sees the annular eclipse, while the person at* P *sees a partial eclipse.*

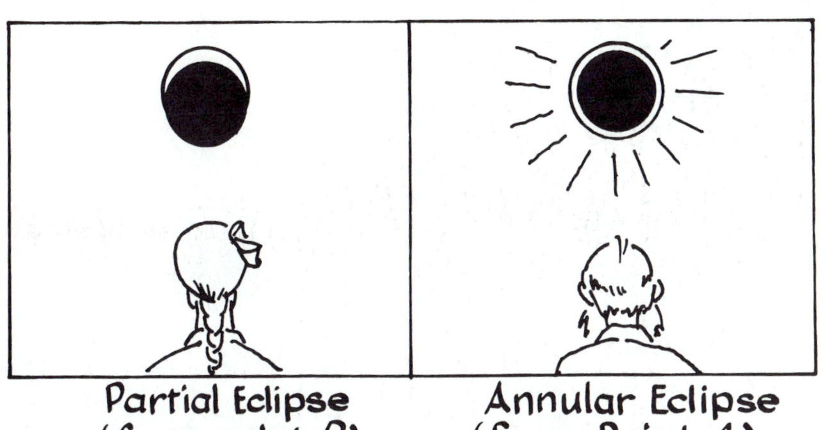

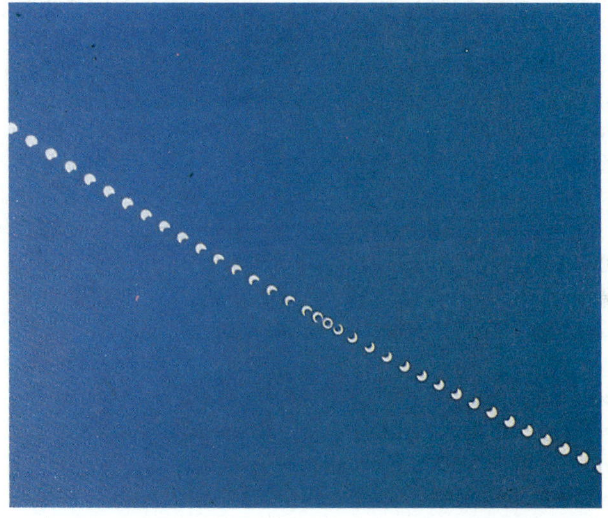

Figure 6–23. *During an annular eclipse, we can see the entire ring—annulus—of the Sun around the Moon.*

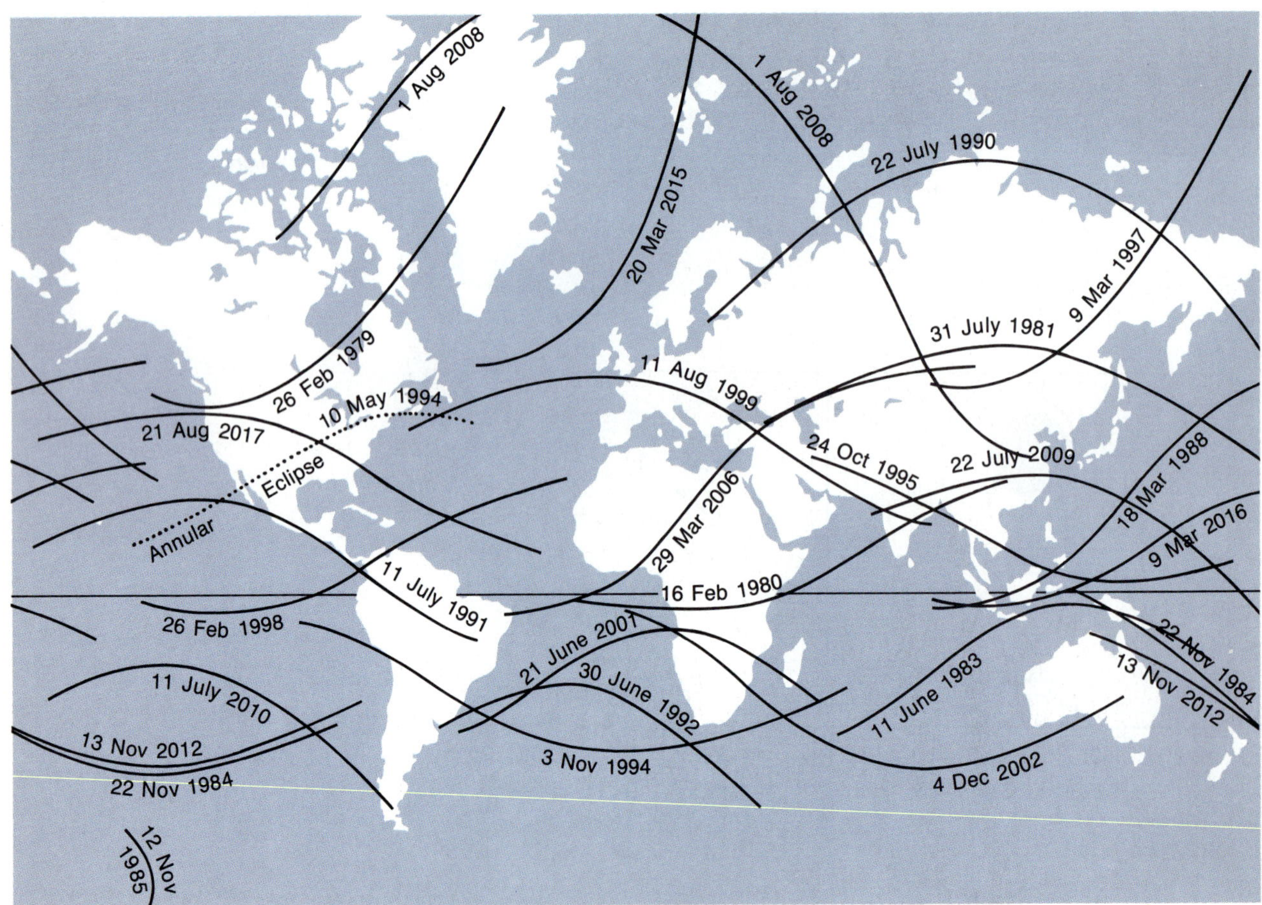

Figure 6–24. *The map shows the path of the moon's shadow during total solar eclipses through 2017. The annular eclipse of 1994 is shown as a dotted line.*

THE CENTER OF MASS

In the last chapter, we saw that Newton explained the orbiting of the Moon around the Earth by using his law of gravitation and his laws of motion. There is something that we omitted, however. One object cannot just stay still while another object orbits around it. To see why, do the Activity "Take Me For a Whirl."

In the Activity you see that if two people hold hands and whirl, they actually turn about a point between them. This point is called the ***center of mass*** of the two people. (The center of mass is sometimes called the center of gravity; for practical purposes the two terms mean the same thing.) As a child, you probably played on a see-saw (or teeter-totter) with someone much heavier than yourself. Recall that the larger person had to sit closer to the pivot point in order to balance. In fact, the two must sit so that their center of mass is at the pivot point of the see-saw if they are to balance on it.

Center of mass: The average location of the various masses in a system, weighted according to how far each is from that point.

This is the point that moves according to Newton's first law of motion when the entire system moves.

Figure 6–25 shows a large and small ball connected by a rod and held by a string tied to the rod. The two balls balance because the string is connected at the center of mass of the objects. If this contraption were thrown into the air with a spinning motion—like a baton—it would spin around that point, the center of mass. The Earth and Moon are somewhat like the two balls on the rod but instead of there being a rod to hold them near one another, there is the force of gravity.

Up until now we have spoken of the Moon orbiting the Earth as if the Earth stayed still and the Moon moved around it. In fact, the two objects revolve around their common center of mass. To determine where this point is located, we use the fact that the Earth's mass is 81 times that of the Moon. Think again of a large and small person on a see-saw. Suppose one person weighs 50 pounds and the other 150 pounds. If one person is three times heavier than the other, the lighter person must sit three times farther from the pivot of the see-saw. Similarly, since the Moon's mass is 1/81 of the Earth's, the Moon is 81 times farther from the center of mass of the Earth-Moon system than the Earth is. This means that the center of mass is about 5000 kilometers from Earth's center and 380,000 kilometers from the Moon. This point is inside the Earth. Figure 6–26 illustrates that it is about 1700 kilometers—1000 miles—below the surface.

Historically, it was the relationship between the distances of Earth and Moon from the system's center of mass that allowed us to calculate the mass of the Moon. In the discussion above, it was stated that if we know that the Earth is 81 times more massive than the Moon, we can locate the center of mass. In actual practice, the logic proceeded the other way. It was by determining the location of the center of mass that the Moon's mass was calculated. Today we know the mass of the Moon more accurately by observing its gravitational effect on space probes that have flown by it but until the space age, the method above was the most accurate. We will see in Chapter 14 that the center of mass method is what allows us to calculate the masses of stars.

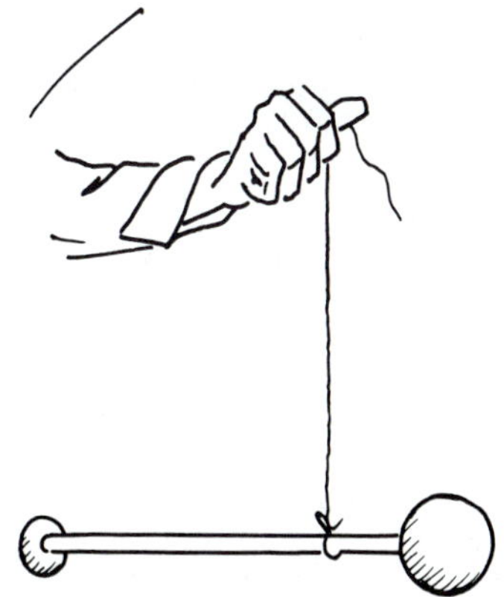

Figure 6–25. *The two balls at the end of the rod balance at the center of mass of the device, where the string is connected.*

Barycenter: The center of mass of two astronomical objects revolving around one another.

The location of the center of mass of the Earth-Moon system was determined by observing parallax of nearby planets due to the Earth's motion as the Moon went around. When observing Mars, the maximum parallax due to this motion is 17 arcseconds.

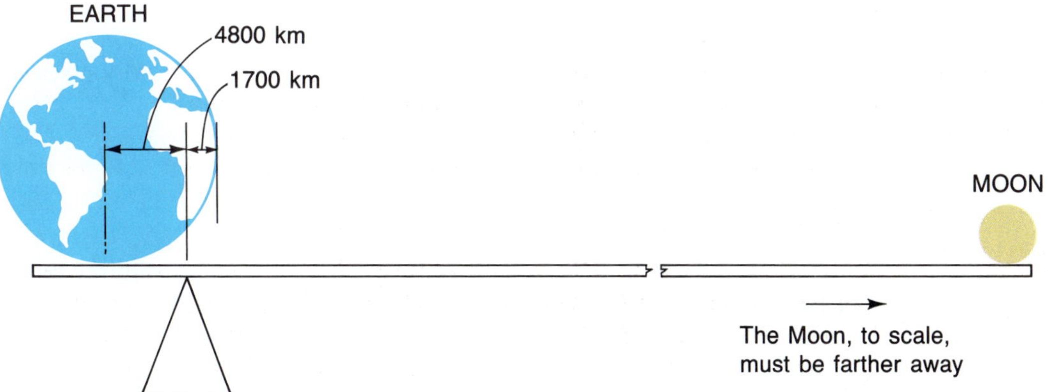

Figure 6–26. *The center of mass of the Earth-Moon system is 4800 kilometers from the center of the Earth. If a model of the system were constructed to scale to sit on a weightless board, the construction would balance as shown.*

ACTIVITY

Take Me for a Whirl

IN THIS ACTIVITY YOU will simulate two objects orbiting each other. Start by imagining that you are the Earth and that this book is the Moon. Hold the book at arm's length and whirl it around you in a horizontal circle. Notice that except for the fact that you are turning around, your body stays in about the same place while you are turning.

The real Moon has much more mass in comparison to the Earth than the book has in comparison to you, and now we want to show what happens when the object simulating the Moon is more massive. You'll need help for this—get another person.

Face the other person while holding each other's hands. Then start whirling around each other as in a child's dance. Notice that in this case you don't stand still; the two of you whirl around a point between you. If both of you are the same mass, the point will be halfway between you. If a smaller or a larger person is available, try it with that person. In this case, the two of you will again be whirling around a point between you, but that point will be much closer to the more massive person (Figure T6–6).

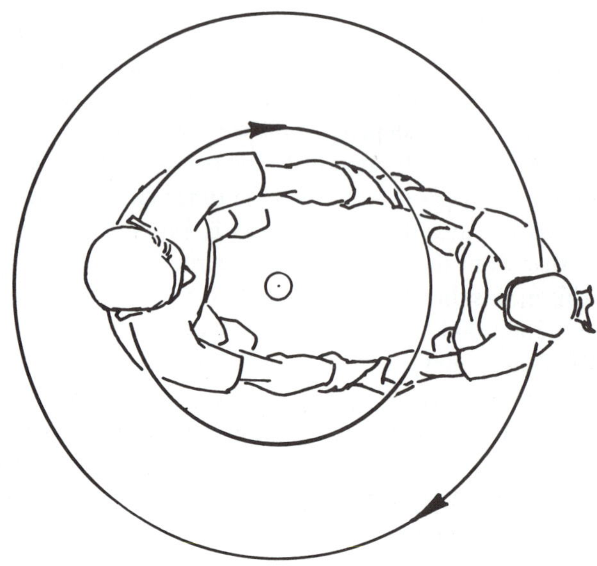

Figure T6–6. *When a large and small person hold hands and whirl, the point around which they rotate is closer to the larger person.*

We will use the concept of the center of mass of the Earth-Moon system to discuss two ideas: a revision of Kepler's Laws and the tides.

KEPLER'S LAWS REVISITED

According to Kepler's laws, the Moon follows an elliptic path with the center of the Earth at one focus. This is not quite true, however. The focus of the ellipse is not at the center of the Earth, but is at the common center of mass of the Earth-Moon system. We have seen that this point is fairly close to Earth's center, however, so Kepler's laws are not wrong by far. In fact, the center of the Earth also moves in an elliptic orbit with that center of mass at one focus.

When we discussed the motion of planets, we stated that according to Kepler's laws they move in elliptic paths around the Sun. There are two cor-

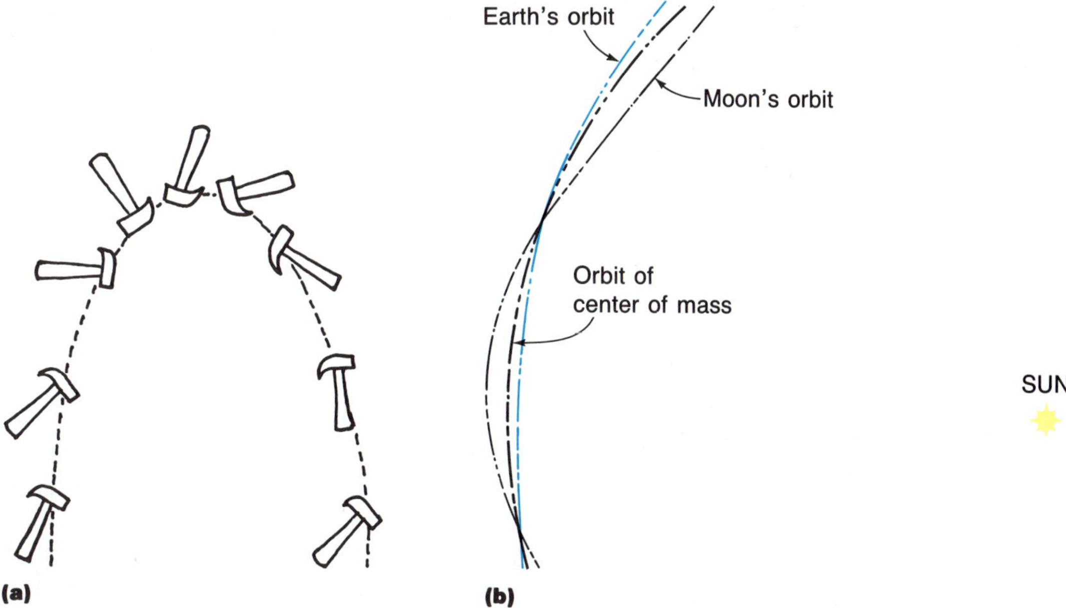

Figure 6–27. *(a) The center of mass of a thrown hammer follows a smooth path as the hammer rotates around that point. (b) Likewise, it is the center of mass of the Earth-Moon system that orbits the Sun in an elliptic path. (By holding a straightedge up to the paths drawn, you can see the Earth and Moon are always curving toward the Sun during the orbit.)*

rections we must make to this. Consider the planet Mercury orbiting the Sun. Corresponding to the case of the Earth and Moon, Mercury and the Sun orbit their common center of mass. The Sun is so very much more massive than Mercury, however, that the correction that must be made here is infinitesimally small. This is also true in the case of the Earth. Although the Sun and Earth actually orbit their system's center of mass, that point is so very close to the center of the Sun that in practice we can say simply that the Earth orbits the Sun.

The Earth has a Moon, however, and the Earth and Moon must be considered a system. In this case it is not the Earth that follows an elliptic path around the Sun, but it is the center of mass of the Earth-Moon system that follows that path. Just as when one throws a hammer with a spinning motion, the center of mass of the hammer follows a regular path—Figure 6–27(a)—so the center of mass of a planet and its satellite is the point that follows Kepler's elliptic path. (See Figure 6–27(b).) Kepler's law of ellipses, therefore, must be corrected as to the point which follows the ellipse.

Kepler's third law must also be changed slightly from what we used in the last chapter. Recall there that we stated that the period P of a planet's orbit is related to its average distance from the Sun, d, by the equation

$$\frac{d^3}{P^2} = KM$$

It was stated that M is the mass of the central object—the Sun in the case of the solar system. In fact, M is the total mass of the two orbiting objects, so that the equation might be written as follows:

Kepler's third law revised this way allows us to determine the masses of moons of other planets, as we will see in a later chapter.

$$\frac{d^3}{P^2} = K(m_1 + m_2)$$

d = average radius of orbit
P = period of orbit
K = constant which depends on units
m_1 = mass of one object
m_2 = mass of the other object

Since the mass of any planet is much smaller than that of the Sun, the equation is fairly accurate when the Sun's mass is used instead of the total mass. (If it weren't, Kepler would never have discovered the relationship.)

Finally, we will describe another slight correction that Newton made to Kepler's laws. Recall that the law of gravity states that *every* object attracts every other one. This includes two planets, like the Earth and Venus. In fact, there is a force between these two planets and when they are nearest to each other, the force changes the path of each enough that the change can be detected. This causes the Earth's (and Venus's) paths to be changed from perfect ellipses. The change is very small, so small that it could not be detected in Newton's time, but it exists.

So we see that the laws of Kepler, like most laws and theories in science, are approximations. For Kepler, they fit his data perfectly. But as more accurate and different observations became available, Kepler's laws had to be modified and improved. Kepler's laws as originally stated are accurate enough to justify their use in very many instances, but they help remind us that scientific theories are always tentative and are subject to revision.

THE TIDES

If you live near the seashore or visit it often, we don't need to describe the tides to you. But "landlubbers" don't often get a chance to experience the tides. Figure 6–28 shows a seashore at low and high tide; the difference in depth of water is obvious.

Most locations on the Earth experience a high tide about every 12 hours and 25 minutes and a low tide midway between high tides. Thus on most days there are two high tides and two low tides. Recall that Newton realized that the force of gravity between the Earth and an object is not really a single force exerted by the entire Earth but is the result of all the forces of gravitational attraction between the object and each part of the Earth—each little mass within the Earth. This idea applies to the force of gravity between the Earth and the Moon. The Moon exerts gravitational force not on the Earth as a whole but on each individual part of the Earth. We will consider only three little parts of the Earth. Figure 6–29 uses arrows to indicate the force of gravity between the Moon and a small mass at three different locations within the Earth. The mass on the side of the Earth nearest the Moon feels the greatest lunar gravi-

Figure 6–28. *Those of us who don't often get to the coast may not be familiar with the phenomenon of tides. Observe how much deeper the water is at high tide.*

tational force since it is closest to the Moon. The mass at the center of the Earth feels less force toward the Moon and the mass at the far side of the Earth feels the least force.

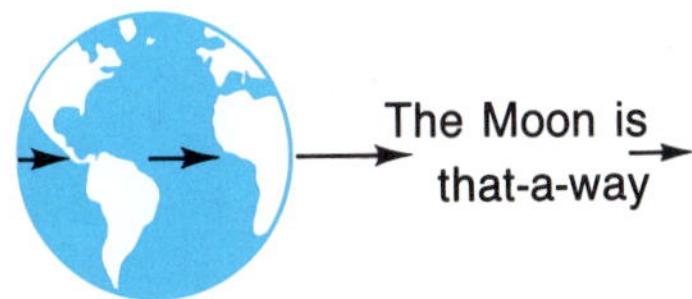

Figure 6–29. *The gravitational force exerted by the Moon on a given amount of mass of Earth is greatest on the side nearest the Moon, less on the mass at the center of the Earth, and least on the mass on the side farthest from the Moon. Arrows here represent the forces.*

Now each of these three parts of the Earth responds to the gravitational force toward the Moon but the part closest to the Moon feels a greater force than the other two. There is more force toward the Moon on one kilogram of mass here than there is on one kilogram of mass at Earth's center, so the mass at the surface feels pulled away from the center. Water covers most of the Earth and since it is liquid and free to flow, some water flows to that area under the Moon. The water becomes deeper at that point and it causes a high tide there.

The high tide on the other side of the Earth is caused by the fact that the center of the Earth feels a greater force toward the Moon than water on that side so the main body of the Earth is pulled away from that water, making another high tide there.

A "differential" pull means that it is different at different points on the Earth.

This differential gravitational pull on the various parts of the Earth results in two areas of the Earth experiencing high tides. The water that went to making those high tides has been pulled away from other parts of the Earth and so a low tide lies midway around the Earth between the areas of high tides.

As you might suspect, the actual tidal phenomena is much more complicated than the above. The Earth is rotating and its land masses disturb the flow of water. The interplay between the shape of the shoreline, the depth of the water, and the location of the Moon all play a part in determining exactly when high and low tides occur at a particular location on Earth and just how high and how low those tides are.

In addition, the Sun causes tides on the Earth. Because the *difference* in the Sun's gravitational pull on opposite sides of the Earth is not very great, these

Spring tide: The greatest difference between high and low tide, occurring about twice a month when the lunar and solar tides correspond.

Neap tide: The least difference between high and low tide, occurring when the solar tide partly cancels the lunar tide.

tides are small and are not noticed independent of the Moon's tides. When they correspond to the Moon's tides (near the times of new and full Moon), however, we see extreme tides on Earth. On the other hand, when the Sun's tides are opposite the Moon's (near quarter Moon), they tend to cancel the Moon's and this causes the change from low to high tide to be less than normal.

Rotation and Revolution of the Moon

Recall from the last chapter that the period of the Moon's rotation exactly matches its period of revolution. This results in it keeping the same face toward Earth at all times. The effect is caused by tidal forces. To understand it, recall that as the Earth rotates, it interacts with tides in the water. If there were no friction between the solid Earth and its oceans, one area of high tide would stay exactly under the Moon and another would be on the opposite side of the Earth. The land masses, however, exert forces on the water as the Earth rotates, causing the point of highest tide to be pushed from directly under the Moon (and directly opposite the Moon). Figure 6–30 illustrates this.

Notice the next time that you are at the seashore that high tide occurs before the Moon is highest in the sky.

If the land masses exert a force on the tidal bulges as the Earth turns, it follows from Newton's third law that the bulges exert an equal and opposite force on the Earth. This results in the Earth's rotation being slowed down because of the ***tidal friction*** exerted on it. The Earth is indeed turning more slowly today than it was years ago. Our days are very slightly longer than were Shakespeare's.

Tidal friction: Friction forces that result from tides on a rotating object.

There are also tides on the dry land. In this case, the rocks and dirt actually stretch to allow the surface to rise and fall. The maximum dry-land tide is about 9 inches. That is, the surface is about 9 inches farther from Earth's center at high tide than at low tide.

Just as the Moon causes tides on Earth, the Earth causes Moon tides. These are similar to the tides on the solid Earth. And just as tides on Earth are causing it to change its speed of rotation, the tides on the Moon have resulted in a change in its speed of rotation. At one time in the past, the Moon must have

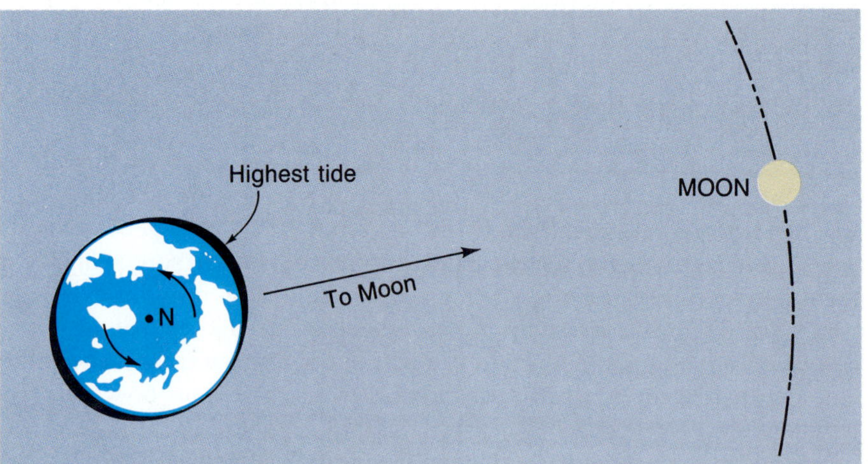

Figure 6–30. *The Earth's rotation tends to drag the tides along with it, so that a high tide is not directly under the Moon but instead is farther to the east.*

had a rotation period different from its revolution period. Through millions of years the tides have slowed the Moon's rotation until it now keeps its same face toward the Earth and tidal friction is very small. (It would be zero if the Moon's orbit were perfectly circular.)

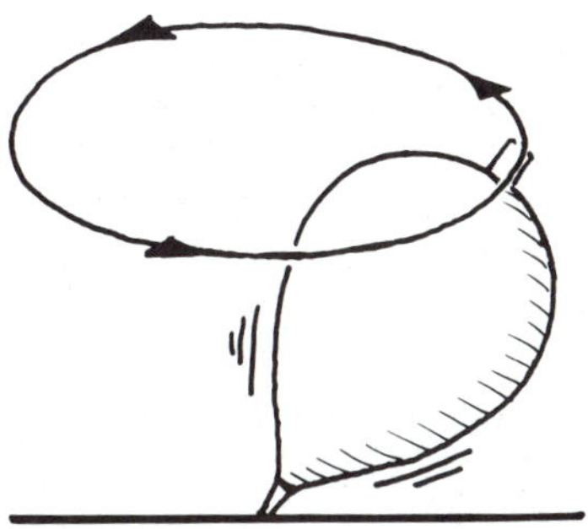

Figure 6–31. *As a child's top spins, it wobbles, or precesses.*

Precession of the Earth

The Earth spins on its axis. Think of what happens when you spin a child's top on a smooth table. The axis of the top does not stay in the same orientation (unless you were able to begin the rotation around a perfectly vertical axis). Instead, the top wobbles around, as indicated in Figure 6–31. What causes the wobble? The mathematics to describe this effect is far from simple, but it boils down to the fact that the top has a tendency to fall over, and its rotation prohibits this from happening. Instead, the top wobbles around, keeping the same angle with the table's surface until friction slows it down. Any time a spinning object feels a force trying to change the orientation of its axis, it will wobble. The wobble is called ***precession***.

Precession: The conical shifting of the axis of a rotating object.

It might seem that the Earth would not precess because there is nothing below it trying to pull it over. There is, however, a force on the Earth tending to change the orientation of its axis of rotation. Refer to Figure 6–32. The Earth drawn there is not spherical, but is "out of round," as is the real Earth, although the real Earth is much more spherical than the one in the figure. The Earth's diameter is about 26 miles greater across the equator than from pole to pole. This is caused simply by the fact that it is spinning. (Imagine what would happen if you made a ball of Jell-O® and got it spinning. The Earth is not Jell-O®, but it is still flexible enough to show the effect.)

Finally, recall that the Moon exerts a gravitational force on each particle of

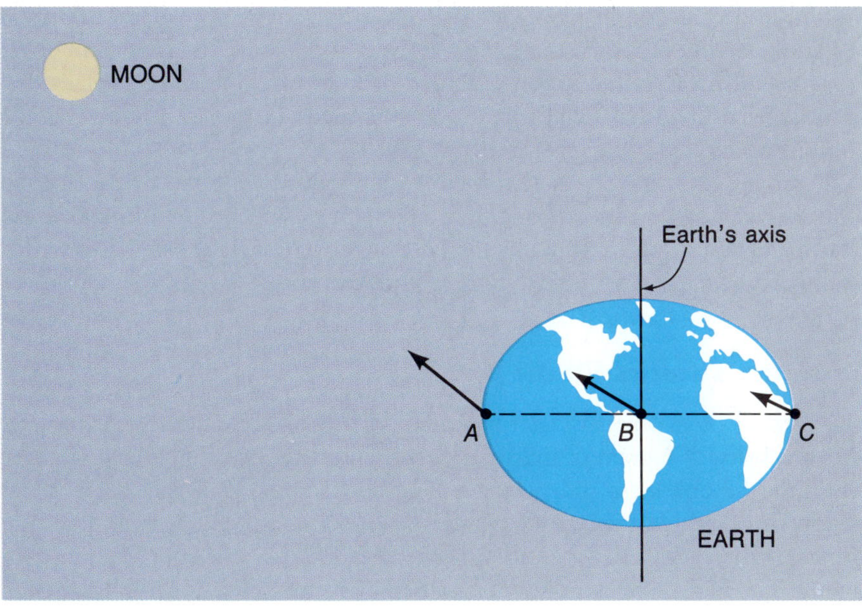

Figure 6–32. *The Earth is not a perfect sphere and the Moon exerts a greater gravitational force on its nearer side, therefore tending to twist the Earth into a different orientation. This causes the Earth to precess.*

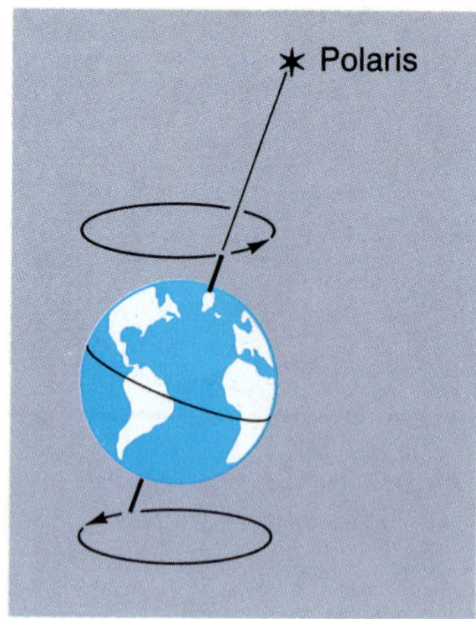

Figure 6–33. *As the Earth spins, its axis precesses with a period of about 26,000 years, pointing to different "pole stars" over the centuries.*

the Earth. The arrows in Figure 6–32 illustrate this force at three places. Since point A is closer to the Moon than point C, the Moon exerts a greater force on a particle at A. This results in an overall force seeking to change the tilt of the spinning Earth's axis. So we have what is needed to cause precession. The Earth's axis does indeed precess, although very slowly. While the child's top may complete a precession in about a second, the Earth requires about 26,000 years! (See Figure 6–33.)

What effect does this have on what we see in the sky? Right now, Polaris is the closest bright star to the Earth's north celestial pole. But it will not remain so forever. The pole will gradually change, until about 12,000 years from now the star Vega will be our "north star."

A corresponding effect is that the position of the vernal equinox changes as the centuries pass. We discussed how this affects the practice of astrology in a previous chapter.

THEORIES OF THE ORIGIN OF THE MOON

Until recently, there have been three theories of the origin of the Moon, called the *capture,* the *double planet,* and the *fission* theories. We will briefly describe each of these theories and then look at the evidence to see which fits the data best.

Because this theory holds that the Earth and Moon had the same origins, it is sometimes called the sister theory.

The double planet theory was suggested in the early 1800s and is the oldest. It holds that as the Earth formed from a spinning disk of material, not all of that material coalesced to form the Earth. A smaller part of it was left orbiting the Earth and formed into the Moon. In Chapter 15 we will discuss theories of the origin of the solar system and will see that this idea is entirely consistent with those theories.

This theory is called the daughter theory.

In 1878, astronomer Sir George Howard Darwin, son of the biologist Charles Darwin, proposed that the Moon was once part of the Earth and broke (or *fissioned*) from it due to forces caused by a fast rotation and by solar tides. The large basin of the Pacific Ocean was proposed as the place from which the Moon was ejected in this fission hypothesis.

Finally, the wife theory.

Early in this century, another theory was proposed. It holds that the Moon was originally a separate astronomical object that happened to come near the Earth and that it was captured by the Earth's gravitational field so that it settled into orbit as the Moon. This is the "capture theory."

How the Theories Fit the Data

Density: The ratio of mass to volume. *For example, if you take a 1 cm cube of metal (1 cm^3 of metal) and determine that its mass is 5 g, then its density is 5 g/cm^3.*

The ***density*** of an object is determined both by its mass and its volume. (Water has a density of 1 gram of mass per cubic centimeter of volume.) As we have seen, the Moon's mass and size have long been known and from these values it can be calculated that the density of the Moon is 3.3 grams per cubic centimeter. The Earth's density, on the other hand, is 5.5 grams per cubic centimeter. The Earth's greater density is caused by its iron core, for iron has a high density. The material in the Earth's crust, on the other hand, has a density close to that of the Moon.

Based simply on comparing densities, the double planet theory seems to be ruled out, for if the Moon formed along with the Earth its density should match Earth's. If the capture theory is correct, however, it would certainly be possible for the Moon to have a density less than the Earth's, for the two objects would have little in common. In fact, it would be a coincidence that the Moon's density matches that of the Earth's crust. The other theory—the fission theory—seems to fit the density data best, for if the Moon came from the Earth's outer layer it would have a density similar to that layer.

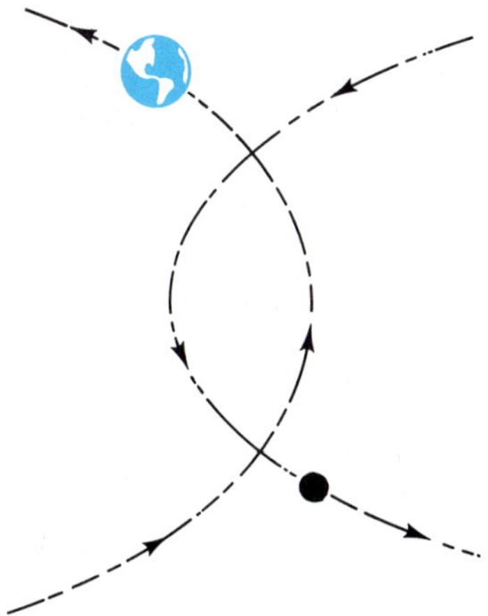

Figure 6–34. *If the Earth and another object (in this figure, a smaller one) had a near-collision, each object's path would be altered but they would not begin to orbit one another.*

There is a problem with the fission theory, however. Astronomers have difficulty explaining how an object as massive as the Moon might have been pulled out of—or thrown off—the Earth. No satisfactory mechanism for this event has been proposed. In addition, the Moon does not orbit in the plane of the Earth's equator as it would if it were ejected from a spinning Earth.

There are also problems with the capture theory. If one astronomical object comes close to another, each of their paths will be changed by the gravitational force between them (Figure 6–34), but one will not capture the other unless there is contact between the two or unless a third object is involved, so that the interaction of the three objects results in one of them being slowed down to an orbital speed. Such a near-collision between three objects seems highly unlikely.

Although we have known the density of the Moon for a long time, its chemical composition was not well known until the Apollo program brought back some 850 pounds of soil and rock samples. This new data brought new problems for the three theories. In many ways, the chemical composition of the Moon is similar to that of the Earth's crust, for both have about the same proportions of some of the major elements: silicon, magnesium, iron, and manganese. The Moon, however, has far smaller proportions of easily vaporized (***volatile***) substances than does the Earth. The Moon has much higher proportions of nonvolatiles (such as aluminum and titanium), which require a very high temperature to vaporize, than does the Earth's crust. The differences in chemical composition, along with the Moon's lack of an iron core, seem to rule out both the fission theory and the double planet theory.

Volatile: Capable of being vaporized at a relatively low temperature.

Thus none of the theories for the origin of the Moon seemed to fit all of the data. The question remained, "How did the Moon get there?"

The Large Impact Theory

In the 1970s a new theory was proposed by A. G. W. Cameron and William Ward of Harvard. They proposed that early in the Earth's history, it was struck at a glancing angle by a large object, that the impact resulted in a fusion of the two objects, and that material was thrown off of the two to form the present Moon. Computer simulation of such a collision shows that if the impacting object has a mass nearly as great as Mars, heat resulting from the collision would vaporize material and eject enough of it into orbit to account for the mass of the Moon, once the material coalesced in its orbit around the Earth. The ***large***

CLOSE UP

Lunar Effects on Humans

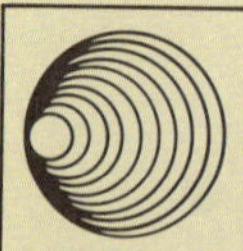

THE MOON IS SOMETIMES given responsibility for much more than the tides. It has been linked with things ranging from human birthrates to insanity to werewolves. We won't have much to say about werewolves except to point out that they seem to be found only in movies and books. Is it possible, though, that the Moon may have strange effects on us humans? Yes, it is conceivable, but we must look at the claims made and the evidence for those claims before concluding anything.

Perhaps it is your experience that the full Moon makes you feel romantic. Rather than some mysterious Moon force causing this, however, the effect is probably due to the attractiveness of the moonlit night (and perhaps of your companion).

The word *lunacy* comes from a belief that insanity can be caused by the Moon. Undoubtedly, if a person firmly believes that exposure to the light of the full Moon will cause him to become insane, there is a better chance that it will happen. Likewise, if a person believes that a thunderstorm will drive her mad, a rumble of thunder may well bring it about. These, however, are psychological effects rather than anything mysterious. There is no quantitative evidence whatsoever that the Moon itself causes insanity.

From time to time newspapers report that more babies were born in some hospital during a certain cycle of the Moon. How do we explain this? We explain it by asking if it is due to anything more than chance. Keep in mind that people remember unusual occurrences and tend to forget everyday things. A gambler utters a certain phrase and the dice come up right, rewarding him with a big pot. From then on, he uses that phrase when rolling his dice. We remember and report to others the unusual things which happen. The birthrate in a hospital—or an entire city—fluctuates due to any number of factors, some of them completely random. To test the hypothesis that the cycle of the Moon affects the birthrate, we must test a large number of people over a long time. Such statistical tests have been done, and they consistently find that there is no correlation between birthrates and Moon phases. It doesn't happen.*

---

*See "The Moon and the Maternity Ward," *Skeptical Inquirer* 3, (Summer 1979).

Large impact theory: A theory that holds that the Moon formed as the result of an impact with a large object upon the Earth. *Might we call this the "battered husband theory"?*

impact theory, as it is called, is able to explain both the similarities and the differences in the compositions of the Earth and the Moon.

At a conference in October 1984, in Kona, Hawaii, called to discuss the origin of the Moon, the large impact theory was the major topic of discussion. Since that conference a consensus has begun to be developed among astronomers that this theory indeed fits the data better than the other three. As with all new theories, this one will be tested against both existing data and new data as the years pass. Although it will probably have to undergo modification, it appears that we may finally have found the answer to the age-old question of the origin of the Moon.

Crime rate is another occurrence sometimes associated with Moon phases. This effect may indeed exist, and it should come as no surprise. Thieves surely prefer dark nights to those lit by a full Moon. Nothing mysterious here.

Could some effect on humans be caused by gravity? Well, the Moon certainly does exert a gravitational force on our bodies. The amount of the force does not depend upon the phase of the Moon, however. In addition, the force is only 3 millionths of the force exerted on our bodies by the Earth, and there is no reason to believe that our body can detect it. After all, our weight changes much more than that when we go from sea level to a high mountain, and we can't feel that difference. (Enhanced athletic performances at high altitudes in events such as the long jump are due more to the thinner air than the decreased gravitational pull on the athletes' bodies.)

What about possible tidal effects on our bodies? The Moon causes tides on the Earth because of the *difference* in its gravitational pull on one side of the Earth and the other. Could it cause some such effect on a person? The two sides of the Earth are 8000 miles apart, while a person is about six feet tall. Therefore, to imagine that there is a meaningful tidal effect in that distance is farfetched, to say the least.

Is it impossible, then, that the Moon may have some effect on us that we do not understand? No, it is not impossible. But we should believe such claims only when there is convincing evidence for them. Extraordinary claims require extraordinary evidence and at best can only be considered as tentative hypotheses.

When someone tries to blame the Moon, planets, or stars for a human event, perhaps we should ask who is trying to evade responsibility for the event. One thing we learn from science is that we humans are responsible for our own destiny. To blame or credit the stars is simply a copout.

CONCLUSION

We began our look at the Earth-Moon system by examining how the distance and size of the Moon were first measured. We will see in future chapters that the ideas behind these measurement techniques are useful far beyond the solar system. We saw that solar and lunar eclipses are each caused by particular alignments of the Sun, Earth, and Moon but that the effects as experienced by humans are far different. Finally, we applied Newton's law of gravitation to motions in the solar system and were required to make slight modifications of Kepler's laws. These modifications will later be important to us in measuring properties of distant stars and galaxies.

RECALL QUESTIONS

1. Define angular size and zenith.
2. When we see an unfamiliar object at a distance, how can we judge its size?
3. How great is the angular size of the Moon? How does this compare to the angular size of the Sun?
4. Explain how the diameter of the Earth was measured by Eratosthenes.
5. How great is the diameter of the Earth? How far is the Moon from Earth?
6. How does the Moon compare to the Earth in diameter?
7. Describe two effects that cause the Moon to appear different sizes at different times. (Hint: See a Close Up for one.)
8. Define umbra and penumbra.
9. Total (and annular) solar eclipses occur more frequently than do total lunar eclipses. Why, then, will you probably observe many more lunar eclipses than solar eclipses during your lifetime?
10. Describe the appearance of a totally eclipsed Moon. Why is it not completely hidden from view?
11. Discuss the danger involved in viewing a solar eclipse and describe four ways to safely view such an eclipse. (Hint: See the Close Up, "Observing a Solar Eclipse.")
12. Why might a partial solar eclipse go unnoticed by most people?
13. Describe some unusual effects that take place during a total solar eclipse.
14. Why are some solar eclipses annular rather than total?
15. Explain why there are two areas of high tide on Earth rather than one.
16. How was the mass of the Moon (compared to the Earth's mass) first determined?
17. At what phase of the Moon does a lunar eclipse occur? A solar eclipse? The highest high tide?
18. Name and describe four theories for the origin of the Moon. Which best fits present data?

QUESTIONS TO PONDER

1. The angular size of your finger held at arm's length is about 2 degrees. On some clear night use this fact to determine the angular distance between the two pointer stars of the Big Dipper. (See the star charts at the back of the book to help you find the dipper.)
2. If you observe the Moon with first one eye and then the other, do you detect a parallax shift against the stars? Why or why not?
3. Explain why the choice of an equatorial location to measure the distance to the Moon makes the measurement simpler.
4. Why must a larger baseline be used to measure the distance to stars by the method of triangulation?
5. When people in Chicago see a total lunar eclipse, what type of eclipse do people one-fourth of the way around the Earth see? (Assume that the Moon is visible in both skies.)
6. The Sun also causes a tidal effect on the Earth. The force of gravity of the Sun on the Earth is greater than the force of gravity of the Moon on the Earth. Why, then, are Sun tides lower than Moon tides?
7. Explain why we never see the back side of the Moon.
8. Is it just coincidence that the Moon's periods of revolution and rotation are the same? If not, explain the cause of this.
9. Does the Earth orbit the Sun in an exact ellipse? Discuss.
10. Why does the duration of a lunar eclipse depend upon the Moon's distance from Earth?
11. In Figure 6–27(b), why don't the Earth and Moon collide when their orbits cross?
12. Why does the time between successive high tides average 12 hours, 25 minutes (and not exactly 12 hours)?
13. Over the years, under what conditions would the very highest tides occur? (An extremely high tide occurred in January 1987.)

CALCULATIONS

1. Compare the distances reached by the umbral shadows of the Moon and the Earth. Compare the diameters of the two bodies. Discuss any relationship you see.
2. Assume that under the conditions described for Eratosthenes' measurement of the Earth's size, the Sun was *exactly* 7 degrees from the zenith at Alexandria. Now at a time when the Moon was directly above the well in Syene, would it have been exactly 7 degrees from the zenith in Alexandria? Why or why not? If not, would its angular distance from the zenith have been greater or less than 7 degrees?
3. If Eratosthenes had observed the Sun to be 14 degrees from the zenith at Alexandria (instead of 7 degrees), would his value for the Earth's circumference have been greater, smaller, or the same?

CHAPTER 7

A controversial theory proposed in 1986 by a University of Iowa physicist, Louis Frank, may be gaining some support. The theory holds that rather than the Earth's water being part of the original planet, most of it has fallen onto the Earth from space during the 4.5 billion years of the Earth's existence. Dr. Frank further proposes that the process is continuing. He hypothesizes that some 10 million icy comets—snowballs from space—strike the Earth each year and that they average about 100 tons each. They have a diameter of about 30 feet and are coated with black hydrocarbons. High in the atmosphere they break up and their ices join the water vapor in the atmosphere; eventually to fall as rain. They add about an inch of water over the entire Earth's surface every 10,000 years.

Astronomers are still very skeptical about the idea but at the Kitt Peak Observatory a new technique was used to search for the fast-moving snowballs and the results agree with predictions made by Frank's hypothesis. At this time, a number of other measurements have been made that correspond to the predictions.

The hypothesis is still in its infancy, and—like many such proposals—it may fall flat on its face. On the other hand, astronomy books of the next decade may present Louis Frank's idea as one that has been verified and is universally accepted.

A Planetary Overview and a Look at Earth

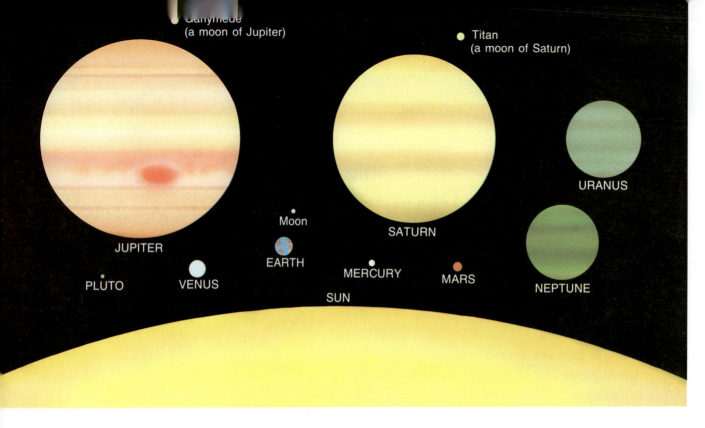

Figure 7–1. *This shows the Sun, planets, and a few of the large moons drawn to scale.*

Figure 7–1 shows the planets as disks, but they are actually spheres. In trying to imagine their comparative sizes, think of them that way.

WE HAVE SEEN IN previous chapters that there are patterns among the planets of our solar system. For example, Kepler's third law tells us of the relationship between a planet's distance from the Sun and its period of revolution. Before beginning our study of individual planets, we will look at other patterns of similarities and differences among the planets.

Figure 7–2. *If the Sun were the size of a basketball, the Earth would be the size of the head of a shirtpin.*

DISTANCES IN THE SOLAR SYSTEM

To say that the Sun is the largest object in the solar system is a gross understatement. We can almost say that the Sun *is* the solar system, so great are its size and mass compared to the other objects. Figure 7–1 shows the planets drawn to scale. At the bottom of the drawing, you see the partial disk of the Sun. The Sun is so large that if it had been drawn as a complete circle fitting the page, many of the planets would have been nearly too small to see. The Sun's diameter is about 1,390,000 kilometers, while the Earth's diameter is about 13,000 kilometers. Thus the diameter of the Sun is more than 100 times that of Earth. To better picture this, consider the Sun to be an object the size of a basketball; a sphere about one foot in diameter. On this scale the Earth

Table 7–1 Average Distances of the Planets from the Sun

Object	Distance from the Sun (AU)	On Our Scale	
		Diameter	*Distance from Sun*
Sun	—	1 foot	—
Mercury	.39	0.04 inch	45 feet
Venus	.72	0.1 inch	80 feet
Earth	1.0	0.1 inch	110 feet
Mars	1.52	0.05 inch	170 feet
Jupiter	5.20	1 inch	200 yards
Saturn	9.54	0.9 inch	350 yards
Uranus	19.19	0.4 inch	0.4 mile
Neptune	30.06	0.4 inch	0.6 mile
Pluto	39.44	0.02 inch	0.8 mile

would be a BB, about the size of the head of a shirtpin, an eighth of an inch in diameter (Figure 7–2).

As we discuss the sizes of solar system objects and the distances between them, try to form a mental picture of the relative distances rather than just memorize the values.

Jupiter, the largest planet, is much larger than the Earth. Its diameter, in fact, is about 11 times that of the Earth. On our scale, with the Sun being a basketball, Jupiter would have a diameter of about 1¼ inches, which is about the size of a ping-pong ball. It's still not much compared to the Sun.

Pluto is the smallest planet, with a diameter about one-fifth that of Earth. In our scale model it would be a grain of sand, less than one-thirty-second inch across! Appendix F lists the actual sizes of the planets along with their sizes compared to the Sun and the Earth.

Now let us consider the distances between the solar system objects. Table 7–1 shows the average distance of each of the planets from the Sun in astronomical units and according to our model. To continue the model in which the Sun is a basketball, we might put the basketball at one end of a tennis court. A BB at the opposite end of the tennis court would be the Earth. Nearly two football fields away is the ping-pong ball Jupiter. Pluto would be a grain of sand a mile away! Between these objects we put nothing—or at least almost nothing. There are only the other planets, all smaller than Jupiter, and the *asteroids,* even smaller objects (to be discussed in the next section).

Asteroid: Any of the thousands of minor planets that orbit the Sun.

Figure 7–3 is the solar system drawn to scale. The planets are all there, stretched out in a straight line from the Sun, which is at the left end. Pluto, 37 billion miles from the Sun, is at the right. To see anything at all on this scale you must look carefully, because even the giant Sun is just a little speck at left and the planets are invisible on this scale. The solar system is mostly empty space.

CLOSE UP

The Titius-Bode Law

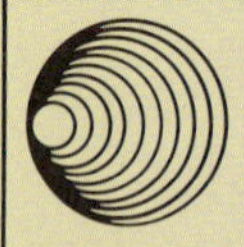

As pointed out in Chapter 2, the relative distances to the planets were known in Copernicus's time. We saw that Kepler used this data to formulate his third law. From the time of Copernicus, people have wondered why the planets are at the distances they are. Is there any pattern to the distances? Kepler thought he had found one when he used the five regular geometric solids to separate the orbits of the six planets. Kepler's geometric rules gave an approximate fit, but even if they had fit perfectly with the six planets known in Kepler's time, the discovery that there were more planets beyond Saturn (but no more regular geometric solids) destroyed any credibility left in his scheme.

In 1766 a German astronomer named Johann Titius found a mathematical relationship for the distances from the Sun to the various planets. The rule was publicized by Johann Bode, the director of the Berlin Observatory, in 1772, and is known today as the Titius-Bode Law or simply Bode's Law. Table T7–1 illustrates how the law works. Column 1 shows a series of numbers starting with zero, jumping to three, and then doubling in value thereafter. Column 2 was obtained by adding 4 to each of those values. Finally, to get column 3, we divide each of the column 2 values by ten. Now compare these figures to the measured distances of each of the planets from the Sun.

The table shows that the Titius-Bode Law fits fairly well, except that there is a gap: No planet is found at 2.8 AU from the Sun. Remember that fitting the data is the first criterion of a good theory. The Titius-Bode Law is too restricted in its application to deserve to be called a theory, but let's take a look at how the Titius-Bode Law fits the other criteria, anyway.

A second criterion for a good theory is that the theory should make predictions that would allow the law to be proven incorrect. The law seems to predict that there should be a planet between Mars and Jupiter, and, further, that the planet should be 2.8 AU from the Sun. In addition, the law predicts that if other planets are found beyond Saturn, the next one will be about 19.6 AU from the Sun. It fits this criterion well;

Table T7–1 Planetary Distances by the Titius-Bode Law Compared to Today's Values

A series of Numbers	Add 4	Divide by 10	Today's Measured Distance (AU)	Planet
0	4	.4	.39	Mercury
3	7	.7	.72	Venus
6	10	1.0	1.0	Earth
12	16	1.6	1.52	Mars
24	28	2.8		
48	52	5.2	5.20	Jupiter
96	100	10.0	9.54	Saturn

it is capable of making specific predictions and being proven wrong.

The third criterion concerns neatness and simplicity. In a sense, the Titius-Bode Law is simple, for it involves a regular pattern. The pattern is not followed throughout, however. Notice that in the first column we do not double the number every time; we start by jumping from zero to three. Thus it is not self-consistent and cannot be called "neat."

Bode published the law in 1772. Nine years later—in 1781—the planet Uranus was discovered by William Herschel in England. Its distance from the Sun was 19.2 AU. The Titius-Bode Law predicted 19.6 AU. A close fit!

With this confirmation of the validity of the Titius-Bode Law, a group of German astronomers (who called themselves the Celestial Police) divided the zodiac into a number of regions, planning to assign a specific region to each of a number of astronomers in order to systematically search for the missing planet at 2.8 AU. The searchers did not find it, however. Instead, a monk who was working on a different project discovered the largest of the asteroids at the distance predicted for Bode's missing planet. (See the Historical Note in this chapter.) Although it was first thought that this was the missing planet, the discovery within a few years of other objects at about the same distance, made it obvious that things were not this simple. The Titius-Bode Law could not account for the large number of planets between Mars and Jupiter.

Two more planets have been discovered since the discovery of the asteroids. How well do they fit the Titius-Bode Law? Table T7–2 shows the complete comparison. Notice that Neptune does not fit the prediction at all but that Pluto comes fairly close. So here we see that the law no longer fits all of the data and therefore it does not satisfy the first criterion as well as before these planets were discovered.

Today the Titius-Bode Law is considered little more than a curiosity rather than a scientific law.* There are a number of reasons for this: The law is not accurate even for the planets it fits, it does not fit Neptune at all, and it is not self-consistent in its structure. Surely, though, it is not just coincidence that it fits even in its limited way. We will see in Chapter 15 that theories that have been proposed for the formation of the solar system account for the fact that the more distant a planet is from the Sun, the farther it is from other planets. Still, the fact that the Titius-Bode Law fits as well as it does must be considered a coincidence. We may be able to judge its significance better when, at some future date, we are able to observe planetary spacing around other stars.

Table T7–2 How Celestial Objects Fit the Titius-Bode Law

Bode's Law Prediction	Today's Measured Distance (AU)	Object
4	.39	Mercury
.7	.72	Venus
1.0	1.0	Earth
1.6	1.52	Mars
2.8	2.8	Ceres
5.2	5.20	Jupiter
10.0	9.54	Saturn
19.6	19.19	Uranus
	30.06	Neptune
38.8	39.44	Pluto

*Relationships such as the Titius-Bode Law are said to be *empirical*. This means that they are found to work, but they are not related to any theoretical framework; we don't know why they work. The Titius-Bode Law isn't a particularly good empirical law, however, for reasons noted above.

Figure 7–3. *This is the solar system drawn to scale, with the Sun at the left and Pluto at the right. On such a small scale, only the Sun is visible—the little dot at left.*

THE ASTEROIDS

The accompanying Close Up discusses a rule relating these distances.

Figure 7–4 is similar to the previous one except that it indicates the locations of the planets by showing their orbits. Notice that there is somewhat of a pattern in their separations in that as one moves farther from the Sun, the orbits of planets are farther apart. Notice also the sudden gap between Mars and Jupiter. Kepler at one time proposed that a small planet might exist there.

On the first night of the nineteenth century, a Sicilian monk named Giuseppe Piazzi discovered an object in the sky that was not on his sky chart. The story of the discovery of this object—named Ceres for the historic Goddess of Sicily—is told in the nearby Historical Note. Calculations of the object's orbit yielded an average distance of 2.8 AU from the Sun, falling between Mars and Jupiter. It seemed that the missing planet had been discovered!

Two years later, another object was found orbiting the Sun at about the same distance. Then a third was found in 1804. By 1890, more than 300 of these objects had been found. We know the objects today as asteroids. Ceres is the largest of them and has a diameter of about 1000 km, or 600 miles. Thus its diameter is only 40 percent of that of Pluto, the smallest of the planets. It—and perhaps another 100,000 smaller asteroids—orbit the Sun between Mars and Jupiter (Figure 7–5). In Chapter 10 we will discuss them, along with other asteroids that orbit outside the main asteroid belt.

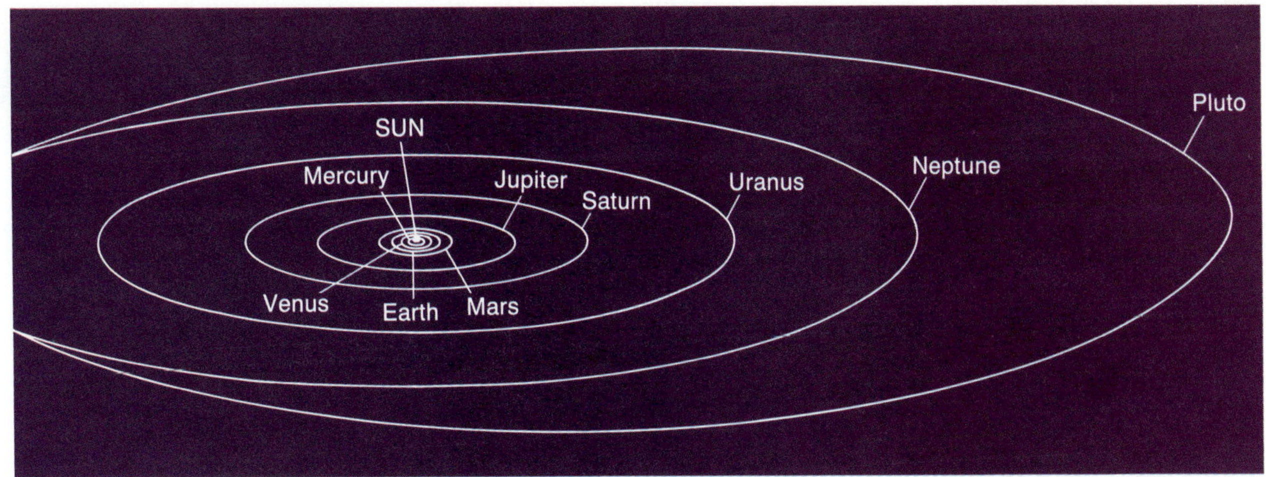

Figure 7–4. *The orbits of the planets are shown here at their relative average distances from the Sun.*

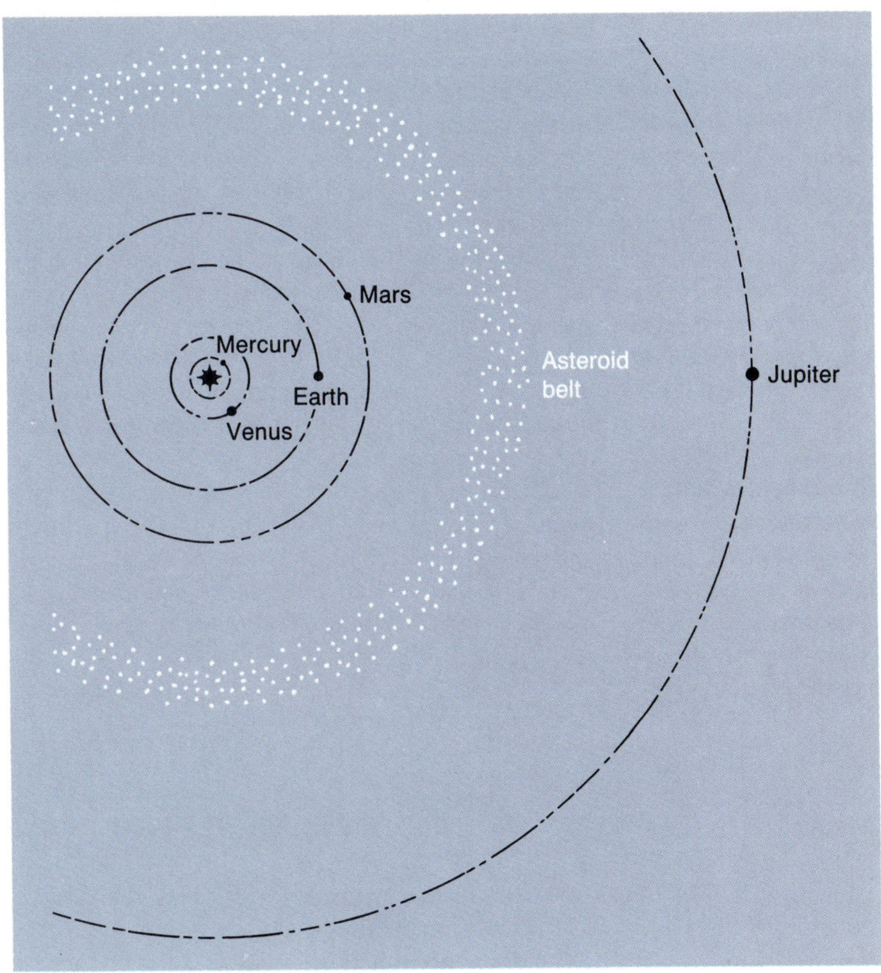

Figure 7–5. *The first asteroids that were discovered (and most that exist) orbit the Sun between Mars and Jupiter.*

HISTORICAL NOTE

The Discovery of the Asteroids

THE GAP BETWEEN MARS and Jupiter had been noticed before the Titius-Bode Law was proposed. Kepler at one time suggested that a small planet might exist there. Then the Titius-Bode law seemed to predict a planet in the gap, and it led Francis von Zach, a German baron, to plan a systematic search for the planet. Giuseppe Piazzi, a Sicilian astronomer and monk, was one of the astronomers who had been chosen to search in one of the sectors into which von Zach had divided the sky. Before he was notified where he was to search, however, Piazzi discovered what he first thought was an uncharted star in Taurus. The object he discovered (on January 1, 1801) was very dim, far too dim to see with the naked eye. Continuing to observe it, he saw that it moved among the stars, and by January 24 he decided that he had discovered a comet. He wrote two other astronomers (including Bode) of his discovery, but on February 11 he became sick and was unable to continue his observations. By the time Bode and the other astronomer received their letters (in late March), the object was too near the Sun to be observed.

Bode was convinced that the hypothesized new planet had been discovered, but he also realized that it would not be visible again until that fall and by that time it would have moved so much that astronomers would not be able to find it again. This was because relatively few observations had been made of its position; not enough for the mathematicians of the time to calculate the object's orbit. Fortunately, a young mathematician named Karl Frederick Gauss, one of the greatest mathematicians ever, had recently worked out a new method of calculating orbits. He worked on the project for months and was able to predict some December positions for the object. On December 31, the last day of the year in which it was seen on the first day, von Zach—the man who had sought to organize the original search—rediscovered it.

The elation over finding the predicted planet did not last long, however, for another "planet" was found in nearly the same orbit about a year later. Its discoverer, Heinrich Olbers, was looking for Ceres when he discovered another object that moved. He sent the results of a few night's observations to Gauss and the mathematician calculated its orbit. The object was given the name Pallas, and a new classification of celestial objects had been found: the new objects that were called *asteroids*.

By 1890 about 300 asteroids had been found using the tedious method of searching the skies and comparing the observations to star charts, looking for uncharted objects. In 1891 a new method was introduced: A time exposure photograph of a small portion of the sky was taken and the photo was searched for any tiny streak. The streak (Figure T7–1) would be caused by any object that did not move along with the stars. These objects were then watched very closely and their orbits determined. Using such methods, well over 3,000 asteroids are now known and named, and it is predicted that some 100,000 asteroids are visible in our largest telescopes.

MEASURING DISTANCES IN THE SOLAR SYSTEM

We saw that Copernicus was able to use geometry to calculate the relative distances to the planets. That is, he was able to calculate that Mars is 1.5 AU from the Sun although he could not determine the value of an astronomical

Figure T7–1. *The two streaks on the time exposure photo are caused by the motion of two asteroids as the camera follows the stars' apparent motions across the sky.*

unit. Today we can measure the distances to planets using radar. We bounce radar signals from a planet and measure the time required for the signal to reach the planet and bounce back. Then, knowing that radar travels at the speed of light, 3×10^5 km/sec, we can calculate the distance to the planet.

The radar signal is typically a burst of 400 kilowatts of power, but the returning signal is only 10^{-21} watt.

EXAMPLE Suppose that a radar signal is bounced from Mars. The signal returns to Earth 22 minutes after being transmitted. How far away is Mars?

Solution First, we realize that 22 minutes is the time for the signal to reach Mars and return to Earth. So a one-way trip requires 11 minutes. Now let's change 11 minutes to seconds (since our radar speed is given in km/sec).

$$11 \text{ min} \times \frac{60 \text{ seconds}}{1 \text{ minute}} = 660 \text{ seconds}$$

Now,

$$\text{velocity} = \frac{\text{distance}}{\text{time}}$$

So

$$\begin{aligned} \text{distance} &= (\text{velocity}) \times (\text{time}) \\ &= (3.0 \times 10^5 \text{ km/sec}) \times (660 \text{ sec}) \\ &= 2.0 \times 10^8 \text{ km.} \end{aligned}$$

Since Mars is about 1.5 AU from the Sun, the distance from Earth to Mars varies from about 0.5 AU to 2.5 AU.

To check that this is a reasonable answer, recall that one astronomical unit is 1.5×10^8 kilometers. So our calculated distance is 1⅓ AU. Since the orbit of Mars is 1.5 AU from the Sun, we see that at some point in its orbit it is possible for Mars to be at our calculated distance from Earth.

Try One Yourself. When Venus is at its closest to Earth, it requires about 4.8 minutes for a radar signal to travel to Venus and back. What is the distance to Venus? Convert the answer to astronomical units and check it with the correct distance given in Table 7.1.

When the nearest planet, Venus, is closest to Earth, a radar signal still requires nearly 5 minutes to get there and back. The great distances in the solar system become more obvious when we consider that if such a signal could be emitted in New York City and reflected from something in Washington, D.C., only 0.002 seconds would be required for the round trip.

Calculating Planetary Sizes

In Chapter 6 we used the following "small angle formula" to calculate the size of the Moon:

$$\text{Width} = \frac{(\text{angular size in degrees}) \times (\text{distance})}{57.3 \text{ degrees}}$$

If you are interested in where this equation comes from, and of other uses for it, check the Close Up in Chapter 14.

The Moon has an angular size of about ½ degree when viewed from Earth. The planets, however, have much smaller angular sizes, on the order of arcseconds rather than degrees. Thus it will be convenient to rewrite this equation for use with arcseconds. Since 57.3 degrees is about 206,000 arcseconds, we obtain

$$\text{Width} = \frac{(\text{angular size in arcseconds}) \times (\text{distance})}{206{,}000 \text{ arcseconds}}$$

The units of width and distance will be the same.

EXAMPLE

When Saturn is at its closest approach to Earth, it is about 1.28×10^9 kilometers from us. At this distance, its angular size is 19 arcseconds. Calculate Saturn's diameter.

Solution

Since 19 arcseconds is certainly a small angle, we can use the small angle formula here:

$$\text{Width} = \frac{(\text{angular size}) \times (\text{distance})}{206{,}000 \text{ arcseconds}}$$

$$= \frac{(19 \text{ arcseconds}) \times (1.28 \times 10^9 \text{ km})}{206{,}000 \text{ arcseconds}}$$

$$= 1.2 \times 10^5 \text{ km}$$

This checks with Saturn's diameter as found in Appendix F. (Note that since the angular size was given to only two significant figures, it was appropriate to express the answer this way.)

Try One Yourself. When Jupiter is 6.3×10^8 kilometers from Earth, its angular size is observed to be 47 arcseconds. What is the diameter of Jupiter?

MEASURING THE MASS OF A SOLAR SYSTEM OBJECT

How do we know the masses of the Sun and planets? To answer this we must return to Kepler's third law, which relates each planet's distance from the Sun to its period of revolution. We saw in Chapter 6 that Newton's formulation of

Kepler's third law was more complete than the original statement by Kepler. The equation for Newton's version of the third law is

Recall that the average distance of a planet from the Sun is equal to the semimajor axis of its elliptical path.

$$\frac{d^3}{P^2} = K\,(m_1 + m_2)$$

d = average distance to Sun
P = period of planet's orbit
K = constant, value depends on units of other quantities
m_1 = mass of one orbiting object
m_2 = mass of the other object

Since the mass of even the largest planet, Jupiter, is less than 0.001 times the mass of the Sun, the sum of the two masses on the right is essentially equal to the mass of the Sun and is therefore the same for each planet. Thus for objects in orbit around the Sun we can write the equation as

$$\frac{d^3}{P^2} \cong KM$$

d = average distance to Sun
P = period of planet's orbit
K = constant, depending on units
M = mass of the Sun

($\cong$ means "approximately equal to")

Kepler did not realize that the mass of the Sun was involved in the equation and he simply stated that the ratio on the left of the equation is the same for each planet. The equation above tells us, however, that if the Sun had less mass, an object at the Earth's distance would have a greater period than the Earth does. (If M is less, P must be greater.) This should make sense from what you know about the law of gravity. Imagine that the mass of the Sun could be magically reduced. It would then exert less gravitational force on the Earth and the Earth, under these circumstances, would have to have a lower speed or it would not stay in its (nearly) circular orbit (Figure 7–6).

Notice that Newton's statement does not actually conflict with Kepler's until great accuracy is demanded. Since the Sun's mass is constant, the value on the right side of the equation is very nearly the same for each of the planets, just as Kepler said. Newton's statement reduces to Kepler's when data is used that is no more precise than Kepler had. Newton's statement of the law, however, allows us to calculate something else—the mass of the Sun. All we need to know in order to do this is the semimajor axis of one planet's elliptical orbit and that planet's period of revolution around the Sun.

There is even more value to the equation. Kepler's third law, as completed by Newton, applies to any system of orbiting objects. Recall that Galileo had compared Jupiter's system of moons to the solar system. Use of the equation here lets us calculate the mass of Jupiter, which is the central object in this case. As we will see, every planet except Mercury and Venus has at least one natural satellite. Thus all we need to know in order to calculate the mass of one of those planets are the distance and period of revolution of at least one of its satellites.

What about Mercury and Venus, which have no moons? Their masses have been calculated on a few occasions by observing their effects on the orbits of

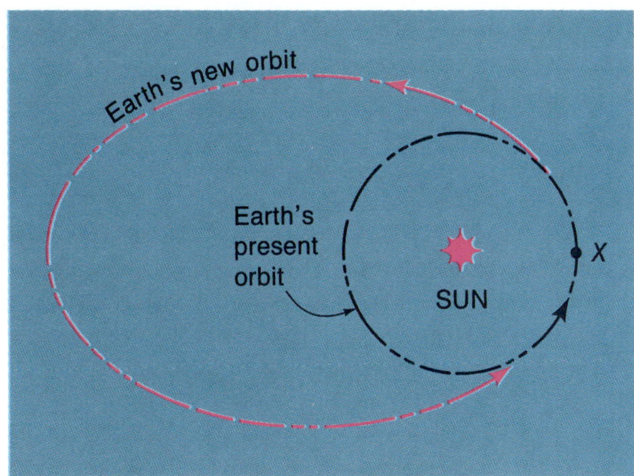

Figure 7–6. *If, when the Earth reached point X in its orbit, the mass of the Sun magically decreased, the Earth would have a speed too great for its present orbit.*

passing asteroids and comets. No asteriod or comet has passed close enough to provide highly accurate data, however, and thus the accuracy of the calculations was limited until space probes flew by these planets. If a space probe is put into orbit around a planet, the equation above applies to it and the equation allows us to calculate the mass of the planet. In practice, the space probe does not actually have to be put into orbit. By analyzing how the gravitational force of the planet changes the direction and speed of a probe during a flyby, the planet's mass can be calculated, although by a more complicated method than the equation we have used.

In the next section, we show how the mass of Saturn is calculated using Newton's version of Kepler's third law.

Saturn's Mass by the Revised Kepler's Third Law

For units of measure in Kepler's third law, we can use astronomical units, years, and solar masses. If we do this, the constant K is equal to 1. (We know this because in the case of the Earth, each term in the equation has a value equal to 1.)

EXAMPLE

Calculate the mass of Saturn using the fact that its satellite Titan orbits it at a distance of 1,222,000 kilometers and that the period of Titan's orbit is 15.9 days.

Solution

First, we change the units from kilometers and days to AUs and years:

$$1{,}222{,}000 \text{ km} \times \frac{1 \text{ AU}}{1.5 \times 10^8 \text{ km}} = 0.0081 \text{ AU}$$

$$15.9 \text{ days} \times \frac{1 \text{ year}}{365 \text{ days}} = 0.044 \text{ year}$$

Now we put these values into the equation:

$$\frac{d^3}{P^2} = K(m_1 + m_2)$$

$$\frac{(0.0081)^3}{(0.044)^2} = 1(m_1 + m_2)$$

$$0.00028 = m_1 + m_2$$

This tells us that the total mass of Saturn and Titan is 0.00028 of the Sun's mass (since m is expressed in solar masses). The mass of the Sun is 2.0×10^{30} kilograms. So Saturn and Titan have a total mass of 5.6×10^{26} kilograms.

Now how do we know how much of this mass is Saturn and how much is Titan? We can observe the motions of Saturn and Titan to see how much each moves as they revolve around their common center of mass. Such observations show us that Titan causes very little motion of Saturn, so that we can conclude that Titan's mass is very much less than Saturn's and we can consider the value above to be Saturn's mass. (Titan's mass, it turns out, is only about 0.0002 of Saturn's.)

Titan's mass is 0.022 of the mass of the Earth.

We look at the tables to check our calculated value against the one accepted by astronomers. Table 7–4 tells us that the mass of Saturn is 5.68×10^{26} kilograms—very close to our calculated value!

Try One Yourself. Deimos is one of Mars's two tiny moons. Its average distance from the center of its orbit is 23,500 km and its period is 1.26 days. From this, calculate the mass of Mars. Check your answer in Table 7–4 or Appendix F.

Mass Distribution in the Solar System

When we consider the masses of the objects that make up the solar system, we must be impressed by the fact that the Sun makes up almost the entire system. Table 7–2 shows the masses by percentages of the total; the Sun's mass is almost 99.9 percent of the total. Jupiter makes up most of the rest, having more mass than the remainder of the planets combined.

PLANETARY MOTIONS

Figure 7–7 illustrates the paths of the planets around the Sun. They are all ellipses, as Kepler had written. Most are very nearly circular but Pluto's orbit is elliptical enough that it overlaps the orbit of Neptune. In 1979 Pluto moved to the location in its orbit where it is inside Neptune's orbit. Until 1999 it will remain closer to the Sun than Neptune is. So until 1999, you should answer the appropriate trivia question by listing Neptune as the planet most distant from the Sun.

Table 7–2 Percentages of the Total Mass of the Solar System

Object	Percentage of Solar Systems' Mass
Sun	99.85
Jupiter	.095
Other planets	.039
Satellites of planets	.00005
Comets	.01 (?)
Asteriods, etc.	.0000005 (?)

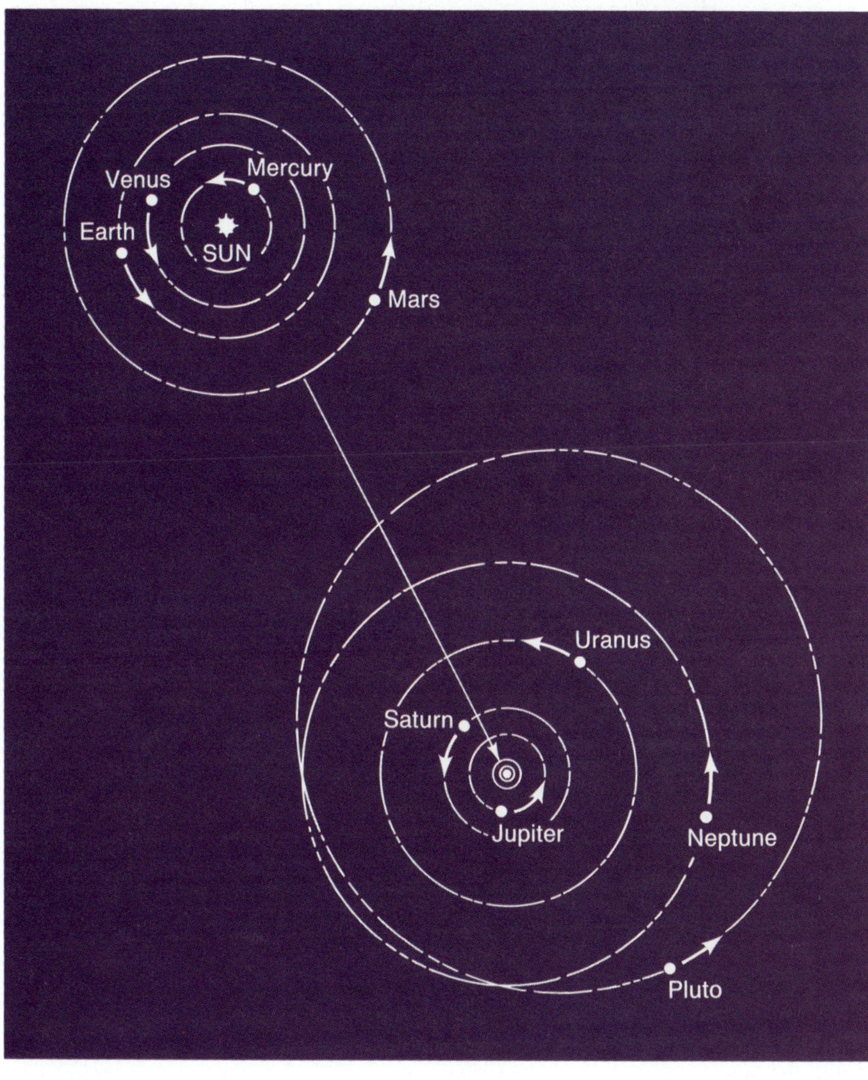

Figure 7–7. *The orbits of the planets are ellipses, according to Kepler's first law, but most are very nearly circular. The obvious exception is Pluto, which for about 20 years of its 248-year period is closer to the Sun than is Neptune.*

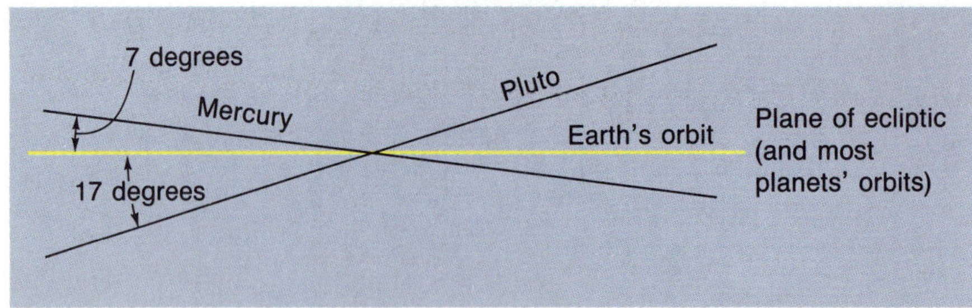

Figure 7–8. *The orbits of most of the planets are in the same plane as the Earth's (the ecliptic), but Mercury's plane is inclined at 7 degrees and Pluto's at 17 degrees.*

Inclination (of a planet's orbit): The angle between the plane of a planet's orbit and the ecliptic plane.

The elliptical paths of all of the planets are very nearly in the same plane. This means that we are able to draw them on a piece of paper without having to foreshorten their paths. Figure 7–8 illustrates the angles between the planes of the various planets' orbits and the plane of the Earth's orbit. Notice that Pluto's orbit is the most "out of line." We will see that Pluto is unusual in a number of other ways.

Figure 7–7 indicates that each of the planets revolves around the Sun in the same direction. When viewed from far above the Earth's north pole, this direction is counterclockwise. We saw in previous chapters that when viewed from this perspective, the Earth rotates on its axis in this same counterclockwise direction and that the Moon also orbits the Earth in a counterclockwise direction. We might ask if this pattern holds elsewhere in the solar system. Yes, in most cases. All of the planets except Venus and Uranus rotate in a counterclockwise direction as seen from above the Earth's north pole.

Most of the other planets have natural satellites revolving around them, just as the Earth does. The direction of revolution of most of these satellites is also counterclockwise, although there are a number of exceptions.

The fact that the planets have their orbits in basically the same plane, that they all orbit in the same direction, that most of them rotate in that same direction, and that most of their satellites revolve in that direction cannot be simple coincidence. We will recall these similarities when we discuss theories concerning the formation of the solar system; we must be sure that the theories explain these properties.

CLASSIFYING THE PLANETS

Jove was another name for the Roman god Jupiter.

As we look at the properties of individual planets, we will see that they divide easily into two groups. It is convenient to group the four innermost planets—Mercury, Venus, Earth, and Mars—together in one group, which we call the terrestrial planets because of their similarity to Earth ("terra"). The next four planets—Jupiter, Saturn, Uranus, and Neptune—are called the Jovian planets because of their similarity to Jupiter. We have already seen that Pluto is unusual in a number of ways. Indeed, it does not fit well into either of the two categories of planets, although it is sometimes classified as a terrestrial planet. As we look at some of the properties of the various planets, we will see even more cases of Pluto fitting neither category.

Table 7–3 Diameters of the Planets

Planet	Diameter		
	Miles	*Kilometers*	*Comparison to Earth*
Mercury	3030	4880	0.382
Venus	7520	12,100	0.951
Earth	7930	12,760	1.000
Mars	4220	6790	0.531
Jupiter	88,900	143,000	11.2
Saturn	74,600	120,000	9.40
Uranus	32,000	52,000	4.08
Neptune	30,000	48,400	3.88
Pluto	1400	2260	0.19

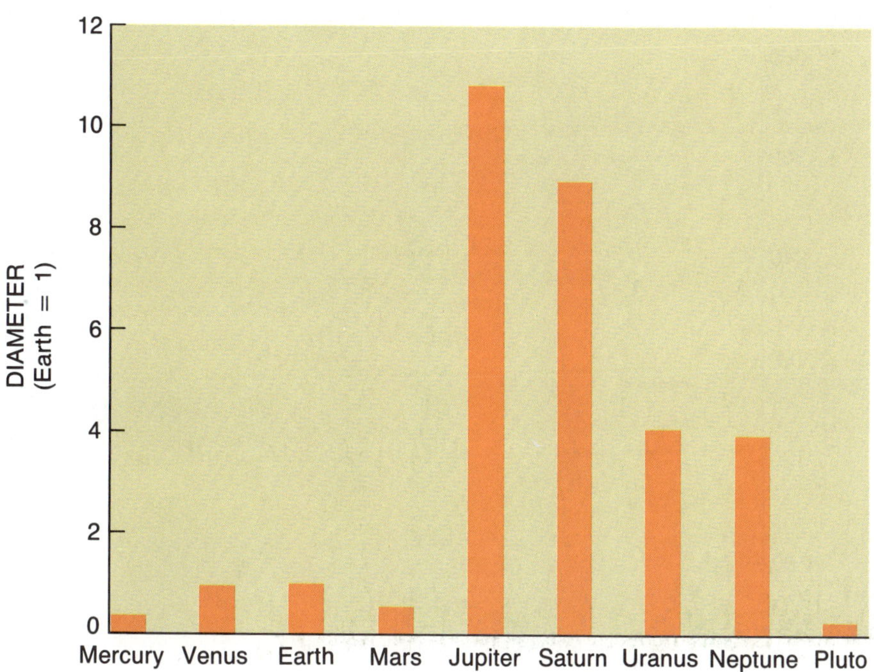

Figure 7–9. *A plot of planetary diameters (Earth's diameter equals one here) makes obvious the distinction between terrestrial and Jovian planets.*

Size

Table 7–3 shows the diameters of the planets, in kilometers, in miles, and as compared to Earth. Figure 7–9 compares their diameters on a graph. Notice that although the four terrestrial planets differ quite a bit from one another, they are all much smaller than the Jovian planets. Pluto, although it is out there beyond the Jovian planets, has a size more like the terrestrials. In fact, it is the smallest of the planets.

Table 7–4 The Masses and Densities of the Planets

Planet	Mass		Density
	*10^{24} kg**	*(Earth = 1)*	*gm/cm³*
Mercury	0.333	0.055	5.43
Venus	4.87	0.815	5.24
Earth	5.974	1.000	5.517
Mars	0.642	0.107	3.93
Jupiter	1900	318.1	1.32
Saturn	568	95.12	0.70
Uranus	86.9	14.54	1.25
Neptune	103	17.2	1.77
Pluto	0.01	0.002	2.0

*To get the mass in kilograms, multiply the value given by 10^{24}.

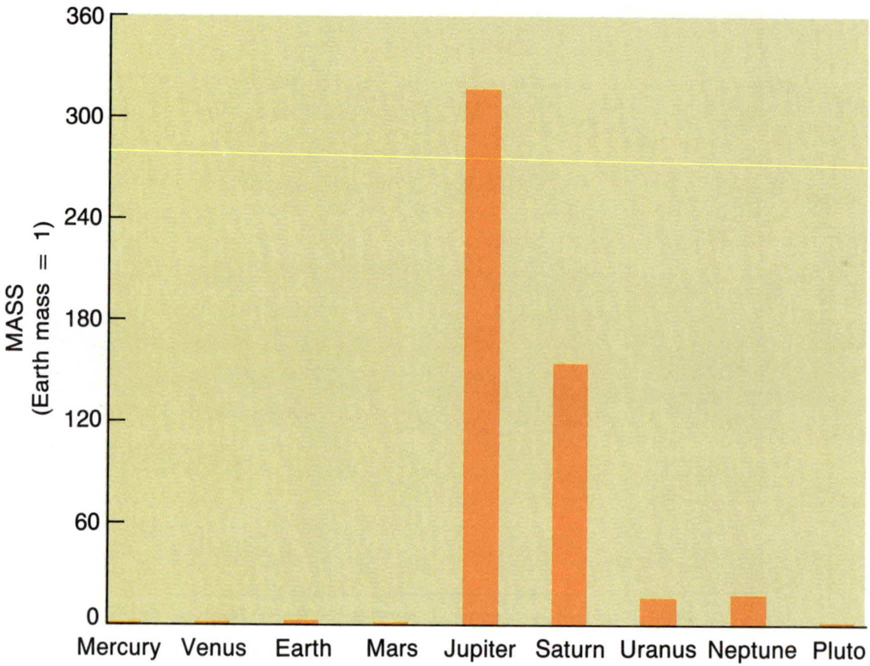

Figure 7–10. *When we plot the masses of the planets, we see that most of their total mass is in Jupiter. On this graph we have made Earth's mass equal to one.*

Mass and Density

When we look at the masses of the planets, we see even bigger differences between the terrestrial and Jovian planets. Look at the planetary masses listed in Table 7–4 and shown graphically in Figure 7–10.

Many people have a tendency to skip over tables and graphs. You are not expected to memorize the values in the tables, but a few minutes looking at

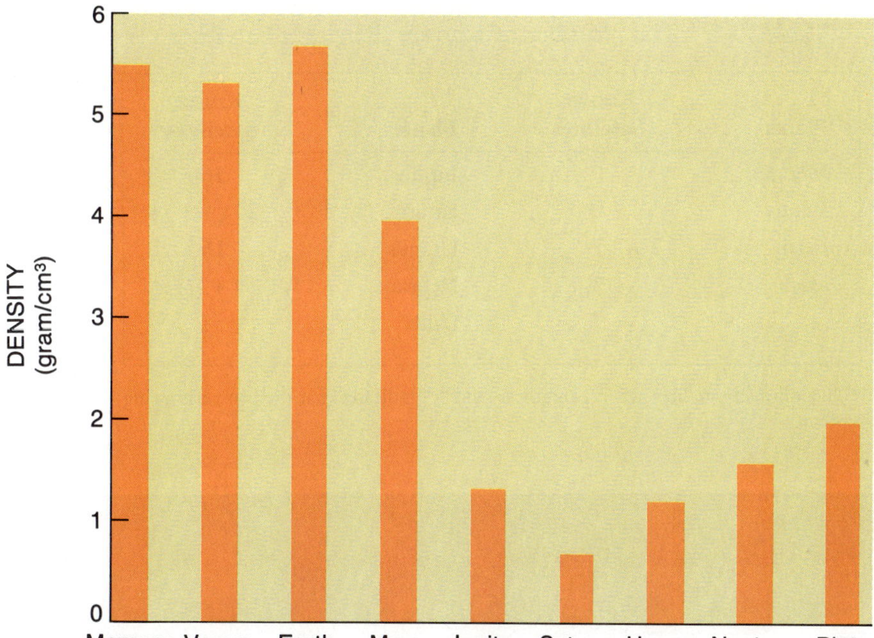

Figure 7–11. *The average density of the terrestrial planets is significantly greater than that of the Jovian planets.*

patterns and thinking of their meaning will yield much knowledge about the solar system. Study the column in Table 7–4 that gives the masses of the planets in terms of Earth's mass. Notice the tremendous difference between the two classifications of planets. Earth is the most massive of the terrestrial planets, but the least massive Jovian planet has more than 14 times the mass of Earth. Again Jupiter stands out as the giant.

The density column of the table (and Figure 7–11) show another difference between the terrestrial planets and the Jovian planets. The terrestrials are more dense. This is because they are primarily solid, rocky objects, while the Jovians are composed primarily of liquid. It used to be common to call Jovian planets "gas planets," but now we know that they actually contain much more liquid than gas.

Satellites and Rings

Table 7–5 shows the number of natural satellites of each planet. Notice that although there is no obvious pattern, the Jovian planets have more satellites. More details of the planetary satellites are found in Appendix H and in discussions of the planets in future chapters.

The planet Neptune is suspected to have rings, but we will not list them as definite until the Voyager II space probe passes near it in August 1989. Figure 7–12 is a photo of the rings of Saturn taken by Voyager II as it passed by the planet. A planetary ring is simply planet-orbiting debris ranging in size from a fraction of a centimeter to several meters. Motions of particles within the rings are extremely complex due largely to gravitational interactions with nearby planetary satellites. In a sense, each of the particles of a ring may be

Table 7–5 The Number of Known Planetary Satellites

Planet	Known Satellites	Planet	Known Satellites*
Mercury	0	Jupiter	16 r
Venus	0	Saturn	20 (+3?) r
Earth	1	Uranus	15 r
Mars	2	Neptune	2 (+1?) r?
		Pluto	1

*Unconfirmed satellites are in parenthesis and "r" indicates that the plant has a ring system.

Figure 7–12. *This is a color-enhanced photo of Saturn's rings taken by Voyager II. A computer was used to emphasize slight color differences between various parts of the rings.*

considered a satellite of the planet, but if we do so, counting satellites becomes meaningless. So we will continue to speak only of the larger "moons" as being planetary satellites. Again, notice the difference between terrestrial planets and Jovians in the number of their satellites.

Rotations

In discussing the rotation speeds of the planets we must distinguish between a solar day and a sidereal day. The ***solar day*** is defined as the time between successive passages of the Sun across the ***meridian,*** so that the length of a solar day on Earth is 24 hours. This is not the same amount of time that the Earth takes to complete one rotation, however. With respect to the stars, the Earth rotates once around in 23 hours, 56 minutes.

Refer to Figure 7–13 to see why there is a difference between the solar day and the ***sidereal day,*** which is the amount of time required for an object to complete one rotation with respect to the stars. In that figure we have put a mark on the Earth to designate a particular location there. When the Earth is at location *X* in the figure, this mark is directly under the Sun. When the

Solar day: The amount of time that passes between successive passages of the Sun across the meridian.

Meridian: An imaginary line that runs from north to south, passing through the observer's zenith.

Sidereal day: The amount of time which passes between successive passages of a given star across the meridian.

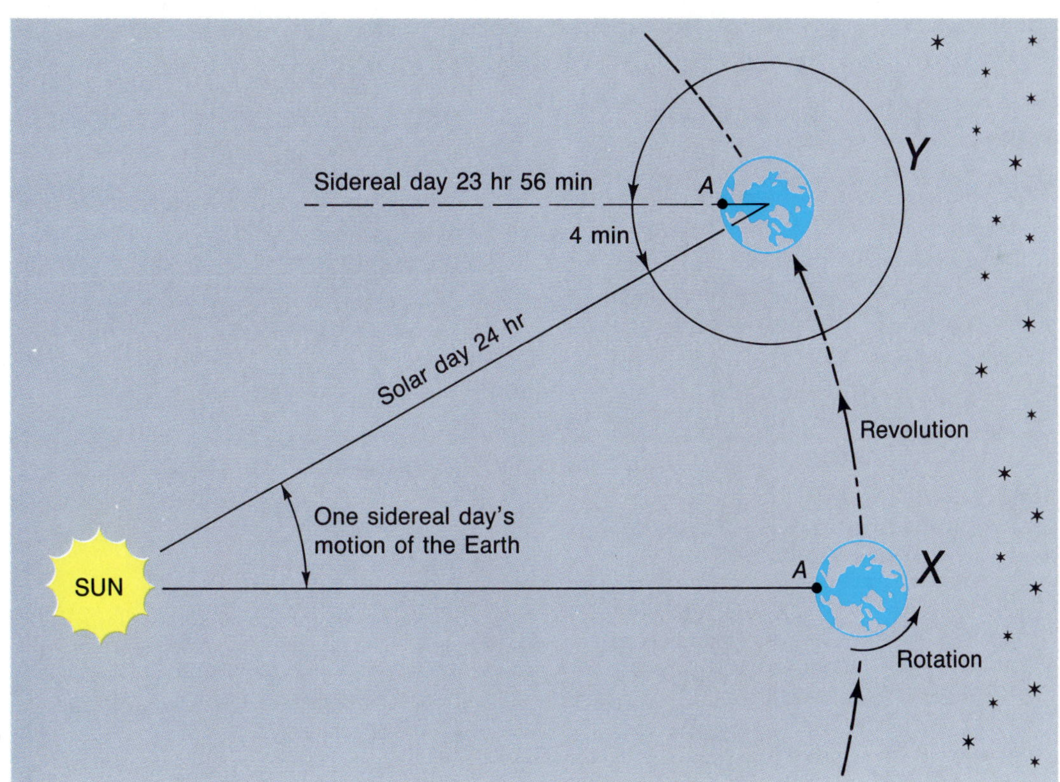

Figure 7–13. *When the Earth has rotated once around with respect to the stars, it has not completed a rotation with respect to the Sun. Thus the sidereal day is shorter than the solar day.*

Table 7–6 The Sidereal Periods of the Planets

Planet	Rotation Period
Mercury	58.65 days
Venus	243 days
Earth	25 h 56 m 04 s
Mars	24 h 37 m 22 s
Jupiter	9 h 51 m
Saturn	10 h 39 m
Uranus	17 h 14 m
Neptune	18 h 12 m
Pluto	6.39 days

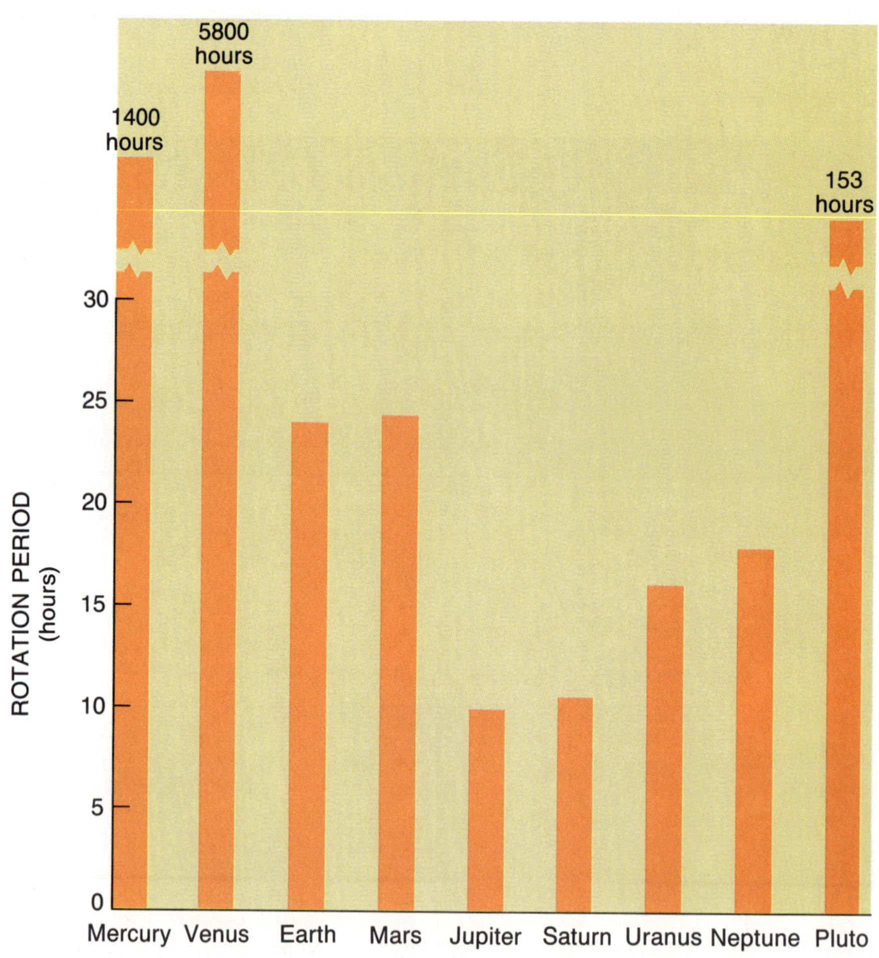

Figure 7–14. *The rotation periods of Mercury, Venus, and Pluto are so much greater than the other planets that their bars go far off the graph.*

Earth has turned so that point A is again under the sun, one solar day will have passed. The Earth moves around the Sun as it rotates, however; moving about 1/365 of the way around the Sun in one day. In the figure we have exaggerated how far it moves so that you can see the effect. Notice that when the Earth has rotated once with respect to the stars, moving from X to Y, the marked spot is not directly under the Sun. The Earth's sidereal period is 23 hours and 56 minutes, but the Earth has to rotate for another 4 minutes to complete a solar day.

Table 7–6 and Figure 7–14 show the sidereal periods of rotation of the planets. While it is true that there are great differences of rotation periods among the terrestrials, notice that all of the Jovian planets rotate faster than the fastest of the terrestrials.

ECCENTRICITIES OF THE PLANETS' ORBITS

When Johannes Kepler determined that the orbit of Mars was an ellipse, he used Tycho Brahe's data concerning that planet's positions among the stars. Those observed positions, though, depended not only upon the motion of Mars but also upon the motion of Earth, from which the measurements were being made. Fortunately for Kepler the Earth's orbit is very circular, having low **eccentricity.** This made his calculations much easier. The eccentricities of planets' (and Ceres's) orbits are given in Table 7–7. Notice how Pluto and Mercury differ so much from the other planets in eccentricity. There is no pattern of differences between terrestrials and Jovians, however.

Eccentricity (of an ellipse): The ratio of the distance between foci to the length of the major axis. (See Figure 7–15.) *Thus a circle has zero eccentricity.*

Table 7–7 Eccentricities of the Orbits of the Planets and Ceres

Planet	Eccentricity of Orbit
Mercury	0.206
Venus	0.007
Earth	0.017
Mars	0.093
(Ceres)	0.077
Jupiter	0.048
Saturn	0.056
Uranus	0.047
Neptune	0.009
Pluto	0.250

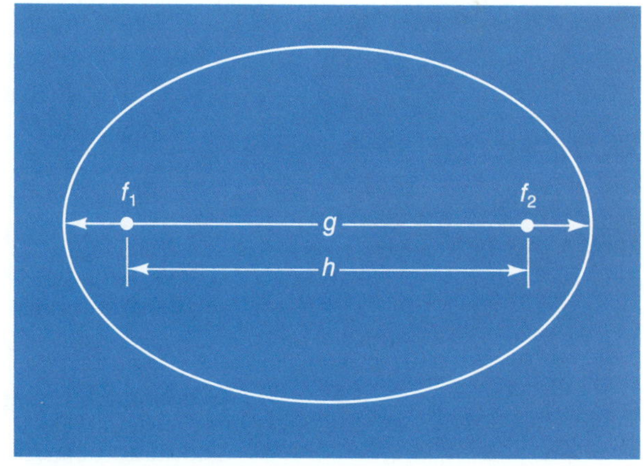

Figure 7–15. *The points* f_1 *and* f_2 *are the foci of the ellipse. The eccentricity is calculated by dividing the distance between the foci* (h) *by the length of the major axis* (g). *Half of the major axis is called the semimajor axis and is what we have previously referred to as the average distance from the planet to the Sun.*

PLANETARY ATMOSPHERES AND ESCAPE VELOCITY

Long before people visited the Moon we knew that it contained no air and no water. This had been predicted by applying Newton's law of gravity, and the same law can also be applied to make predictions concerning planetary atmospheres. To see the connection between the law of gravity and an object's lack of atmosphere, we will first discuss jumping off the Earth. This discussion leads to an idea that will help us understand not only why some planets have no atmosphere, but—in Chapter 16—what a black hole is.

We will start by imagining an Earth with no air. On such an Earth, if we throw something upwards it will not be slowed by air friction. It will still feel the effect of gravity, however, and it will slow down, stop, and then fall back to Earth. So we throw it harder. It rises farther, and as it gets higher, the force of gravity on it is less. Thus its rate of slowing—its deceleration—is less at greater heights. Could we throw the object fast enough so that gravity never stops it and brings it back down? The answer is yes. We can calculate from the laws of motion and gravitation that the speed needed is about 11 kilometers per second (7 miles per second). An object fired upward from Earth at this speed or greater will continue to rise, slowing down all the time, but never stopping. We call this speed the ***escape velocity*** from Earth.

Escape velocity: The minimum velocity an object must have in order to escape the gravitational attraction of an object such as a planet.

When we consider that the speed of sound is only about 0.3 kilometer per second, we see that the escape velocity from Earth is a tremendous speed. The reason that we imagined an Earth without air friction is that in practice, if we fired an object from the surface at 11 kilometers/second, it would be slowed—and probably destroyed—by air friction. In the space program we have fired objects into space with a velocity exceeding escape velocity; every space probe that has been sent to other planets has completely escaped the Earth. The reason that air friction does not destroy them is that rockets carried them above the atmosphere before increasing their speed to escape velocity.

The escape velocity of an astronomical object depends on the gravitational force at the object's surface (or from whatever height we are launching the object). Recall that the gravitational force at the surface of the Moon is only ⅙ of that at the surface of Earth. A 120-pound astronaut weighs only about 20 pounds on the Moon. The escape velocity from the Moon is therefore less than that from Earth. It is only about 2.5 kilometers per second—eight times the speed of sound in air at the Earth's surface.

Phobos, a natural satellite of Mars, is tiny by astronomical standards and it has an escape velocity of only about 30 miles per hour. You can't jump this fast, but if you have a good arm you could throw a ball from Phobos fast enough that it would never return.

To see what escape velocity has to do with the question of the atmosphere of an astronomical object, we must briefly discuss the nature of a gas.

Gases and Escape Velocity

The states of matter in our normal experience are three: solid, liquid, and gas. Some understanding of the gaseous phase is necessary to understand planetary atmospheres. To picture a gas, picture a swarm of bees buzzing around a hive. The bees represent individual molecules of a gas. This analogy is faulty in a few ways, however. First, compared to their size, the molecules of a gas are much, much farther apart than are the bees. Second, molecules move in straight lines until they collide. Then they bounce off and move straight again. Finally, remember that molecules are moving through completely empty space. There is nothing between them. Nothing.

There are three additional things we must keep in mind about the molecules of a gas:

The energy of motion of the molecules is what determines the temperature of an object.

1. As gas molecules bounce around, different ones have different speeds at any given time. Some will be moving fast and some slow. In this sense, they are like the bees.
2. The average speed of the molecules depends on the temperature of the gas. Gases at higher temperature have faster-moving molecules.
3. At the same temperature, less massive molecules have greater speed. For example, since a molecule of oxygen has less mass than one of carbon dioxide, if we have a gas that is a mixture of oxygen and carbon dioxide, the oxygen will, on the average, be moving faster.

Now let's consider the atmosphere of the Earth. Our atmosphere is held near the Earth by gravitational forces. Consider a molecule in the upper part of our atmosphere. At great heights above Earth the atmosphere has a low density; we say that it is "thin." That means that the molecules are much farther apart than down here at the surface. Suppose that at some instant a particular molecule up there happens to be moving away from Earth. There are very few other molecules around, so a collision is unlikely and our molecule acts just like any other object moving away from Earth. The force of gravity slows it down. Whether the molecule will return to Earth or escape depends upon how the speed of the molecule compares to the Earth's escape velocity. If the molecule's speed is greater than escape speed, the molecule is gone, never to return to Earth.

Notice that we are making no distinction here between the terms speed *and* velocity.

It is obvious from the fact that we have an atmosphere on Earth that the velocities reached by molecules of the air do not exceed escape velocity. Recall, however, that molecules of lower mass have greater speeds. Hydrogen molecules have less mass than those of any other element. It is therefore no coincidence that there is essentially no hydrogen in the Earth's atmosphere: The temperature of the upper atmosphere is great enough for hydrogen molecules to escape. Any hydrogen that is released into the Earth's atmosphere eventually will be lost. The chemical element hydrogen does not exist alone in our atmosphere, but only as part of more massive molecules. A molecule of water vapor, for example, contains hydrogen.

We have stated that the escape velocity from the surface of the Moon is about 2.5 kilometers per second. At the temperatures reached on the sunlit side of the Moon, all but the most massive gases attain speeds greater than this and therefore the Moon has essentially no atmosphere. The Apollo astronauts were not surprised to find no air to breathe when they landed on the Moon.

The Atmospheres of the Planets

The average speed of a particular type of molecule depends upon the temperature of the gas, but at any given time some molecules will be travelling faster than average. This means that although the average speed of a particular gas may be less than the escape velocity from a planet, the gas may still gradually escape because the speed of a small fraction of its molecules exceeds the escape velocity. Because of this, we must use a multiple of the average speed in considering whether or not a gas will escape from a planet. Theory shows that rather than consider the average speed, the value we should use is ten times the average speed of the molecules of the gas in order that the planet be able to retain that gas for billions of years.

Figure 7–16 is a graph of the average speeds of various molecules versus their temperature. The dashed lines represent ten times the average molecular speed. Each planet, as well as some planetary satellites, are plotted on the graph at their respective temperatures and escape velocities. A planet can retain a gas if the planet lies *above* the dashed line for that gas. Find the Moon on the graph and notice that every gas has an average molecular speed that will allow it to escape. The planet Mercury is similar. On the other hand, only hydrogen and helium escape from Earth and Venus; the four Jovian planets retain all of their gases.

Table 7–8 Characteristics of the Jovian and Terrestrial Planets

Terrestrials	Jovians
Near the Sun	Far from the Sun
Small	Large
Mostly solid	Mostly liquid and gas
Low mass	Great mass
Slow rotation	Fast rotation
No rings	Rings
High density	Low density
Thin atmosphere	Dense atmosphere
Few moons	Many moons

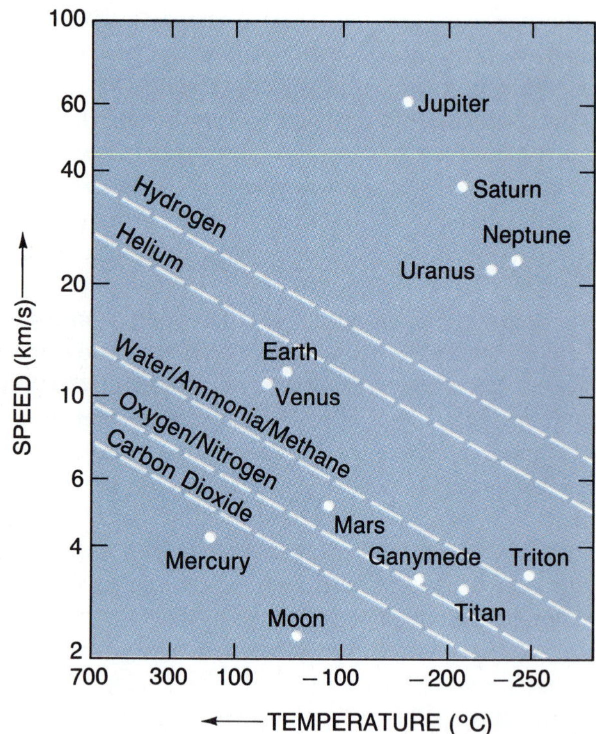

Figure 7–16. *This graph shows how the speeds of various gases depend upon their temperatures. The dashed lines represent ten times the gases' average speeds. The planets and some planetary satellites are marked at their corresponding temperatures and escape velocities.*

Because of the differences in gravitational force, we find that the atmospheres of terrestrial planets differ from those of Jovian planets. The terrestrial planets have much thinner atmospheres.

Table 7–8 summarizes the differences we have discussed between the terrestrial and Jovian planets.

EARTH

In the next few chapters we will discuss the individual objects that make up our solar system. Before we do so, however, we should look at the one with which it is natural for us to make comparisons—our home planet.

The Earth's Atmosphere

Figure 7–17 shows the Earth's atmosphere drawn to scale around the planet. Compared to the size of the Earth, the atmosphere reaches a very small distance above the surface. The atmosphere gets thinner and thinner as one gets farther from Earth and so it is impossible to put a definite boundary on it, but when one reaches a distance of 100 or 150 kilometers above the surface, the atmosphere is essentially nonexistent.

Our atmosphere consists of approximately 80 percent nitrogen and 20 percent oxygen. Other constituents such as water vapor, carbon dioxide, and ozone make up a very small percentage of the atmosphere even though they are very important to life on Earth. We will discuss the atmospheric importance of carbon dioxide when we discuss the planet Venus in the next chapter.

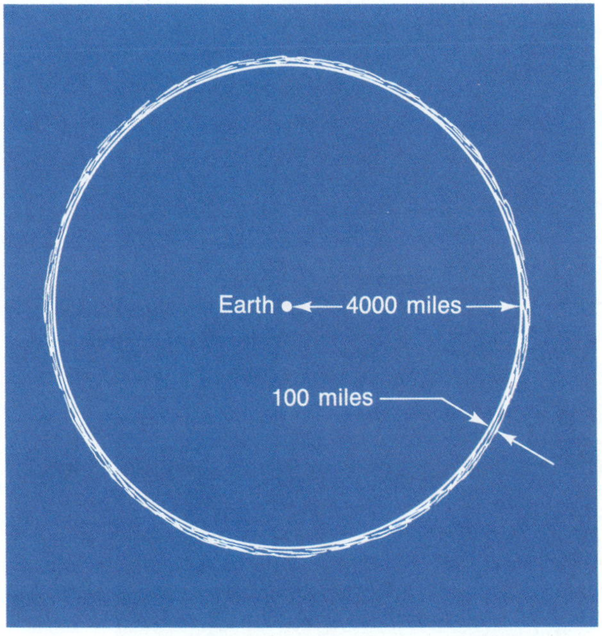

Figure 7–17. *The radius of the Earth is about 4000 miles. The atmosphere has no definite limit, but at 100 miles height it is very thin. A height of 100 miles is represented to scale in this drawing.*

The Earth's Interior

We learn about the Earth's interior primarily by detecting two types of waves that result from earthquakes. These waves travel through the Earth and by analyzing their times of travel from distant earthquakes, geologists can deduce some properties of the materials deep within the Earth.

Crust (of the Earth): The thin, outermost layer of the Earth.

Mantle (of the Earth): The thick, solid layer between the crust and the core of the Earth.

Core (of the Earth): The central part of the Earth, probably consisting of a solid inner core surrounded by a liquid outer core.

Differentiation: The sinking of denser materials toward the center of planets or other objects.

The interior of the Earth is made up of three layers, as shown in Figure 7–18. The ***crust,*** which is the outer layer, extends to a depth of less than 100 kilometers and is made up of the common rocks with which we are familiar here on the surface. The density of the crust is about 2.5 to 3 gm/cm^3.

The ***mantle*** extends nearly halfway to the center of the Earth, about 2900 km below the surface. This layer, although it is solid, is able to flow very slowly when steady pressures are exerted on it but it will crack and move suddenly under extreme sudden pressures. The mantle is more dense (3 to 9 gm/cm^3) than the crust and therefore the crust is able to float on it.

The ***core*** of the Earth seems to be divided into two parts, a liquid outer core and a solid inner core. The core is even more dense than the mantle, ranging from 9 gm/cm^3 to 13 gm/cm^3. Because of its high density, the core of the Earth is thought to be made up primarily of iron and nickel, the most common heavy elements. As is shown in Table 7–9, the average density of the Earth is 5.52 gm/cm^3, or more than 5 times the density of water.

The pattern of increasing density of materials within the Earth tells us something of the Earth's past, for such ***differentiation*** could only have come about when the Earth was in a molten state, when the heavier elements would have sunk through the less dense layers. The molten state must have resulted from heating during the formation of the planet and from energy released by radioactive elements early in the Earth's history.

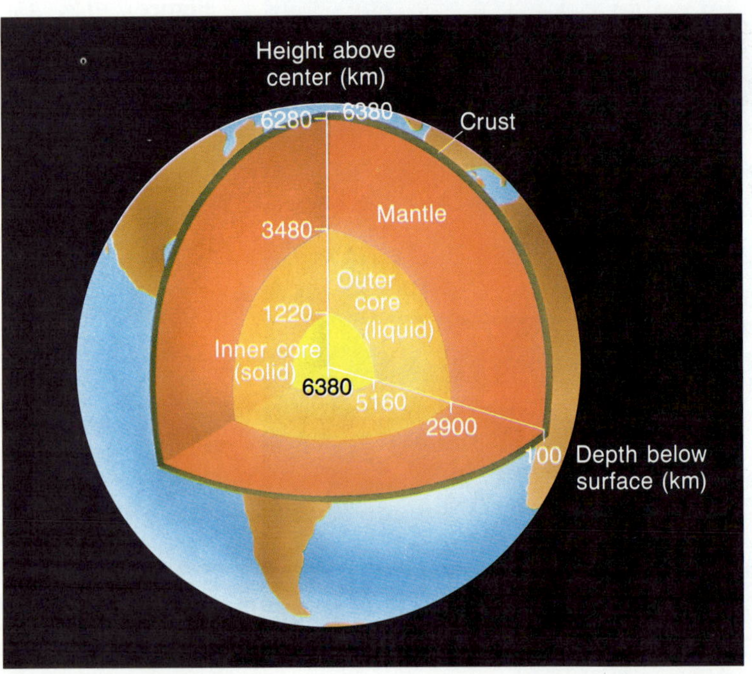

Figure 7–18. *The interior of the Earth, showing its primary layers.*

Table 7–9 The Planet Earth

Diameter	12,760 km
Mass	5.97×10^{24} kg
Density	5.52 gm/cm^3
Surface gravity	9.8 m/s/s
Escape velocity	11.18 km/s
Sidereal rotation period	23.934 hours
Solar day	24.0 hours
Albedo	0.31
Tilt of equator to orbital plane	23.44°
Orbit	
Semimajor axis	1.496×10^8 km
Eccentricity	0.0167
Sidereal period	365.24 days
Moons:	1

The Earth's Magnetic Field

If a piece of paper is laid over a magnet and filings of iron are sprinkled on the paper, a pattern such as that shown in Figure 7–19(b) results. The area around a magnet where a magnetic force is felt is called a ***magnetic field.*** In a magnetic field a small magnet aligns with the field as indicated in Figure 7–20.

Magnetic field: A region of space where magnetic forces can be detected.

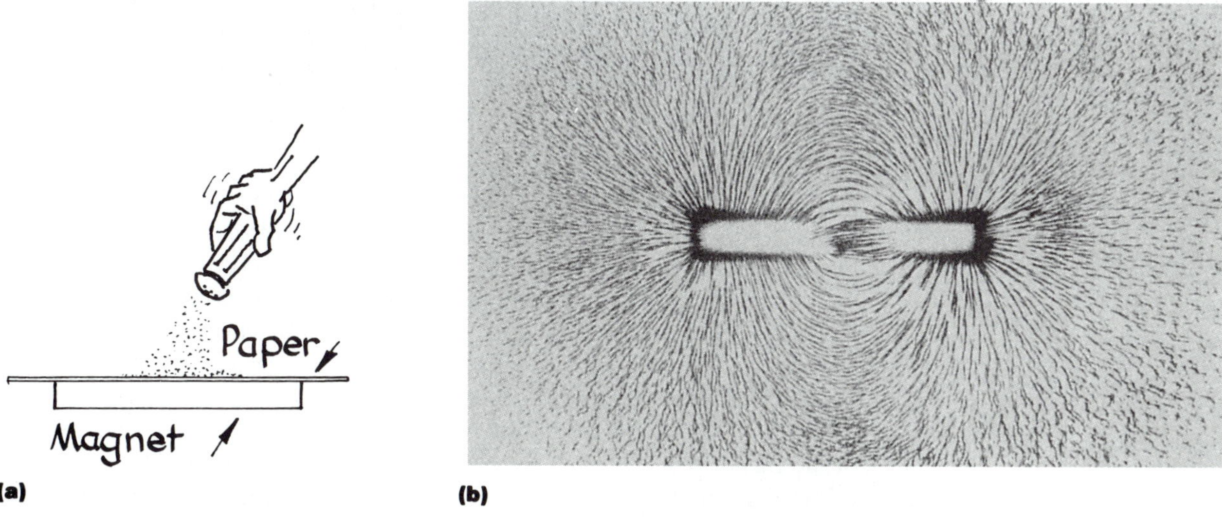

Figure 7–19. *(a) An edge-on view of iron filings being sprinkled onto a piece of paper that covers a bar magnet. (b) The resulting magnetic field pattern made by the iron filings. The outline of the magnet is obvious, but notice also the pattern the filings form beyond the magnet.*

Figure 7–20. *The arrows in the circles represent small magnets that are suspended so that they are free to rotate in the magnetic field of the large magnet.*

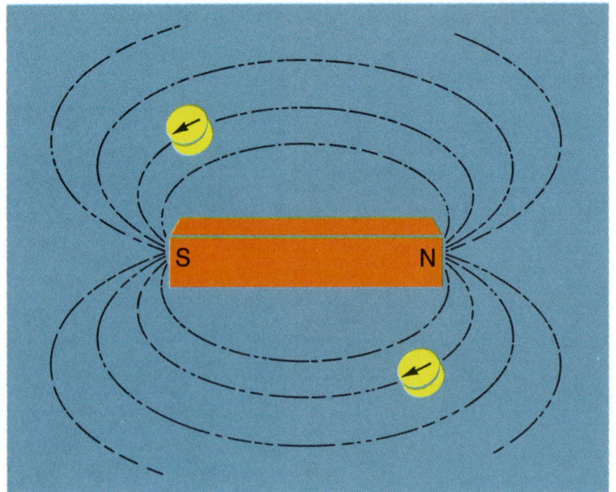

Figure 7–21. *The magnetic field of the Earth is similar to that of a bar magnet.*

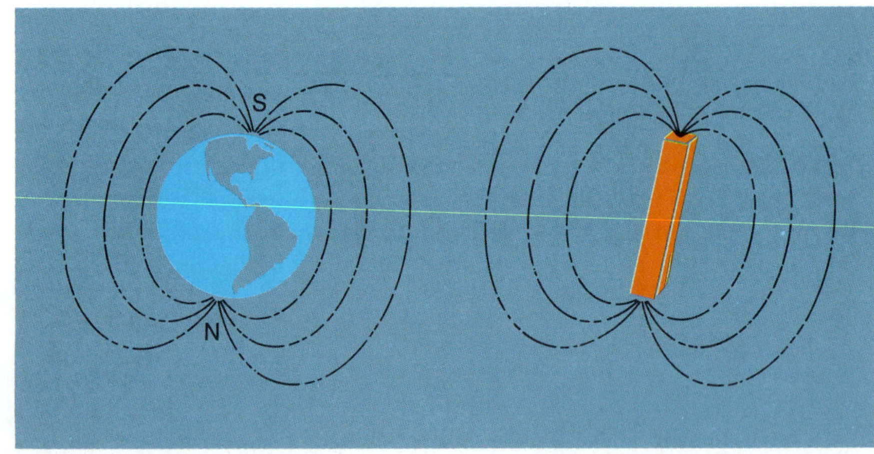

A magnetic compass is simply a magnet that is free to rotate so that it can align with a magnetic field.

When a magnet is suspended so that it is free to rotate near the Earth, it always aligns so that opposite ends point in specific directions, indicating that the magnet is aligning with a magnetic field that is related to the Earth. By plotting the direction of the magnetic field at various places on and off the Earth's surface, we can determine that the Earth's magnetic field has a shape similar to that of a bar magnet, as shown in Figure 7–21. Notice that the poles of the Earth's magnetic field—those points toward which the magnetic field converges—are located near (but not exactly at) the poles of the rotation axis of the Earth.

One might picture the Earth as containing within it a bar magnet that is not quite aligned with its spin axis. We know that the magnetic field of the Earth is not caused by a bar magnet inside, but the origin of the field is not completely understood. Magnetic fields (even those near a magnet) are caused by moving electric charges, and it is thought that the Earth's field is the result

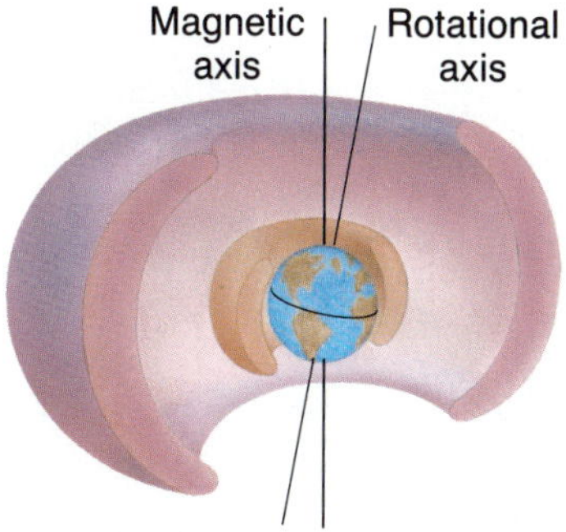

Figure 7–22. *The Van Allen belts are regions where the magnetic field of the Earth traps charged particles. They surround the Earth except near the poles.*

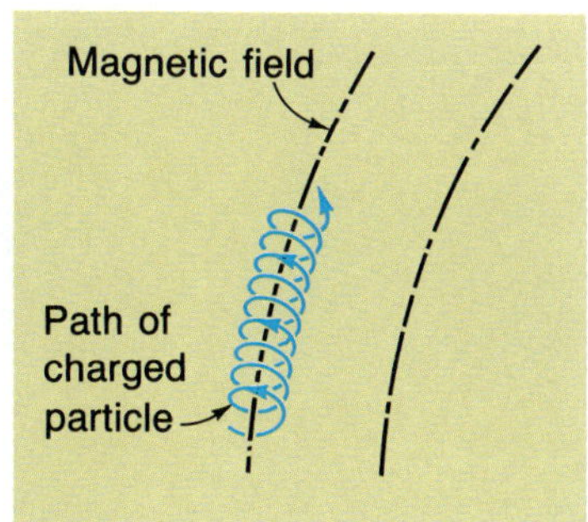

Figure 7–23. *Charged particles are forced to move in spirals around the lines of magnetic field. This causes them to become trapped in the Earth's field.*

of the motion of charges within the Earth's rotating liquid core. The locations of the Earth's magnetic poles are known to have wandered during the geological ages and the magnetic field has even undergone complete reversals. At least part of the effect is due to movements of the Earth's crust with respect to the interior. It has been hypothesized that a major cause for the reversals was a change in the Earth's rotation rate that resulted from bombardment with meteoroids. Many questions remain before hypotheses concerning pole wandering can be confirmed or denied.

You can produce a magnetic field by causing a current (moving electric charge) in a coiled wire.

One effect of the Earth's magnetic field became the first scientific discovery of the space age. Early spacecraft discovered the existence of electrically charged particles swarming in doughnut-shaped regions high above the Earth's surface (Figure 7–22). The cause for these "Van Allen belts" is well understood. Just as an electric current (moving electric charges) causes a magnetic field, a magnetic field exerts a force on charged particles. When charged particles enter a magnetic field, they are forced to move in a spiral around the lines of the field as shown in Figure 7–23. In the case of the Van Allen belts, the charged particles (protons and electrons) come primarily from the Sun and they have been captured by the magnetic field of the Earth.

We see the effect of the particles trapped within the Earth's magnetic field in the ***auroras*** that are seen in the skies near the north and south poles of the Earth (Figure 7–24). These beautiful displays of light are the result of disturbances of the Earth's magnetic field that cause some of the particles to follow the magnetic field lines down to the atmosphere, where they collide with atoms of the air and cause it to glow. We will discuss the reason for such disturbances when we study the Sun in Chapter 12.

Aurora: Light radiated in the upper atmosphere due to impact from charged particles.

The Earth from Space

We Earthlings have learned much about neighboring planets by sending spacecraft by them, by putting craft into orbit around them, and by sending landers

Figure 7–24. *An aurora is caused by charged particles trapped in the Earth's magnetic field striking atoms in the upper atmosphere.*

Figure 7–25. *It might be possible for a viewer with a large telescope on Mars to see signs of the Great Wall of China.*

to their surfaces. Let us consider what could be learned about the Earth if we lived on another planet and used similar technology to study the Earth.

The atmosphere of the Earth and its cloud layer hinder detailed viewing of the surface from afar. Inhabitants of nearby planets, however, could detect differences between our continents and oceans and they could see the Earth's white polar caps (as we see those on Mars). In order to determine what makes up the continents, oceans, and polar caps, our neighbors could analyze the spectrum of light reflected from these areas. They would be able to determine that most of our planet is covered by water, and that our polar caps are frozen water. In addition, they would see our changing cloud cover and determine that it, too, is made up of water.

Could they detect signs of life? Probably not. Although we think of ourselves as important, we have not changed our planet in a way that would be obvious to an observer on another planet or moon. The only object we have constructed that might possibly be visible is the Great Wall of China (Figure 7–25), not at all a modern accomplishment as we measure time. The only sign of life they are likely to detect from afar is our electromagnetic noise; i.e., radio and TV broadcasts, as described in the accompanying Close Up.

As an alien spacecraft approached Earth and went into orbit around it, it would obtain views such as those in Figure 7–26. Now its occupants might be able to see signs of life. Perhaps their first visible evidence of life would come when they viewed the dark side of our planet, for urban areas would be visible at night because of the wasted light that escapes from them into space (Figure 7–27).

CLOSE UP

ET Life I—Stray Radio Signals

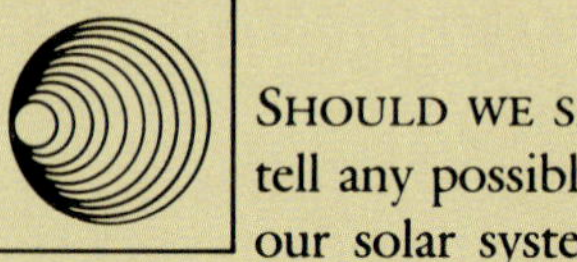

SHOULD WE SEND SIGNALS into space to tell any possible intelligent beings beyond our solar system that we exist? Astronomers have used our largest radio telescope to send a message to regions of space that we think likely to be populated with intelligent beings. We will put off a discussion of this message until Chapter 13, when we will discuss radio telescopes (which normally are used to detect radio signals from inanimate objects in space).

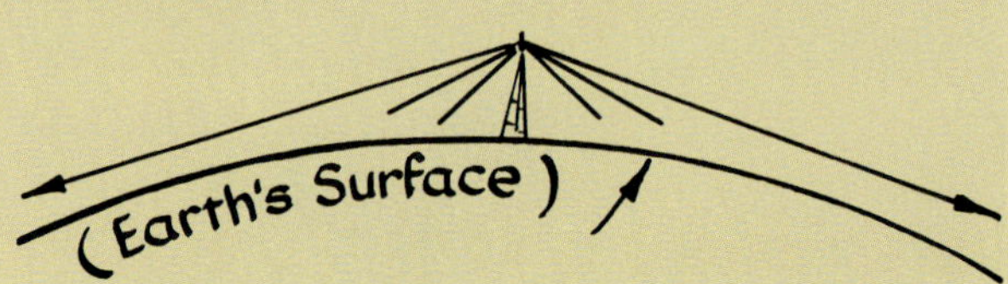

Figure T7–2. *Radio and TV signals are leaked into space from our antennas.*

The other way that we signal our existence is accidental, and it occurs daily. We started such signalling more than a half century ago, when we started broadcasting radio programs here on Earth, and as the number and power of our radio (and then television) stations increased, so did the power of the signals that leaked out into space. Figure T7–2 shows how such signals leak into space. Each transmitter of radio or television signals wastes some of its power transmitting to space. The strength of these stray signals is enough that they could be detected by beings on a planet circling a nearby star if those beings had a system of large radio telescopes. However, the signals that leak from Earth are so weak and the signal from a given station would be received for such a short time by an antenna elsewhere in space, that there is probably no way that they could be unscrambled. Even if they could, their interpretation by extraterrestrials would be extremely difficult.

This does not mean, however, that any beings which may receive the signals could not use them to learn a lot about our planet. To start with, they would detect that the strength of the signals changes regularly every 24 hours and would likely conclude that this was caused by a 24-hour rotation rate of our planet. In addition, they could determine something about our approximate technological population distribution from the pattern of the intensity changes. Here's how: Figure T7–3 shows the approximate density of radio and TV broadcasting antennas on our globe. As the earth turns, the strength of radio waves leaking into space would vary depending upon which area of the earth was leaking its signals toward the extraterrestrial radio receiver. By analyzing the changing pattern of intensity of waves from Earth, the distant scientists could ascertain that technology is very unevenly distributed around our planet.

If the hypothetical extraterrestrials have been watching our part of space for some time, they would know that we are a very young civilization in technological terms, for we only recently invented radio.

If intelligent beings are out there listening in on us, they are not going to be able to watch our evening news and they couldn't watch our situation comedies or our soaps. (Would they conclude from these that intelligent life exists on Earth?) They could learn something about us, nonetheless.

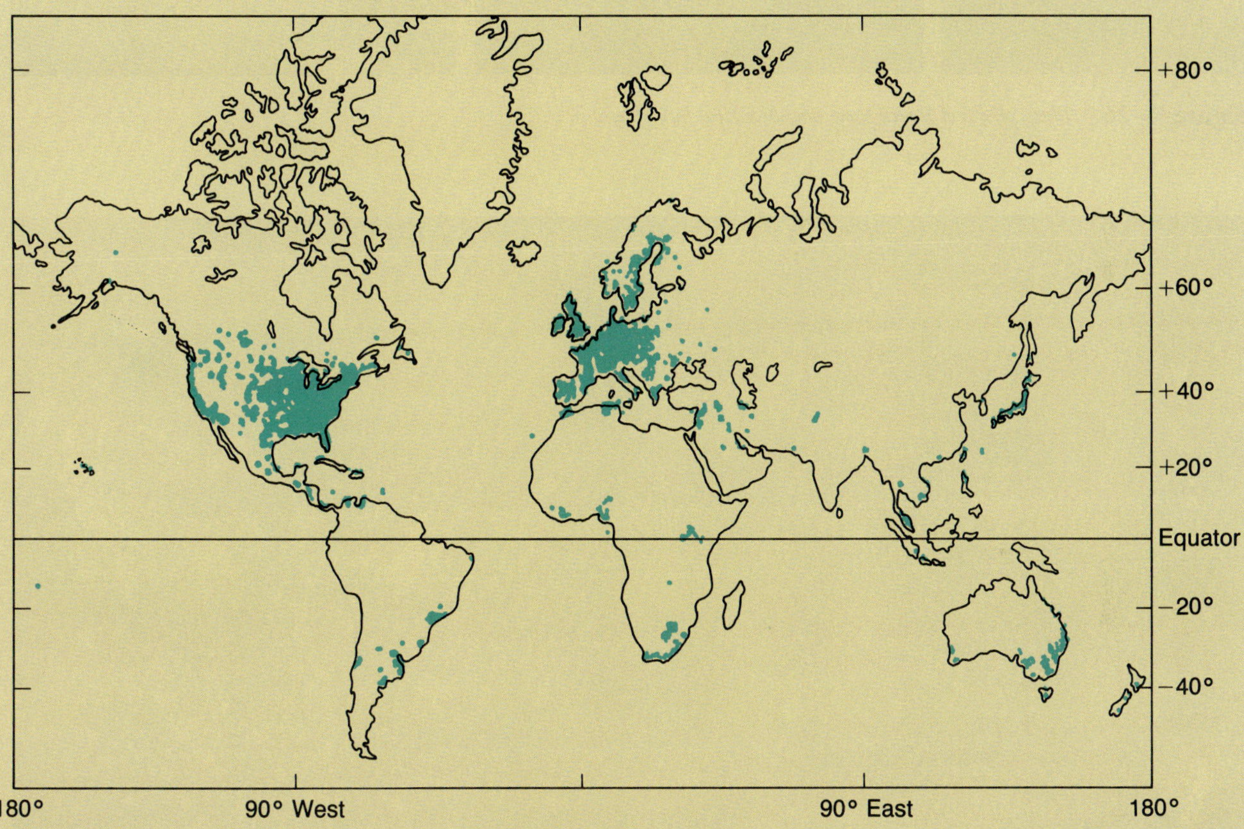

Figure T7–3. *This map shows the distribution of the more powerful television transmitters on Earth. Since they tend to be located near centers of population and wealth, they are not evenly distributed.*

Figure 7–26. *These photos of Earth show how beautiful our planet is.*

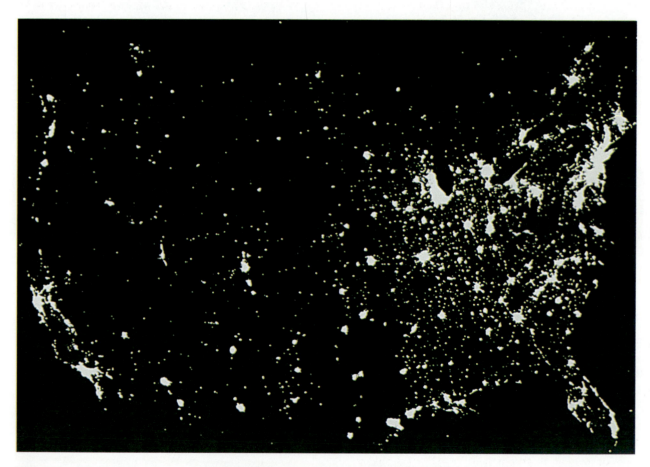

Figure 7–27. *This is a composite of a number of photos showing the U.S. from space at night.*

If landers were sent to Earth to look for life, their ease in finding it would obviously depend upon where they landed. But if they were equipped like our Viking missions that landed on Mars to search for life, they would be capable not only of photographing our large animals, but they would be able to detect organic molecules in the soil (or ice) no matter where they landed.

CONCLUSION

In this chapter we have seen that our solar system contains a myriad of objects, vastly different from one another. We have seen similarities, however, and have been able to classify the planets into two categories: terrestrial and Jovian. We have looked at some of the characteristics that prohibit us from listing the (normally) most-distant planet, Pluto, in either category.

One of the most important topics discussed during our overview of the solar system was how we measure some of the properties of the objects circling the Sun, including distances between them, their sizes, and their masses.

Our next step will be a more detailed description of each of the planets as well as the more minor objects within the solar system—comets, meteoroids, and asteroids.

RECALL QUESTIONS

1. In a previous chapter we saw that Kepler's Third Law relates a planet's period to its distance from the Sun. This law was expanded by Isaac Newton to include what other quantity?
2. How does the Sun's mass compare to the total mass of all other objects in the solar system?
3. How does the Sun's size compare to the sizes of the planets?
4. What is the largest planet, and how does it compare in size and mass to the Earth?
5. What was the first celestial object discovered after Bode's Law was proposed? Did it fit predictions made by the law? (Hint: See the Close Up.)
6. What is the name of the largest asteroid, who discovered it, and in what century was it discovered?
7. How do we know the masses of the planets?
8. Define density.
9. Distinguish between a sidereal day and a solar day. Which is longer on Earth?
10. There are common features regarding the directions of rotation and revolution of the planets. Describe them. Which planets are unusual as to their direction of rotation (spin)?
11. Name the terrestrial planets and the Jovian planets. Why are the latter called *Jovian?*
12. In what ways are the terrestrial planets similar to one another but different from the Jovians?
13. What are the two primary constituents of the Earth's atmosphere?
14. How does the eccentricity of the Earth's orbit compare to that of other planets?
15. How does the Earth compare to the other planets in density?
16. What is meant when we say that the escape velocity from the Earth is 7 miles per second?
17. What two factors determine the speed of the molecules of a gas?
18. Explain why hydrogen escapes from the Earth's atmosphere but carbon dioxide does not.
19. Does the Moon have an atmosphere? Discuss.
20. Show on a sketch the relative positions and sizes of the Earth's core, mantle, and crust.
21. What is a magnetic field, and how can one be detected?
22. What are the Van Allen belts and what causes them?

QUESTIONS TO PONDER

1. State Bode's Law and discuss how well it fits the three criteria of a good scientific theory. (Hint: See the Close Up.)
2. Write a description of the reports that might be sent back by extraterrestrials as they view Earth through the window of their approaching space ship.
3. Suppose two planets were found to have natural satellites at identical distances from their centers. The period of the satellite of planet *X*, though, is found to be greater than that of planet *Y*. Which planet has the greater mass? Explain your reasoning.
4. If the period of the satellite of planet *X* in question 3 is three times that of planet *Y*, how much more massive is one planet that the other?
5. Why is the mass of each planet much greater than the mass of any of its satellites?
6. Kepler's third law allows us to calculate the total mass of a planet and one of its satellites. How can we tell what fraction of the total mass is the planet and what fraction is the moon?
7. Is the escape velocity from Earth the same whether we are considering a point just above the atmosphere or a point higher up? Explain.

CALCULATIONS

1. The eccentricity of Pluto's orbit is 0.25. Use a drawing of an ellipse to explain what this means quantitatively.
2. If another planet were found beyond Pluto, how far would Bode's law predict it to be from the Sun (assuming it to be the next planet in his scheme)?
3. Io is one of Jupiter's moons first seen by Galileo. Its mass is very much less than Jupiter's. Io is 422,000 km from Jupiter and orbits it every 1.77 days. Use this data to calculate the mass of Jupiter, and then check your answer in Table 7–4.
4. A Martian astronomer calculates (from Perkel's Third Law, perhaps) that the total mass of the Earth/Moon system is 6.05×10^{24} kg. Observing the motion of the two bodies, the Martian determines that the center of mass of the two is 0.0122 of the way between their centers (measured from the Earth). What does the Martian calculate as the mass of the Moon? Check the Martian's answer in Table 7–4.
5. At a time when Jupiter is 9.0×10^8 km from Earth, its angular size is 33 seconds of arc. Use this data to calculate the diameter of Jupiter, and then check your answer in Appendix F.
6. Use the appendices to find the diameter of Mars and the size of its orbit. Assume a circular orbit for Mars and calculate its distance from Earth (in kilometers) at closest approach. Finally, calculate its angular size at that distance.
7. Suppose that a radar signal is bounced from Jupiter. It returns to Earth 100 minutes after being sent. How far away is Jupiter at the time of the measurement?

CHAPTER 8

Dramatic rendition of the Soviet Phobos spacecraft.

Phobos, one of the moons of Mars, would be an interesting place to visit. A person who weighs 130 pounds on Earth would weigh only about a half pound on tiny Phobos. Assuming that the visitor is wearing a very flexible space suit, she would be able to jump up and not come down for 15 seconds! The escape velocity from Phobos is about 12 meters/second, about 25 miles per hour. A ballplayer could throw a baseball completely off this moon, never to return. Throwing the ball over the horizon at a lower speed would result in the ball going into orbit, and since the distance around Phobos is about 90 kilometers (50 miles), such a ball would take a couple of hours to complete its orbit. You might imagine quite a strange baseball game.

The Terrestrial Planets

WE WILL BEGIN OUR trip to the planets with the planet closest to the Sun and will then proceed outward. As we discuss each planet in turn, you should compare that planet with Earth and with planets previously discussed. Concentrate on remembering patterns and comparisons between planets rather than memorizing numerical facts.

MERCURY

In legend, Mercury was the god of roads and travel and the messenger of the Roman gods. He wore winged sandals and delivered his messages with godlike speed. It is for this speedster that the fastest planet of the solar system is named.

Mercury As Seen from Earth

Although Mercury was one of the planets known to the ancients, it is the naked-eye planet least seen by people on Earth. Even Copernicus supposedly lamented near the end of his life that he had not seen it. The reason Mercury is so hard to see is that it is so close to the Sun that it can be seen with the naked eye only either shortly after sunset in the western sky or shortly before sunrise in the eastern sky (see Figure 8–1.)

When Mercury is viewed by telescope we are able to see that it exhibits phases like Venus and the Moon. Surface detail is difficult to discern, however; primarily because when Mercury is near the horizon, its light is passing through so much of the Earth's atmosphere. Figure 8–2 illustrates this situation. The best telescopic views of Mercury actually are made when it is high overhead during the day, but even then, surface features are not well-defined and the telescope shows only that Mercury's surface contains bright and dark areas. Early in this century there were reports indicating a cratered surface but this idea was not well accepted. Details of its surface features were not seen clearly until the planet was visited by space probes.

Mercury's Surface Compared to the Moon's

In November 1973, Mariner 10 left Earth on a mission to Mercury. After passing near Venus, it swept by Mercury in March 1974, taking numerous photos. Its orbit was then adjusted so that after passing Mercury, it orbited the Sun with a period just twice that of Mercury's, coming back near Mercury once during each of its orbits. Figure 8–3 indicates Mariner 10's orbit. Mariner's systems lasted long enough to get photos on two passes of Mercury after the first. In all, Mariner 10 provided us with more than 4000 photos of Mercury.

Figure 8–4(a) shows a mosaic of Mariner's views of Mercury and part (b) of the figure shows some detail of part of the planet's surface. Notice its similarity to our Moon; both are covered with impact craters produced by debris from space. We find, however, that the walls of the craters of Mercury are less steep than those of the Moon so that Mercury's craters are less prominent than the

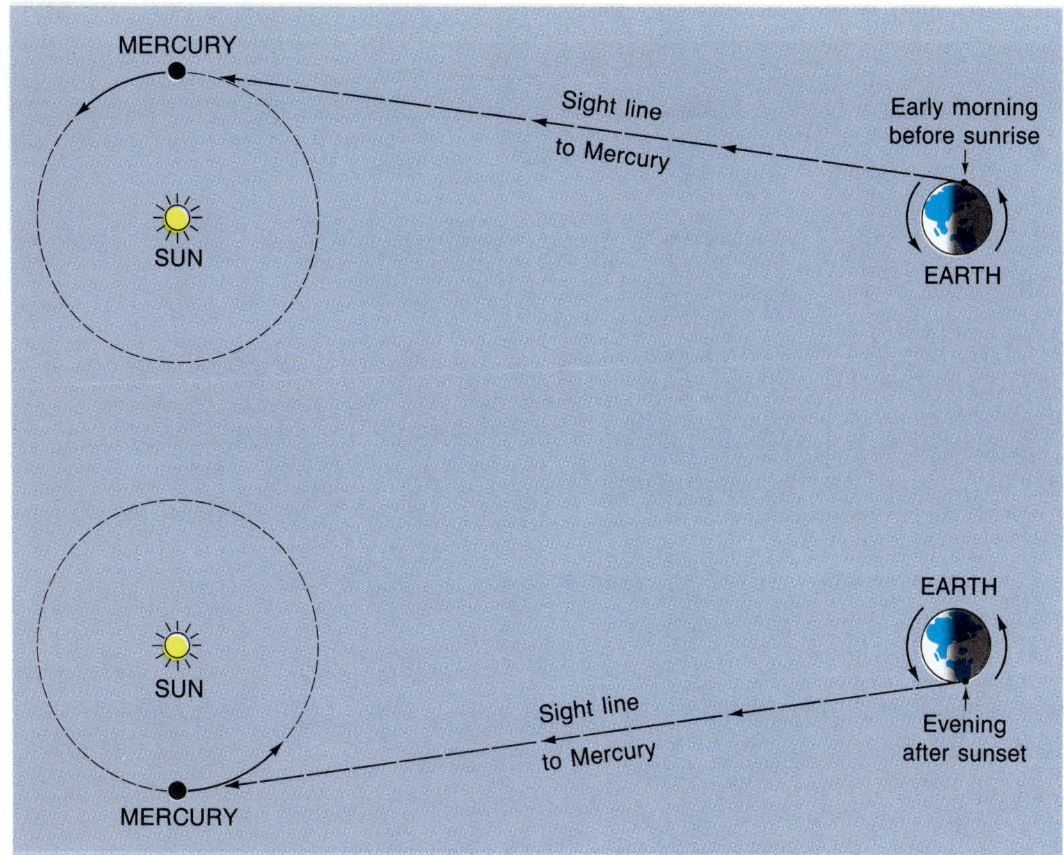

Figure 8–1. *Mercury can only be seen from Earth either shortly before sunrise (top), or shortly after sunset (bottom).*

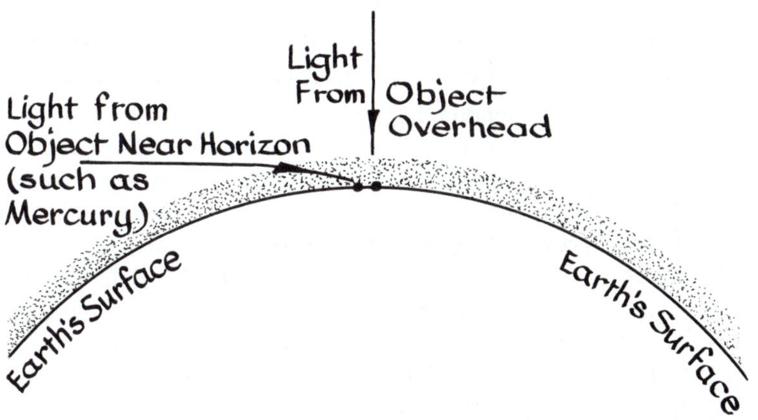

Figure 8–2. *The thickness of Earth's atmosphere is exaggerated here, but notice that light from an object near the horizon must pass through more atmosphere than light from an overhead object. (The bending of light entering the atmosphere will be discussed in Chapter 11 and is not important here.)*

Moon's (Figure 8–5). This would be expected, because Mercury's surface gravity is about twice that of the Moon and loose material will not stack as steeply under the greater gravitational force.

Like the craters on the Moon, most (or all) of Mercury's craters were formed by impacts with infalling objects—meteorites.

In comparing photos of Mercury and the Moon, notice that ray patterns of material ejected from craters are less extensive on Mercury than on the Moon.

ACTIVITY

Viewing Mercury, Venus, and Mars

TABLE T8–1 SHOWS THE dates between 1989 and 1994 when Mercury will be situated for best viewing. Between these dates Mercury rises at least 1½ hours before sunrise or sets at least 1½ hours after sunset. The longer the rising or setting time is separated from sunrise or sunset, the easier it is to find Mercury in the sky. Mercury is highest in the sky about midway between each of the pairs of dates and the longer the time between the pair of dates, the better the viewing will be at the midpoint. For example, in 1992, the June/July period provides a better viewing opportunity than the March dates, and Mercury will be highest about June 27.

Within about 45 minutes of sunrise and sunset, the sky is so bright that Mercury is particularly difficult to see with the naked eye. Thus, if you are viewing in the morning, you should begin your search at least an hour before sunrise. In the evenings, there is no need to begin searching before about 45 minutes after sunset. Mercury is never easy to find for the first time, and you must have a very clear sky and a low horizon. Look low in the sky in the area lit by the sun, searching for what appears to be a dim star.

Table T8–2 shows favorable dates for viewing Venus in the morning and evening sky. Between each pair of dates shown, Venus rises at least one hour before the sun or sets at least one hour after. The dates of Venus's greatest brilliance is shown, but between the dates indicated, Venus should be easy to find any time the sky is clear. Venus is bright enough that you can find it as close as 20 minutes to sunrise or sunset.

Table T8–3 shows some rising and setting times for Mars. Here is an example of how to use the table: Suppose you wish to know where to find Mars in May of 1990. Since it was rising at the beginning of morning twilight the previous December and will rise at midnight in August, you can figure that it will rise around 2:00 or 3:00 A.M. in May. Thus you must look for it in the early morning in the eastern sky.

Table T8–1 Favorable Dates for Viewing Mercury

Year	Evening Viewings	Morning Viewings
1989	Apr. 21–May 10	Oct. 9–Oct. 12
1990	Apr. 6–Apr. 19	Jan. 21–Jan. 30
1991	Mar. 24–Mar. 30	Jan. 3–Jan. 18 Dec. 17–Jan. 5
1992	Mar. 6–Mar. 10 June 19–July 4	Dec. 1–Dec. 19
1993	June 1–June 22	Nov. 16–Nov. 29
1994	May 15–June 8	July 14–July 28

As you look at Mars, observe its color compared to nearby stars. It should be obvious why Mars is called the red planet.

If a telescope is available to you, use it to observe the planets. You should be able to see the phases of Venus (and perhaps Mercury). The color of Mars appears even more obvious in a telescope and when Mars is at opposition and closest to Earth (when it is rising at sunset), some observers report that they can see the white polar caps even in a small telescope.

Helpful advice is available for the amateur astronomer in a number of monthly magazines, particularly *Astronomy* and *Sky and Telescope*. These include diagrams of planetary locations and hints on viewing. The two listed here are highly recommended. In addition, helpful planetary information, including positions during each month of the year, can be found in reputable almanacs such as *The World Almanac and Book of Facts*.

Table T8–2 Favorable Dates for Viewing Venus

Year	Range of Dates	Morning or Evening	Date of Greatest Brilliance
1989+	May 25–Jan. 10	E	Dec. 14
1990	Jan. 24–Sep. 13	M	Feb. 20
1991	Jan. 1–July 29	E	July 16
1991+	Sep. 1–Mar. 4	M	Sep. 28
1992+	Sep. 13–Mar. 25	E	Feb. 24
1993	Apr. 10–Nov. 25	M	May 6
1994	Mar. 10–Oct. 10	E	Sep. 30
1994+	Nov. 12–June 30	M	Dec. 6

Table T8–3 Selected Rising and Setting Times for Mars

Date	Description*
Early Dec. 1989	Rising at beginning of morning twilight
Early Aug. 1990	Rising at midnight
Late Nov. 1990	Rising at sunset, setting at sunrise
Mid May 1991	Setting at midnight
Mid July 1991	Setting at end of evening twilight
Mid Feb. 1992	Rising at beginning of morning twilight
Early Sept. 1992	Rising at midnight
Early Jan. 1993	Rising at sunset, setting at sunrise
Early June 1993	Setting at midnight
Mid Aug. 1993	Setting at end of evening twilight
Early June 1994	Rising at beginning of morning twilight
Early Nov. 1994	Rising at midnight

*When midnight is indicated, the reference is to standard time (rather than daylight saving time).

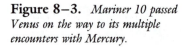

Figure 8–3. *Mariner 10 passed Venus on the way to its multiple encounters with Mercury.*

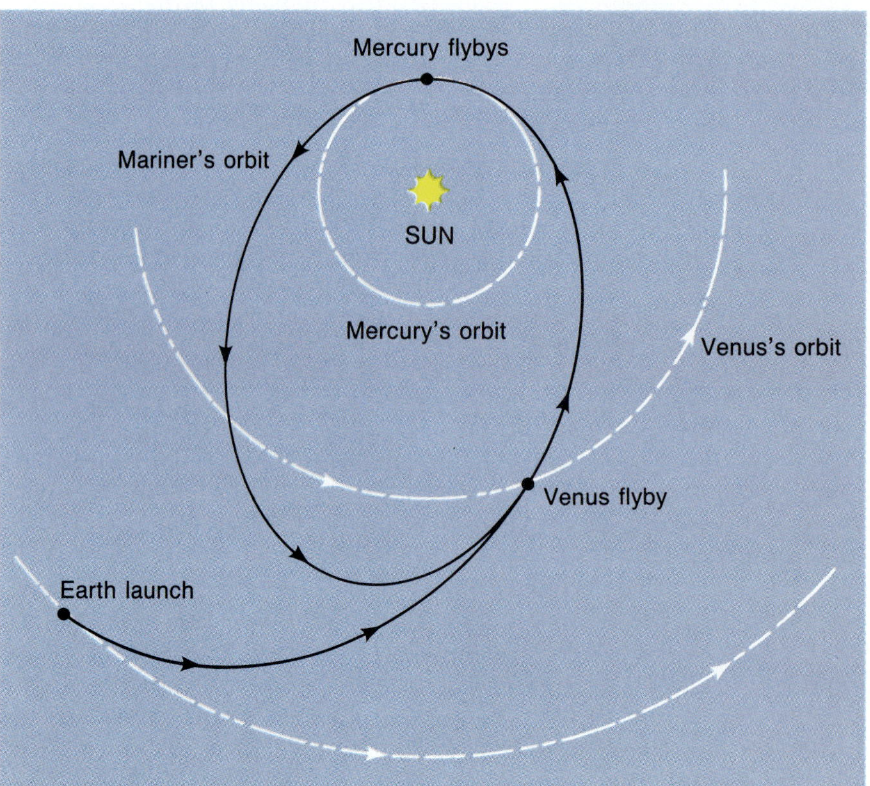

(a)

(b)

Figure 8–4. *(a) A mosaic of photos of Mercury taken by Mariner 10. (b) Some of the craters of Mercury.*

When craters were formed by meteorite impacts, the greater gravitational force on the surface of Mercury resulted in material not being thrown as far from the craters.

Notice also that Mercury lacks the large mare we see on the Moon. Instead, its craters are fairly evenly spread across the surface, separated by smooth plains. This difference between the two surfaces is related to differences in the rate at which the objects cooled after formation. As we will see when we discuss the formation of the solar system, the terrestrial planets and the Moon began as balls of molten rock. As they cooled, a solid crust formed on their surfaces. Debris falling from space struck the crust with enough energy to penetrate it and let molten material (lava) from below flow up through the break. The results of these lava flows are obvious on the Moon, where large maria were formed from the lava that welled from beneath the surface.

A larger planet cools more slowly than a small one, however; so Mercury's crust formed slowly. For a long time after the crust started forming, meteoritic debris was still able to penetrate it, allowing lava to flow out and obliterate older craters. This resulted in the plains we see between the craters.

Another difference between Mercury's surface and the Moon's is the great number of long cliffs (called ***scarps***) that are found on Mercury (Figure 8–6). These are found at many locations around the planet and suggest that after its crust hardened during its formation, Mercury shrank a little, causing the cliffs to be formed.

On Mercury's surface, Mariner photos revealed a large impact crater named the Caloris Basin (Figure 8–7). It consists of a number of concentric rings

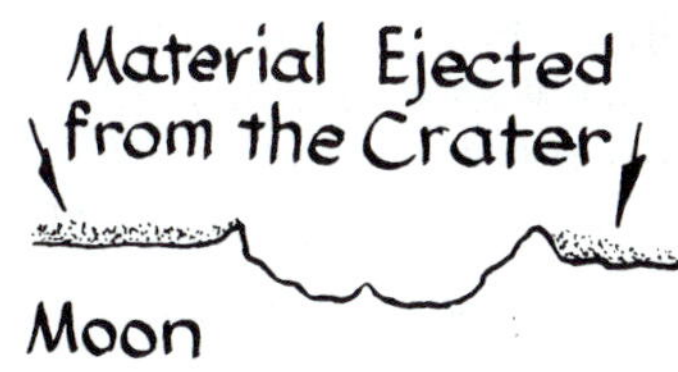

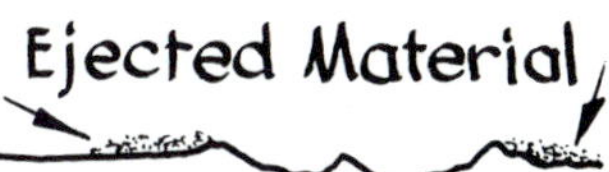

Figure 8–5. *Craters on the moon have steeper walls than those on Mercury and because of the greater surface gravity on Mercury, material ejected from the crater does not travel as far.*

Scarps: Cliffs in a line. They are found on Mercury, Earth, Mars, and the Moon.

Figure 8–6. *Near the horizon of Mercury in this photo, one can see a line of cliffs called scarps. These are much more extensive on Mercury than on the Moon.*

Figure 8–7. *The center of the Caloris Basin lies just out of the photo toward the left. Notice the ring pattern that was formed by the impact of a large object (now buried under the surface). The basin is about 1400 kilometers (850 miles) in diameter.*

somewhat like those of Mare Orientale on the Moon (Figure 8–8), although the Mercurian feature is larger. Both "bull's eyes" were caused by large objects striking the surfaces; the impacts resulting in shock waves that caused the rings of cliffs. On the opposite side of Mercury from Caloris Basin, we find a jumbled, wavy area strange enough that it has been dubbed "weird terrain." This was probably the outcome of the shock wave that was sent around the planet as a result of the Caloris Basin impact. When the waves met on the other side of the planet, they caused a permanent disruption of the surface there. Similar, although less obvious, features are seen on the Moon opposite major craters. We will see another consequence of the Caloris Basin impact when we discuss Mercury's motions.

Figure 8–9 is a drawing of part of Mercury's surface. Notice that its features have been named after prominent artists, composers, and writers. Various Earthly cultures are well mixed on Mercury. Observe that the crater Mark Twain is close to that named for Ts'ao Chan, and Dostoevskij is just a little south of Milton.

Data from Mariner confirmed that Mercury has negligible atmosphere, as was expected considering the escape velocity from Mercury's surface and the high daytime temperatures caused by its proximity to the Sun. Notice in Figure 8–10 that although the escape velocity from the surface of Mercury is greater than from the Moon, Mercury's temperature is higher and therefore none of the gases shown are expected to be found on Mercury.

Figure 8–8. *Mare Orientale, on the Moon, is similar to the Caloris Basin, but is only about 1000 kilometers (620 miles) in diameter.*

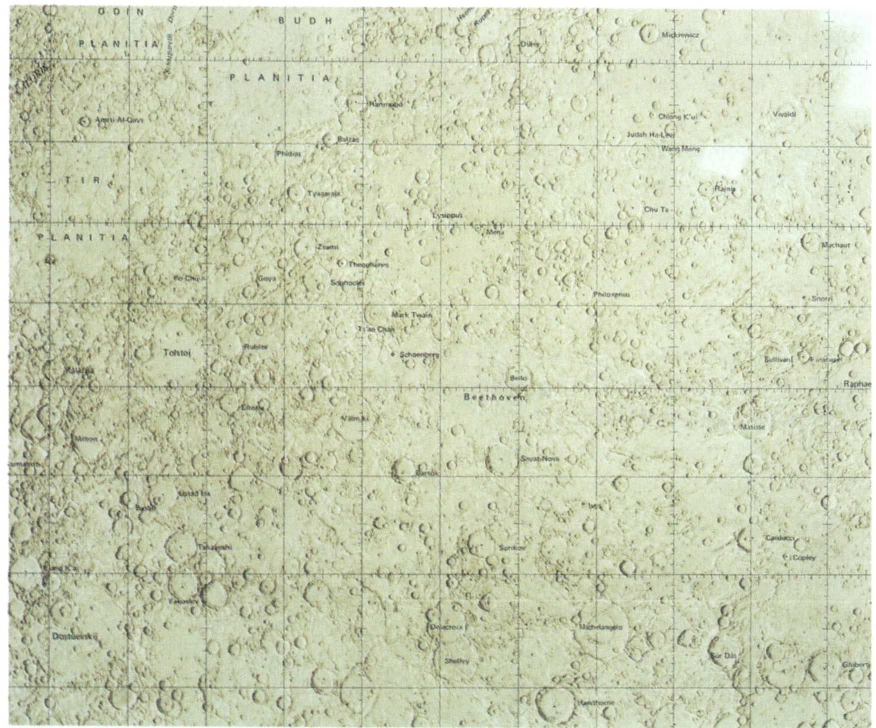

Figure 8–9. *Part of Mercury's southern hemisphere, with some features named.*

Figure 8–10. *Both Mercury and the Moon lie below the dashed lines representing escape velocities of the various gases, showing that the gases can escape from these two objects.*

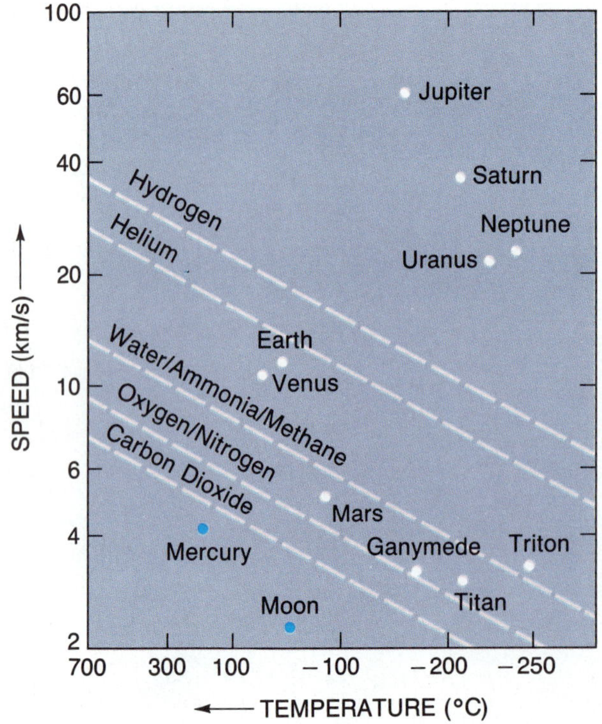

Size, Mass, and Density

4880 km = 3030 miles.

Mercury is the second smallest planet, exceeding only Pluto in size. In fact, there are two natural satellites in the solar system that are larger than tiny Mercury. (Figure 8–11 compares Mercury with some other solar system objects.) With a diameter of 4880 kilometers, Mercury's total surface area is only slightly greater than that of the Atlantic Ocean.

The mass of Mercury, accurately determined from its gravitational pull on the Mariner 10 probe, is only 0.055 of the mass of the Earth. Its average density is 5.43 times the density of water, slightly less than the Earth's 5.52.

The density of an astronomical object is determined not only by the material of which it is composed but by how much that material is compressed by the gravitational field of the object. For example, an object on the surface of Mercury feels less gravitational pull toward the planet than it would on Earth. Thus if Mercury were made up of the same elements as the Earth, matter below the surface would be less compressed by material above and Mercury would have considerable less density than the Earth does. The fact that Mercury's average density is only slightly less than Earth's means that Mercury has a higher concentration of the heavy elements than does the Earth. Evidence from Mariner indicates, however, that rocks on the surface are similar to Earth rocks so this higher-density material must be below the surface. Later we will see more evidence for the nature of the matter in Mercury's core.

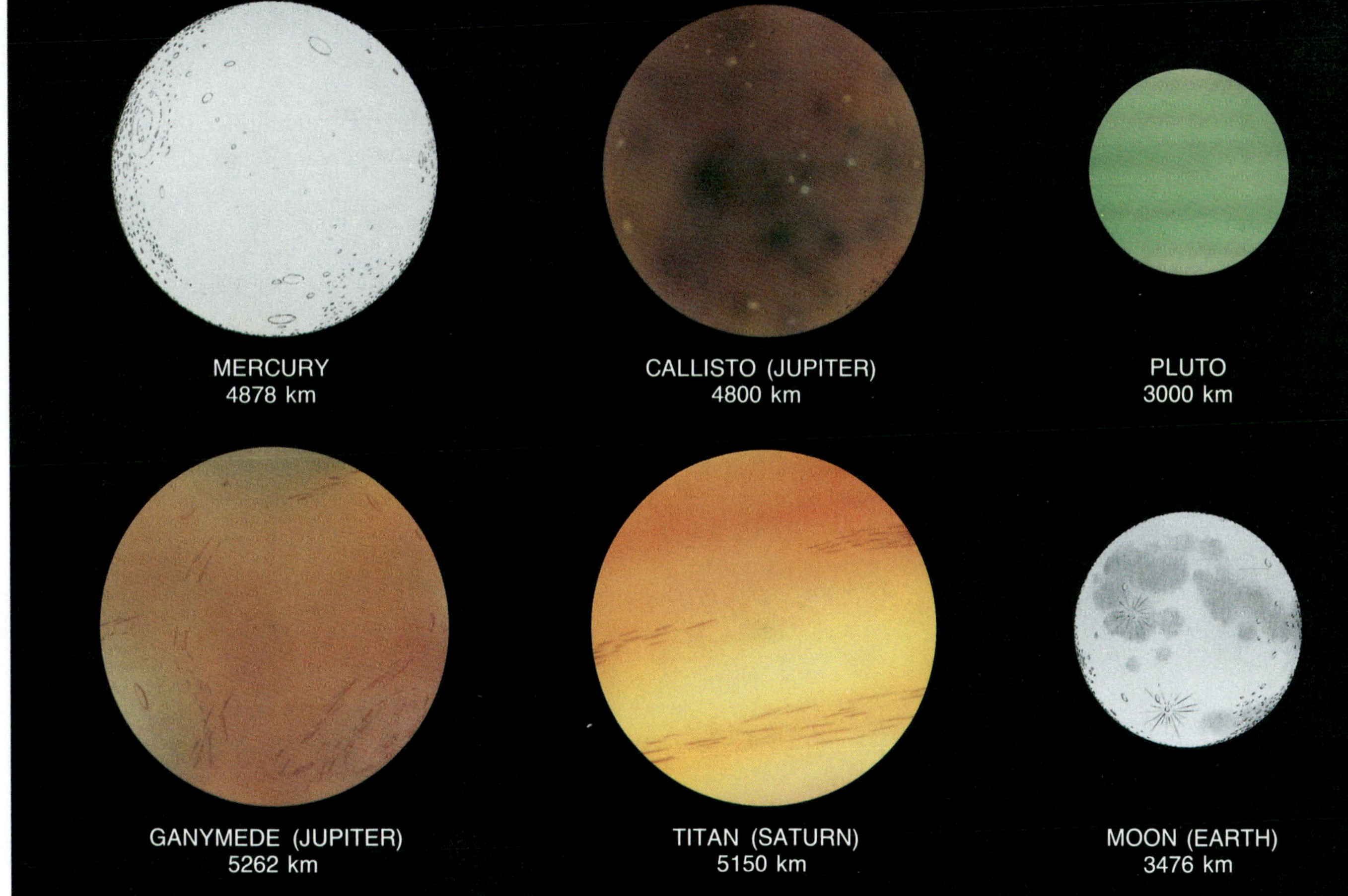

Figure 8–11. *The size of Mercury is compared to some other solar system objects. Pluto is smaller than Mercury, but one of Jupiter's and one of Saturn's moons is larger. (In the case of moons, the parent planet is shown in parentheses.)*

Calculating the Density of a Planet A knowledge of the average density of a planet or other astronomical object is important if we wish to determine the internal make-up of that object. We will illustrate the calculation of the density of a planet with an example.

EXAMPLE

Calculate the average density of Mercury given that its diameter is 4878 kilometers and its mass is 3.30×10^{23} kilograms.

Solution

Density is defined as mass divided by volume, so we will first express Mercury's diameter in meters and then calculate its volume.

$$4878 \text{ km} \times \frac{1000 \text{ m}}{1 \text{ km}} = 4{,}878{,}000 \text{ m} = 4.878 \times 10^6 \text{ m}$$

We assume that Mercury is a sphere and use the equation for the volume of a sphere:

$$\text{volume} = (4/3)\ \pi\ (\text{radius})^3$$

The radius is half the diameter. Substituting this, we obtain

$$\begin{aligned}\text{volume} &= (4/3)\ \pi\ (2.439 \times 10^6\ \text{m})^3 \\ &= 6.078 \times 10^{19}\ \text{m}^3\end{aligned}$$

A regular card table has a volume—including the space under it—of about one cubic meter.

Note that m^3 means cubic meters. We express it as m^3 because that is consistent algebraically: Cubing the symbol m yields m^3. The value we calculated for Mercury's volume is much too large for us to imagine in terms of everyday objects, of course.

Now we calculate the density:

$$\begin{aligned}\text{density} &= \frac{\text{mass}}{\text{volume}} \\ &= \frac{3.30 \times 10^{23}\ \text{kg}}{6.08 \times 10^{19}\ \text{m}^3} \\ &= 5.43 \times 10^3\ \text{kg/m}^3\end{aligned}$$

This corresponds to a density of 5.43 grams per cubic centimeter.

Try One Yourself. Use Appendix F to find the mass and diameter of Mars, and from this information, calculate its average density.

Mercury's Motions

Being the closest of the planets to the Sun, Mercury circles the Sun in less time (88 days) than any other planet and moves faster in its orbit than any other (average speed: 48 kilometers/second). Again, except for Pluto, Mercury's orbit is the most eccentric of the planets. Its distance from the Sun varies from 47 million kilometers to 71 million kilometers.

48 km/s is equivalent to 110,000 mi/hr.

Because of Mercury's proximity to the Sun and its elongated orbit, it exhibits an interesting rotation rate. Recall that because of tidal effects, the Moon keeps its same face toward the Earth. Astronomers once thought that Mercury likewise points the same face toward the Sun at all times. Radar observations, however, indicate that this is not the case. They show that Mercury rotates on its axis once every 58.65 Earth days, which is precisely two thirds of its orbital period of 87.97 days. This means that the planet rotates exactly 1½ times for every time it goes around the Sun.

Why would this pattern occur? The answer is that Mercury is not perfectly balanced; one side is more massive than the other. Because of this, the Sun

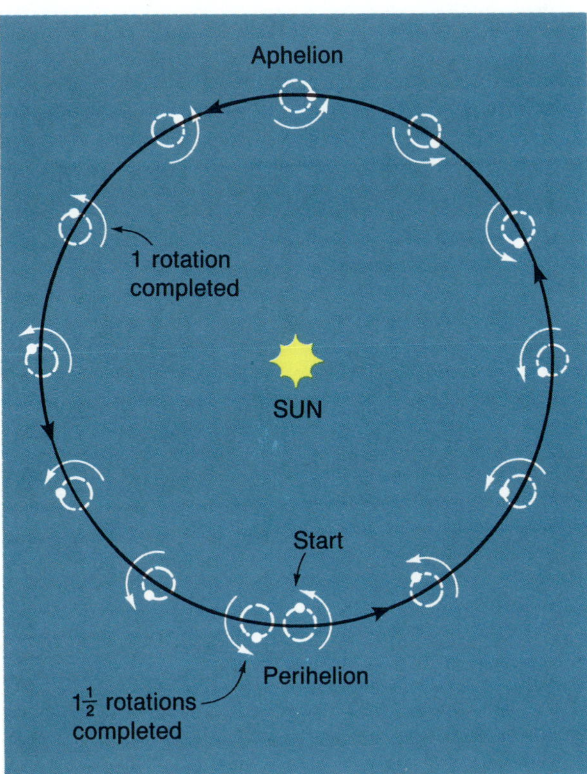

Figure 8–12. *Mercury's rotation period is such that a point on its surface that is under the Sun at one perihelion position is opposite to the Sun on the next.*

exerts an uneven force on the planet, especially when it is closest to the Sun (at ***perihelion***). After countless revolutions this has resulted in the rotational period of the planet being coupled with its revolution period. Figure 8–12 shows this situation, showing a dark spot on Mercury's surface in order to make it obvious that opposite sides of the planet face the Sun at each perihelion passage.

Perihelion: The point in its orbit when a planet (or other object) is closest to the Sun.

Aphelion: The point in its orbit when a planet (or other object) is farthest from the Sun.

The point marked on Mercury's surface in Figure 8–12 could represent Caloris Basin, for that Mercurian feature falls directly under the Sun at every other perihelion position. This is unlikely to be a coincidence. It is hypothesized that the object that hit Mercury's surface and created the basin must have been made of dense material and the object remains under the planet's surface, causing Mercury to be lopsided and therefore to have its periods of rotation and revolution coupled as they are.

The Latin word calor *means* heat. *The unit for measuring heat is the* calorie.

Mercury's solar day is far different from its sidereal day. Refer again to Figure 8–12 to see why. Notice that after two-thirds of an orbit, Mercury has completed one rotation with respect to the distant stars. This does not mean that it has completed one solar day, however. Mercury does not complete one rotation with respect to the Sun until 176 Earth days have passed. To see this, notice that if a person were at the marked point when the planet is at the position marked "start," the Sun would be at the highest point in the person's sky. As time passed, the Sun would get lower, finally setting when the planet has completed half a revolution. When Mercury reaches "start" again, it is midnight for that person. It takes 88 days for Mercury to complete its orbit

Table 8–1 The Planet Mercury

	Value	Compared to Earth
Diameter	4878 km	0.38
Mass	3.302×10^{23} kg	0.055
Density	5.43 gm/cm^3	0.98
Surface gravity	3.7 m/s^2	0.38
Escape velocity	4.25 km/s	0.39
Sidereal rotation period	58.65 days	58.65
Solar day	176 days	176
Surface temperature	−150°C to +450°C	
Albedo:	0.06	0.2
Tilt of equator to orbital plane	0	—
Orbit		
Semimajor axis	5.79×10^7 km	0.387
Eccentricity	0.2056	12.3
Inclination to ecliptic	7.0°	
Sidereal period	87.97 days	0.24
Moons	0	

and that constitutes only half of a day for the person on the planet. A solar day on Mercury is therefore 176 Earth days (or two Mercurian years)!

Mercury's Interior and Magnetic Field

Before the Mariner 10 mission, Mercury was thought to have no magnetic field. To see why, recall that the Earth's field is caused by the rotation of the planet and its metallic core. Mercury's rotation rate was thought to be too slow to produce a magnetic field.

Mariner 10 did detect a magnetic field on Mercury, although it was not very strong; about 1 percent as strong as the Earth's field. The field appears to be shaped like the Earth's, with magnetic poles nearly aligned with the planet's spin axis. The presence of the magnetic field leads us to conclude not only that Mercury has a metallic core but that the core occupies a larger fraction of the planet than does the Earth's core; perhaps extending two-thirds of the way to its surface (Figure 8–13). This conclusion also is supported by Mercury's high density.

Temperatures on Mercury

Imagine the Earth with no atmosphere to shield us from the Sun and with a daylight period lasting 88 of our days. This is the situation on Mercury. Tem-

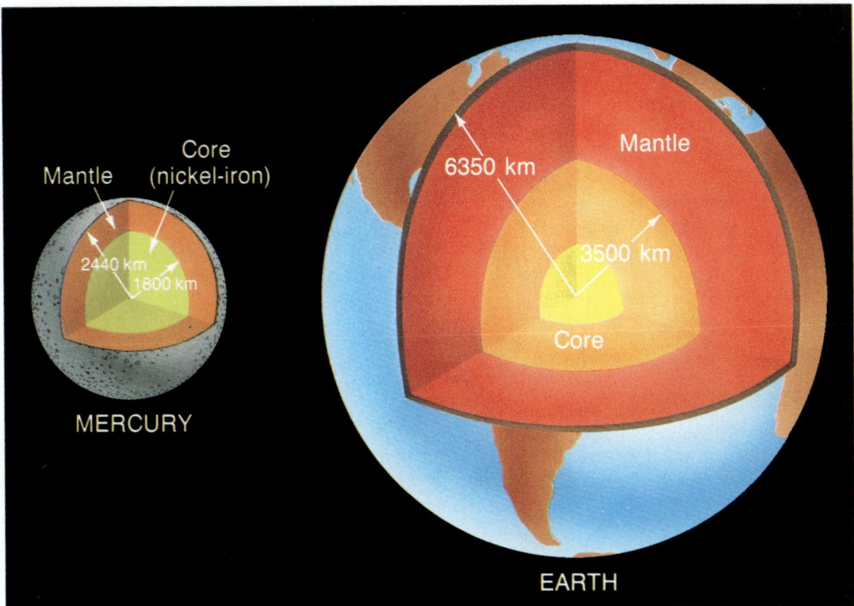

Figure 8–13. *Mercury's core occupies a greater portion of the planet's volume than does Earth's core.*

peratures there reach as high as about 450 degrees Celsius (C). The element lead melts at about 330 degrees C, so lead would be molten at midday on Mercury.

450 degrees C is about 800 degrees F.

On the other hand, the nighttime side of Mercury also has no atmosphere to hold in the heat during its long nights. It therefore gets as cold as −150 degrees C (which is −250 degrees F). Mercury's temperature variations are much greater than any other planet.

VENUS

Venus is never seen farther than 47 degrees from the Sun and so—like the case of Mercury—it is only visible either in the evening sky after sunset or in the morning sky before sunrise. Except for the Sun and Moon, however, Venus can be the brightest object in our sky. Seeing such a bright object above the horizon has fooled many people into thinking they were seeing something else. During World War II, pilots of fighter planes sometimes shot at Venus thinking that it was an enemy plane. Today it is common for reports of UFOs to increase when Venus is at its brightest in the sky.

Venus is such a beautiful sight that it is little wonder that the ancients named it after the goddess of beauty and love. We'll see that as a possible home for transplanted Earthlings, however, it is not such a beautiful planet after all.

Size, Mass, and Density

Venus has historically been known as the sister of Earth. It is similar to Earth in many ways: Its diameter is 95 percent of Earth's, its mass is 82 percent of Earth's, and of all the planets, its orbit is located closest to us.

Figure 8–14. *Venus is at greatest brilliance when its elongation is 39 degrees, which occurs about 36 days before or after (shown here) its new phase.*

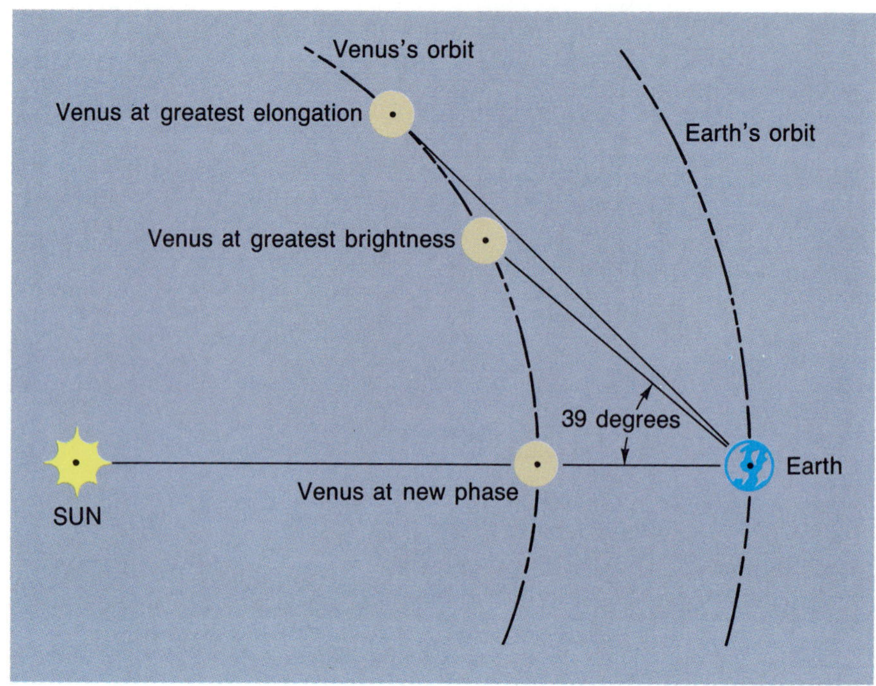

From the values for its diameter and mass we can calculate that Venus has a density 5.24 times that of water; not much different from Earth's 5.52. It would have a slightly higher density if its material were compressed as much as Earth material. Soviet spacecraft that landed on Venus have shown that its surface rocks are of about the same composition as Earth rocks and we therefore conclude that in order to have a density as high as it does, Venus must have a very dense interior and is probably differentiated, with a metallic core.

Venus As Seen from Earth

Venus's phases can be seen even in a small telescope. Try it.

Galileo was the first to observe the phases of Venus. He argued that the fact that it exhibits a gibbous phase proves that it orbits the Sun and not the Earth. Although its phases are not visible to the naked eye, they have a definite effect on how bright Venus appears to us. When Venus is between Earth and Sun, it is closer to us than any other planet gets. It is then in a new phase, however, and cannot be seen from Earth. Just before and just after the new phase, its crescent is so small that to the naked eye the planet appears dim. On the other hand, when it is nearly full and we see almost its entire disk, it is on the other side of the Sun from us and is so far away that it appears dim again. Venus appears brightest at an intermediate position, when it shows less than a full face but when it is fairly close to Earth (Figure 8–14). This position occurs when Venus is 39 degrees from the Sun; about 36 days before and after its new phase. Table T8–2 in the Activity box shows when Venus will be at its brightest over the next few years.

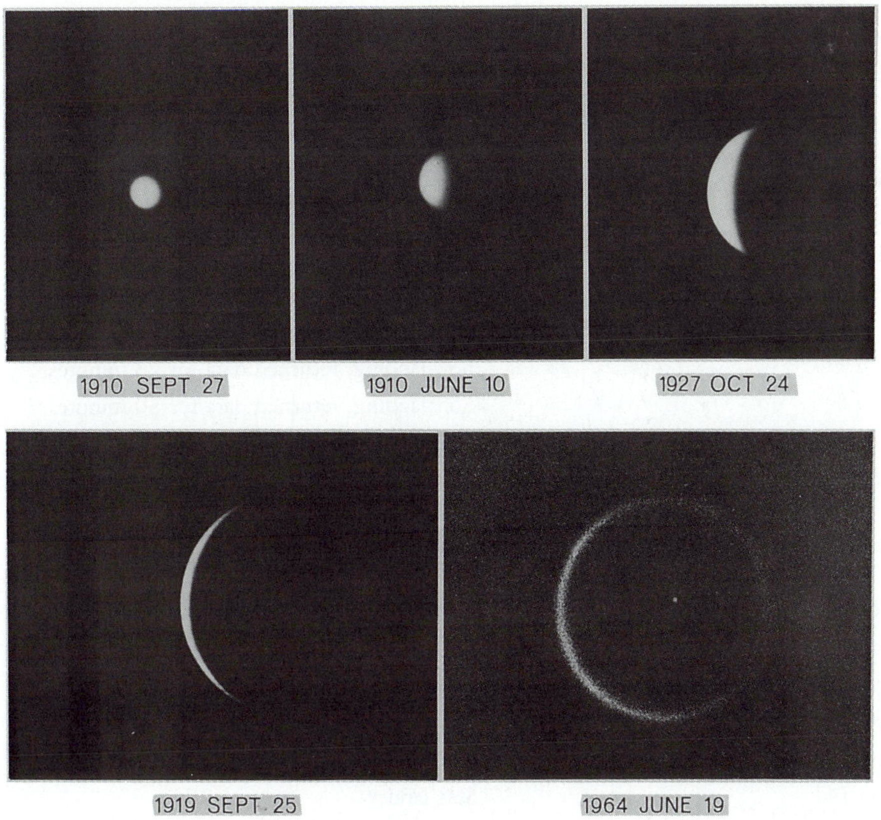

1910 SEPT 27

1910 JUNE 10

1927 OCT 24

1919 SEPT 25

1964 JUNE 19

Figure 8–15. *Various phases of Venus.*

In a telescope, the angular diameter of Venus varies from 10 seconds of arc when it is most distant to 64 seconds of arc when it is closest. Figure 8–15 contains photos of the planet in its various phases and shows the great change in its angular diameter as seen from Earth.

Venus's Motions

Venus orbits the Sun in a very circular orbit—more circular than any other planet. Its period is 225 days, resulting in a very nearly constant orbital speed of about 35 kilometers/second.

35 km/s is 78,000 mi/hr.

The surface of Venus is not visible from Earth, for the planet is covered by dense clouds. Radar, however, allows astronomers to learn about nearby planets by bouncing radio waves from them. Radar can penetrate Venus's cloud cover and since 1961 we have been bouncing radar signals from its surface. This is how we first learned about its rotation rate and its surface features.

The use of radar to study Venus told us that Venus's rotation is very unusual. The planet rotates *backwards* compared to most other rotations in the solar system. That is, if we view the solar system from above the Earth's north pole, we see that all planets circle the Sun in a counterclockwise manner and that all the planets except Venus and Uranus rotate in this same direction. Venus's

Table 8–2 Missions to Venus

Launch Date	Spacecraft	Description
1962	Mariner 2 (U.S.)	Flyby, 35,000 km from surface
1967	Venera 4 (USSR)	Crushed in descent
1967	Mariner 5	Flyby, 4100 km
1969	Venera 5	Crushed in descent
1969	Venera 6	Crushed in descent
1970	Venera 7	Soft landing, returned data for 23 minutes
1972	Venera 8	Soft landing, returned data for 50 minutes
1973	Mariner 10	Flyby, 5700 km
1975	Venera 9	Orbiter plus soft lander, operated 53 minutes
1975	Venera 10	Orbiter plus soft lander, operated 65 minutes
1978	Pioneer Venus 1 (U.S.)	Orbited as low as 150 km from surface
1978	Pioneer Venus 2	Four hard landers
1978	Venera 11	Flyby (25,000 km) plus soft lander
1978	Venera 12	Flyby (25,000 km) plus soft lander
1981	Venera 13	Soft lander
1981	Venera 14	Soft lander
1984	Venera 15 Vega 1*	Soft lander returned data for 21 minutes, balloon in atmosphere operated 46 hours
1985	Venera 16 Vega 2*	Lander analyzed soil samples, balloon returned data for 46 hours

*The Venera 15 and 16 spacecraft were renamed after their Venus encounter, and they continued to Halley's comet.

rotation is very slow; its sidereal rotation period being 243 days. Its period of revolution around the Sun, however, is only 225 days. These rotation and revolution rates result in the solar day on Venus being about 117 Earth days. So a day spent on Venus would be quite unusual. The Sun would rise in the west, stay up for about 58 Earth days, and set in the east.

Inferior conjunction: The configuration of a planet when it is as nearly between the Earth and the Sun as it can get during a given orbit.

The rotation rate of Venus is such that the same side of it faces the Earth at each ***inferior conjunction***. It is possible that this is simply a coincidence, but it is much more likely that this is another example of gravitational coupling such as we have seen both for the Moon and for Mercury. In those cases, the object's rate of rotation is linked to its period of revolution by gravitational forces caused by the object around which it rotates—the Earth and Sun respectively. In the case of Venus, however, gravitational forces between Venus and the Earth have caused the rotation period of Venus to adjust so that its same face is always turned to the Earth when that planet is closest to us.

Figure 8–16. *This photo of the surface of Venus was taken by Venera 13. The object at bottom right is part of the lander.*

Exploration of the surface of Venus by radar has yielded much knowledge about our sister planet, but it was when we sent space probes to orbit Venus and to land on its surface that we fully realized how different this planet is from ours.

Visits to Venus

Every 1.6 years, Venus and Earth are in favorable positions for launching space probes toward Venus. The U.S. has launched a number of such probes but the USSR has been more active in searching out the mysteries of our planetary neighbor, having soft-landed a number of craft on the planet. Table 8–2 highlights the visits that have been made to Venus and Figure 8–16 shows a photo of Venus's surface taken by a Soviet Venera spacecraft.

Synodic period: The time interval between successive similar configurations of a planet-Sun-Earth (for example, successive inferior conjunctions). *Venus's synodic period is 1.6 years.*

No magnetic field has been detected on Venus and based on the sensitivity of the instruments that have searched for it, if one does exist, its strength must be less than 1/10,000 of Earth's. Because of Venus's slow rotation we would not expect Venus to have a strong magnetic field, although according to present theories of the cause of planetary magnetic fields it should be strong enough for us to detect. Recall that Earth's field is known to have reversed its direction in the geologic past. Perhaps the magnetic field of Venus is now in the process of reversing, which would explain why it is so weak.

At the time of writing, the next scheduled U.S. mission to Venus is the Magellan Venus-orbiting radar mapper scheduled for launch in April 1989.

The Atmosphere of Venus

The length of a day on Venus is not the only feature that makes that planet a poor candidate for the title of Earth's sister. The Venera landers confirmed what we had begun to learn about the unusual atmosphere of Venus.

The Earth's atmosphere is composed of nearly 80 percent nitrogen and 20 percent oxygen, with small amounts of water, carbon dioxide, and ozone. Venus, on the other hand, has an atmosphere made up of about 96 percent carbon dioxide, 3.5 percent nitrogen, and small amounts of water, sulfuric acid, and hydrochloric acid—the same acids that are in car batteries! In fact, the clouds we see on Venus are made up in large part of sulfuric acid droplets. Venus is indeed inhospitable.

350 km/hr is 225 mi/hr.

The upper atmosphere of Venus is very windy. The winds there reach speeds up to 350 kilometers/hour. The wind blows in the direction of rotation of the planet, but remember that the planet is rotating very slowly. As one moves lower in the atmosphere, the wind velocity decreases until it is nearly zero at the surface. (Even at low velocities, however, the high density of the atmosphere near the surface would produce hurricane-like forces.) We do not fully understand the causes of Venus's wind patterns. Undoubtedly the slow motion of the sun across Venus's surface (because of its slow rotation rate) is the basic cause but the details are not known. Further study of the weather on Venus would not only help us to know that planet better but would increase our understanding of weather patterns on Earth.

The gas of Venus's atmosphere is so thick that if we could survive its hazards, we could strap on wings and fly.

As the space probes descended through the atmosphere of Venus, it was not only the acidic atmosphere and the great winds that presented a hazard. They were also subjected to tremendous pressures, for the atmospheric pressure on the surface of Venus is about 90 times that on the surface of Earth. As if the acid atmosphere and high pressure were not enough, Venus is also inhospitable because of its high temperatures: about 460 degrees C (850 degrees F) near the surface.

A Theory Explaining Venus/Earth Differences

We stated that both Earth and Venus have small amounts of water vapor in their atmospheres. Venus's atmospheric water is about all the water that exists on the planet, however. Why is this planet, so close to the size of the Earth, so different from Earth?

Terrestrial planets get much of their atmospheres from gases released from their interiors; primarily through volcanic action. Volcanoes on Earth release large amounts of both water vapor and carbon dioxide and it is likely that this is also the case for volcanoes on Venus. Both planets are theorized to have once had water on their surfaces and substantial amounts of carbon dioxide in their atmospheres. Most of the carbon dioxide on Earth is now dissolved in the

Table 8–3 The Planet Venus

	Value	Compared to Earth
Diameter	12,104 km	0.95
Mass	4.869×10^{24} kg	0.82
Density	5.24 gm/cm^3	0.95
Surface gravity	89.8 m/s^2	0.90
Escape velocity	10.36 km/s	0.93
Sidereal rotation period	243 days	243
Solar day	117 days	117
Surface temperature	+460°C	
Albedo	0.75	2.4
Tilt of equator to orbital plane	178°	7.6
Orbit		
Semimajor axis	1.095×10^8 km	0.723
Eccentricity	0.068	4.1
Inclination to ecliptic	3.39°	
Sidereal period	243.0 days	0.67
Moons	0	

oceans and in rocks such as limestone, which is found in the oceans. Venus, however, would have had a higher surface temperature due to its position nearer the Sun and this is thought to have evaporated water from its surface, reducing the amount of carbon dioxide that would be dissolved. The clouds of water vapor, combined with large amounts of carbon dioxide in Venus's atmosphere, then resulted in an effect we experience only to a mild degree here on Earth: the greenhouse effect.

Did you ever notice that here on Earth, our coldest nights occur when the sky is clear? When we have cloudy skies at night, the difference between day and night temperatures is minimized. The reason for this is that in order for the Earth's surface to cool down at night, it must radiate into space a portion of the heat it gained during the day. The primary heat loss occurs in the form of infrared radiation. Infrared radiation, however, does not readily pass through water vapor, so the clouds reflect the infrared waves back to Earth. Thus clouds form a sort of blanket for the Earth.

Early in Earth's and Venus's history, both planets experienced this blanketing effect but because of Venus's higher temperature, it had more water vapor than did Earth. In addition, it had large amounts of carbon dioxide in its atmosphere. This gas is also opaque to infrared radiation. Visible light, on the other hand, is not blocked by carbon dioxide. Thus although most of the visible light from the Sun was reflected by the clouds of Venus, some of it passed through the atmosphere and was absorbed by the surface, causing the

Albedo: The fraction of incident sunlight that an object reflects.

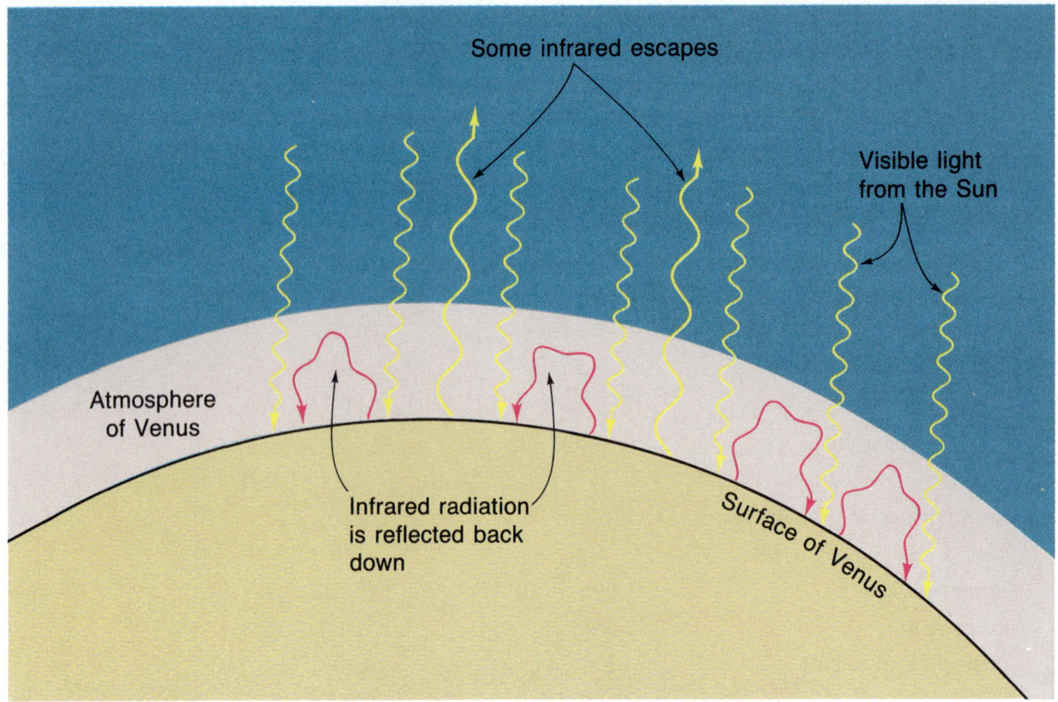

Figure 8–17. *Some visible light penetrates the atmosphere of Venus and reaches the surface, heating it. Most of the infrared radiation emitted by the surface, however, is reflected back by the atmosphere, causing Venus to be very hot.*

surface to heat up. Meanwhile, high in Venus's atmosphere, ultraviolet radiation from the Sun was breaking water molecules into their constituents: hydrogen and oxygen. The hydrogen escaped from the planet and the oxygen combined with other elements in the atmosphere. This left the atmosphere with large amounts of carbon dioxide, continuing to trap infrared radiation.

The hotter Venus's surface got, the more infrared radiation it emitted. Because of the carbon dioxide in the atmosphere, though, much of this radiation could not escape. Thus the planet continued to heat up. The high surface temperature on Venus baked further carbon dioxide out of the surface rocks. So a chain reaction resulted: the more carbon dioxide that was released, the hotter the surface became and the more carbon dioxide was released.

The planet did not continue heating forever, of course. The atmosphere was not *completely* opaque to infrared. A small fraction of the infrared radiation that hit it was remitted into space and that energy was lost to the planet. The amount of solar energy being absorbed by Venus was constant, while the amount being released depended upon the temperature of the surface. Thus Venus continued to heat up until the amount of radiant energy leaving it was equal to the amount striking it. At this point an equilibrium condition was reached. Figure 8–17 illustrates the situation.

Venus continues to have high temperatures due to trapped infrared radiation. Such heating is called the ***greenhouse effect*** because it is somewhat similar

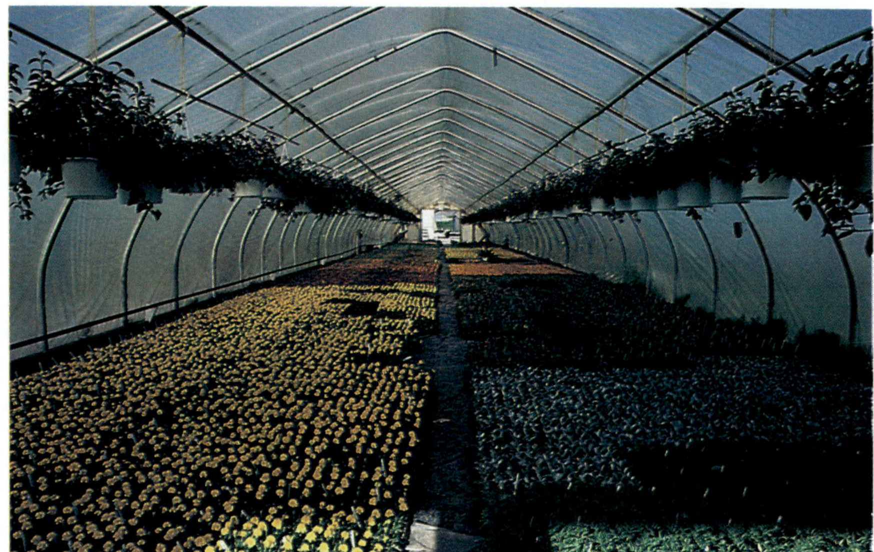

Figure 8–18. *Although visible light can penetrate the glass of a greenhouse, most infrared radiation leaving the surface inside is reflected back by the glass, resulting in inside temperatures being high.*

to what happens in greenhouses on Earth (Figure 8–18). Sunlight penetrates the glass of a greenhouse's roof and walls and is absorbed by the plants and soil within. These surfaces then emit infrared radiation. Glass, like water vapor and carbon dioxide, is a poor transmitter of infrared and so the air in a greenhouse is kept warmer than the outside air. This is not the entire story in the case of a greenhouse, however. A very important factor in the heating of greenhouses is the fact that the walls prohibit the warm air inside from mixing with the cooler outside air. This trapping of the air is not a factor on a planet and so the heating of Venus and of a greenhouse are not truly similar.

Greenhouse effect: The effect by which infrared radiation is trapped within a planet's atmosphere by reflection from particles (for example, carbon dioxide) within that atmosphere.

If our theory of Venus's greenhouse heating is correct, the major differences in conditions today on Venus and Earth are due to a fairly slight difference in temperature far back in their histories. This difference was magnified by the greenhouse effect until conditions on Venus are now often associated with Hell. This scenario is what causes many scientists to fear the release of large amounts of carbon dioxide into Earth's atmosphere by the burning of fuels. It is feared that a fairly slight increase in Earth's temperature could set off a runaway greenhouse effect here, destroying Earth as we know it.

Climatologists report that, worldwide, the six warmest years in the last 100 have occurred during the 1980s. Working with scientists from other fields, including astronomy, they are trying to determine whether this represents a chance fluctuation or whether it is due to the greenhouse effect. If the greenhouse effect is responsible, significant changes in our lifestyles may be necessary in order to reverse the process.

The Surface of Venus

USSR Venus landers have shown us that the surface of Venus is rock-strewn, at least at the landing sites. In addition, they showed that rocks in some locations

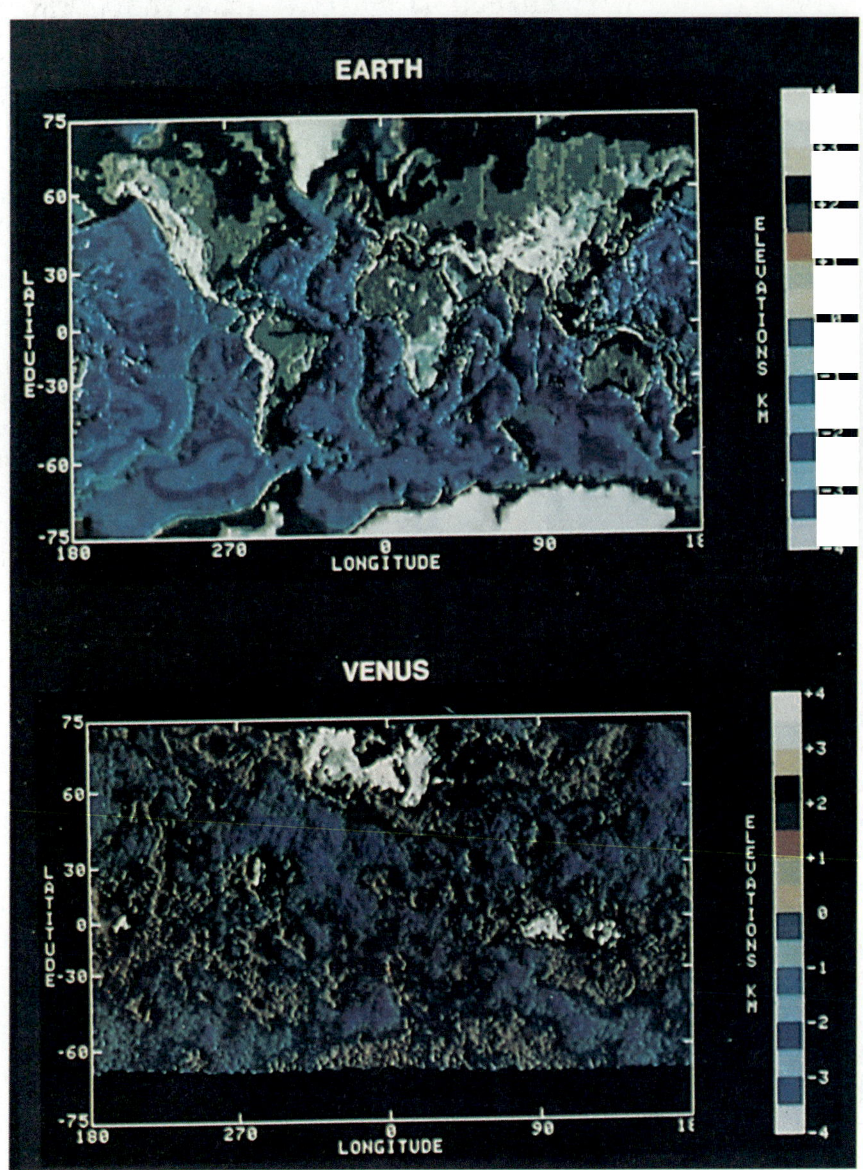

Figure 8–19. *A relief map of Venus, made from the Pioneer Venus radar survey, is compared to one of Earth.*

are more weathered than in others, indicating that they have been exposed on the surface for a much longer time. The sharp-edged rocks were a surprise to scientists studying Venus until it was realized that the wind is fairly calm at the surface. If there had been great winds at the surface, the dense gas would have caused considerable dust movement and erosion of the rocks.

We indicated earlier that we receive information concerning Venus's surface by using radar sent from Earth. Additional information was obtained by Pioneer Venus 1, which orbited the planet in 1978. It bounced radar signals from the surface and used the reflected waves to map much of the planet. Just as radar in airport control towers is used to tell the distance to airplanes, Pioneer Venus's radar was used to determine distances to points on the surface of Venus. Thus the Venus orbiter enabled us to map the heights of various features on Venus.

About two-thirds of the surface of Venus is covered by rolling hills with craters here and there. Highlands occupy less than 10 percent of the surface and lower-lying areas make up the rest. Figure 8–19 is a map of its surface, along with a similar map of Earth. Colors on each map represent similar elevations on each planet. On the Venus map, the area of high elevation at upper center is Ishtar Terra, the most prominent highland area. It is comparable to the U.S. in area and its highest mountain, Montes Maxwell, is about 12 kilometers (7.5 miles) above the average rolling surface. (Mt. Everest, by comparison, rises only about 9 kilometers above sea level.)

Montes Maxwell was named after James Clerk Maxwell, who did very important pioneering work concerning the electromagnetic spectrum.

All of Venus, including the rolling plains that make up most of its surface, is dryer than the driest desert on Earth. Earth is covered primarily by rolling water, and Venus by rolling desert. Venus is Dry. Hot. It has tremendous pressure. Venus is uncomfortable!

As we leave Venus and move outward from the Sun, we next encounter Earth. This planet is so important that it was included in the last chapter, and so it will be skipped here. That makes Mars next.

MARS

The ancient Greeks called the fourth planet from the Sun Ares, naming it after their mythical god of war. Our names for the planets, however, come from their Roman names, and indeed, in Roman mythology the god of war is Mars. Why is this planet associated with warfare? Perhaps because of the association of its red color with blood. Mars is indeed red. Later in this text we will refer to red stars, but the red stars that are visible to the naked eye are not very red at all compared to the planet of war.

Size, Mass, and Density

Mars is a small planet, closer in size to our Moon than to the Earth. Figure 8–20 compares Mars to Mercury, Venus, Earth, and the Moon. Mars falls between Mercury and Venus in size, having a diameter about half of Earth's. This makes the surface area of Mars about one-fourth that of Earth. The volume of Mars, then, is about one-eighth of Earth's ($\frac{1}{2}^3$). Now recall that both Mercury and Venus have densities comparable to Earth's. If Mars follows this pattern its mass should be about one-eighth that of Earth, which would make the ratio of its mass to its volume about like Earth's. The mass of Mars, however, is only one-tenth of Earth's and its density is only 3.95 times that of water, or less than ¾ of Earth's density.

Area depends on the square of the diameter and $\frac{1}{2}^2$ is ¼.

Mars's Motions

Mars orbits the Sun at an average distance of 1.524 AUs (about 228 million kilometers) from the Sun. Its orbit, however, is fairly eccentric and its distance

The eccentricity of Mars's orbit is 0.093, while Earth's is 0.017.

Figure 8–20. *This mosaic of photos shows the objects with their correct relative sizes.*

Figure 8–21. *Mars is closer to the Sun during summer in its southern hemisphere. In this drawing both the tilt of Mars's axis and the change in its distance from the Sun is exaggerated.*

You should be able to figure out why the two periods are nearly the same for Mars.

from the Sun varies from about 210 million kilometers to 250 million kilometers. From Kepler's Third Law we can calculate that Mars requires 1.88 years to complete its orbit around the Sun.

We saw that a solar day on Mercury was 176 Earth days, and although Venus was supposed to be Earth's sister, a day there lasts 117 Earth days. In this sense, it is Mars that should be called Earth's sibling for it has a sidereal rotation period of 24 hours, 37 minutes and its day is 24 hours, 40 minutes.

The equator of Mars is tilted 25.2 degrees with respect to its orbital plane. This is very close to Earth's 23.4 degrees. Since the tilt of a planet's axis causes opposite seasons in the planet's two hemispheres, we expect such seasons on Mars. There is another feature of Mars's motion that affects its seasons, however. The eccentricity of Mars's orbit causes it to be much closer to the Sun at some times of its year than at others. It turns out that Mars is 17 percent closer to the Sun during the northern hemisphere's winter than during its summer (see Figure 8–21). Being closer to the Sun in winter and farther in summer means that seasonal temperature variations are moderated in the northern hemisphere.

In the southern hemisphere, the reverse is true. Mars is closer to the Sun during summertime, and farther from the Sun in winter. Thus the southern hemisphere experiences greater seasonal temperature shifts than the northern hemisphere.

The same effect occurs for Earth, which is closest to the Sun in January (wintertime in the northern hemisphere). The Earth's orbit is so circular, though, that we are only about 3.3 percent closer in at one time than the other (rather than 17 percent). In addition, water absorbs and releases great amounts of heat as it changes temperature and there is more water in the Earth's southern hemisphere than in its northern. This tends to cancel out the small effect caused by a changing distance from the Sun.

Mars As Seen from Earth

Mars is the only other planet with surface features that can be seen from Earth. The amount of surface detail that is visible, however, depends on a number of factors. First, the surface of Mars is often obscured by dust storms on that planet, which will be discussed later. In addition, surface visibility varies greatly depending on Mars's distance from Earth.

We had to say "other" because Earth's surface can be seen from Earth.

As you might expect, Mars is best seen when it is directly opposite the Sun in the sky. When a planet is in such a position, 180 degrees from the Sun, it is closest to Earth and is said to be in ***opposition***. This happens about every 2.2 years for Mars. But since Mars's orbit is so eccentric, the distance from Earth to Mars at opposition might be as little as 55 million kilometers or as great as 100 million kilometers. The opposition of 1988 was a close one and the planets were separated by only 52 million kilometers (Figure 8–22).

Opposition: The configuration of a planet when it is opposite to the sun in our sky. That is, the objects are aligned as follows: Sun-Earth-planet.

In Figure 8–23 you can see dark and light areas on the surface of Mars. The light areas have a red color and many observers report that the darker areas appear somewhat green. These markings were observed as early as 1660 and the rotation rate of Mars was determined from their motion. In addition, you can see a white cap at the pole of the planet. Observing this cap as Mars orbits the Sun, you can see it diminish in size as that pole faces the Sun and then grow again when that pole faces away from the Sun. This effect is similar to the Arctic and Antarctic areas on Earth. The tilt of Mars's axis is very similar to Earth's and we are observing these poles as they experience the Martian summer and winter.

Syzygy: A straight line arrangement of three celestial objects. *Word-game players like this word.*

Other seasonal changes can be observed on Mars. Large parts of the planet change periodically from a dark to a light color and back, depending on the position of the planet in its orbit. Since the darker areas appear green (to some observers, at least), this color change led to speculation that there is vegetation on the planet and that it changes color in response to seasonal growth. If there is vegetable life, could there be animal life on Mars? What about intelligent life, then? Before exploring the possibility of life on Mars, we will consider Mars's two tiny moons.

The reason for this reported green color is probably because green is the complementary color to red. Stare at something red for a little while and then close your eyes. You see green.

The Moons of Mars

Mars has two natural satellites named Phobos and Deimos. Both are small and irregularly shaped. Phobos, for example, is 28 kilometers across its longest dimension, but it is 23 kilometers and 20 kilometers across other dimensions (17 × 14 × 12 miles). Thus it is shaped like a potato. Deimos is even smaller

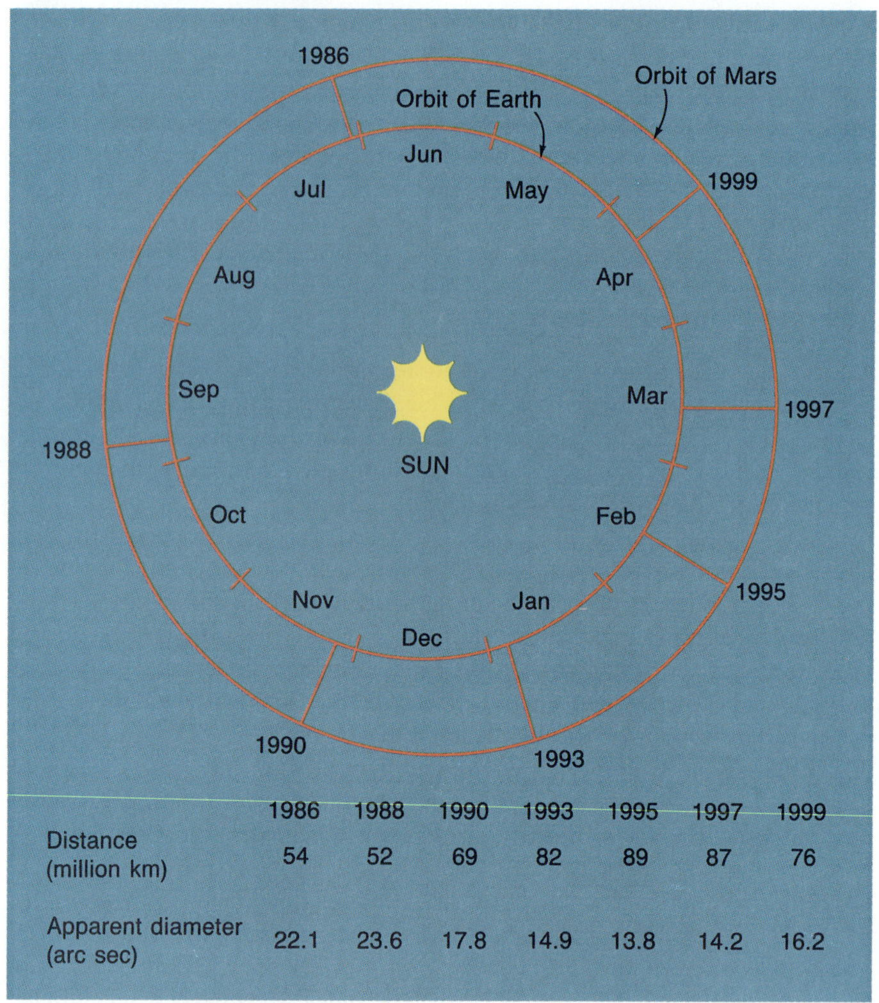

	1986	1988	1990	1993	1995	1997	1999
Distance (million km)	54	52	69	82	89	87	76
Apparent diameter (arc sec)	22.1	23.6	17.8	14.9	13.8	14.2	16.2

Figure 8–22. *Oppositions of Mars occur during the years shown. The differences in distances are exaggerated in the drawing.*

and is similarly shaped. Figure 8–24 shows photos of the two satellites taken by Viking I and compares them in size to Manhattan Island, New York. Notice the prominent cratering of their surfaces.

The surfaces of both moons of Mars are very dark, similar to that of many asteroids. Their densities are 1.9 and 2.1 times that of water (for Phobos and Deimos respectively), also similar to the densities of the rocky asteroids. Such similarities lead us to believe that these satellites are captured asteroids rather than objects that were formed in orbit around Mars during the formation of the solar system. A capture like this could have taken place as the asteroid passed very close to Mars and was either slowed by friction with the Martian atmosphere, by collision with a smaller asteroid, or by gravitational pull from another asteroid. Such an event may not seem likely, but given the number of asteroids which have passed by Mars over billions of years it is certainly a possibility.

The Martian moons are not only small, but they orbit very close to the planet and have short periods of revolution, about 0.3 days for Phobos and 1.3

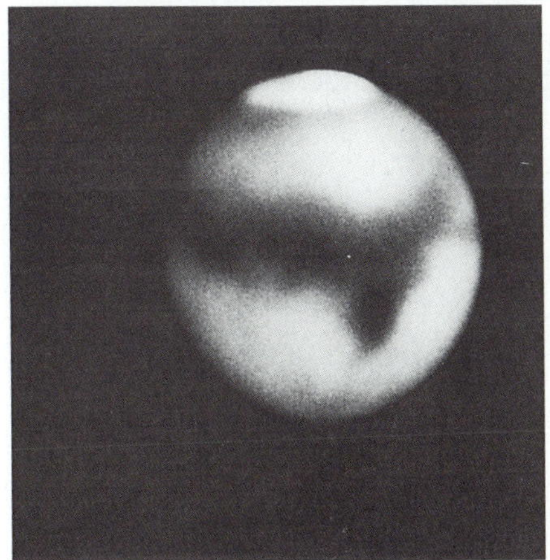
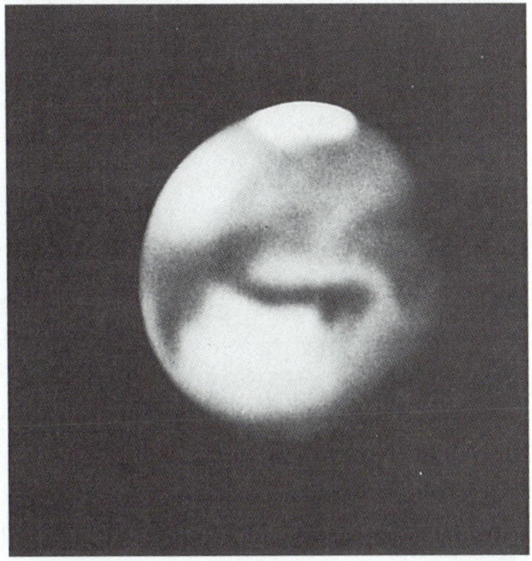

Figure 8–23. *These telescopic views of Mars were taken during the 1971 opposition and are as good as can be obtained from Earth. Notice the white polar cap.*

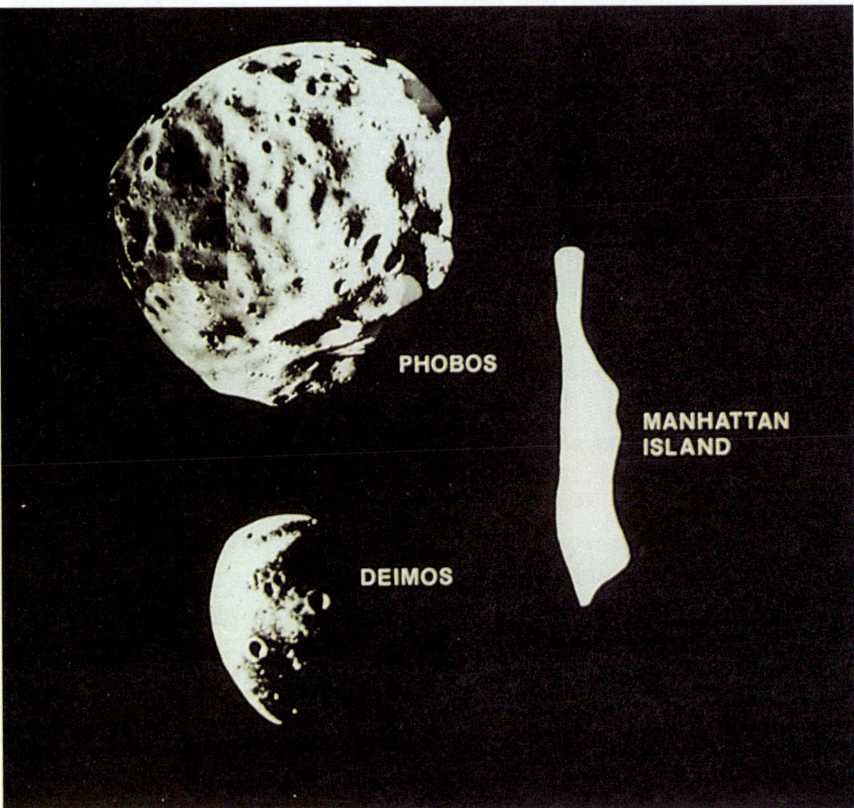

Figure 8–24. *Phobos and Deimos compared in size to Manhattan Island.*

days for Deimos. Both revolve in the same direction as most other solar system objects, counterclockwise as seen from north of Earth.

The counterclockwise motion of our Earthly Moon results in it moving slowly eastward among the stars as viewed from Earth. As we watch the Moon during a single night, however, this eastward motion is overwhelmed by its

Figure 8–25. *Although the stars would appear to move toward the west when viewed from Mars (just as from Earth), Phobos would move across the sky toward the east and Deimos would hover, moving slowly toward the west.*

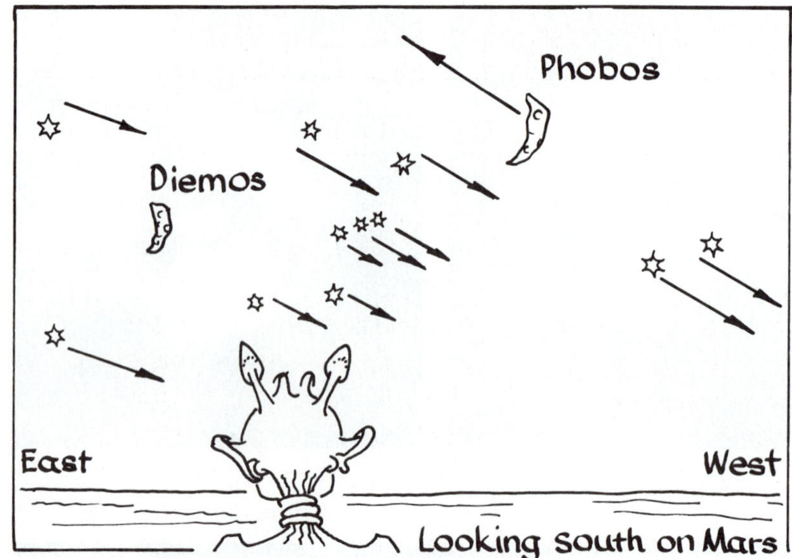

apparent westward motion that results from the rotation of the Earth. The eastward motion is far from obvious to the casual observer during one night. Things would be quite different for a Martian. Recall that Mars rotates with a period of about 24½ hours in a counterclockwise direction. Phobos takes only about 8 hours to circle Mars, however. The result is that an observer on Mars would see stars move across the sky just as we do on Earth, but that person would see Phobos rise in the west and move across the sky toward the east! (See Figure 8–25.) Deimos would hover over one's head for long periods of time, moving only very slowly toward the west.

Spherical or Nonspherical? Table 8–4 presents data on Mars. Notice that the diameter of Mars is slightly greater across the planet's equator than it is pole-to-pole. We will discuss the reason for this in the next chapter, when we discuss Jupiter (which is even more "out of round"). We have seen here that Phobos and Deimos also are not spherical. Before explaining why they are not, we should first explain why one might expect celestial objects to be spherical in the first place.

What is the shape of a raindrop? Raindrops are difficult to see, and most people think that they look somewhat like Figure 8–26(a); having a tail. In reality they are much more like the drop shown in part (b) of that figure; they are spherical. And they are spherical for basically the same reason that planets tend to be spherical. Forces of attraction between the molecules of a drop of water pull the molecules together. They are pulled as close as possible so that the surface of the drop has the least possible area, and the area is least when the drop is in the shape of a sphere. Consider what would be the situation if a drop of water were a cube. In this case, a molecule at the corner of the cube could get closer to the center of the drop by moving to the center of one face of the cube. So the drop would not remain a cube.

Table 8–4 The Planet Mars

	Value	Compared to Earth
Equatorial diameter	6794 km	0.53
Polar diameter	6759 km	0.53
Mass	6.4191×10^{23} kg	0.11
Density	3.93 gm/cm^3	0.71
Surface gravity	3.71 m/s^2	0.379
Escape velocity	5.02 km/s	0.45
Sidereal rotation period	24.623 hours	1.03
Solar day	24.67 hours	1.03
Surface temperature	−140°C to +20°C	
Albedo	0.15	0.5
Tilt of equator to orbital plane	25.20°	1.08
Orbit		
Semimajor axis	2.28×10^8 km	1.524
Eccentricity	0.0934	5.6
Inclination to ecliptic	1.85°	
Sidereal period	1.881 years	1.881
Moons	2	

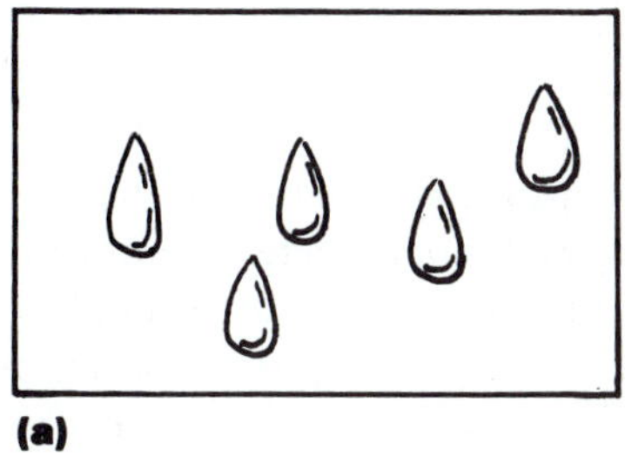
(a)

(b)

Figure 8–26. *(a) Falling raindrops are often drawn like this. (b) A photo of an actual raindrop (splashing into a pool) shows that it is fairly spherical in shape.*

In the case of a planet it is the force of gravity which tends to pull its parts together, rather than the cohesive forces between molecules of a water drop. The result is the same as with a drop of water, however; a planet takes the shape that will give it the smallest possible surface area. Earthly mountains, which are formed by disturbances within the Earth (such as volcanic action), are forever being disturbed by wind and rain so that their parts can be pulled closer to the center, making the planet a more perfect sphere.

HISTORICAL NOTE

The Discovery of the Martian Moons

WE POINTED OUT IN Chapter 3 that Johannes Kepler was somewhat of a mystic and that he practiced numerology. One of the patterns he saw in the heavens concerned the numbers of satellites of the various planets. Mercury and Venus have none, Earth one, and Jupiter 4 (known at that time). Between Earth and Jupiter was Mars, and Kepler proposed that Mars must have two moons if it is to fit the pattern. Scientists don't use logic like this today, but that was Kepler's way. From the time of Kepler, popular thought held that Mars must have two moons circling it. The idea appears a number of times in literature, most notably in Jonathan Swift's *Gulliver's Travels*, where Swift fills in such details as the moons' size (small) and orbital periods (10 hours and 21.5 hours).

There was no evidence for the existence of any satellites of Mars until 1877 when Asaph Hall, the astronomer of the newly constructed U.S. Naval Observatory near Washington, D.C., decided to search very near the planet for moons. To do so, he used a disk that blocked the planet's brightness from his view. There is a controversy about how quickly he found the moons, but Carl Sagan reports that he was unsuccessful for the first few nights and was about to quit when his wife encouraged him to continue searching. He did find the two moons and named them Phobos (Fear) and Deimos (Terror), the names of the horses that pulled the chariot of the Greek god of war.

In his book, *The Cosmic Connection*, published before Viking sent back photos of the moons, Sagan states that a feature on one of the moons should be named after Mrs. Hall. Look at the Figure T8–1, a photo of Phobos. The large crater is about 10 kilometers across and is named Stickney, Angelina Hall's maiden name, and the name proposed by Sagan. Another large crater on Phobos is named for her husband.

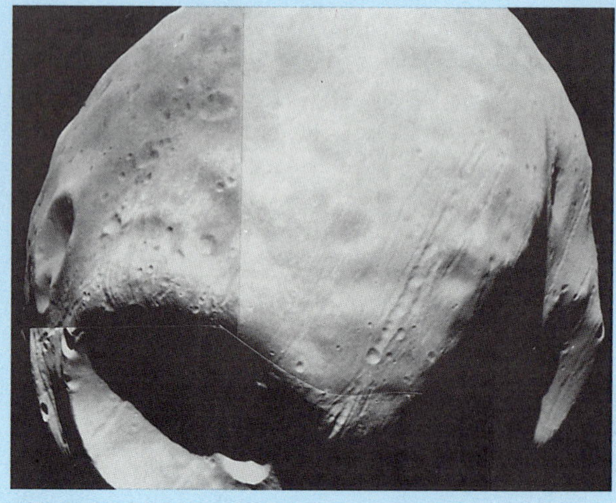

Figure T8–1. *The large, prominent crater is named Stickney, the maiden name of Mrs. Hall, the wife of the discoverer of Mars's moons.*

The actual orbital periods of the two moons are 7.7 hours and 30.3 hours. Although Swift had predicted 10 hours and 21.5 hours, he hit close enough to the actual times that some people believe that he had special knowledge of some sort. Actually, it was reasonable for him to assume that the moons, if they exist, must be close to the planet and must be small. Otherwise they would already have been discovered. And if they are close to the planet, Kepler's laws tell us that their periods of revolution must be short. There is no reason to hypothesize strange explanations for Swift's guesses. He just knew about Kepler's work.

Figure 8–27. *The Lowell Observatory in Flagstaff, Arizona, where Percival Lowell searched for life on Mars. The observatory is still used for planetary studies.*

The same tendency exists in the case of planetary satellites but the gravitational force is too small to produce a spherical shape for these small objects. On the surface of Mars's two moons, the gravitational force toward the center is much too small to pull the satellite into a sphere. The strength of the rock of which the moon is made resists the weak gravitational force that tends to reshape it.

Most asteroids are likewise nonspherical.

Life on Mars

Speculation about life on other planets—particularly Mars—is nothing new. The famous mathematician Frederick Gauss was so convinced that intelligent life existed on Mars that in 1802 he proposed that a huge sign be marked in the snows of Siberia to signal the beings there.

During the favorable (close) opposition of Mars in 1863, Father Angelo Secchi, an Italian astronomer, observed Mars and drew a colored map of Mars's surface. The map showed some lines which Father Secchi called *canali*, an Italian word best translated as *channels*. Then in 1877 the director of the Rome Observatory, Father Giovanni Schiaparelli, observed Mars and drew a more elaborate map showing many *canali*. In translation to English, however, Schiaparelli's *canali* became *canals*, rather than channels, giving the implication of artificial waterways. This struck a responsive chord with the public and Schiaparelli's findings were taken as confirmation of an intelligent race of beings on Mars.

In 1894 an American astronomer, Percival Lowell (1855–1916), had an observatory built on a hill near Flagstaff, Arizona to concentrate on the study

Figure 8–28. *Percival Lowell's chair is kept in the observatory for historical reasons even though it takes up valuable space.*

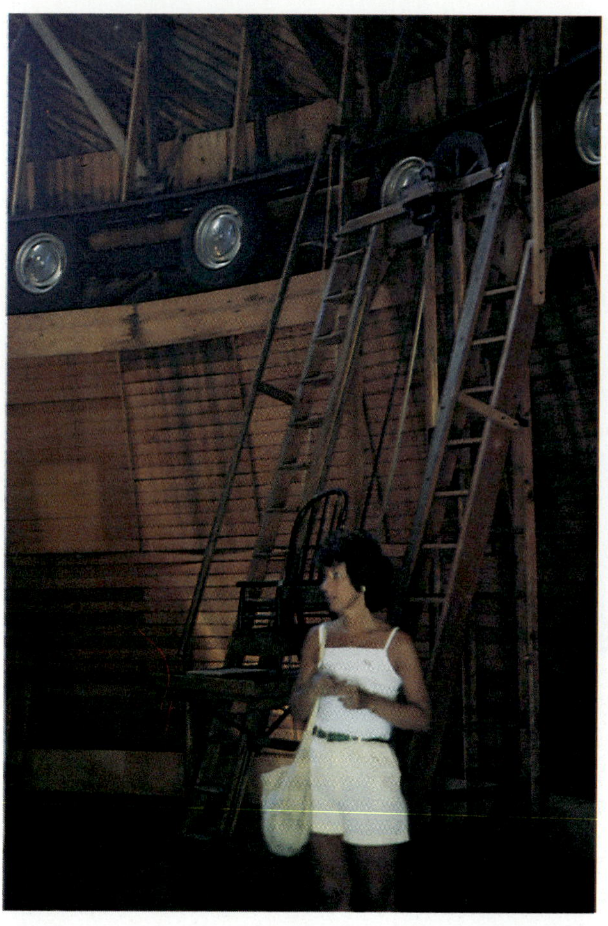

Lowell's observatory (Figure 8–27) is on Mars Hill, now within the city of Flagstaff. The observatory is still engaged in active research, financed largely by the Lowell family. It is open at times to the public. Figure 8–28 shows the chair and lift system that Lowell used.

The Wall Street Journal, in December 1907, said that one of the most important events of the year was the proof that there was intelligent life on Mars.

of Mars and its intelligent life. Lowell's drawings of Mars include as many as 500 canals (Figure 8–29). Recall that white polar caps are easily visible on Mars and that they change in size according to the season. Lowell believed that the canals were built to transport water from the polar caps to the dryer parts of the planet.

Other astronomers failed to see canals; even some using telescopes much larger than Lowell's and Schiaparelli's. In 1894, as Lowell was completing his observatory, the astronomer Edward Barnard said, "to save my soul I can't believe in the canals as Schiaparelli draws them." Vesto Slipher, Lowell's assistant, believed that he saw craters on Mars but he never published his views.

Then in 1898, H. G. Wells wrote *The War of the Worlds*, describing a fictional invasion of Earth by Martians. When Orson Welles dramatized this novel in a very realistic radio broadcast in 1938, he made it appear that a radio music show was being interrupted frequently to report on the invasion as it occurred. Many people did not hear (or forgot about) the announcement at the beginning of the show that it was a dramatization and there was widespread panic, especially in New Jersey, where the invaders were supposed to be landing.

Figure 8–29. *This map of Mars was drawn by Percival Lowell in 1907.*

Invasion and Its Results

A series of Martian invasions did start in the late 1960s and it is still in progress. The invasion, however, is proceeding in the opposite direction than H. G. Wells envisioned and it is a peaceful endeavor, although competition between major powers is certainly a factor in this invasion. In 1965 Mariner 4 passed by Mars, sending back 21 pictures of the surface. These pictures ended speculation about canals, for none were observed. The planet was seen to be covered by deserts and craters. Other questions were raised, however, for the pictures confirmed that major dust storms are common on Mars and yet the surface contains numerous craters; craters that should have been worn down by the constant pounding of wind and dust. We know now that the reason that erosion does not occur very quickly is that the atmosphere of Mars is extremely thin, so that the pressure at the surface is about 1/200 of the air pressure at Earth's surface. Dust stirred up by this thin atmosphere must then be extremely fine; nothing like a sandstorm in one of our deserts.

Mariner 4 was followed in 1969 by Mariners 6 and 7 and in 1971 by Mariner 9. The latter spacecraft went into orbit around Mars. It arrived while

CLOSE UP

ET Life II—Viking's Search for Life

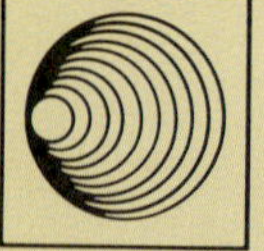

In the late summer of 1975, Viking I and Viking II were launched from Cape Kennedy (now Cape Canaveral) toward Mars. After travelling to Mars, Viking 1 separated into two parts, one part staying in orbit and another descending to a soft landing on the planet. Viking 1 landed on Mars on July 22, 1976, the seventh anniversary of our first step on the moon. Then in August, Viking 2 landed nearly on the other side of the planet.

One of the primary purposes of the Viking mission was to look for signs of life on our neighboring planet. Various tests were conducted toward this end. The first was made by a television camera that scanned the area for any signs of plant or animal life or even for footprints. There was really not much expectation that such large life forms existed on Mars and so the negative results were not particularly disappointing. The other tests, however, were more sophisticated and were designed to detect less obvious life forms.

To perform these experiments, each lander contained an arm with a scoop at its end (Figure T8–2) so that it could retrieve some soil from the surface. The first of these experiments was called the "labelled release experiment." It was performed by taking about a teaspoon of soil and dampening it with a rich nutrient that should have been absorbed by any organism in the soil. Some of the carbon atoms in this nutrient were carbon 14, a radioactive form of carbon. It was thought that after an organism absorbed nutrients the organism should release carbon to the air, including some of the radioactive carbon. Such radioactivity can be easily detected.

To try to insure that a positive result from this experiment would be due to biological processes rather than chemical processes, other soil samples were heated to 300 degrees F before being fed the nutrient. This was done in order to first sterilize the soil and kill any living matter that might be in it. In this way, the investigators could see whether the unheated soil acted differently from that which had been heated to kill life forms. (In a procedure like this, the sample that was sterilized is called the *control*. Such a procedure allows a comparison to be made and is common in all branches of science.)

When the labelled release experiment was tried on the unheated soil by both Viking I and II, the radioactivity of the air increased quickly. In the case of the control, the change in radioactivity was much less. Thus the experiment seemed to indicate that life was present. But there was a problem: The radioactivity showed up *too* quickly, faster than should occur if it had been caused by an organism absorbing and releasing the carbon. And when more nutrient was added, there was no further increase in the radioactivity of the air. This argued *against* the presence of life in the soil.

Although the initial results of the experiment caused some excitement among the researchers monitoring the results from Earth, the final conclusion was that the release of carbon was caused by simple chemical reactions with elements in the soil and did not involve any biology. Recall that the surface of Mars is thought to have plentiful oxygen; enough to rust the iron. A hydrogen peroxide molecule is a water molecule with

a great dust storm was occurring and there was fear that its batteries would be depleted before the dust settled and the surface could be seen. Fortunately, the storm ended and much was learned about the surface, particularly about the spectacular volcanoes and canyons.

The largest of the volcanoes has been named Olympus Mons. Figure an extra oxygen atom. It seems likely that hydrogen peroxide is present in the soil and that this chemical reacted with the carbon to produce carbon dioxide, which was then detected by the radiation monitors. Once the reaction used up the hydrogen peroxide from the Martian soil, the reaction stopped and no more carbon was released. The reason that the control sample did not show the activity was that the heat broke down the hydrogen peroxide before it was exposed to the nutrient containing carbon.

Another experiment aboard Viking was a test for respiration. Here, the soil sample was put in a container with inert gases (those which don't react chemically) and a nutrient was added. Finally, the gas was examined for changes in its chemical composition. If the living organism released any gases, they would be detected. When the experiment was performed, no more new gases appeared than would be expected from normal chemical reactions.

A common procedure in laboratories on Earth is to use a *mass spectrometer* to search for very small quantities of given chemicals in a sample. The Viking landers contained mass spectrometers capable of finding organic molecules even if they made up only a few billionths of the sample. All life on Earth contains these very large molecules that use carbon as their foundation. No such molecules were found.

We have learned a great deal about the surface of Mars since the original Vikings landed there. This knowledge could be used to improve the experiments on a future mission and avoid the confusion created by the hydrogen peroxide reactions. Is there life on Mars? Before we landed, most scientists doubted it. The experiment was judged worth trying, however, for if we had found life, the knowledge we could have gained by examining a life form different from our own would have been tremendous.

Figure T8–2. *Viking I's sampler scoop is at the end of the arm extending from right.*

The Vikings did not find life, but on the other hand they did not show absolutely that life does not exist on Mars. We still do not know the answer to the question, but we must admit that the chances appear slimmer now than they were before the Viking experiments.

8–30 compares it to two of the largest mountains on Earth. Its height of 15 miles is twice that of our largest mountain and its base would cover the entire state of Arizona. Figure 8–31 is a photo of the mountain and its surroundings. Around its base are gigantic cliffs. What a sight this mountain must be from the surface!

The base of Olympus Mons has a diameter of 400 miles!

Figure 8–30. *Olympus Mons compared to Mount Everest and Hawaii.*

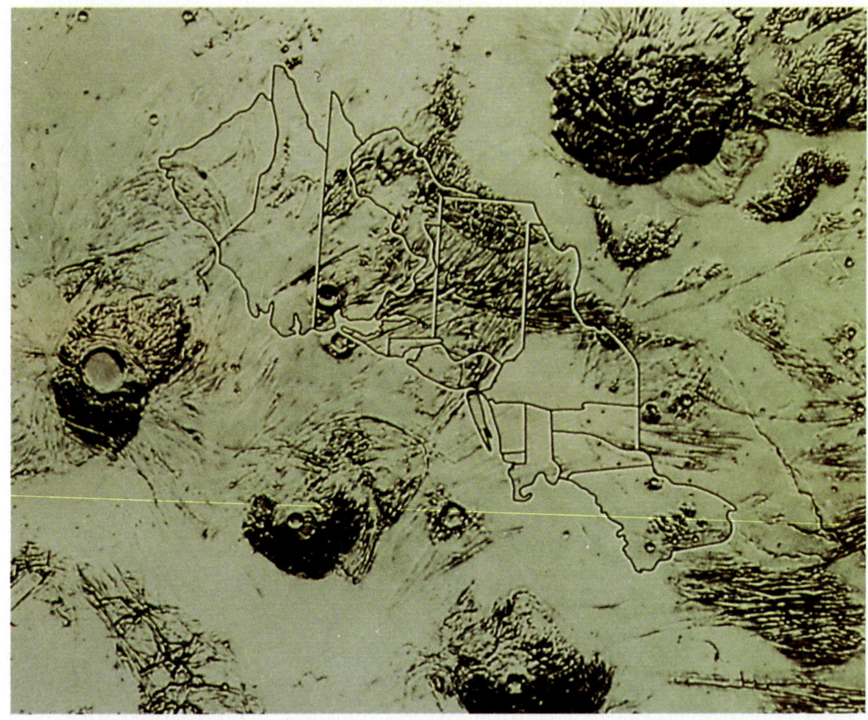

Figure 8–31. *This is a photo of Olympus Mons and its three companion volcanoes, with northeastern U.S. overlaid to show the comparative sizes.*

Venus also has at least one mountain larger than any Earthly mountain. Why do these planets have such tall volcanoes? On all three planets, volcanoes are formed above hot spots that lie deep within the planets. From time to time, material wells up from these hot spots, spilling lava out and building up the volcanoes. The reason that Mars and Venus have larger mountains than Earth has to do with motions of the crusts of the planets. On Earth, the crust is divided into plates that move across the underlying material. This means that the crust moves across a given hot spot and the volcanic action slowly shifts along the surface. (This is why we find a number of Earthly volcanoes in a straight line, with only one being active.) Apparently, the crusts of Venus and Mars are not as mobile as Earth's, and the same part of the crust stays above a particular hot spot. This causes the volcano to grow higher and higher.

Stretching away from the area of Olympus Mons is a canyon that more than matches the volcano. If placed on Earth, *Valles Marineris* (named for the Mariner spacecraft) would stretch all the way across the United States, as shown

(a)

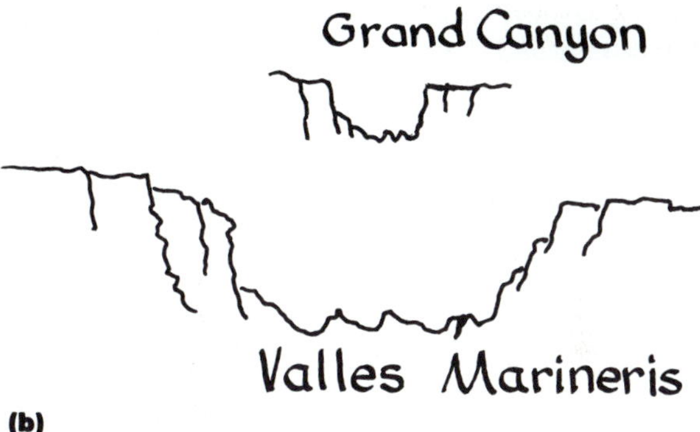

(b)

Figure 8–32. *(a) Valles Marineris would stretch nearly across the U.S. (b) In width and depth, also, it dwarfs the Grand Canyon.*

in Figure 8–32(a). Its comparison to the Grand Canyon is shown in part (b) of the figure.

On the seventh anniversary of humans' first landing on the Moon, Viking I landed on a cratered plain of Mars and took the first closeup photo of the surface (Figure 8–33). Figure 8–34 shows the landscape near Viking I in approximately true color. The red color we see from Earth is real!

We'll see the reason for the red color later.

The Viking landers confirmed something that had been seen by Mariner 9: It appears that running water was once common on Mars. Figure 8–35 shows dry riverbeds that look very similar to the arroyos found in our desert southwest. Arroyos are formed on Earth by infrequent flows of large amounts of water from rainstorms. At present, there can be no rainfall on Mars, however, because atmospheric pressure on Mars is much too low for liquid water to exist. The Martian arroyos are so similar to those found on Earth that we conclude that at one time there must have been liquid water flowing on Mars. Where would the water have gone? The answer lies, at least partly, in those white polar caps.

The polar caps of Mars consist of two parts: a water ice base that is covered during the winter by frozen carbon dioxide. In summer the carbon dioxide evaporates, leaving behind the ice. The temperature does not rise high enough on Mars to melt that ice. If it did, there would only be enough water to cover the surface to a depth of less than 0.1 millimeter; much, much less than the amount of water on Earth.

Other water seems to be hidden in permafrost below the surface of Mars. We have no way at present of knowing how much water is there and whether

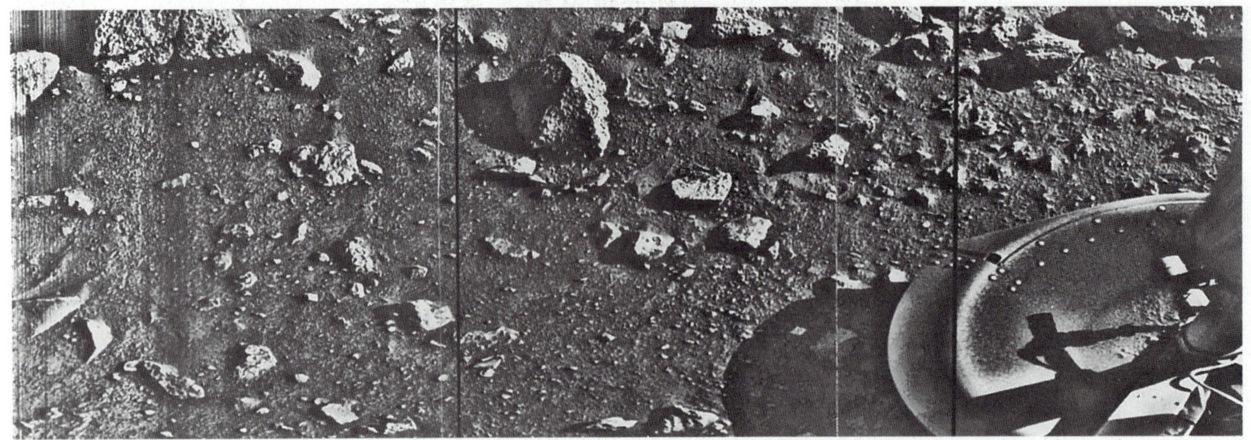

Figure 8–33. *The first photo taken from the surface of Mars (taken just minutes after landing) shows the footpad of Viking I. The large rock in the center is about four inches across.*

Figure 8–34. *The horizon is about 2 miles away in this photo. Color photos were made by combining three photos, each made with a different color filter on the camera.*

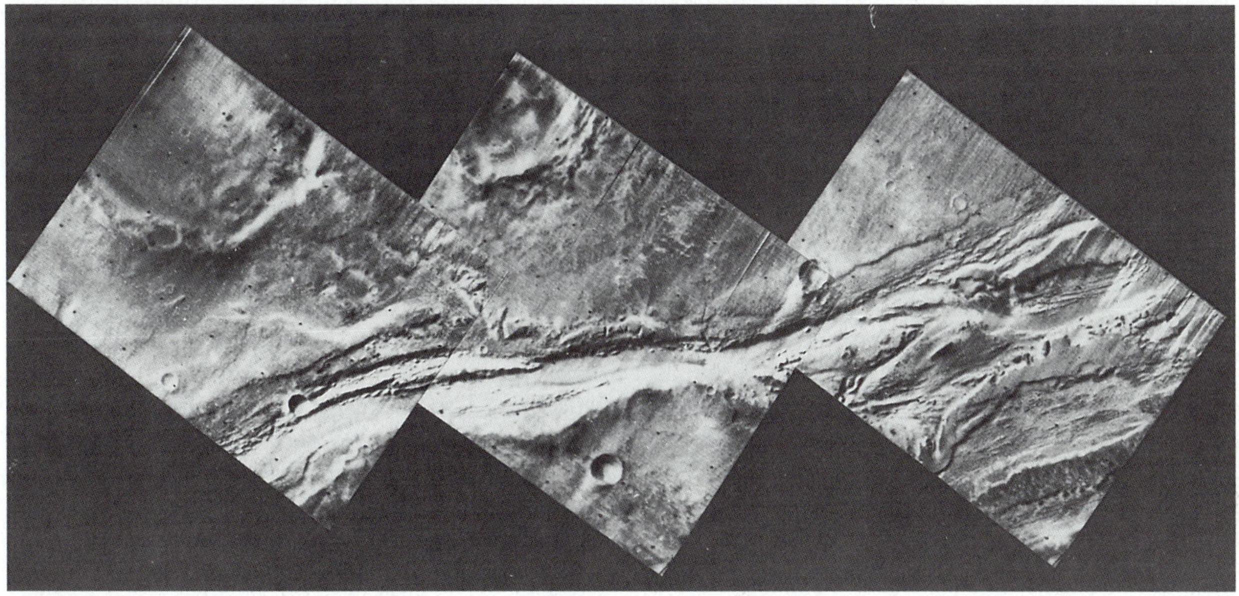

Figure 8–35. *What appear to be dry riverbeds on Mars lead us to believe that water once flowed on the surface.*

this water and the water in the polar caps is enough to have provided the moisture for a warmer, more hospitable Mars at some time in its history. An alternative hypothesis is that the riverbeds were formed in brief, cataclysmic periods when ice was melted by meteorite impacts or volcanic heating.

Surface Conditions

Near the Martian equator, noontime temperatures reach as high as 20 degrees C, a comfortable temperature for humans. At night, however, the temperature at the same location might drop to − 140 degrees C. The reason for this extreme difference in temperature is that the atmosphere is so thin. As we saw in the case of Venus and Earth, a planet's atmosphere shields it from the Sun during the day and serves as a blanket at night by reflecting back some infrared radiation toward the surface. The amount of shielding and blanketing is determined by the amount and type of atmosphere. The atmosphere of Mars is 95 percent carbon dioxide and one might suppose that Mars would have a greenhouse effect like Venus does. The low atmospheric pressure at the surface of Mars means that there is simply too little atmosphere of any type to significantly moderate the temperature on Mars.

20 degrees C equals 68 degrees F and − 140 degrees C equals − 220 degrees F.

The escape velocity from Mars is 5 km/s, less than half of Earth's escape velocity. Even though Mars is colder than Earth, almost all of the water vapor, methane, and ammonia have escaped the atmosphere along with the less massive gases.

Refer again to Figure 8–10.

Another difference between Mars and Earth: Earth has a layer of ozone that prohibits most of the Sun's ultraviolet radiation from reaching the surface.

On Mars, there is no ozone so the Sun's intense ultraviolet radiation passes through the atmosphere and breaks up water vapor into hydrogen and oxygen. The hydrogen escapes and the oxygen enters into chemical reactions with other elements.

One of the elements with which the oxygen combines is iron. The surface of Mars is rich with iron, and when oxygen and iron react we get a compound that is usually a nuisance to us on Earth: rust. It is this rust that gives Mars its characteristic red color. The surface is rusty!

From the Viking landers we also learned what causes the seasonal color variations. The fine grains that make up the dust storms are of a lighter color than the underlying surface. In the springtime, this dust is stripped away by the wind, exposing the darker surface.

So there is no vegetation on Mars. In fact, the Viking landers found no signs of any vegetation at all. No animal life. No life even of microscopic size. The Viking landers scooped up some of the Martian soil to analyze its chemical composition and to look for signs of microbial life. They found no organic chemicals whatsoever in its soil, meaning that they found no evidence for life ever existing on Mars. It is important to see that this is not the same as saying that they found evidence that life never existed on Mars. No current evidence for life was found. Perhaps our experiments were too limited in what they were searching for, or perhaps we simply need to look in other locations. We are forced to conclude, however, that there is very little chance that life now exists on Mars.

Today we have a far different idea of the red planet than did people of only a few decades ago. We have learned much of Mars's exciting terrain, with gigantic volcanos, long and deep canyons, dry riverbeds, and planet-wide dust storms.

We learned enough from the Viking missions to make further missions even more useful. This is the way things work in science. Each good experiment not only answers some of the questions being asked, but it indicates what questions should be asked the next time.

WHY EXPLORE?

We have learned more about our planetary neighbors in the last quarter century than was learned in all of our previous history. But there are those who question the value of using our resources to send spacecraft to the planets. There are two answers to this.

First, seeking knowledge about our universe is one of the things that separates humans from any other known creature. It is our nature to explore, just as it is our nature to enjoy music and art. If we never achieve one practical benefit from planetary exploration, such exploration would be valuable simply because seeking new knowledge is one of our highest aspirations. The amount of money that should be spent on such endeavors might be a point for argument, but when we compare the total amount of money spent on the basic sciences to what is spent on warfare, we see that the money spent on science is trivial.

Second, there are many practical benefits of planetary exploration. Few would argue that knowledge of Earth is without practical benefit, but many people do not realize that a study of other planets is a valuable source of knowledge about our Earth. Without a study of other planets, we would be severely limited in what we would know about our own planet. For example, it is by learning more about the greenhouse effect on Venus and the lack of such an effect on Mars that we will find out how serious is the threat of a runaway greenhouse effect here on Earth. We cannot afford to experiment with our planet, but an examination of other planets provides us with just such an experiment. By studying weather systems of other planets, we learn about our own. Was there once life on Mars? If so, why does it not seem to be there now? Could the same thing happen on Earth? Surely such knowledge is of practical benefit.

CONCLUSION

Throughout history, people have wondered about the planets. With the advent of telescopes, particularly the larger ones, we began to learn something about their surfaces. It was the planetary visits of the space age, however, that made us think of these planets as *places*, rather than as celestial objects. We now have detailed maps of their surfaces, and landers have invaded Venus and Mars to begin exploration.

As we learn more about the other terrestrial planets we are finding that in many ways they are similar, for they share a common history of formation, but we also find major differences. Some of the differences exist because the planets differ in size and mass. These factors determine whether a planet will retain an atmosphere, and the atmosphere of a planet determines how infalling meteorites affect its surface. Other differences, to be discussed further when we discuss theories of the formation of the solar system, occur simply because of the planets' different distances from the Sun.

In the next chapter, we move out to the Jovian planets. Each of these giants has its own story to tell. Each forms another piece of the puzzle of how the solar system—including the inhabitants of its third planet—came to be.

RECALL QUESTIONS

1. Which is the smallest of the terrestrial planets?
2. Which planet is closest to the Earth in size? In rotation period?
3. Why is the surface of Mercury difficult to see from Earth?
4. What is unusual about the revolution and rotation rates of Mercury and why has this occurred?
5. How was the most accurate value for Mercury's mass obtained?
6. Explain why Mercury's surface experiences such great extremes of temperature.
7. In what ways is Venus the sister of Earth? In what ways is it different?
8. What is the name of the series of space probes sent by the U.S. to Venus? By the USSR? Which nation has landed probes on the surface of Venus? On Mars?
9. What is the greenhouse effect? Why is the effect not an exact analogy to a greenhouse on Earth?
10. Describe the surfaces of Mercury, Venus, and Mars.
11. How does the tilt of Mars's axis compare to Earth's?
12. What is meant by a planet being in opposition?

13. Of what are Mars's polar caps composed?
14. Why can't liquid water exist on Mars? What is the evidence that liquid water once flowed on Mars?
15. How do the daily temperature changes on Mars compare to those on Earth? Why does this difference exist?
16. Explain why large celestial objects tend to be spherical and how some small ones avoid that fate.

QUESTIONS TO PONDER

1. What would be the temperature effect on a planet if its atmosphere reflected most of the visible light striking it but let infrared radiation pass through?
2. Explain why we can say that Viking did not prove that life does not exist on Mars.
3. Why does the inverse square law of radiation only apply when the source of light is small compared to the distance from it? (Hint: Imagine the change in illumination caused by standing three feet from an entire wall of light bulbs and then moving twice as far from the wall.)
4. A planet that rotates in the same direction in which it orbits has a longer solar day than sidereal day. Give one example of this and explain why it occurs.
5. At this time, the U.S. has one more mission planned to Mars during this century; an orbiter called the Mars Geoscience/Climatology Orbiter, or MGCO, that is scheduled for launch in 1992. Report on the status of this mission.
6. Mars has no ozone layer protecting its surface from ultraviolet radiation. Our industrial society may be destroying the ozone layer here on Earth. Report on the latest research concerning depletion of the Earth's ozone layer.
7. Report on the status of (or results of) the Magellan mission to Venus.
8. It has been suggested that we may be able to "terraform" Venus and/or Mars so that they are inhabitable by humans. Assuming that we have (or will have) the technology to do so, what do you think of the idea?

CALCULATIONS

1. Venus has a diameter of 12,000 kilometers and a mass of 4.87×10^{24} kilograms. Calculate its volume and average density. How does its density compare to that of water?
2. If a planet has a diameter one-third of Earth's, how does its surface area compare to Earth's? Its volume?
3. Jupiter's diameter is 11 times Earth's diameter. Compare the volume of Jupiter to the volume of Earth.
4. Show that a density of 1 gram/cm^3 corresponds to 1000 km/m^3.

CHAPTER 9

A painting depicting the Galileo spacecraft over one of Jupiter's moons, Io.

In 1979 the last Voyager spacecraft passed by Jupiter. The next major exploration of Jupiter and its moons is scheduled for launch in November 1989. The spacecraft, called Galileo, will follow an interesting path to Jupiter. Instead of being launched outward, *away from the Sun, it will begin its trip by moving toward the Sun and it will pass by Venus in February 1990. Its path is shown in Figure 9–18. The gravitational pull of Venus will cause it to increase its speed. Then, after a year in space, it will visit its next planet—Earth—passing by at an altitude of 3600 kilometers and again increasing its speed. Next it will move into the asteroid belt, where it will spend a year (and perhaps provide us with our first close look at an asteroid). It still will not have enough speed to get to Jupiter and so it will loop back in toward the Sun, passing by the Earth once again in December 1992, three years after launch. This Earth flyby, however, will give it the speed it needs to reach Jupiter; an event now scheduled for December 7, 1995.*

As Galileo nears Jupiter, it will launch a probe into the planet's atmosphere. Falling quickly toward Jupiter, the probe is expected to relay data back to its mother ship for about an hour before it is destroyed in the dense Jovian atmosphere. The Galileo orbiter will then continue for two years to orbit among Jupiter's moons, sending us information on this miniature system and the planet at its center.

The Jovian Planets

WE NOW TURN OUR attention to the planets of the outer solar system—the Jovian planets. (We will save Pluto for the next chapter.) Although we found that Mercury, Venus, and Mars are strange worlds for us Earthlings, we will see that the outer planets are even stranger. As we continue our journey out from the Sun we will take these planets in order. There is still a lot we do not know about our neighboring worlds and you will see that as we get farther and farther from our home base, we know less and less about the planets there.

JUPITER

The size of a planet is not apparent to the naked eye but long before the invention of the telescope, Copernicus had deduced that Jupiter was larger than Venus in spite of the fact that Venus at its brightest is brighter than Jupiter. He concluded this from the two planets' relative brightnesses and distances; figuring that since Jupiter is so much farther away, it would have to be much larger to appear so bright if it shined only by reflected light. Galileo observed the planets' angular sizes with a telescope and was able to determine that Jupiter was indeed larger, for he could use their angular sizes and relative distances to calculate their relative sizes. Today we know that Jupiter is the largest object in the solar system besides the Sun. It is appropriate, then, that it is named after the king of the Roman gods.

Copernicus simply assumed that each planet reflects the same percentage of light that strikes it.

Jupiter Seen from Earth

Jupiter's mass is calculated by observing the radii of the orbits of its moons and their periods of revolution, as we saw in Chapter 7. And Jupiter is massive! It has more than twice the combined mass of all the other planets, their moons, and the asteroids. When we compare Jupiter to the Earth we find that Jupiter is 318 times more massive than our little planet.

In size, Jupiter is even more remarkable. Figure 9–1 shows the relative sizes of Jupiter and the Earth. Jupiter's diameter is about 11 times that of the Earth, making its volume about 1300 times that of the Earth.

If Jupiter's mass is 318 times the Earth's and its volume is 1300 times the Earth's, then its density is $^{318}/_{1300}$, or only about one-fourth as much as the Earth! While the Earth has an average density about twice as great as a common rock, Jupiter's density is just slightly more than that of water (1.3 times as dense as water). This means that Jupiter must be composed of more light elements than is the Earth.

You might have a tendency to picture Jupiter as a lumbering, slow-moving giant. Indeed, it is 5.2 astronomical units from the Sun and takes nearly 12 years to circle the Sun. It moves around in its orbit at about 500 miles a minute, slightly less than half the Earth's orbital speed. Its rotation is not slow, however. This gigantic planet spins on its axis once every 9 hours and 50 minutes. This is the first Jovian planet we have studied, and we will find that fast rotation is common for Jovians. They have rotational velocities much greater than the terrestrial planets.

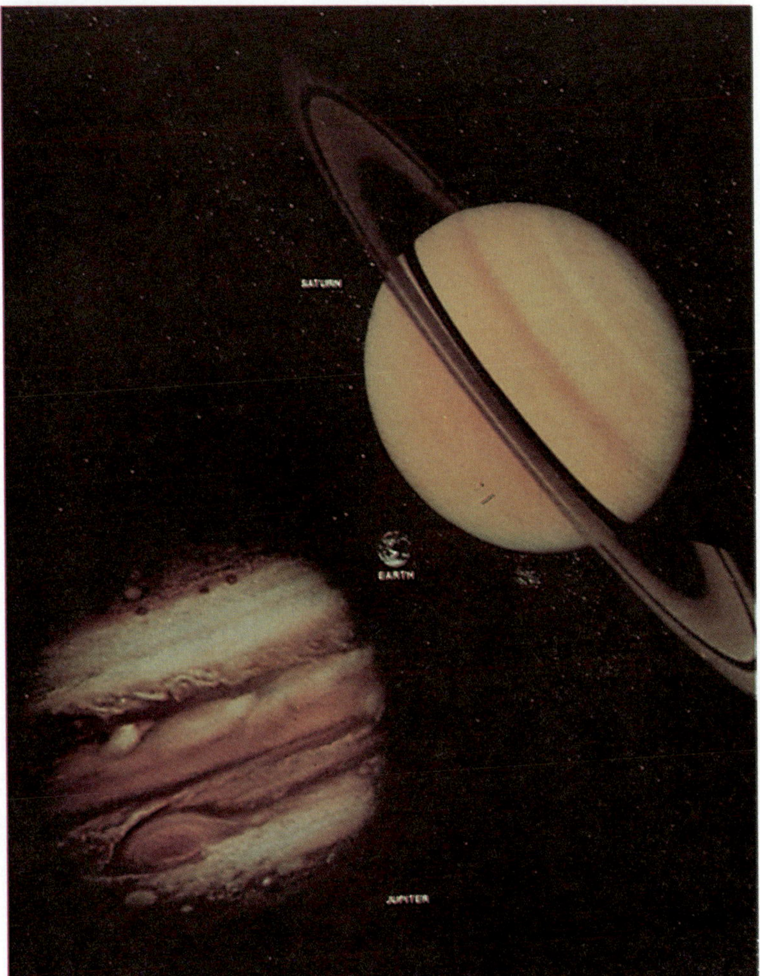

Figure 9–1. *Jupiter and Saturn dwarf our Earth.*

Things are not quite this simple, however. Through even a small telescope we can see that Jupiter has dark and light belts parallel to its equator. Figure 9–2 is a photo taken with a large Earth-based telescope and shows obvious parallel bands around the planet. Observations of these show that the ones near the equator rotate slightly faster than those nearer the poles. The band at Jupiter's equator has a rotation period of 9 hours, 50 minutes. Bands closest to the poles complete one rotation in 9 hours, 56 minutes. This "spreading" of the rotation, or ***differential rotation,*** indicates that the visible surface of Jupiter is not solid.

Differential rotation: Rotation of an object in which different parts have different periods of rotation.

Another thing about Jupiter that can be observed fairly easily is that it is not spherical—it is "out of round." If you use a ruler to measure both the polar diameter and the equatorial diameter of Jupiter in Figure 9–2, you will find that the equatorial diameter is definitely greater. In fact, it is 6 percent greater than the polar diameter. We saw in the last chapter that Mars also is nonspherical but not nearly to this degree. These planets are said to be ***oblate,*** and the reason lies in their rotation.

Oblateness: A measure of the "flatness" of a planet, calculated by dividing the difference between the largest and smallest diameter by the largest diameter. See Figure 9–3.

$$\text{obl} = \frac{(d_{\text{large}} - d_{\text{small}})}{d_{\text{large}}}$$

Figure 9–2. *The belts around Jupiter are easily visible in this photo taken from Earth. The red spot is called—imaginatively—"the red spot."*

Figure 9–3. *Oblateness is calculated by subtracting the length of the short diameter from the long and the dividing by the long diameter.*

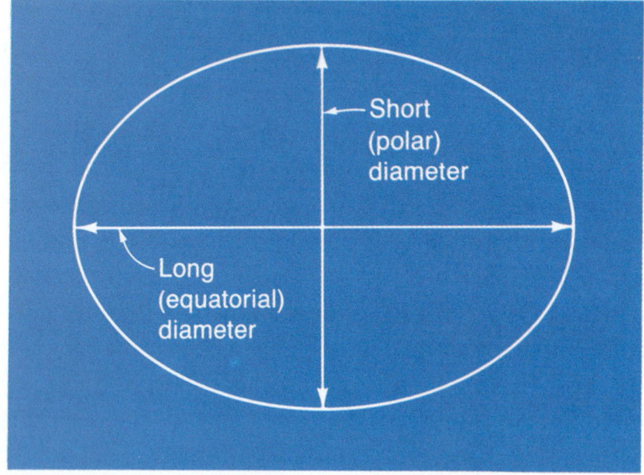

Figure 9–4 shows an arrangement of circular, flexible metal pieces on a rod that can be made to rotate. Figure 9–5(a) is a time exposure of the device rotating slowly and part (b) shows it spinning faster. Notice that when it spins fast, the metal bands flex outward, giving the object an oblate appearance. This distortion is similar to what occurs in the case of planets. It is due to the fact

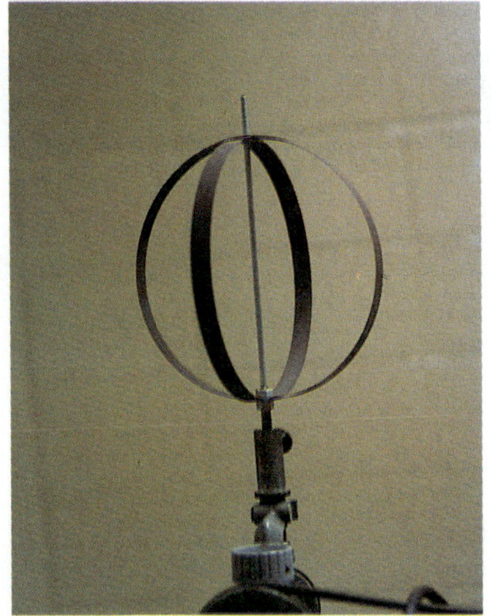

Figure 9–4. *This device consists of two hoops mounted on a rod that can be spun at a rapid rate.*

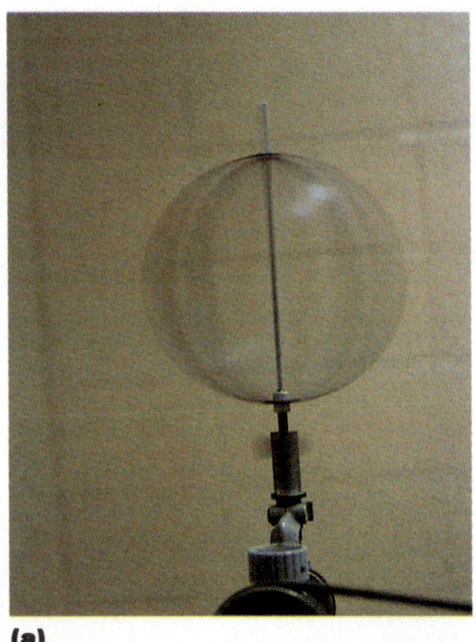

(a)

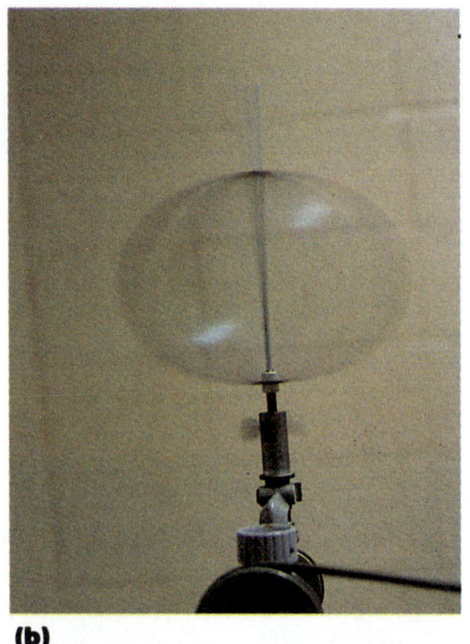

(b)

Figure 9–5. *(a) Here, the device of the last figure is shown in slow rotation. (b) When the hoops are made to rotate faster they distort, resulting in the oblate shape seen here.*

that the object distorts as it overcomes the inertial tendency of its surface to continue in a straight line rather than move in a circle around the axis of spin.

Jupiter from Space

In December 1973 Pioneer 10 flew to within 81,000 miles of the surface of Jupiter. Then in December 1974 its twin, Pioneer 11, came within 30,000 miles of the surface. These two craft sent back numerous black and white photos of the surface. In addition, the craft contained instruments to detect charged particles, radiation from the planet, and the magnetic field of Jupiter. After passing Jupiter, Pioneer 11 continued on to Saturn; also giving us our first close view of that planet. Both of these spacecraft are now on their way out of the solar system, to glide endlessly through space.

By using color filters while taking black and white photos, color images were reconstructed.

The knowledge gained from the Pioneer missions helped in the design of the next generation of spacecraft sent to look at the outer planets. Most of the photos of Jupiter you see today were taken by the two Voyager missions to Jupiter. Voyager 1 and 2 flew by Jupiter in 1979 and sent back some 33,000 photos, as well as much other data concerning Jupiter and its moons. Figure 9–6 shows the entire planet photographed by Voyager.

The Red Spot

It was pointed out in the last chapter that if we knew more about weather on Venus, we would know more about weather on our own planet. Jupiter may provide an even better study of weather systems than does Venus, for weather

ET Life III—Letters to Extraterrestrials

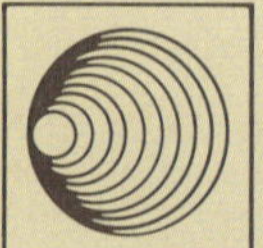

Pioneer Plaques

In late 1971 it was pointed out to Dr. Carl Sagan that the trajectories of the Pioneer 10 and Pioneer 11 spacecraft would take them out of the solar system and that it might be possible to include on them a message to extraterrestrials. Sagan called NASA authorities and within three weeks got approval to put a plaque on both Pioneer 10, which was scheduled to be launched the following March; and Pioneer 11, which was to be launched a year later. Carl and his wife Linda, along with Dr. Frank Drake, designed the message, which was etched on a 6 × 9-inch gold-anodized aluminum plaque. Figure T9–1 shows their design.

Some of the message on the plaque is easily recognizable (to humans, at least.) The umbrella-shaped object behind the man and woman is an outline of the Pioneer spacecraft drawn to scale with the people. Thus the finder can determine the size of a human. (The man on this scale is 5 feet 9½ inches tall.)

The two circles at upper left represent hydrogen atoms emitting radiation of a particular wavelength. The length of this wave is the basis for the binary numbers at far right. These numbers serve as a second way of showing the height of the woman, giving her height in multiplies of that wavelength. In addition, the binary numbers below the sketch of the solar system give its dimensions. The Pioneer spacecraft is shown leaving the third planet and swinging by the fifth and sixth.

The spidery-looking feature at left shows the directions from the solar system to pulsars, pulsating sources of waves we detect in deep space (and which we will study in a later chapter). The binary numbers along the lines represent the period of the pulses expressed in terms of the period of the wave from hydrogen. By analyzing the directions and periods of the pulsars, the finders could tell where our solar system is located in the Milky Way. In addition, since pulsars slow down with time, they could tell when the craft was launched.

Could extraterrestrials interpret the message? Curiously enough, since hydrogen atoms and pulsars are

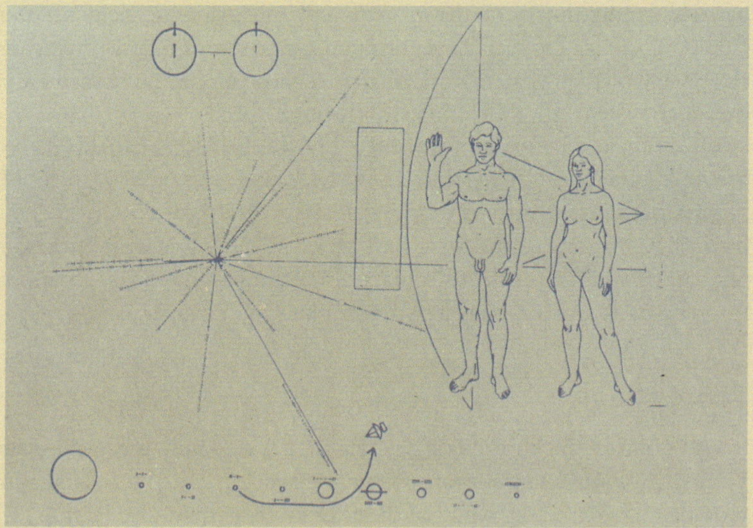

Figure T9–1. *This plaque aboard Pioneer 10 and 11 carried a message to extraterrestrials—and to Earthlings.*

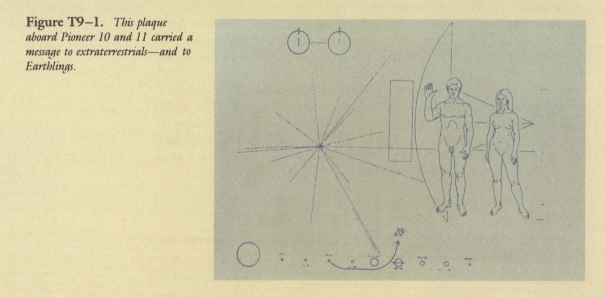

common to all creatures of the universe, it is thought that they would be more likely to interpret the pulsar sketch than the human figures, which we recognize immediately.

When news of this plaque was released, there were outcries against sending such "pornography" into space. Real Carl Sagan's *Cosmic Connection, An Extraterrestial Perspective* (New York, NY: Doubleday, 1980) for an interesting account of reactions to the plaque. Sagan says that the plaque was as much a message to *us* as it was to extraterrestrials.

Voyager Records

In 1977, Voyagers 1 and 2 were launched into space to rendevous with Jupiter and Saturn. Plans were later adjusted to allow Voyager 2 to continue to Uranus and Neptune. Finally, both spacecraft would pass out of the solar system to drift through interstellar space. With more time available to plan a message, a group of people that included the three who prepared the Pioneer plaque plus Ann Druyan, Timothy Ferris, and Jon Lomberg, worked to put a much more complete message aboard Voyagers 1 and 2.

The messages on Voyager are contained on two copper records, designed to be played at 16⅔ revolutions per minute. Instructions for playing the records are included in pictures on the cover. On the records are 90 minutes of the world's greatest music, 118 pictures, and greetings in nearly 60 languages (including one nonhuman message from a humpback whale).

The music on the Voyager includes parts of compositions from Bach (*Brandenburg Concerto No. 2* and *The Well-Tempered Clavier*), Beethoven (Symphony No. 5, String Quartet No. 13 in B-flat), and Mozart (*The Magic Flute*). In addition it includes Chuck Berry's "Johnny B. Goode," Australian Aborigine songs, a Peruvian wedding song, and numerous other selections from the cultures of our world.

The records contain pictures as well as sound. The pictures include many photos of scenes of nature here on Earth, of the Earth from space, of the Earth as changed by humans, of people in various activities, and of the biology of humans. Figure T9–2 shows a few of them. It is interesting to note that because of the criticism of the plaque sent aboard Pioneer, NASA refused to allow pictures of naked humans to be included. Enough information is included to make it clear that we are not egg-layers, but would an extraterrestrial be confused by the lack of detail in this regard? The record does list the senators and members of Congress who were on the important committees involved with space exploration. Perhaps that will tell extraterrestrials something about us.

Will the Plaques Be Found?

The messages were not put aboard the Pioneer and Voyager craft with high hopes of them ever being found. They were sent in much the same manner that a child puts a message in a bottle and throws it into the sea. In fact, the space messages have even less chance than the child's bottle of being found. Even if the galaxy abounds in intelligent life, the emptiness of space and the slow speed of the spacecraft make the chances slim indeed. The soonest any of the craft will pass within two light years of a star is 40,000 years!

The messages are really a symbol of hope, a shout that "we are here." As stated in *Murmurs of Earth*—which tells the story of the Voyager records—after the Earth is a charred cinder the messages will continue to travel through space, "preserving a murmur of an ancient civilization that once flourished—perhaps before moving on to greater deeds and other worlds—on the distant planet Earth."*

---

*From *Murmurs of Earth* by Carl Sagan and F. Drake, A. Druyan, J. Ferris, J. Lomberg, L. Sagan (New York, NY: Ballantine Books, 1978).

(Continues)

CLOSE UP

Figure T9–2. *A few of the photos carried by the records aboard the Voyager spacecraft.*

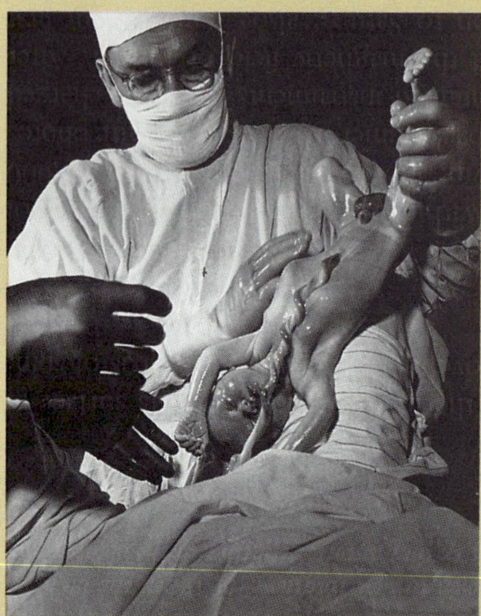

patterns in the upper atmosphere of Jupiter are far removed from any solid surface Jupiter may have and therefore are almost unaffected by complications produced by surface irregularities. Jupiter's weather system should therefore be simpler in nature and allow us to learn what happens when surface features are not a major factor. (The practice of examining a simple system in order to form a working hypothesis about a complicated one is common in science.)

Figure 9–6. *This photo of Jupiter was taken from Voyager 1 at a distance of 40 million kilometers from the planet.*

The fact that surface features have little effect on Jupiter's upper atmosphere is probably what allows Jupiter's weather patterns to last for such long periods. The prime example of this is the giant red spot on Jupiter (seen on Jupiter in Figure 9–6). This spot was seen as early as the mid–1600s, so it has lasted for more than 300 years. The spot is about 50,000 kilometers (30,000 miles) in length and nearly 15,000 kilometers across. When we realize that the Earth is about 13,000 kilometers in diameter, we can appreciate the immensity of this feature. From time to time over the centuries, the red spot has faded in intensity but it has never disappeared completely.

Data from Jupiter indicates that the red spot is a high-pressure storm system similar to the much smaller Earthly systems reported by your local weather reporter. High-pressure systems in Earth's northern hemisphere rotate in a clockwise direction and in the southern hemisphere they rotate in the opposite direction. The red spot is in Jupiter's southern hemisphere and indeed it rotates counterclockwise—with a period of about six days. Figure 9–7 is a close-up of part of the spot; we can see the swirling currents at its edges. In addition, the figure shows the size of Earth relative to the red spot.

The spot is actually more gray than red and is difficult to see in a small telescope.

The Composition of Jupiter

The terrestrial planets are all similar to the Earth in that most of their substance is hard, rocky material like that of the Earth. Jupiter is different. It is composed of about 75 percent hydrogen and 24 percent helium, with small amounts of water, methane (CH_4), ammonia (NH_3), and a few other compounds. This composition is closer to the composition of the Sun and other stars than it is to the terrestrial planets.

This is composition by mass. Counting atoms, it would be about 90 percent hydrogen and 10 percent helium.

The colors seen in Jupiter's upper atmosphere apparently are the result of chemical reactions induced by sunlight and/or by lightning in its atmosphere. The light areas around the planet are high-pressure areas where gas is rising from inside Jupiter and the dark belts are low-pressure regions where the gas descends. The rapid rotation of Jupiter moves the regions around the planet so that it has a belted appearance.

As might be expected from the fact that the planet has differential rotation, measurements made by Voyager show that neighboring bands of Jupiter's atmosphere are moving at different speeds around the planet. As a result of these different speeds, there is considerable swirling at the boundaries between bands. The stormy nature of these swirls is obvious in the view of the belts shown in Figure 9–8.

The gaseous atmosphere of Jupiter is fairly thin; only a few thousand miles in depth. Naturally, as you get lower in the atmosphere, the pressure is greater (because the gas at lower elevation supports the gas above). On Jupiter, the pressure soon gets so great that the hydrogen no longer acts as a gas but starts

Figure 9–7. *Jupiter's red spot (at right) is seen with the great turbulence around it. Earth is superimposed at the same scale to help us appreciate the size of the spot.*

acting like a liquid. The molecules are forced so close together that the hydrogen acquires a liquid nature; it becomes liquid hydrogen. On the Earth we have a gaseous atmosphere above a liquid ocean. On Jupiter, however, there is no such boundary between the liquid and the gas. At the top of the atmosphere, we would classify the hydrogen as a gas. Much lower, we would call it a liquid. In between, the gas is more and more dense as we get deeper and deeper in the atmosphere. If we could travel inward from the gaseous region, we would find ourselves in thicker and thicker gas until we finally decided that we were no longer in a gas but were in a liquid.

About 10,000 miles below the clouds there is another change in Jupiter's composition, caused by the increasingly greater pressures and higher temperatures. At this depth, temperatures are so high that collisions between atoms result in the atoms being broken into separate protons and electrons. We call this state *liquid metallic hydrogen*. Hydrogen in this form cannot be produced

Figure 9–8. *This photo of the belts of Jupiter shows the swirling that results from the different speeds of adjacent belts. Io, one of Jupiter's moons, is seen in front of the planet.*

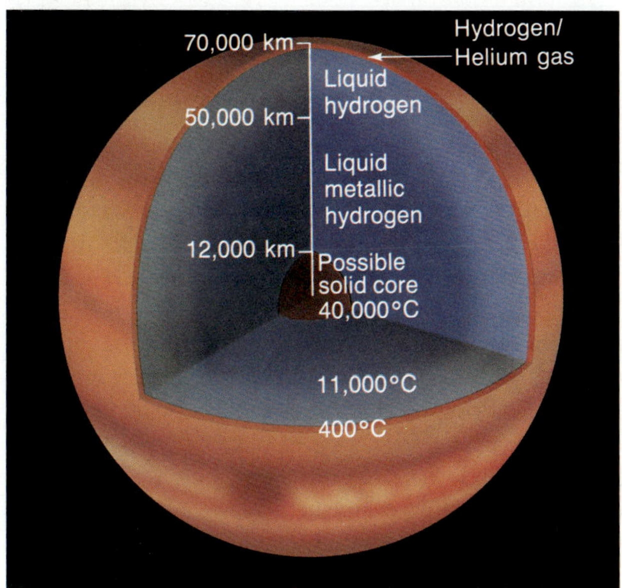

Figure 9–9. *The interior of Jupiter largely consists of liquid metallic hydrogen with (probably) a solid core.*

on Earth because we cannot create such a combination of high temperature and high pressure, but it is predicted by nuclear theories. Liquid metallic hydrogen has properties far different from normal hydrogen, including the fact that it is a good conductor of electricity. (This is why it is termed *metallic.*) As indicated by Figure 9–9, most of the planet is made up of this state of matter. Our knowledge of the properties of matter under the extreme pressures that exist at

Table 9–1 The Planet Jupiter

	Value	Compared to Earth
Equatorial diameter	142,800 km	11.2
Polar diameter	134,200 km	10.6
Mass	1.899×10^{27} kg	318
Density	1.32 gm/cm^3	0.24
Surface gravity	26.4 m/s^2	2.69
Escape velocity	59.6 km/sec	5.3
Sidereal rotation period	9.842 hours	0.41
Albedo	0.34	1.1
Tilt of equator to orbital plane	3.12°	0.13
Orbit		
Semimajor axis	7.78×10^8 km	5.203
Eccentricity	0.0485	2.9
Inclination to ecliptic	1.3°	
Sidereal period	11.86 years	11.86
Moons	16	

these depths is very limited because we are unable to reach such pressures in our laboratories.

Figure 9–9 shows a rocky core at the center of Jupiter. We have no direct evidence that such a core exists but the core is hypothesized on the basis that there must have been heavy elements in the original material of which Jupiter is made. These heavy elements exist on the moons of Jupiter and there is good reason to think that they also exist on the planet. If so, they would have sunk to the center of Jupiter and formed a solid core. The size of the solid portion of the planet is unknown but it is thought to have a diameter perhaps as small as the Earth's or as much as a few times the Earth's. In any case, it makes up a very small portion of the entire planet.

Jupiter has a very strong magnetic field, ten times as strong as Earth's. We saw in the previous chapter that in order for a planet to have a magnetic field, two conditions are necessary: rotation and electrically conductive material within the planet. The terrestrial planets appear to have iron (or nickel-iron) cores that are good electrical conductors. Jupiter does not depend upon its core to produce its magnetic field, for the liquid metallic hydrogen that makes up much of the planet is responsible—along with Jupiter's fast rotation—for its strong magnetic field.

Magnetosphere: The volume of space in which the motion of charged particles is controlled by the magnetic field of the planet rather than by the solar wind.

Like the Earth's magnetic field, Jupiter's field deflects the solar wind around the planet as well as trapping some of the charged particles of the wind in belts around it. Jupiter's ***magnetosphere,*** however, is about 100 times larger than Earth's and the charged particles trapped within its radiation belts move at such

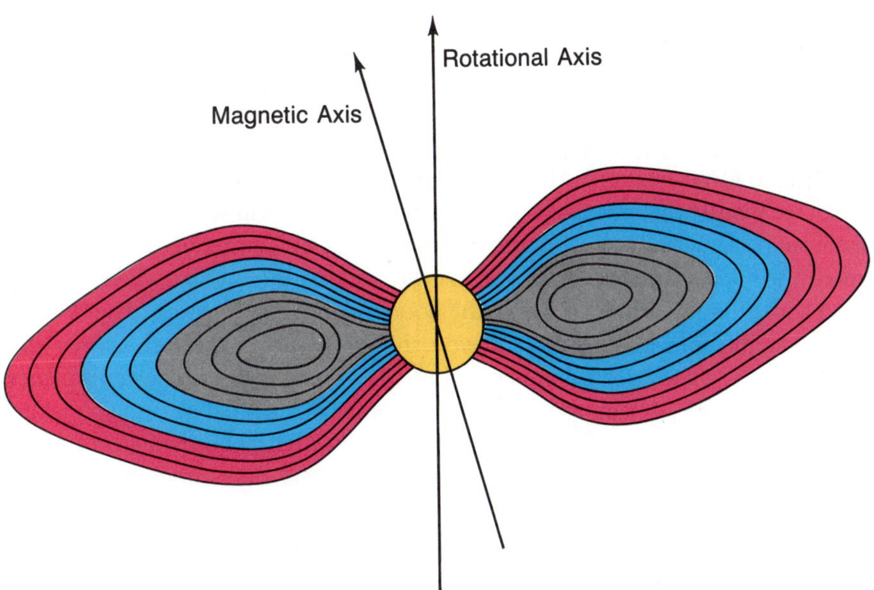

Figure 9–10. *Jupiter's radiation belts are flattened by the planet's rapid rotation. Compare them to the Van Allen radiation belts in Chapter 7.*

great speeds that they interfere with electronic circuits on spacecraft that pass through them. Figure 9–10 illustrates the radiation belts on Jupiter.

Energy from Jupiter

There remain a number of puzzles concerning Jupiter besides the composition of its interior. One puzzle results from the fact that Jupiter emits more energy than it absorbs from the Sun. This was known before the Pioneer and Voyager missions and those missions only confirmed the observations. Let's explain why this is a problem.

When we speak of solar radiation, we include—along with visible light—ultraviolet, infrared, and all other types of radiation. (See Chapter 11.)

We would expect the radiation coming from a planet to be simply the sum of the solar radiation that is reflected from the planet and the infrared radiation that is emitted as the planet re-emits the absorbed solar radiation. We don't expect a planet to have an energy source of its own. It is easy to calculate the amount of solar energy striking any planet. To do so, all we have to know is how much energy strikes a square mile here on the Earth, use the inverse square law to calculate how much strikes a square mile on the planet in question and then multiply this amount by the cross-sectional area of the planet. When the energy striking a planet is then compared to the energy reflected and emitted from that planet, we expect to find that the energy coming from the planet is the same as the energy absorbed. If it isn't, the planet must either be heating up as it gains energy, or cooling as it loses energy.

Recall that in discussing Venus, we said that the greenhouse effect results in the planet being hot enough that it emits as much energy as it absorbs.

Jupiter, though, does not behave as expected; it emits almost 2.5 times more energy than it absorbs. There are three possible ways for it to do this. One possibility is that it may have an internal energy source such as chemical reactions or a source of radioactivity within it. There is no reason to believe that there is much energy produced in Jupiter by either of these methods, but it was once hypothesized that Jupiter acts like a miniature star, with significant nuclear fusion reactions going on within it. (Nuclear fusion reactions provide the energy of the stars.) Further calculations showed that Jupiter is not massive enough and does not have sufficient internal pressures or temperatures to support fusion reactions, however.

A second possibility is that Jupiter is continuing to shrink in size. (Later, when we discuss the formation of the solar system, we will see that planets form from much larger clouds of matter.) If Jupiter is still shrinking, the source of energy would be gravity. A simple Earthly case of this occurs whenever you drop something. The object strikes the floor and heats up slightly. (The heating is very slight, and to experience it you should probably cheat and *throw* the object down to the floor a number of times.) As we will see later, when a planet is forming, material being pulled inward by gravity is heated up and so gravity serves as an energy source. This has been hypothesized as occurring on Jupiter and calculations shows that it is possible that gravity does supply some—but not all—of the extra energy from Jupiter.

Do not get the idea that Jupiter gets significantly cooler from year to year. The planet is so large that the extra energy emitted from it corresponds to an extremely small temperature change.

Finally, it is now thought that most of the extra energy from Jupiter is leftover energy from its formation. That is, when Jupiter was formed it heated up because of gravitational energy (as did all the planets) and it is still in the process of cooling. Its immense size has served to insulate its interior so that it cools slowly and this cooling continues even today.

Jupiter's Family of Moons

Jupiter's 16 moons (at last count) can be divided into two groups: Most are small objects and probably are captured asteroids like the Martian moons but the largest four (those discovered by Galileo) are far different from the others and each has a story to tell. Most of our knowledge of Jupiter's system comes from the two Voyager missions and the photos in this section of the text were taken by those missions.

Figure 9–11 shows the relative sizes of the four Galilean moons, along with the four terrrestrial planets, Earth's Moon, and Saturn's largest moon.

The volcanoes on Io might be more appropriately called geysers, for we usually associate Earthly volcanoes with mountains and the eruptions on Io do not come from mountain tops.

The Galilean moon nearest Jupiter is Io. Figure 9–12 shows its mottled surface; Io is a strange sight. Its yellow orange color is caused by the element sulfur, which covers its surface. The biggest surprise to astronomers came when Linda Moribito of NASA, while examining Voyager photos, discovered an active volcano on Io. Further examination showed that Io has many active volcanoes or geysers. There were eight or nine active volcanoes photographed by the two Voyager craft. (See Figure 9–13.) The circles and dark spots you see in the photo of Io are not impact craters; they are the results of volcanoes. In fact, the volcanoes explain the lack of impact craters, for the surface of Io constantly is being changed by the release of hot sulfur from below the surface.

We must ask what is the source of the energy for these volcanoes. In the case of geysers and volcanoes on Earth, the energy source is primarily internal heat retained from the Earth's formation and energy released by radioactive materials within the Earth. But a small object cannot retain heat nearly as long as a large one and Io would have lost any heat that resulted from its formation.

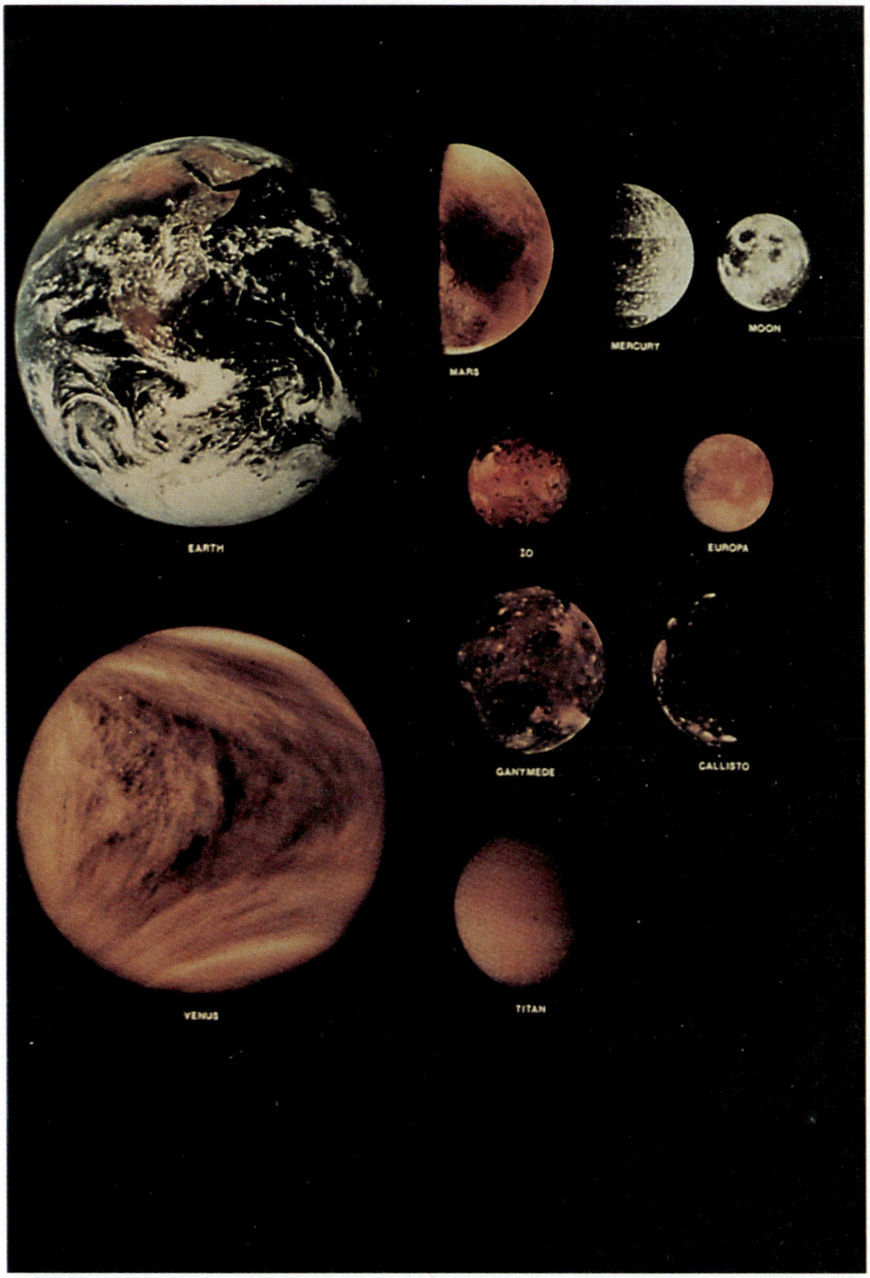

Figure 9–11. *Compare the sizes of the Galilean moons to other objects we have studied.*

Radioactivity cannot account for the tremendous volcanic activity on that moon; rather, it is tidal forces that produce the heat within Io. These tides result from the passing of Jupiter's other moons, particularly the large Galilean moons. As they move by, they are thought to pull Io's surface inward and outward by as much as 100 meters and this constant flexing of the moon results in heat from friction. We see that heat being released in the activity of the volcanoes.

An interesting example of the confirmation of a scientific prediction occurred here, because just three days prior to Voyager's flyby, S. Peale published a paper predicting that tidal heating of Io would result in vulcanism.

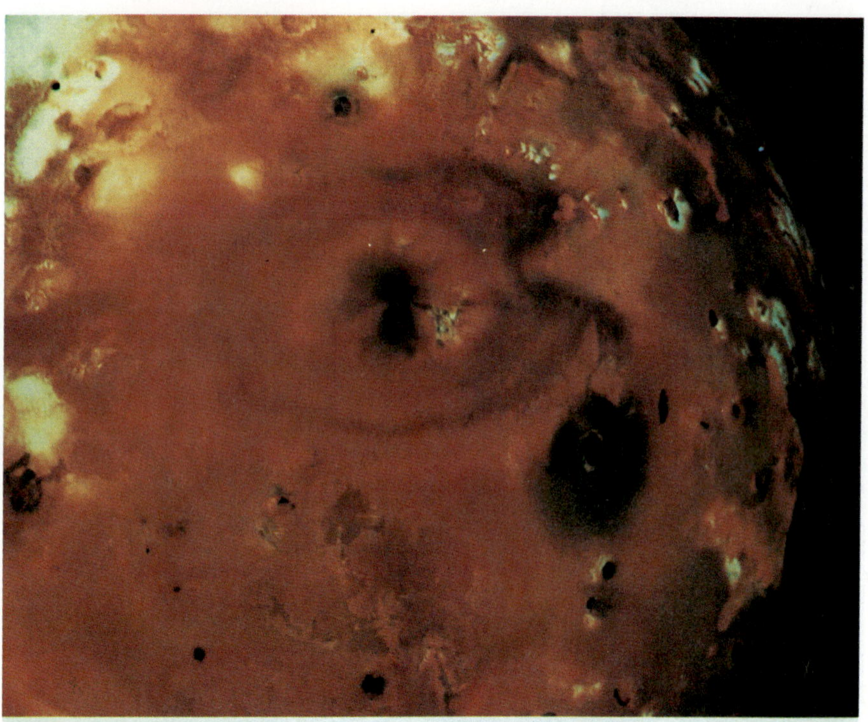

Figure 9–12. *Io, the innermost of the Galilean satellites, has a number of active volcanoes on its sulphur surface.*

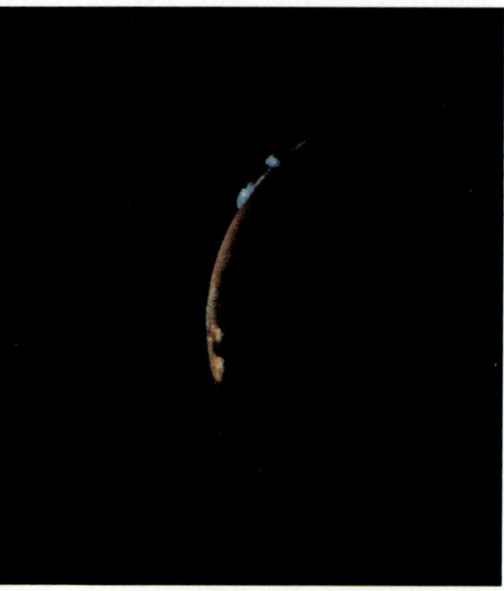

Figure 9–13. *This photo shows one of Io's volcanoes erupting at the upper left.*

Io is surrounded by a halo of sodium atoms and some of the sodium is swept away from Io by Jupiter's magnetic field, resulting in a faint ring of sodium near Io's orbit.

The mass of Io has been calculated based on its gravitational attraction for passing spacecraft and its density is calculated to be about 3.5 times that of water. This leads us to conclude that the satellite is composed mostly of rock, with a relatively shallow layer of sulfur on its surface. No water is found on Io; Io is a strange, volcanic desert.

Europa, 1.5 times as far from Jupiter as Io, presents a far different picture to us. Its surface is not sulfur, but something more familiar to us—ice. Figure 9–14 shows its cracked-billiard-ball appearance. Europa's density is slightly less than Io's but again we must conclude that it is mostly rock covered by an ocean of frozen water.

Note the lack of craters on Europa. Only three were found in Voyager photos. Again, this does not indicate that meteoroids have not struck Europa but rather that its surface is active so that craters do not remain visible for long. Although it is farther from Jupiter than is Io, Europa also experiences tidal heating. There appears to be at least one volcano on Europa, although what appears to be a volcano may simply be water spraying out through a crack in the moon's crust; forced out by pressure resulting from tidal flexing. The cracks in Europa's surface are hypothesized to be another result of this flexing.

Ganymede is the largest moon in the solar system; it is larger than the planet Mercury. On its surface (Figure 9–15) we see ice, but notice how different Ganymede is from Europa. There are craters, indicating that its surface is not

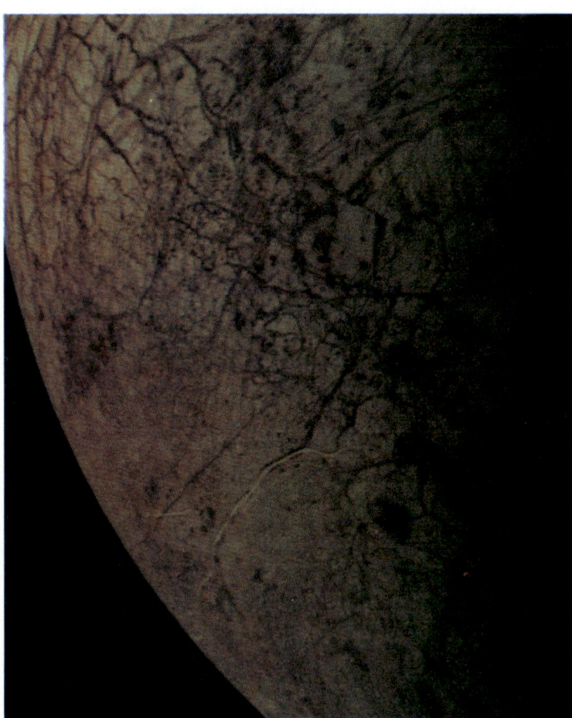

Figure 9–14. *The lack of craters on Europa indicates that the surface is active.*

as active. Its ice is also darker than Europa's. This darker color is the dust from meteorite falls that spreads across the surface of the satellite. On Europa, the surface is constantly being refreshed with water from below, causing the meteorite dust to be spread through a much deeper layer of the satellite. The less active surface of Ganymede, on the other hand, leaves the dust on the surface. Notice, though, that there are light-colored streaks on Ganymede. These are where cracks have formed and icy slush from below has welled up to fill the cracks.

The outermost of the Galilean moons is Callisto, Figure 9–16. Now we see a more familiar cratered surface. At its greater distance from Jupiter, Callisto experiences little tidal heating and has a very inactive surface. The newer craters are the whitest, showing where a meteorite impact has brought clean ice to the surface. The large white crater seen near the top of the photo is named Valhalla and is the largest impact crater known in the solar system. It has bull's eye rings around it that are cracks in the surface caused by the impact.

As one moves from the innermost Galilean moon to the outermost, patterns of change are obvious. The farther the moon is from Jupiter:

1. the less active the surface,
2. the lower the average density of the moon, and
3. the greater the proportion of water.

When we study the origin of the solar system we must look for an explanation for these patterns, for we have seen similar patterns of change among

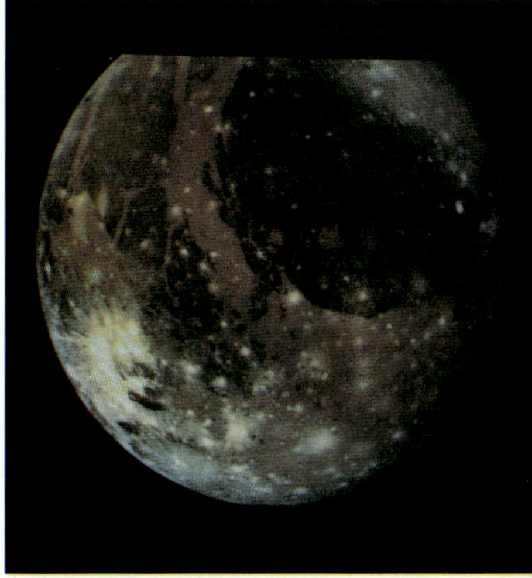

Figure 9–15. *Ganymede is larger than Mercury.*

Numbers 2 and 3 listed here are related, for since water has a lower density than most other substances on the moons, the more water on a moon, the less its density.

Figure 9–16. *Callisto is the Galilean satellite that is farthest from Jupiter and that has the least active surface.*

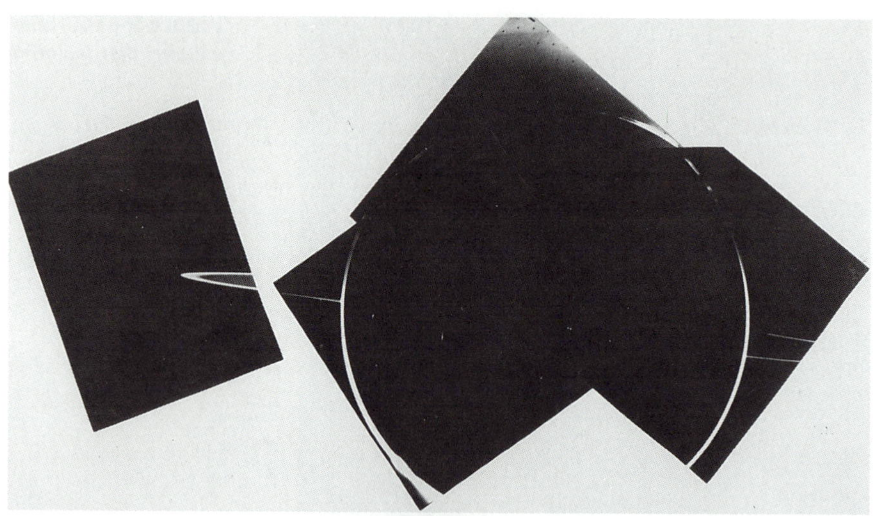

Figure 9–17. *This is a mosaic of photos taken by Voyager 2, showing a ring of Jupiter. The spacecraft was in the shadow of Jupiter, so that the planet is outlined and seen as a circle of sunlight that was scattered by the edges of Jupiter's atmosphere.*

the planets as one moves from Mercury to the outer planets. This confirms Galileo's intuition that the Jupiter system represents the solar system in miniature.

Jupiter's Rings

A surprising discovery by Voyager 1 was that Jupiter has rings. Figure 9–17 shows the rings as they were photographed by the Voyager 2 camera when it was in Jupiter's shadow. The light that reached the camera from the rings was therefore scattered forward by the material of the rings and since large chucks of matter cannot scatter light in this manner, the rings are known to be made up of very tiny particles (as small as particles of cigarette smoke). The photo indicates that the rings are also fairly close to Jupiter; they extend to only about 1.8 planetary radii from the planet's center.

Calculations indicate that the particles of Jupiter's rings could not have been there since the formation of the solar system because radiation pressure from the Sun, as well as forces from Jupiter's strong magnetic field, gradually send some particles down into the planet and others away into space. Therefore the ring is thought to be continually replenished; probably from small moonlets within it or near it. Two of these moonlets, called J15 and J16 (being the fifteenth and sixteenth of Jupiter's moons to be discovered), have been found. We will discuss planetary rings further when we discuss Saturn and its very prominent system of rings.

Future Space Probes to Jupiter

A mission to Jupiter has been planned and ready since 1982. Various problems with launching vehicles (including those resulting from the Challenger accident)

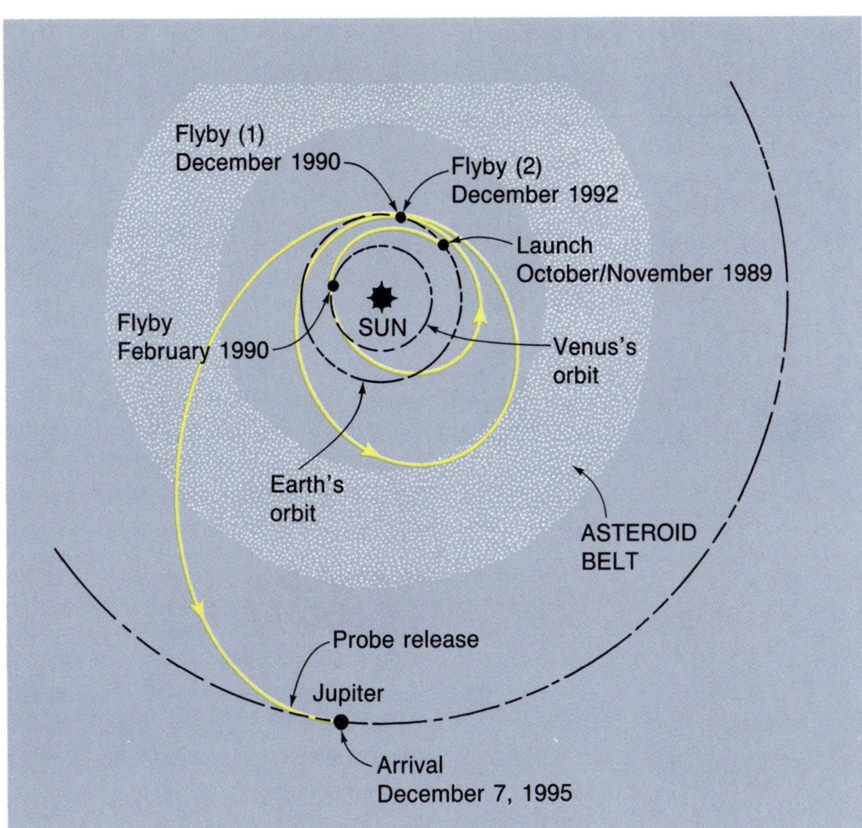

Figure 9–18. *Galileo's planned route to Jupiter. Note that it will fly by Venus once and Earth twice.*

have continued to delay the launch but it is now scheduled for November 1989. Named Galileo to honor the intellectual giant who was the first to make telescopic observations of Jupiter, it will consist of an orbiter that will circle Jupiter for about two years, photographing the planet and its moons and measuring features such as Jupiter's strong magnetic field. In addition, the Galileo craft will have a probe that will descend into Jupiter's atmosphere. This probe will have a heat shield and parachute and before it is destroyed by the intense pressure and heat in the depths of the planet, it should answer many of our questions. And, as with most good experiments, it will undoubtedly raise even more questions than it answers. On the first page of this chapter is a painting of the Galileo probe passing by Io and Figure 9–18 shows its long path to Jupiter.

SATURN

Probably the most impressive object visible in a small telescope is the planet Saturn. Figure 9–19(a) is a photo taken by an Earth-bound telescope. The telescope used by Galileo when he discovered Saturn, however, did not allow him to see the rings nearly so clearly. He called Saturn "the planet with ears" and was unable to explain what appeared to be bumps on opposite sides of it. It was some 50 years later that Christian Huygens recognized that the "ears" were due to rings around the planet.

Huygens (1629–1695) was a Dutch physicist and astronomer who made major advances in the fields of optics and light.

ACTIVITY

Observing Jupiter and Saturn

JUPITER AND SATURN ARE beautiful objects to observe if you have access to a telescope. Even with binoculars you may be able to see the four Galilean moons of Jupiter and the rings of Saturn. Make a careful sketch of the location of Jupiter's moons and then view them a few hours later or the next night. You will be able to see that they have moved. In addition, see if you can tell that Jupiter and Saturn are not perfectly round. Look for dark bands across the disk of Jupiter.

Table T9–1 shows selected rising and setting times for Jupiter and Table T9–2 does the same for Saturn. For dates between those shown you can find the planet as in the following example: Suppose you wish to find Jupiter in February 1990. The table shows the planet setting around sunrise in December and at midnight

Table T9–1 Selected Rising and Setting Times of Jupiter

Date	Description
Late September 1989	Rises at midnight
Late December 1989	Rises at sunset, sets at sunrise
Late April 1990	Sets at midnight
Early June 1990	Sets at end of evening twilight
Mid-August 1990	Rises at beginning of morning twilight
Late October 1990	Rises at midnight
Late January 1991	Rises at sunset, sets at sunrise
Late May 1991	Sets at midnight
Early July 1991	Sets at end of evening twilight
Mid-September 1991	Rises at beginning of morning twilight
Early December 1991	Rises at midnight
Late February 1992	Rises at sunset, sets at sunrise
Early June 1992	Sets at midnight
Late July 1992	Sets at end of evening twilight
Mid-October 1992	Rises at beginning of morning twilight
Mid-January 1993	Rises at midnight
Late March 1993	Rises at sunset, sets at sunrise
Early July 1993	Sets at midnight
Early September 1993	Sets at end of evening twilight
Early November 1993	Rises at beginning of morning twilight
Late February 1994	Rises at midnight
Early May 1994	Rises at sunset, sets at sunrise

in April; since February is about midway between these dates, it should set about 2:00 or 3:00 A.M. in February. Thus you should look for it in the evening sky fairly high in the east. After rising, it will get higher in the sky, cross the meridian, and progress toward the western horizon until it sets. Notice that from early June until mid-August of 1990 Jupiter will be too close to the Sun to be easily visible.

In each table, midnight refers to standard time rather than daylight saving time.

Although these tables should be of help to you in finding the planets, a better plan would be to pick up an issue of either *Astronomy* or *Sky and Telescope*, two monthly magazines that contain maps and hints for viewing the planets and other celestial objects.

Table T9–2 Rising and Setting Times for Saturn, 1990

Date*	Description
Early March	Rises at beginning of morning twilight
Mid-May	Rises at midnight
Mid-July	Rises at sunset, sets at sunrise
Early October	Sets at midnight
Late December	Sets at end of evening twilight

*The chart can be used for three to four years before or after 1990 by subtracting a week for each year before 1990 and adding a week for each year after. For example, in 1994 Saturn will rise at midnight in mid-June.

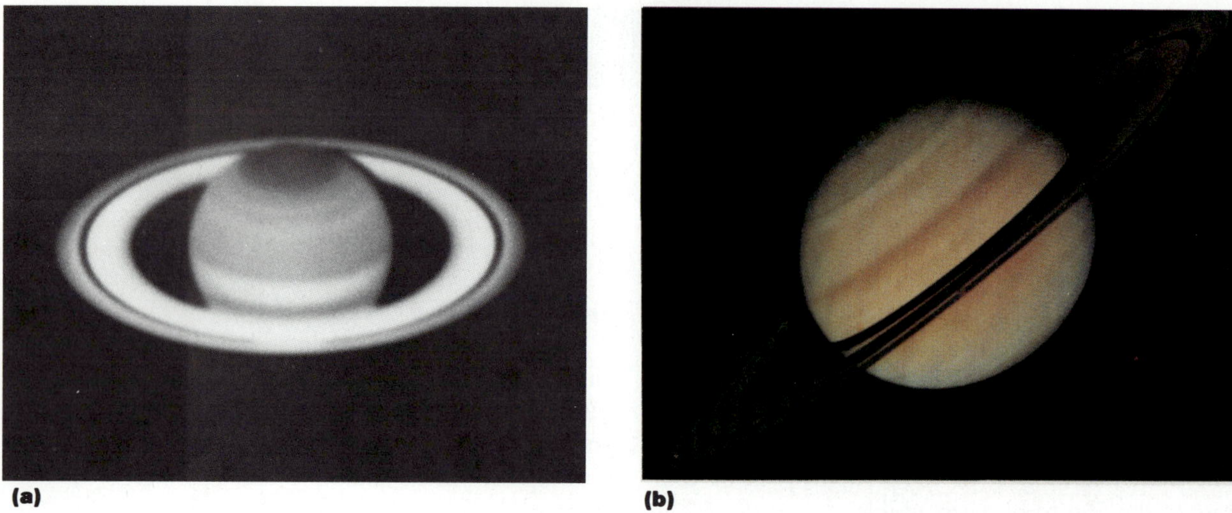
(a) (b)

Figure 9–19. *(a) Saturn as shown by an Earthbound telescope. (b) Saturn as photographed by Voyager.*

Size, Mass, and Density

Saturn has an equatorial diameter of 120,000 kilometers (75,000 miles); not much smaller than Jupiter and more than nine times that of the Earth. Except for its obvious rings, Saturn is in many ways similar in appearance to Jupiter. Notice in Figure 9–19(b) that, like Jupiter, Saturn has bands around it. If the planets are indeed similar, we would suspect that Saturn has a density like Jupiter's but in fact Saturn's density is less. While the Earth has a density like a common rock and Jupiter is slightly more dense than water, Saturn's density is only 0.7 as much as that of water. As we will see later, the low density of Saturn is probably due to a less dense core and a lower percentage of liquid metallic hydrogen.

Motions

Saturn orbits the Sun at 9.5 astronomical units in a fairly circular orbit. Thus its distance from the Earth varies only from about 8.5 AU to 10.5 AU. Its appearance from the Earth varies greatly, but not because of changes in distance. Figure 9–20 shows a number of views of Saturn taken at various times during half of its 29.5 year period of revolution. To see the reason for the change in appearance of its rings, think of Saturn revolving at its great distance from the Earth and Sun as shown in Figure 9–21. The rings of Saturn are in the plane of the planet's equator and this plane is tilted 27 degrees with respect to the planet's orbital plane. The planet keeps this same tilt while moving around the Sun (and the Earth), so when it is viewed from the Earth we see the rings at different orientations at different times.

Saturn's northern hemisphere was tilted most toward Earth in 1988 and its rings will be edge-on to us in 1996.

When we consider the rotation of Saturn, we find two more similarities to Jupiter. The various belts of Saturn's atmosphere rotate at different rates, just

Figure 9–20. *Photos of Saturn taken from Earth during half of Saturn's 29-year period.*

Figure 9–21. *Saturn at four points in its orbit, along with its appearance from Earth. At points* X *and* Z *the rings are edge-on to Earth.*

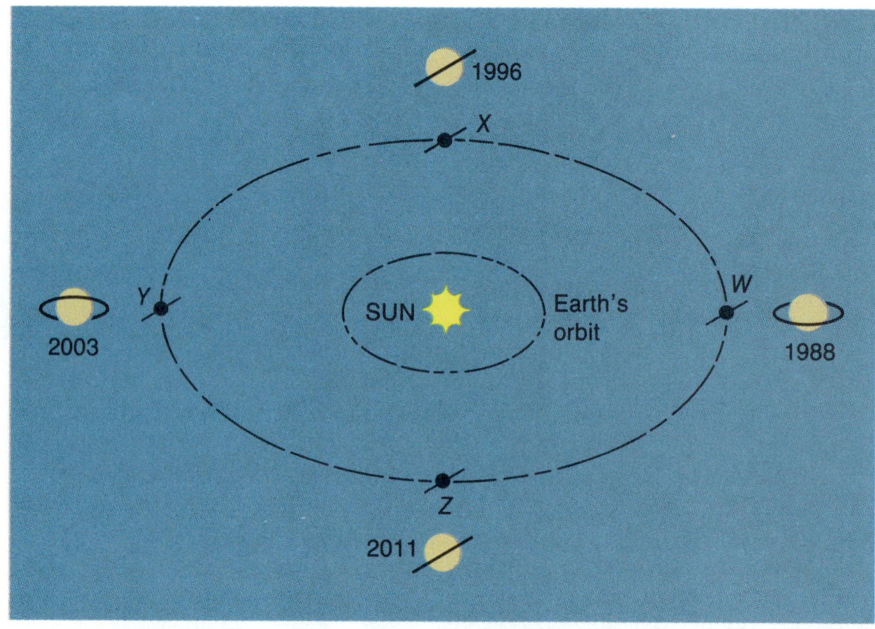

as in the case of Jupiter, and its equatorial rotation rate of 10 hours and 39 minutes is close to Jupiter's 9 hours and 50 minutes.

Saturn's oblateness is 0.098, so that its equatorial diameter is nearly 10 percent greater than its polar diameter.

Recall that Jupiter's oblateness was explained by its high rotation rate. Saturn has a similar rotation rate but is even more oblate. This is because the gravitational field at its surface is weaker than at Jupiter's and since there is less gravitational force tending to keep the planet in a spherical shape, Saturn is very "out of round"; more than any other planet.

Figure 9–22. *Saturn's interior resembles Jupiter's but Saturn contains a smaller percentage of liquid metallic hydrogen.*

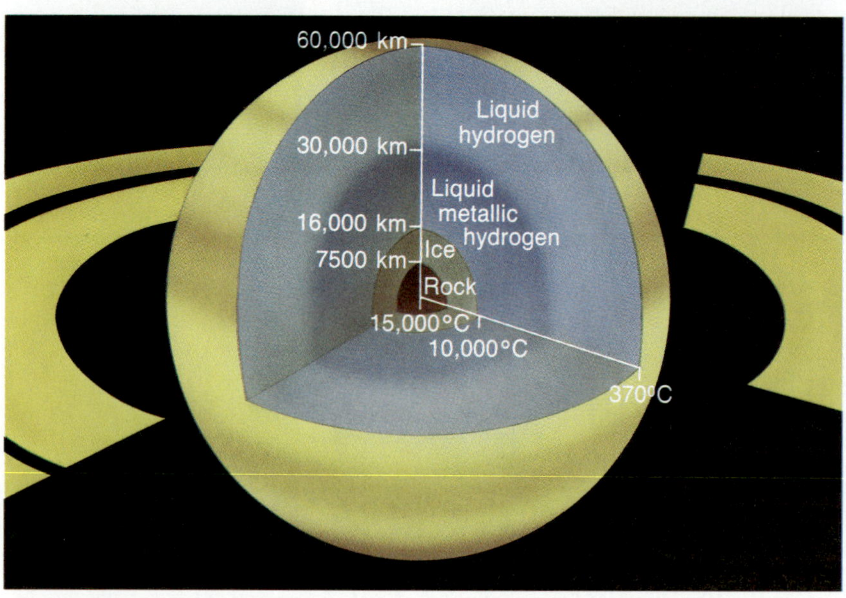

Pioneer, Voyager, and Saturn

As you might suspect, we have learned much about Saturn from space probes. Pioneer 11 passed Saturn in September 1979, Voyager 1 about a year later, and Voyager 2 a year after that. (Pioneer 10 swung around Jupiter in such a manner that its path did not take it to Saturn.) Knowledge gained by each of the first two probes was used to guide experimenters in decisions concerning the following ones.

The composition of Saturn is similar to that of Jupiter but Saturn has a slightly higher percentage of hydrogen and a corresponding lower percentage of helium. Saturn is colder, however, and this inhibits the chemical reactions that give Jupiter's atmosphere its varied colors.

Recall that liquid metallic hydrogen can exist only at very high pressure. One must descend deeper into Saturn than into Jupiter to find such pressure and therefore Saturn has less of this conducting material (and consequently a weaker magnetic field) than Jupiter. Figure 9–22 shows a cross-section of Saturn indicating that probably it has a solid dense core like Jupiter.

Another similarity between Saturn and Jupiter is found in the fact that Saturn also radiates more energy than it absorbs; about 2.2 times as much. As in the case of Jupiter, the source of this energy may be residual heat from its formation although other hypotheses have been proposed to explain this phenomenon.

As Pioneer and Voyager flew by Saturn, it was not just the planet itself on which they focused their instruments. The moons of Saturn and its beautiful rings were of special interest.

Table 9–2 The Planet Saturn

	Value	Compared to Earth
Equatorial diameter	120,000 km	9.4
Polar diameter	108,000 km	8.5
Mass	5.684×10^{26} kg	95.0
Density	0.70 gm/cm^3	0.13
Surface gravity	11.7 m/s^2	1.19
Escape velocity	35.6 km/sec	3.2
Sidereal rotation period	10.665 hours	0.43
Albedo	0.34	1.1
Tilt of equator to orbital plane	26.73°	1.1
Orbit		
Semimajor axis	1.427×10^9 km	9.555
Eccentricity	0.0556	3.3
Inclination to ecliptic	2.49°	
Sidereal period	29.46 years	29.46
Moons	23	

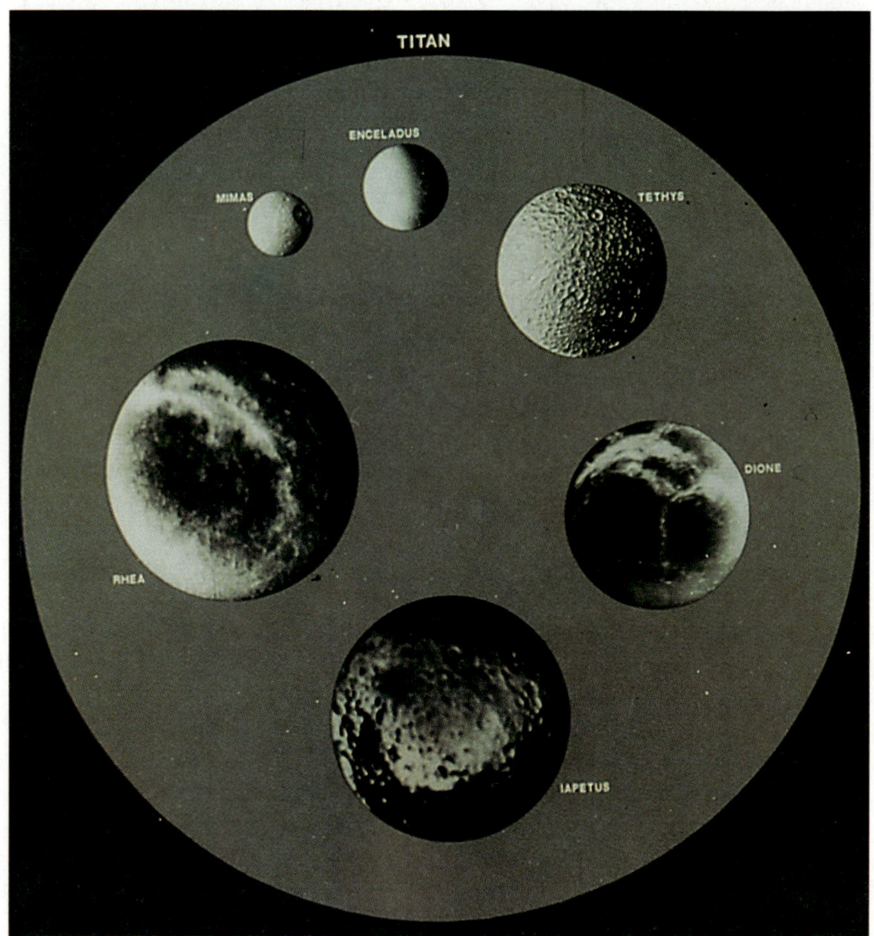

Figure 9–23. *Saturn's seven largest moons. The surface of giant Titan, shown behind the rest, is hidden by its dense clouds.*

Titan

Saturn has some 23 moons, most of which consist of dirty ice like we see on some of Jupiter's moons. Although the Voyager photo of Mimas and its prominent crater (Figure 9–23) is a curious sight and each of Saturn's moons is unique, we will discuss only Titan, the largest. Refer to Appendix H for details concerning the moons of Saturn as well as other moons of the solar system.

Titan may turn out to be the most interesting moon in our system but our knowledge of its surface is limited by the haze that covers it. What makes it unique is that it has an atmosphere composed of about 85 percent nitrogen, 12 percent argon, and 3 percent methane. When sunlight strikes methane, it can cause the formation of large organic molecules. If this is happening on Titan, these molecules must drift down to the surface, making a soup many meters thick on the surface. This raises the question of whether life might have formed from these organic molecules.

Titan is only slightly larger than Mercury, which has no appreciable atmosphere. Titan, however, has an atmosphere more dense than Earth's. How can this be? Recall that an object can retain an atmosphere only if the escape

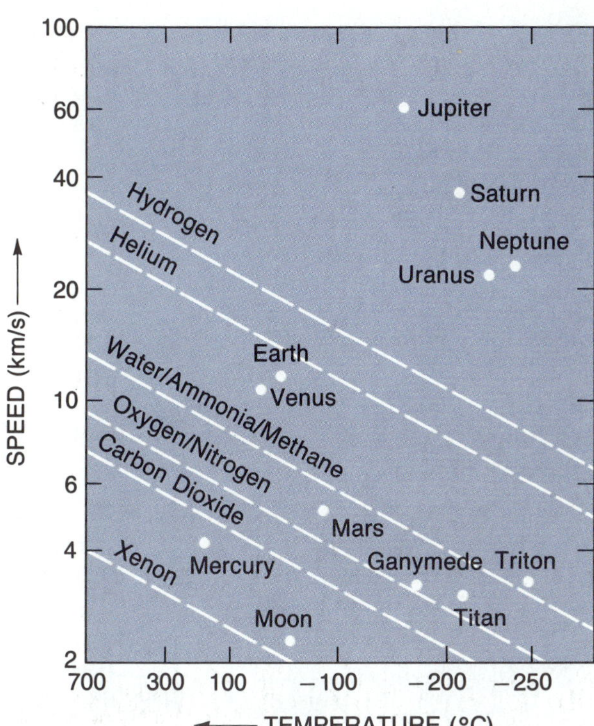

Figure 9–24. *This is a repeat of a graph shown in previous chapters, but some moons are included here. The graph predicts that Titan should retain the more massive gas molecules.*

velocity from its surface is greater than the speed of gaseous molecules there. Figure 9–24 shows a number of planetary moons. Notice that the escape velocity from Titan is somewhat less than from Mercury. The difference between the two objects that allows Titan to retain an atmosphere, however, is their temperatures. On Mercury's Sun side, we find temperatures up to 450 degrees C, but the temperature of Titan's surface is only about −220 degrees C. Thus the molecular speed of the gas molecules on Titan is less than the escape velocity from the planet and it has retained its atmosphere.

Planetary Rings

Notice in Figure 9–20 that the rings of Saturn are not visible when they are edge-on to us. This is because they are very thin; only about a mile across. The rings, however, are not solid sheets but are made up of small particles of water ice or perhaps rocky particles coated with ice. The great amount of empty space between the particles in the rings means that if they were compressed, their thickness would be reduced from a mile to about a yard.

Each of the particles that make up the rings of Saturn revolves around the planet according to Kepler's laws. Thus, particles nearer the planet move faster than those farther out. Each of the particles is, in a sense, a separate satellite of Saturn.

Three distinct rings are visible from Earth and long before the advent of space flight they were named (from outer to inner) rings A, B, and C. Photos from space indicate that there are many more rings than this, as is evident in

Figure 9–25. *Voyager photos revealed the rings of Saturn to be very complex.*

Figure 9–25. A number of reasons have been given for the existence of the spaces between the rings and it is now evident that different spaces have different causes. Some explanations are quite complicated, but others are much simpler.

One of the spaces that is easily explained is the one between rings A and B, known as *Cassini's division* after the French Astronomer G. D. Cassini. To explain this space between rings we must compare the motion of a particle in the Cassini division to the motion of Mimas, one of Saturn's moons. A particle in Cassini's division would orbit the planet with a period of 11 hours, 17½ minutes. This is just half of the orbital period of Mimas. Thus any particle found in the division will be at the same place in its orbit each time Mimas passes near it. Each time this happens Mimas exerts a slight gravitational tug on the particle, deflecting it slightly from its path. If Mimas did this at random times during the orbiting of the particle, the gravitational effect would be random and the particle's path would not be severely affected. Because of the synchronous relationship between the periods, however, the tug is repeated regularly and the overall effect is to pull the particle out of its orbit. This is what is happening to particles that drift into Cassini's division and it is the reason the division exists.

If such synchronous gravitational tugs were the only mechanisms at work determining the structure of Saturn's rings, Cassini's division might be completely clear of particles but in fact it is not. There are many other forces at work, including gravitational forces from other moons and particles as well as electromagnetic forces from Saturn's magnetic field.

Other features of the rings are explained by the existence of small moons orbiting near the rings. These satellites, called *shepherd moons,* cause some of the rings to keep their shape. Shepherd moons are discussed in a Close Up in this chapter. Still other features of the rings seem to be similar to spiral wave patterns seen in galaxies. Various hypotheses have been used to explain the various features of the rings but no comprehensive theory has yet been proposed.

Why Rings? The origin of the material of the rings is not well understood. It was once thought to be material left over from the formation of Saturn and its moons, but data from Voyager seems to suggest that the material was captured from a passing comet. In any case, we do understand the reason that the rings have not formed into one or more moons. To see why, we should review the effect of tides.

Recall how the gravitational force between the Earth and the Moon causes tides on both the Earth and the Moon. It is not simply the existence of a gravitational that which causes the tides, however; but the fact that there is a *difference* in the amount of gravitational force exerted on masses that are at difference places on the Earth (or the Moon). We considered tides in Chapter 6 but we will now look at them in another way; focusing here on Kepler's laws.

According to Kepler's third law—expanded to apply to planets—if two objects are orbiting a planet, the object closer to the planet must move faster in its orbit if it is to stay in that orbit. Figure 9–26 illustrates a moon in a circular orbit around a planet but in the figure we have divided the moon into three parts, each at a different distance from the planet. Kepler's law tells us that if the three parts of our moon were not connected, part *X* would have to move faster. If it were to move at the same speed as part *Y*, it would fall inward toward the planet. Likewise, part *Z* should travel more slowly than *Y* and if it were to travel at the same speed as *Y*, it would fly out of *Y*'s orbit. This, in fact, is just another way to look at the cause of tides, for each part of the moon tends to do just what we have described.

Kepler's third law was stated as $P^2 = Cd^3$. *The velocity of something moving in a circle with period* P *is given by* $v = 2\pi d/P$. *So* $P = 2\pi d/v$. *Putting this in Kepler's law, we get* $v^2 = 4\pi/Cd$. *Thus, the greater the distance, the less the speed.*

Why doesn't a moon fly apart because of this effect—the tidal force? The answer, of course, is that the moon has its own forces holding it together. A rock on its surface is pulled by the force of gravity toward the center of the moon. Artificial satellites, on the other hand, are held together by the strength of the materials of which they are made.

It is important to see that it is the *relative difference* in distance from the planet that causes the effect. If a moon is far from its planet, the outside edge may be only a fraction of a percent farther from the planet than is the center of the moon. The same moon located close to its planet will have its outer edge

Shepherd Moons

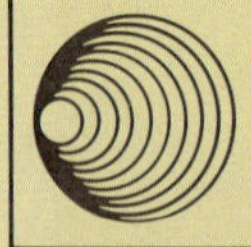

WHEN THE RINGS OF Uranus were discovered in 1977, they were seen to be very narrow. (Voyager 2 data later showed that only two of the ten rings had widths of more than a dozen kilometers.) Thus these rings are different from the rings of Saturn, most of which are fairly wide. In addition, the edges of Uranus's rings are well-defined rather than fading gradually from maximum density inside the ring to nothing beyond its edge. Peter Goldreich of the California Institute of Technology and Scott Tremaine of Princeton University proposed a solution; hypothesizing that a pair of moons, one on each side of a ring, could explain why the ring stays in a narrow band rather than spreading out over space.

Figure T9–3 illustrates their proposal. The two moons orbit just inside and just outside the narrow ring. Kepler's third law tells us that the moon closer to the planet moves faster than the particles of the ring. As that moon catches and passes the particles on the inside portion of the ring, gravitational pull from it increases the particles' energy just a little, causing them to move outward in their orbit. (This energy comes from the moon and therefore the moon loses energy and moves inward a bit. The moon is so much more massive than the particles, however, that the change in its path is almost insignificant.)

In an opposite manner, as particles on the outside edge of the ring pass the more distant moon, they lose energy and move inward. These effects are very small on each particle; much less than is indicated by the figure. Nonetheless, Goldreich and Tremaine showed that the effect is significant enough that as the moons continued to orbit near the rings, they force the particles together; acting as shepherds for the flock of particles in the ring.

The scientific community awaited Voyager's passing by the rings of Uranus to see if it found such moons. The answer came earlier than expected. Voyager 2 at that time had not yet passed Saturn. When it did, it photographed Saturn's rings and on each side of Saturn's narrow F ring was a moon (Figure T9–4). Calculations showed that the two moons near the F ring would indeed perform the function of shepherds for that ring. The F ring shepherds, however, are large enough that they disturb the ring in an additional way; causing strands in it (Figure T9–5).

Figure T9–3. *As the inner moon passes the inner portion of the ring, it forces ring particles outward. The outer moon forces particles inward. The effect is greatly exaggerated here.*

In January 1986, Voyager 2 passed Uranus and returned a remarkable photo (Figure T9–6) of the epsilon ring of Uranus along with two shepherds. Other rings around Uranus exist without the aid of shepherd moons and so although Goldreich and Tremaine were right in their prediction of the action of shepherd moons, this mechanism is only a part of what is happening in planetary ring systems.

Inside the Roche Limit?

The shepherd moons of Saturn's F ring and those orbiting Uranus are within the Roche limit of each planet. How can they exist there without being pulled apart by gravitational forces? There are a number of answers to this question. First, it is an oversimplification to refer to a single Roche limit. For a smaller moon, the limit is closer to the planet than it is for a

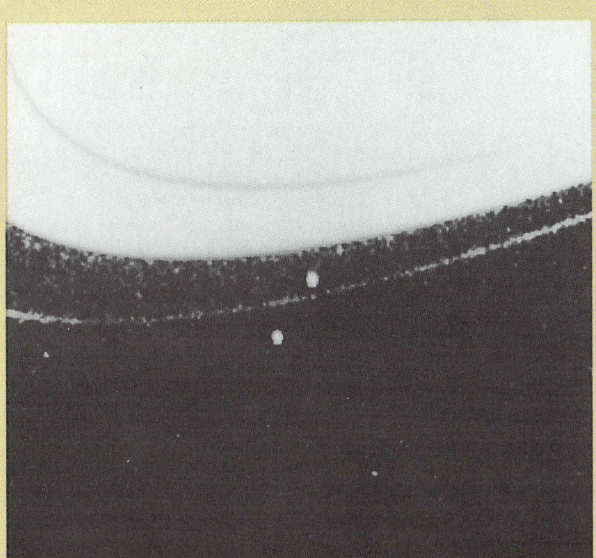

Figure T9–4. ***Voyager photographed these shepherd moons near the F ring of Saturn.***

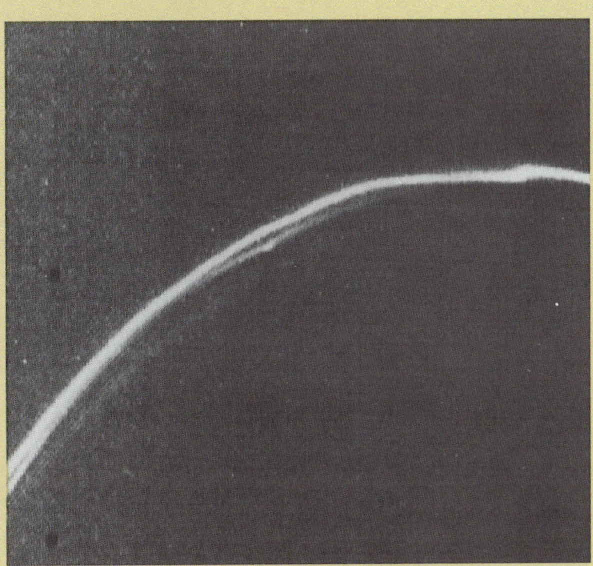

Figure T9–5. ***Strands in Saturn's F ring illustrate that the mechanics of planetary rings is very complicated.***

larger moon. The shepherds are small moons. In addition, the position of the limit depends upon the density of the moon. But most important, the Roche limit concerns only moons that are held together by gravitational forces. Obviously the Space Shuttle orbits inside the Roche limit of the Earth. But the Shuttle is not held together by gravitational forces; it is held together by the strength of the metal of which it is constructed. In like manner, a single large rock can exist inside the Roche limit, held together by cohesive forces within the material. Each shepherd moon may well be a single large rock rather than being made of smaller particles held together by gravity.

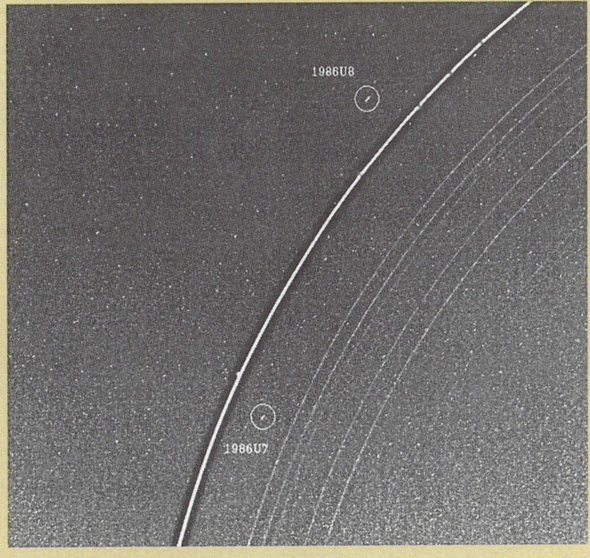

Figure T9–6. ***Moons 1987U7 and 1987U8 were photographed shepherding a ring of Uranus.***

Figure 9–26. *The moon is divided into three parts for analysis. Part X is closer to the planet than the rest of the moon and would have to move faster if it were a separate satellite. If there were no attractive force toward the rest of the planet at Xs present speed, it would fall inward and begin an eccentric orbit.*

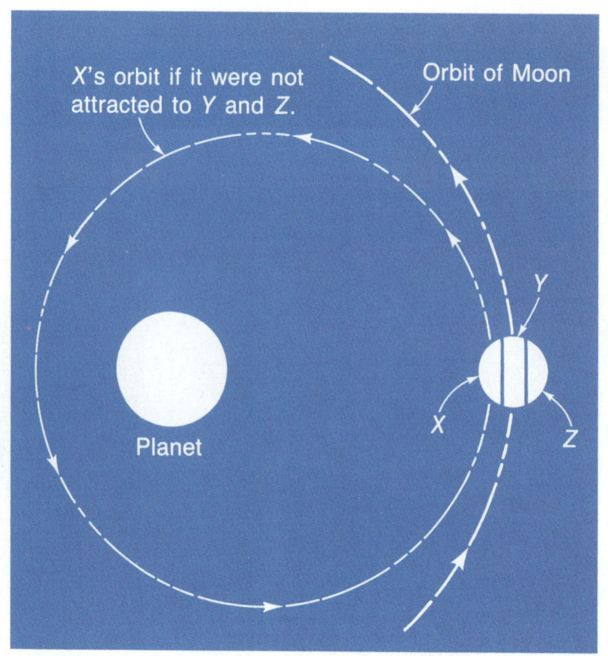

Roche limit: The minimum radius at which a satellite (held together by gravitational forces) may orbit without being broken apart by tidal forces.

Artificial satellites are held together by the strength of their materials and so the Roche limit does not apply to them.

a greater fraction of the distance from the planet than is its center. This causes the effect of tides to be greater if a moon is nearer a planet. There is a critical distance, called the ***Roche limit***, inside which the tidal force on a moon will be greater than the moon's own gravitational force. Inside this distance a large moon will be unable to hold itself together.

Now we come to Saturn. The rings of Saturn are inside this Roche limit. Thus the gravitational force between any small particles within the rings is less than the tidal force that tends to pull them apart. And so no moons are formed from the particles.

Figure 9–27 gives an idea of typical particle sizes and their distribution in the rings. The actual particles are irregular in shape rather than the spheres shown in the computer drawing, however. It is interesting to note that if all of the material of the rings of Saturn were formed into a single moon, the moon would be about the mass of Janus (one of the smallest of Saturn's moons) and only one millionth the mass of the Earth's Moon.

URANUS

Uranus (YOOR-uh-nuss) was unknown to the ancients even though they undoubtedly saw it, for it is barely visible to the naked eye even under perfect viewing conditions. It was plotted on star charts made by telescope as early as 1690. In anything but large, high-resolution telescopes, however, Uranus appears only as a speck of light rather than a disk. This, coupled with the fact that it moves very slowly, caused it to go unnoticed as a planet until 1781 when English astronomer William Herschel noticed that this particular "star" seemed

Figure 9–27. *This computer simulation shows the particles in Saturn's A ring as spheres (which is not correct), but the sizes and spacings shown are correct compared to the size of the person.*

to have a size. He thought at first that it was a comet but after calculations showed that the planet's orbit was nearly circular, Herschel realized that he had discovered a new planet.

Herschel proposed that the new planet be named "Georgium Sidus" (the Georgian star) after King George III, but the name didn't take. Johann Bode (of Bode's Law fame) suggested "Uranus," after the mythical Greek god of the sky.

The size of Uranus is difficult to determine from the Earth because its disk is so indistinct in a telescope (Figure 9–28). Recall that the size of an astronomical object can be determined from its distance and angular size. If the image is indistinct, the angular size cannot be determined accurately, of course. The first reliable value for Uranus's diameter came from a telescope aboard a high-altitude balloon operated by Princeton University. This telescope was lifted by the balloon to an altitude of 15 miles; above most of the distortion caused by the Earth's atmosphere.

An improved determination of the diameter of Uranus was made in 1977 when Uranus was observed passing in front of a star. Such an event, when a moon or planet eclipses a star, is called an ***occultation*** and provides us with a valuable measuring method. Since the speed of Uranus in its orbit was known, all that had to be measured during the occultation in order to calculate Uranus's diameter was the time during which the star was occulted. In fact, various observatories on the Earth were able to observe slightly different occultations as indicated by Figure 9–29. This served as a check on Uranus's exact path as well as its size.

Occultation: The passing of one astronomical object in front of another.

The diameter of Uranus is 52,000 kilometers; about half the diameter of Saturn and one-third that of Jupiter. Long before Voyager 2 passed Uranus in

52,000 km is 32,000 miles.

HISTORICAL NOTE

William Herschel, Musician/Astronomer

WILLIAM HERSCHEL, WHO HAS been called the greatest observational astronomer ever and the Father of Stellar Astronomy, was a musician. At least that was his training and that is what provided his living for a great part of his life. His father Isaac was an oboeist in the band of the Hanoverian Foot Guards. (Hanover was a province in the former state of Prussia in Germany.) Armies at that time were accompanied by a band, the members of which did not participate in the actual fighting but had the function of providing fanfare and spirit. Friederich Wilhelm Herschel, who adopted the name William when he moved to England, was one of Isaac's five children. All of the children were raised as musicians and William learned to play the violin as soon as he could hold the small one his father obtained for him. When he was 14 he joined his father and older brother in the military band.

The semimilitary life was not for him, however, so he and his brother requested a discharge and just after William turned 19, they moved to England together. They were almost penniless and without a good command of the English language, but their musical abilities allowed them to get work copying music and tutoring. William served for a brief time as instructor to another military band but spent most of his young adult life teaching, performing, and composing music. Herschel's notes refer to a number of symphonies he composed, but little of his music remains today. It attracted attention in England at the time, however, and Herschel was able to join in the fashionable society of London and Edinburgh.

William's father Isaac had not been highly educated and he was not wealthy, but he was a man with a natural curiosity and an interest in learning. Fortunately, he passed his intellectual enthusiasm on to his children. Thus William the musician spent his spare time becoming self-educated. Besides learning English, he learned Italian and some Latin. It was common to see him reading as he rode horseback across the countryside.

One of the interests that William inherited from his father was the study of astronomy. He began to build his own telescopes and after teaching as many as eight music lessons a day, he would spend the evenings observing the heavens. By 1778 he had built an ex-

1986, the mass of the planet had been calculated by applying Kepler's Third Law to its five known moons. When its density is then calculated, we find that Uranus has a density of 1.25 times that of water, greater than Saturn but slightly less than Jupiter. Taking into account the surface gravity on Uranus (which is less than on Saturn or Jupiter), we can calculate what Uranus's density would be if the planet were made up entirely of hydrogen and helium. Its density is greater than this. It is therefore thought that Uranus must have a dense, rocky core (perhaps the size of the Earth) and that this core may be surrounded by a deep liquid water ocean. Figure 9–30 shows one theoretical model of the cross-section of the planet.

The purpose of the Uranus observations during the occultation of 1977 was to determine the planet's size more accurately. A surprise was in store for the observers, however. A short while before Uranus was expected to occult cellent reflecting telescope about six inches in diameter. This was no easy task, for construction of the mirror involved metal casting.

Herschel's life changed after March 13, 1781. On that night he saw a hazy patch in the sky and upon observing it over a few nights, he saw that it moved. What he thought was a comet turned out to be the first planet to be discovered in recorded history.

It is interesting that William Herschel was not immediately accepted by the scientific community. Herschel's telescopes were of such quality that he was able to see things in the heavens that other astronomers couldn't, and many did not believe his claims about the details of what he was seeing. There seems to have been some professional jealousy concerning this musician who, in his writings in astronomy, did not use the jargon accepted by the profession. Herschel's skills as a telescope maker and as an observer won out, however, and his discovery of Uranus could not be denied. He was soon recognized and celebrated by scientific societies. Within a year after hc discovered Uranus, he received a pension from King George III so that he could devote full time to astronomy, his former hobby.

While he was still devoting most of his time to music, Herschel had brought his sister Caroline to live with him. (She served as a vocalist at some of his performances.) Now that he was an astronomer, she became a very important assistant and an accomplished astronomer herself. Together they made the first thorough study of stars and nebulae (faint, hazy objects not understood at the time). One important discovery they made was the existence of double stars—stars in orbit about one another. It was after his death that the orbits of these stars were analyzed in enough details to determine that the stars obey Newton's laws of motion and gravitation and therefore that these laws are valid not only here in our solar system but even among the distant stars.

Herschel continued serious observing until he was nearly 70 and he died at the age of 84. After his death, Caroline wrote a large catalog of his observations and received many honors for her own astronomical knowledge.

William Herschel did not marry until the age of 49. Four years later he had a son, John, who himself became a noted astronomer.

the star, the light from the star blinked five times. Then after the occultation, similarly timed blinks were observed but in the opposite order. The conclusion was reached that Uranus has at least five rings that are invisible from the Earth but that are dense enough to obscure the light from a star behind them. By comparing the results obtained by a number of observatories, details of the rings' orientations and sizes could be determined. Figure 9–31(a) shows the apparent path of the star past the rings for the five observatories. Use of the same method during an occultation in 1978 revealed a total of nine rings.

The rings of Uranus cannot be seen from the Earth because they reflect only about 5 percent of the sunlight that hits them. The rings must be made of material as dark as soot. By comparison, Saturn's rings reflect 80 percent of the light that hits them.

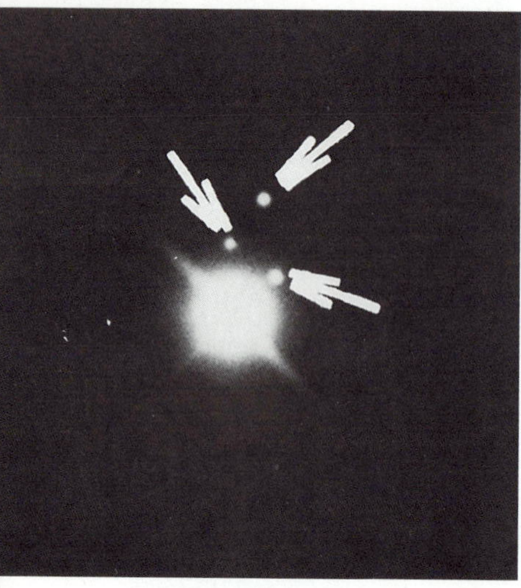

Figure 9–28. *Uranus as seen from Earth, with three of its moons.*

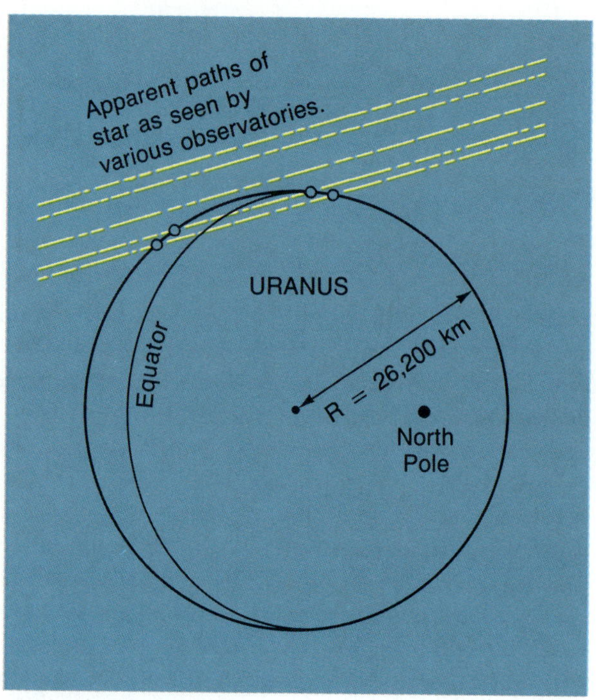

Figure 9–29. *As Uranus was observed to pass by a star, five observatories at different locations on Earth saw the planet at different places with respect to the star. The drawing shows the path of the star from each observatory as it appears to move by Uranus.*

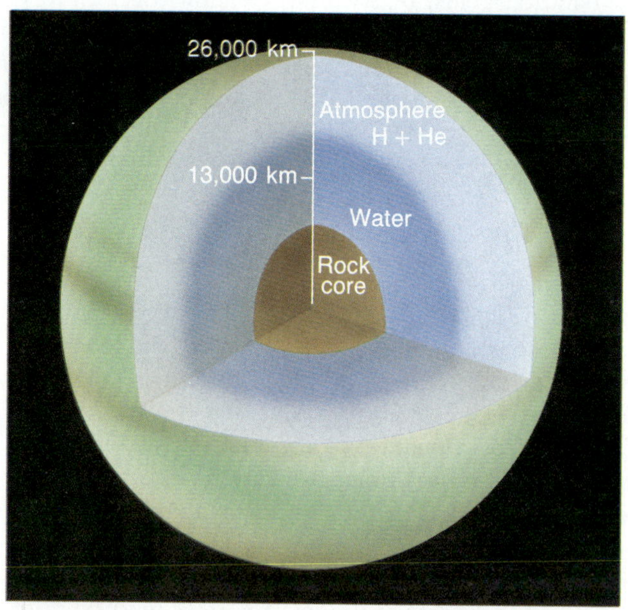

Figure 9–30. *A model of Uranus's interior.*

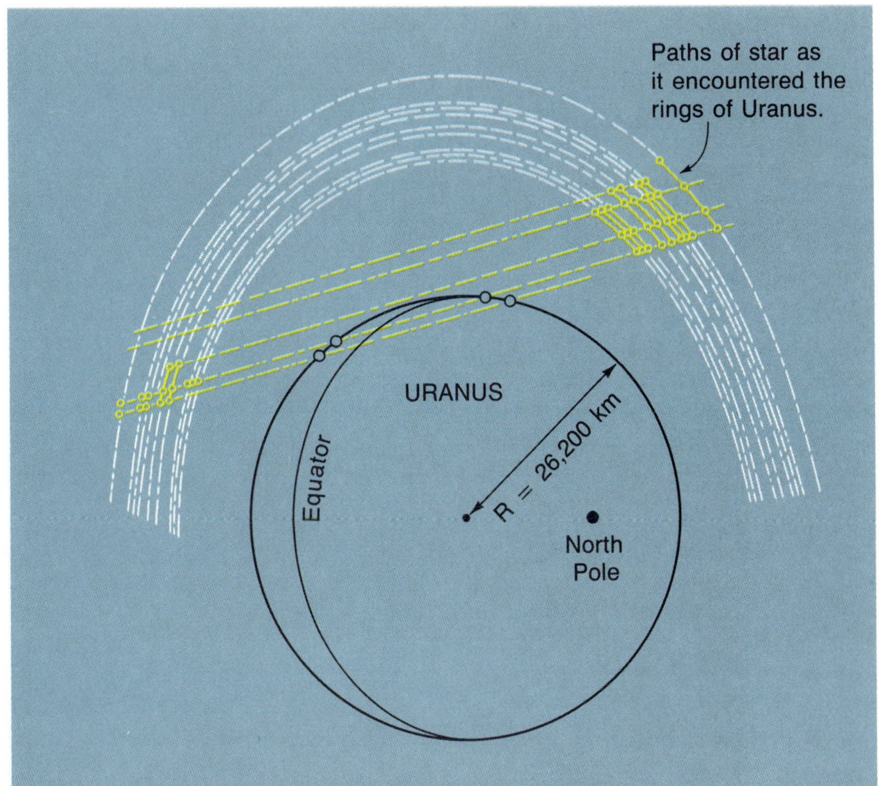

(a)

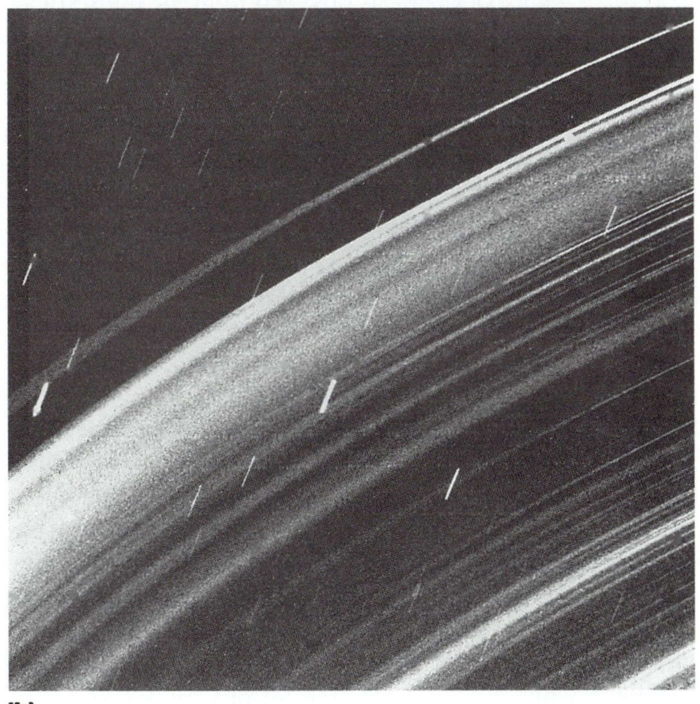

(b)

Figure 9–31. *(a) As the five observatories prepared to observe Uranus's occultation of the star, the star appeared to blink on and off, revealing the rings of Uranus. (b) Voyager's photograph of the rings revealed great numbers of ringlets.*

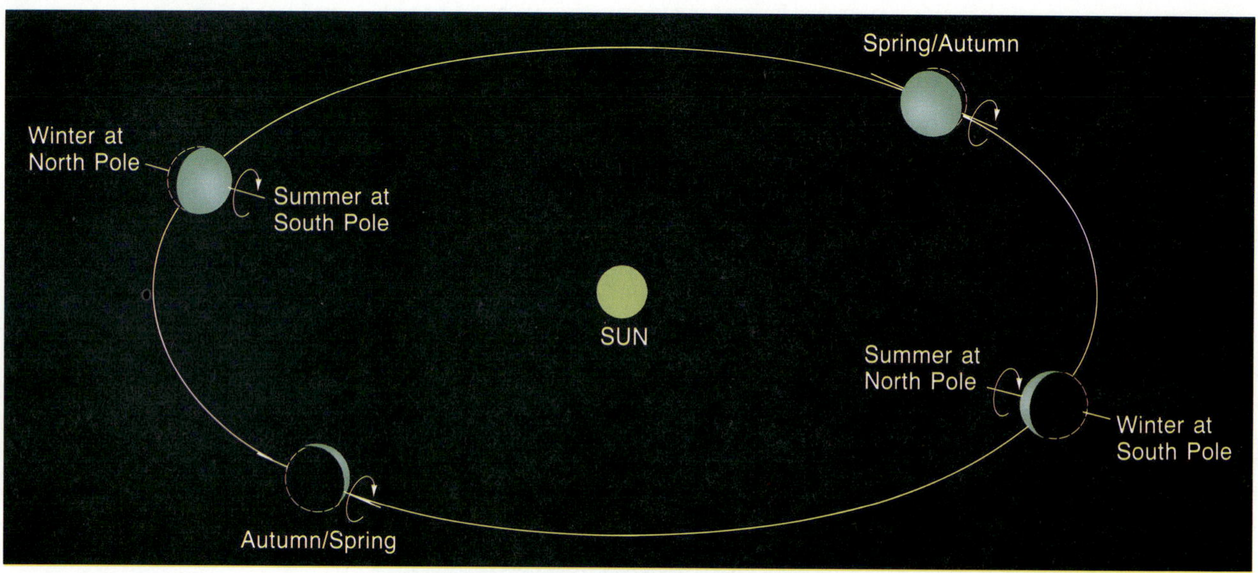

Figure 9–32. *Uranus's axis is tilted so that its poles point nearly directly at the Sun at times.*

Uranus's Orientation and Motion

Recall that Venus is unusual in that it rotates in a backwards direction: clockwise as observed from the north. Except for Venus, every other planet we have studied rotates normally; counterclockwise as observed from the north. In addition, all planets we have studied, including Venus, have their equatorial plane tilted less than 30 degrees to their plane of revolution around the Sun. Here is where Uranus is unique. Its equatorial plane is tilted nearly 90 degrees to its plane of revolution. This means that during its 84-year orbit, its north pole at one time points almost directly to the Sun and at another time faces nearly away from the Sun. Figure 9–32 illustrates this. Remember that the tilt of a planet's axis is what causes the planet to have seasons. One might think that such a large tilt of Uranus's axis would have a great effect on its weather. Data from Voyager indicates that Uranus has a fairly uniform temperature over its entire surface (about −200 degrees C), however; indicating that the atmosphere is continually stirred up, with winds moving from one hemisphere to another.

On Earth, there is little air exchange between the northern and southern hemispheres.

We said Uranus's tilt was nearly 90 degrees. In fact, tables list its tilt as 98 degrees. Isn't this the same as an 82-degree tilt? Not really. To show the difference, we first need to define what we mean by the north pole of a planet. After all, only the Earth has its axis oriented so that Polaris, the north star, is above its north pole. To decide which pole of a planet is to be called north, we can imagine standing on the planet facing one of its poles. If when we do this the rotation of the planet carries us around to our right, we are facing the north pole. The north pole of Uranus, as defined in this manner, lies on the other side of the plane of the ecliptic from the Earth's north pole. Figure 9–33 illustrates this difference. Thus we usually list the tilt of Uranus's axis as 98 degrees rather than 82 degrees. The fact that the angle is greater than 90 degrees

Another way to define the north pole of a planet: Grab the planet with your right hand in a manner such that your fingers point in the direction of its rotation. Your thumb will then point to the north pole.

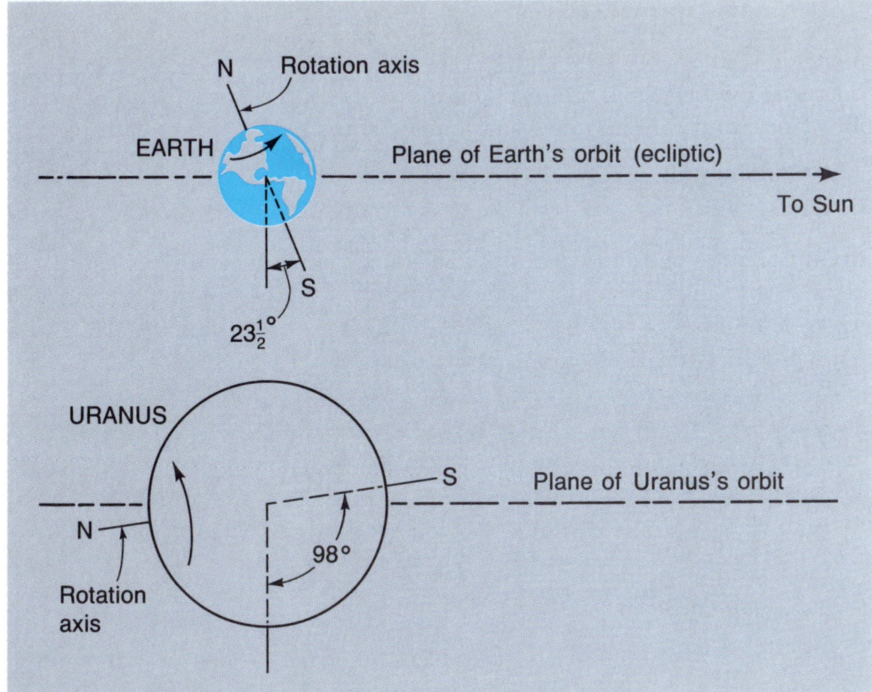

Figure 9–33. *An arrow on Earth's surface indicates its rotation. Notice that Uranus's rotation arrow points in the same direction around its pole as does Earth's arrow.*

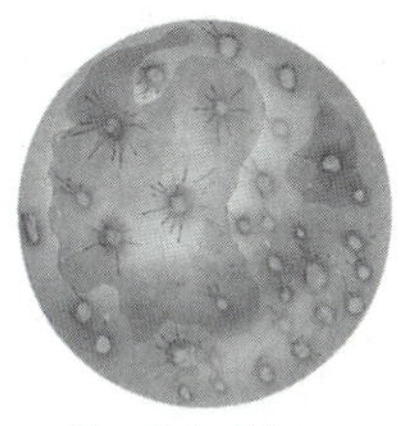

Figure 9–34. *Some of Uranus's moons compared in size to Earth's.*

means that, like Venus, Uranus has a backward rotation. The fact that the angle is so close to 90 degrees makes Uranus unique.

Features of Uranus's surface are not visible from the Earth, so until Voyager approached Uranus, there was uncertainty about its period of rotation. Voyager photos told us that, like Jupiter and Saturn, there are various bands of clouds on Uranus and that these bands rotate with different periods; from 16 hours at the equator to nearly 28 hours at the poles. Uranus has about the same chemical makeup as Jupiter, about three-fourths hydrogen and most of the remainder helium.

Uranus's magnetic field is comparable to Saturn's. Probably it originates in electric currents within the planet's layer of water (Figure 9–30). The magnetic field is unusual in that its axis is tilted 55 degrees with respect to the planet's rotation axis. No other planet has nearly this large an angle between the two axes.

The corresponding angle for Earth is 12 degrees.

Five moons of Uranus (Figure 9–34) were known before Voyager and now we know of ten more. All are dark, low-density, icy worlds. The innermost, Miranda, is perhaps the strangest looking object in the solar system. Figure 9–35 shows its varied terrain. You may wonder who has been farming it. Because of its racetrack-like grooves, one feature has been named Circus Maximus after the arena of ancient Rome. One possible explanation for the strange features of Miranda is that it experienced a tremendous collision with another object and was broken into pieces that later fell back together.

Table 9–3 The Planet Uranus

	Value	Compared to Earth
Equatorial diameter	52,000 km	4.1
Polar diameter	49,000 km	3.9
Mass	8.698×10^{25} kg	14.5
Density	1.25 gm/cm^3	0.23
Surface gravity	9.1 m/s^2	0.93
Escape velocity	21.1 km/sec	1.9
Sidereal rotation period	16 to 28 hours	
Albedo	0.34	1.1
Tilt of equator to orbital plane	97.86°	
Orbit		
Semimajor axis	2.870×10^9 km	19.19
Eccentricity	0.0472	2.8
Inclination to ecliptic	0.77°	
Sidereal period	84.01 years	84.01
Moons	15	

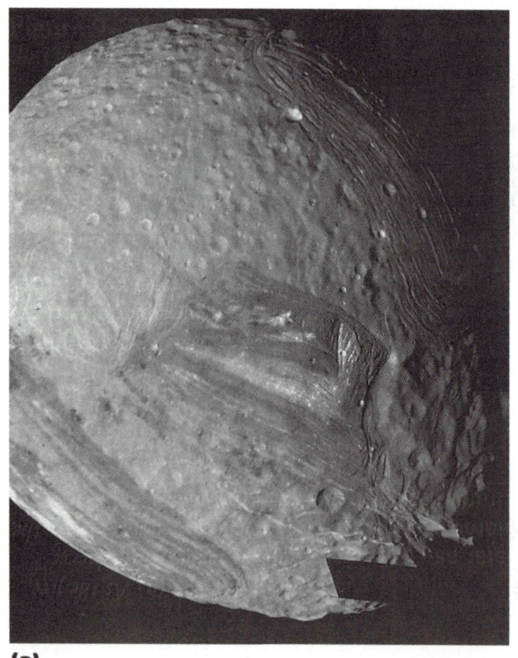

(a)

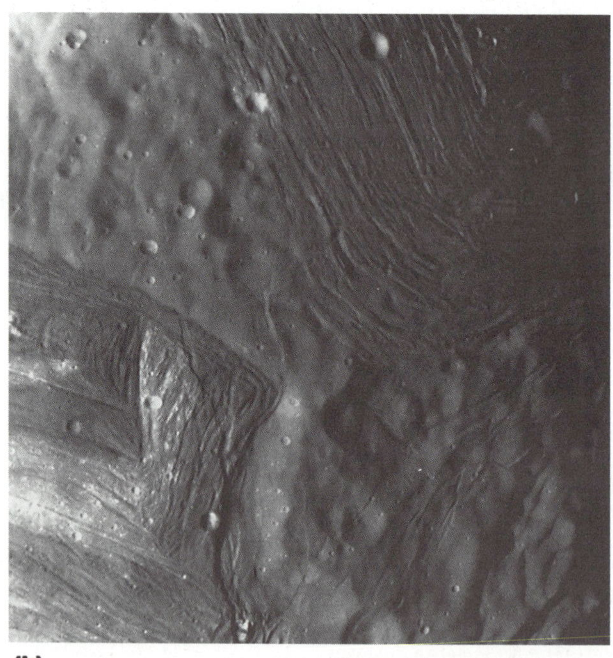

(b)

Figure 9–35. *(a) Miranda. (b) A close-up showing Circus Maximus (upper right) and the Chevron (lower left).*

Two of Uranus's moons are shepherd moons (see Close Up), keeping one of the rings in formation. Material in the rings is very sparse, however; so much so that there is more material in Cassini's division of Saturn's system than in all of the rings of Uranus! Nine rings had been observed (by occultations) from the Earth; only one more was found by Voyager. The ten rings, however, contain hundreds of smaller ringlets, as seen in Figure 9–31(b).

NEPTUNE

Uranus and Neptune are similar in many ways. The first similarity is size; Neptune is very nearly the same size as Uranus (Figure 9–36). We saw that Uranus appears only as a small, indistinct disk in a telescope so when we consider that Neptune is half again as far away from the Earth, it is no surprise that Neptune appears even more indistinct from the Earth, even in the largest telescopes.

Although their sizes are similar, Neptune is more massive than Uranus for it has a greater density—about 1.8 times that of water. It is the most dense of the Jovian planets (but still quite a bit less dense than any terrestrial planet.)

The composition of Neptune seems to be very close to that of Uranus. Since Neptune is more dense than Uranus, it is probable that it, too, has a solid core; perhaps larger than Uranus's.

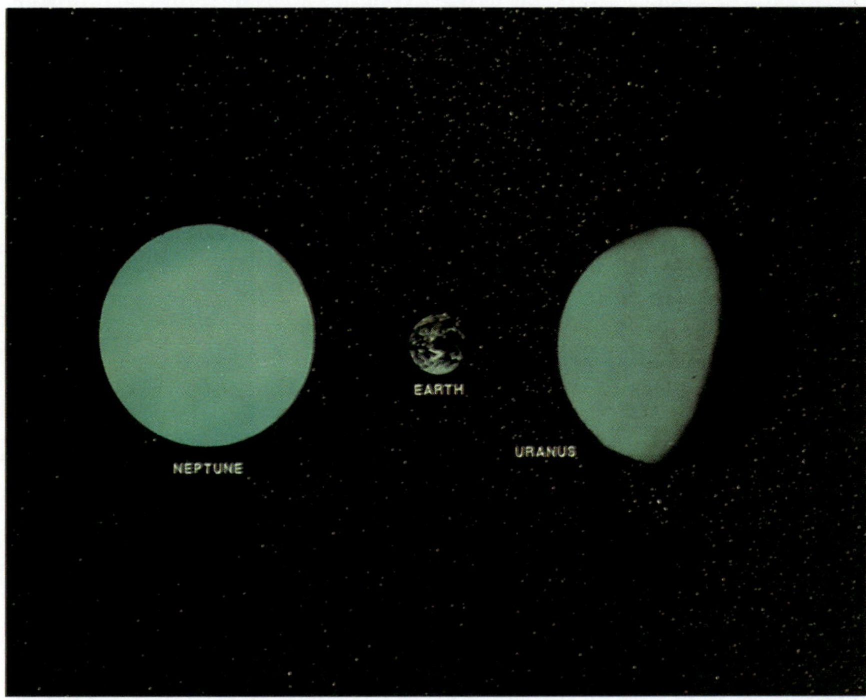

Figure 9–36. *Uranus, Neptune, and Earth are shown to scale.*

CLOSE UP

The Discovery of Neptune

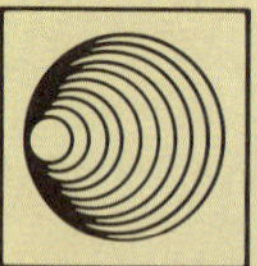

THE DISCOVERY OF NEPTUNE is especially interesting because it reveals some very human aspects of science.

After Uranus was discovered in 1781, astronomers examined old star charts and found that Uranus had been plotted on charts as far back as 1690. Those who had plotted it had mistaken it for just another star but the positions they marked allowed later astronomers to calculate the orbit of the planet. As time went on, however, it was clear that Uranus was not following the orbit predicted by Newton's laws. Astronomers took into account the gravitational pull from Saturn and Jupiter that caused Uranus's path to deviate slightly from a perfect ellipse, but Uranus still did not follow the expected path. By the early 1800s it was obvious that something was wrong. Even taking into account the gravitational effect of other known planets, the deviation of its orbit reached two minutes of arc by 1844. Two minutes of arc may not seem much, for two stars that far apart would be difficult to distinguish with the naked eye. But it was enough that astronomers could not ignore it.

Two possibilities existed: Either Newton's laws were in error or an unknown planet was disturbing the orbit or Uranus. Most astronomers felt that the latter possibility was more likely but there were those who thought that perhaps the law of gravity acts differently at the great distances involved.

In 1845, John C. Adams, a 26-year-old British mathematician, took up the problem of calculating the position of the hypothesized planet. He realized that he would be able to calculate the unknown planet's location from the fact that Uranus was first ahead of, and then lagged behind, its predicted positions.

In late September 1845 Adams took his results to Sir George Airy, the British Astronomer Royal, telling him that the planet would be found near a certain position in the constellation Aquarius. Adams first had problems contacting Airy. He must have felt like a student trying to find his advisor, for the first time he tried to deliver his results in person Airy was out of the country. Another time, Airy was out of his office. A third time, he was at dinner and his butler would not let him be disturbed. Adams left messages but they never reached Airy.

Airy was one of the astronomers who doubted the existence of the new planet, thinking that Newton's laws were inexact at the distance of Uranus. When

Does Neptune have rings? We might expect so, since the other Jovian planets have ring systems of various types. Following the discovery of Uranus's rings, whenever Neptune has occulted a star attempts have been made to find rings. Results are inconclusive. Observations made in the summer of 1984 indicated that it does have rings but that the rings are not continuous. They seem to be more like arcs than rings, as illustrated in Figure 9–37. The data indicate that the arcs are similar in density to Uranus's rings and there appear to be two separate ones—each only 8 to 20 kilometers wide. Both were found outside the Roche limit and so their existence would indeed present us with a

Airy finally got one of Adams's messages, he therefore refused to use valuable time on the large telescope for what he thought would be a fruitless search. The youth and inexperience of Adams also seems to have contributed to Airy's skepticism.

During the same period, the French mathematician Joseph Leverrier was making similar calculations of Uranus's orbit. In November 1845 Leverrier published the first part of his calculations in a scientific journal. Then in June 1846 he published his final results. Although neither knew of the other's work, Adams's and Leverrier's predicted positions for the planet were within one degree of one another.

Similar to Adams's experience with his own countrymen, Leverrier was unable to get French astronomers to search for the planet. Finally, on September 23, 1846—one year after Adams had first requested a search—Leverrier asked Johann Galle in Berlin to look for the planet. Galle quickly convinced the director of the observatory where he worked of the worthiness of the project and the new planet was discovered on the very first night of searching.

Credit for scientific discovery goes to the first person who documents his or her results. (Today such documentation normally takes the form of publication in a scientific journal.) There is little doubt, however, that if Airy had used Adams's calculations to find Neptune, the prediction would be credited to Adams. For years, controversy swirled about the priority of prediction. Recall from your study of history that England and France are historical rivals. So in this case, not only was the reputation of the two mathematicians involved but also national prestige. (Similar disagreements have arisen in this century concerning scientific discoveries made in the U.S. and the USSR.) Today the two men are given joint credit and we find the story of the discovery interesting enough to retell it in nearly every astronomy text.

mystery. Perhaps the particles of the arcs are prevented from coming together to form a moon by a phenomenon similar to that of shepherd moons.

Voyager 2 will pass Neptune in August 1989. In fact, according to present plans, it will pass closer to Neptune than any Voyager craft did to any other planet. Astronomers look forward to Voyager's help in solving some of the mysteries concerning this distant planet. They particularly want to learn more about its rings or arcs.

We know of two moons of Neptune, but since Voyager 2 will not pass this planet until 1989, we can show no photos of their surfaces.

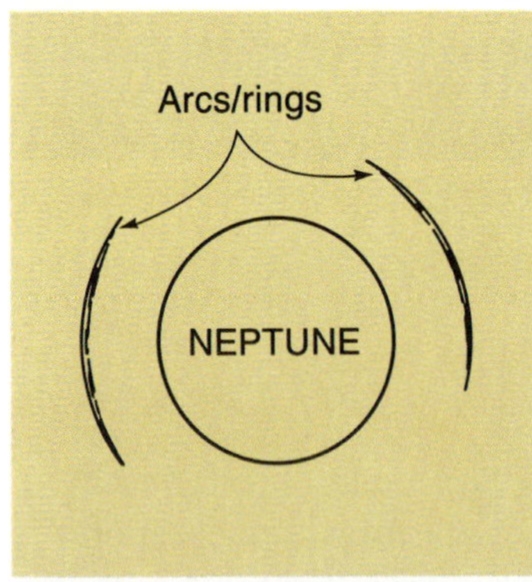

Figure 9–37. *The rings of Neptune seem to be discontinuous and are therefore thought of as arcs instead of rings. Voyager data will either confirm or refute this tentative observation.*

Table 9–4 The Planet Neptune

	Value	Compared to Earth
Equatorial diameter	48,400 km	3.8
Polar diameter	47,400 km	3.7
Mass	1.028×10^{26} kg	17.2
Density	1.77 gm/cm^3	0.32
Surface gravity	12.0 m/s^2	1.22
Escape velocity	24.6 km/sec	2.2
Sidereal rotation period	18 to 20 hours	
Albedo	0.29	0.94
Tilt of equator to orbital plane	29.56°	1.3
Orbit		
Semimajor axis	4.497×10^9 km	30.06
Eccentricity	0.0086	0.5
Inclination to ecliptic	1.77°	
Sidereal period	164.79 years	164.79
Moons	2 (+?)	

CONCLUSION

The Jovian planets certainly are different from the worlds of the inner solar system. Compared to the terrestrial planets, the Jovians are larger, more massive, more fluid (rather than solid), less dense, and rotate faster. We find that their many moons are vastly different from one another; the differences caused at least partly by their distances from the Sun and from their planet.

As we learn more about the Jovian planets, we are even more amazed at these beautiful giants. We see great similarities between them but we also see great differences. Jupiter, Saturn, Uranus, and Neptune are similar enough that we classify them into a single group but each is an individual world with its own history.

We have one more planet to discuss—Pluto. This planet is such an enigma that we placed it in the next chapter, along with various small objects that are part of our solar system.

There are many mysteries remaining in our study of the solar system. These mysteries are what make its study so interesting. If we were to run out of mysteries, science would cease to exist and we would have lost one of the pursuits that makes life so exciting.

RECALL QUESTIONS

1. Explain how Copernicus was able to estimate the relative sizes of Jupiter and Venus from a knowledge of their relative distances from the Earth and Sun.
2. We say that Jupiter has a rotation rate of 9 hours, 50 minutes. What is it on the planet which has this rotation rate?

3. What is the Galileo probe?
4. Starting at the cloudtops and descending downward, describe Jupiter's interior. What leads us to think that Jupiter has a solid core?
5. What causes the swirling between the bands we see in Jupiter's atmosphere?
6. Explain why, as the years go by, Saturn appears so different when viewed from the Earth.
7. Describe the rings of Saturn and explain the Roche limit.
8. How was it determined that Uranus was a planet rather than a comet?
9. Describe two ways the diameter of a planet can be measured from the Earth.
10. Why were the rings of Uranus not observed directly from the Earth? How were they observed?
11. What is unusual about the orientation of Uranus?
12. How do the densities of the Jovian planets compare to the densities of the terrestrial planets?
13. What causes the heating of the inner Galilean moons that results in volcanism? On which moons is this most prevalent?

QUESTIONS TO PONDER

1. Report on the latest findings by the Voyager mission concerning Neptune, its moons, and its rings.
2. Discuss the problem of Jupiter's energy balance and list possible explanations.
3. If there have been developments regarding the Galileo probe since this writing, look them up and describe them.
4. Many artificial satellites orbit the Earth very close to it, far inside the Roche limit. Why are they not pulled apart? (It is not simply their small size that prevents this. Planned space stations are fairly large.)
6. How does knowing the composition of the surface of a planet allow us to determine its size?
6. Does one pole of every planet point to Polaris? What is meant by the direction *north*? What defines which pole of a planet is its north pole?
7. A series of concentric rings can be seen on Callisto. Name two other occurrences of this phenomenon in the solar system.
8. How can it be that Titan has an atmosphere, while Mercury (only slightly smaller) has almost none?

CALCULATIONS

1. If an object had a diameter equal to four times the Earth's diameter, how would its volume compare to the Earth's?
2. If an object had a mass equal to three Earth masses and a volume five times that of the Earth, how would its density compare to the Earth's?
3. The Earth receives energy from the Sun at the rate of 1.4 kilowatts per square meter. Jupiter is 5 AU from the Sun. At what rate does it receive energy from the Sun?
4. Suppose a planet has an angular diameter of 11 arc-seconds when it is 7.5 AU from the Earth. Use the small angle formula to calculate its diameter.

CHAPTER 10

Woodcut titled *Meteoric Shower of Nov. 13, 1833.* Artist unknown. This woodcut was originally published in *Bible Readings for the Home Circle,* Review and Herald Publishing Company, 1889.

***I**n **Our First Century** R. M. Devens described the Leonid meteor shower of November 13, 1833. The woodcut (perhaps by K. Jausin; its origin is uncertain) depicts this event, which Devens called one of the 100 great and memorable events in American history. Devens wrote, "During the three hours of its continuance, the day of judgment was believed to be waiting for sunrise, and, long after the shower had ceased, the morbid and superstitious still were impressed with the idea that the final day was at least only a week ahead. Impromtu meetings for prayer were held in many places, and many other scenes of religious devotion, or terror, or abandonment of worldly affairs, transpired, under the influence of fear occasioned by so sudden and awful a display."*

From David W. Hughes, "A Mysterious Woodcut," *Sky and Telescope,* September 1987.

Pluto and Solar System Debris

WE DEVOTED A CHAPTER to the four terrestrial planets and another to the four Jovian planets. That leaves only Pluto. This planet was not included among the others because it is so different from them that it fits in neither of the above categories. We will discuss it here before moving on to the debris of the solar system: the meteoroids, comets, and asteroids.

THE DISCOVERY OF PLUTO

The fact that astronomers were able to predict the existence and location of Neptune based on irregularities in the orbit of Uranus led them to try the method once more. (See the Close Up, "The Discovery of Neptune," in Chapter 9.) Analysis of the orbital data of Uranus led to the conclusion that although the gravitational pull from Neptune accounted for about 98 percent of the variation from Uranus's expected orbit, there still were unexplained irregularities. A number of astronomers used the data to predict a ninth planet. The one whose work led to success was Percival Lowell, a successful businessman-turned-astronomer. Lowell had built an observatory near Flagstaff, Arizona, in 1894 and in 1905 he made his prediction of the existence of the new planet. He used the Lowell Observatory telescope to search for the disk of "Planet X" until he died in 1916, but he had no success.

This is the same Percival Lowell who thought he had found evidence for life on Mars.

In the 1920s a new photographic telescope was donated by Percival Lowell's brother and was installed at Lowell Observatory. On April 1, 1929, Clyde W. Tombaugh started a new search for the predicted planet. Tombaugh, however, used a different method of searching. Instead of looking for a small, faint disk, he concentrated on looking for the motion a planet must exhibit.

Tombaugh searched for the moving planet with the aid of an instrument called a *blink comparator*. This instrument is essentially a microscope containing a mirror that can be flipped quickly so that the observer can alternately look at two different photographs of the sky. The astronomer takes photos of the same area of the sky a few days apart. The two photos are then arranged in the comparator so that the stars in each photo appear at the same place as he or she changes from viewing one to viewing the other. If a moving object such as a planet is in the photos, it will appear to jump from one spot to another as the astronomer changes views.

Searching for an object in this manner is very tedious work as the comparator is scanned slowly over one pair of photographs after another. The work was especially difficult in Tombaugh's search because the predicted position of Planet X was in Gemini, which is near the Milky Way and therefore in a region with a great number of faint stars. This resulted in some 300,000 star images in each single photograph. Nevertheless, Tombaugh was successful.

On February 18, 1930, more than ten months after he began, Tombaugh detected a difference indicated by the arrows in the two photos in Figure 10–1. Consider that in searching for the planet, Tombaugh had to search not just this one pair of photos but numerous pairs. (Of course, he had the blink comparator.)

After the discovery of Pluto was confirmed by other astronomers, Tombaugh went to college enrolling as a freshman at the University of Kansas.

The new planet was announced by Tombaugh on March 13—the 75th anniversary of Lowell's birth and the 149th anniversary of the discovery of

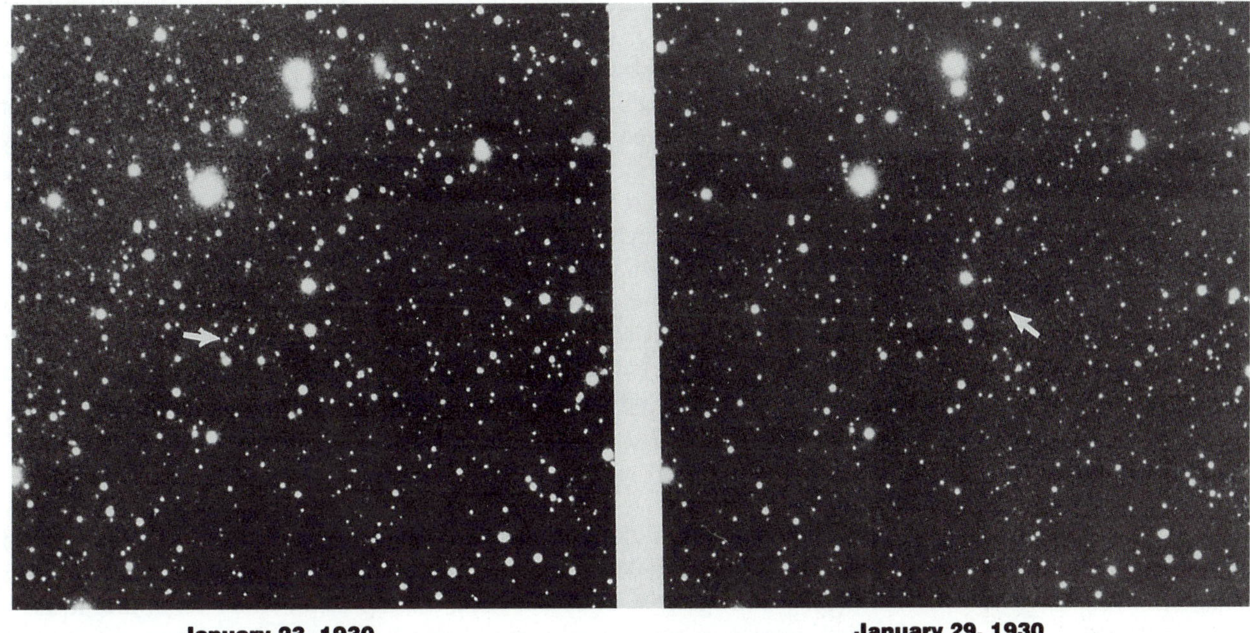

Figure 10–1. *These are the two photos by which Pluto was discovered. We have indicated the change in position of Pluto.*

Uranus. It was named Pluto after the mythical Greek god of the underworld (and also for Percival Lowell, since the first two letters of the name are his initials).

Pluto was discovered nearly 6 degrees from where Lowell had predicted it and it soon became clear that its size was so small that its mass must also be small; too small to cause the irregularities that had been seen in Uranus's orbit. More recent analysis of the orbital data used by Lowell and his contemporaries indicates that the irregularities they perceived were simply due to the limited accuracy of the data. This is an example of the importance of knowing the amount of uncertainty involved in a measurement. The irregularities noted by Lowell were not caused by another planet at all but were variations due to limited accuracy of data.

Thus we must conclude that Pluto was discovered by accident. If Lowell's calculations and predictions had not been made, however, the search would not have been carried on so diligently and the planet's discovery probably would not have been made until much later.

Pluto As a Planet

We know little about Pluto because of its great distance from us. It has a very eccentric orbit, so that although its average distance from the Sun is about 40 AU, its distance ranges from 30 AU to about 50 AU. Pluto's orbit is more eccentric than any other planet. Recall from Chapter 7 that there is another

Pluto's orbital eccentricity is 0.25, compared to Mercury's 0.21 and Earth's 0.017.

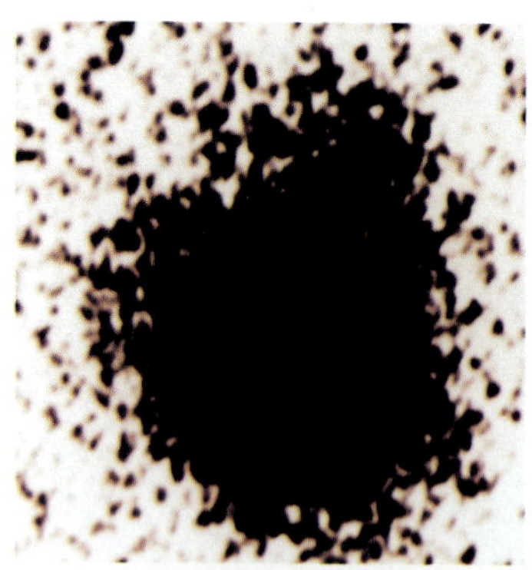

Figure 10–2. *This is the best photo we have of Pluto and Charon. This is a negative, so that the bright objects are dark here.*

unusual aspect of Pluto's orbit: it is tilted 17 degrees to the ecliptic, while no other planet's orbit is tilted more than 7 degrees.

The size of Pluto can be calculated from its angular size and distance, but the planet is so small and its image so fuzzy that such a calculation does not tell us much. In 1965, astronomers watched as Pluto's motion carried it very near a star. The fact that it failed to occult the star allowed them to calculate that its diameter must be less than 5500 km (3400 miles). In 1976, an analysis of the spectrum of light (Chapter 11) from Pluto led to the conclusion that its surface was primarily frozen methane. Knowing the fraction of light reflected from frozen methane and the amount of light that reaches us from Pluto gives us another way to calculate its size. By this method we learn that Pluto must be less than 3300 km in diameter. More recent measurements make it even smaller; only 2300 km in diameter.

Charon and the Mass of Pluto

Since no features of Pluto's surface are visible from the Earth, its rotation period has been difficult to determine. In 1956 it was observed that the planet changes slightly in brightness every 6.4 days, making astronomers think that it has a dark area on its surface and that its period of rotation is 6.4 days.

In 1978, James W. Christy of the U.S. Naval Observatory was analyzing data from its Flagstaff location and noticed that Pluto seemed to have a bump on one side. Figure 10–2 shows this bump and indicates the limited resolution of photos of this distant, tiny planet. Continued observation showed the bump to move from one side to another and it was concluded that Pluto has a moon. The moon was named Charon after the mythical boatman who ferried souls to the underworld to be judged by Pluto and also after Christy's wife, Charlene.

Charon orbits Pluto every 6.4 days and is responsible for the periodic change in brightness that was observed. Astronomers are confident that tidal friction between the two objects has resulted in them becoming gravitationally locked so that Pluto's and Charon's periods of rotation are each 6.4 days; the same as their periods of revolution.

In order to account for the perceived irregularities in Uranus's orbit, Lowell had predicted a mass for Pluto of about 6 Earth masses. We see that it has much, much less mass and that its discovery was indeed an accident.

From the distance between Pluto and Charon and their period of revolution, Kepler's third law allows us to calculate their total mass. Then comparing the light received from each and assuming that they have similar reflecting surfaces, we can conclude that Pluto's mass is less than 0.002 of the Earth's and Charon's mass is about one-tenth of that. With a diameter of about 2300 km, this yields a density for Pluto nearly twice that of water; indicating that it probably contains rocky matter along with its ices and frozen gases. Pluto is so distant, however, that we must keep in mind that the values above are extremely uncertain and may be in error by as much as a factor of two. Figure 10–3 shows Pluto and Charon drawn to scale according to our best data.

Voyager will not pass near Pluto and its moon, Charon, and there are no plans for missions to this remote system.

A Former Moon of Neptune?

For a number of reasons, including its unusual orbit (which intersects that of Neptune) and its small size (which is so much less than any Jovian planet),

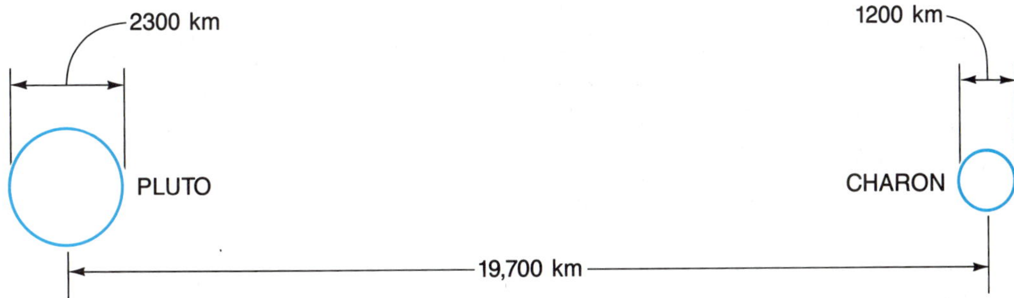

Figure 10–3. *Pluto and Charon drawn to scale.*

some astronomers have proposed that Pluto was once a moon of Neptune. There are moons of Jupiter, Saturn, and the Earth that are larger than Pluto. Perhaps a collision or near passage of some unknown celestial object caused Pluto to be ejected from the Neptunian system. Such a collision would be highly unlikely but not impossible.

Although their orbits intersect on a two-dimensional drawing, the two objects don't actually cross paths.

The discovery of Charon, however, has made this hypothesis seem even more unlikely. How could Pluto have gotten a moon for itself if it were once a moon of Neptune? Again, this would not be impossible but such a double set of coincidences reduces its likelihood. Nonetheless, the possibility cannot be completely ruled out for if an event is possible, it may have happened.

A more recent proposal is that Pluto and Charon should be classed as asteroids rather than a planet and moon. A number of asteroids are known to exist as gravitationally-bound pairs. The suggestion indicates how arbitrary our classification of celestial objects is. Nature is unified; we divide objects within it into discrete groups only to help our understanding.

SOLAR SYSTEM DEBRIS

One dictionary defines debris as "an accumulation of fragments of rock"—not far from what we will be describing. It could be said that the solar system is made up of one large object (the Sun), a number of medium-sized objects (the planets), and debris. We have discussed the medium-sized objects (as well as some smaller moons) in the last two chapters and we will discuss the large object in a later chapter. Now we will look at the debris. The material of which the debris is composed is not basically different from that which makes up the planets. Rather it is material which, when the solar system was being formed, did not become part of the Sun or the planets. The debris comes in a number of forms, including asteroids, meteoroids, comets, and dust.

ASTEROIDS

Since the first night of the nineteenth century, when Father Giuseppe Piazzi discovered Ceres, we have determined accurately the orbits of some 3000 asteroids. Figure 10–4 is a photo showing the motion of an asteroid across the

CLOSE UP

You Can Name an Asteroid

NAME AN ASTEROID? SOUNDS like a magazine advertisement. But the fact is that only about 3000 of the perhaps 100,000 asteroids that appear on photos (and thus have a diameter of about 1 kilometer or greater) have been named. Why? Because in order to name an asteroid, you must know its orbit accurately. You determine its orbit by taking photos of it and perhaps searching photos taken by other astronomers who took them with an interest in other objects. After you have determined the orbit, you must wait until the asteroid has completed at least one more cycle around the Sun in order to check your predicted orbit.

If you have indeed calculated a reliable orbit, you will be given the privilege of naming the asteroid. The official name of the asteroid will include not only the name you give, but a number indicating the order of discovery. For example, 1 Ceres and 2 Pallas. These first asteroids were named after gods of Roman and Greek mythology. Names from mythology quickly ran out and today almost any name is acceptable. Asteroid number 1001 is Gaussia, named after the mathematician who discovered the method of calculating orbits that made possible the determination of the orbit of Ceres. As you can verify in the internationally accepted catalog of asteroids, the Soviet *Ephemerides of Minor Planets,* asteroid number 1814 is named for Johann Bach, 1815 for Beethoven, and 2309 is named Mr. Spock for the Star Trek character. But you could name yours after whomever you want.

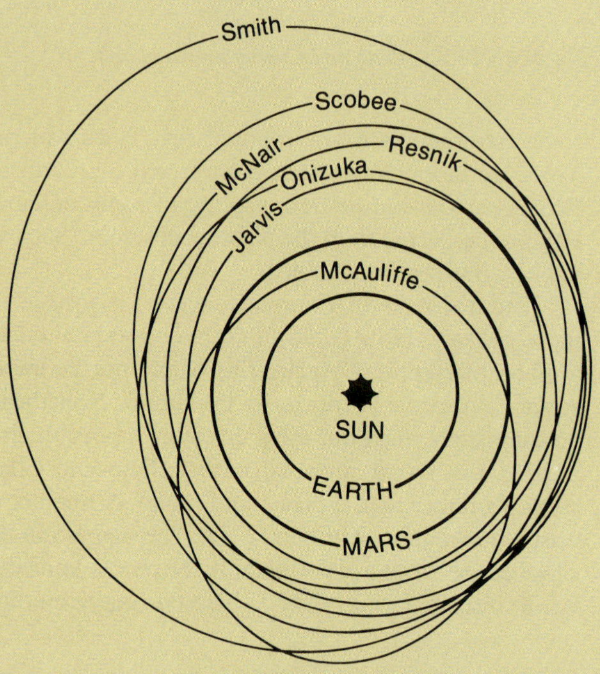

Figure T10–1. *The orbits of the asteroids named after each of the seven Challenger astronauts.*

Some of the more recently named asteroids honor the seven astronauts lost in the Challenger explosion in January 1986. Figure T10–1 shows the approximate orbits of these seven asteroids.

background of stars; it is from photos like this that most asteroids are discovered and their orbits analyzed. Approximate orbits are known for another 6000. It is estimated that about 100,000 different asteroids appear on today's photos of the heavens. These objects traditionally have been called *minor planets,* indicating their stature as objects that orbit the Sun.

Only when the orbit of an asteroid is determined well enough that its

Figure 10–4. *The telescope was following the stars during this time exposure. The asteroid Eros appears as a short streak.*

location can be predicted on succeeding oppositions do we give it a name. Since calculations of asteroid orbits are no trivial task and usually are of no great value to astronomical knowledge, most of the asteroids go nameless.

Ceres is by far the largest of the asteroids, being greater than 1000 km (600 miles) in diameter. In fact, Ceres makes up about 30 percent of the entire mass of all asteroids. Two others (Pallas and Vesta) have diameters greater than 300 kilometers, about 30 more are between 200 and 300 km in diameter, and about 100 are larger than 100 km. All the rest are smaller than that.

The masses of the three largest asteroids are calculated by observing the perturbations they cause on smaller asteroids passing nearby. Masses of other asteroids are estimated based on their sizes and expected densities. But how are their sizes known? Although they are too small to form disks of measurable size in our telescopes, there are two other methods that allow us to determine their sizes.

Measuring the Size of an Asteroid

One way to know the size of an asteroid is to observe it as it moves across in front of a star. By observing such an occultation from various places on Earth, we can get fairly accurate values for the size of the asteroid. (This is the method we described in the previous chapter for measuring Uranus's size.)

A more accurate method, especially for smaller asteroids, relies on measuring the total amount of radiation received from the asteroid. Asteroids do not reflect 100 percent of the light that strikes them, of course. They absorb some and reflect the rest. The key idea, though, is that they must re-emit every bit of radiation they absorb. (This re-emitted radiation is in the form of infrared radiation.) If an object absorbed more energy than it emitted, it would necessarily increase in temperature. As we discussed when describing the greenhouse effect on Venus, a temperature increase would result in an increase in infrared radiation emitted until equilibrium is reached. On the other hand, an object cannot emit more radiation than it absorbs unless it has an internal source of energy. (We saw that this is the case for Jupiter and perhaps some other Jovian planets.)

Do not associate the word radiation *only with nuclear radiation. The word refers to electromagnetic radiation (photons) and to movement of charged particles.*

Table 10–1 The Planet Pluto

	Value	Compared to Earth
Diameter	2300 km	0.2
Mass	10^{22} kg	0.002
Density	2 g/cm^3	0.35
Surface gravity	0.5 m/s^2	0.05
Escape velocity	1 km/sec	0.1
Sidereal rotation period	6.39 days	6.4
Albedo	0.5	1.3
Tilt of equator to orbital plane	118°?	
Orbit		
Semimajor axis	5.900×10^9 km	39.44
Eccentricity	0.250	15
Inclination to ecliptic	17.2°	
Sidereal period	248.5 years	248.5
Moons	1	

Thus step one in measuring the size of an asteroid is to measure the radiation we receive from it and then use the inverse square law to calculate the total energy the asteroid is reflecting and emitting. This is equal to the total energy the asteroid is receiving from the Sun.

Step two in the process is to calculate how much solar energy strikes one square meter of the asteroid's surface. We are able to do this because by knowing the orbit of the asteroid, we know its distance from the Sun and we can use the inverse square law and a knowledge of the amount of energy emitted from the Sun to calculate how much solar radiation strikes a square meter of the asteroid's surface.

So now we know both the amount of radiation striking a square meter of its surface and the total amount of radiation hitting it. It is a straightforward matter to calculate how many square meters make up its surface.

The Orbits of Asteroids

The direction of motion of all of the asteroids is, like that of the planets, counterclockwise as viewed from north of the solar system. Most of their planes of revolution are near the plane of the ecliptic, although we know of one with a plane of revolution 64 degrees from Earth's (Figure 10–5).

Asteroid belt: The region between Mars and Jupiter where most asteroids orbit.

Most of the asteroids orbit the Sun at distances from 2.2 to 3.3 astronomical units (in what is called the ***asteroid belt***). This corresponds, by Kepler's third

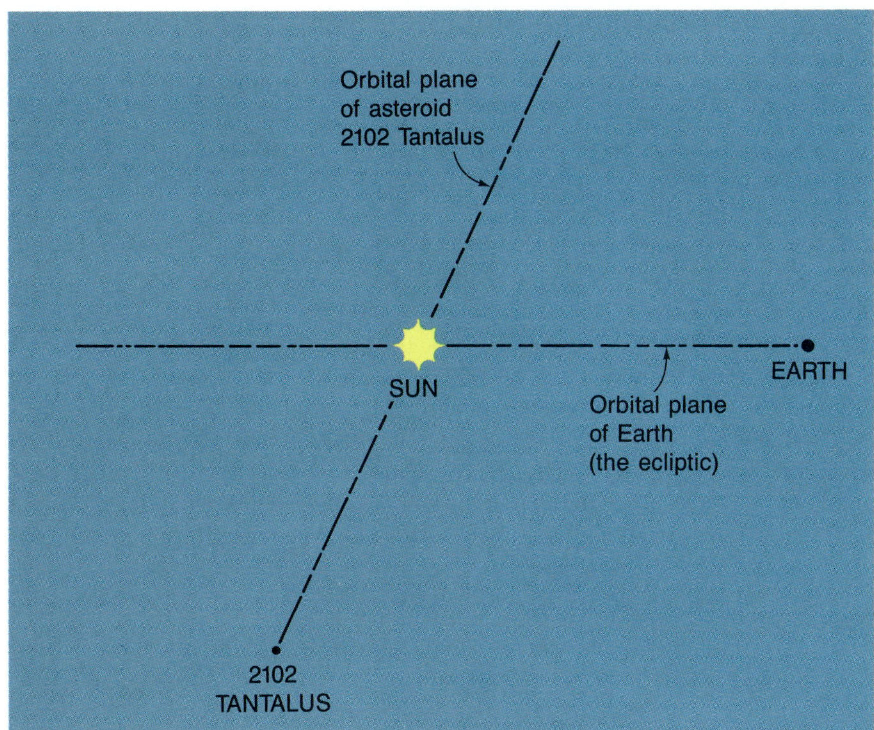

Figure 10–5. *Most asteroids orbit the sun in planes close to the ecliptic but the orbital plane of the asteroid 2102 Tantulus is at 64 degrees inclination to the ecliptic.*

law, to periods of from 3.3 to 6 years. Most of the asteroids in the asteroid belt have fairly circular orbits, although not as circular as the orbits of the major planets.

We must not think of the asteroid belt as being crowded. Thc asteroid belt is nothing like the swarms of rocks through which spacecraft ducked and dove in the popular video game of some years ago. When the Pioneer and Voyager spacecraft passed through the asteroid belt, they certainly did not have to make evasive maneuvers to avoid asteroids.

If the 100,000 asteroids that appear in photographs were in exactly the same plane and were spread evenly across the asteroid belt, neighboring asteroids would be separated by 2 million kilometers; more than a million miles. In reality, the asteroids are not regularly spaced in the asteroid belt but, on the other hand, they are not all in the same plane. So two million kilometers is a low estimate for the average distance between the 100,000 asteroids large enough to appear on photos. There are many more asteroids smaller than this; but even so, the asteroids present no major hazard to spacecraft. Compared to the popular picture of a crowded asteroid belt, the belt is almost empty. To illustrate this, imagine that we make a scale model in which the largest asteroid—Ceres—is the size of a grain of sand. In this case, a normal size asteroid would be one meter away, and would be too small to be seen (Figure 10–6).

Not all of the asteroids orbit the Sun between Mars and Jupiter. About 50 with diameters larger than 1 kilometer are known to have orbits eccentric enough

CLOSE UP

Chaos Theory

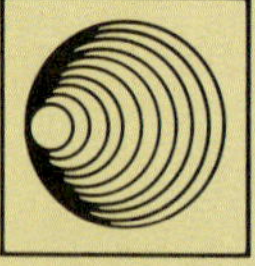

MANY PHENOMENA IN NATURE are very complicated; they involve numerous objects and complex motions. Think of the path taken by smoke as it climbs from the end of a cigarette. The motions of the myriad particles in the ring system of a planet or the motions of the asteroids in the asteroid belt are other examples. Sometimes a single large object that is acted on by a great number of other objects exhibits what seems to be unusual motions. An example of this is Hyperion, a moon of Saturn that tumbles in erratic motions as it orbits the planet. Scientists historically have assumed that if they could analyze the multitude of forces on these objects, their seemingly random motion could be understood using Newton's laws but it appeared that the complete understanding of some complicated systems was beyond our capability.

A new method of analysis, called *chaos theory,* is finding success in investigating some of these systems. The origin of the theory's name comes from its ability to explain seemingly chaotic situations. The theory works by finding patterns within a system's complexity using the power of computers to find the patterns. The theory is still in its infancy but people working with it predict that someday it will be considered one of the great theories of science. Time will tell.

Apollo asteroids: Asteroids that cross the Earth's orbit.

that they cross the Earth's orbit. Some of these have passed fairly near the Earth in recent history. In 1986 the asteroid Icarus passed 6 million kilometers (4 million miles) from our planet. The recent record, however, is the asteroid Hermes, which passed within 900,000 kilometers of Earth on October 30, 1937. This is slightly more than twice the distance of the Moon from Earth. When we discuss meteors, we will describe what happens when a large asteroid hits the Earth.

Kirkwood's Gaps As the orbits of more and more asteroids were calculated during the nineteenth century, it became clear that they are not evenly distributed across the asteroid belt. At certain distances from the Sun, there are gaps in the belt. Figure 10–7 illustrates the situation. One very prominent gap occurs at 2.50 AU and another at 3.28 AU. Daniel Kirkwood, an American astronomer, first explained these in 1866. He noticed that 3.28 AU corresponds to a period of revolution of 5.93 years, and that this is just half of Jupiter's period of 11.86 years. An asteroid at 2.50 AU would orbit the Sun in 3.95 years, which is one-third of Jupiter's period. We saw a similar situation in the last chapter in the explanation for Cassini's division in Saturn's rings. In both cases, synchronous tugs from a large object on smaller particles gradually move those particles out of their orbits and result in gaps.

Jupiter has cleared out other regions of the asteroid belt where it exerts a regular gravitational pull on objects that may once have been there. Besides the

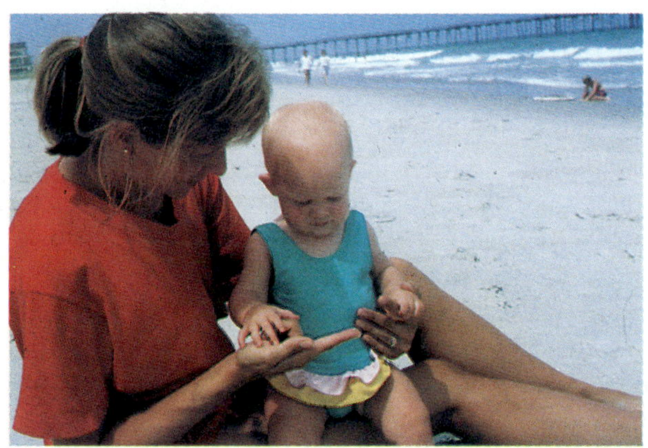

Figure 10–6. *If Ceres, which has a diameter of about 1000 kilometers, were the size of a grain of sand, its nearest neighbor would typically be a meter away but it would be too small to see, as is the baby's grain of sand here.*

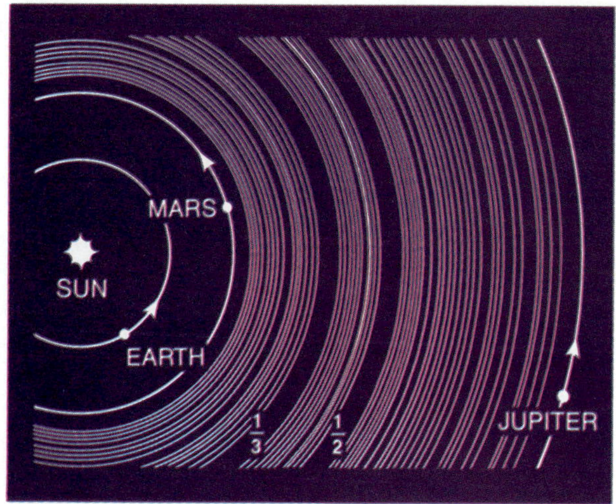

Figure 10–7. *Kirkwood gaps appear where asteroids would have periods of one-third and one-half of Jupiter's.*

gaps at locations corresponding to one-third and one-half of Jupiter's period, there are also major gaps corresponding to two-fifths and three-fifths of the period of the giant planet.

The Origin of Asteroids

It was once thought that asteroids are the remains of the explosion of a planet. This theory has been abandoned today for two reasons. First, there is no known mechanism by which a planet could explode. Second, if all of the asteroids were combined into one object, that object would be only about 1500 kilometers in diameter; much less than our 3500 kilometer-diameter Moon. Such a small object does not fit the pattern of planetary sizes.

It is considered much more probable that the asteroids are simply primordial material that never formed into a planet. There is a good reason that the material in the region of the asteroid belt did not form into a planet; the reason is Jupiter. The gravitational pull from that planet causes a continual stirring effect on the objects in the asteroid belt. This prevents the small gravitational forces that exist between the asteroids from pulling them together to form a larger object.

Even today, there is evidence that Jupiter causes collisions between asteroids that result in their fragmentation. In fact, some of the asteroids that have orbits outside the main belt are thought to have resulted from such collisions. Analyses of the orbits of such asteroids have shown that there are groups ("families") whose orbits, when traced backwards, indicate that they were once together. They apparently broke apart as the result of a collision with another asteroid caused by the gravitational force of Jupiter.

We will return to a brief discussion of asteroids when we take up the question of the origin of meteoroids later in the chapter. Now we turn to another category of solar system debris—comets.

COMETS

One of the most spectacular astronomical sights available to the naked eye is a comet. Figure 10–8 is a photo of Comet West, which made its appearance in 1976. Unless a person has seen a comet, it is common to assume from a photo such as Figure 10–8 that the comet streaks across the sky in the direction opposite to its tail. Perhaps from seeing streaks behind fast-moving characters in cartoons, we assume that the comet in the figure is moving downward at great speed. In fact, if you see a comet in the sky, you observe no rapid motion at all. Unless you observe carefully, the comet will appear to stay in the same place among the stars, having only the motion across our sky caused by the rotation of the Earth. (It does move among the stars, of course, and its motion can easily be seen over a few days.) Don't confuse a comet with a meteor, which *does* streak across the sky—and which will be discussed later.

Figure 10–8. *Comet West, discovered by Richard West of European Southern Observatory in 1976.*

Until the investigations of comets by Tycho Brahe (the same person who took such careful data on Mars used later by Kepler), it was common opinion that comets were a phenomenon in the Earth's atmosphere. When a bright comet appeared in 1577, Tycho took measurements of it from a number of locations and was unable to detect the parallax that would be associated with an object in our atmosphere. In addition, he could not detect any parallax of the object due to the rotation of the Earth. (See Figure 10–9.) He concluded from his observations that the comet must be at least three times as far away as the Moon (for which he could observe parallax) and that the comet is an object that orbits the Sun.

It is interesting to note that although Johannes Kepler discovered that planets move in elliptical orbits, his conclusion (from a study of a bright comet of 1607) was that comets move in straight lines through the solar system.

Kepler was nearly right, for an elongated ellipse is almost a straight line except at the ends.

Comet Orbits—Isaac Newton and Edmund Halley

Isaac Newton proposed that comets orbit the Sun according to his law of universal gravitation and laws of motion just as do the planets. He concluded that since they were visible from Earth for only short periods of time (typically a few months), their orbits were very eccentric; very elongated.

Edmund Halley was a friend of Newton. In fact, he was the person who talked Newton into publishing the *Principia,* who personally financed the book, and who was the only person for whom Newton expressed appreciation in the book. Halley used Newton's methods, his own observations, and descriptions

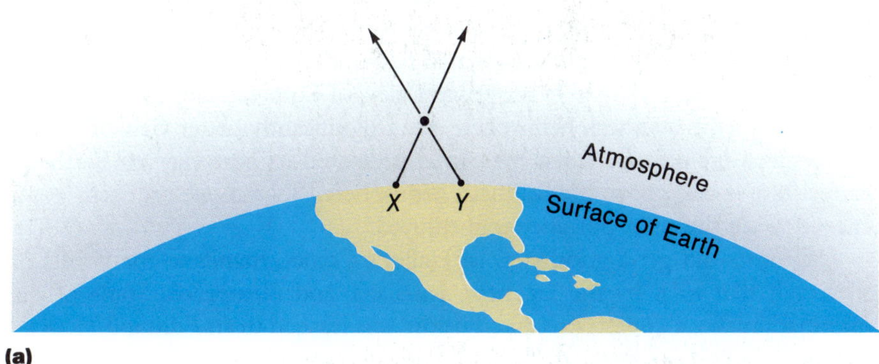

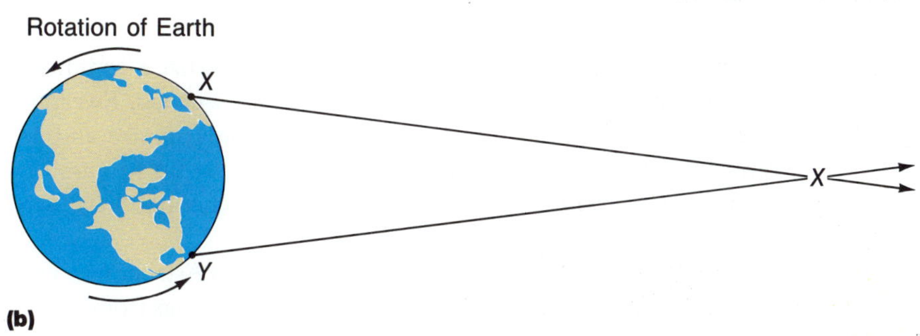

Figure 10–9. *(a) An object in the atmosphere will be seen by observers* X *and* Y *(who are at different locations) to be among different stars. (b) An object as near as three times the Moon's distance will exhibit parallax detectable with the naked eye as the Earth turns.*

of previous comet sightings to calculate the orbits of a number of comets. He noticed that the orbits of comets of 1531, 1607, and 1682 were very similar and suspected that they might be the same comet, but he was at first confused by the fact that the time that had elapsed between one appearance of the comet and the next was not always the same. When he realized that the comet's path would be changed slightly by the gravitational pull of a planet—particularly Jupiter or Saturn—when the comet passed nearby, he hypothesized that these three comets were in fact three appearances of the same comet. He predicted that the comet would reappear in 1758.

On Christmas night of 1758, the comet was sighted. In honor of its predictor, it was named Halley's Comet. The brilliant Edmund Halley, however, had died sixteen years earlier at the age of 85 and was unable to see his hypothesis verified. By investigating reports of comets in literature, we can now trace Halley's Comet (or *Comet Halley*) back as far as the year 239 B.C.

Today we discover about a dozen comets each year. Of the nearly 1000 whose orbits have been calculated, about 100 have a period of revolution around the Sun of less than 200 years. Most have extremely long periods of up to a million years.

The planes of revolution of comets are not limited to the plane of the planets but are randomly oriented. Comets sweep past the Sun from all directions.

Halley's is the comet best known to the general public, but it is just one of many comets that appear in our sky. Its fame is probably due not only to the story of its discovery but to the fact that it is the brightest short-period comet. Its period closely matches the human lifetime, so most of us get only one chance to see this comet. This is why the 1985/1986 return was particularly disappointing for observers in the northern hemisphere. Figure 10–10 helps explain why relatively few people saw the comet on that pass. Notice that when its tail was longest, the Earth was situated so that the comet's tail did not stretch across the sky. This resulted in the poorest view of Comet Halley in 2000 years. People in Earth's southern hemisphere got a significantly better view of it than northerners did because when the comet presented its best view to Earth, in early 1986, it was above the southern hemisphere. But no one on Earth got a chance to see the sight our ancestors enjoyed.

Scientists did get a good view of Halley's Comet, thanks to the flotilla of spacecraft sent to it by the Japanese, Russians, and Europeans. Some of the information about comets presented below was learned from Edmund Halley's comet.

The Nature of Comets

Coma (KOH-mah): The part of a comet's head made up of diffuse gas and dust.

Nucleus (of comet): The solid chunk of a comet, located in the head.

Tail (of comet): The gas and/or dust swept away from a comet's head.

Figure 10–11 illustrates the composition of a typical comet. What is commonly called the head of the comet consists of the ***coma*** and the ***nucleus.*** Sweeping away from the head is the comet's ***tail,*** which varies greatly in size and appearance from comet to comet. The head of a comet may be up to a million kilometers in diameter; almost as large as the Sun. Some comet tails have been as long as an astronomical unit.

The basic model of a comet nucleus proposed by Harvard Astronomer Fred L. Whipple in 1950 had to be modified only slightly by the findings from Comet

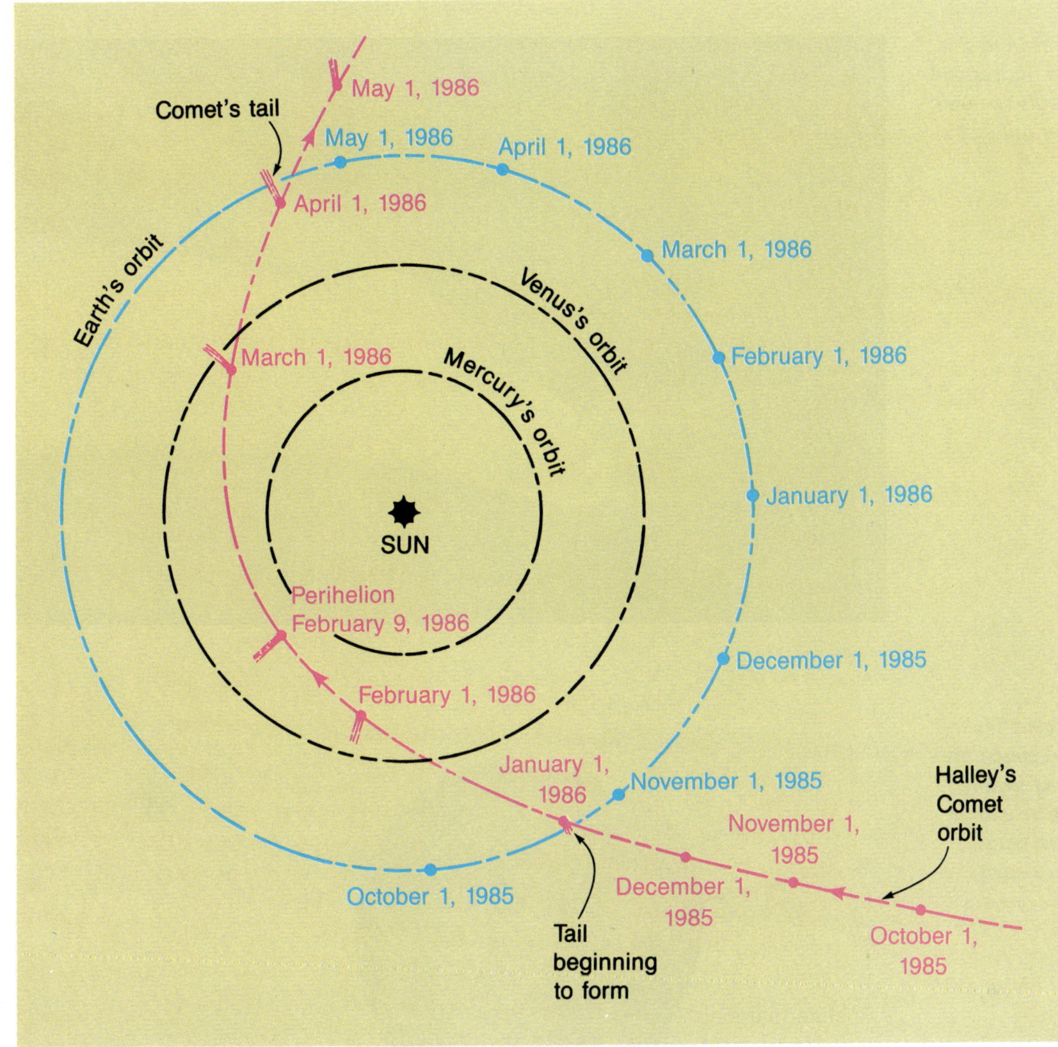

Figure 10–10. *Halley's Comet was closest to Earth in spring 1986, when it was above the southern hemisphere.*

Halley. Whipple proposed that the nucleus resembles a dirty snowball made up of water ice, frozen carbon dioxide, a few other frozen substances, and small solid grains (the "dirt").

According to Whipple's model, when the comet is far from the Sun the nucleus—a few kilometers in diameter—is all there is. As the comet approaches the Sun its surface becomes warmer, causing the materials to melt and vaporize. The coma, he proposed, consists of the molecules of these materials that hover around the nucleus, held there by the small gravitational force toward the nucleus. The dust that is known to exist in the coma comes from "dirt" that was mixed with the "snow," he said.

Observations of Comet Halley in 1986 revealed that the surface of the nucleus is very black and that it has cavities. The vaporized material from inside

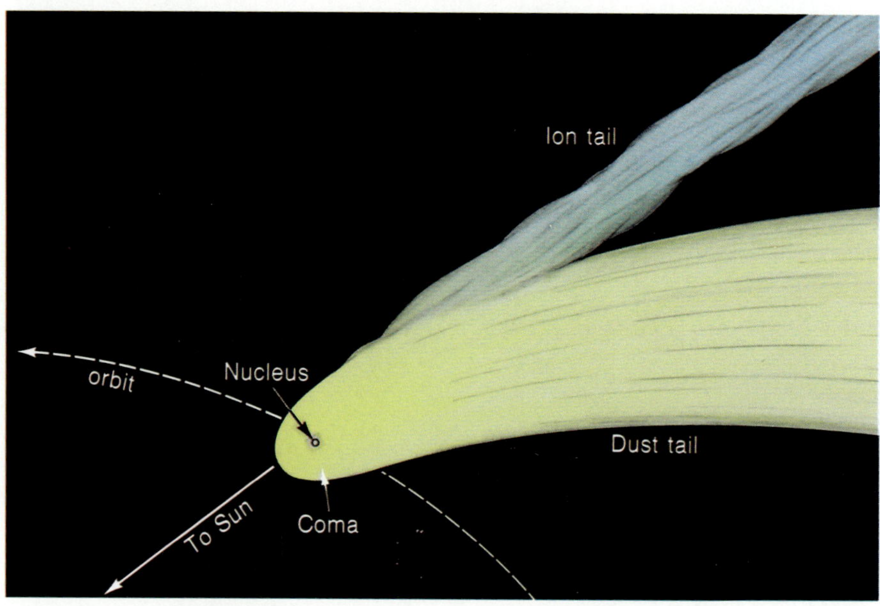

Figure 10–11. *The head of a comet is made up of its nucleus and coma. There are normally two tails; a relatively straight ion tail and a curved dust tail.*

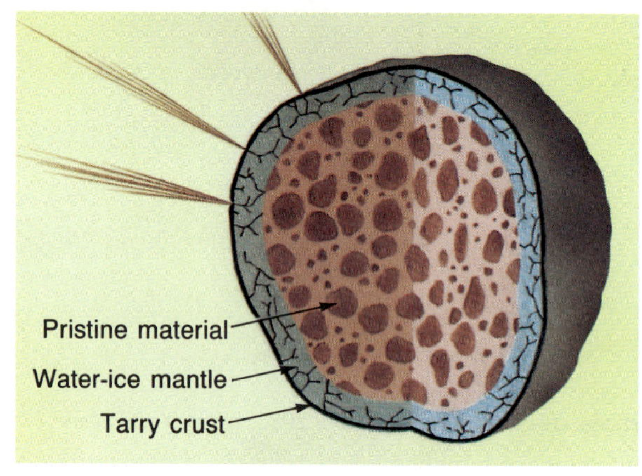

Figure 10–12. *Comet Halley (and presumably other comets) are made up of a mixture of ice and dust surrounded by an ice mantle and covered with a dark crust. Material from inside is ejected through the crust as the comet is heated by the Sun.*

spews forth from these holes in the surface crust. Figure 10–12 shows a cross-sectional view of the nucleus of Comet Halley. Even before the flybys, it had been observed that material is ejected from a comet's nucleus in jets rather than being uniformly evaporated from the surface as originally proposed in Whipple's model.

In miles, this is 10 mi. × 4.5 mi. × 5 mi.

The nucleus of Comet Halley is probably typical of other comets. Its shape has been described as like an avocado; about 16 kilometers in length, 7.5 kilometers in width, and 8 kilometers across its other dimension.

The coma and tail of a comet are mostly empty space. Giotto, the space probe sent to Comet Halley by the European Space Agency, flew to within about 600 kilometers (350 miles) of the nucleus and measured the density of material in the coma. It found that the coma is billions of times less dense than

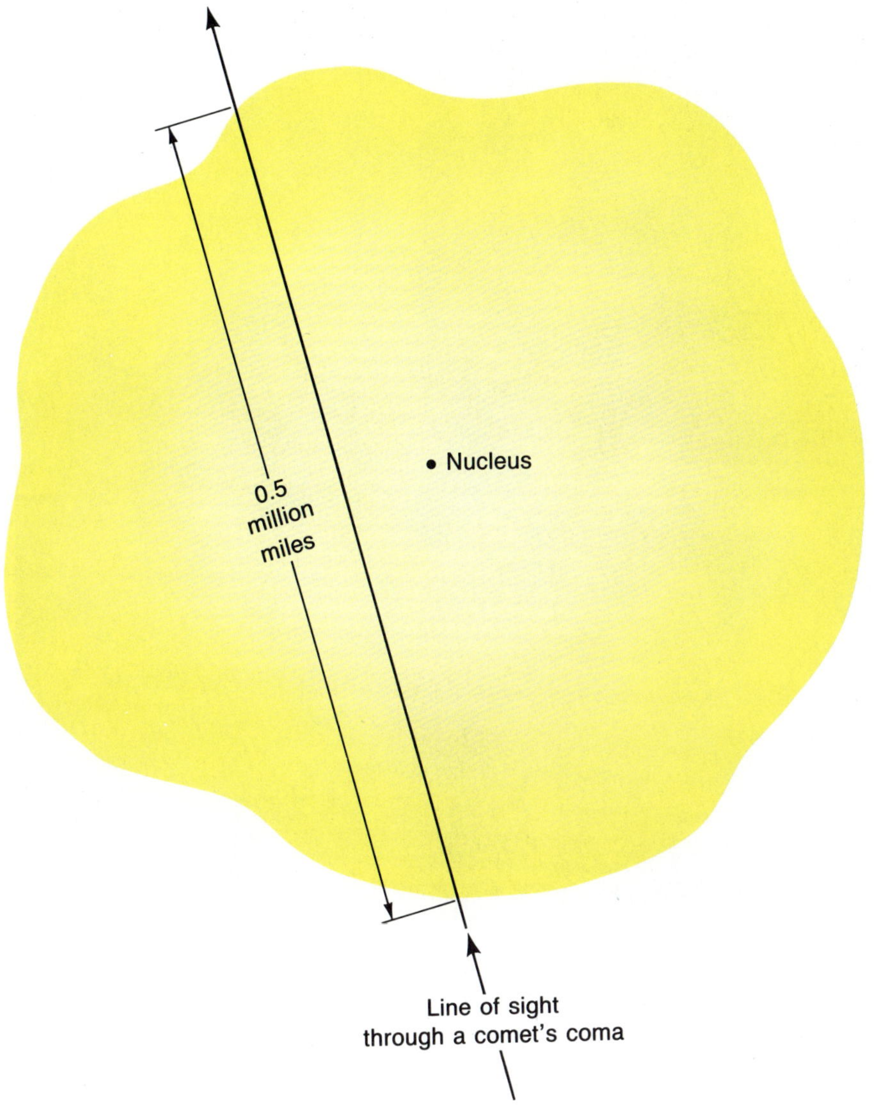

Figure 10–13. *Although the density of material in a comet's coma is low, its large size means that there is quite a bit of material along any line of sight that passes through it.*

the atmosphere of Earth at sea level. The entire mass of a typical comet is less than one billionth the mass of the Earth, and more than 99.99 percent of this mass is in the nucleus. The reason that we can see the coma at all is that its molecules and dust particles reflect sunlight to us and ultraviolet light from the Sun causes the molecules to fluoresce in much the same manner as a "black light" poster glows in ultraviolet light. There are very few molecules and dust particles in any particular volume of the coma but the coma is extremely large, so that there is a lot of material along any particular line of sight (Figure 10–13).

Figure 10–14 is a photo taken from the Giotto spacecraft. Notice the jets of material coming primarily from one side of the nucleus; the side facing the Sun. It seems that what is happening is that the Sun is warming the dark crust

The spacecraft Giotto was named after the artist who included Comet Halley as the star of Bethlehem in his fresco of the nativity scene. He had just seen a return of the comet a few years earlier, in 1301.

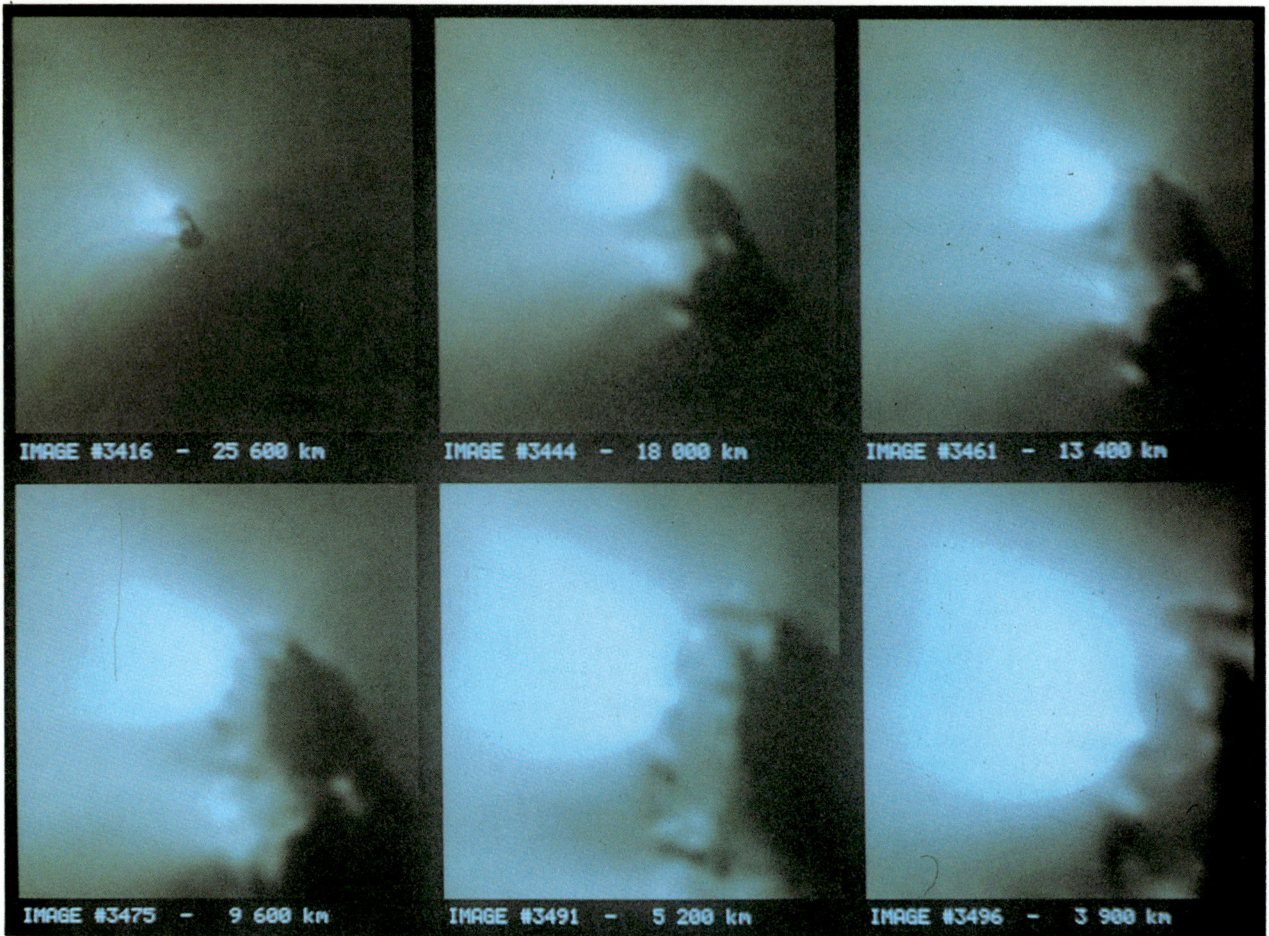

Figure 10–14. *These are photos of Comet Halley taken at various distances by the Halley Multicolour Camera on board the European Space Agency's Giotto spacecraft. The jets of material come from the sunlit side of the nucleus.*

of the comet, causing the ices below to vaporize and spew out through openings in the crust. Comet Halley rotates with a period of about 53 hours and as one side rotates away from the Sun, the jets on that side shut down, while jets on the side turning toward the Sun become activated.

Comet Tails

As indicated earlier, the tail of a comet does not necessarily follow the head through space. Rather, a comet's tail always points away from the Sun. Figure 10–15 shows a comet near the horizon at sunset. Notice the location of the tail relative to the Sun. If the tail points away from the Sun, this means that

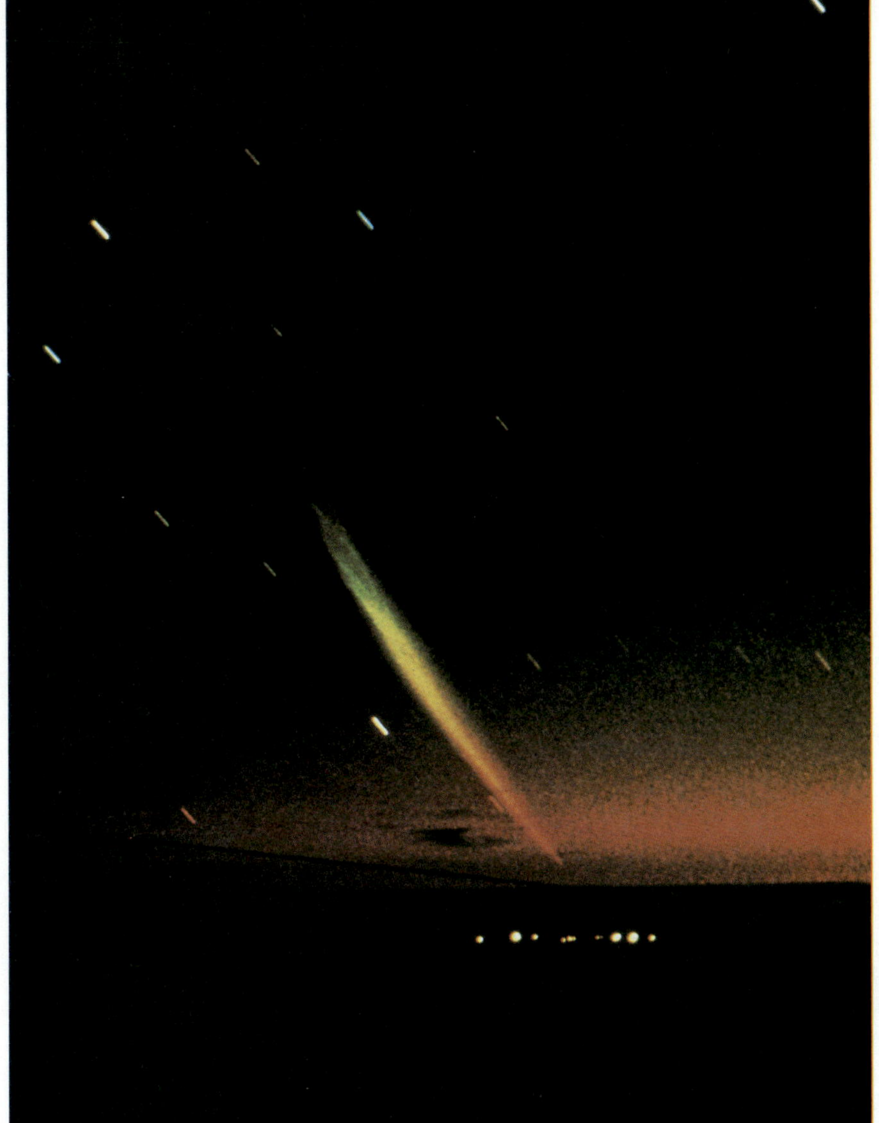

Figure 10–15. *Comet Ikeya-Seki was near the horizon after sunset when this photo was taken in 1965.*

after a comet has passed the Sun, its tail actually *leads* it through space (Figure 10–16). This takes some explaining.

First, comets have two tails although one or both may be very small and they may change greatly as time passes. Figure 10–17 shows four photos of a 1957 comet. Notice that one tail is very straight while the other curves away from the comet. The straight tail consists of charged molecules (***ions***) being swept away from the comet by the solar wind—nuclear particles that are always being emitted by the Sun. These molecules move away at a great speed and form a straight tail.

Ion: A charged atom resulting from the atom's loss or gain of an electron.

The curved tail is caused by the grains of dust in the coma being pushed away by the weak pressure of solar radiation. This radiation pressure is not detectable by us in our everyday life but light does indeed exert a tiny force on

Figure 10–16. *In general, a comet's tails point away from the Sun. Notice that the tails are largest shortly after the comet has passed the Sun.*

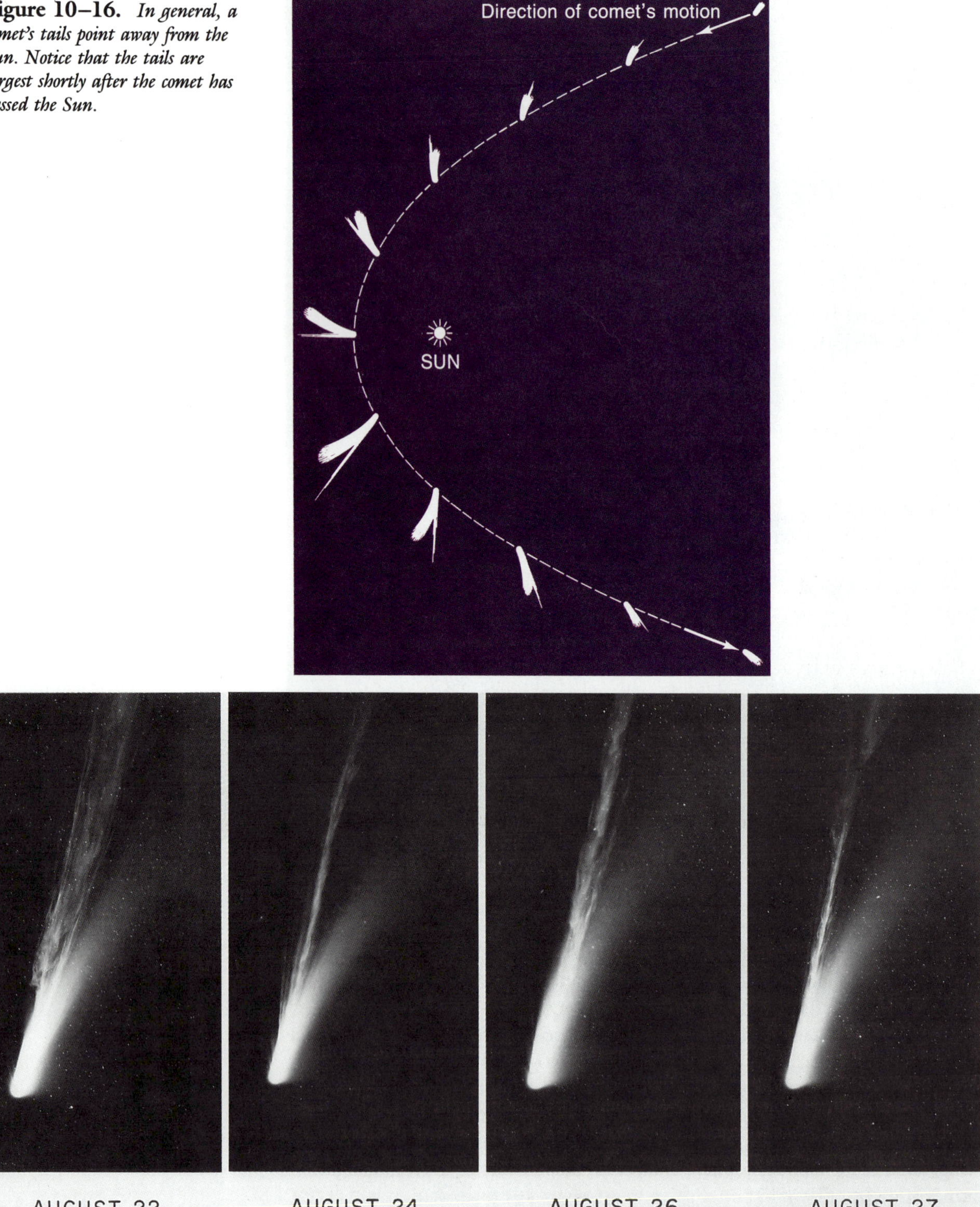

Figure 10–17. *The two tails of Comet Mrkos are obvious in these photos, taken on different dates during its appearance in 1957.*

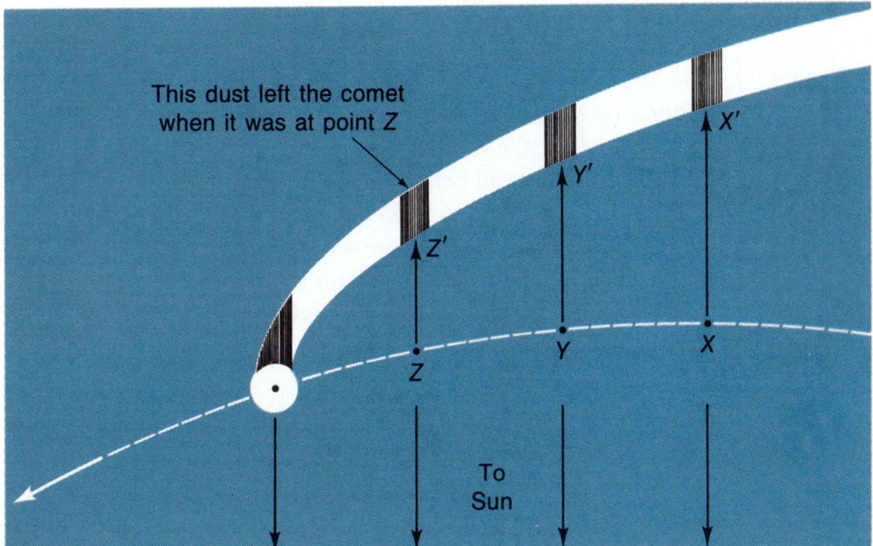

Figure 10–18. *Dust that was blown off the comet when it was at point X is now at X'. Dust blown off at Y is at Y'. This causes the dust tail to be curved.*

objects it strikes. In the weak gravitational field of the comet's nucleus, the force is great enough to move dust particles away from the head. They move much more slowly than the molecules that form the ion tail, however, and Figure 10–18 shows why this results in the dust tail being curved. Dust particles that left the comet when it was at location *X* have reached *X'*, those that left at *Y* are at *Y'*, and so on.

Sometimes only one tail of a comet is visible because one is behind the other as seen from Earth. Some comets, on the other hand, are not as dusty as others and do not contain a prominent dust tail.

In general, the tails of comets point away from the Sun regardless of the motion of the comet nucleus because the agents responsible for them (the solar wind and light) are emitted radially from the Sun.

Comet Death

If a comet loses material as its gas and dust are pushed off to form its tails, comets must have limited lifetimes. Comet Halley spews about 25 to 30 tons of its mass through its jets each second when it is close to the Sun. This sounds like a great deal, but it means that the nucleus loses less than 1 percent of its mass on each pass of the Sun. Comet Halley is now on its way toward the outer portion of its elliptical orbit and since its jets are now inactive, it is no longer losing mass. (It will continue out past Neptune's orbit before returning to our region of the solar system in 2061.)

Comets do gradually lose their ices, however. It is thought that some comets finally evaporate away all of their nucleus so that they just fizzle out. Other comets, after they lose all of their volatile materials, become simple chunks of rock and probably would be classified as asteroids. A third way that comets die is to fall into the Sun. A comet that comes close to the Sun is slowed by the Sun's atmosphere and after a number of passes, it will be slowed enough that it falls into the Sun.

THE OORT CLOUD

Thus far we have been considering only short-period comets—those with periods of a few hundred years. Most comets have much longer periods and approach us from far beyond the orbits of the planets. About a dozen comets are discovered each year; almost all being long period comets.

In 1950 the Dutch astronomer Jan Oort proposed that in a region of space that lies from 10,000 AU to 100,000 AU from the Sun there exist great numbers of comets that orbit the Sun (Figure 10–19). This shell of comets surrounding the solar system has become known as the ***Oort cloud.*** Naturally, the Oort cloud is too far away for its comets—which are simply nuclei at that distance—to be visible from Earth.

Oort cloud: The theorized spherical shell lying between 10,000 and 100,000 AU from the sun containing billions of comet nuclei.

The comets of the Oort cloud can be divided into two categories. First, many of the comets we observe are in elliptical orbits with extremely long periods. They obey Kepler's second law, however; this means that they move at extremely low speeds when they are far from the Sun. If this is true, they spend the overwhelming majority of the time well beyond planetary distances. For example, a comet with a period of 500,000 years would spend about 499,999 of those years beyond the orbits of the planets. Thus many of the comets in the Oort cloud are there simply in accordance with Kepler's laws.

Second, there must be many comets in the Oort cloud whose orbits never approach the inner solar system. This would not necessarily mean circular orbits, for a comet could vary from 10,000 Au to 100,000 AU from the Sun and still remain in the Oort cloud. From time to time one of the comets of this category must pass near another comet, causing it to change its path. This may result in a more circular path or it may result in it being moved into an orbit that takes

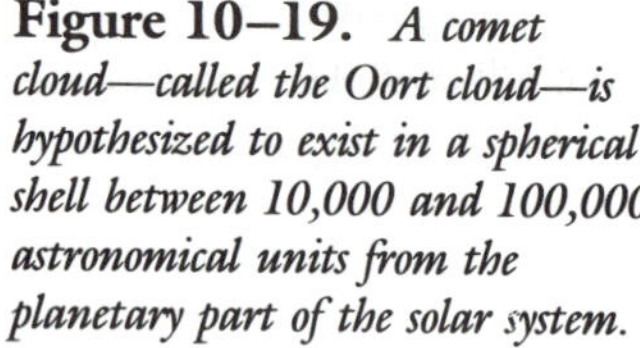

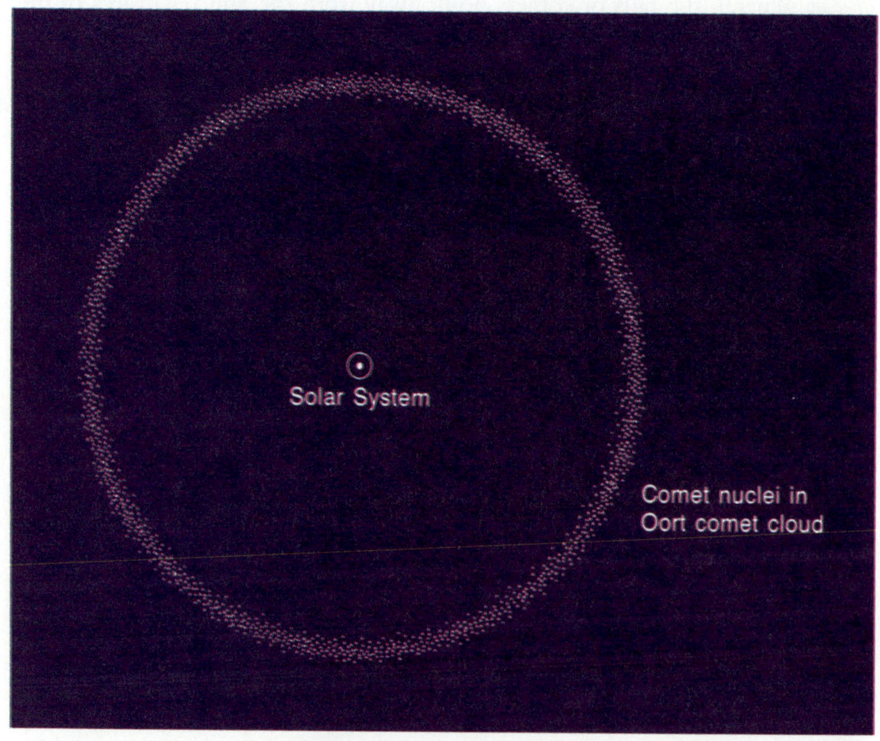

Figure 10–19. *A comet cloud—called the Oort cloud—is hypothesized to exist in a spherical shell between 10,000 and 100,000 astronomical units from the planetary part of the solar system.*

it toward the inner solar system. In addition, although the outer Oort cloud stretches only about one-third of the way to what is now the nearest star, every few million years a star passes closer to the Sun than this and gravitational forces from the star cause changes in the orbits of comets. Some comets are believed to be deflected into the inner solar system by this method also. These comets from the Oort cloud become long-period comets.

The Origin of Short-Period Comets

The age of the solar system is about 5 billion years. Why have those 100 comets with periods of less than 200 years not evaporated away their ices so that they are no longer observed? The answer must be that these short-period comets are relative newcomers to the inner solar system. We must ask where they come from and an obvious hypothesis is that long-period comets sometimes become short-period comets. How might this occur?

If the path of a comet depended solely on its gravitational attraction toward the Sun, there would be no way in which its orbit could be changed so that it became a short-period comet. Other objects (particularly the massive Jupiter) affect its orbit, however. The combined effects of the gravitational forces from Jupiter and the Sun can cause one of these comets to change its orbit so that it then becomes a short-period comet. Figure 10–20 illustrates one way for

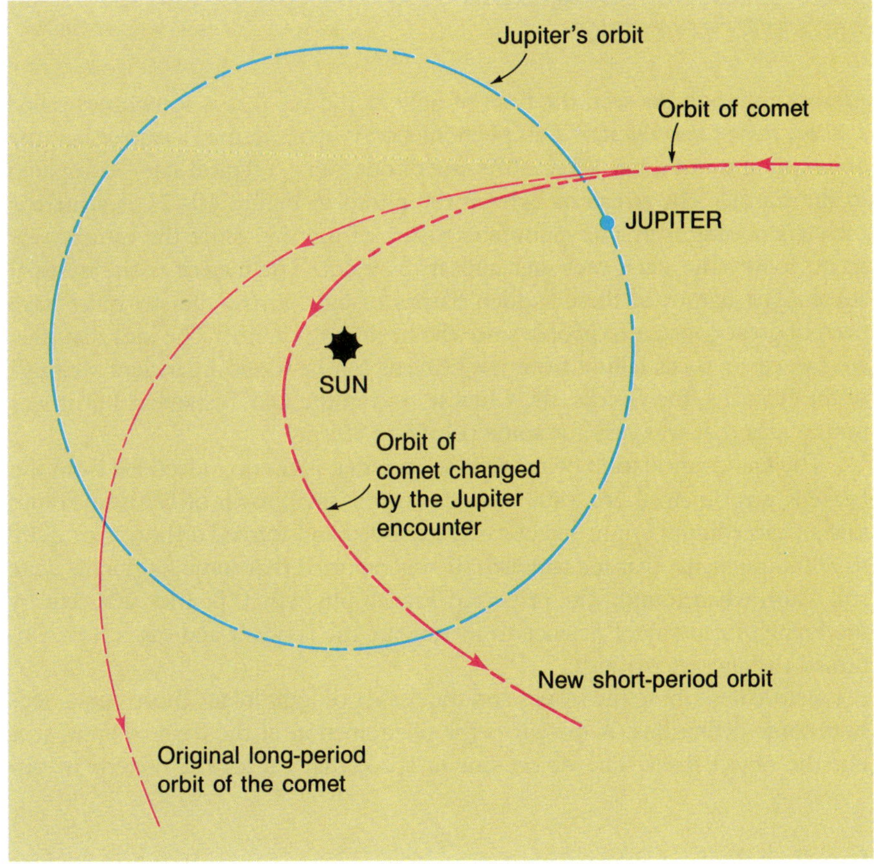

Figure 10–20. *Gravitational attraction to a planet—Jupiter, here—can cause a comet to alter its orbit; perhaps changing a long-period comet to a short-period one.*

this to occur. All of the short-period comets were captured in the inner solar system by such a mechanism. On the other hand, the Sun and a planet can have the effect of changing the orbit of a comet so that it leaves the solar system and the Oort cloud entirely.

The large range of estimates illustrates how little we know of the Oort Cloud.

Most comet orbits are either elliptical or parabolic, not hyperbolic. The latter would indicate that the comet is not bound to the Sun.

Astronomers have no idea how many comets are in the Oort cloud. Estimates range from a million to a trillion. We must remember that the Oort cloud is simply an hypothesis to explain three observations. First, new comets are continually joining the inner solar system. Second, long-period comets pass the Sun from all directions. Their observed motions cause us to conclude that they have their origin in a regularly distributed cloud around the sun. Finally, the orbits of the vast majority of comets indicate that they are gravitationally bound to the Sun.

The word *cloud* in reference to comets orbiting the sun can be misleading. When you think of an Earthly cloud, you may think of a volume fairly crowded with water droplets. But the Oort cloud is far from crowded. We pointed out that the asteroid belt is mostly empty space, but it is crowded compared to the Oort cloud. If the Oort cloud contains a trillion comets—a high estimate—there would still be an average distance of 16 AU (1.5 billion miles) between comets. A future interstellar traveller stranded in the middle of the Oort cloud with a telescope probably would not be able to find a comet nucleus.

METEORS, METEOROIDS, AND METEORITES

Meteor: The phenomenon of a streak in the sky caused by the burning of a rock or dust particle as it falls.

Almost everyone has seen the flash of light in the sky that is sometimes called a *falling star* or *shooting star*. This phenomenon is better termed a ***meteor*** because obviously it is not a star. (Recall the size of our Sun—a typical star—compared to the Earth.) The streak of light in the photo in Figure 10–21 is a meteor the camera caught in a 15-minute exposure of the sky. Since the camera was set to follow the stars, each star appears as a dot. The meteor is the unusual streak. The nature of these sudden flashes of light across the sky must have been of great concern to people since the beginning of time. The idea that they are caused by rocks falling from the heavens can be found in ancient writings of the Romans, the Greeks, the Chinese, and in the Old Testament but it was not an idea that was easy for some people to accept.

The first confirmation by modern science that rocks do indeed fall from the heavens was on April 26, 1803. Citizens of the small town of L'Aigle, France saw an exceedingly bright meteor that exploded and formed a shower of 2000 to 3000 fragments that fell to Earth. It was reported that some fragments were still warm when found. The respected French physicist J. B. Biot was sent by the French Academy of Science to investigate the incident and his report confirmed the ancient writings.

Before describing the stones and the flashes of light in any more detail, let's state some definitions. A meteor is the phenomenon of the flash of light; it is not the object itself. The object out in space that causes the meteor in our

Figure 10–21. *The streak of light is a meteor.*

atmosphere is called a ***meteoroid.*** Finally, if the object survives its fall through the atmosphere and lands on Earth, it is called a ***meteorite.***

Meteoroid: An interplanetary chunk of matter smaller than an asteroid.

Meteorite: An interplanetary chunk of matter after it has hit a planet or moon.

Fireball: An extremely bright meteor. More than 10,000 fireballs can be seen each day over the Earth.

Meteors

Most meteors are very dim. You might see one out of the corner of your eye and wonder if you saw something or not. Others, however, are very bright and might cause a long streak across the sky. In fact, the brightest are brighter than a full Moon. The light is the result of the meteoroid burning itself up as it passes through our atmosphere. As the stone enters the atmosphere, it experiences friction caused by the molecules of air. This heats up the object as well as the air it is passing through. At first glance, this may seem odd to you if you have held your hand out of a car window and experienced the *cooling* effect of the air striking your hand. The difference is in the speeds of the objects—your hand and the meteoroid. A meteoroid's speed is typically 50 km/s (100,000 miles/hr). Thus the air molecules are striking the object at tremendous speeds. This causes the surface of the object to heat up until it vaporizes, streaming its atoms in its wake. These hot atoms, along with the similarly heated air, glow like the gas in a fluorescent lamp and present us with the phenomenon known as a meteor.

It is difficult for an individual observer to estimate the height or speed of a meteor but if two observers at different locations each record the same meteor on a photo, they can use triangulation to calculate these quantities. By means of such measurements it has been determined that the typical meteor begins to glow at a height of about 130 kilometers (80 miles) and burns out at about 80 kilometers (50 miles). The speed of a meteoroid as it enters the Earth's atmosphere might be anywhere from about 10 km/s to about 70 km/s.

These speeds correspond to 20,000 mi/hr and 150,000 mi/hr.

Meteoroids

Most meteoroids vaporize completely in the atmosphere and never reach the Earth's surface. The energy source that produces the light we see is simply the motion of the meteoroid. By calculating how much light is emitted from the meteoroid as it burns up and by estimating how much of its original energy of motion changes to light (rather than heat), we can calculate the mass of the original meteoroid. Most meteors are produced by meteoroids with masses from a few milligrams (a grain of sand) to a few grams (a marble-size rock).

The "energy of motion" referred to here is called the kinetic energy *of the meteoroid.*

Under ideal viewing conditions, a person looking for meteors can see an average of five to eight meteors per hour. Since a meteor can be seen only if it is within 150 to 200 kilometers of the viewer, we can calculate that over the entire Earth there must be about 25,000,000 meteors a day bright enough to be seen by the naked eye. The number visible in telescopes would be hundreds of billions. If meteors average a fraction of a gram in mass, this means that about 100 tons of material from space strikes the Earth each day!

If a meteor trail is recorded from more than one location, its path can be determined accurately enough to calculate the path of the original meteoroid before it was slowed by the atmosphere. In this way we find that, as expected, most meteoroids are simply tiny particles orbiting the Sun. Meteoroids differ from asteroids in that most asteroids orbit the Sun close to the plane of the ecliptic, while small meteoroids do not suffer this limitation; their orbit around the Sun may be in any orientation. (See Figure 10–22.) A majority have very eccentric orbits rather than the nearly circular orbits of most asteroids.

This does not rule out the asteroid belt as the origin of these meteoroids, however. It is thought that many small meteoroids are debris from collisions between asteroids. Such collisions would break the asteroids into a number of smaller pieces, including pieces as small as sand grains. These tiny pieces would come out of the collision in all directions and move in elliptical orbits at all orientations with the ecliptic.

Many meteors, however, are from a source other than the asteroids—they are due to material evaporated from a comet's nucleus and then blown off the comet by the action of the Sun. This leads us to a discussion of meteor showers.

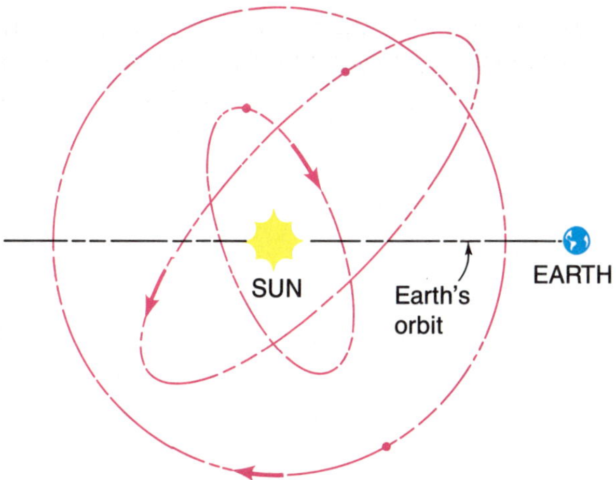

Figure 10–22. *Small meteoroids strike the Earth from all directions, indicating that their orbits are not confined to the ecliptic. Three of countless meteoroid orbits are shown here.*

Figure 10–23. *This photo shows the Leonid meteor shower of November 1966. The streaks radiating from one area of the sky are all part of the shower.*

Meteor Showers

On some nights, we see many more meteors than normal and if we observe closely, we see that there is a pattern to the directions of the meteors. Figure 10–23 is a time exposure showing the phenomenon. The streaks are meteors. Notice that they seem to point to (or rather, *from)* one point in the sky. This phenomenon is called a ***meteor shower*** and is caused by the Earth passing through a swarm of small meteoroids. Figure 10–24 illustrates this. A cluster of tiny particles is shown in the Earth's path. These tiny meteoroids cause the shower.

Meteor shower: The phenomenon of a large group of meteors seeming to come from a particular area of the celestial sphere.

Radiant (of a meteor shower): The point in the sky from which the meteors of a shower appear to radiate.

To see why the meteor shower seems to originate in a certain constellation, think of the Earth moving through space. When it encounters the swarm of meteoroids the Earth is moving in a particular direction; toward some constellation in the sky. The meteor trail seems to originate in that constellation or close to it. The reason that the shower may not appear to come exactly from the constellation toward which the Earth is moving is that the swarm itself has a speed and therefore its direction of hitting the Earth is determined by a combination of the Earth's and the swarm's direction of motion.

Meteor showers are named after the constellation from which they seem to come. The constellation behind the meteor shower in Figure 10–23 is Leo, and the shower is known as the Leonid meteor shower. The meteor shower about to occur in Figure 10–24 will likewise be a Leonid shower.

The particles that cause a meteor shower strike the Earth along nearly parallel paths yet they appear to diverge from a point. The reason for this apparent contradiction is the same as the reason that parallel strips of a long, straight interstate highway seem to diverge as they near a viewer standing in the median (Figure 10–25). The particles forming the shower are coming into the Earth's

ACTIVITY

Observing Meteors

To observe meteors, you should have a clear sky away from city lights, so your first step is to get out into the country. Second, you must avoid a night with a bright Moon that will light up the sky so that you will have difficulty viewing. Finally, the best time for observing is after midnight; preferably from around 2 A.M. until the beginning of morning twilight.

The reason for choosing a time after midnight can be seen by referring to Figure T10–2. In this figure, the Sun (not shown) is to the left and the Earth is moving toward the top of the figure. Thus the half of the Earth lined in red is on the leading edge as the Earth moves through space. Now consider a car driving through a rain shower and you see why most meteoroids hit the Earth on this leading side. The analogy is not quite exact because different meteoroids have very different motions through space, while raindrops all fall in about the same direction. Meteoroids are able to catch up with the Earth from behind and hit the trailing half of the Earth. Still, many more strike the Earth's leading edge as the Earth sweeps them up.

Referring again to the figure, you see that it is midnight for a person standing at point *A* and that the Sun is rising for a person at *B*. Meteors, then, are better observed in the early morning hours.

Once you have found your observing location, you must wait for your eyes to adapt to the darkness. Unless you have chosen a night of a heavy meteor shower, you must be patient. Lay back and relax. (Did you bring a lawn chair?) Patience should reward you with the streak of a meteor across the sky. It is likely that you will see the meteor in some direction other than where you are looking; out of the corner of your eye. There is a good reason for this: the part of your retina that sees things out of your direct line of sight is more sensitive to motion and to changing light. An experienced observer takes advantage of this effect when looking, for example, for the motion of a satellite in the sky. The observer looks to the side of where the object is expected to be seen.

The best advice to the person looking for meteors is to pick a night when a meteor shower is expected. Table 10–2 lists the dates of the most prominent annual meteor showers.

atmosphere along parallel paths but as they near us, they seem to be spreading out.

For centuries we have known that some meteor showers repeat regularly each year but the origin of the showers was not known. Then in 1866 it was shown that the particles that cause the Perseid meteor shower (which occurs around August 12 and appears to originate in the constellation Perseus) have almost exactly the same orbital path that an 1862 comet had. It was concluded that the meteoroids of the shower were simply particles that had long ago come from the comet and formed its tail. Those particles were in orbit about the Sun when their paths intersected the Earth's. Today we are able to associate most of the major annual meteor showers with some comet. Table 10–2 lists the major showers and their associated comets.

The table also indicates the average number of meteors expected to be seen during the maximum of the shower. The intensity of some showers changes

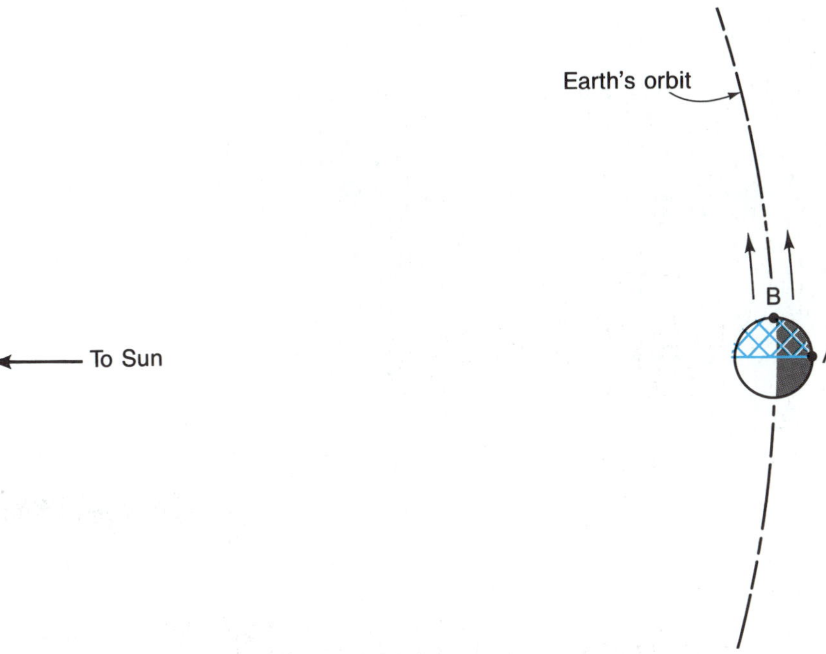

Figure T10–2. *The Earth is moving upward in the drawing and it strikes most meteoroids on its leading edge. Those striking it from behind must catch up. It is midnight at point A and sunrise at point B.*

greatly year to year, however. Figure 10–26 shows why this occurs. In part (a), the particles that had been blown from the comet have become spread fairly evenly around the comet's orbit and so each time the Earth passes through this orbit, about the same number of particles strike our planet.

In part (b) of the figure, however, the particles are clumped in one region of the comet's orbit. If the Earth and this swarm happen to meet, we see much greater meteor activity. The Leonid showers are irregular in intensity from year to year. In 1833, 1866, and again in 1899 the Earth passed through a major swarm of particles in the comet's orbit and nearly 100,000 meteors could be seen in an hour. The maxima of the Leonid showers occur about every 33 years, which is just the period of the comet from which they originated. As the years go by, the particles spread out more and more along their orbit, but the display was spectacular in November 1966, when Figure 10–23 was taken. We are therefore looking forward to another beautiful display around 1999.

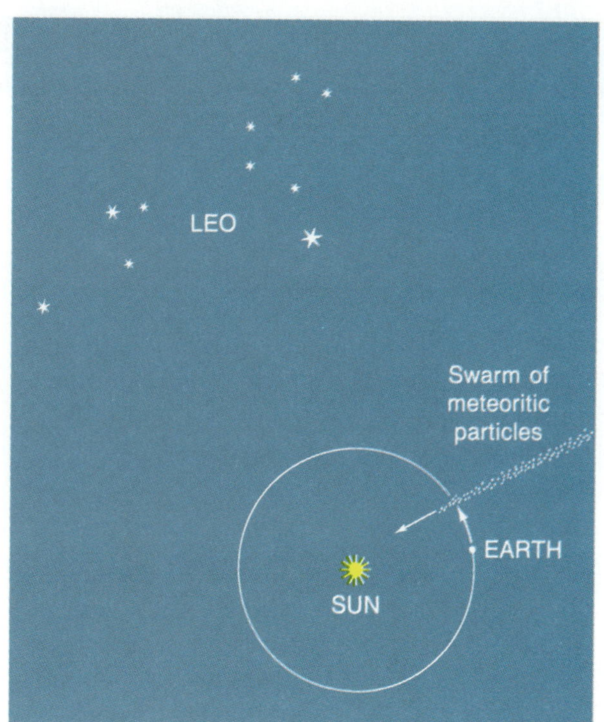

Figure 10–24. *The Earth here is moving toward the constellation Leo and is about to encounter a swarm of meteoritic particles.*

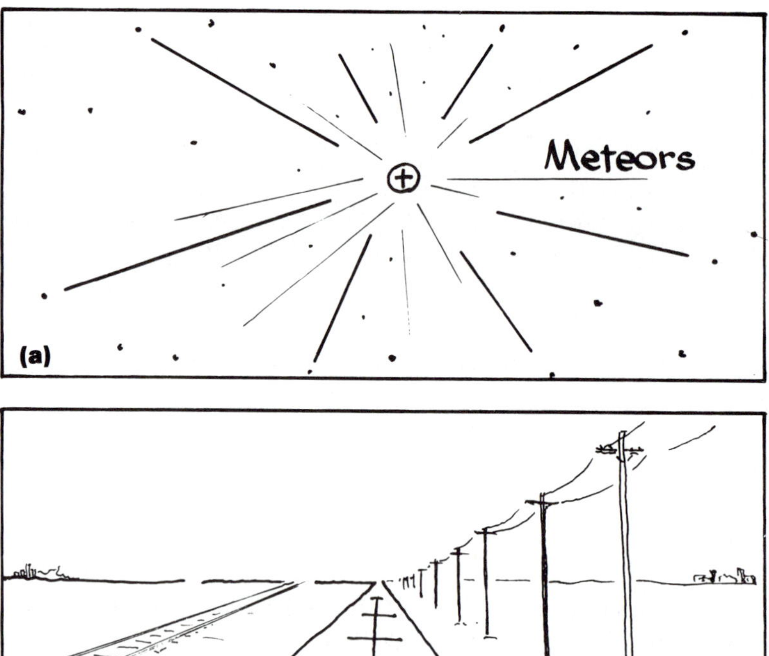

Figure 10–25. *(a) Meteors in a meteor shower seem to come from a single point in the sky. (b) The divergence of the meteors is the same phenomenon seen by a person standing on a long straight highway. The edges of the road seem to diverge as they come toward the observer.*

Table 10–2 Some Major Meteor Showers

Shower	Date of Maximum*	Associated Comet	Expected Hourly Rate
Quadrantids	Jan. 3	??	50
Lyrids	April 21	Comet 1861 I	15
Eta Aquarids	May 4	Comet Halley	20
Delta Aquarids	July 30	??	20
Perseids	Aug. 12	Comet 1862 III	50
Draconids	Oct. 9	Comet Giacobini-Zinner	15
Orionids	Oct. 20	Comet Halley	20
Taurids	Oct. 31	Comet Encke	15
Leonids	Nov. 16	Comet 1866 I	15
Geminids	Dec. 13	??	50

*The dates given are for the approximate date of maximum and may vary slightly from year to year.

Figure 10–26. *In part (a) the particles are spread around the entire orbit but in (b) they are clustered in one area.*

METEORITES

Meteorites are classified in three categories: iron meteorites (or *irons*), which are made up of 80 to 90 percent iron (most of the rest being nickel); stony meteorites (*stones*), which are just what their name implies but often contain flakes of iron and/or nickel; and *stony irons,* about half iron and half rock. Figure 10–27 shows a number of different meteorites.

About 90 percent of all meteorites are stones, but if you find a meteorite, it will likely be an iron. The reason for this apparent paradox is that stony

Figure 10–27. *(a) This iron meteorite has become magnetized. (b) The McKinney Meteorite, found in Texas, is a stony meteorite. (c) This 14-ton iron is the largest meteorite ever found in the U.S. After he found it in 1902, a farmer and his son moved it three-fourths of a mile and charged people 25 cents to view it.*

meteorites are very similar to regular Earthly stones, especially after the meteorites have weathered for a few years. Thus they are difficult to recognize. Irons, however, are found by people using metal detectors and are different in appearance from regular Earthly rocks.

How do we know, then, that most meteorites are stones? Meteorites can be found in the Antarctic, where they become exposed when snow is blown from them. Here, where there are no natural rocks to cause confusion, 90 percent of the meteorites we find are stones.

The largest meteorite ever found, the Hoba meteorite, weighs about 65 tons and is in Namibia, in southwestern Africa. Iron meteorites such as this are much stronger than stones and therefore do not break up in the atmosphere as do stones. The Hoba meteorite (so called after the name of the farm on which it lies) barely disturbed the surrounding surface of the Earth, apparently because it entered the atmosphere at a very small angle and was greatly slowed down before reaching the surface.

You may be unable to travel to Africa to see the largest meteorite but the second largest is in captivity in New York City. The 34-ton giant was hauled from Greenland in the 1890s by the explorer Robert Peary and is now in the Hall of Meteorites of the American Museum of Natural History.

Meteorite Hits—Craters and Near Misses

What happens when a meteorite strikes the Earth? Quite naturally, it makes a hole. You can confirm that such a hole will be much larger than the meteorite by throwing a rock into mud to simulate a meteorite strike. You will see that a crater much larger than the rock is produced. Figure 10–28 is a photo of the most prominent impact crater on Earth, Meteor Crater near Winslow, Arizona. The crater is in the desert about 40 miles east of Flagstaff. The crater, nearly a mile across, is 180 meters deep and has a rim rising 45 meters above the surrounding desert ground. To appreciate the size of the crater, look at the lower left corner of Figure 10–28 and find the two-lane road leading to the parking lot and guest building.

The crater is 600 feet deep and rises 150 feet above the surrounding ground.

Some 25 tons of iron meteorite fragments have been found around the crater; some as far as four miles away. The meteorite that formed Meteor Crater is estimated to have had a total mass of 300 million kilograms (300,000 tons) and to have been about 45 meters across. It struck about 25,000 years ago at a speed of about 11 kilometers/second (25,000 miles/hour), releasing an energy equivalent to 1000 Hiroshima bombs. If you are ever in northern Arizona it is worth a trip to see this crater, which is only about 100 miles from the Grand Canyon.

Many other meteorite craters have been found on Earth but most are far less impressive than Meteor Crater because weather has worn them down or they are under the sea. Figure 10–29 shows the locations of craters caused by meteorite impacts. Meteor Crater is not only the most recently formed major crater but it is well preserved because of its location in the dry Arizona desert.

In more recent times, a meteoroid that never reached the ground created quite a furor in the United States. On August 10, 1972, a fireball passed about 60 kilometers (40 miles) above the western U.S. moving northward to Canada. It was visible long enough to be photographed and was even recorded on movie film by a tourist. The meteoroid that caused it is estimated to have weighed as much as 1000 tons. Fortunately, it hit the Earth's atmosphere at a glancing angle and its motion was not slowed enough by the thin atmosphere to allow

This fireball is still cited by UFO enthusiasts as evidence for visits from extraterrestrials.

Figure 10–28. *This view of Meteor Crater shows the guesthouse and entrance road at lower left.*

it to be pulled to Earth. Figure 10–30 shows the trail it left above the Tetons in Wyoming. After skimming the atmosphere, it continued back into space. If it had hit the atmosphere a few kilometers lower, it would have landed somewhere in Canada, producing a crater about 100 meters in diameter.

One might speculate on whether, if this object had hit the surface of the Earth, the resulting explosion might have been misinterpreted as a nuclear attack from an enemy. What would have been the response? What if such a meteorite had struck near a major city in the U.S. or USSR? Could a relatively small meteorite cause the population of Earth to self-destruct?

It is estimated that a meteorite larger than 1 kilometer in diameter strikes the Earth on the average once every 3 or 4 million years. A hit by a 1 kilometer-diameter meteorite would be equivalent to a 3000-megaton bomb; producing a crater 10 kilometers in diameter. Such an explosion would have effects far beyond the area of impact because it could send enough dust into the atmosphere

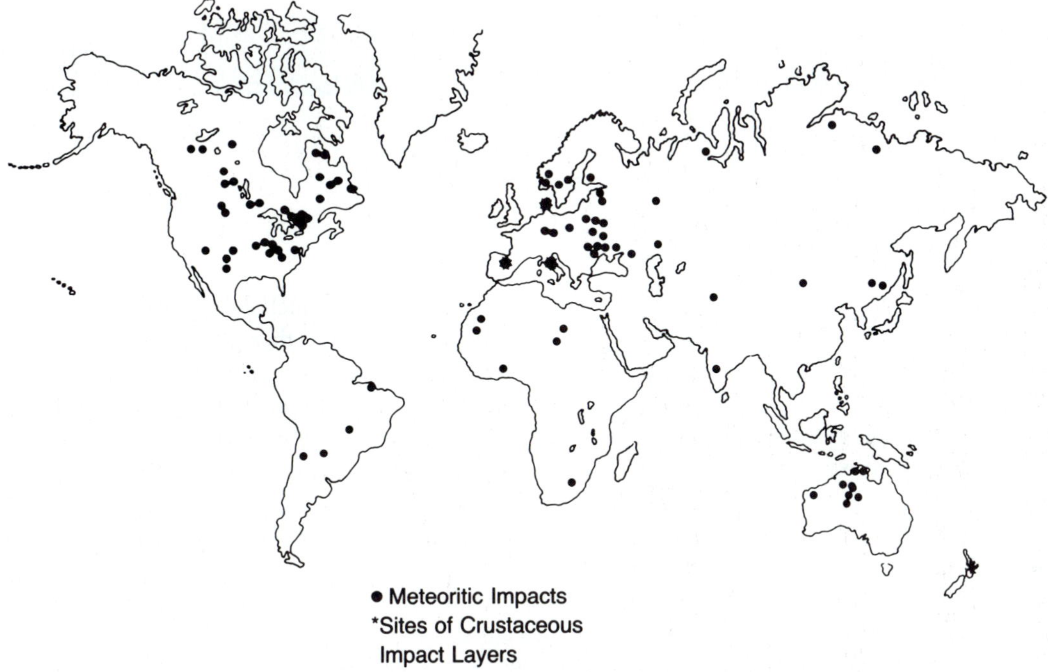

Figure 10–29. *This map shows the locations of major impact craters. Few are well-preserved enough to be obvious to a person at the site.*

to block out a significant amount of sunlight. An hypothesis put forward nearly a decade ago suggests that such a meteor strike was responsible for the extinction of the dinosaurs. (See the accompanying Close Up.)

How likely is it that a person could be hit by one of the tiny meteorites that regularly strike Earth? Based on the number of meteorites that reach the ground, on the population of the world, and on the size of people we can estimate that there is about one chance in 20 of a person somewhere in the world being hit by a meteorite in any year. There is one known case of this happening, at least if you count bounces. On November 30, 1954 a meteorite came through the roof of a house in Sylacauga, Alabama, and hit a woman on the bounce, severely bruising her hip.

The chances of being hit by a meteorite are very much less than being struck by lightning. And if it happens you will become famous in astronomy books.

The Tunguska Event

Something spectacular fell in Tunguska, Siberia, on June 30, 1908. The fireball was bright enough to be seen in daylight and the sound that resulted from its explosion some 10 kilometers above the Earth was heard as far away as 1000 kilometers (600 miles). The explosion, which was equivalent to about a 50-megaton bomb, knocked down trees as far as 30 kilometers away and a man 80 kilometers away was knocked down by the shock wave from the explosion. The photo Figure 10–31 was taken some 20 years later.

CLOSE UP

Meteors and Dinosaurs

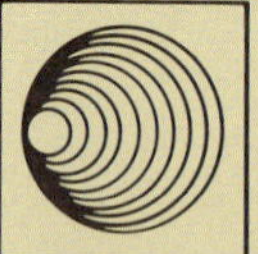

SIXTY FIVE MILLION YEARS ago, dinosaurs became extinct. In fact, nearly 75 percent of all species on Earth became extinct during the same short time period (as geological time is measured). The reason for this mass extinction has been debated by scientists for some time and a number of different catastrophes have been hypothesized as the cause for the extinctions.

In 1980 astronomy entered the picture. In that year, Walter Alveraz, Luis Alveraz (Walter's father), Frank Asaro, and Helen Michel proposed a solution for which they had real evidence. They had found, first in the Apennine Mountains of Italy and then at a number of sites around the Earth, a layer of clay containing significant amounts of chemical elements normally rare in the Earth's crust. Figure T10–3 shows the layer. The elements iridium, platinum, and osmium were found in quantities many times greater than normally found on Earth. How could this layer have been deposited? Evidence comes from the fact that these elements are much more abundant in meteorites than in the crust of the Earth. Perhaps a giant meteorite—an asteroid—struck the Earth, exploded, and sent its debris high into the atmosphere to fall to Earth and form the layer. To account for the thickness of the layer found around the world, the meteorite must have been about 10 kilometers in diameter; not an unusual size for an asteroid. The debris, which also would have included earthly material pulverized by the impact, would spread as a cloud around the Earth and then gradually settle to form the layer found by the Alveraz team.

But would the explosion have wiped out entire species all around the Earth? No, not directly. But consider what the effect a giant cloud of dust would have on vegetation on Earth. Such a cloud could remain in the atmosphere for more than a year, darkening the Earth below. Much vegetation would die, along with many animal species that depended on vegetation for food. Dinosaurs are included in this category as well as many other species. But not small creatures such as rodents that could forage for seeds and nuts. And, indeed, the fossil record shows that such creatures did survive this period.

One test of this asteroid hypothesis is to find out how long ago the clay layer was formed; to see if it is 65 million years old. If you have studied geology, you know that such dating of past events is a common procedure. The result? The clay was deposited about 63 million years ago, remarkably close to the value given for the mass extinction as shown by the fossil record.

Another question arises: Where is the crater produced by such a large meteorite? It would be expected to be about 100 kilometers in diameter. Since we find no such giant crater, the obvious answer is under the ocean. After all, two-thirds of the Earth is covered by water so the chances are two out of three that a meteorite would fall in the ocean. If it did, the water would not prevent such a large object from producing a crater at the bottom of the ocean and sending up the hypothesized cloud of dust. We observe no such crater beneath the sea but details of the ocean depths are still largely unknown and ocean currents would have eroded such a crater so that it would not be obvious.

Since 1980, when the Alveraz team presented their data, their hypothesis has become more and more accepted. (In fact, it is usually referred to as a *theory,* in accordance with our discussion in the first few chapters, where we say that an idea should be called a theory only after it has gained some acceptance.)

A Speculative Hypothesis

Some scientists claim to find evidence that there has been a regular cycle of extinctions, with about 26 million years between. If this cycle does indeed exist, could astronomical events have been responsible? At least one hypothesis has been presented.

This hypothesis proposes that the extinctions are not caused by a single asteroid, but rather by collisions between the Earth and a great number of comets. (This would also explain the lack of a large meteor crater from 65 million years ago.) Now all we need is a mechanism that would cause a great number of comets to fall to Earth over a short period of time and to have such an event to occur in a regular cycle.

When we discuss stars in more detail in later chapters, we will find that about half the stars in the heavens exist not as single stars but rather as two (or more) stars orbiting one another. Suppose the Sun is not alone but is part of a system of two stars in orbit one another with a period of 26 million years. If this is the case, and if the other star has a very elliptical orbit so that it comes close to the Oort cloud at its closest approach to the Sun, it could disrupt the cloud; it would send comets in all directions, some toward the inner solar system where the Earth is innocently orbiting. This would mean that once every 26 million years the Earth would be showered by comets, explaining the cycle of mass extinctions.

The fact that we observe no such star means that the star must either be a very small, dim one, or else that it is no longer in orbit around the Sun. The undetected star, called "The Death Star" and "Nemesis," has been the subject of several searches but has not been found.

Figure T10–3. *This rock in Petriccio, Italy, contains a clay layer hypothesized to be due to an impact (or impacts) of asteroids.*

This hypothesis has been the subject of strenuous debate among scientists. It is highly questionable because it requires a great number of coincidences and requires an orbit for Nemesis that carries it so far from the Sun that other stars would probably break up the pair.

So where do we stand? It appears that the asteroidal explanation for the mass extinction of 65 million years ago is on fairly firm footing. On the question of the 26 million year cycle of extinctions, many scientists argue that there is not sufficient evidence that such a cycle even exists and that the cause proposed for the supposed cycle are just too imaginative to be considered as anything more than an interesting conjecture.

Figure 10–30. *Although this trail of light was reported by some people as being due to a UFO, it is actually a large meteor.*

Figure 10–31. *Something hit Siberia in 1908 and caused the destruction still evident in this photo taken 20 years later.*

No meteoritic material has been found at the site, leading some to speculate that the explosion was caused by a UFO rather than a natural meteor. One hypothesis is that the object was the nucleus of a small, icy comet, perhaps 50 meters in diameter, which exploded. Another—very speculative—hypothesis that has been proposed is that a miniature black hole struck the Earth to cause the explosion. The black hole could have passed completely through the Earth and come out through the ocean on the other side.

Although we do not know the cause of whatever happened that day in Siberia, there is no evidence to cause us to think that it involved extraterrestrial beings. Scientists always look first for explanations involving phenomena that have been observed in other circumstances or for which there is established theory. Meteorites, comets, and even black holes fit these criteria.

This would not have been a black hole produced by a supernova (Chapter 16), but one produced in the big bang (Chapter 18).

CONCLUSION

The solar system is indeed a collection of diverse objects. As our space probes venture out to study it in more detail, we are continually surprised by the beautiful diversity we find. We have seen in the last few chapters the great differences that exist not only between planets, asteroids, comets, and meteors, and satellites; but even between objects within these various categories. On the other hand we cannot help but be impressed by the similarities we see; even between objects we place in different categories.

In Chapter 12 we will see that the objects in our solar system were formed at about the same time and that their formation and differentiation were determined by the conditions that existed at the time. We are beginning to understand why they are different and why they are the same but we have a long way to go. That is what makes study of the solar system exciting.

RECALL QUESTIONS

1. What is a blink comparator and how is it used?
2. List two ways in which Pluto's orbit is unusual compared to the other planets.
3. We cannot determine the size of Pluto from the size of its image in a telescope. How, then, can we know its size?
4. What did the discovery of Charon allow us to calculate about Pluto?
5. Why have we not named all of the asteroids that have been observed?
6. How are the masses of asteroids determined?
7. What causes the Kirkwood gaps?
8. How were the Pioneer and Voyager spacecraft guided through the asteroid belt so that they were not wiped out in a collision?
9. Why did the asteroids not form into a planet between Mars and Jupiter?
10. Why do we think that it is unlikely that the asteroids are the remains of a planet that exploded?
11. Describe the three main parts of a comet. Include approximate sizes of each part for a typical comet.
12. Describe the modified "dirty snowball" model of a comet's nucleus.
13. What causes a comet to have two tails? Why are they different shapes?
14. What is the Oort cloud?
15. Why was Halley's Comet so unspectacular in 1985/86? What part of Earth received the best view?
16. We cannot hope to see the Oort cloud. What makes us think that it exists?

17. Explain why the passing air heats a meteoroid, while a wind cools a person.
18. When you see a meteor, what is the approximate probable size of the particle producing it?
19. What causes a meteor shower and why do the meteors appear to come from just one part of the sky?
20. Name and describe the three main types of meteorites. Which is most common? Which is easier to find? Why?

QUESTIONS TO PONDER

1. Any measurement, by the very nature of measurement, has limited accuracy. Discuss the implications of this in the case of the discovery of Pluto.
2. Describe two methods of measuring the sizes of small asteroids.
3. Figure 10–10 shows the tail of Halley's Comet being longest *after* the comet had passed closest to the Sun. Why is the tail not longest when the comet is nearest the Sun?
4. If a comet is seen in the west shortly after sunset, in what direction will its tail point: toward or away from the horizon? In what direction will it point if it is seen in the east in the morning sky?
5. Find out about a meteor crater in your part of the country or the world. The geology department of your college is a suggested source of information.
6. What effect on the motion of a comet would you expect to result from the jetting of material from its nucleus?
7. Compare the size and shape of Phobos to that of the nucleus of Halley's Comet.
8. It was argued that asteroids must emit the same amount of radiation which they receive. How do we know that they are not still releasing heat that resulted from their formation early in the solar system's history?

CALCULATIONS

1. There is a Kirkwood gap at 2.82 AU from the Sun. Calculate the period of revolution of a particle if it were in this gap and determine the ratio of this period to Jupiter's period.
2. Halley's Comet has a period of about 76 years. How long is its semimajor axis?

CHAPTER 11

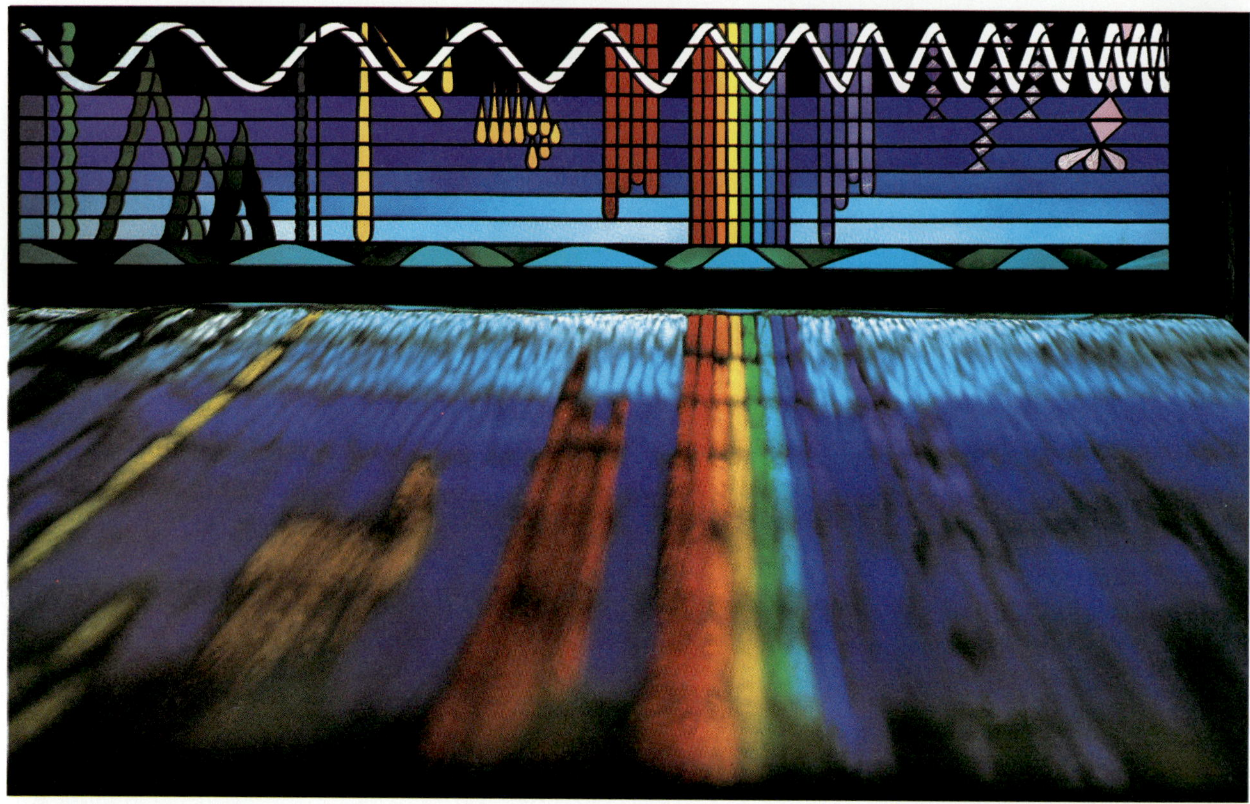

The sky's electromagnetic spectrum appears in stained glass at the Smithsonian Institute.

The Rama committee was still manageably small, though doubtless that would soon be rectified. His six colleagues—the [United Planets] representatives for Mercury, Earth, Luna, Ganymede, Titan and Triton—were all present in the flesh. They had to be; electronic diplomacy was not possible over solar system distances. Some elder statesmen, accustomed to the instantaneous communications which Earth had long taken for granted, had never reconciled themselves to the fact that radio waves took minutes, or even hours, to journey across the gulfs between the planets. 'Can't you scientists do something about it?' they had been heard to complain bitterly, when told that face-to-face conversation was impossible between Earth and any of its remoter children. Only the Moon had that barely acceptable one-and-a-half-second delay—with all the political and psychological consequences which it implied. Because of this fact of astronomical life, the Moon—and only *the Moon—would always be a suburb of Earth.*

(Quoted from Arthur C. Clarke, *Rendezvous with Rama,* Harcourt Brace Jovanovich, 1973, pp. 27, 28.)

Light and the Electromagnetic Spectrum

A STUDY OF ASTRONOMY is engrossing because we learn about the fantastic objects that make up our universe. But perhaps just as interesting is the study of *how* we know what we know. How do we know that our Sun is a star like the thousands of others we see in the sky at night? How do we know what the stars are made of? How do we know their sizes? How can we measure a star's mass?

In fact, how can we learn anything about objects which we cannot hold, feel, weigh, and experiment with? By examining the radiation that they emit. This radiation, including not only visible light but many other types of radiation, carries to us a tremendous amount of information about the objects that it comes from. Before we can understand how astronomers use radiation to answer questions about the stars, we must learn something about the radiation itself.

In this chapter we will study the nature of light and learn how we measure three major properties of stars: Their temperatures, their compositions (that is, the chemical elements of which they are made), and their speeds relative to the Earth. This is quite a lot to learn about the stars, but it is just the beginning.

THE WAVE NATURE OF LIGHT

Spectrum: The order of colors or wavelengths produced when light is dispersed.

Figure 11–1 shows a beam of white light passing through a prism. The emerging light is separated into colors; into a ***spectrum.*** Isaac Newton showed that the prism does not add color to the light but rather that color is already contained in white light and that the prism merely separates the light into its colors. Analysis of the spectrum of light from stars is extremely important in astronomy. What is it about light that allows it to be separated into colors and what is different about the various colors of light which result?

Wave Motion in General

Wavelength: The distance from a point on a wave to the next corresponding point.

Light acts like a wave. Figure 11–2 is a simplified drawing of a wave. As indicated in the figure, the distance between successive peaks (crests) of the wave is called the ***wavelength*** of the wave. The wave in the drawing has a wavelength of 2¼ inches; waves you make by dipping your hand in a swimming pool might have a wavelength of about this value.

A wave does not sit still, however. Figure 11–3 represents a movie film taken of a water wave as it passes a fisherman's floating cork. Notice first that the wavelength of the wave remains the same when the wave moves and second, that the wave causes the fisherman's cork to move up and down as the wave passes. This indicates that the water itself moves up and down rather than along the direction of the wave's motion. As the wave travels along the surface, the water's motion is primarily in the vertical direction and not along the direction of the wave.

Frequency: The number of repetitions per unit time.

Now suppose that the fisherman counts the number of times the cork moves up and down in one minute. He finds that the cork moves through a complete cycle 30 times each minute. We say that the ***frequency*** of the cork's motion is 30 cycles per minute and that therefore the frequency of the wave is 30 cycles/minute.

Figure 11–1. *A prism separates white light into its component colors.*

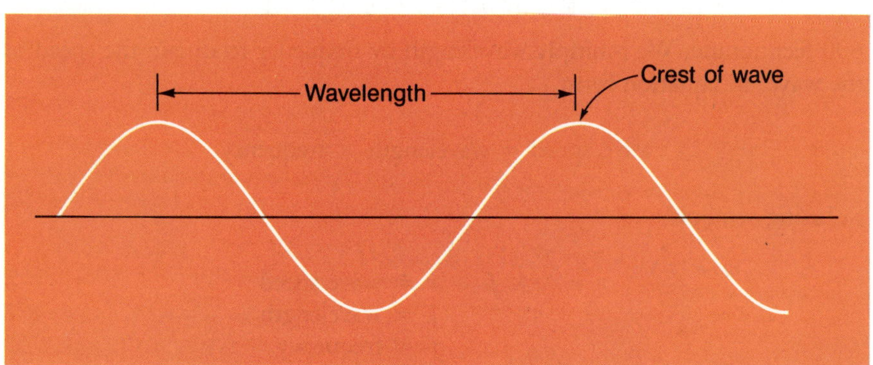

Figure 11–2. *Wavelength is the distance between successive crests of a wave.*

Suppose the fisherman measures the wavelength of the waves to be 20 feet. Can he use the fact that the wavelength is 20 feet and the frequency is 30 cycles/minute to calculate the speed with which the wave is moving? Before reading on, think about the situation and see if you can figure out how to calculate the wave speed.

Each wave, from crest to crest, is 20 feet long and 30 of these waves pass the fisherman each minute. This must mean that the waves move at a speed of

Figure 11–3. *A movie film of a wave passing under a fisherman at the end of a long dock. The wave causes the cork to bob up and down.*

600 feet/minute. We multiply wavelength by frequency to obtain the speed of the wave. In equation form,

$$\text{wave speed} = \text{wavelength} \times \text{frequency}.$$

or, using symbols,

λ is the lower-case Greek lambda.

$$v = \lambda \times f \qquad \begin{aligned} v &= \text{wave speed} \\ \lambda &= \text{wavelength} \\ f &= \text{frequency} \end{aligned}$$

This equation applies not only to water waves, but to all types of waves, including light and sound waves. Let's show an example of its use in the case of sound.

EXAMPLE Sound travels at a speed of about 335 meters/second (1100 feet/second) in air. What is the wavelength of a sound that has a frequency of 262 cycles/second?

Solution

Using the equation,

$$\text{wave speed} = \text{wavelength} \times \text{frequency}$$
$$335 \text{ m/s} = \lambda \times 262 \text{ cycles/s}$$

We now solve the equation for the wavelength and do the calculation.

$$\lambda = \frac{(335 \text{ m/s})}{262 \text{ cycles/s}}$$
$$= 1.28 \text{ meters}$$

So the length of the sound wave is about 1¼ meters.

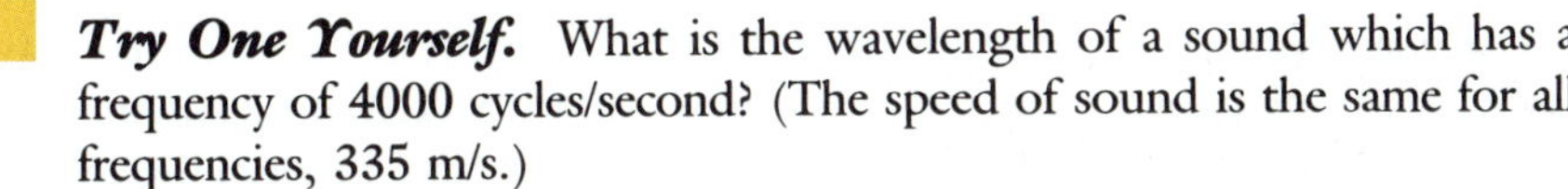

Try One Yourself. What is the wavelength of a sound which has a frequency of 4000 cycles/second? (The speed of sound is the same for all frequencies, 335 m/s.)

Hertz (abbreviated Hz): The unit of frequency equal to one cycle per second.

If you did the suggested exercise, you saw that a sound of higher frequency has a shorter wavelength, since the speeds are the same. The reason we used sound in the example rather than light is that sound waves have frequencies, velocities, and wavelengths within our everyday experience, whereas light waves don't. Now we will return to light and the spectrum produced when white light shines through the prism.

Light As a Wave

White light is made up of light of many different wavelengths. All wavelengths, however, travel at the same speed in a vacuum (and interstellar space is essentially a vacuum). That speed is tremendous, 3.00×10^8 meters/second, which is 300,000 kilometers/second or 186,000 miles/second. If a light beam could be made to travel around the Earth, it would circle the globe seven times in one second.

Light is extremely fast and its wavelengths are unimaginably short. The wavelength of the reddest of red light is about 7×10^{-7} meters, or 0.0000007 meters. The wavelength decreases as one moves across the spectrum from red to violet and the wavelength at the violet end of the spectrum is about 4×10^{-7} meters.

We found that in describing the solar system, meters and kilometers are not convenient units with which to describe distances and so we use astronomical units (AUs) there. In describing the wavelengths of light, a meter is much too long to be convenient; another unit is used in science—the ***nanometer.*** One nanometer (abbreviated *nm)* is 10^{-9} meters. So the shortest violet wavelength and the longest red wavelength are 400 nm and 700 nm respectively.

Nanometer (abbreviated nm): A unit of length equal to 10^{-9} meter.

The same equation relates the speed, wavelength, and frequency of light as relates those terms for other types of waves: speed = wavelength × frequency.

ACTIVITY

Making Waves

THIS IS AN EASY activity, but one that can help you understand wave motion. You need a large container of water; perhaps a bathtub. The amount of water you put in is not critical as long as you make it a few inches deep.

Now dip your finger into the water at about the center of the tub and watch a ripple spread out from your finger. The ripple is a small wave. Notice how long it takes the ripple to reach one end of the tub. Now hit the water harder with your finger; making a bigger, deeper wave. Does this wave take a different amount of time to reach the end? Do not be fooled into seeing only what you expect to see. Perhaps you can find a stopwatch to use to measure the time of the wave's travel to the end. I think you will see that a shallow wave and a deep wave travel at the same speed. Light works this way, too. Light of low intensity travels just as fast as the brightest light.

Next, dip your finger regularly, making a long train of waves. Make a note of your estimate of the wavelength—the distance between crests of the waves. Now dip your finger with a greater frequency. Not harder, but more often. Is the wavelength different? Dip less frequently. What happens to the wavelength? Does the effect you see make sense?

To see that it makes sense, notice again that the waves of short wavelength travel at the same speed as the waves of long wavelength. Wavelength does not affect wave speed. Thus when your finger forms a wave crest, the crest moves off at a fixed speed. If you make the next crest quickly, the distance between the crests—the wavelength—will be short. If you take more time before making the next crest, the first will travel farther and the wavelength will be greater.

This is just what we expect from the equation which relates speed, frequency, and wavelength. If the speed is the same for all of the waves, the one of greater frequency will have a shorter wavelength. It works the same way for light.

Finally, as you dip your finger at a regular frequency, move the finger slowly toward one end of the tub. Can you see any difference in the wave pattern in front of and behind your finger? This may not be easy to see, but it illustrates a very important phenomenon in astronomy. We'll discuss it—the Doppler effect—later in this chapter.

It is probably not worth memorizing these frequencies of light waves but it is handy to remember that the wavelength of visible light ranges from 400 nm (violet) to 700 nm (red).

When the equation is used with the velocity of light to calculate frequencies of light waves, extremely high frequencies are obtained: 400 nm corresponds to 7.5×10^{14} cycles/second and 700 nm corresponds to 4.3×10^{14} cycles/second. The frequency of visible light is tremendously large; larger than we can imagine.

It is important to note that the color changes continuously as we move across the spectrum of visible light so that there are not just a few colors in the spectrum; rather, there are as many as we care to name. In Figure 11–4 there is sufficient difference between the colors at points *A* and *D* that we have a different name for each. We would probably call the colors at points *A* and *C* both red, however; even though we can tell that they differ from one another. Likewise, the wavelengths of the light at point *A* and *B* are also different, even though we may not be able to distinguish a difference in color between those points.

D C B A

Figure 11–4. *Every different point along the spectrum is a different wavelength of light; even points* A *and* B, *which may appear to be the same color.*

The wavelength of a particular light determines its color, but since color is so subjective and since people are unable to distinguish between two very similar colors, scientists describe light by referring to wavelength rather than color. They might mention the color in some cases, but this is in order that we might better picture the situation. Remember that light can be described more accurately than by simply calling it "red" or "green."

Another important point: As we describe the light coming from astronomical objects, we will sometimes refer to its wavelength and sometimes to its frequency. Remember that if you know one you can calculate the other.

Invisible Electromagnetic Radiation

Can a light wave have a wavelength longer than 700 nm? Yes, although such a wave is not visible to us. The waves we see—visible light—are just a small part of a great range of waves that make up the ***electromagnetic spectrum.*** Waves longer than 700 nm (the approximate limit of red) are called *infrared* waves. Figure 11–5 shows the entire electromagnetic spectrum. Notice that the infrared region of the spectrum goes from 700 nm at the border of visible light up to about 10^{-4} m, which is a tenth of a millimeter or 100,000 nm. Electromagnetic waves longer than that are called *radio* waves.

Electromagnetic spectrum: The entire array of electromagnetic waves.

Infrared, radio, ultraviolet, X rays, and gamma rays are defined by their frequency and/or wavelength, as shown in Figure 11–5.

Going the other way from visible light (toward shorter wavelengths), we first encounter *ultraviolet* waves, then *X-rays* and *gamma rays.*

It is important to emphasize that all of these types of waves (or rays, as certain portions of the spectrum are known) are essentially the same phenomenon. They differ in wavelength and this causes them to differ in some of their other properties. For example, visible light is just that—visible. Ultraviolet is invisible to us but it kills living cells. In addition, it is the ultraviolet radiation emitted by the Sun that causes our skin to tan or burn. Infrared is perceived by us as heat radiation. And yes, the radio waves in the spectrum are the same radio waves we use for transmission of messages on Earth. They are handy for carrying messages containing sound and pictures (in the case of TV) for a number of reasons, including the fact that they pass through clouds and bend around obstacles.

Again, it is important to realize that all of these various waves are electromagnetic waves, just as is visible light. We give the various regions different names because of their properties and the uses we have for them. Nature does not see the spectrum as seven or eight or ten different regions; all of these regions are simply electromagnetic waves.

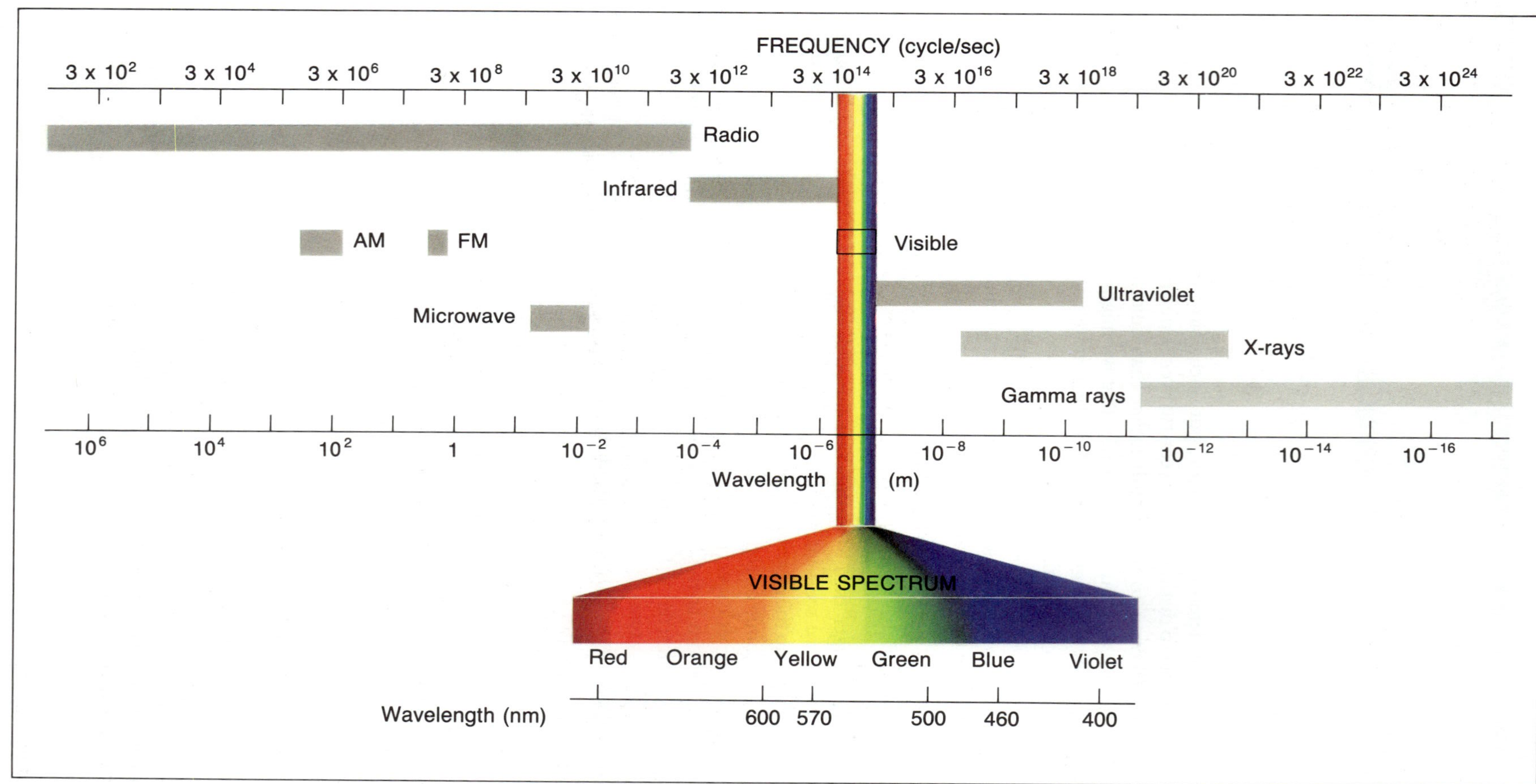

Figure 11–5. *The electromagnetic spectrum is divided into a number of regions, depending upon the properties of the radiation. Region boundaries are not well defined. Notice the small portion of the spectrum occupied by visible light.*

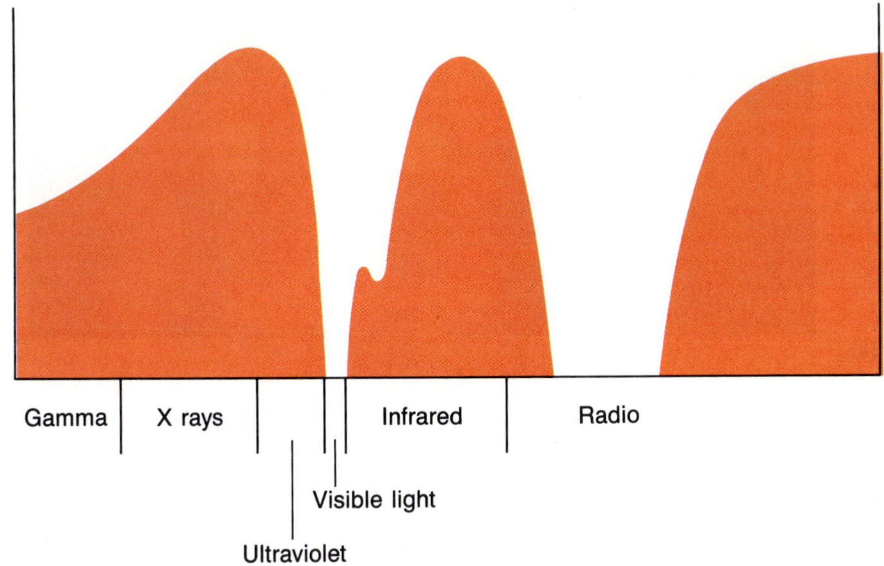

Figure 11–6. *The height of the colored region indicates the relative amount of radiation of a given wavelength blocked by the atmosphere from reaching Earth's surface. There are two regions of the spectrum to which the atmosphere is transparent: visible light and part of the radio region.*

Reception of Electromagnetic Waves from Space The electromagnetic spectrum is important to astronomers because waves of all the different regions of the spectrum are emitted by celestial objects. Notice that visible light is a very small fraction of the entire spectrum. We humans think of it as being the important part, but this only shows our limited outlook. Astronomers learn a great deal from the invisible radiation emitted by objects in the heavens.

One of our problems with this invisible radiation is that most of it does not pass well through air so it does not reach the surface of the Earth. Our air is transparent to visible light and to part of the radio spectrum but most of the rest of the electromagnetic spectrum is blocked to some degree. Figure 11–6 is a chart showing the relative transparency of the atmosphere to various regions of the spectrum. Where the colored region is highest, the least amount of radiation gets through. Notice that not much ultraviolet radiation penetrates to the surface. This is fortunate, for recall that this radiation damages living cells.

Astronomers refer to windows *in the atmosphere, saying that there is a visual window and a radio window. This means that our atmosphere allows radiation in these two regions of the spectrum to penetrate to the surface.*

Astronomers, however, wish to detect and examine these nonpenetrating radiations from space. They accomplish this by putting detectors on balloons to send them high into the atmosphere or on artificial satellites to get them completely above the atmosphere. This will be covered in a later chapter.

THE COLORS OF PLANETS AND STARS

Now we'll return to a discussion of the visible spectrum and talk about how the spectrum from an object is analyzed in order to determine some properties of heavenly bodies. We can divide this analysis into two different categories. First, we look at the overall spectrum of light from the object. We typically call this the color of the object but remember that it actually involves examining

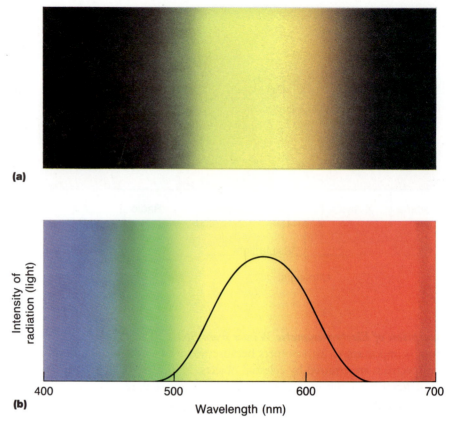

(a)

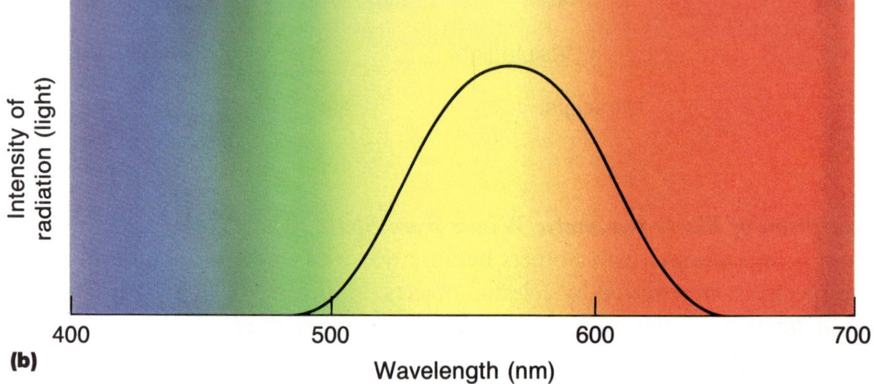

(b)

Figure 11–7 *(a) If light reflected from the lemons were sent through a prism to reveal its spectrum, we would see that mostly yellow light is reflected by them. (b) This graph indicates the relative intensity of the various wavelengths of light from the lemons.*

the actual wavelengths of light rather than just the color. The second way, to be discussed later in this chapter, involves examining individual parts of the spectrum. Let's look at the color aspect.

Color from Reflection—The Colors of Planets

When you see a spectrum spread out before you—on a screen perhaps—you see a particular color coming from a given location on the spectrum. This is because the wavelength associated with that color is coming to your eye from that spot. The color we see in most objects is not of a single wavelength, however. Figure 11–7(a) might be the spectrum of light from some lemons. The spectrum contains very many different wavelengths of light, but notice that there is no violet or blue light in the spectrum and that the center of the spectrum is indeed in the yellow. Part (b) of the figure shows a graph that indicates the intensity of light of each wavelength. Where the graph is higher, the light of the corresponding color is brighter.

The lemons have the spectrum illustrated in the figure because when white

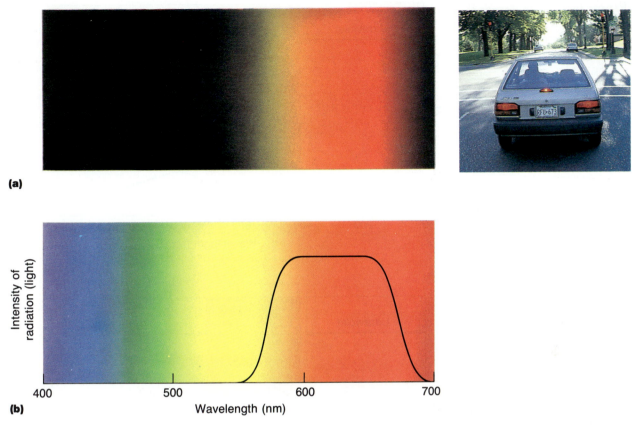

(a)

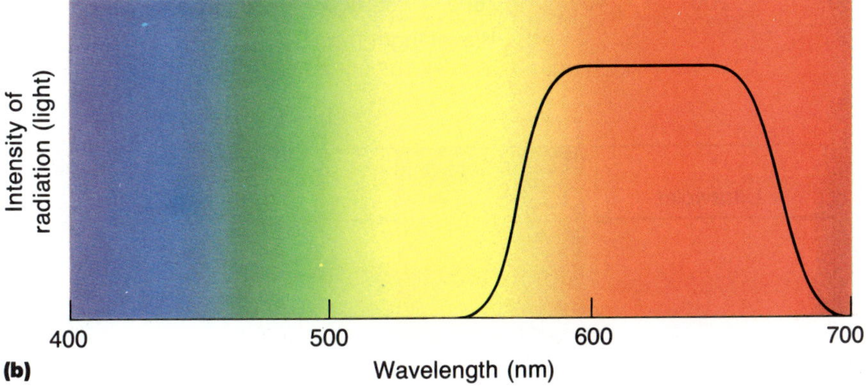

(b)

Figure 11–8. *The spectrum of light from the tail light of this particular car.*

light strikes them, they absorb some of the wavelengths of the white light; reflecting only those wavelengths we see in their spectrum (the wavelengths centered on the color yellow). This is what determines their color.

The planets have their color because of a process like that described for the lemons. The rusty red color of Mars occurs because the material on its surface absorbs some of the wavelengths of sunlight; reflecting a combination of wavelengths that looks rusty red to us.

Figure 11–8 represents the color spectrum and intensity/wavelength graph of the light from the red tail lights of a car. The spectrum includes not only the many wavelengths of the red part of the spectrum but also some orange. In this case, the bulb inside the tail light emits white light but part of that light is absorbed by the plastic cover. The light that gets through the cover has the spectrum indicated in the figure.

The colors of the Sun and other stars is determined in part by a process somewhat like the tail light of the car, for light from a star is produced by emission within the star and some is absorbed as it passes through the outer layers of the star. We will look at this in more detail later.

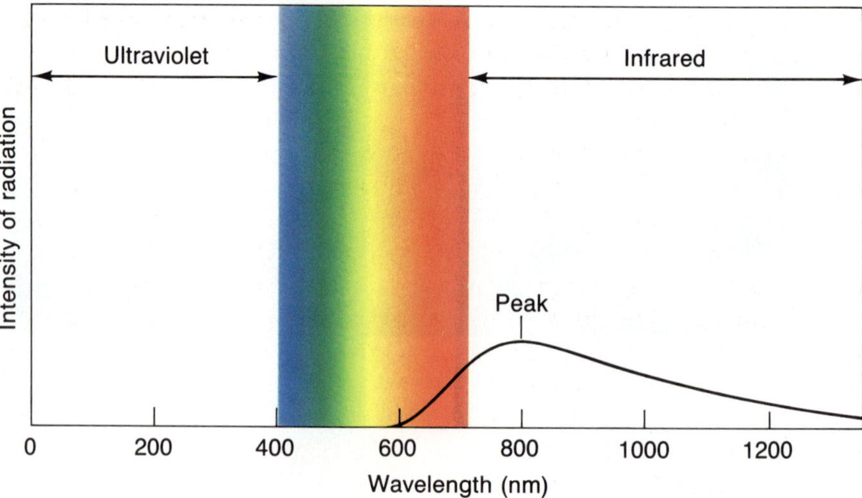

Figure 11–9. *The spectrum of a red-hot stove burner shows that it emits more infrared radiation than visible light.*

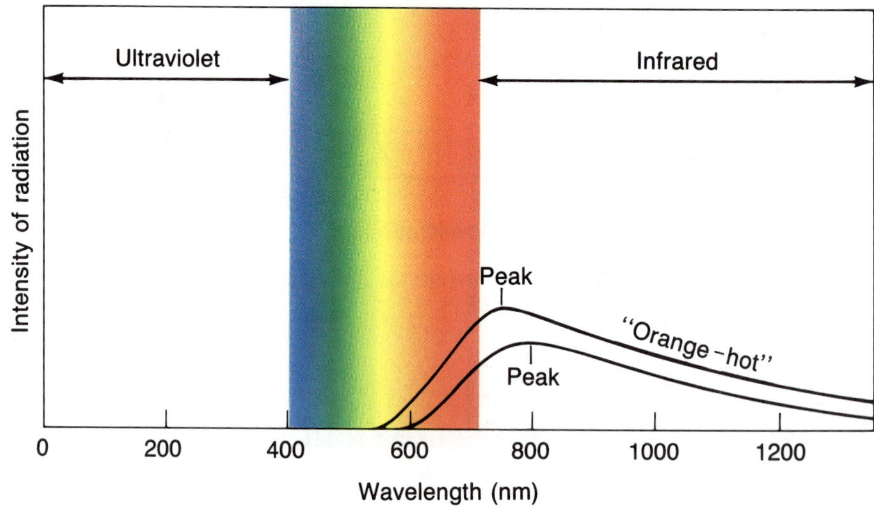

Figure 11–10. *The line representing an orange-hot burner emits more radiation of all wavelengths and its peak is at a shorter wavelength than the peak for the cooler burner.*

Color As a Measure of Temperature

We want you to imagine a dark room because in a well-lit room, the lamp and stove element reflect light and you can see them even if they are turned off.

The light emitted by the Sun and other stars can be compared to light coming from a light bulb or from the element of an electric stove in an otherwise dark room.

Consider what happens when you turn the burner of an electric stove to a low setting. It glows a dull red. Figure 11–9 is an intensity/wavelength graph of the stove burner glowing dull red. Notice that on this graph we have included not only the visible portion of the spectrum but quite a lot of the infrared. Recall that we experience infrared radiation as heat. The stove burner is emitting quite a bit of infrared radiation. (If you don't believe this, put your hand near it and feel the heat it radiates.) In fact, the graph indicates that more infrared radiation is being emitted than visible radiation and the graph reaches a peak in the infrared portion of the spectrum. Some red light is emitted, but almost no light from the center and violet end of the spectrum.

Astronomers usually call the intensity/wavelength graph of a star its thermal spectrum.

Now turn up the heat on the stove. The burner begins to take on an orange glow. Figure 11–10 is an intensity/wavelength graph of this burner, with the line for the red burner still included. Compare the two lines. First, more radiation of all wavelengths is emitted in the case of the orange burner. This should correspond to your experience, for you can feel that there is more infrared being emitted and you can see that there is more light coming from the burner. Second, the peak of the graph has moved over toward the visible portion of the spectrum; toward the shorter wavelengths. (Remember, infrared radiation has longer wavelengths than visible light.) Finally, the graph does not taper off on the left until somewhere near the middle of the visible spectrum.

This is about as far as you can go with an electric stove burner. If you have a lamp available with a dimmer switch on it, you can take it to the next step; making it glow yellow-hot. Figure 11–11(a) includes the graph of such a lamp.

Finally, suppose you light up the lamp all the way. It gets white-hot. It appears white because it is emitting light about equally in all visible regions of the spectrum. As Figure 11–11(b) shows, its intensity/wavelength graph peaks near the center of the visible portion of the spectrum.

As indicated by the sequence above, the intensity/wavelength graph of an object emitting electromagnetic radiation can be used to detect its temperature. The graph of Figure 11–12 shows three lines that represent widely different temperatures, including one hotter than any previously shown. The temperature corresponding to each line is shown.

If you want to research the theory explaining the relationship between wavelength and intensity of radiation, look up blackbody radiation *and* Wein's Law *in a general physics text.*

The peak of the intensity/wavelength curve for a light-emitting object always falls at a wavelength that depends upon the object's temperature. This is important to us in astronomy because by plotting this graph for a star we can know the temperature of its surface. The intensity/wavelength graph provides astronomers with an extremely valuable tool in their attempts to understand stars.

It is curious that when we call something "red hot" we generally mean that it is extremely hot. "Orange hot," however, is hotter than red hot. Even hotter than that is "yellow hot" and then "white hot." The temperatures indicated in Figure 11–12 are typical of temperatures of stars, and the line that reaches its peak at the shortest wavelength on the graph represents the hottest star of the three. Notice that a star of this temperature emits more violet and blue light than it does the longer wavelengths. We say that it is a blue star. We often refer to stars by color; this is a quick way to indicate their temperature. A white star is hotter than a red star. In practice, of course, a red star does not appear red like a Christmas tree bulb but it definitely has a red tint. If you have the opportunity to use a telescope to observe some pairs of closely spaced stars you can see this color difference easily but to the unpracticed naked eye color differences between stars are not at all obvious. The important point is that by examining the thermal spectrum of a star, we can determine the temperature of the surface of the star without ever visiting it!

Beta Cygni (named Alberio) is the second brightest star in Cygnus and is a good example of a closely spaced pair of stars with obvious color differences.

Figure 11–11. *The spectra for the yellow-hot and white-hot objects continue the trend seen previously. The white-hot object emits radiation across the visible spectrum.*

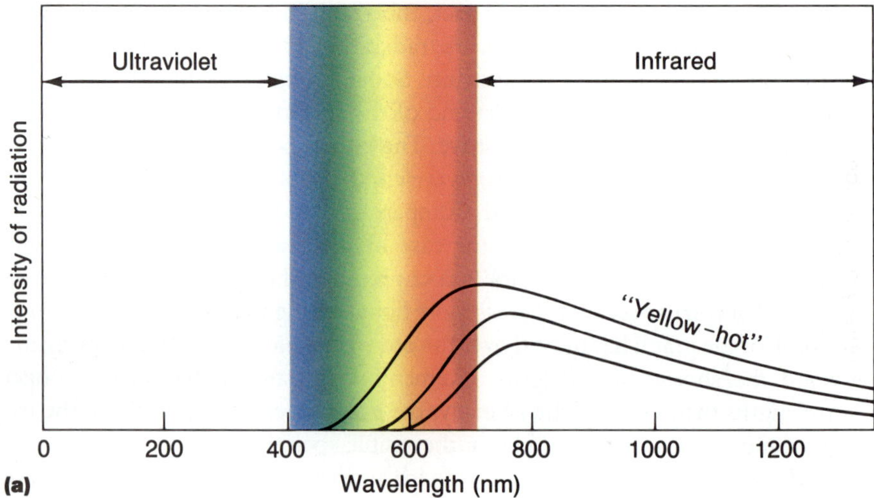

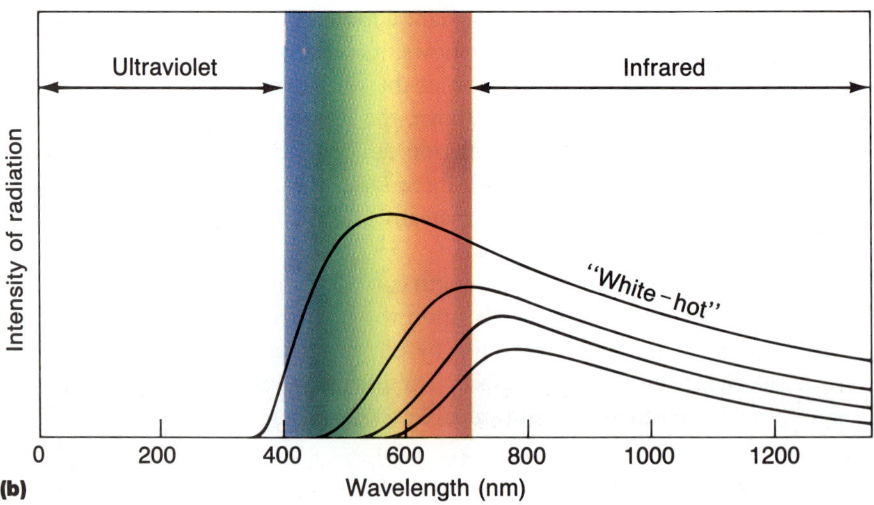

SPECTRA EXAMINED CLOSE UP

Continuous spectrum: A spectrum containing an entire range of wavelengths rather than separate, discrete wavelengths.

The spectrum of visible light described above and shown in Figures 11–1 and 11–4 is called a ***continuous spectrum.*** Such a spectrum is produced by heating a solid object (in this case the filament of a lamp) until it is hot enough to emit visible light. Not all spectra are of this type but before we explain other types of spectra, including those from stars, we must look at how light is emitted from atoms.

The Bohr Atom

If mercury gas is heated until it emits light and that light is sent through a prism, the spectrum of Figure 11–13(a) results. This is certainly not a continuous

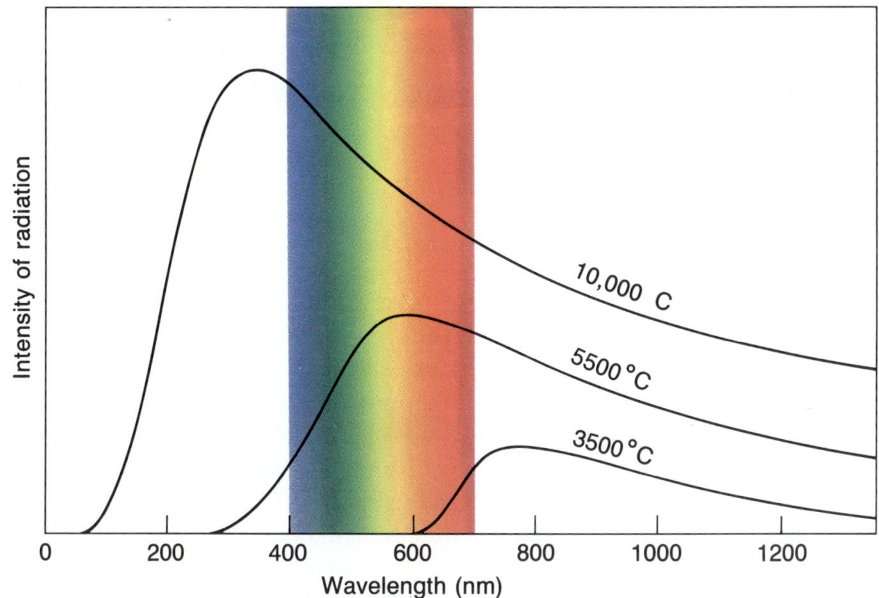

Figure 11–12. *The graph is extended to temperatures found in stars. The 10,000 degree object would be called "blue hot," for it emits more radiation in the blue-violet region of the spectrum than in the other regions.*

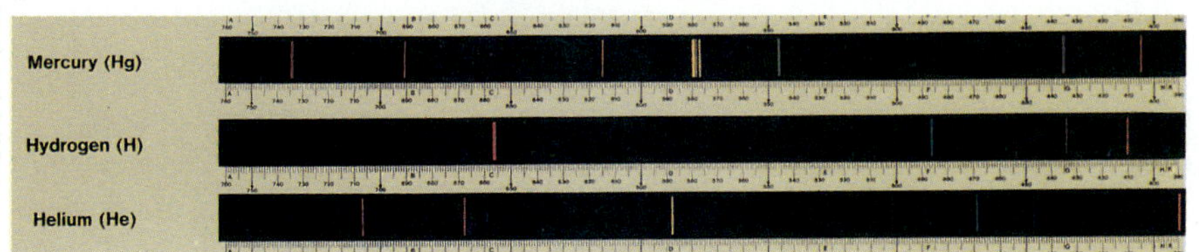

Figure 11–13. *Mercury, hydrogen, and helium gases each emit specific wavelengths of light when heated. This applies not only to visible light but to infrared and ultraviolet.*

spectrum. Figure 11–13(b) shows the spectrum produced by heated hydrogen gas and Figure 11–13(c) is from helium gas. Heated gasses do not produce continuous spectra; but rather spectra consisting of distinct lines, each line representing one wavelength of light. Thus a heated gas produces only a certain number of distinct wavelengths of light rather than a continuous mixture of wavelengths like we see from a hot solid object.

In 1913 the Danish scientist Niels Bohr presented a model of the atom that explains how the various types of spectra are produced. His model describes the atom as having a ***nucleus*** with a positive electrical charge, circled by electrons with a negative electrical charge (Figure 11–14). Positive and negative electrical charges attract one another and this electrical force holds the electrons in orbit around the nucleus.

The ***Bohr atom,*** as Niels Bohr's model is called, is based on three new ideas, or postulates:

1. *Electrons can orbit the nucleus only with certain specific energies.* In order to imagine the different energies of an electron, we think of it in orbit around the nucleus; each energy corresponding to a specific orbital distance. Thus we speak of "allowed" orbits for the electrons. The element hydrogen has only one

The element mercury is a silvery liquid at everyday temperatures. Its spectrum is produced by heating it until it forms a gas and then until it becomes hot enough to emit light. Common street lights (the bluish-colored ones) contain mercury, as can be seen by looking at a spectrum of their light.

Nucleus (of atom): The central, massive part of an atom.

Bohr atom: The model of the atom proposed by Niels Bohr, containing electrons in orbit around a central nucleus and explaining the emission of light.

HISTORICAL NOTE

Niels Bohr

In 1964 George P. Thomson, son of the discoverer of the electron and one of the leading physicists of this century in his own right, said of Niels Bohr: "He led science through the most fundamental change of attitude it has made since Galileo and Newton, by the greatness of his intellect and the wisdom of his judgements." In spite of tributes like this, most students know Bohr only through a few pages of a textbook which discusses the Bohr atom. Who was this giant who is compared to Galileo and Newton?

Niels Bohr was born in 1885 into a very cultured Danish home. His father was a physiologist and university professor and both of his parents read at least four languages and were lovers of art and music. His father's interest in science led Niels to that subject and Niels was the kind of person who gave maximum effort in whatever he tried. For example, he and his brother Harald were famous throughout Denmark as soccer players. (Harald even made the Danish Olympic team in 1908.)

Niels was married in 1910. He and his wife Margarethe had six sons and a number of grandchildren. He was very family-oriented and greatly enjoyed his children. Margarethe and Niels remained devoted to one another until Niels's death in 1962. One son, Aage, followed him in the study of physics and the two worked together on a number of projects (and rode motorcycles together—one of Niels' hobbies).

In 1922 Niels Bohr was awarded the Nobel Prize in physics "for his services in the investigation of the structure of atoms and of the radiation emanating from them." In his acceptance speech he emphasized the limitations of his theory and he seems to have been more aware of its limitations than other scientists who worked with the theory.

One major project on which Niels and Aage Bohr worked was the development of the atomic bomb. Niels's mother was Jewish and after Hitler's army overran Denmark, he and his family were forced to flee their native land to avoid arrest. Niels and Aage came to the United States and helped with the Manhattan Project (the code name for the bomb development effort).

Although Bohr was convinced that the atomic bomb was necessary to stop Adolf Hitler, his main concern was not its development but its control after the war. He was one of the few influential people of the time who foresaw the postwar competition between the U.S. and USSR in nuclear weaponry. He was convinced that the U.S. and Great Britain, who were working jointly on the bomb, should inform Russia of its development (without giving details of the methods used.) He argued that scientists all over the world understood the basic theory of the bomb anyway and that if the U.S. developed it and dropped it without telling the Soviets, it would cause suspicion and distrust on the part of Russia, who was our ally at the time.

Bohr was able to arrange a visit with Winston Churchill, the prime minister of Great Britain. He hoped to convince the British leader of the dangers of a postwar nuclear arms race. The meeting lasted only a half hour, however. Churchill told the person who had introduced Bohr, "I did not like the man when you showed him to me, with his hair all over his head." Bohr said simply, "we did not speak the same language." Churchill seems to be one of the few people

Figure T11–1. *Niels Bohr and his sons.*

who were not impressed with Niels Bohr. (In fact, Churchill wanted him put under intelligence surveillance.)

Bohr later spoke for an hour and a half with President Franklin Roosevelt about the same issue. He thought that he had convinced the president, but after Roosevelt met with Churchill, Roosevelt agreed not to tell the Russians about the weapon.

As was learned later, a Russian spy had been working in the U.S. on the Manhattan Project and Russia knew more about the bomb than Bohr was proposing to tell anyway. We now know that suspicion and distrust did indeed grow between the two powers, and, as Bohr had predicted, the USSR quickly developed its own atomic bomb and then the hydrogen bomb. If Bohr's advice had been taken, might today's world be safer?

Bohr's contribution to science goes far deeper than the development of the planetary model of the atom, as important as that is. His philosophical ideas on the nature of physical theory is perhaps his greatest contribution, but its discussion would require much more room than we have here.

Niels Bohr was very interested in people and was greatly admired by all with whom he worked. (His code name in the Manhattan Project was Nicholas Baker, but his fellow scientists, who were mostly younger, called him "Uncle Nick.") He was a very soft-spoken, gentle man and his influence on scientific thinking will last for ages to come.

Figure 11–14. *The Bohr atom pictures electrons in circular orbits about a central nucleus.*

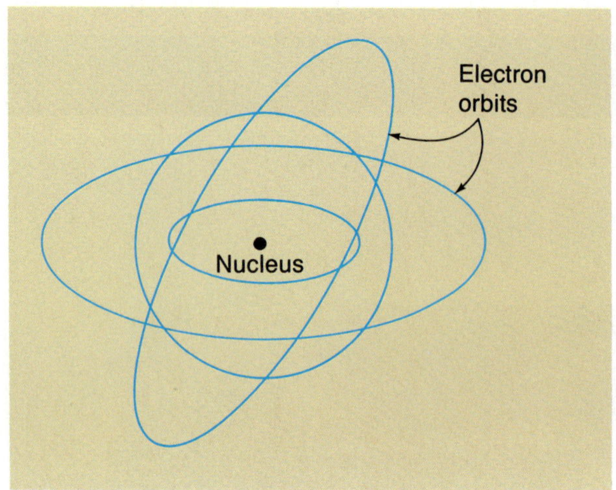

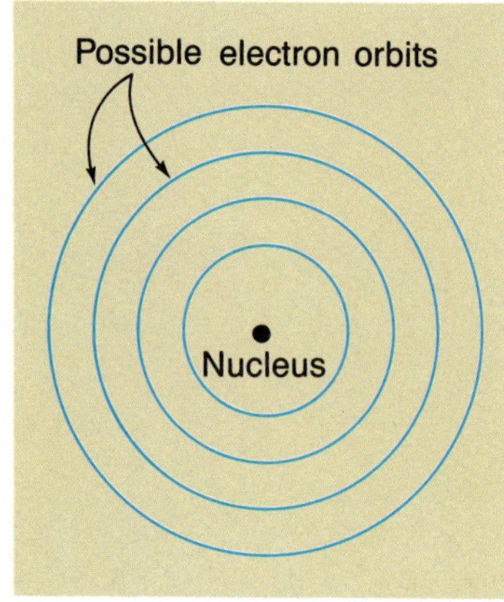

Figure 11–15. *There are only certain discrete energy levels available to the hydrogen atom, represented in the Bohr atom by discrete orbits for its electron.*

electron but many possible energy levels and therefore many allowed orbits. Figure 11–15 represents the hydrogen atom, showing orbits this electron might have. The point of Bohr's first postulate is that the electron can have only these particular energies and therefore these particular orbits. This is far different from the case of the solar system, where there are no limitations on possible positions of planetary orbits. There is no law of nature, for example, that would prohibit Earth from orbiting the Sun at a greater distance (which would mean that the Sun/Earth system would have greater energy). Bohr proposed the revolutionary idea that there is such a law for electrons. Nature only allows certain electron orbits.

2. *An electron can move from one energy level to another, changing the energy of the atom.* In terms of electron orbits, when energy is added to an atom the electron moves farther from the nucleus. On the other hand, the atom loses energy when an electron moves from an outer orbit to an inner orbit. This lost energy comes out of the atom in the form of electromagnetic radiation.

In our previous discussion of waves and light we assumed that light acts as a simple long wave, similar to the waves you can make in a swimming pool. Theoretical work by Albert Einstein, Niels Bohr, and others showed that light is more complicated than this. The Bohr model holds that light is emitted not as continuous waves, but in tiny bursts of energy; each burst being emitted when an electron moves to an orbit closer to the nucleus. According to the Bohr model, light is emitted when an electron falls from an outer to an inner orbit. (We say "falls" because an electron is attracted toward the nucleus in a manner analogous to the way we are attracted to Earth and it seems natural to think of the electron's move inward as a fall.) These tiny bursts of electromagnetic energy are called ***photons*** of radiation.

You might think of the energy/distance relationship as follows: An electron "feels" a force of attraction toward the nucleus and to pull it farther away requires energy.

Photon: The smallest possible amount of electromagnetic energy of a particular wavelength.

The applicable equation is $E = hf$, *where* E *is the energy of the photon,* h *is a constant (called* Planck's constant), *and* f *is the frequency of the light.*

3. *The energy of a photon determines the frequency (or wavelength) of light that is associated with the photon.* The greater the energy of the photon, the greater the frequency of light, and vice versa. Thus a photon of violet light has more energy than a photon of red light. (Recall that violet light has a greater frequency than red light.)

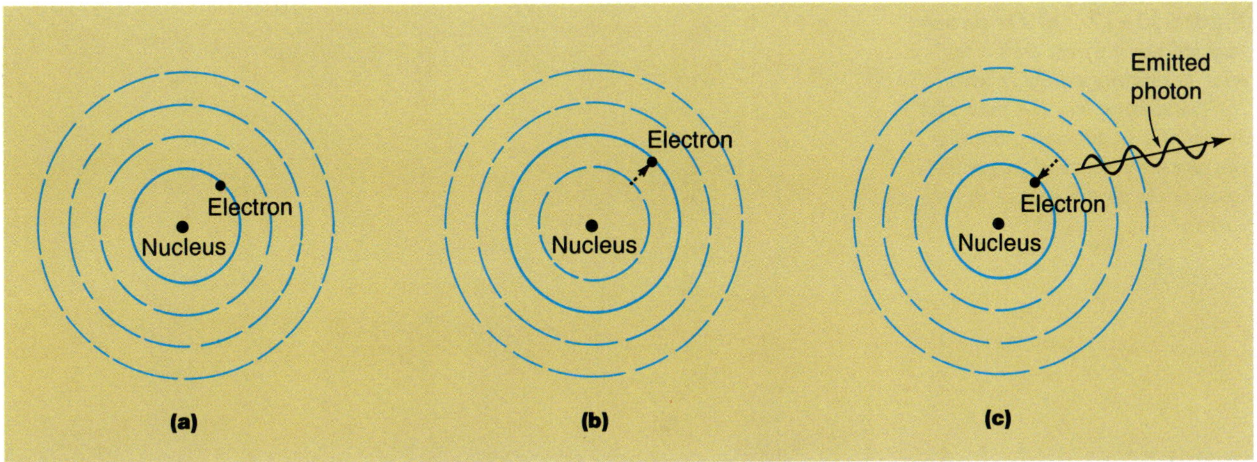

Figure 11–16 *(a) The electron is in the lowest possible energy state—the lowest orbit. (b) When the atom gains energy (perhaps by collision with another atom), the electron jumps up to a higher orbit. (c) The electron then quickly falls down to its original orbit, emitting energy in the form of a photon of electromagnetic radiation as it does so.*

Emission Spectra

We can use the Bohr model of the atom to explain why only certain wavelengths are seen in the spectrum of light emitted by a hot gas. In its normal, lowest-energy state, the electron of a hydrogen atom is in its lowest possible orbit as indicated in Figure 11–16(a). If this atom is given energy (perhaps by an electric current or by collisions with other atoms), the electron might jump to a more distant orbit. Figure 11–16(b) illustrates this. An atom will not stay in its energized state long. Quickly the electron falls down to a lower orbit; emitting a photon as it does, shown in Figure 11–16(c). The energy of this photon is exactly equal to the energy difference between the two orbits. Finally, since the energy of the photon determines the frequency of the radiation, the radiation coming from this atom must be of thc corresponding frequency (and color).

The lowest energy state of an atom is usually called the ground state *and the energized states are called* excited states.

We have described one atom emitting one photon. In an actual lamp that contains hot hydrogen gas, there are countless atoms being given energy and countless atoms emitting photons as they lose energy. Different atoms will have different amounts of energy, depending upon the energy they have absorbed (from a collision with another atom, for example). If a particular atom's energy corresponds to the electron being in the third orbit, the atom might release its energy in a single step as shown in Figure 11–17(a), or it might release it a step at a time as shown in part (b) of the figure.

Suppose there is enough energy available to cause electrons to move to the fourth orbit. Figure 11–18 shows all of the possible falls an electron might take to get back to the lowest orbit. Each of these falls corresponds to a certain specific energy and therefore to a certain specific frequency of emitted radiation. The electrons of some atoms will fall by some paths and the electrons of other atoms will fall by other paths. The result is that radiation of a number of different frequencies will be emitted from the entire group of atoms. Not all frequencies will be emitted, however—just the ones that correspond to the electron jumps.

Hence, the spectrum from a heated gas is not a continuous spectrum. It contains only certain definite frequencies. We call such a spectrum an ***emission spectrum.*** An emission spectrum has a few individual bright lines instead of a continuous band of colors.

Emission spectrum: A spectrum made up of discrete wavelengths rather than a continuous band of wavelengths.

Figure 11–17 *(a) The electron may fall directly from orbit 3 to orbit 1, emitting a single photon. (b) The electron may fall from orbit 3 to orbit 1 in two steps, emitting two photons whose total energy is equal to the photon emitted in part (a).*

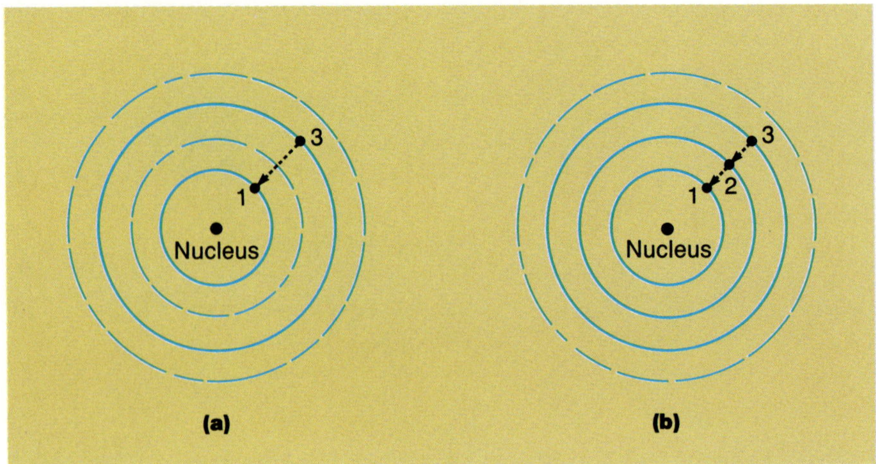

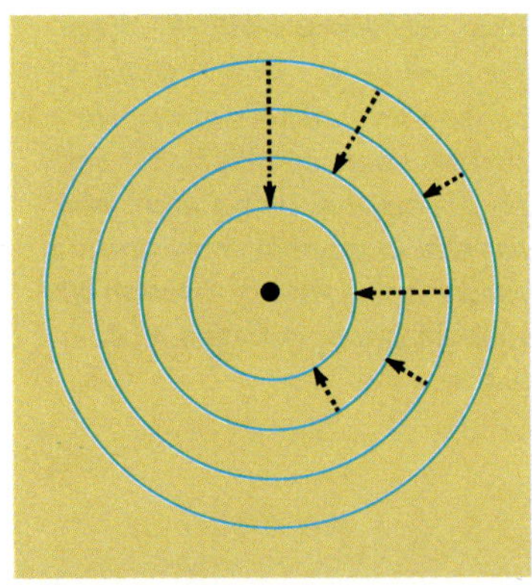

Figure 11–18. *These are the possible paths an electron may take to get from orbit 4 to orbit 1. Each fall corresponds to a different energy change and therefore to a photon of different frequency.*

Refer back to Figure 11–13, which shows the emission spectra of hydrogen, helium, and mercury. The three spectra are different. This is because the allowed energy levels of the atoms are different for each chemical element. No two chemical elements have the same set of energy levels and thus no two chemical elements have the same emission spectrum. This provides us with a tremendously valuable method of identifying elements, since each has a unique spectral "fingerprint." This process has a number of important applications here on Earth but since in this book we are more interested in the stars, we'll look at a stellar application.

Absorption Spectra—The Stars

Figure 11–19 is a spectrum of the light from the Sun. It appears *almost* like a continuous spectrum, but a number of wavelengths of its light are missing! The explanation for this is also based on the Bohr atom and is fairly easy to understand.

Photosphere: The region of the Sun from which visible radiation is emitted. We will discuss the Sun in the next chapter.

From the visible surface (the ***photosphere***) of the Sun, a continuous spectrum is emitted like the white light spectrum we saw in Figure 11–4 that was produced from the light of a heated solid object. Even though the Sun is, in most ways, more like a gas than a solid, it produces a continuous spectrum rather than an emission spectrum. This is because as atoms are pushed together their energy levels are broadened (as if slightly different orbits are allowed). As atoms become more and more tightly pressed together, their energy levels begin to overlap so that a full range of possible orbital energies are possible. Thus an entire range of photon energies is emitted by the atoms and instead of separate, distinct spectral lines appearing in the spectrum, an entire range of frequencies appears.

Before the light from the Sun gets to us on Earth, it must pass through the relatively cooler atmosphere of the Sun as well as through the atmosphere of the Earth. The Sun does indeed have an atmosphere and as we will see in a later chapter, its atmosphere is much denser and much deeper than Earth's. As the light passes through these gases, atoms of the gases absorb some of it. This

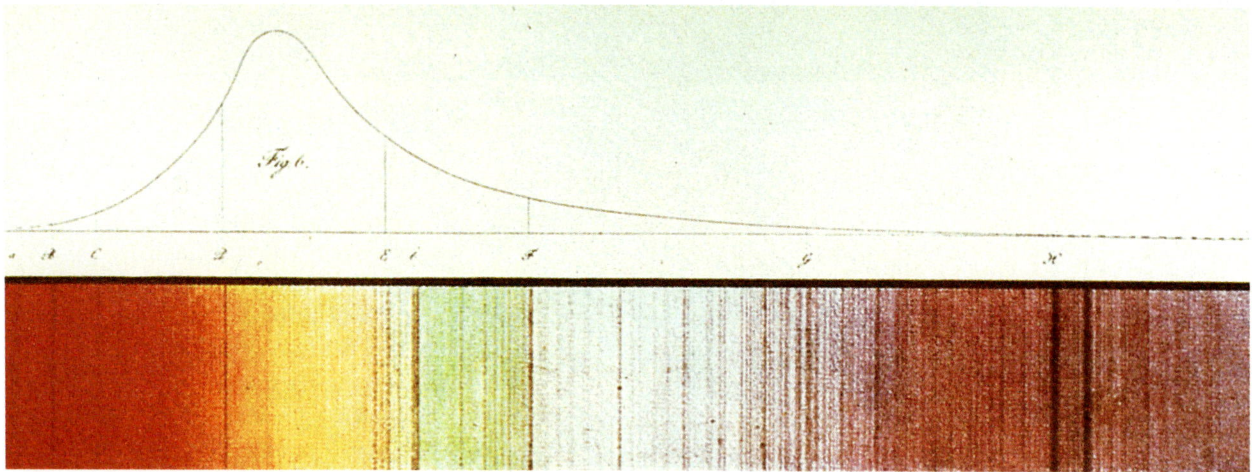

Figure 11–19. *The Sun's spectrum may appear at first to be a continuous spectrum, but when magnified, we see dark lines across it where specific wavelengths do not reach us.*

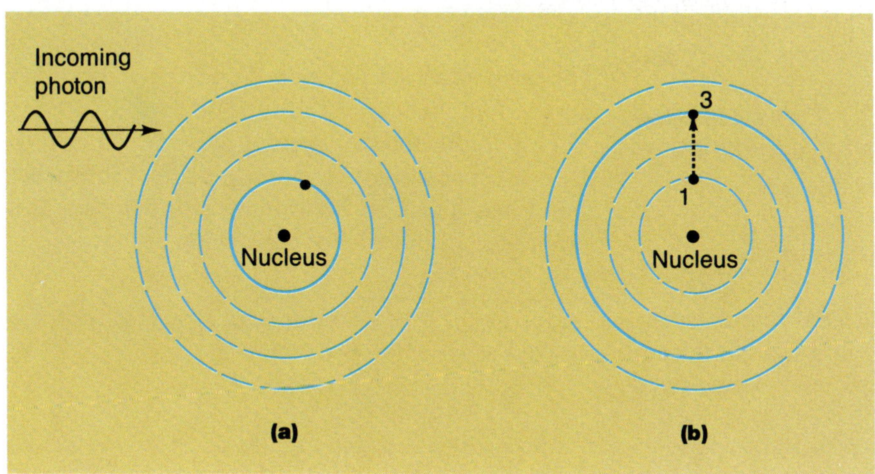

Figure 11–20. *Absorption of the photon results in the atom gaining energy, represented by the electron moving to an outer orbit (the third orbit, in this example).*

absorption of energy raises the atom's energy level but since only certain specific energy levels are possible, only certain amounts of energy can be absorbed by the atom. This results in the reverse of what we had before: Instead of an atom emitting a photon as it releases energy, it absorbs a photon as it absorbs energy. And just as photons of certain energies are emitted by a hot gas, photons of the same energies are absorbed by the cooler gas of the atmosphere. Figure 11–20 illustrates the absorption process in terms of electron orbits.

If white light (a continuous spectrum) is passed through hydrogen gas that is too cool to emit light, it will absorb some frequencies of this light—the same frequencies it would emit as an emission spectrum. Figure 11–21(a) is the emission spectrum of hydrogen we saw before. Figure 11–21(b) shows the

HISTORICAL NOTE

Mystery: The Discovery of Helium

IN 1868 A FRENCH astronomer, Pierre Jules Janssen, went to India to observe a total solar eclipse and to study the emission lines in the part of the atmosphere just above the surface. One bright yellow line appeared that was particularly mysterious. The metal sodium has a bright yellow spectral line and at first Janssen thought that the line he saw was the sodium line. Its wavelength, however, did not match that of sodium's yellow line and Janssen could not reproduce the new spectral line in his laboratory.

Sir Norman Lockyer, an English astronomer who also had observed the new line during the eclipse, hypothesized that the line was due to an element not yet found on Earth and he named it ***helium*** after the Greed word *helios,* meaning "Sun." Reports from Janssen and Lockyer of the discovery of the new element arrived at the Paris Academy of Science within a few minutes of each other and helium is listed as a joint discovery by the two.

For more than 26 years, helium was considered an hypothesized element. In 1895, an American mineralogist found that an unusual gas was released by a certain mineral when it was treated with acid. He reported that the gas was nitrogen, but other scientists disagreed. Sir Norman Lockyer was sent a sample of the gas and upon seeing its spectrum, said, "the glorious yellow effulgence . . . was a sight to see."

Another scientist to examine the gas, Sir William Ramsey, wrote to his wife:

> There is argon in the gas; but there was a magnificent yellow line, brilliantly bright, not coincident with, but very close to, the sodium yellow line. I was puzzled, but began to smell a rat.*

The rat Ramsey smelled turned out to be none other than helium, the element of the Sun.

That the method of spectral analysis was able to discover an element on the Sun a quarter-century before it was found on Earth testifies to the value and importance of this tool of astronomy.

*Tilden, William A., Sir William Ramsey, Memorials of His Life and Work. *London: Macmillan, 1918.*

Absorption spectrum: A spectrum which is continuous except for certain discrete wavelengths.

absorption spectrum that results when white light is passed through cool hydrogen gas. Notice that the dark lines of this spectrum correspond exactly to the bright lines of the emission spectrum.

You might object that after the cool gas has absorbed radiation, it must re-emit it. Shouldn't this then cancel out the absorption? No. As shown in Figure 11–22, the re-emitted light is sent out in all directions. So certain frequencies of the light that was originally coming toward the Earth are scattered by the atmosphere of the Sun. This results in less light of those frequencies reaching us and an absorption spectrum is observed.

The spectrum of the light that is re-emitted is an emission spectrum. During a total eclipse of the Sun, light from the main body of the Sun is blocked out and astronomers can see the light emitted by the Sun's atmosphere and examine

Figure 11–21. *Part (b) shows the absorption spectrum of hydrogen. Notice that the absorbed wavelengths are the same as the emission lines in part (a).*

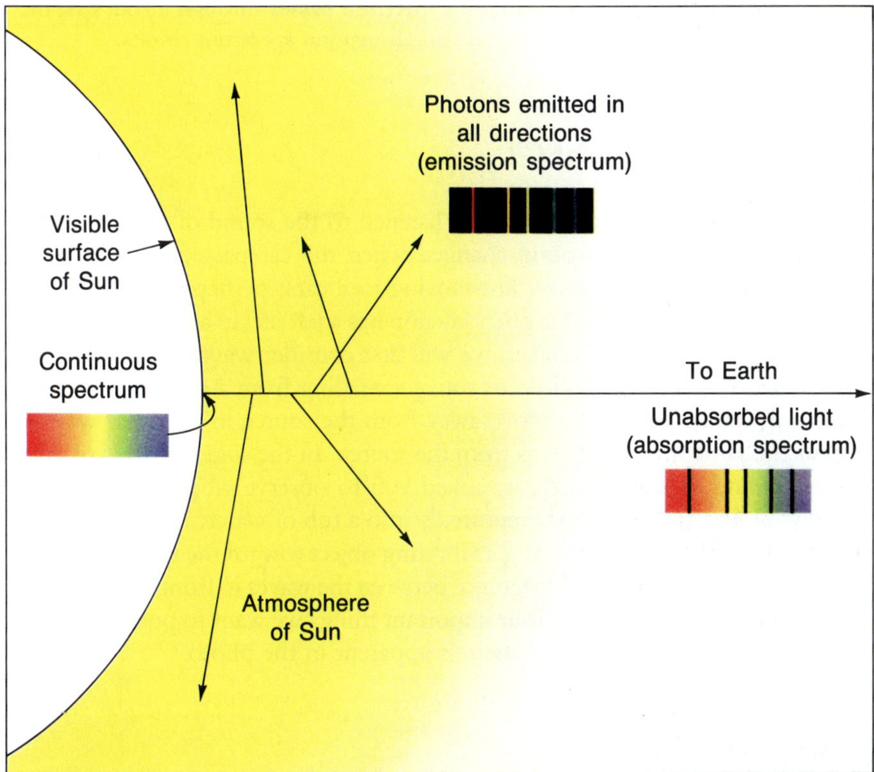

Figure 11–22. *Consider a ray of light leaving a point on the Sun and moving toward the Earth. As it passes through the Sun's atmosphere, some wavelengths are absorbed and then re-emitted in random directions. This results in the solar spectrum being an absorption spectrum. Light from the Sun's atmosphere is an emission spectrum.*

its emission spectrum. It is by examining the emission spectrum from the gas near the Sun that astronomers discovered helium. (See the accompanying Historical Note.)

In the case of the Sun or another star, there are a number of chemical elements in its atmosphere (that is, different kinds of atoms, each with its own pattern of electron orbits and spectral lines). The white light passing through this gas has many frequencies absorbed that correspond to the various chemical elements of the gas. By examining the complicated absorption spectrum that results, we are able to deduce what elements are in the star's atmosphere. Thus we answer a question which, just a century ago, was thought to be unanswerable. We now know what the surface layers of stars are made of! As you might appreciate from the complexity of the Sun's spectrum (look back at Figure

Gustav Kirchhoff (1824–1887) was a German physicist and astronomer whose primary work was in the field of spectroscopy—the study of the spectrum.

The elements in the Earth's atmosphere also absorb radiation, the frequencies of which depend upon what elements are in our atmosphere. But we know what those elements are and they can be taken into account.

11–19), the analysis is fairly complicated, but it is now a common one in astronomy.

The different types of spectra and their causes can be summarized in what are called Kirchoff's laws of spectral analysis:

Kirchhoff's Laws: 1. A hot, dense glowing object (a solid or a dense gas) emits a continuous spectrum.
2. A hot, low-density gas emits light of only certain wavelengths—an emission spectrum.
3. When light having a continuous spectrum passes through a cool gas, the gas absorbs some wavelengths and an absorption spectrum results.

THE DOPPLER EFFECT

Have you ever stood near a road and listened to the sound of a car as it sped by you? Recall how the sound changed when the car passed. The change is especially noticeable for noisy, fast-moving race cars; perhaps you have heard it on televised auto races. This phenomenon has a parallel in astronomy—a very important one. To understand it, we will first consider waves on water.

Figure 11–23(a) is a photo of waves spreading from a disturbance on the surface of water. The waves move away from the source in a regular way and appear the same in all directions from the source. In the suggested activity near the beginning of this chapter, we asked you to observe what happens if you move your finger as you dip it repeatedly into a tub of water. Figure 11–23(b) was made in this manner, moving a vibrating object toward the right as it makes waves on water. Look at the difference between the waves in front of and behind the moving source. There are four important things we want to point out about this case, although only one of them is apparent in the photo.

(a)

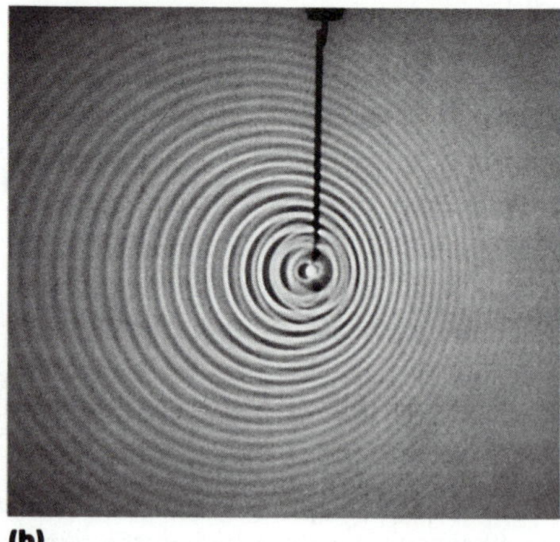
(b)

Figure 11–23. *Waves leaving a moving source (b) are compressed in front of the source and stretched behind it.*

1. Even though the source of the waves is moving, the waves still travel at the same speed in all directions. The motion of a source does not push or pull the waves. It just disturbs the liquid and the disturbance moves away at a speed that depends only on the characteristics of the liquid—water, in this case.

2. The wavelength of the waves in front of the moving source are shorter than they would be if the source were stationary and the wavelengths behind the moving source are longer. Recall that where the wavelength is shorter, the frequency is higher; and where the wavelength is longer, the frequency is lower. This means that if there were corks on the water, the corks in front would bounce up and down with a greater frequency than if the source were not moving and corks behind the moving source would bounce with a lower frequency.

This is the same effect that causes the sound of a car to change as the car passes. The car emits sound just as the vibrating object produces water waves. When you are in front of the car, you are in the region of shorter wavelength and higher frequency. In the case of sound, high frequency means high pitch. After the car has passed its sound waves are stretched in wavelength, causing you to hear a lower pitch.

This effect, which we see here both in water waves and in sound waves, is called the ***Doppler effect,*** named after Christian Doppler, the man who first explained it. Before we apply it to light waves and astronomy, let's continue the list of important things to know about it.

Doppler effect: The observed change in wavelength from a source moving toward or away from the observer.

3. The sound does indeed get louder as a car approaches (and the water waves get higher as the vibrator approaches a bobbing cork), but this is not what the Doppler effect is about. The Doppler effect refers only to the *change in frequency* (and therefore wavelength) of the wave.

4. The frequency does not get higher and higher as the source approaches at a uniform speed. Rather, the frequency is observed to be constantly higher than normal as the source approaches and constantly lower than normal as it recedes.

The Doppler Effect in Astronomy

We do not receive water waves and sound waves from the stars, but only electromagnetic waves. However, the Doppler effect also works for electromagnetic waves. This means that if an object is coming toward us, the light we receive from it will have a shorter than normal wavelength and one moving away will have a longer than normal wavelength. Have you noticed this as you watched a car coming toward you and then going away? No. The reason you can't observe this is because the amount of the shortening and lengthening of the wave depends on the speed of the object *compared to the speed of the wave.*

For water waves, which move fairly slowly, you observe the Doppler effect even for a slow-moving wave source. In the case of sound, which has a speed much greater than water waves, you notice the Doppler effect only for fairly

Figure 11–24. *If a spaceship could move fast enough, a green light on it would appear red to the person being left behind.*

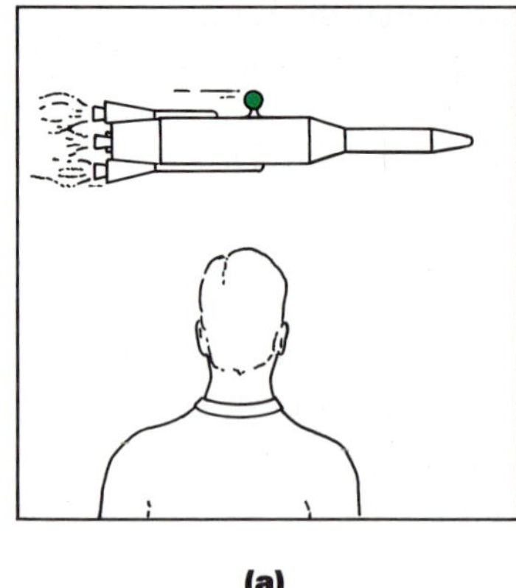

(a)

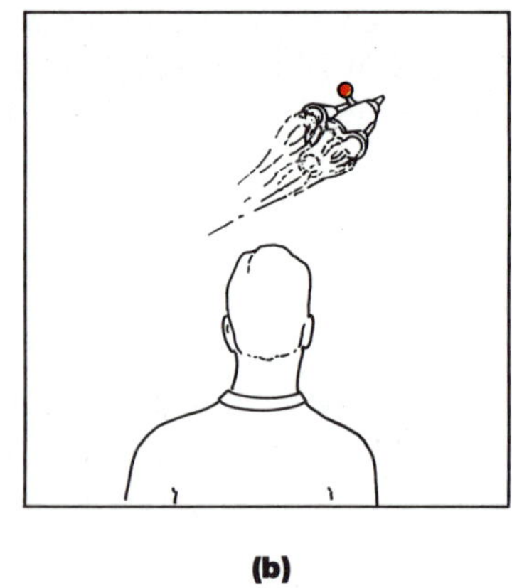

(b)

fast objects. A car going 70 miles per hour is moving at about 10 percent of the speed of sound.

Redshift: A change in wavelength toward longer wavelengths.

Blueshift: A change in wavelength toward shorter wavelengths.

In the case of light, you don't perceive the Doppler effect for a car traveling at 70 miles per hour because it is moving at only one ten-millionth the speed of light. To describe the Doppler effect for light, an example is sometimes given of a spaceship with a green light on it. If the spaceship is moving away from us at a very great speed, the light we see from the lamp is stretched in wavelength so that it may appear red, as shown in Figure 11–24. The light is ***redshifted.*** If the spaceship is approaching, the light is ***blueshifted.***

The spaceship example does illustrate the Doppler effect, but it is very misleading in an important respect: Except for very distant galaxies, objects in the heavens do not move with speeds great enough to actually change their color appreciably. The redshift or blueshift caused by the Doppler effect is very small in most cases. If the spectra of stars were continuous spectra, there would be few cases in which the Doppler effect could be used to detect motion. It is the spectral lines (usually the absorption lines) that make the Doppler effect such a powerful tool.

As an example of how we use absorption lines to detect the Doppler effect for stars, imagine that we record the spectrum of hydrogen gas in the laboratory. Figure 11–25(a) shows the spectrum. Figure 11–25(b) represents the spectrum of a star having only hydrogen in its atmosphere (which is unrealistic, but this is a simplified example). Notice that the absorption lines in the star's spectrum do not align exactly with the emission lines of the laboratory spectrum. They are shifted slightly toward the red. This indicates that the star is moving away from us.

There are three major differences between our example and a measurement of a real star: First, there are many more spectral lines in a real star's spectrum. (Look back at Figure 11–19, the solar spectrum.) Second, the Doppler shift is almost always much smaller than that indicated in the example. When we later show photos of actual spectra, they will usually be a magnified portion of a small part of the visible spectrum. Third, astronomers do not normally use color film in recording the spectrum. This may seem odd, but recall that color is not easy to describe accurately, while wavelength is. The wavelengths of absorption

Figure 11–25 *(a) Emission spectrum of hydrogen when the source is stationary with respect to the observer. (b) Absorption spectrum of hydrogen for a receding source. Note the shift of all lines toward redder colors.*

lines can be measured very accurately and compared to the lines from a laboratory spectrum to determine the existence of and the precise amount of the Doppler shift.

The Doppler Effect As a Measurement Technique

Thus far we have described the Doppler effect as being a method for detecting whether a star is moving toward or away from us, but it is more powerful than this. From measurements of the *amount* of the shifting of the spectral lines, we can determine the ***radial velocity*** of the star relative to Earth. The radial velocity is the star's velocity toward or away from us, and must be distinguished from its ***tangential velocity,*** which is its velocity across our line of sight. Figure 11–26 distinguishes the two velocities.

Radial velocity: Velocity along the line of sight, toward or away from the observer.

Tangential velocity: Velocity perpendicular to the line of sight.

The equation that allows us to calculate radial velocity from Doppler shift data is as follows:

$$\frac{\Delta\lambda}{\lambda} = \frac{v}{c}$$

$\Delta\lambda$ = wavelength difference
λ = wavelength for stationary source
v = velocity of object
c = velocity of light (in a vacuum)

Figure 11–26. *The red car has a radial (to-and-fro) velocity with respect to the observer, while the green car has a tangential (side-to-side) velocity.*

This equation applies if the object is moving much slower than the speed of light, as is the case except for very distant galaxies. For these galaxies a slightly more complex equation is needed as Einstein's special relativity theory becomes important.

Here λ is the wavelength of the spectral line from the stationary laboratory source, $\Delta\lambda$ is the difference in wavelength between the star's spectrum and the laboratory spectrum, v is the velocity of the object, and c is the speed of light. If we solve this equation for what we are usually calculating, the velocity of the object, we obtain

$$v = c\,(\Delta\lambda/\lambda)$$

To illustrate how easy this equation is to apply, let's do an example.

EXAMPLE The wavelength of one of the most prominent spectral lines of hydrogen is 656.285 nm. (This is in the red portion of the spectrum.) In the spectrum of Regulus (the brightest star in the constellation Leo), this line is observed to appear longer by 0.0077 nm. Calculate the speed of Regulus relative to Earth and determine whether it is moving toward or away from us.

Solution First, notice that the data given says that the wavelength of the line in the spectrum of Regulus is *longer* by 0.0077 nm. This means that the star is moving *away* from us. To determine its speed, we will substitute the given values in the equation which relates Doppler shift and speed.

Notice that since $\Delta\lambda$ was stated to two significant figures, the answer was likewise stated that way.

$$\begin{aligned} v &= c\,(\Delta\lambda/\lambda) \\ &= (3.0 \times 10^8 \text{ m/s})\,(0.0077 \text{ nm}/656.285 \text{ nm}) \\ &= 3500 \text{ m/s} \\ &= 3.5 \text{ km/s} \end{aligned}$$

The annual variation in the Doppler shift provides evidence for the Earth's motion around the Sun. The evidence was not available, of course, when the question was controversial.

So Regulus is moving away from Earth at a speed of about 3.5 kilometers/second. This is about 2 miles/second, or 8000 miles/hour; a typical speed for nearby stars. Since the Earth's speed in its orbit around the Sun is about 30 km/s, the Earth's motion would have to be taken into account in making the measurement. Our calculation assumed that the Earth was between the Sun and Regulus so that it had no motion toward or away from that star. Figure 11–27 illustrates this.

Try One Yourself. The nearest star to the Sun that is visible to the naked eye is Alpha Centauri. With Earth's motion removed, the 656.285 nm line of hydrogen has a wavelength of 656.237 nm in Alpha Centauri's spectrum. Calculate the radial velocity of this star relative to the Sun and tell whether it is moving toward or away from the Sun.

It may seem strange but because of the sensitivity of the Doppler effect, radial velocity is much more easy to detect and measure than tangential velocity.

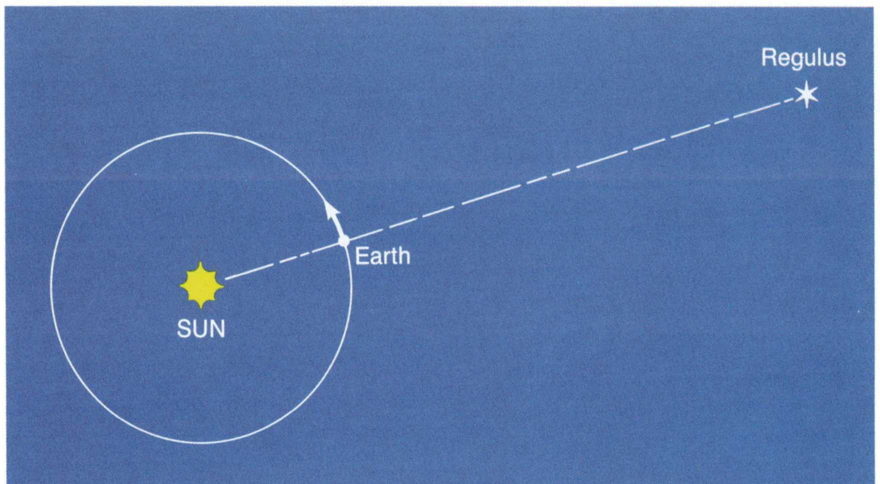

Figure 11–27. *When Earth is between the Sun and a particular star, its motion is perpendicular to a line to that star and thus it has no radial velocity with respect to the star.*

Regulus is moving with a speed of "only" 8 km/s away from us. Though this speed is great by everyday standards, if Regulus also has this speed across our line of sight we are not able to detect it because the star would take a very long time at this rate to move a measurable angle across the sky (because of its great distance).

Other Doppler Effect Measurements

The use of the Doppler effect is not limited to measuring the speeds of stars. Other measurements include the following:

1. Measuring the rotation rate of the Sun. Galileo was the first to observe sunspots on the Sun and he reported that they move across the Sun and thus provide a method to measure the Sun's rotation rate. The Doppler effect gives a second method. The measurement is done by examining the light from opposite sides of the Sun. Light from the side moving toward us is blueshifted and light from the other side is redshifted (see Figure 11–28).

2. In a similar manner, the rotation rates of planets (and the rings of Saturn) can be measured. The fact that the light in this case has been reflected

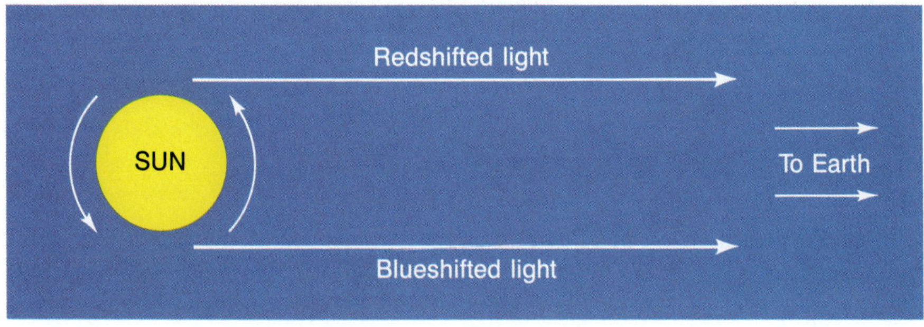

Figure 11–28. *Light from the edge of the Sun rotating toward Earth is blueshifted, and light from the opposite edge is redshifted.*

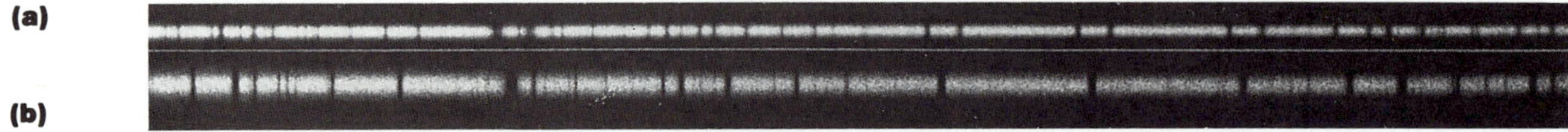

Figure 11–29. *Two spectra, taken at different times, of what appears as a single star. In the top spectra, one star is moving toward Earth and the other way, while in the bottom, the two stars are moving tangentially with respect to Earth.*

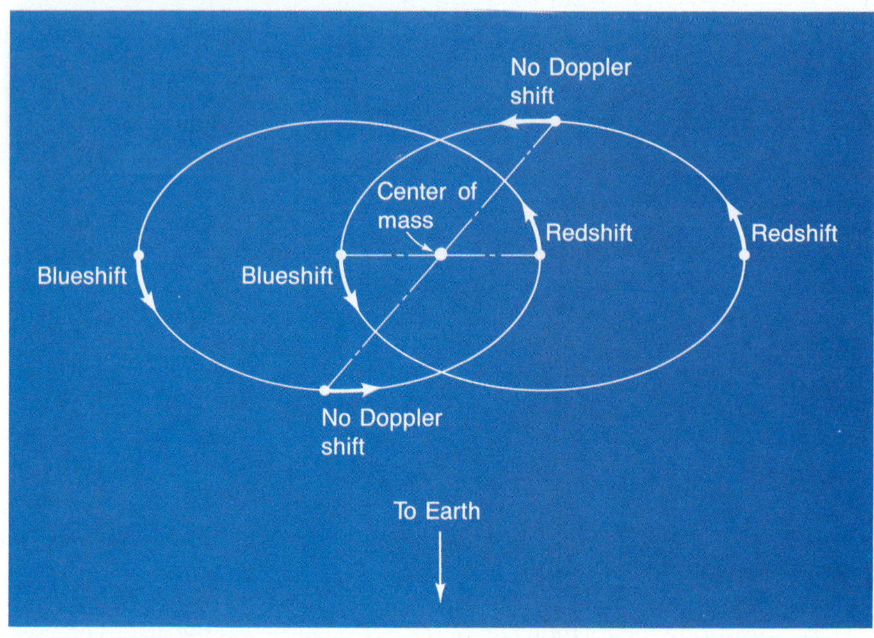

Figure 11–30. *The spectra of the last figure were produced by a binary system of two stars revolving around one another. The stars move in elliptical orbits, always being on opposite sides of the center of mass from one another.*

Police radar catches speeders by bouncing waves from their cars and detecting the Doppler shift in the reflected waves.

by the planet (or rings) rather than emitted by it does not matter; it is still shifted by the Doppler effect. In fact, the rotations of Mercury and Venus were first revealed by reflected radar waves.

3. Figures 11–29(a) and (b) show two spectra of what appears in a telescope to be a single star. The spectra were recorded at two different times. The explanation for the splitting of the spectral lines is that this is not a single star at all but a pair of stars revolving around one another. Figure 11–30 shows the reason why we sometimes see one line and sometimes two. Such ***binary stars*** are important to us in our quest to learn more about the stars and we will return to a study of them later.

Binary star: A pair of stars gravitationally bound so that they orbit one another.

RELATIVE OR "REAL" SPEED?

The speed measured by the Doppler effect is the speed of the object *relative to the speed of the Earth.* As mentioned earlier, we must take into account the Earth's speed in any calculation made with the Doppler effect. You may argue that even if we do this, we are only measuring the speed of the object with respect to the Sun. What about the object's *real* speed? There is no such thing.

The reason for this last statement is that *all* speeds are relative to something. When you say that a car is moving at 50 miles/hour, you mean "relative to the surface of the Earth." When you walk up the aisle of a moving plane, you have one speed relative to the seated passengers, another relative to the Earth, another relative to the Moon, another to the Sun, etc.

All motion is relative; all nonmotion, too. When we say that something is at rest, we usually mean that it is at rest relative to the Earth. (Not always, however. You might tell your little brother to sit still in the back seat of your car even though the car is moving. In this case you are asking him to sit still relative to the car.) It is meaningless to say that something is absolutely at rest.

So there is no loss of meaning to our finding the velocity of an object relative to the Earth and then correcting for the Earth's motion around the Sun. All motion is relative.

This understanding of the relativity of motion is called Galilean relativity *or* Newtonian relativity. *Einstein's theory of relativity (Appendix A) goes much further than this.*

CONCLUSION

We have seen the spectra of stars used in two very different manners. First we treated the spectra as continuous spectra and examined their intensity/wavelength graph. By measuring the wavelength of the maximum intensity of radiation, we can determine the temperature of a star's surface. In doing this analysis, we ignore the absorption lines. These lines, remember, are very narrow; their presence does not change the overall pattern of the intensity/wavelength graph.

Second, we looked at the absorption lines within the spectrum. These lines allow us to determine the chemical composition of stars and, when we use the Doppler effect, the shift in the lines allows us to calculate the radial speeds of stars relative to Earth as well as the speeds of rotation and revolution of celestial objects.

We have begun our study of the stars with a look at how three very important properties of stars are measured: temperature, composition, and motion. In future chapters we will see how other properties are measured. Our goal is to find out what stars are, how they behave, how they come to be, and what finally becomes of them.

RECALL QUESTIONS

1. Define wavelength and frequency in the case of a wave.
2. Suppose you produce two different waves in a swimming pool. One has a frequency of 2 cycles/second, and the other has a frequency of 3 cycles/second. Which has the greater wavelength?
3. Which has the higher frequency, light of 400 nm or light of 450 nm?
4. Approximately what is the least and greatest wavelength of visible light?
5. Name seven regions of the electromagnetic spectrum in order from longest to shortest wavelength. (Do not list the colors of visible light as separate regions of the spectrum.)
6. Which of the following is not an electromagnetic wave: radio, infrared, visible, sound, ultraviolet?
7. Which parts of the electromagnetic spectrum penetrate the atmosphere?
8. Define the following terms: electron, nucleus, photon.

9. State the three postulates upon which the Bohr atom is based.
10. How is the energy of the photon related to the frequency of the light?
11. Why do the various elements each have different emission spectra?
12. Name the three different types of spectra and explain how each is produced.
13. Explain how we know what elements are in the atmospheres of the stars.
14. Carefully explain what the Doppler effect is. In the case of light waves, does the Doppler effect tell us anything about the intensity of the light in front of and behind a moving light source?
15. Does measurement by the Doppler effect tell us the full speed of a star? Explain.
16. Distinguish between the way the spectrum is used to determine the temperature of a star and the way it is used to determine the star's composition and/or motion toward or away from us.
17. Suppose someone argues that the measurement provided by the Doppler effect is not really meaningful because it tells us only the star's motion relative to Earth and not its real motion. How do you answer this?
18. List three different types of motions that can be measured by the Doppler effect.

QUESTIONS TO PONDER

1. In the case of a wave, the meanings of the terms *frequency* and *velocity* are often confused. Distinguish between the two, using water waves to give an example.
2. It is especially easy to get sunburn when skiing high in the mountains. How does the high altitude contribute to causing sunburn?
3. What does an object do to the white light that strikes it to give the object its color? How does this differ from the color of an object which emits its own light?
4. Five different stars are described by the following colors: white, red, orange, blue, yellow. List the stars in order from coolest to hottest.
5. Draw an intensity/wavelength graph for a red star and another for a white star. Describe two ways in which they differ.
6. Compare the idea of allowed electron orbits in the Bohr atom with Bode's rule for planetary orbits.
7. Some stars appear redder than others. Is this the result of the Doppler effect? If so, explain. If not, explain why not.
8. A particular emission line of hydrogen gas (a green one) in the laboratory has a wavelength of 486.1 nm. If the spectrum of a certain galaxy shows this wavelength at 486.3 nm, is the galaxy moving toward or away from us?
9. Look up and report on the use of the Doppler effect in police radar and in measurement of speeds in sports (particularly baseball).

CALCULATIONS

1. If the speed of a particular water wave is 12 meters/second and its wavelength is 3 meters, what is its frequency?
2. The speed of sound in air is about 335 m/s. What is the wavelength of a sound wave with a frequency of 256 cycles/second?
3. Express 500 nm in meters.
4. The 656.285 nm line of hydrogen is measured to be 656.305 nm in the spectrum of a certain star. Is this star approaching or receding from the Earth? Calculate its radial speed relative to the Earth.
5. If a certain star is moving away from Earth at 25 km/s, what will be the measured wavelength of a spectral line which has a wavelength of 500,000 nm for a stationary source?

CHAPTER 12

SOLAR ECLIPSE EXPEDITION TO INDONESIA
MARCH 11 – 20, 1988

Tour Features:
- $2\frac{1}{2}$ Minutes of Totality.
- Tour of exotic Bali Island.
- Java's Borobudur Temple.
- Southern Sky Star Parties.
- Eclipse lectures and discussions.
- Fully escorted.

Price Includes:
- Roundtrip airfare from Los Angeles.
- 7 nights deluxe hotel accommodations.
- American Breakfast daily.
- All ground transportation and transfers.
- Land tours in Bali, Java.

PRICE = $1690.00 PER PERSON, DBL. OCC.

For more information, contact:

Joel K. Harris
Twilight Tours
3316 W. Chandler Boulevard
Burbank, CA 91505
(818) 841-8245

The Sun

THE SUN, THE CELESTIAL object of most importance to life on Earth, is just an ordinary star. The cosmic importance of the Sun is limited to the fact that it is the king of the planetary system in which we live; many other stars are bigger and brighter, many are more interesting and unusual, and most will far outlive the Sun.

We will begin this chapter with a brief overview of the major properties of the Sun. Then we will investigate the source of the Sun's energy and consider various theories for the production of its tremendous energy. Finally, we will describe the Sun in more detail; starting at its center, where energy production takes place, and proceeding outward.

SOLAR DATA

As viewed from Earth, the Sun has an average angular diameter of 31′ 59″, just barely less than 32 minutes or arc. By taking 1.50×10^8 kilometers as the average distance from Earth to Sun and by using either the method shown in the accompanying Activity or the relationship between angular size, distance, and actual size (see Chapter 6), we can calculate that the diameter of the Sun is 1.39×10^6 kilometers. Thus the Sun's diameter is about 110 times the Earth's and about 10 times Jupiter's. Figure 12–2 illustrates the great size of the Sun.

From Kepler's third law (as revised by Newton) we calculate that the mass of the Sun is 1.99×10^{30} kilograms. This is more than 300,000 times the mass of the Earth! The Sun's average density, then, is 1.41 gm/cm^3; about the same as the density of Jupiter. This is just the first of a number of similarities we will see between the king of the solar system and the king's subjects, the planets.

When Galileo first used a telescope to view the Sun nearly 400 years ago, he observed dark spots moving across its face. Figure 12–3(b) is a series of

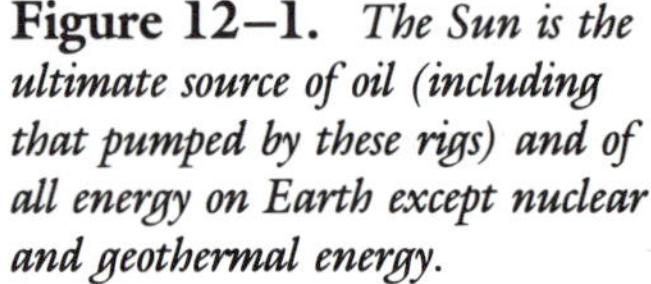

Figure 12–1. *The Sun is the ultimate source of oil (including that pumped by these rigs) and of all energy on Earth except nuclear and geothermal energy.*

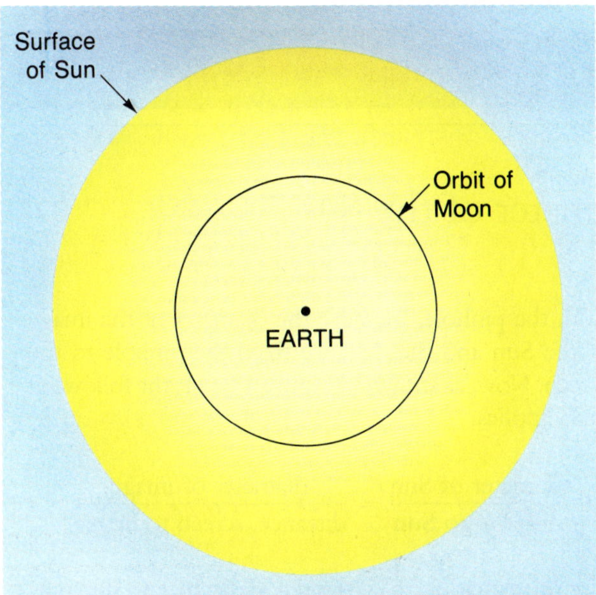

Figure 12–2. *If the Earth were at the center of the Sun, the Moon would orbit about half way to the Sun's surface.*

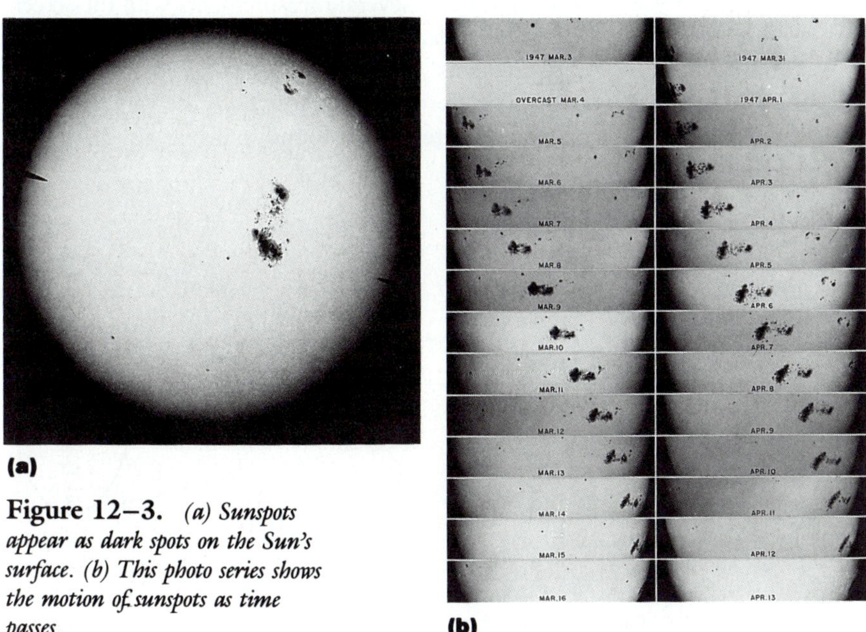

Figure 12–3. *(a) Sunspots appear as dark spots on the Sun's surface. (b) This photo series shows the motion of sunspots as time passes.*

photos of the motion of such ***sunspots.*** Galileo concluded from sunspot motions that the Sun rotates with a period of more than a month. Today we know that the Sun exhibits differential rotation; rotating with a period of 25.4 days at its equator and nearly 40 days near its poles. Recall that the equatorial regions of Jupiter and Saturn also rotate faster than their polar regions.

Sunspot: A region of the photosphere that is temporarily cool and dark compared to surrounding regions.

The gas planets are nonspherical, so we might ask if the Sun is also out of round. The answer is no. This is easily explained, however, for the Sun rotates

ACTIVITY

Measuring the Diameter of the Sun

WITH SIMPLE EQUIPMENT ONE can measure the size of the Sun with a fair degree of accuracy. All that is needed is a sunny day, a piece of cardboard, a ruler, and the knowledge that the Sun is 150,000,000 kilometers from Earth. In a Close Up in Chapter 6 we discussed observing an eclipse by pinhole projection; we will use pinhole projection here also.

Punch a small hole (perhaps one-eighth inch) in a piece of cardboard and hold the cardboard so that the Sun shines through the hole onto a surface behind it (Refer to Figure T6–4.) You may have to adjust the size of the hole to get an image bright enough to see clearly and you might use additional cardboard to shield your screen from reflected sunlight.

Being sure that the screen is perpendicular to a line from the pinhole, measure the diameter of the image of the Sun and the distance from the pinhole to the screen. Now, as shown in Figure T12–1, the following ratio applies:

$$\frac{\text{diameter of Sun}}{\text{distance to Sun}} = \frac{\text{diameter of image}}{\text{distance, screen to image}}$$

Use this equation to calculate the distance to the Sun.

In order to get a feel for the accuracy of your measurement, make several measurements with the screen at different distances. How closely do your various measurements agree? What is the largest source of error in this procedure? How does the value you obtained compare to that found in Table 12–1?

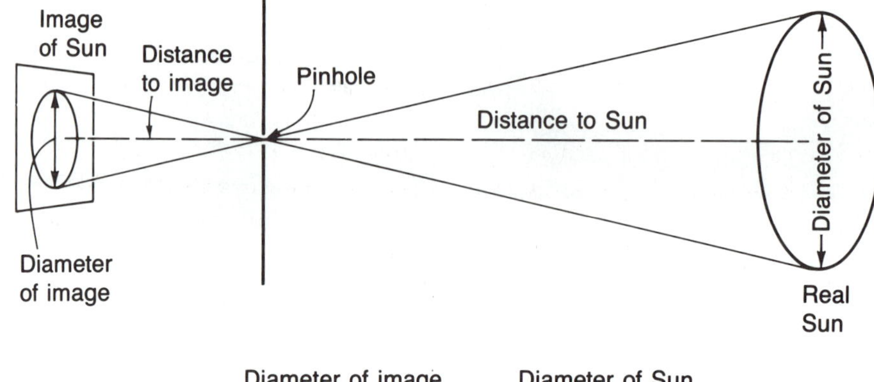

$$\frac{\text{Diameter of image}}{\text{Screen to image distance}} = \frac{\text{Diameter of Sun}}{\text{Distance to Sun}}$$

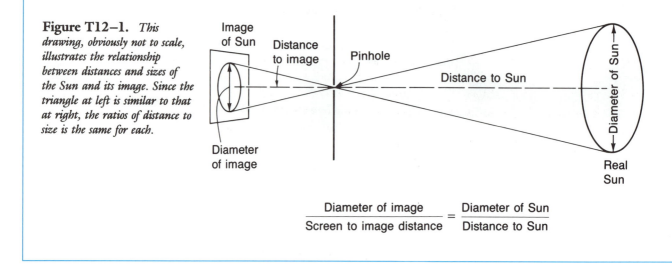

Figure T12–1. *This drawing, obviously not to scale, illustrates the relationship between distances and sizes of the Sun and its image. Since the triangle at left is similar to that at right, the ratios of distance to size is the same for each.*

Today the rotation of the Sun can be measured by observing the Doppler shift of light from opposite sides of the Sun.

much slower than the Jovian planets. While the effect of rotation tends to distort the shape of an object, its gravitational field tends to produce a spherical shape and the Sun's rotation is simply too slow to produce a measurable equatorial bulge.

Table 12–1 Solar Data

	Value	Compared to Earth
Diameter	1,392,530 km	109
Mass	1.989×10^{30} kg	330,000
Density	1.41 gm/cm^3	0.26
Surface gravity	270 m/s^2	28
Escape velocity	617 km/s	55
Surface temperature	6000°C	
Luminosity	3.9×10^{26} watts	
Tilt of equator to ecliptic	7.25°	
Rotation Period		
Equator	25.38 days	
40° latitude	28.0 days	
80° latitude	36.4 days	

SOLAR ENERGY

The Sun emits energy in all portions of the electromagnetic spectrum. A valuable piece of information about the Sun is the rate at which it emits its energy—its ***luminosity*** or total power output. Fortunately, this is not too difficult to measure and calculate.

Luminosity: The rate at which electromagnetic energy is being emitted.

We start by measuring the rate at which solar energy strikes the Earth's atmosphere. This determination was made long ago by measuring the amount of energy from the Sun that strikes an area on the Earth's surface and then correcting for the energy absorbed by the atmosphere. Today it is done most accurately from satellites above the atmosphere. Measurements show that solar energy strikes the upper atmosphere of the Earth at the rate of 1380 watts/m^2. The following example shows how this can be used to calculate the Sun's luminosity; the "wattage" of the Sun.

EXAMPLE

We will calculate the energy output of the Sun in Watts, given that solar energy strikes the Earth at the rate of 1380 watts/m^2 and that the Sun is 1.50×10^8 kilometers from Earth. First, since we have expressed the area on Earth in square meters, we should also express the Earth-Sun distance in meters.

$$1.50 \times 10^8 \text{ km} \times \frac{1000 \text{ m}}{1 \text{ km}} = 1.50 \times 10^{11} \text{ m}$$

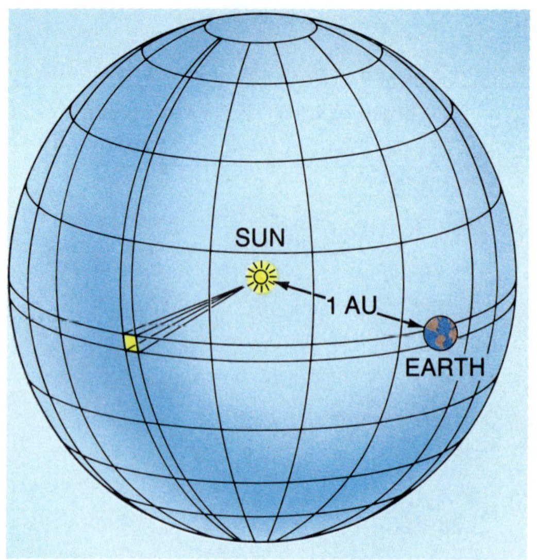

Figure 12–4. *An imaginary sphere is shown drawn around the Sun at the distance of the Earth. It is a simple matter to calculate the area of such a sphere and to determine the solar energy that strikes each square meter of it.*

Imagine a sphere around the Sun at the Earth's distance. We must calculate how many square meters are on that surface. (See Figure 12–4.) To do this, use the equation for the area of a sphere, with 1.5×10^{11} meters being the radius of the sphere in the calculation.

$$\begin{aligned} \text{Area of sphere} &= 4\,\pi\,r^2 \qquad (r = \text{radius of the sphere}) \\ &= 4\,(3.14)\,(1.50 \times 10^{11}\ \text{m})^2 \\ &= 2.83 \times 10^{23}\ \text{m}^2 \end{aligned}$$

Each of these square meters receives 1380 watts of solar power. Therefore,

$$\begin{aligned} \text{Total solar power} &= 1380\ \text{watts/m}^2 \times 2.83 \times 10^{23}\ \text{m}^2 \\ &= 3.9 \times 10^{26}\ \text{watts} \end{aligned}$$

Try One Yourself. How many watts of power strike one square meter of Mars's surface? (Mars is 2.3×10^{11} meters from the Sun.) Solve this by using the total solar power calculated in the example and the area of a sphere at Mars's distance.

As seen in the example, the Sun's luminosity in 3.9×10^{26} watts. When we consider that this amount of power is enough to keep 5 trillion trillion household light bulbs burning, we see that it is truly an awesome amount of power.

The example above illustrates nicely why the inverse square law applies to radiation from the Sun (or any distant object). The area of a sphere that is centered on the Sun depends upon the square of the sphere's radius. Thus a sphere twice the distance from the Sun has an area four times as great, and only one-fourth as much energy strikes a square meter on the surface of this more distant sphere.

The Source of the Sun's Energy

It is estimated that a 1 percent change in solar luminosity would result in a temperature change on Earth of 1 or 2 Celsius degrees (about 2 to 4 Fahrenheit degrees). When we consider that the last major ice age on Earth resulted from a temperature decrease that averaged only 5 Celsius degrees across the planet, we see how critical it is that the Sun maintain a uniform rate of energy production. What is the source of this energy that must have remained constant far into the past?

Prior to this century, a number of hypotheses had been suggested to explain the source of the Sun's energy. All have now been rejected, based on additional data. For example, it was proposed that chemical reactions (such as the burning of a fuel) are the source of the Sun's energy. We now know that this cannot be the case, simply because if the Sun were made of a fuel such as coal or oil it would burn out in a few centuries at the rate that it is releasing energy.

In the mid–nineteenth century, Hermann von Helmholtz and William Lord Kelvin proposed that the source of the Sun's energy is a very slow gravitational contraction. We will see later how such a contraction produces energy. Their calculations showed that at the Sun's present rate of energy production, gravitational contraction could have produced the energy by a reduction in the size of the Sun so slight that it would not have been enough to notice in recorded history. The rate of energy production would not have changed appreciably over the past 100 million years. Their theory seemed to be a good one; it fit the available data.

Then in this century geologists learned that the Earth's age is not a few hundred million years but rather a few ***billion*** years; ten times longer. The contraction theory had to be abandoned and the source of the Sun's energy was again an open question.

The conversion between mass and energy is what the equation $E = mc^2$ *is about.* E *stands for the amount of energy that can be created from a certain amount of mass,* m. *The* c *is the conversion factor and has a value equal to the speed of light.*

In the first decade of this century, Albert Einstein proposed, as part of his special theory of relativity, that mass and energy are interconvertible. That is, one can be changed into another. Late in the 1920s it was hypothesized that this process could be the source of energy in the Sun. Then during the 1930s, physicists worked out the theory that today explains how the Sun has produced

Figure 12–5. *If the Sun's energy production decreased only a few percent, we could have another ice age.*

its tremendous power for the past 4 to 5 billion years and how it will continue this production for another similar period of time.

Solar Nuclear Reactions

Proton: The massive, positively charged particle in the nucleus of an atom.

Neutron: The massive nuclear particle with no electric charge.

Do not confuse the terms proton *and* photon*; they are very different. Refer to Chapter 11 for a description of photons.*

Recall that the Bohr model of the atom proposed that the atom consists of a nucleus surrounded by orbiting electrons (Figure 12–6). That nucleus will be our focus now. An atom's nucleus makes up almost 99.99 percent of the mass of the atom and consists of two kinds of particles: ***protons*** and ***neutrons***. Protons have a positive electrical charge and neutrons have no electrical charge. The number of protons in the nucleus determines what element the atom is. For example, if the nucleus of an atom contains one proton, that atom is necessarily an atom of hydrogen. If it contains two protons, it is helium. If six, carbon. If ninety-two, uranium. A nucleus of hydrogen, on the other hand, is not limited to a specific number of neutrons. Although most hydrogen nuclei contain no neutrons, some have one neutron and a few, two.

It is important to distinguish nuclear reactions from chemical reactions. The latter are those with which we are familiar in our everyday life and involve the combining of a number of atoms into a molecule or the separating of a molecule into individual atoms or into smaller molecules. When we burn paper, for example, the paper's carbon atoms combine with oxygen atoms from the air; producing carbon dioxide molecules. The nuclei of the individual atoms do not come into play. Figure 12–7 illustrates this and emphasizes that the forces between atoms involve only forces exerted by the electrons of one atom on the electrons of another, and not forces exerted by the nuclei on one another.

Fusion (nuclear): The combining of two nuclei to form one.

Nuclear reactions, on the other hand, involve forces between nuclear particles; orbiting electrons are not part of these reactions. There are many types of nuclear reactions but only one will be of interest to us: the ***fusion*** reaction. In a nuclear fusion reaction, two nuclei combine to form a larger nucleus. They "fuse."

The primary source of energy in the Sun (and in all stars during most of their lifetimes) is a series of nuclear fusion reactions in which four hydrogen nuclei are fused to form one helium nucleus. In the process, a small fraction of the mass of the nucleii is changed into energy. This is where Einstein's theory comes into play. Let's look at the process.

Most hydrogen nuclei consist simply of one proton. Prior to the fusion reaction, we have four hydrogen nuclei and after the reaction, there is one helium nucleus. Let us subtract the mass of one helium nucleus (6.6466×10^{-27} kg) from the mass of four hydrogen nuclei (each having a mass of 1.6736×10^{-27} kilogram):

$$\begin{array}{r} \text{mass of 4 hydrogen nuclei} = 6.6942 \times 10^{-27}\ \text{kg} \\ \underline{\text{mass of 1 helium nucleus} = 6.6466 \times 10^{-27}\ \text{kg}} \\ \text{difference} = 0.0476 \times 10^{-27}\ \text{kg} \end{array}$$

The difference between the mass of the original matter and the resulting matter is very small, not only in terms of the actual amount (less than 10^{-28} kilogram) but also in that the lost mass is only 7 tenths of one percent of the

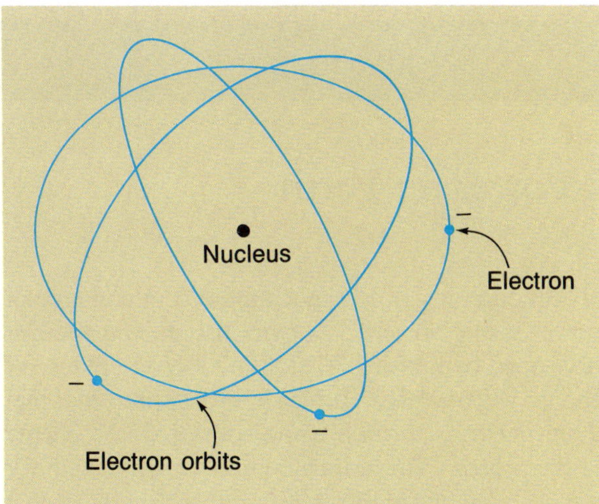

Figure 12–6. *The Bohr atom, with electrons (that have a negative charge) circling the positive nucleus. If the atom were this size, the nucleus would still be too small to see.*

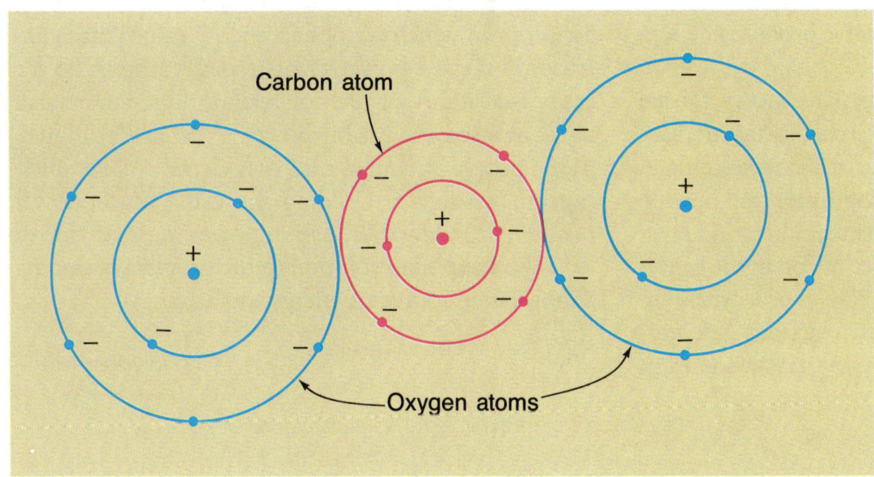

Figure 12–7. *A carbon dioxide molecule (like all molecules) is held together by bonds between the electrons of its two oxygen atoms and one carbon atom. Nuclear forces do not come into play in the bonding.*

original. Not even 1 percent of the mass is changed into energy. In fact the energy produced by one single fusion is only enough to lift a mosquito a fraction of a millimeter. In order to produce the Sun's output of 3.9×10^{26} watts, a total of 4.3×10^{9} kilograms—nearly 5 million tons—of matter must be converted into energy each second. This is about 10^{38} nuclear reactions, involving the transformation of some 55 billion kilograms of hydrogen to 54.6 billion kilograms of helium. Although this is a tremendous amount of matter by human standards, it is almost insignificant when compared to a Sun's total mass. At its present rate of consumption of hydrogen, it would take close to 100 billion years to convert all the Sun's hydrogen to helium.

In practice, the process by which hydrogen is converted into helium incorporates three steps. The reactions start with a fusion of two protons (hydrogen nuclei) and end with production of a helium nucleus containing two

CLOSE UP

Fission and Fusion Power on Earth

THE DREAM OF UNLIMITED energy has been with us at least since the beginning of the industrial age and wonder at the tremendous power of the Sun existed long before that. We know now that nuclear fusion reactions are the source of the Sun's energy and just a few decades ago humans harnessed this energy (if *harnessed* is an appropriate word here) in the hydrogen bomb. The bomb's name comes from the fact that it uses a form of hydrogen (deuterium) as its fuel. The nuclear reaction in the H-bomb is similar to that in the Sun; it produces helium from hydrogen.

Peaceful uses of fusion power have not yet been developed, although much research has taken place over the last four decades and is still in progress. A hint at the problems of controlling fusion can be seen by considering the tremendous temperatures and pressures that are necessary to produce fusion in the center of stars. On the other hand, there is essentially an unlimited supply of fuel for such reactions—deuterium—and therefore controlled energy production from fusion is a very attractive goal.

Although fusion is a much more common type of power in the universe, its twin, fission, was the first developed by humans. Fission involves the release of energy when a large nucleus is broken into two medium nuclei, with mass being converted into energy in the process. The atomic bomb (poorly named, for it uses the energy of the nucleus rather than the energy of the entire atom) was the first application of fission power. Since the development of the A-bomb, we have learned to control this reaction and fission is the reaction from which we obtain energy to produce electricity in nuclear power plants today. Contrary to the ready availability of fuel for fusion, the uranium that must be used for fission power is definitely limited. Many people, scientists and nonscientists alike, question the wisdom of building and using fission power plants; but in any case, fission power must be viewed as only a temporary solution to the problem of finding a long-range source of energy on Earth.

protons and two neutrons. During the process two other smaller nuclear particles are produced, as well as a gamma ray. The solar fusion reactions are presented in more detail in Table 12–2 and Figure 12–8.

Deuterium: A hydrogen nucleus that contains one neutron and one proton.

Positron: A positively charged electron emitted from the nucleus in some nuclear reactions.

Neutrino: An elementary particle that has little or no rest mass and no charge but carries energy from a nuclear reaction.

Look at the first reaction in the table. Here, two hydrogen nuclei fuse to form the nucleus of another type of hydrogen, called ***deuterium,*** which has a neutron in its nucleus along with the proton. In addition, two other particles are formed and these two particles fly away from the reaction at great speeds. One of them, the positive electron, agitates nearby nuclei as it moves among them, causing these nuclei to increase in speed. If nuclei increase in speed, however, the effect we observe is a rise in temperature. It is by means of this process that the energy from fusion becomes thermal energy. The other particle, the ***neutrino,*** escapes from the Sun and does not cause significant heating within the Sun. We discuss this particle in a Close Up in this chapter.

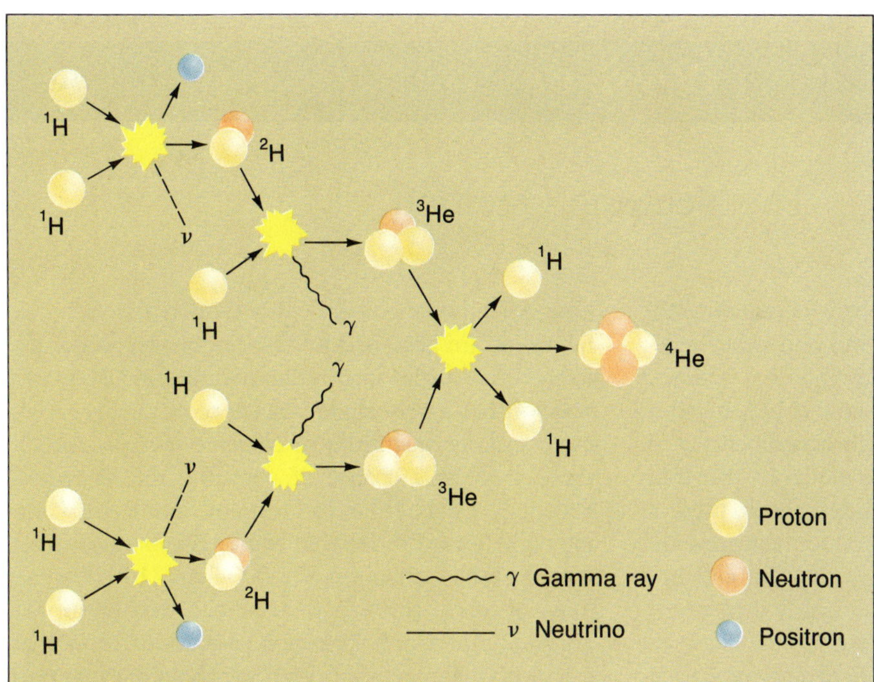

Figure 12–8. *The proton-proton chain begins with four protons and ends with a helium nucleus. The four protons are shown at left combining in separate reactions to produce two deuterium nuclei (each with a proton and a neutron).*

Proton-proton chain: The series of nuclear reactions that begins with four protons and ends with a helium nucleus.

Table 12–2 The Proton-Proton Chain

Reaction	Explanation
$^1_1H + ^1_1H \rightarrow$ $^2_1H + e^+ + \nu$	Two protons combine to produce a deuterium nucleus (^2_1H; the 2 indicating the total number of protons and neutrons in the nucleus), a positron (e^+, which is a positive electron), and a neutrino (ν).
$^2_1H + ^1_1H \rightarrow$ $^3_2He + \gamma$	A deuterium nucleus joins with another proton to produce a helium nucleus (^3_2He, containing two protons and one neutron) and a gamma ray (γ).
$^3_2He + ^3_2He \rightarrow$ $^4_2He + 2\,^1_1H$	Two helium nuclei fuse to form the common type of helium (^4_2He) and two protons.

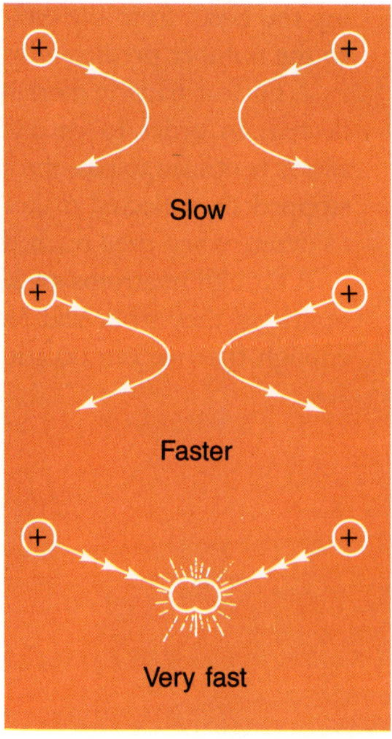

Figure 12–9. *Nuclei moving toward one another at too slow a speed will be repelled because of their positive charges but if they are moving fast enough, electrical repulsion will not be strong enough to prevent them from colliding and fusing.*

Hydrogen, of course, exists throughout the world. Yet it does not, on its own, fuse into helium. The reason for this is that all nuclei have the same type electric charge (positive) and therefore they repel one another. This electrical repulsion force acts over great distances, at least compared to the very short distances over which nuclear forces act. This means that the particles are unable to get close enough together for the attractive nuclear forces to take over unless they happen to be moving toward one another at a great speed (Figure 12–9). Since the temperature of a gas determines the speed of its particles, the particles of hot hydrogen are more likely to fuse than those of cool hydrogen. Fusion occurs only in high-temperature matter.

CLOSE UP

The Neutrino Question

WE OBSERVE DIRECTLY ONLY only the outer layers of the Sun and other stars. What we know of their interiors is almost completely theoretical; achieved now largely by computer modeling. Several years ago experiments were devised in an attempt to test the model by detecting neutrinos from the Sun's core. Nuclear theory predicts that the Sun's fusion reactions produce neutrinos and that these neutrinos pass through the Sun and into space at the speed of light (or very nearly that speed). A number of groups of experimenters around the world are attempting to see if the quantity of neutrinos reaching the Earth from the center of the Sun corresponds to the number predicted by theory.

A nuclear particle must be stopped (or at least deflected) in order to detect it and the property that allows neutrinos to pass through the Sun without being stopped or deflected makes them extremely difficult to detect when they reach the Earth.

Although neutrinos are seldom involved in nuclear reactions, they strike and react with nuclei frequently enough that we can occasionally detect the reaction. One detection method depends upon the fact that neutrinos are absorbed by chlorine nuclei and transform those nuclei into radioactive argon. Even this reaction has a very slight chance of occurring, however; so large quantities of chlorine must be used if there is to be a chance of detecting the neutrinos. Raymond Davis, Jr. and a group of researchers working with him at Brookhaven National Laboratory have placed a tank containing 378,000 liters (100,000 gallons) of liquid cleaning fluid, C_2Cl_4, deep in an old gold mine in South Dakota. Calculations show that even with that much chlorine, only about one neutrino should react each day to produce a radioactive argon atom. Still, the researchers know that if the argon is produced, their equipment is sensitive enough to detect the argon after many days.

Davis' group has tried the experiment many times but they can detect only about one-third of the number of neutrinos predicted by the theory. When their results were first reported in the late 1970s, astronomers and nuclear physicists were shocked and began looking for explanations. There are several possible causes for

In addition, we have seen that a great number of fusions of hydrogen nuclei must occur each second in order to produce the Sun's power. Thus we know that the density of matter must be extremely high in the region of the Sun where fusion takes place. To see how and where this occurs, we will investigate the internal structure of the Sun.

THE SUN'S INTERIOR

Conditions inside the Sun must be theorized based upon what we see near the surface and what we know about the behavior of matter at the temperatures and pressures which are necessary to sustain fusion reactions. We know that at temperatures as high as the Sun's, matter must exist as a gas rather than a solid the discrepancy. The first suspect in a case like this is always the experimental apparatus. The apparatus and procedure have been checked and rechecked, however, and the fault does not lie there.

It is possible that our theory concerning the detection of neutrinos is in error. Neutrinos are of various types and only one type is detected by this experiment; the type theorized to be produced in the Sun. Could it be possible that neutrinos change from one type to another during their eight-minute flight from the Sun? This would mean that the problem lies in our understanding of neutrinos rather than in our theory of solar energy production.

Of course we may be wrong about the nuclear reactions within the Sun. The error may be a relatively slight one; concerning, perhaps, the various types of neutrinos that are produced. Then again, we must recognize the possibility that there is a major error in our theory of energy production within stars. A recent suggestion is that a previously unknown particle is carrying off some of the energy thought to have been carried by neutrinos. (The two names that have been suggested for this particle are interesting: *cosmion*, an important-sounding term, and *Weakly Interacting Massive Particle*—WIMP—a more whimsical name.)

Another interesting possibility exists. The prediction of the number of neutrinos being produced is based on the energy we receive from the Sun. Recall, however, that photons of electromagnetic radiation require some 10 million years to reach us after their production in the core of the Sun. Thus the energy we receive had its beginning millions of years ago. But neutrinos reach us in only eight minutes. Could it be that the Sun is now in a period of low energy production? If so, this won't have an effect on Earth until thousands or millions of years from now, when it will cause another great ice age.

Experiments using another element, gallium, are now underway in Europe and the USSR in hopes of solving the mystery. (The latter is another of the numerous U.S.–USSR cooperative ventures in science.) As of now, we cling to our theory of energy production within the Sun, hoping either to discover the missing neutrinos or to determine why they are missing.

or a liquid and this fact makes the analysis easier, for gases are much simpler than liquids or solids. As we will discuss later, the temperature on the surface of the Sun is about 6000 degrees C and it increases greatly below the surface. At these temperatures, solids and liquids cannot exist and most electrons are stripped away from their nuclei. The result is that most of the material of the Sun's interior consists of free nuclei and free electrons. The behavior of this material, however, is similar to that of a simple gas and so we must first study properties of gases. The properties of importance to us are temperature, pressure, and ***particle density***.

Particle density: The number of separate atomic and/or nuclear particles per unit of volume.

Pressure, Temperature, and Density

When you blow up a balloon, the gas inside exerts a pressure on the rubber of the balloon and supports it against its tendency to contract. The pressure exerted

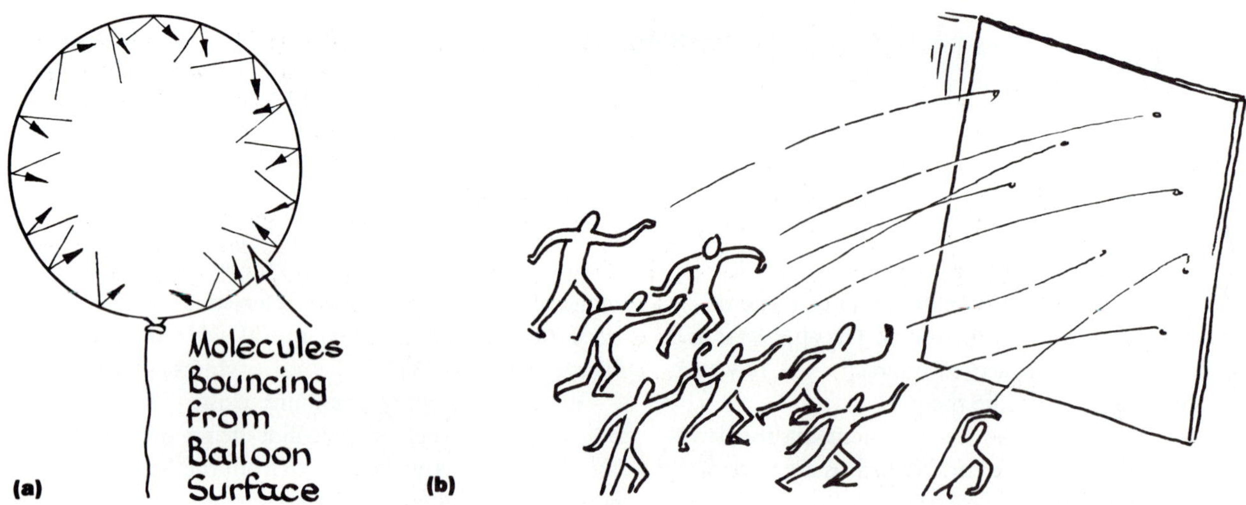

Figure 12–10. *The walls of a balloon are held out by numerous collisions by molecules inside the balloon in the same way that a great number of tiny sand-throwers could exert a force on a board and topple it backwards.*

by the gas is the result of collisions of the individual molecules of the gas with the rubber surface. Figure 12–10(a) illustrates this. Each molecule, as it strikes the rubber wall and rebounds, exerts a tiny force on the wall. Although we cannot detect each individual bounce and the force produced, the overall force exerted by the gas is simply the total of all of these individual impulses. To see this, imagine tiny grains of sand being fired at a board by a great number of sand-throwers as in Figure 12–10(b). The force exerted on the board by a single grain of sand might seem negligible but the overall result could be a force great enough to cause the board to move.

Pressure: The force per unit of area.

Notice that in the discussion above we sometimes spoke of force and other times of pressure. ***Pressure*** is defined as the amount of force exerted per unit area and might be expressed as pounds per square inch. So when we think of a single grain of sand or a single atom rebounding from something, it is more natural to speak of the force exerted by the particle. On the other hand, when we think of many sand grains or many atoms striking over a great area, we speak of the force exerted on each unit of area, or the pressure exerted.

What determines the pressure of a gas, then? There are two factors: the speed of the molecules of the gas and their density. To see that each of these factors is important, think again of the molecules of gas in the balloon. If we heat this gas, causing the molecules to move faster, each collision with the inside surface of the balloon will be more violent; exerting more force on the wall of the balloon. In addition, a greater speed results in more collisions. For these two reasons, the total force exerted on one square centimeter of the wall will be greater; the pressure exerted by the gas will be greater. On the other hand, if we cool the gas, it will exert less pressure. Conclusion: the pressure exerted by a gas is directly related to the gas's temperature.

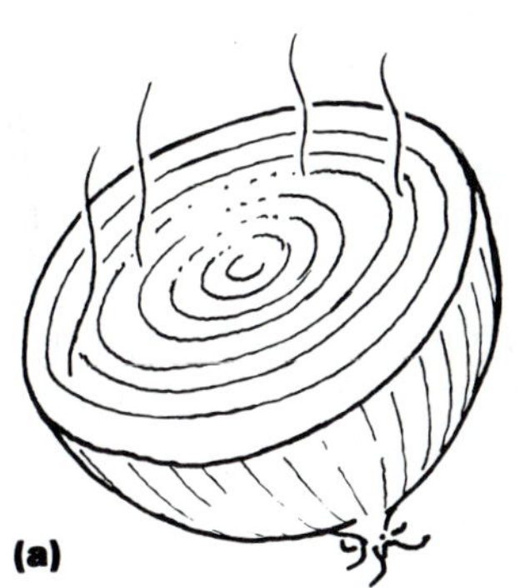

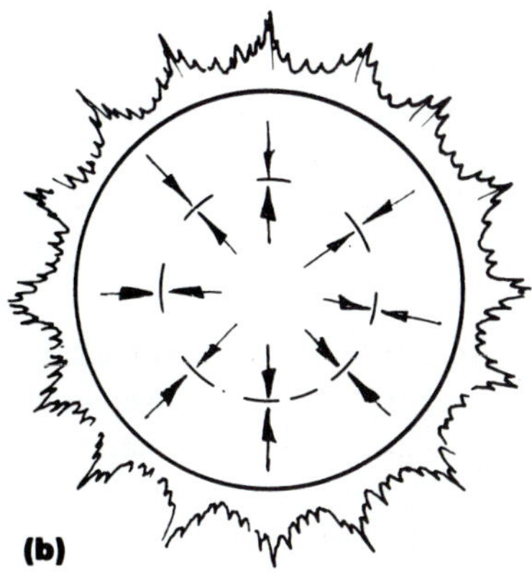

Figure 12–11. *(a) One can think of the Sun as consisting of multiple layers like those of an onion, except that there is no distinct boundary between the Sun's layers.*
(b) Within the Sun, there must be the same pressure upward on any layer as there is downward.

To appreciate the effect of the particle density on pressure, imagine that twice as much gas is somehow put into the balloon without allowing it to expand. If this is done, twice as many molecules will be striking the inside walls of the balloon, exerting twice as much pressure. This, of course, is what causes the balloon to expand when more gas is added. Conclusion: the pressure exerted by a gas is directly related to the density of the gas.

Pressure, density, and temperature are interrelated. If one changes, one or both of the others must change. Now we turn back to the Sun and consider how these factors determine the character of the Sun's interior.

Hydrostatic Equilibrium

The Sun is held together by gravitational force. The force of gravity holds the solar material to the Sun just as the force of gravity holds the atmosphere of the Earth near its surface. In this respect, the Sun is merely a big ball of gas. Our Earth's atmosphere is more dense near the surface; not simply because the force of gravity is greater there than higher up but also because the pressure exerted by the gas above compacts the lower layers. Gases lower in the atmosphere have to support the gases above. The same applies to the Sun. At any particular depth below the Sun's surface, the pressure of the gas at that point must be enough to support the gas above. Thus it is convenient to think of the Sun as having layers, like the various layers of an onion as shown in Figure 12–11(a). Keep in mind, however, that in the Sun there are no distinct boundaries between layers but rather a continuous change as we move toward or away from the center.

Since the Sun is in a state of equilibrium (that is, neither noticeably contracting nor expanding), the pressure downward on any thin layer must be equal to the pressure exerted upward on that layer as shown in figure 12–11(b). This allows us to calculate the pressure at any depth below the surface, for knowing the total mass of the Sun we can calculate the weight of gas above any particular layer. Knowing this, we can calculate the pressure that is needed within the layer in order to support the gas above.

Hydrostatic equilibrium: A balance between the pressures in a planet's atmosphere or within a star.

The equilibrium conditions in the Sun are known as ***hydrostatic equilibrium***. The name is almost self-explanatory: "hydro" refers to the fluid, gaseous state. This is basically just a more complex case of the situation we discussed with the inflated balloon. In that case, the stretched rubber holds the air inside in a compressed state. As long as the outward pressure of the compressed air inside is enough to support the inward pressure of the rubber, equilibrium is maintained.

Since the gas at the center of the Sun is supporting the weight of the gas all the way out to the surface, we should expect great pressures at the center. In fact, the pressure there is calculated to be 1.3×10^9 times that on the surface of the Earth. This tremendous pressure pushes protons close enough together that hydrogen fusion can take place. Only near the center of the Sun is the density of hydrogen great enough to support fusion. The solar core, where fusion is taking place, extends out to perhaps 10 percent of the radius of the Sun.

The fusion reaction in the core provides a heat source that obviously must be taken into account when calculating conditions within the Sun. Recall that when a gas is heated, it tends to expand. A balloon expands when its temperature rises but then it stabilizes at a (different) equilibrium condition. Likewise, the Sun exists in a state of equilibrium, with the force of gravity balanced by forces tending to expand the gas.

To see how hydrostatic equilibrium works, imagine that we could somehow compress the Sun artificially. If this occurred, the pressure within the Sun would increase. The fusion rate would then increase, pushing the Sun back out to another equilibrium position. When in a later chapter we discuss the life cycle of stars we will see that once the energy production of a star slows and the core cools, contraction of the core begins. As long as energy production is stable, however, the Sun remains in equilibrium.

To see what happens to the energy produced in the core of the Sun, we must look at energy transport within the Sun.

Energy Transport

We observe that energy is radiated from the surface of the Sun. The fusion reactions in the Sun, however, occur at its core, where the temperatures and pressures are the greatest. The heat energy produced must then be transported out to the surface.

There are three possible methods by which heat can be transferred from the center of the Sun outward; the same three processes which occur here on Earth and, indeed, everywhere in the universe. They are conduction, convention, and radiation. We'll discuss each in turn.

Conduction (of heat): The transfer of heat in a solid by collisions between atoms and/or molecules.

If you put one end of a spoon on the burner (or in the flame) of a kitchen stove and hold the other end, you'll feel your end of the spoon gradually getting warmer. Heat is being transferred by vibration through the metallic crystal structure of the spoon. The method of transfer here is called ***conduction***. Imagine the atoms near the end of the spoon in the fire. As that end heats, the atoms

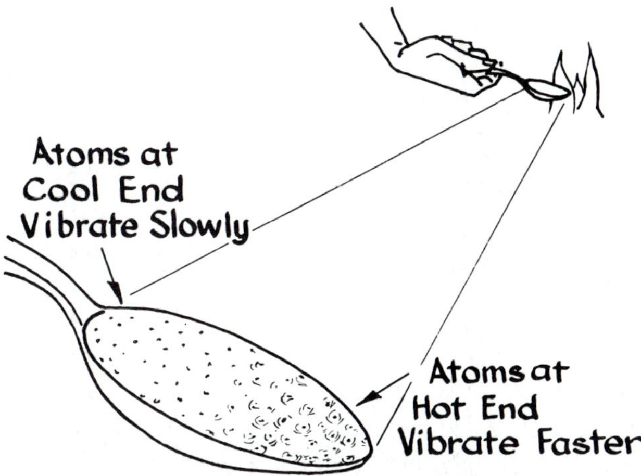

Figure 12–12. *The fast-vibrating atoms at the end of the spoon in the fire cause atoms next to them to vibrate faster. This continues until atoms at the far end are vibrating fast, meaning that that end also is hot.*

Figure 12–13. *Hot air rises from the burner, warming the hand. This is an example of convection of heat.*

vibrate at greater speeds (see Figure 12–12). These atoms exert forces on adjacent atoms of the metal, however, and cause those atoms to pick up the vibration. Gradually, the increased vibration spreads up the spoon until the atoms at the other end are also vibrating more rapidly than they were. Notice that in heat conduction, atoms do not move from one region to another but vibrational energy—heat—is transferred.

Conduction requires that the particles of the substance be in close contact, as are the atoms in a solid. In a star, this is not the case, except in some extremely dense stars, and so conduction is not a significant factor in transporting heat from within the Sun.

Convection: The transfer of heat in a gas or liquid by means of the motion of the material.

Convection occurs when the atoms of a warm fluid (liquid or gas) move from one place to another. Put your hand about a foot above a hot stove burner. You will feel hot air rising from the burner, as illustrated in Figure 12–13. This takes place because heated air is less dense than cooler air and therefore the hot, less dense air rises. The result is that heat is transferred upward from the stove by the motion of the hot gas. In forced-air central heating/cooling systems the motion of the hot (or cool) air is caused by fans.

In a star, convection between adjacent layers is significant only when the temperature difference is great compared to the pressure difference. In the case of our Sun, this condition is met only in the region within about 150,000 kilometers of the surface (Figure 12–14). In this region, convection constantly mixes the solar material as hot gas rises and cooler gas descends. Deeper within the Sun, convection is almost inconsequential and mixing does not occur to a great degree.

Figure 12–14. *This illustrates the relative thickness of the three heat-transport zones of the Sun.*

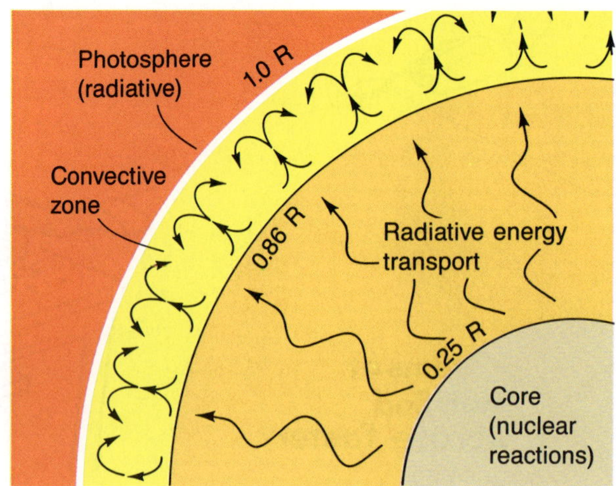

Figure 12–15. *A photon is absorbed and re-emitted numerous times as it travels from the Sun's core. Each re-emission is in a random direction.*

Radiation (of heat): The transfer of heat by infrared waves.

The final method of heat transfer is by ***radiation.*** If you hold the palm of your hand exposed to the stove burner, you can tell that your hand is being heated by another method other than rising hot air. To emphasize this, hold your hand off to the side, where it is not in the stream of hot air, and you will feel heat radiated to your hand. Radiation of energy occurs in all portions of the electromagnetic spectrum, depending only upon whether the hot object emits the particular wavelength of radiation and on whether the receiving object absorbs that radiation or not. The air of your kitchen, for example, is transparent to most electromagnetic radiation produced by the stove burner so it does not absorb the radiation and is not heated by it directly. Your hand, however, being opaque to the radiation, absorbs it and is heated by it.

Inside the Sun, and most stars, radiation is the principle means of heat transport. If the Sun were transparent, the electromagnetic radiation produced in the core would travel outward at the speed of light and reach the surface in about 2 seconds. In actuality the material of the Sun is nearly opaque and so the radiation that is emitted travels only a small distance before being absorbed. It is then re-emitted, absorbed, re-emitted, etc.; with a typical distance between successive absorptions of 1 centimeter (Figure 12–15). The reemissions occur in random directions and the result is that energy that began perhaps as a gamma ray photon from the fusion of two protons travels a very circuitous path and may take on the order of millions of years to reach the surface. This seems an impossibly long time for something travelling at the speed of light but we must keep in mind both the multitude of absorptions and re-emissions taking place and the extreme length of the roundabout path taken by the energy.

Figure 12–14 shows the various portions of the interior of the Sun: the core, the zone where radiation is the primary energy transporter, and the zone where convection is most important. Once the energy of the Sun reaches the surface, it is again radiated outward and it is this radiation that reaches Earth; primarily in the form of ultraviolet, visible, and infrared radiation. Before it reaches Earth, however, it must pass through the solar atmosphere; a region much larger than the main body of the Sun itself.

THE SOLAR ATMOSPHERE

The Sun has no surface in the sense that the Earth does. When we speak of the surface of the Sun, we are speaking of the part of the Sun from which we receive visible light, the *photosphere*. We divide the region of the Sun beyond the photosphere into two parts: the chromosphere and the corona. We will discuss the three portions of the Sun's atmosphere in turn.

The Photosphere

Although the photosphere is a very thin layer (a maximum of about 500 kilometers thick), it does have a thickness. This means that some of the light we receive from the Sun comes from one depth within the photosphere and other light comes from other depths. In saying that the photosphere is 500 kilometers thick, we are saying that we can see to that depth. When we look at an edge (the **limb**) of the Sun, we see to a lesser depth because we are seeing the Sun at a grazing angle. Figure 12–16 shows this effect, which is important to us in that it allows us to analyze light from different depths within the photosphere and therefore to determine the temperature at different depths. We learn that the photosphere varies in temperatures from about 8000 degrees C at its deepest to about 4000 degrees C near the outer edge. Overall, the light we receive from the photosphere is representative of an object about 6000 degrees C. An intensity/wavelength graph of the radiation from the Sun (Figure 12–17) peaks near the center of the visible spectrum.

Limb (of the Sun or Moon): The apparent edge of the object as seen in the sky.

The fact that the solar spectrum peaks near the center of the visible region is not just a coincidence, for our eyes evolved so as to use the available electromagnetic radiation efficiently.

The pressure of the outer photosphere (calculated as we did for inner layers, from knowing the gravitational force there and the amount of material above each layer) is only about 0.01 the pressure at the surface of the Earth. Knowing the temperatures and gas pressures of the photosphere, we can calculate the density of particles there. We find that the density of the matter there is only 0.001 of the density of air at sea level on Earth (even though the gravitational field there is 28 times what it is at the surface of the Earth).

Figure 12–16. *From the center of the disk of the Sun, we receive light that originated at a greater depth than the rays we receive from near the edge.*

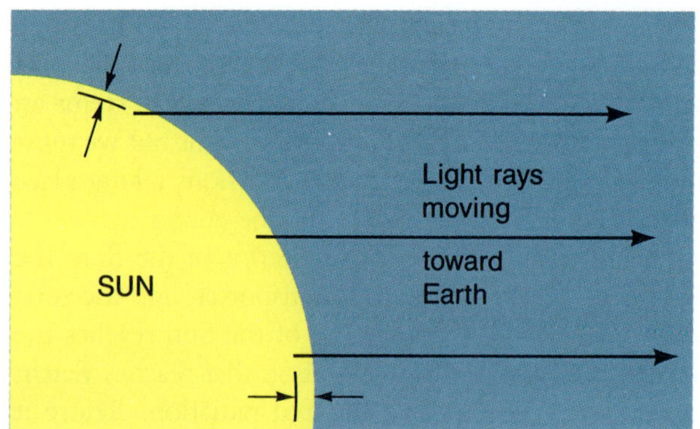

Figure 12–17. *The intensity/wavelength graph of light from the Sun reaches a peak in about the center of the visible portion of the spectrum.*

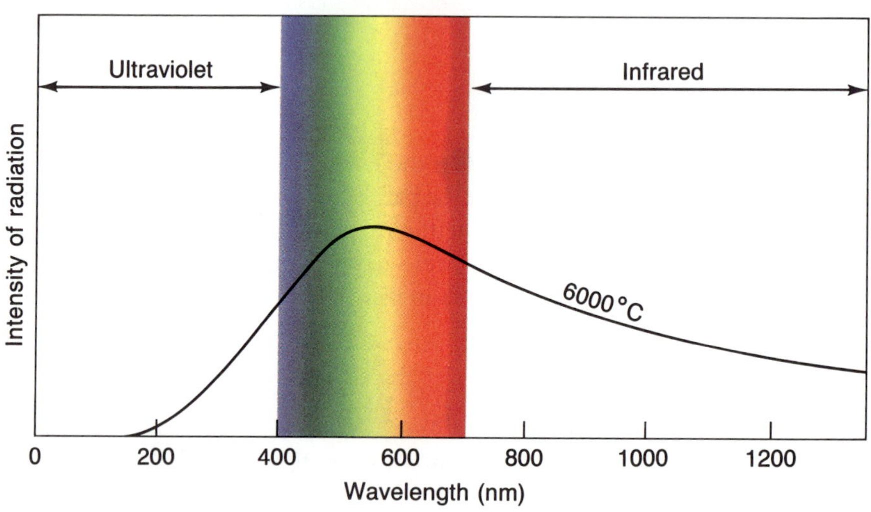

Granulation: Division of the Sun's surface into small convection cells.

These are percentages by mass, not by number of atoms.

Figure 12–18 is a photograph of part of the photosphere. There are irregularly shaped bright areas surrounded by darker areas. Recall from our discussion in the last chapter of the intensity/wavelength diagram that a hotter object emits more radiation than a cooler object of the same size. The brighter areas of the sun are brighter simply because they are hotter. This ***granulation*** of the Sun's surface is the result of convection. Granules are areas where hot material (the light areas) is rising from below and then descending (the dark surroundings). Figure 12–19 illustrates the effect and demonstrates that the photosphere is a boiling, churning region.

Using methods described in the last chapter, we can determine the chemical composition of the photosphere. We find that the mass of the photosphere is about 79 percent hydrogen. The remaining 21 percent consists of some 60 elements. All of these elements are known on Earth and we find them in about the same proportions on Earth that we do in the Sun's atmosphere (with a few exceptions). The exceptions are of two types: Elements such as helium that have masses so low that they would have escaped Earth if they were once here in abundance, and elements we find on Earth but whose characteristic spectra are

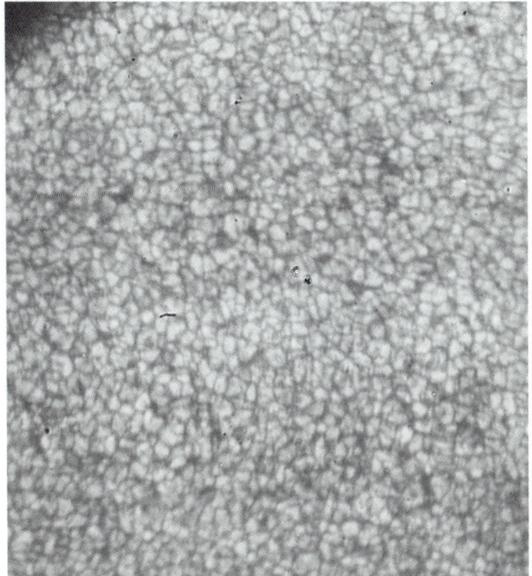

Figure 12–18. *This photograph of the Sun's surface was taken from a high-altitude balloon to eliminate effects of atmospheric blurring. The granules are typically 800 to 1200 kilometers across.*

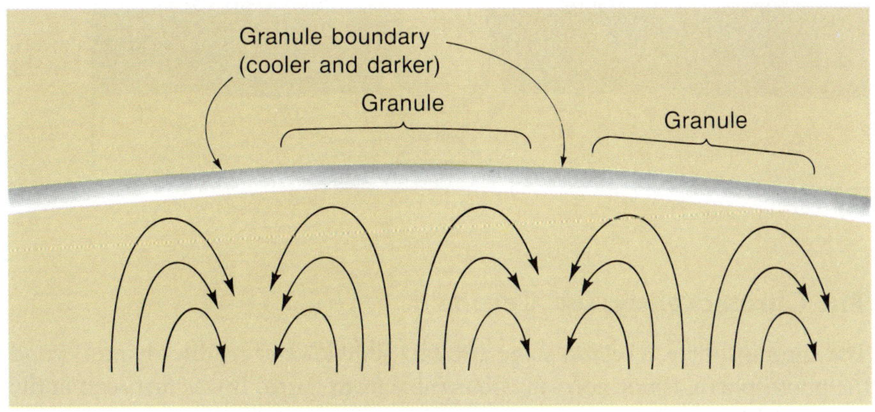

Figure 12–19. *Granules are seen where hot material from below the photosphere rises. Where it descends after cooling slightly, we see the darker edges of the granules.*

such that they would not be detectable in the solar spectrum if the elements are as rare in the Sun as they are on Earth. The similarity of composition between astronomical objects as different as Earth and the Sun is an observation that will be of importance to us in Chapter 15 when discussing the formation of the solar system.

Notice that it is the composition of the Sun's atmosphere—not the entire Sun—that we deduce from the solar spectrum. From our knowledge of nuclear fusion we know that helium must be more abundant in the core of the Sun than in the atmosphere, where it comprises about 20 percent of the mass. Overall, the Sun is theorized to be made up of about 75 percent hydrogen, with most of the rest being helium; leaving only about 2 percent for the remainder of the elements.

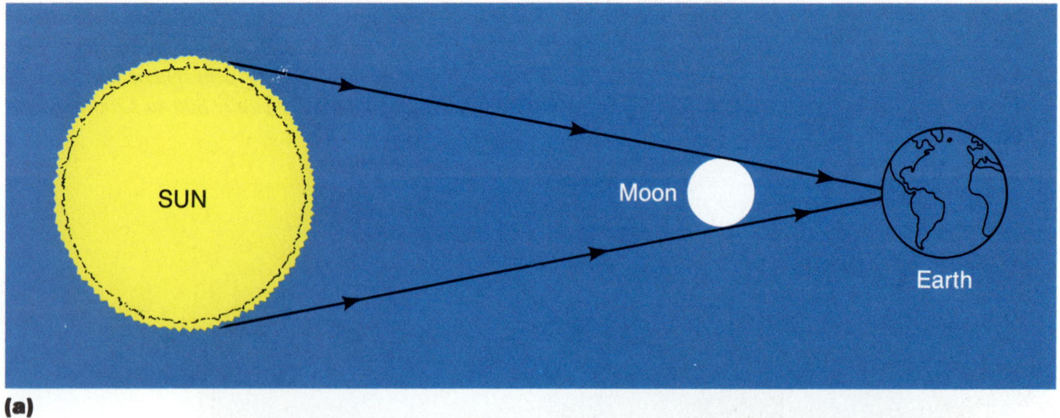

(a)

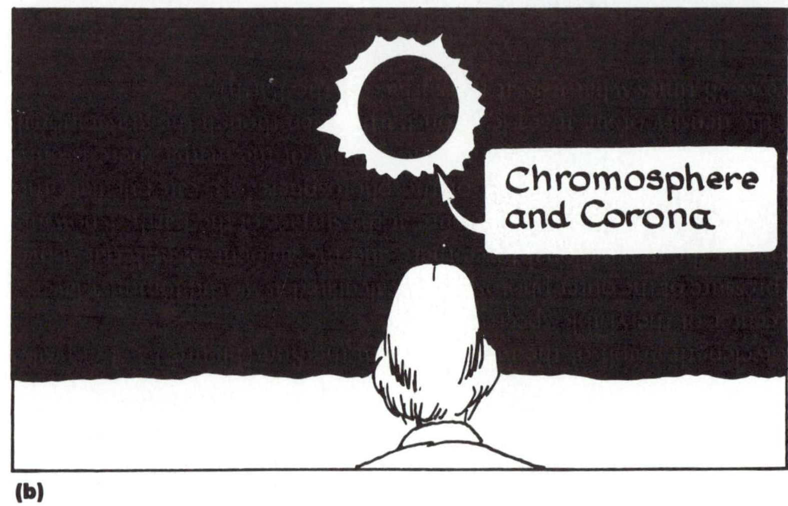

(b)

Figure 12–20. *The chromosphere is seen against the dark background of space during a solar eclipse.*

The Chromosphere and Corona

The chromosphere, a region some 2000 to 3000 kilometers thick lying beyond the photosphere, is not normally observable from Earth. It was first seen in the seventeenth century during a solar eclipse. It appears as a bright red flash, lasting only a few seconds, when the Moon has just covered the photosphere. During solar eclipses from 1842 to 1868 it was examined in more detail. Its spectrum was observed to be a bright line (or emission) spectrum because in viewing it we are seeing light from a hot gas with the dark sky behind it (Figure 12–20). Because the chromosphere is so much dimmer than the photosphere it is only observed at the time of an eclipse, when the brighter portions of the Sun are blocked out.

A telescope designed to photograph the atmosphere of the Sun (when there is no eclipse) is called a coronagraph.

Today the chromosphere and the region beyond it, the corona, can be observed by the use of a telescope that produces an artificial eclipse of sorts. With this instrument we can observe the Sun's atmosphere at various depths. We learn that as one moves outward from the photosphere, the temperature increases instead of diminishing as we would expect. It is as high as 100,000 degrees in the outer portions of the chromosphere and continues to increase

Figure 12–21. *The solar corona during the eclipse of February 16, 1980.*

beyond the chromosphere into the corona, where it may reach millions of degrees.

It might seem that these regions would be extremely bright because of their high temperature. The reason that they are dim is because they have so little density; there is hardly any matter there to glow. The corona's density is less than one billionth that of the Earth's atmosphere.

The reason for the high temperatures within the chromosphere and corona is not well understood. The prevailing theory has been that the high temperature is the result of sound waves that are produced within the convective regions of the Sun and that intensify as they pass outward until they are absorbed in the chromosphere and corona, heating these regions. Recent data, however, calls this explanation into question and many astronomers now believe that the heating is caused by an interaction between the Sun's magnetic field and its differential rotation. The reason for the high chromosphere and corona temperature is just one of the unanswered questions astronomers have about the Sun.

The word corona *comes from a Greek word for crown.*

The corona, a region extending for millions of kilometers from the Sun, has been observed during total solar eclipses for centuries although many people used to claim that it was just an optical illusion caused by the sudden dimming of light as the Sun is eclipsed. Figure 12–21 shows its extremely irregular appearance.

Figure 12–22. *This photo, taken from space, shows the obvious loop shape of some prominences.*

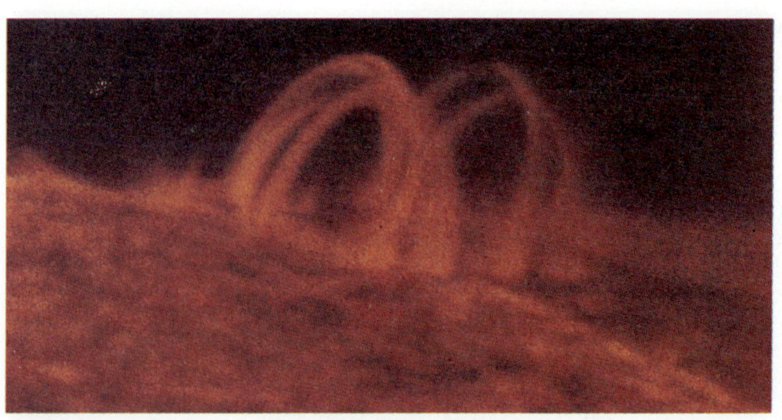

Figure 12–23. *This sequence of photos, taken from space, shows how this particular prominence progressed as charged particles were pushed from the Sun by its magnetic field.*

Figures 12–22 and 12–23 show spectacular occurrences in the Sun's atmosphere. These are ***prominences,*** the ejection of solar material up into the chromosphere and corona. Some of these are relatively slow-moving and remain fairly stable for as long as a few days. They may reach as high as thousands of kilometers above the photosphere. Some move much more quickly, ejecting material from the Sun at speeds up to 1500 km/s and reaching heights of nearly a million kilometers. Prominences are often associated with sunspots and the solar activity cycle to be discussed in the next section.

When we divide the Sun into regions, we must remember that the boundaries between the regions are artificial ones for we have named the various regions and distinguished between them on the basis of certain selected properties. For example, we consider that the process of heat transport is an important one and we therefore talk of a radiative zone and a convective zone. If we emphasized another property, we might not make a division between these two parts of the Sun at all. In addition, remember that although in our drawings the boundaries between various regions appear sharp, in actual practice they are much less well defined. This is especially true when we speak of the outer limits of the corona, where the coronal material becomes the solar wind.

Figure 12–24. *Observing the area near the Sun reveals the solar wind. The color seen here is not real, of course, but indicates different intensities of the wind at the time the photo was taken.*

The Solar Wind

The ***solar wind*** (Figure 12–24) is a continuous outflow of charged particles from the Sun, mostly in the form of protons and electrons. These particles stream through space, taking mass away from the Sun. Near the Earth, the solar wind normally travels at about 400 km/s and has a density of from 2 to 10 particles per cubic centimeter. Recall that one effect of the solar wind is that it causes comet tails to point away from the Sun as its particles sweep comet material along with them.

Perhaps a more dramatic effect of the solar wind is the auroras seen near the poles of the Earth. Auroras result from the solar wind being trapped by the magnetic field of the Earth. Where the Earth's magnetic field lines converge toward the surface of the Earth, the electrically charged particles strike the molecules of the upper atmosphere and cause them to emit the beautiful, eerie glow we call an aurora.

Prominence: The projection of solar material beyond the disk of the Sun.

Solar wind: The flow of nuclear particles from the Sun.

SUNSPOTS AND THE SOLAR ACTIVITY CYCLE

Observations of dark spots on the Sun were reported by the Chinese as early as the fifth century B.C. It is sometimes possible to see very large ones with the naked eye if the Sun is viewed when it is very near the horizon. Europeans did not report sunspots until Galileo saw them with his telescope; perhaps because the Europeans did not have observers as astute as the Chinese or perhaps—after Aristotelian thought was adopted—because Aristotle had proclaimed that the Sun was flawless. (We tend not to see what we disbelieve.)

ACTIVITY

Observing Sunspots

SUNSPOTS WERE FIRST OBSERVED by the naked eye, as described in the text, but such a method is not recommended. You would probably have to search the setting Sun for long periods of time over many years before you saw your first sunspot and staring even at the setting Sun might damage your eyesight.

A more realistic way to view sunspots is with a telescope. *Do not, however, look directly at the Sun with a telescope*. You can obtain a solar filter to put over the front of the telescope but it is an even better to use the telescope to project an image of the Sun on a screen, as described below.

First, a caution: The intensity of sunlight is so great that you run a risk of damaging your telescope. One way to decrease this risk is to put over the objective lens a piece of cardboard with a hole cut in it smaller than the lens. Tape it down so it won't fall off and it will block out some of the light. If your telescope has a finderscope, you should cover it by taping a piece of cardboard (without a hole) over its objective lens. This will not only prevent eye damage to someone who, out of habit, looks through it, but it will block the Sun from burning out the finderscope's crosshairs.

In using a telescope to view the Sun, set it up by first focusing on a distant object (NOT THE SUN). Then pull the eyepiece out just slightly. Now is the time to cover the finderscope and partially cover the objective. Point the telescope exactly at the Sun. This is not as easy as it sounds, and the best way is to move it until the shadow of the telescope tube is smallest. DON'T LOOK THROUGH THE FINDER-

In the late eighteenth century, Alexander Wilson hypothesized that sunspots were places where we were seeing through the outer surface of the Sun and into a cooler interior. William Herschel, the discoverer of Uranus (Historical Note, Chapter 9), even thought that the interior of the Sun might be cool enough to support life. Today's spectroscopic measurement of solar temperatures reveals that sunspots are indeed about 1500 degrees cooler than the surrounding photosphere. They are far from cool and dark, however; for we receive about as much light from an average sunspot—as wide, perhaps, as twice the diameter of the Earth—as we do from the full Moon, even though we are about 400 times farther from the Sun than from the Moon. Sunspots are temporary phenomena, lasting anywhere from a few hours to a few months.

The explanation for sunspots involves the magnetic field of the Sun, which can be measured using a technique discovered late in the last century. The magnetic field of an object causes the splitting of each emission line of the object's spectrum into two or more lines and the strength of the magnetic field can be determined from the extent of the splitting. The splitting can be measured in the spectrum of light from individual parts of the Sun and is an important tool in studying the Sun.

The splitting of spectral lines by strong magnetic fields is called the Zeeman Effect, *after the Dutch physicist who discovered it.*

Sunspots often appear in pairs, aligned in an east-west direction. Early in this century it was found that the magnetic field in a sunspot is about 1000

SCOPE. When this is done, you should be able to see a spot on a screen held behind the eyepiece. Focus by moving the screen and eyepiece to various positions until you get the view you want. A cardboard shield around the telescope will shadow your image from direct rays and improve your view. If you have a star diagonal (which reflects the image off to the side), use it. Trace the image of the Sun and any spots you see. Then repeat your observation in a day or two and look for motion of the sunspots.

Figure T12–2 shows a small telescope being used to project the Sun's image during an eclipse.

Have fun, but be careful.

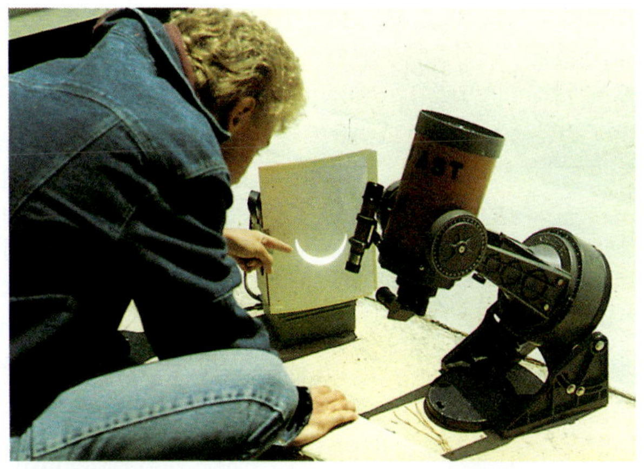

Figure T12–2. *The telescope is being used to project an image of the Sun during an eclipse. This image is much clearer than one obtained by pinhole projection.*

times as strong as the magnetic field of the surrounding photosphere. In addition, we find that sunspot pairs have opposite magnetic polarities, one being north and the other south.

Sometimes the Sun contains a great number of sunspots (Figure 12–25), and sometimes few or no sunspots are seen. In 1851, Heinrich Schwabe, a German chemist and amateur astronomer, discovered that there is a fairly regular cycle of change in the number of sunspots and that the cycle lasts about 11 years. He found that although individual spots do not last long, about the same number are found on the Sun at any one part of its cycle. Figure 12–26 is a graph showing the number of sunspots during the last 400 years (although data is sparse in the early 1600s). It shows a cycle that varies somewhat in period but that averages about 11 years.

Sunspots have an effect on our life on Earth that is not yet well understood. Notice on Figure 12–26 that during the last half of the 1600s and the first decade of the 1700s there were very few sunspots observed, even though the telescope had been invented and was being used to look for sunspots. In addition, reports of eclipses during those years report a very calm corona and chromosphere. Coincident with their period of very little solar activity, northern Europe experienced record low temperatures and unusual weather occurred all over the Earth. Recent data indicates that there is a positive correlation between

The extended sunspot minimum of the late 1600's is called the Maunder Minimum, *after E. W. Maunder, who developed the butterfly diagram (Figure 12–26).*

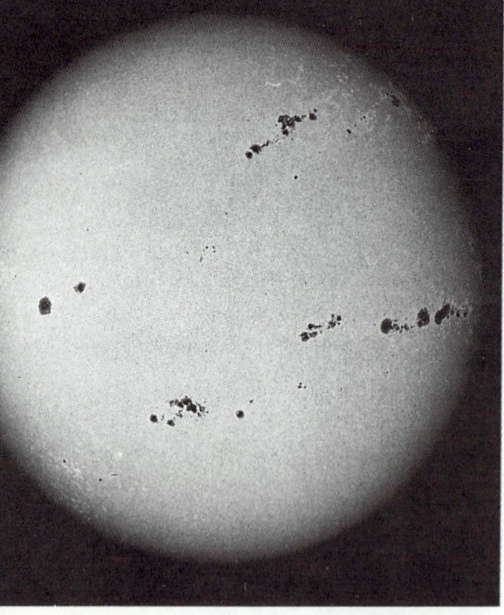

Figure 12–25. *At the time of this photo, December 1957, the Sun was near a sunspot maximum.*

sunspot activity and the Sun's luminosity. That is, the Sun increases in brightness when there are more sunspots. The reason for this is not known, but the correlation might explain the "Little Ice Age" in northern Europe.

In addition; severe droughts occurred on the Great Plains of the United States in the years 1842, 1866, 1890, 1912, 1934, 1953, and 1976. Find these dates on the sunspot graph and you will see that they correspond nicely to sunspot minima. The next sunspot minimum is expected in 1996 or 1997. Should we expect a drought, or are past correspondences only coincidence?

A Model for the Sunspot Cycle

At a sunspot maximum, most spots occur about 35 degrees north or south of the equator. (Figure 12–25 shows the Sun at about sunspot maximum.) Then, as the cycle progresses the spots are seen closer and closer to the equator. By the time they reach the equator, the cycle is at a minimum and new spots are beginning to form again at greater latitudes. If we plot the location of the spots as time goes by, we get a pattern shown in Figure 12–27, called a *butterfly diagram* for obvious reasons. It is important to point out that a given sunspot does not move from high to lower latitudes. (A sunspot's lifetime is much shorter than the 11 year solar cycle.) Instead, the diagram tells us that as time passes and old sunspots die out, new ones form closer to the Sun's equator.

A further interesting feature of the 11-year cycle is that during one cycle, the easternmost sunspot of each pair in a given hemisphere of the Sun is a north magnetic pole and the western sunspot a south pole. (In the other hemisphere, the opposite is true.) Then during the next cycle the pattern reverses, with the eastern sunspot in that hemisphere being a south pole. From one cycle to another, the general magnetic field of the Sun also reverses. Thus the entire magnetic cycle of the Sun has a 22-year period rather than an 11-year one.

A modern hypothesis explains the existence of sunspots and their 11-year cycle as being due to patterns of magnetic field lines within the interior of the Sun. It is thought that groups of these lines form "tubes" threading through

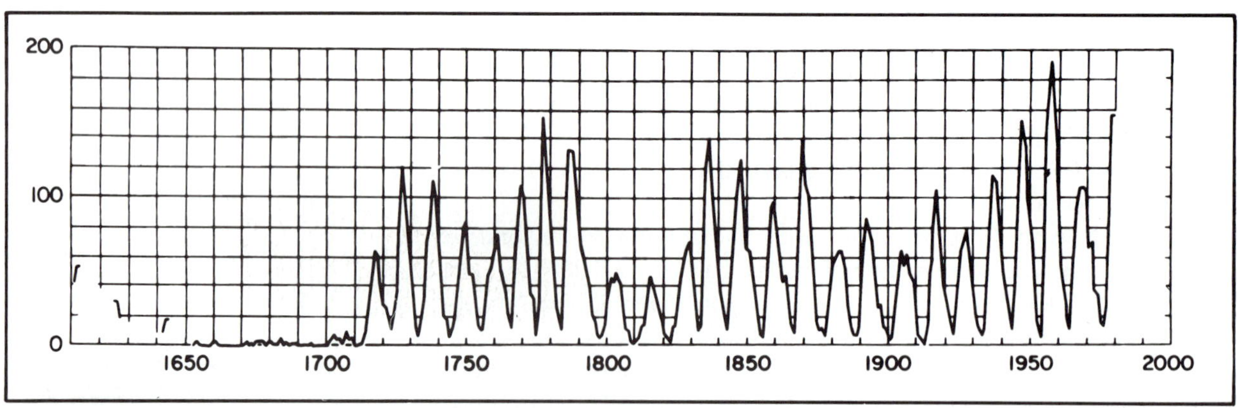

Figure 12–26. *The number of sunspots varies with a period of about 11 years but there are great variations in the intensity of that maximum.*

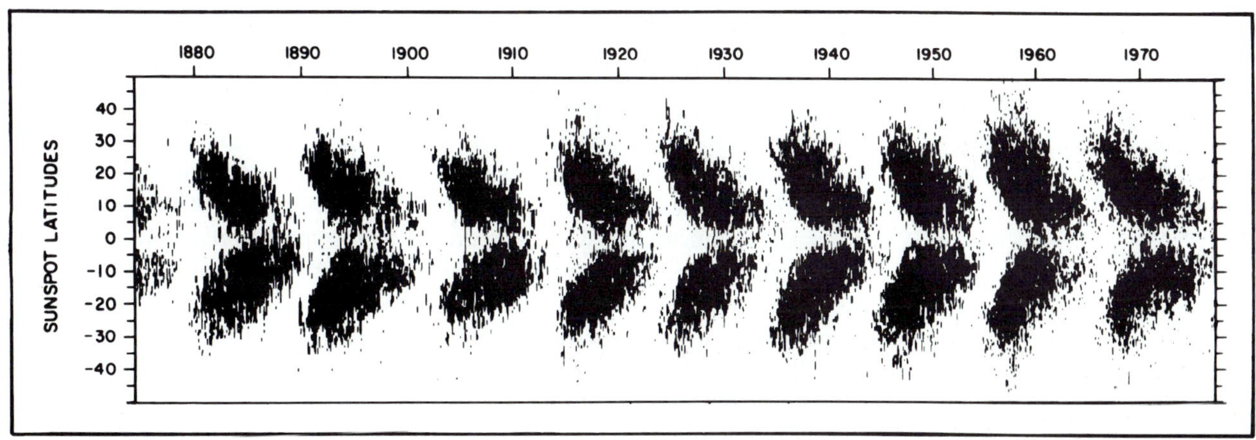

Figure 12–27. *The previous figure plotted the number of sunspots as time passed. This one, the butterfly diagram, shows their location on the Sun.*

the Sun. When the tubes first form, they are relatively straight and buried deep within the Sun as shown in Figure 12–28(a). The differential rotation of the Sun, however, causes the lines to wrap around the Sun as shown progressively in Figures 12–28(b) and (c). As the tubes become more and more twisted around the Sun, they are forced to the surface. When they break through we see a pair of sunspots, one with a north magnetic pole and one with a south magnetic pole. This breaking through the surface causes the lines to weaken and die out at the same time as more lines are forming deep within the Sun. Detailed analysis shows that such a chain of events would cause the magnetic field direction in the tubes (and therefore the magnetic field of the entire Sun) to reverse in direction when new lines are formed.

This model, though perhaps hard to imagine, does fit the observations. As with any new hypothesis, we should expect that as new observations are made it may well have to be adjusted and refined or even discarded.

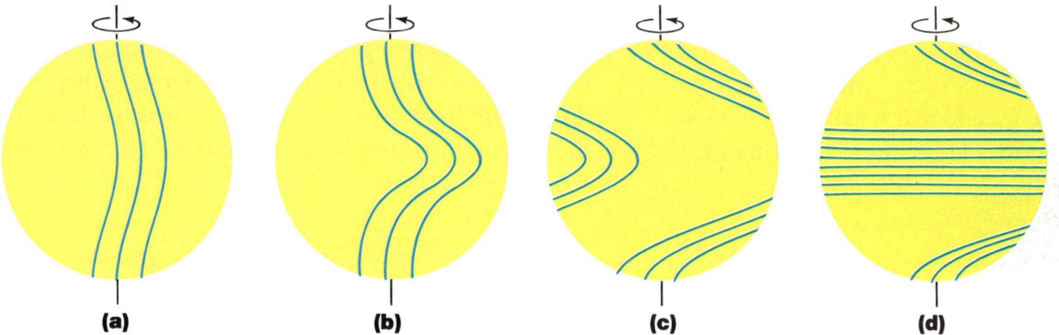

Figure 12–28. *It has been proposed that tubes of magnetic field lines form just below the Sun's surface. The Sun's faster rotation near its equator then twists the tubes around the Sun until they force one another through the surface.*

Solar Flares

Solar flare: An explosion near or at the Sun's surface, seen as an increase in activity such as prominences.

The turbulent magnetic field of the Sun is what is responsible for the prominences discussed earlier, giving prominences the unique shapes such as those in Figures 12–22 and 12–23. It also causes the colossal flareups called ***solar flares*** that normally occur during sunspot maxima. Lasting from a few minutes to a few hours, a solar flare can release the equivalent energy of a few thousand of our largest nuclear weapons. According to the model presented in the last section, these flares occur when a great number of twisted tubes of magnetic field lines release their energy at once through the photosphere. Extremely energetic particles are blasted out of the Sun, reaching the Earth in about three days. They are responsible not only for spectacular auroras but also for disruptions of Earthly radio transmissions.

The radio disruption occurs when particularly energetic particles of the solar wind strike a layer of the Earth's atmosphere called the ionosphere. The ionosphere plays a part in radio transmission because it reflects radio waves back down to the surface of Earth. Normally, the Earth's magnetic field prohibits particles of the solar wind from reaching the ionosphere by deflecting and trapping them, but the high-energy particles in the solar wind are able to penetrate to the ionosphere. When the ionosphere is disrupted by the solar wind, we may experience static in radio reception or even complete loss of signal.

In our discussion of the formation of the solar system, we will see that the solar wind had a major function in determining the nature of today's solar system.

Do solar flares have other effects on the Earth and on the lives of those of us on this planet? As we learn more about the Sun, we may find that what seem like small quirks in the Sun's behavior are actually of major importance to life on Earth.

CONCLUSION

The importance of the Sun to life on Earth is obvious. To astronomers, the Sun takes on even more importance because of all of the stars in the sky it is by far the closest. Since astronomers must understand stars if they are to understand the workings of the universe, the Sun becomes critical in such a study. As we learn more about the Sun, we find that it may be even more important in our everyday life than previously thought; its various activities altering our weather through the years and therefore our food supply.

Previous chapters of the text have been devoted to a study of the planets and in this chapter we have explored the workings of the Sun. We will once again return to the solar system in Chapter 15 when we discuss the formation of stars, for it was along with the formation of the Sun, an ordinary star, that our solar system began.

RECALL QUESTIONS

1. What is meant when it is said the Sun has "differential" rotation?
2. Name two methods of detecting solar rotation.
3. Describe some evidence that shows that the source of solar energy cannot be chemical reactions.

4. Distinguish between chemical and nuclear reactions, giving an example of each.
5. What is it about the nucleus of an atom that distinguishes one atom from another?
6. Name the chemical element that is consumed and the element that is produced in the Sun. What produces the energy when the change occurs?
7. In what physical state is most of the material in the interior of the Sun?
8. Define and explain *hydrostatic equilibrium.*
9. How do we know how great are the pressures at certain depths below the surface of the Sun?
10. List the three methods of heat transfer, giving an example of each. What method(s) is important in which region(s) of the Sun?
11. If all electromagnetic radiation travels at the speed of light, how can it be that radiated energy takes so long to get from the center of the Sun to the surface?
12. Describe the photosphere as to thickness and temperature.
13. If the chromosphere and corona are so hot, why are they not brighter than the photosphere?
14. According to present theory, what causes sunspots?
15. Name two Earthly phenomena caused by the solar wind.

QUESTIONS TO PONDER

1. Do some library research to report the estimated effect on sea level of an increase of a few degrees in Earth's average temperature. Would this have much effect on any seashore with which you are personally familiar?
2. If the magnetic field of the Sun reverses every 22 years, what is meant when we refer to the "northern" hemisphere of the Sun?
3. Explain how energy produced by mass-decrease becomes heat within the Sun.
4. Distinguish between force and pressure as defined in science, and give an example of units in which each can be expressed.
5. Describe the relationship between pressure, density of particules, and temperature in a gas.
6. How would the size of the Sun change if the rate of fusion reactions increased?
7. What method of heat transport results in a room being warmer near the ceiling than lower down?
8. Explain how we measure the temperature at various depths in the photosphere.
9. If air is transparent to radiation, how does radiation from a hot stove ever heat a kitchen?
10. If the chromosphere is hotter at greater distances from the Sun, how can heat be transferred *outward* through it?
11. Explain why hot air rises and relate this to why convection does not occur between most layers in the Sun.
12. We see the Sun not as it is now, but as it was eight minutes ago. The energy we detect, however, began millions of years ago. Discuss the implications this has on what we mean by the word "now."

CALCULATIONS

1. Calculate the angular size of a 15,000-kilometer sunspot as seen from Earth.
2. If one kilogram of coal is burned in one second, it will produce 2 million watts of power. How many kilograms would have to be burned each second to produce the Sun's energy output? If the Sun were made of coal, how much time would pass before it would burn out at its present rate of energy production? (In fact, oxygen would be needed for the burning—nearly three times more mass of oxygen than of coal—so only one-fourth of the Sun's mass would be coal.)
3. How many watts of power strike one square meter of Venus' cloudtops? (Venus is 1.1×10^{11} meters from the Sun.)

CHAPTER 13

Rainbows over the Very Large Array of radio telescopes in New Mexico.

As of 1989, Pioneer 10 has been travelling for 10 years and it is now farther from the Sun than the most distant planet. The radio waves that we continue to receive from the spacecraft require more than five hours to travel to Earth.

The signal from Pioneer has at least two uses. First, it sends us data by which we learn more about the properties of space at the limits of the planetary orbits. (We know now that the solar wind extends at least to Pioneer's distance.) Second, NASA is using it to test radio telescopes, for the strength of the signal emitted by the spacecraft is only one watt; about one-twentieth the energy of a burning candle. How long will we be able to detect the weak waves from this craft, now more than 4 billion miles away from Earth?

Telescopes: Windows to the Universe

GALILEO GALILEI WAS THE first to use a telescope to study the heavens systematically. Much was learned before Galileo's time by naked eye observations but Galileo's telescope changed astronomy—and our outlook on the universe—tremendously. As we make our telescopes larger and larger and take them into space, above the distortions of the Earth's atmosphere, we realize that we are just beginning to learn about the mysteries of our universe.

This century has brought on a multitude of telescopic tools and our view of the skies has expanded to parts of the spectrum well beyond visible light. The word *telescope* now includes instruments used to map the sky in all regions of the electromagnetic spectrum.

We will begin this chapter by describing some properties of light that are important to the understanding of visible-light telescopes. We will then discuss the use of such telescopes. Finally, we will look at some of the nonvisible-light telescopes that are so indispensable to modern astronomy. (Further details on the optics of telescopes will be found in Appendix L.)

REFRACTION AND IMAGE FORMATION

The word medium *refers to the material that transmits the light.*

In our discussion of light thus far, we have concentrated on its wave properties. We will see the effect of these properties later in this chapter but for the moment we can forget about them and concentrate on a more obvious phenomenon: the path that light travels. That path is usually very simple, for light travels in a straight line as long as it remains in the same medium. It may, however, change direction upon entering a second medium. Figure 13–1(a) shows a ray of light passing through a wedge of glass. We see that the light travels in a straight line before and after passing through the glass and that it travels in a straight line inside the glass but that the ray bends when it passes through each surface of the glass. Part (b) of that figure shows what happens when the two sides of the wedge are more nearly parallel; the light bends less. The phenomenon

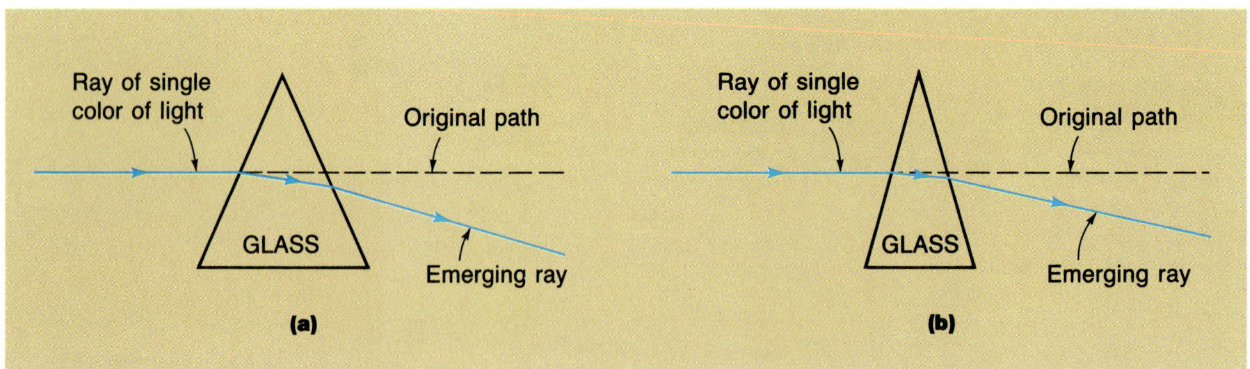

Figure 13–1. *When light passes through a wedge of glass, it is bent from its path. The less the angle of the wedge, the less the bending. (The "wedge" is really a prism, but we are concentrating here on the bending of the light rather than its separation into colors.)*

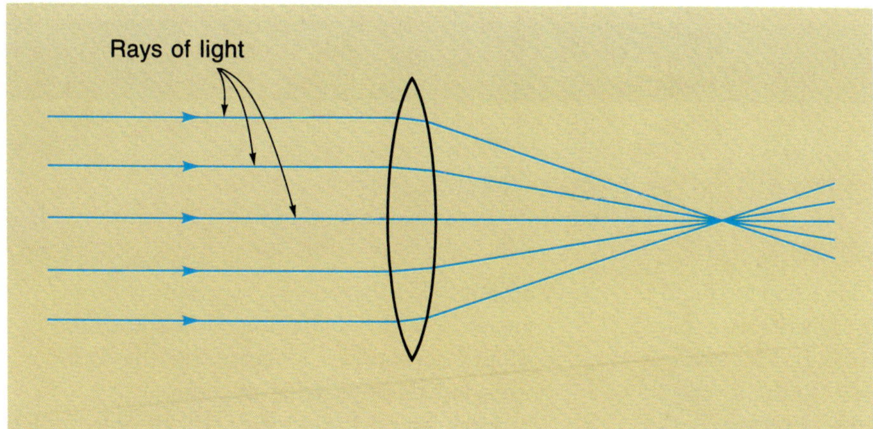

Figure 13–2. *A lens bends incoming rays of light toward a single point.*

of the bending of a wave as it passes from one medium into another is called ***refraction***.

Refraction: The bending of light as it crosses the boundary between two materials.

Figure 13–2 shows a number of rays of light (that came from beyond the left side the page) passing through a lens-shaped piece of glass. The surfaces of a lens have just the right curvature to cause all of the rays of light shown in the figure to pass through the same spot. To see the importance of this, imagine that each of the rays you see in the photo came from a distant star. If we put a piece of paper at the point where the light rays come together, as in Figure 13–3(a), all the light from that star that passes through our lens will come to a single point on the piece of paper. In fact the rays of light coming from other stars will likewise come to a focus on the paper, forming an ***image*** of that area of the sky.

Image: The visual counterpart of an object, formed by refraction or reflection of light from the object.

The ***focal point*** of a lens is that point where light from a very distant point directly in front of the lens comes to a focus. This is the point where the rays converged in Figures 13–2 and 13–3(a).

Focal point (of a converging lens or mirror): The point at which light from a very distant object converges after being reflected or refracted.

The ***focal length*** of the lens is the distance from the lens to the focal point. Depending upon the curvature of their surfaces, different lenses have different focal lengths.

Focal length: The distance from the focal point of a lens or mirror to its center.

THE REFRACTING TELESCOPE

Lenses, of course, are at the heart of the telescope—particularly the refracting telescope. (We will study the reflecting telescope later in this chapter). The simplest use of a telescope (in principle, anyway) is as a lens for a camera. Figure 13–4(a) shows a camera mounted on a small telescope. What might not be obvious in the photo is that the camera's normal lens has been removed. The camera is using the long-focal length lens of the telescope in place of its regular lens. Figure 13–4(b) shows the arrangement. The telescope lens simply replaces the regular camera lens and brings the image to a focus on the film (just as it formed an image on the piece of paper in Figure 13–3). As we will see later, telescopes can be mounted so that they can track the stars across the sky and

CLOSE UP

The Hubble Space Telescope

THE HUBBLE SPACE TELESCOPE (Figure T13–1) is scheduled for launch from a space shuttle in December 1989. Its expected impact on astronomy has been likened to the invention of the telescope. The reason for this optimism lies in the fact that the telescope will be above the blurring effect of the Earth's atmosphere. With its 2.4-meter (94-inch) mirror, it will be the largest astronomical instrument ever placed in orbit and will be able to detect objects 50 times dimmer than the dimmest that can be detected from Earth's surface.

The Space Telescope will have five scientific instruments on board:

1. The Faint-Object Camera will take advantage of the large telescope mirror to photograph extremely faint celestial objects. The instrument will have the ability to accumulate exposures over many of the satellite's orbits to produce an image.
2. The Faint-Object Spectrograph will separate the light from dim objects into its various wavelengths in order to determine the characteristics of those objects.
3. The High-Resolution Spectrograph will operate in the ultraviolet region of the spectrum, enabling astronomers to determine the composition of the interstellar medium.
4. The High-Speed Photometer will be able to detect rapid changes in light intensity from objects over a wide range of wavelengths. It will be able to concentrate either on point sources of light or on the light from a wider field of view.
5. The Wide-Field Camera will help astronomers more accurately determine the distance scale of the universe, thereby testing cosmological theories of the universe as a whole.

The satellite will orbit the Earth at about 600 kilometers (300 miles) and will transmit its data to the Science Institute at John Hopkins University by NASA's existing satellites and ground stations. It will be interesting to see how much this instrument changes astronomy in the next decade.

allow astronomers to take long time exposure photos of the heavens. Long time exposures allow us to photograph objects much fainter than can be seen by simply looking through a telescope. Thus the use of a telescope with a camera is much more important to a professional astronomer than its use for direct viewing.

Objective lens (or objective): The primary light-gathering element—lens or mirror—of a telescope.

Eyepiece: The magnifying lens (or combination of lenses) used to view the image formed by the objective of a telescope.

Figure 13–5(a) shows a small telescope being used for direct observation. The primary lens, the lens through which the light passes first, is called the ***objective lens,*** or simply the ***objective***. This lens brings the light to a focus, shown at point F in Figure 13–5(b). This is the point where the film was located when the telescope was used with a camera. For direct viewing, a second lens, the ***eyepiece,*** is added just beyond the focal point. This lens simply acts as a magnifier to enlarge the image.

(a) (b)

Figure T13–1. *(a) The Hubble Space Telescope will provide resolutions of up to 0.1 arcseconds during its anticipated 15-year lifetime. (b) The successful launch of Discovery on September 30, 1988 provided hope to astronomers that the Hubble Space Telescope will soon be in orbit.*

Chromatic Aberration

Figure 13–6 shows that a prism separates light into its colors. This effect is caused by the fact that different wavelengths of light travel at different speeds in a transparent material. When white light hits the surface of a transparent material at an angle, this results in different wavelengths of light refracting by different amounts. Then when they strike the second surface of the prism, the colors are further separated from one another.

Dispersion: The separation of light into its various wavelengths upon refraction.

In the case of a lens, except for the ray of light that goes straight through the center, rays go through the lens in much the same way that they go through a prism. And like the light going through the prism, the light passing through a lens separates into colors. This causes the lens to have a slightly different focal

Figure 13–3. *(a) Rays of light from a distant star are brought to focus at a single point. (b) Light from a star higher in the sky will be brought to a focus at a different place, lower on the paper.*

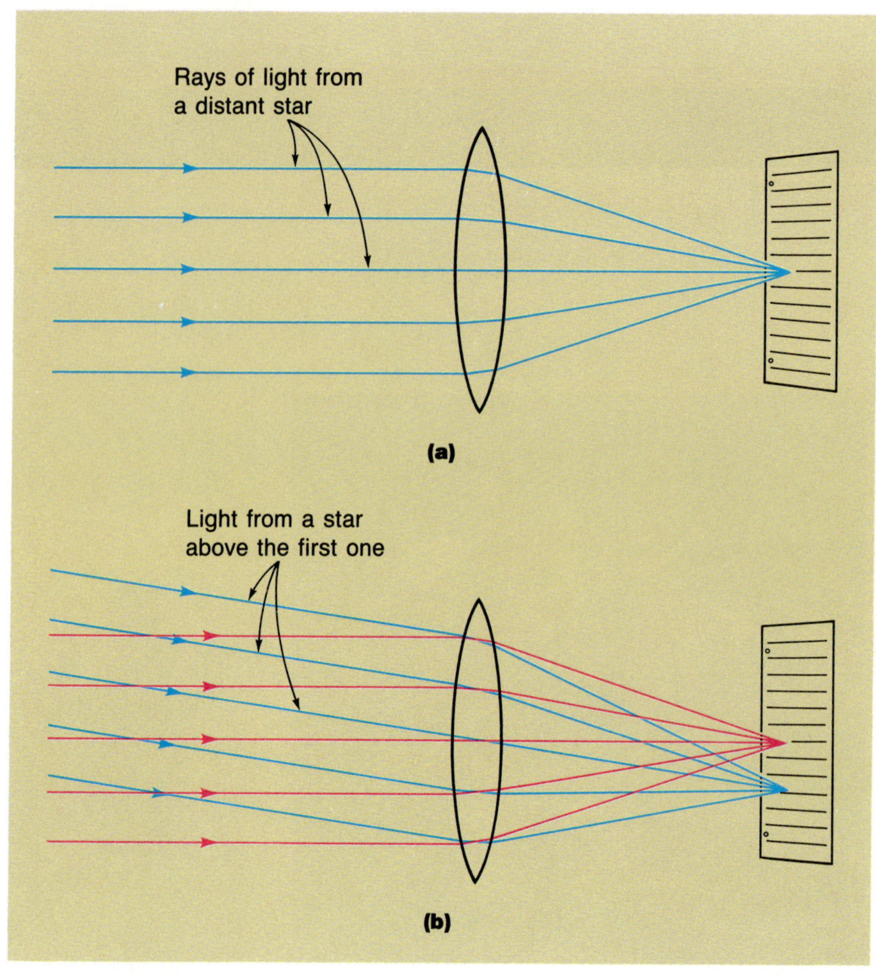

length for each wavelength of light. Figure 13–7(a) exaggerates the effect but illustrates the idea.

The separation of colors by the glass of a lens means that there is no single place where an image is exactly in focus. Figure 13–7(b) shows the situation if the film of a camera is placed at the point where the red light focuses. In this case, the other colors are out of focus and the result will be an image with a fuzzy, bluish edge. This phenomenon, called ***chromatic aberration,*** occurs when a telescope is used for direct viewing as well as when it is used with a camera. It is a problem not just in telescopes but in regular cameras as well. Fortunately it can be corrected, at least in part.

Chromatic aberration: The defect of optical systems that results in light of different colors being focused at different places.

The amount of color separation that occurs when light passes through a lens depends not only on the curvature of the glass but also upon the type of glass. Some kinds of glass separate the colors more than other kinds. This fact is used in telescopes and cameras to correct for chromatic aberration. In all but the cheapest toy-store telescopes, the objectives of refracting telescopes are made

(a)

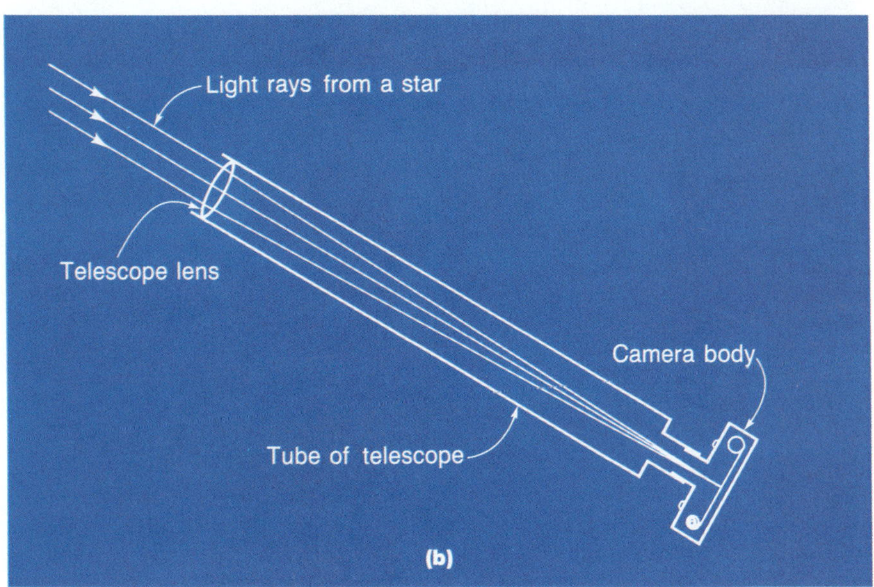

(b)

Figure 13–4. *A simple way to use a telescope for photography is to let the telescope serve as the camera's lens. The image is focused directly on the film by the telescope's main lens.*

up of two lenses instead of one. As shown in Figure 13–8(a), the second lens has reverse curvature from the first. This curvature is not enough to undo all of the converging effect of the first, however, and the light is still brought to a focus. The second lens is made of a different type of glass than the first and although it does not cancel out the bending of the light, it does cancel out most of the color separation as shown in Figure 13–8(b). Such a combination of lenses is called an ***achromatic lens.*** If you have a camera, examine the print around the edge of the lens. It will probably say "achromat," although this

Achromatic lens (or achromat): An optical element that has been corrected so that it is free of chromatic aberration.

Figure 13–5. *When a telescope is used for direct viewing, the eyepiece is used as a magnifier to view the image formed by the objective lens. This image is formed at the focal point* (F) *of the objective.*

(a)

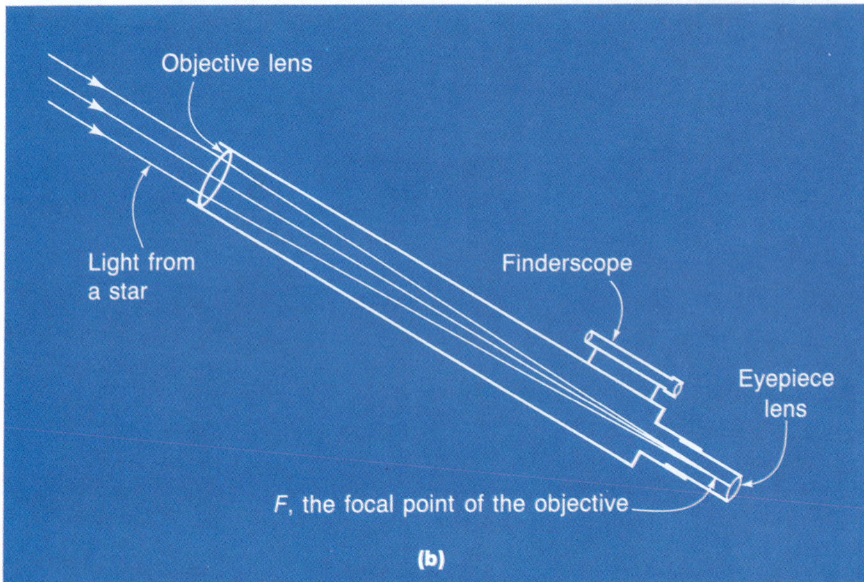

(b)

correction is so common on today's cameras that the lens may not even be marked as such.

Thus far, our examples have been of small telescopes; telescopes that amateur astronomers are likely to have. Research telescopes, however, are much larger. To see why large telescopes are desirable, we must examine the reasons for using a telescope and the various features that make one telescope better than another for a particular use.

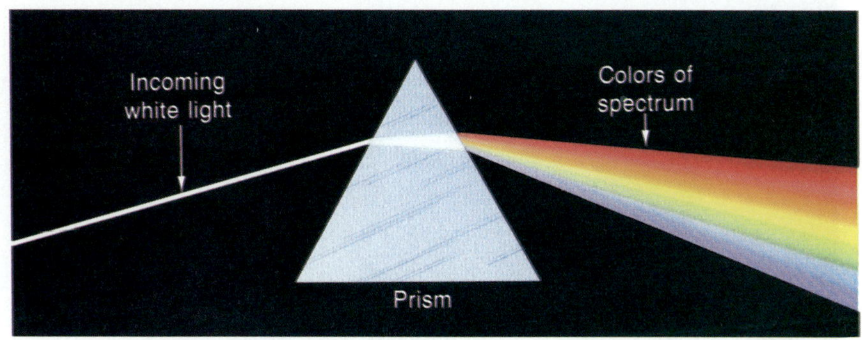

Figure 13–6. *A prism separates white light into its colors by bending different wavelengths of light at different angles.*

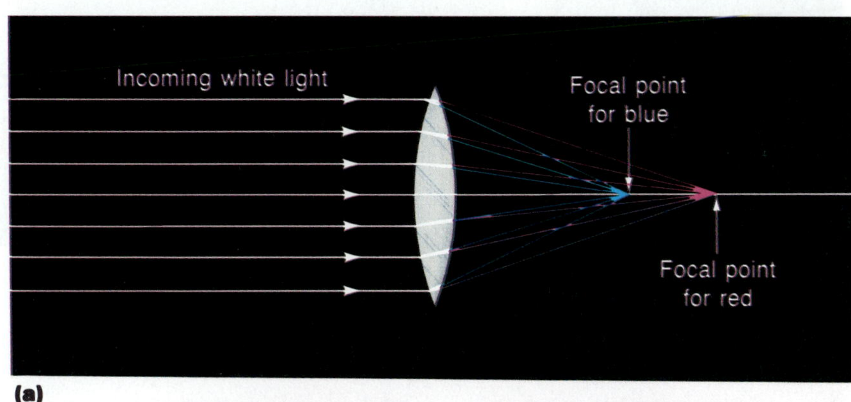

(a)

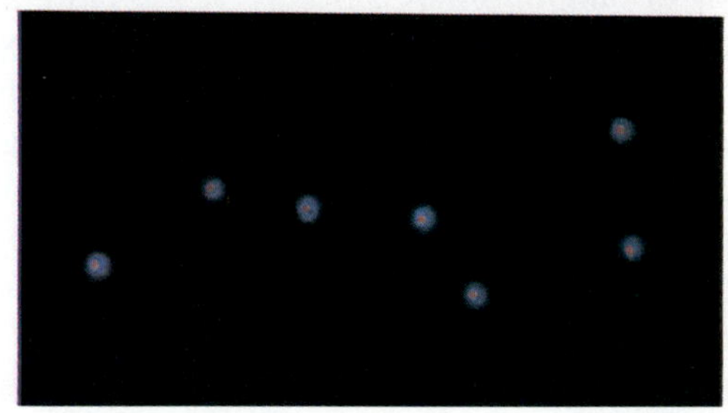

(b)

Figure 13–7. *(a) A lens exhibits a prism effect, separating white light into its colors. It therefore focuses each color at a different place. Only red and blue are shown here. (b) If a lens is not corrected for this defect, an image will have color fringes around it.*

THE POWERS OF A TELESCOPE

When most people think of a telescope's power, they think of magnification. Magnification, however, is only one of three major powers of a telescope and it is the least important. We will start our discussion of powers of a telescope with this least important one, and then we will consider the two other powers: light-gathering power and resolving power. In our discussion, we will again describe small telescopes but remember that the same ideas apply to large telescopes used by professional astronomers. In fact, the reasons for using such large telescopes are related to the powers of telescopes.

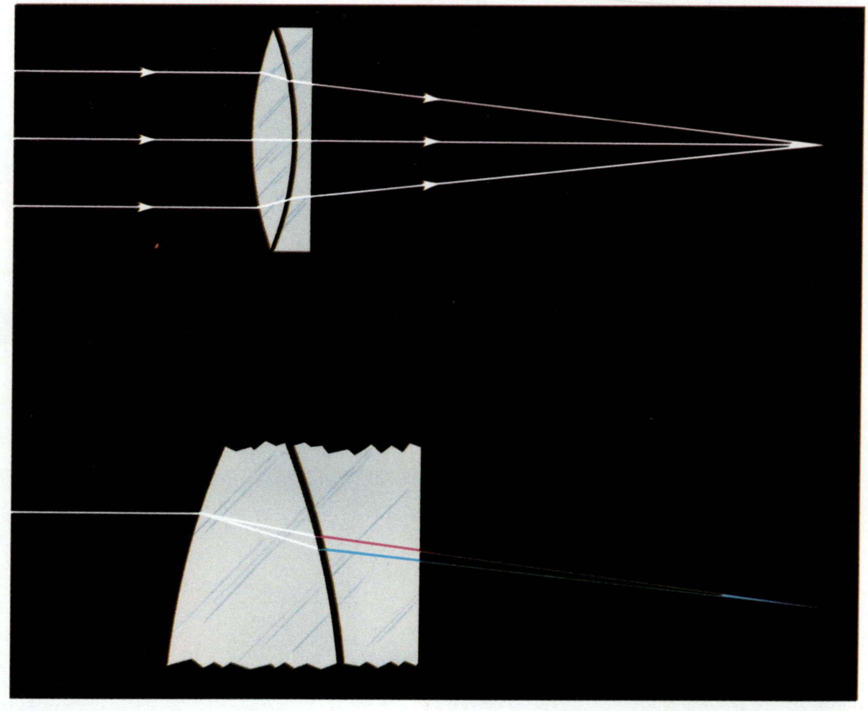

Figure 13–8. *A lens can be corrected for chromatic aberration by the proper combination of lenses of different types of glass. The second lens here brings the separated colors back together at the image.*

Magnifying Power

Magnifying power (or magnification): The ratio of the angular size of an object when it is seen through the instrument to its angular size when seen with the naked eye.

For a telescope (and binoculars and a number of other optical instruments), ***magnifying power*** or ***magnification*** is defined as the ratio of the angular size of an object when it is seen through the instrument to the object's angular size when seen with the naked eye. Recall from Chapter 6 that the angular size of an object is a measure of how large the object looks to us. Figure 13–9(a) and (b) compare the angle made by the Moon seen with the naked eye and then seen through a particular telescope. You might estimate from the angles in the figure that this telescope has a magnification of about six. That is, the telescope magnifies the object six times.

What determines how great the magnification of a telescope will be? As the objective lens forms an image at the focal point, the greater the focal length of that lens, the larger the image. In the case of the eyepiece (being used as a magnifier), the opposite is true: A lens with a shorter focal length results in a more magnified image. Thus for a telescope, the greatest magnification is achieved by having a long-focal-length objective and a short-focal-length eyepiece. We can calculate the magnifying power of a telescope by a simple equation:

$$\text{Magnifying Power} = \frac{\text{focal length of objective}}{\text{focal length of eyepiece}}$$

or, in symbols,

$$M = \frac{F_{\text{obj.}}}{F_{\text{eye.}}}$$

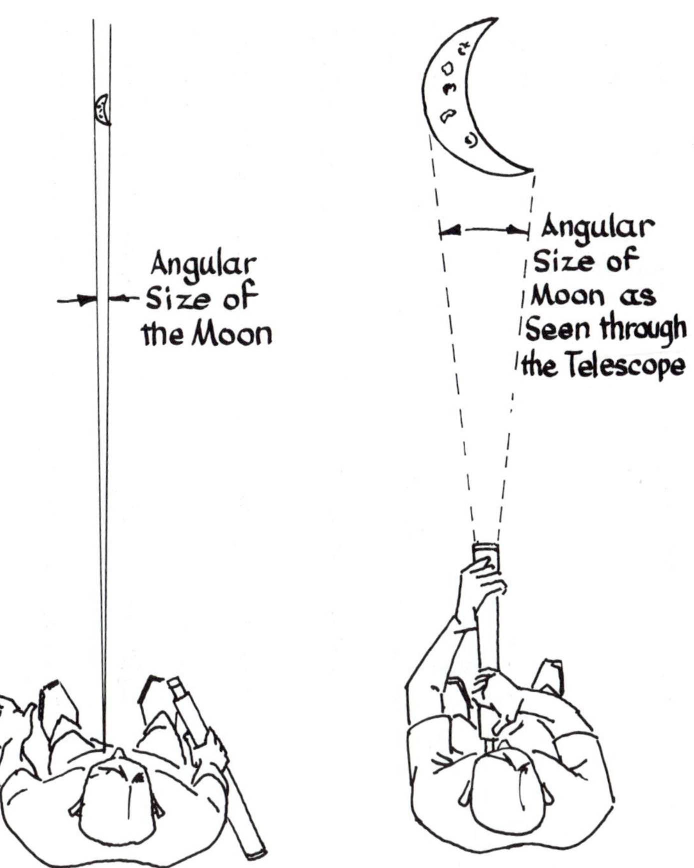

Figure 13–9. *(a) The moon seen by the naked eye. (b) In a telescope, the angular size of the moon is apparently larger. The magnification of the telescope is the ratio of the two angular sizes.*

EXAMPLE

The telescopes used in my introductory astronomy laboratory have objectives with focal lengths of 1250 millimeters. One eyepiece that is used has a focal length of 25 millimeters. What is the magnification produced by the telescope? What is the angular size of the Moon as seen through this telescope using the 25 mm eyepiece? (The naked eye angular size of the Moon from Earth is about ½ degree.)

Solution

To calculate the magnification, we use the equation that relates it to focal lengths,

$$M = \frac{F_{\text{obj.}}}{F_{\text{eye.}}}$$

$$= \frac{1250 \text{ mm}}{25 \text{ mm}}$$

$$= 50$$

Thus the magnifying power of the telescope is 50. We say that the magnification is 50 times, or 50X.

Now, since the Moon's angular size to the naked eye is ½ degree and the telescope magnifies the Moon 50 times, the Moon's angular size in the telescope will be 50 times this, or 25 degrees.

Try One Yourself. What is the magnification produced by a telescope having an objective with focal length of 1.5 meters when it is being used with an eyepiece having a focal length of 12 millimeters? (Hint: In doing the calculation, you must express the two lengths in the same units: either meters or millimeters.)

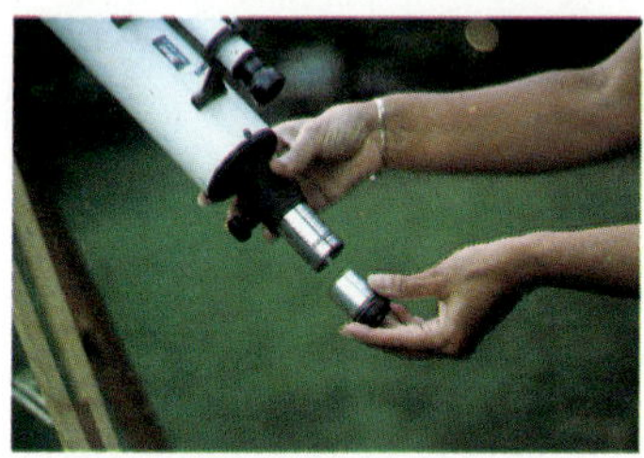

Figure 13–10. *An eyepiece is easily changed in order to change the magnification of the telescope.*

Notice that both in the example and in the suggested problem, we hinted that one might use an eyepiece of a different focal length. Indeed, it is a minor matter to change the eyepiece of a telescope. Figure 13–10 shows someone putting an eyepiece into a small telescope. An eyepiece like this costs relatively little and so it is common to have a number of different eyepieces for a telescope in order to have different magnifications available. The obvious question is, "Why not always use the greatest magnification?" There are a number of answers to this question.

The first answer is that as you increase the magnification, you decrease the ***field of view*** of the telescope. Field of view refers to how much of the object is seen at one time. Figure 13–11 shows the decrease in field of view as the magnifying power increases. This is entirely reasonable, for if the object appears larger not as much of it will be contained within the view of the telescope.

When viewing an object—the Moon, for example—you may wish to see the whole object rather than just a portion of it. If so, you will use an eyepiece with a long focal length; thus producing less magnification.

Field of view: The actual angular width of the scene viewed by an optical instrument.

The other answers to the question of why we do not always use the greatest magnification relate to the other powers of a telescope.

Light-Gathering Power

There is another difference between the three views of Figure 13–11. Notice that the more magnified the image, the darker. To see why this occurs, consider the part of the Moon we see in part (c). In part (a), the light from this part of the Moon was concentrated on a small part of the image but it is more spread out in the more magnified view part (c). That is, the same light that, in part (a), covers only part of the image has to illuminate the entire magnified image of part (c). The image is therefore darker. There are two ways to make it brighter and still retain this magnification. If one is taking a photo, a longer time exposure can be used. When we do this we allow the light's effect to accumulate on the film over a longer time and the film becomes more exposed. The other way is to capture more light from the Moon in the first place. This can only be done by using a larger objective and it brings us to the second power of a telescope: the ***light-gathering power,*** which is often the most important power.

Light-gathering power: A measure of the amount of light collected by an optical instrument.

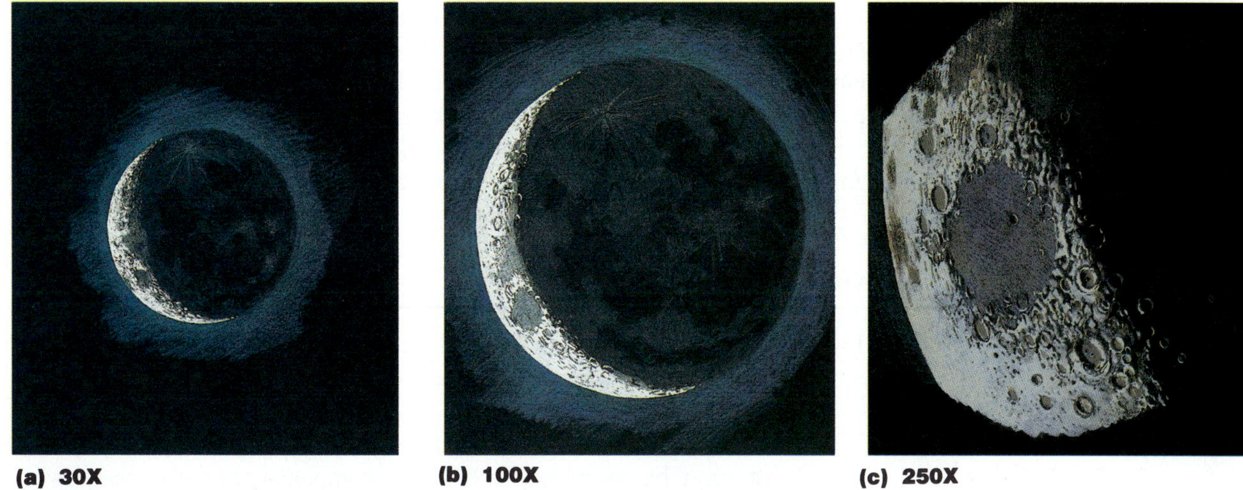

Figure 13–11. *Increasing the magnification decreases the field of view and makes the image darker.*

The light-gathering power of a telescope refers to the amount of light it collects from the object. In the case of observing bright objects like the Moon and nearby planets, the objects reflect so much light to Earth that lack of light is not a problem. Most objects that are observed with a telescope, however, are very faint and in order to get an image we need to capture as much light as possible from them. This is true whether we are using the telescope for photography or for direct viewing.

Light-gathering *power may seem an odd thing to call this power, for a telescope does not* gather *light in the sense of searching it out. It simply captures the light that hits the objective and brings that light to a focus.*

The major way to increase the light gathered by the telescope is to use a telescope with a larger objective. The amount of light that strikes the objective simply depends upon the area of the objective. This is the primary reason it is desirable to have a telescope with a large-diameter objective and why the size of the objective is one of the features specified when discussing a telescope. For example, one might describe a particular telescope as a refractor with a

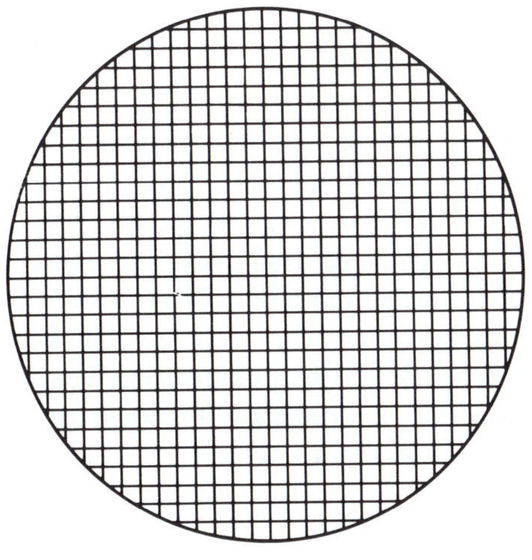

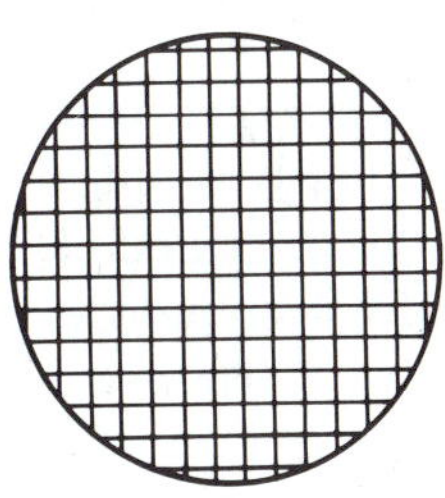

Figure 13–12. *The larger circle has twice the diameter of the smaller. If you count the squares, you will find four times as many in the large circle. Thus a telescope with an objective of twice the diameter has four times the light-gathering power.*

12-centimeter objective of focal length 140 centimeters. Notice, though, that while we state the *diameter* of the objective, it is the *area* that is important. Since the area of a circle is proportional to the square of the diameter, we must be careful in making comparisons of light-gathering power.

To illustrate light-gathering power, let's work an example comparing the light received by two telescopes with objectives of different size.

EXAMPLE How does the light-gathering power of a telescope with a two-inch objective lens compare to that of one with a three-inch objective?

Solution The areas of circles depend upon the squares of their diameters, and so it is the squares that must be compared in order to compare light-gathering power. We'll set up a ratio of the squares of the two diameters:

$$\frac{3^2}{2^2} = \frac{9}{4}$$

$$= 2.25$$

Thus the light-gathering power of the larger telescope is more than twice that of the smaller.

Try One Yourself. If I trade my three-inch telescope for a five-inch model, by how many times do I increase my light-gathering power?

The desire to be able to see fainter and fainter objects in space has led us to make larger and larger telescopes. Before discussing large telescopes, however, we must look at another advantage of size; the third and last power we need to consider.

Resolving Power

One property of light that we assume in our everyday life is that it travels in straight lines (unless it reflects from a mirror or is refracted at a surface). If we could not assume this we would be unsure whether an object we see is in front of us or behind us. Yet there are exceptions to this rule: light does not always travel in straight lines. The exceptions are usually unimportant but look at Figure 13–13. This is a magnified photo of the shadow of a screw. The shadow was made by holding the screw a few meters from a screen and illuminating it with a small, bright light source. Notice that the edges of the shadow are not distinct and that in fact there are light and dark fringes near the edges.

Figure 13–14(a) is a photo of light that hit a screen after passing through a very small hole. Again, we see that the spot of light does not have sharp edges; instead, we see dark and light fringes again.

In both of these cases, the light passing near the edge of the object (the side of the screw of the border of the hole) has "spread out" slightly. Figure 13–14 shows the idea. The effect is small and is seldom seen in everyday life because to see it you need a very bright, small light source and even then you must look carefully. We will not discuss the reason that light acts this way except to point out that water waves behave similarly, as shown in Figure 13–15. Such bending of waves as they pass by the edge of an obstacle is called ***diffraction***.

Diffraction: The spreading of light upon passing the edge of an object.

The amount of diffraction that occurs when light passes through an opening depends upon two things: the wavelength of the light and the size of the opening. The longer the wavelength, the more diffraction; and the larger the opening, the less diffraction.

In a telescope, the objective itself forms the opening through which the light passes. Even for the smallest telescope, this opening is much larger than the hole that was used to produce Figure 13–14(a). Thus, diffraction as extreme as that in the photo will not occur in a telescope. But the effect is there; light

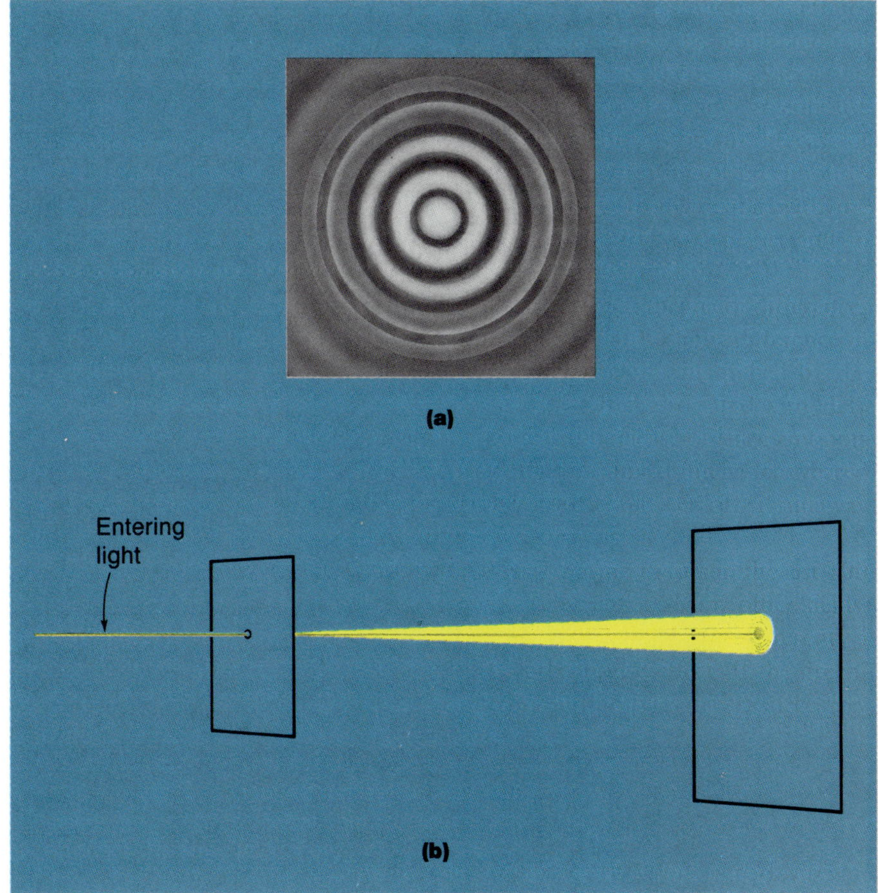

Figure 13–14. *This illustrates how light bent in passing through the hole.*

Figure 13–13. *If an extremely small light source is used, the shadow of an object will have light and dark fringes at its edge. This is due to diffraction of light. (A laser was used to make this photo but laser light is not necessary to produce the effect.)*

Figure 13–15. *Water waves are striking the dam from the left. They spread out after passing through the opening.*

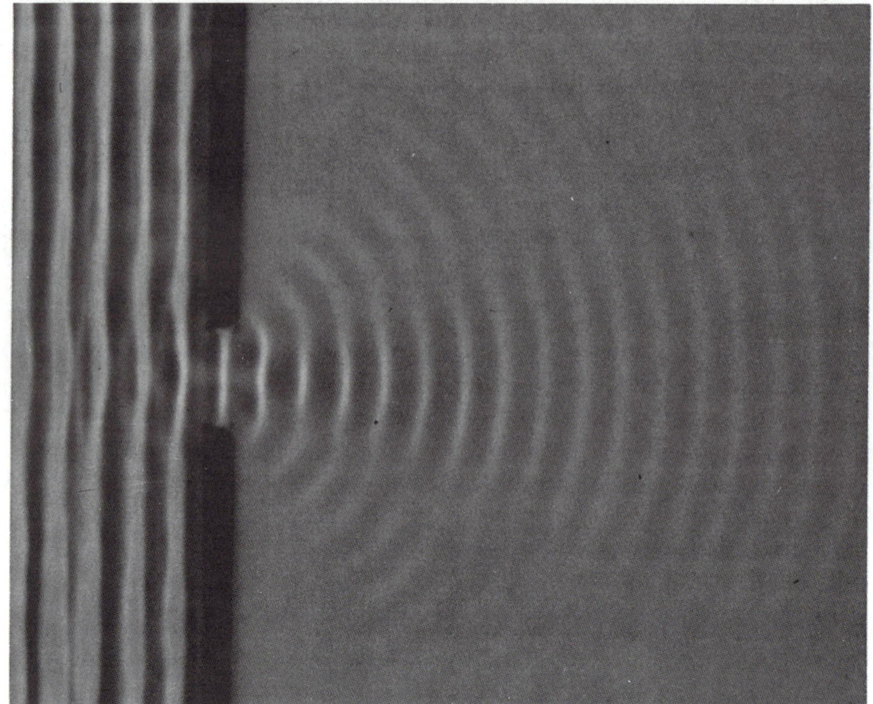

that should bend regularly and accurately according to the laws of refraction fans out a slight amount. This results in the image not being exactly clear. A star that should appear as a single point image is blurred out into a small spot. And when we increase the magnification, we simply make the blurred spot bigger and fainter.

Figure 13–16 represents three views of Saturn taken with different telescopes. The image is clearest in the right view. This is not because some of

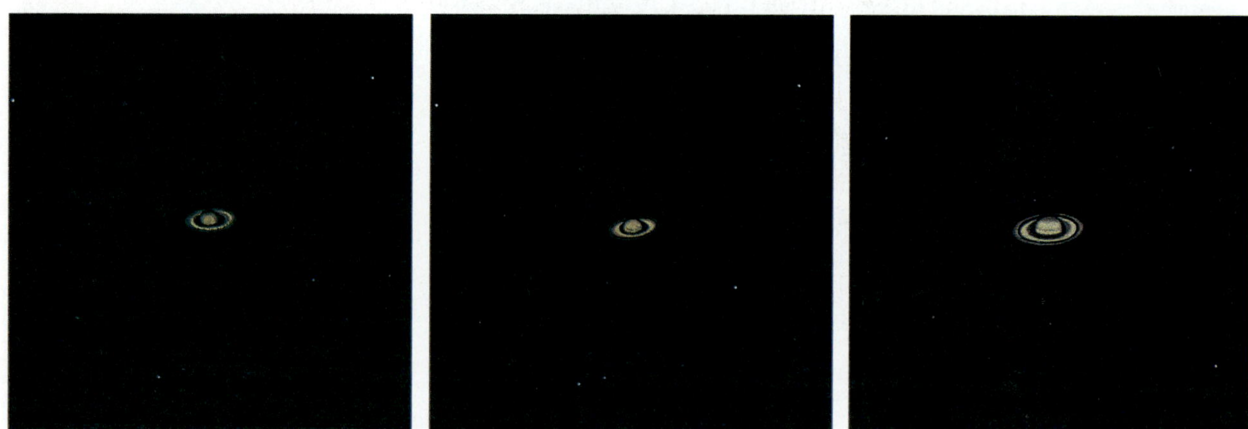

Figure 13–16. *The differences in resolution are obvious in these images of Saturn.*

Figure 13–17. *This star in the handle of the Big Dipper is seen as a single star by many people but can be resolved into two stars by those with better eyesight.*

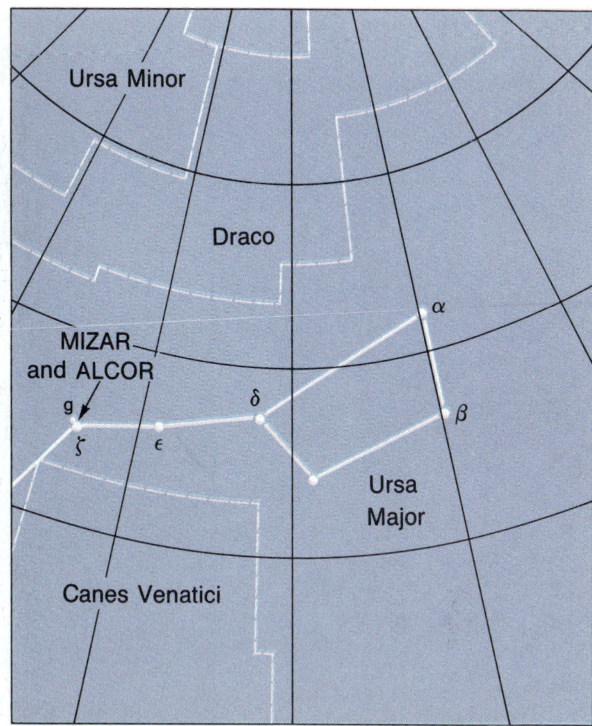

the telescopes were not in focus, but rather because diffraction limited the clarity of the other two images.

Figure 13–17 is a photo of the Big Dipper and indicates what seems to be a single star in the Dipper's handle. If you look at the Dipper in a clear dark sky and if you have fairly good eyes, you can see that this star is not one, but two stars. We say that your eyes, and the clarity of the sky, allow you to *resolve* the pair of stars. Someone with poorer eyesight may be unable to resolve the pair. Now suppose you look at this pair of stars with a small telescope. Figure 13–18 shows what you see. The figure shows three stars; two very close together. In fact, the two stars that are close together were seen as one star when viewed with the naked eye; it was the brighter of the naked-eye pair. The fainter of the naked-eye pair is the third star in the photo, at the opposite side of the field of view. The telescope is able to resolve the group of stars into three.

The brighter star is named Mizar and the dimmer Alcor. American Indians called the stars "the horse and rider."

The ***resolving power*** of an instrument determines the smallest angular separation two stars can have and still be resolved as two by the instrument. Thus resolving power is described in terms of an angle.

Resolving power (or resolution): The measure of the ability of a telescope to see fine details in an object.

What is it about a telescope that determines its resolving power? Naturally, the quality of the lenses is a major factor but even with perfect optical components, the resolving power of a telescope is limited by diffraction. Since less diffraction occurs with a large objective, the maximum resolving power can be achieved by a telescope with a large-diameter objective.

The best human eye has a resolving power of about one minute of angle, or 1/60 degree. A telescope with a 15-centimeter (about six inch) objective will have a maximum resolving power of about one second, or 1/3600 degree. Based

Figure 13–18. *Even a small telescope resolves Mizar into two stars (lower right). Alcor is at upper left.*

on size alone, the largest telescopes should have a resolving power far greater than this but in fact the lack of clarity of the Earth's atmosphere becomes a major factor in limiting the resolution of large telescopes.

This atmospheric turbulence is what causes the twinkling of the stars when they are viewed with the naked eye.

The lack of clarity of the atmosphere is caused by two factors: turbulence of the air and air pollution—either due to modern civilization or simply due to dust. Recall that light is refracted as it passes from one material into another if there is a difference in its speed in the two materials. In fact, light travels at slightly different speeds in air at different temperatures. Our atmosphere always contains some amount of turbulence and this causes air at various temperatures to move across the line of sight of a telescope. This results in the image moving slightly and it places a limit on the resolution of even the largest telescope. Thus even the largest telescopes on Earth have a practical resolving power of about one-half arc second. As we will discuss later, the limit the atmosphere places on resolving power is the primary reason that astronomers look forward to placing a major optical telescope in Earth orbit.

Power Summary

Of the three powers of telescope, magnification is the least important in most applications. We have seen that magnification is very easy to increase for a given telescope, but that there is a limit to the magnification that is practical, both because higher magnifications produce fainter images and because the effects of diffraction and lack of atmospheric clarity limit the sharpness of images at higher magnification.

In addition, we have seen two advantages of large telescopes. First, large telescopes gather more light than small ones; allowing us to see fainter objects. Second, larger telescopes have less diffraction and therefore a greater resolving power (although this advantage is largely neutralized when looking through the Earth's atmosphere).

THE REFLECTING TELESCOPE

As illustrated in Figure 13–19, an inwardly curved mirror will also bring rays of light to a focus. This allows us to use it as the objective of a telescope. Figure 13–20 shows the arrangement devised by Isaac Newton. A small flat mirror is

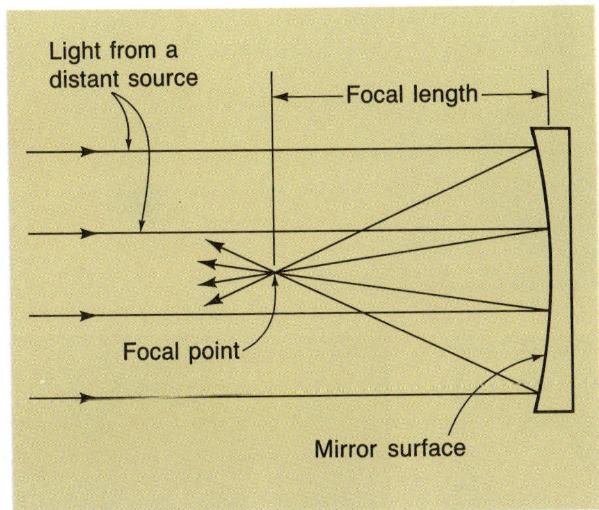

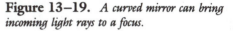

Figure 13–19. *A curved mirror can bring incoming light rays to a focus.*

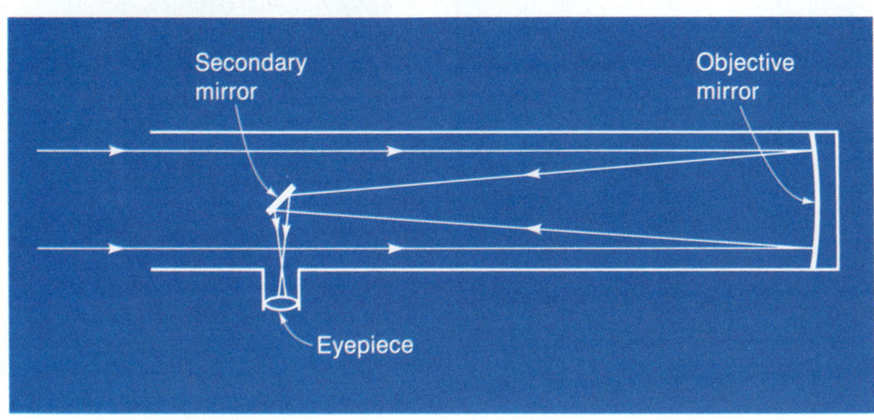

Figure 13–20. *The Newtonian focal arrangement places a small flat mirror in the path of the reflected rays so that they are bounded off to the side and into the eyepiece (or camera or other instrument).*

Figure 13–21. *This is a telescope built by Isaac Newton.*

Figure 13–22. *(a) The objective mirror is at the bottom of the tube and the person is looking into the eyepiece of this Newtonian telescope. (b) The secondary mirror of a Newtonian telescope blocks only a small fraction of the incoming light.*

Figure 13–23. *The curvature of the primary mirror of a real telescope is much smaller than the curvatures of mirrors shown in our diagrams. (This is the 2.4-meter mirror for the Hubble Space Telescope.)*

arranged in front of the objective mirror to deflect the light rays out to the eyepiece or camera body. Figure 13–22(a) shows a person using a six-inch Newtonian reflector. You might be concerned that the secondary mirror blocks some light from hitting the objective. True, but in the telescope shown in the photo, the secondary lens only blocks out a circle about 1.5 inches in diameter. Thus the fraction of the area blocked is only $[(1.5 \text{ in})/(6 \text{ in})]^2$, or 1/16 of the total area receiving light. Figure 13–22(b) indicates the size of the secondary mirror and shows how it is supported.

An inwardly curved mirror is said to be concave, *so the objective mirror of a telescope is concave. A mirror with an outward curvature—such as the passenger-side mirror on many cars—is said to be* convex.

The mirrors of a reflecting telescope are *front-surface mirrors*. This means that the shiny surface (normally aluminum) is on the front rather than the back of the mirror. By looking at the image in the large objective mirror of Figure 13–23 you can see how slight is the curvature of such a mirror. An aluminized surface is fragile and can be easily scratched, so mirrors in your house or purse have the reflecting surface on the back of the glass and this surface is covered with a protective layer.

The largest refractor in existence is the 40-inch diameter telescope at Yerkes Observatory at Williams Bay, Wisconsin (Figure 13–24). Reflecting telescopes

ACTIVITY

Mirrors

USE ANY REGULAR HOUSEHOLD mirror for this activity. A vanity mirror in a dorm room will be fine. First, look at your image in the mirror. This is a different type of image than is produced by the objective mirror of a telescope, for the image in your mirror cannot be focused on a screen. The image in your mirror is not formed by light actually striking the image; but rather the light that bounces from the mirror *seems* to have come from the image. Where is your image located when you look in the mirror? That is, is your image located at the surface of the mirror or behind the mirror? To answer this, you must imagine that you have a twin and that the image you see is your twin. Now where would your twin have to be standing in order to be at the distance of the image in the mirror?

Try to estimate how far behind the mirror your image is located compared to your distance in front of it. This is tough to estimate, but if you have a pocket mirror and a friend, there is a way to make such an estimate: Put the pocket mirror between your index finger and the next finger so that you see the image of your index finger (from about the knuckle out) in the mirror. Now spread the two fingers and move the mirror around until the image of your index finger lines up with your second finger. Ask a friend which finger is closer to the mirror. Using this method, I hope that you can see that the image in a flat mirror is the same distance behind the mirror as the object is in front.

Another way to check this is to use an adjustable camera (one you have to focus) to take your picture in a mirror. Suppose you stand three feet in front of the mirror. At what distance do you focus the camera? Three feet? No; six feet.

This type of image, in which the light rays do not

are made much larger than this, however. Before describing these telescopes, let's explain why reflectors can be made larger than refractors. The reasons include the following:

1. In order to be achromatic, a refractor requires two lenses. This means that there are four surfaces of glass that have to be shaped correctly. A front-surface mirror, on the other hand, has only one critical surface. Since it is extremely important to obtain perfectly shaped surfaces, limiting the number of surfaces greatly simplifies the construction of the objective.

2. It is impossible to completely correct lenses for chromatic aberration. When light reflects from a mirror, however, all wavelengths reflect in exactly the same direction; thus automatically eliminating chromatic aberration problems.

3. Since a reflector's mirror is front-surfaced, the light doesn't pass through the glass of the mirror. Thus the glass does not need to be as perfect as that used for a refractor. It is difficult (and therefore expensive) to make large pieces of glass without tiny air bubbles or other imperfections.

converge to form an image, is called a *virtual* image. This is the type of image produced by a magnifying glass and by a telescope eyepiece. The telescope objective produces a *real* image. (You cannot form a real image with a flat mirror; a *concave* mirror—curved inward—is required.)

To see that the reflecting surface of your everyday mirror is on the back side of the glass, move the tip of a pencil or pen up to the mirror and notice where the image is located when the tip touches the glass of the mirror. Notice that the image of the pencil point does not touch the actual point. There seems to be space between them. This space exists because there is a thickness of glass between the point and the reflecting surface. Thus the pencil point is actually about 1/16 inch in front of the mirror surface. This means that the image is about 1/16 inch behind the surface and that the object and image are therefore about 1/8 inch apart.

Finally, in a brightly lit room, hold the corner of a white sheet of paper against the mirror while you view at a large angle to the mirror. (That is, don't view from directly in front of the mirror.) Can you see a second, fainter, image of the corner of the paper? This is caused by reflection from the front surface of the glass. Notice that this second image *does* seem to contact the actual corner of the paper; the object and image touch since the object is zero distance from the reflecting surface. The existence of this second image for a rear-surface mirror is another reason telescopes use front-surface mirrors.

4. Glass sags. If a large piece of glass is left in the same position for a long period of time, it will change its shape slightly. This is unimportant for the windows in your house but considering the accuracy needed for the surfaces of telescope objectives, it can be a problem. When a large refractor is not in use, it must be stored in different positions each time so that its shape doesn't change over the years. The objective of a reflector is also made of glass, of course but its mirror can be supported anywhere across its back rather than only around its edge as is the case for a lens.

All of the above reasons make a large reflector much less expensive and more practical than a large refractor, and it accounts for the fact that all really large telescopes are reflectors.

Optical Arrangements of Reflectors

A reflecting telescope with an eyepiece arrangement as described above is called a ***Newtonian telescope,*** or is said to have a **Newtonian focus.** The Newtonian

Newtonian focus: The optical arrangement of a reflecting telescope in which a plane mirror is mounted along the axis of the telescope so that it intercepts the light from the primary and reflects it to the side.

Figure 13–24(a). *The 40-inch diameter Yerkes Observatory refractor, photographed in 1921. The telescope tube is nearly 20 feet long, indicating that the focal length of the objective is that long. (Find Albert Einstein in the photo.)*

Cassegrain focus: The optical arrangement of a reflecting telescope in which a mirror is mounted so that it intercepts the light from the primary and reflects it back through a hole in the center of the primary.

focus is a common one for small telescopes. Figure 13–25 shows the arrangement common in large telescopes, the ***Cassegrain focus*** (invented by G. Cassegrain, a French optician who lived at the time of Newton). Notice that the eyepiece or camera body is at the back of the telescope. In this arrangement, the secondary mirror is not a flat mirror but has an outward curvature. The effect of the curved secondary mirror is that an objective that actually has a short focal length can be given a longer effective focal length and thus can be contained in a short telescope.

Equatorial Mounting

The simple fact that the Earth rotates makes it necessary to add another feature to an astronomical telescope. Suppose you are viewing Jupiter and its moons

Figure 13–24(b). *A more recent photo of the Yerkes Observatory telescope.*

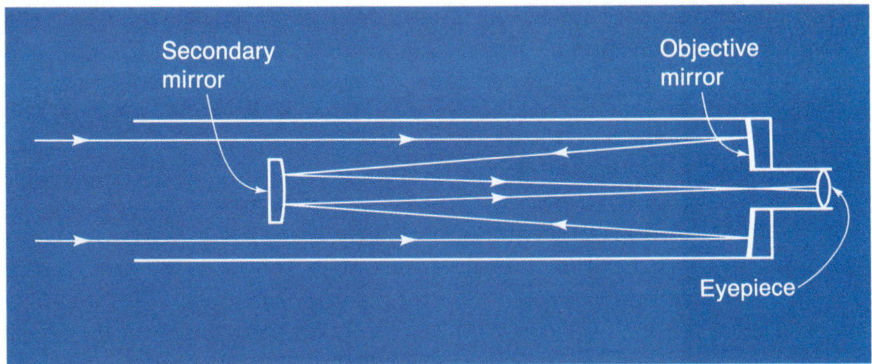

Figure 13–25. *In the Cassegrain focal arrangement, the secondary mirror is curved outward (convex) and the light is reflected back through a hole in the primary mirror.*

Figure 13–26. *A telescope mounted so that is rotation axes are vertical and horizontal must be moved about both axes if it is to track a star across the sky.*

The axis arrangement of the telescope in Figure 13–26 is called an altazimuth mount. *One axis changes the telescope's altitude and the other its* azimuth *(the angle around the horizon measured from north.)*

A telescope mounted in the manner of Figure 13–27 is said to have an equatorial mount, *because one of the axes is aligned with the equator. (The other is parallel to the Earth's axis.)*

with your backyard telescope. Because of the Earth's rotation, you will see Jupiter sliding across your view until it is gone. This motion across the sky is not obvious to the naked eye but the motion is magnified in the telescope and unless you move the telescope to follow Jupiter, the planet will move out of your limited field of view. Figure 13–26 shows a small telescope with two axes about which it can rotate; one horizontal and one vertical. To track a star using this telescope, motion around both axes would be required as shown in the photos.

Figure 13–27(a) shows this telescope mounted so that its two axes are tilted. The user orients this telescope so that when it is in the position in photo (b), it is pointing toward the north celestial pole. Now look at part (c) of the photo and you see that in tracking Jupiter (or a star), the telescope needs to be rotated around only one axis. This simplifies things greatly because a small clock motor (mounted in the base of the telescope) can be used to move the telescope around this axis at the correct speed in order to follow the object across the sky. Using the motor, the telescope tracks the object and keeps it in the field of view.

Some modern telescopes now have simple horizontal/vertical axes of rotation. In order for these telescopes to track a celestial object, a computer is used to turn the telescope around each axis at the proper rate.

LARGE OPTICAL TELESCOPES

Prime focus: The point in a telescope where the light from the objective is focused.

In very large telescopes, observing is often done at the ***prime focus***. This is the point where the light from the objective mirror comes to a focus. Figure 13–28 shows the largest telescope in the United States, the 200-inch (five-meter) Hale telescope on Palomar Mountain. Look for a person in the photo in order to appreciate the size of the instrument. Figure 13–29 shows the observer's cage at the prime focus of the instrument. The cage is located at the top end of the telescope and the astronomer in the cage is carried around with

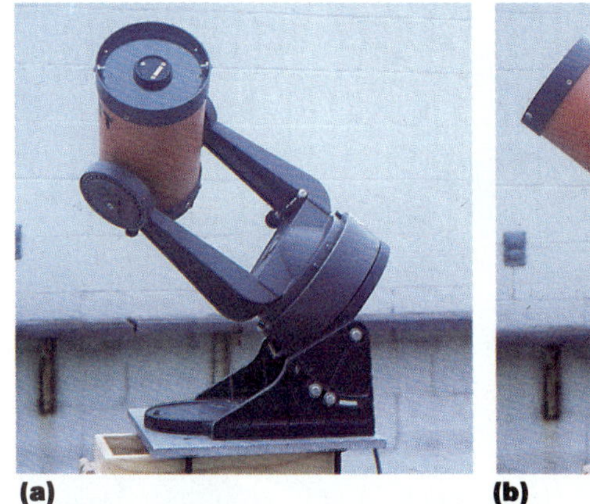

(a)

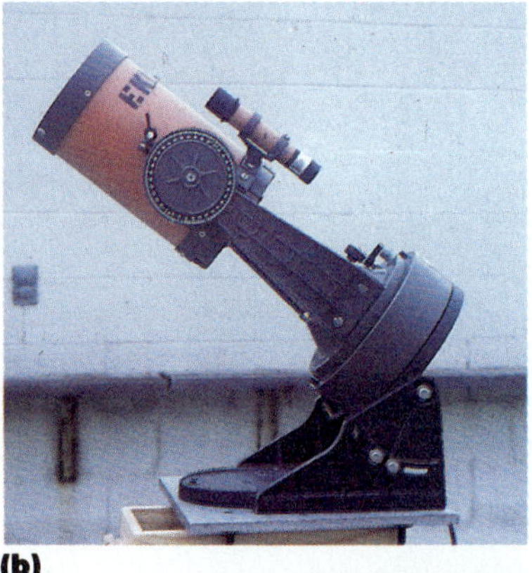

(b)

Figure 13–27. *(a) This telescope is mounted so that one axis is parallel to the Earth's axis. (b) Here the telescope is pointed toward the north celestial pole. (c) As the telescope moves to follow a star, it rotates around only one axis.*

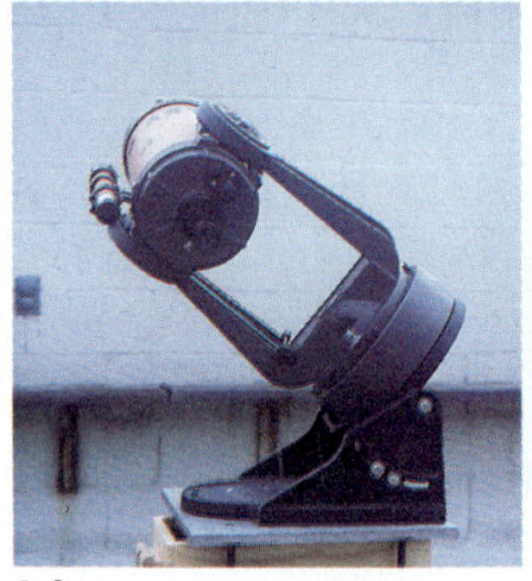

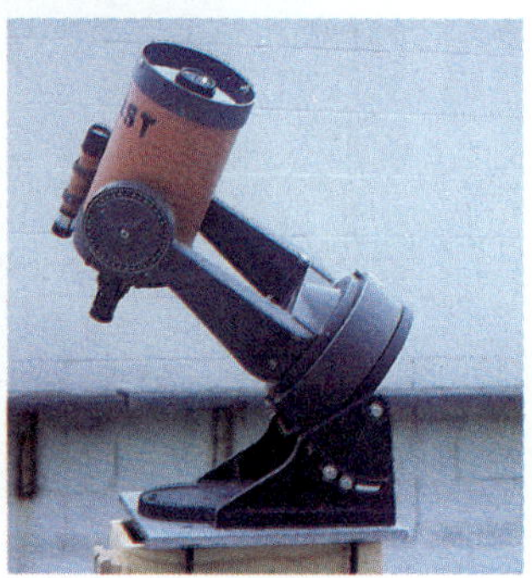

(c)

the telescope as it moves. The observer's cage does block some light to the objective mirror, but only a small fraction of it.

Look at the photo of the Hale telescope and notice how the equatorial mounting was achieved in this case. As this telescope follows an object across the sky, the shiny circular surface at the upper right slides around in its base, forming one of the axes of rotation.

To achieve the best viewing conditions, it is advantageous to locate telescopes high in the mountains in dry, clear climates (Figure 13–30). This eliminates as much as possible the blurring effects of the atmosphere.

The 1990s will be a particularly exciting time for astronomy, for a number of new telescopes are now being planned or are under construction. Figure 13–31 shows a new design that uses multiple mirrors rather than a single one. This telescope, at Mt. Hopkins, Arizona, has six 1.8-meter objective mirrors with secondary mirrors arranged so that they all focus at the same point. The total area of the six mirrors is equivalent to one large objective 4.5 meters in diameter. The problem of aligning six mirrors and keeping them in alignment is a great one, but the cost savings achieved by constructing a number of small mirrors rather than one large one more than makes up for this.

A consortium of western European countries now has plans to use the idea of multiple mirrors to build a telescope much larger than any of today's. The Very Large Telescope, as its designers call it, will actually consist of four telescopes that can be used independently or used together as a single telescope.

Figure 13–28. *The Hale Telescope on Palomar Mountain. To appreciate the size of the telescope, look for the person at the bottom.*

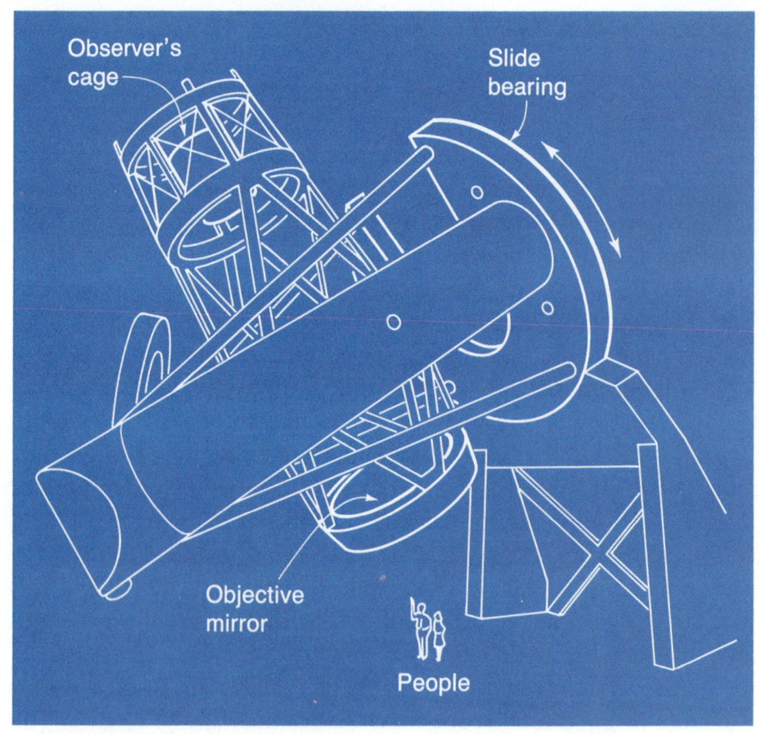

Figure 13–29. *This shows a person in the observer's cage of the Hale telescope.*

The four mirrors (Figure 13–32) will have a total light-gathering power equivalent to one telescope 16 meters in diameter! According to present plans, the telescope will be located high in the mountains of Chile.

The SST and the HST

Telescopes are used for much more than to obtain an image of objects in the sky. In Chapter 11 we saw that a tremendous amount of information is obtained by spectral analysis—the examination of light that has been separated into its colors. In order to obtain data for such an analysis, a ***spectrometer*** is connected to a telescope. This instrument separates light into its colors by use of a prism—or more commonly, a ***diffraction grating***—and produces either a photo of the spectrum or numerical data concerning the intensity of light at various wavelengths.

Spectrometer: An instrument that separates electromagnetic radiation according to wavelength. (A spectrograph is a spectrometer that produces a photograph of the spectrum).

Diffraction grating: A device that uses the wave properties of electromagnetic radiation to separate the radiation into its various wavelengths.

Fiber optics: The transmission of light through small filaments of glass (or plastic).

Penn State University and the University of Texas are building a telescope designed specifically for spectroscopy. Figure 13–33 is a model of the instrument, whose multiple mirrors will have an effective aperture of about eight meters. The Spectroscopic Survey Telescope (SST) will not produce photographic images but instead light will be carried from its focal point (at the top of the support in the figure) to a spectrometer by means of ***fiber optics***.

The SST will cost about 6 million dollars when it is completed in the early 1990s. This is an extremely low cost for a telescope of this size. Part of the savings is accomplished by the use of multiple mirrors but additional money is saved by the fact that the telescope will move only on one axis: the vertical axis.

Figure 13–30. *Fifteen major telescopes of the European Southern Observatory cover the summit of this mountain in the Chilean Andes.*

Figure 13–31. *The six mirrors of the Multiple-Mirror Telescope on Mount Hopkins, Arizona, is equivalent to one 4.5-meter objective.*

Figure 13–32. *Each of the four telescopes of the European Southern Observatory's Very Large Telescope will have an 8-meter reflector. When used together, they will have a light-gathering power equivalent to a single 16-meter telescope.*

Thus it will be able to move only parallel to the horizon. As the Earth rotates, it will be able to scan nearly all of the sky but its mirrors will not track an object across the sky. Instead, it will use a different plan to obtain sufficient light from a particular stellar object. As the object moves across the sky above the mirrors, an assembly will move across the focal plane of the telescope and will be able to collect light from the object for as much as an hour. (In Figure 13–33, the moveable light collector is located in the circle at the top.)

The most exciting telescope, however, must be the Hubble Space Telescope (Figure T13–1). The HST will operate in Earth-orbit, thereby eliminating the blurring effects of the atmosphere. This telescope was originally scheduled for launch in the mid–1980s but problems in the Space Shuttle program have delayed its launch. The Space Telescope will use a 2.4-meter mirror (Figure 13–23) and a number of instruments will be on board, including two spectrometers. (The first Close Up in this chapter describes the telescope's various functions.)

Although the Space Telescope's 2.4-meter aperture is much smaller than the large telescopes on Earth's surface (see Table 13–1), its location above the atmosphere will be able to detect stars as much as 50 times fainter than the faintest seen from Earth. This means that it will be able to see objects seven times farther away. (Recall the inverse square law: if one moves seven times farther away, the light received is 7^2 times dimmer.) In addition, it will have the ability to observe the sky in infrared and ultraviolet regions. Astronomers eagerly anticipate the answers—and new questions—that will be provided by the Space Telescope.

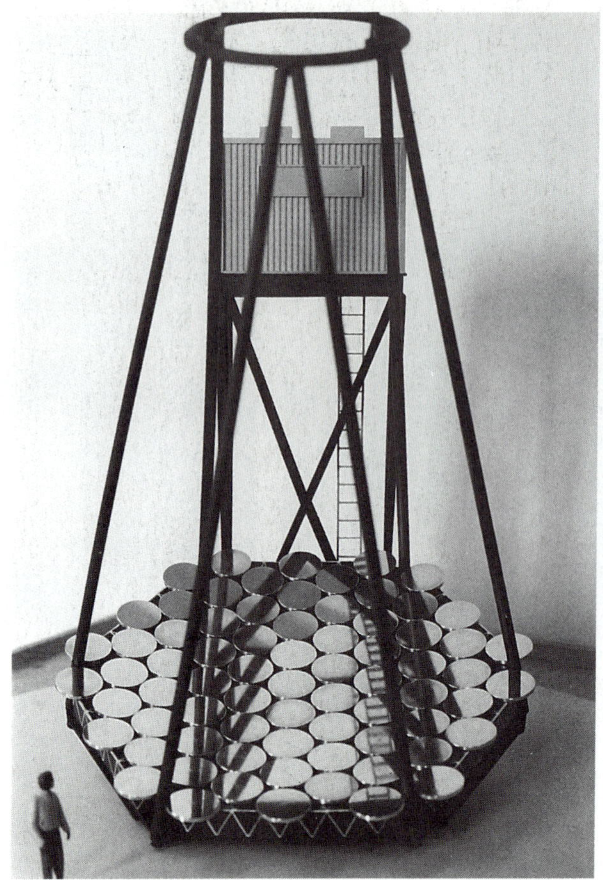

Figure 13–33. *The many 1-meter mirrors of the Spectroscopic Survey Telescope will focus light on a collector (at the top) that will move to track the chosen celestial object. The actual telescope will appear more complex than this model and will have 85 mirrors rather than the 73 shown.*

RADIO TELESCOPES

Thus far we have concentrated on optical telescopes; telescopes that gather visible light. Besides visible light, only one other type of radiation from space penetrates the atmosphere significantly to be studied in detail from the surface: radio waves. In 1931, a scientist working on radio transmission for Bell Laboratories noticed that static received by his antenna originated in the Milky Way. When better radio receivers were designed (largely during World War II), astronomers were able to pinpoint the sources of celestial radio waves and the field of radio astronomy was born.

Two problems arise in examining radio waves from space. First, the intensity of radio waves from space is much less than the intensity of light waves. Second, since the wavelengths of radio waves are a million times greater than the wavelengths of visible light, there is a corresponding decrease in the resolution of images made with radio waves. (Recall that diffraction is greater with longer wavelengths.)

Both of these problems are solved in the same way—by making radio telescopes extremely large. Figure 13–34 shows two radio telescopes at the National Radio Astronomy Observatory at Green Bank, West Virginia; the

Table 13–1 Major Telescopes of the World (2.5 meters or larger)

Observatory	Location	Telescope
European Southern Obs.	Chile	16 m "Very Large Telescope"*
Keck Observatory (California Institute of Technology and University of California)	Mauna Kea, HI	10 m Keck Telescope*
McDonald Obs.	Mount Locke, TX	8 m Spectroscopic Survey Telescope*
Soviet Special Astrophysical Obs.	Caucasus, USSR	6.0 m Bol'shoi Tel. Azimutal'nyi
Mt. Wilson & Las Campanas Obs.	Palomar Mountain, CA	5.0 m George E. Hale Telescope
Smithsonian Observatory	Mount Hopkins, AZ	4.5 m Multiple Mirror Telescope
Royal Greenwich Obs.	Canary Islands, Spain	4.2 m William Herschel Telescope
Cerro Tololo Observatory	Cerro Tololo, Chile	4.0 m
Anglo-Australian Obs.	Siding Spring Mt., Aus.	3.9 m Anglo-Australian Telescope
Kitt Peak National Obs.	Kitt Peak, AZ	3.8 m Mayall Telescope
Royal Observatory Edinburgh	Mauna Kea, HI	3.8 m United Kingdom IR Telescope
Canada-France-Hawaii Obs.	Mauna Kea, HI	3.6 m Canada-France-Hawaii Telescope
European Southern Obs.	Cerro La Silla, Chile	3.6 m
Apache Point Obs.	Sunspot, NM	3.5 m
Max Planck Institute (Bonn)	Calar Alto, Spain	3.5 m
Astronomical Research Corp.	Sacramento Peak, NM	3.5 m*
Lick Observatory	Mount Hamilton, CA	3.0 m Shane Telescope
Mauna Kea Observatory	Mauna Kea, HI	3.0 m NASA Infrared Telescope
McDonald Observatory	Mount Locke, TX	2.7 m
Haute Provence Observatory	Saint Michele, France	2.6 m
Crimean Obs.	Simferopol, USSR	2.6 m Shajn Telescope
Byurakan Astrophysical Obs.	Armenia, USSR	2.6 m
Mt. Wilson and Las Campanas Obs.	Las Campanas, Chile	2.5 m Irenee du Pont Telescope
Royal Greenwich Observatory	Canary Islands, Spain	2.5 m Isaac Newton Telescope
Royal Greenwich Observatory	Canary Islands, Spain	2.5 m Nordic Optical Telescope

*Telescopes under construction or in the planning stage.

smaller one being 45 meters in diameter and the larger 100 meters in diameter. One hundred meters is the length of a football field!

In November 1988 the 100-meter telescope collapsed. Replacing it would cost some 10 million dollars.

Notice that the telescopes do not have shiny reflecting surfaces. In fact, the larger one has a wire mesh surface. This is possible because of a feature of waves that goes hand-in-hand with the diffraction effect: Although longer wavelengths diffract more when going through an opening, they do not require as smooth a surface for reflection. (Have you noticed the poorer reception by your car radio when the car is surrounded by the metal superstructure of a bridge? The bridge reflects some of the radio waves.)

Radio telescopes are similar in principle to the satellite dishes we use to receive TV signals from Earth satellites (Figure 13–35). In each case the reflector

(a)

(b)

Figure 13–34. *Radio telescopes at the National Radio Astronomy Observatory near Green Bank, West Virginia. (a) The 100-meter telescope shown here collapsed in November 1988. (b) A 45-meter telescope.*

Figure 13–35. *This satellite dish works on the same principle as a radio telescope. Radio waves from the satellite are focused onto the radio antenna a few feet in front of the center of the dish.*

simply directs waves to a small radio antenna located at the focal point of the "mirror." You can see the supports for the antenna in the photos.

Satellite dishes point to satellites that are in geostationary orbits; staying at a fixed location above the Earth's equator.

One difference between radio telescopes and a satellite dish is that radio telescopes must be able to move if they are to track a celestial source of radio waves. Imagine the difficulties in constructing these giant structures so that they can be moved accurately. You can see the tracking mechanism on the telescope of Figure 13–34(b).

The world's largest radio telescope, located in Arecibo, Puerto Rico, is not capable of independent movement. Shown in Figure 13–36, it is a telescope constructed by stretching wire mesh across a natural bowl between hills. This

Figure 13–36. *The radio telescope near Aricebo, Puerto Rico is the world's largest. Its radio antenna is suspended above it from three supports.*

telescope is 300 meters (1000 feet) in diameter and scans the sky as it moves along with the Earth. Slight changes in the direction from which it detects radio signals can be achieved by moving its antenna, which hangs from the cables suspended above the bowl.

The image formed by a radio telescope is not a normal photo, of course. The radio telescope simply detects the intensity of radio signals from the area of the sky toward which it is pointed. One way to display and examine the data received is to plot a graph of the intensity of the radiation as the telescope moves across a small portion of the sky. Figure 13–37(a) shows what such a plot might look like. A more complete image of the radio-emitting object can be obtained by scanning the radio telescope back and forth across the celestial object and feeding the data into a computer that is programmed to represent as different colors the various intensities of the radio waves. Figure 13–37(b) was drawn in this manner and indicates the intensity of radio waves from a small portion of the sky.

CLOSE UP

ET Life, Part IV—SETI

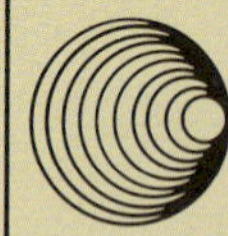

RADIO TELESCOPES ARE DESIGNED and built to detect and measure radio waves coming from objects in space. If, however, there are intelligent beings out there, these telescopes will be useful both in detecting their presence and in communicating with them. In a Close Up in Chapter 15 we will describe various questions that must be answered before we can calculate the likelihood that extraterrestrial intelligence exists and how many life sites there are likely to be. Another way to answer the question of whether intelligent beings exist is to search for signals from such beings. SETI is the acronym for the Search for Extraterrestrial Intelligence.

Radio telescopes have been used on occasion to search the heavens for evidence of radio signals from intelligent beings. Frank Drake, an American astronomer whose name we have encountered before in discussing the question of extraterrestrial life, in 1960 used a telescope of the National Radio Astronomy Observatory to search for signals from two nearby stars. The search, which Drake called Project Ozna, involved looking for unusual patterns in radio signals; different from the signals emitted by inanimate objects such as stars and galaxies.

Since that time a number of other astronomers have done similar searches. In one, done in the mid 1970s, more than 600 nearby stars were watched for about 30 minutes each; again looking for abnormal signals. The larger the telescope, of course, the weaker the signal it will be capable of detecting. Thus the largest radio telescope, the one at Arecibo, has been used by Frank Drake and Carl Sagan. No evidence of signals from intelligent life forms was found in any of the attempts.

The searches thus far have involved relatively few stars and each was observed for a relatively short period of time. The astronomers doing the searches realized that even if intelligent life were plentiful in the universe, the chances of finding it during these short searches were slim, at best. In order to have a realistic chance of finding signals from intelligent beings, a much more extensive effort would be needed.

During the summer of 1971, NASA, in cooperation with the American Society of Engineering Education, funded Project Cyclops, a study by a number of scientists and engineers of the search for extraterrestrial intelligence. The group concluded that such a search was indeed possible with the technology then available. They made recommendations concerning aspects of the search such as which frequencies were most likely for transmissions of messages. (Answering this question involves choosing a frequency that is not absorbed by our atmosphere or by material in interstellar space and that is not emitted strongly by natural objects in space.)

The participants in Project Cyclops recommended the construction of an array of radio telescopes, each 100 meters in diameter, for a comprehensive search to last a few decades. Figure T13–2 shows an artist's conception of the telescope array. It is necessary to look at details of the drawing in order to appreciate the size of each antenna and of the entire array. The Cyclops array would cost billions of dollars and although it has the support of a number of astronomers and interested laymen, it has not been funded. Its supporters admit that there is no guarantee of success in finding extraterrestrial life but they point out the tremendous effect it would have on humanity to find another race of beings in the universe and to learn something of their planet, their biology, their society, their science, etc.

The Project Cyclops group gave us an answer to "Can we?" The rest of us must answer the question "Should we?"

Figure T13–2. *The proposed Cyclops array. Find the building at right center.*

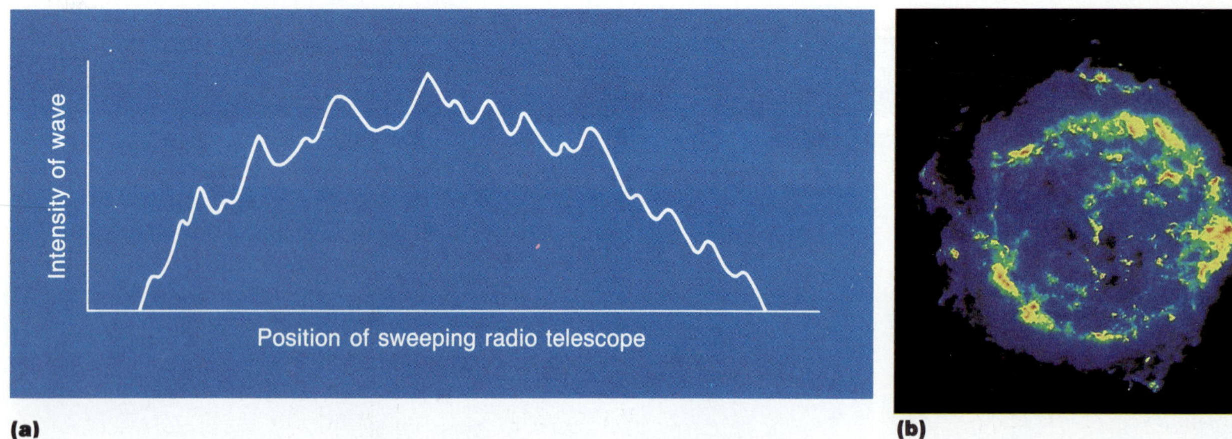

(a) **(b)**

Figure 13–37. *(a) A radio telescope scanning once across a celestial source of radio waves detects the intensity of the radio signal on the line along which it is moving, so that a graph like this might be plotted. (b) If the radio telescope is programmed to repeatedly scan the radio source a computer can plot a contour map of the source, indicating each different intensity with a different color.*

INTERFEROMETRY

Recall that the resolution of a telescope depends upon both the diameter of the telescope and the wavelength of the radiation. The greater the diameter of the telescope, the greater the resolution (assuming atmospheric clarity is not a factor) but the greater the wavelength, the poorer the resolution. Even though radio telescopes are very large, radio waves are so long that the best resolution from a single radio telescope is in the order of a number of arc minutes. This means that a radio source the size of a star would still appear in a radio telescope to be a blur as large as half the diameter of the moon.

The intensity of the signal received by only two spots would be much less than from the entire dish, of course.

The solution to this problem lies in the fact that a giant radio telescope would retain the same resolution if only two portions of its outer surface were being used, as in Figure 13–38(a). Astronomers take advantage of this idea by combining two radio telescopes so that they act as one; the two telescopes substituting, in a sense, for two portions of the outer part of a giant telescope as in Figure 13–38(b). In this way, they are able to obtain resolutions equal to that of a single large telescope.

Using two telescopes to act as one is not a simple matter, however. To understand the problem, refer to Figure 13–39. Radio waves from a distant source are shown striking the dish of a radio telescope. Notice that they are reflected so that a single wave gets to the detector (located at the focal point of the dish) at the same time from all areas of the dish. This feature must be retained in using two radio telescopes as one. Waves from each single dish must be combined in the correct relationship. We say that the waves from the two telescopes "interfere" with one another when they combine and therefore the technique of linking two (or more) telescopes so that they act as one is called

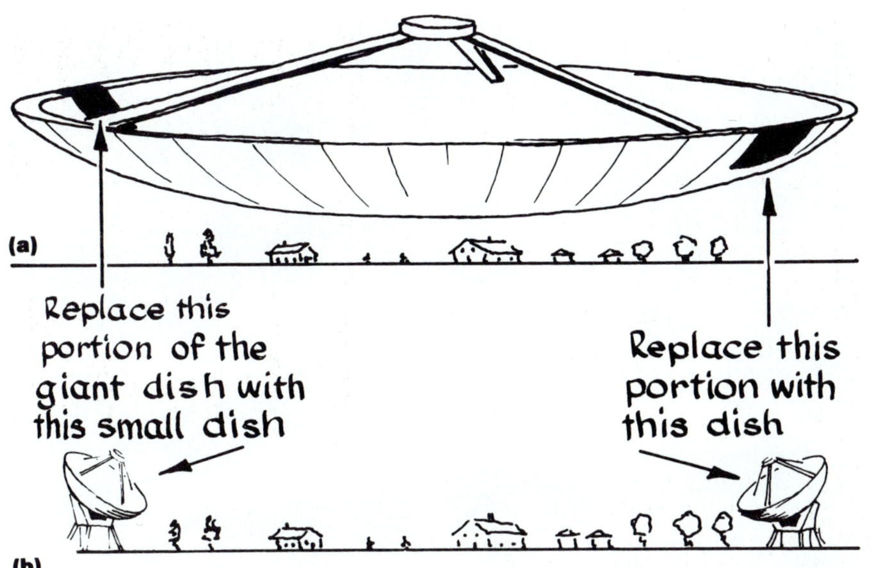

Figure 13–38. *Two small radio dishes (b) can be made to have the same resolution as a large radio telescope (a) that has a diameter equal to the distance between the two small ones. The strength of the signals detected will be much greater in the case of the single large telescope, of course.*

interferometry. Interferometry has become possible using widely separated radio telescope dishes due to today's extremely accurate atomic clocks.

Interferometry: A procedure that allows a number of telescopes to be used as one by taking into account the time at which individual waves from an object strike each telescope.

In the New Mexico desert is an array of radio telescopes used for interferometry. Figure 13–40 is a photo of this Very Large Array. The telescopes ride on a double pair of railroad tracks so that they can be moved and the arrangement changed.

Very accurate atomic clocks allow us to coordinate the waves received by very distant radio telescopes so that they can even be placed far across the Earth from one another, achieving resolutions as high as a small fraction of an arc second. This is better than optical telescopes!

There are plans to launch a radio telescope into space in 1995. This telescope would be used in conjunction with Earth-bound telescopes to provide an even longer baseline for interferometry and thereby provide even greater resolution.

The distance between the telescopes will be even greater than the length of the project's name: Orbital Very Long Baseline Interferometry.

Interferometry must also be employed in the newest optical telescopes. Recall that the Very Large Telescope being planned for Chile is actually four separate telescopes rather than a collection of mirrors that focuses at the same point (like the multiple-mirror telescope on Mount Hopkins). To use the four telescopes as one, the signals from each must be combined using techniques developed for radio telescopes.

DETECTING OTHER ELECTROMAGNETIC RADIATION

The visible and radio regions of the spectrum penetrate Earth's atmosphere well. In addition, some infrared radiation penetrates to the Earth's surface, as you

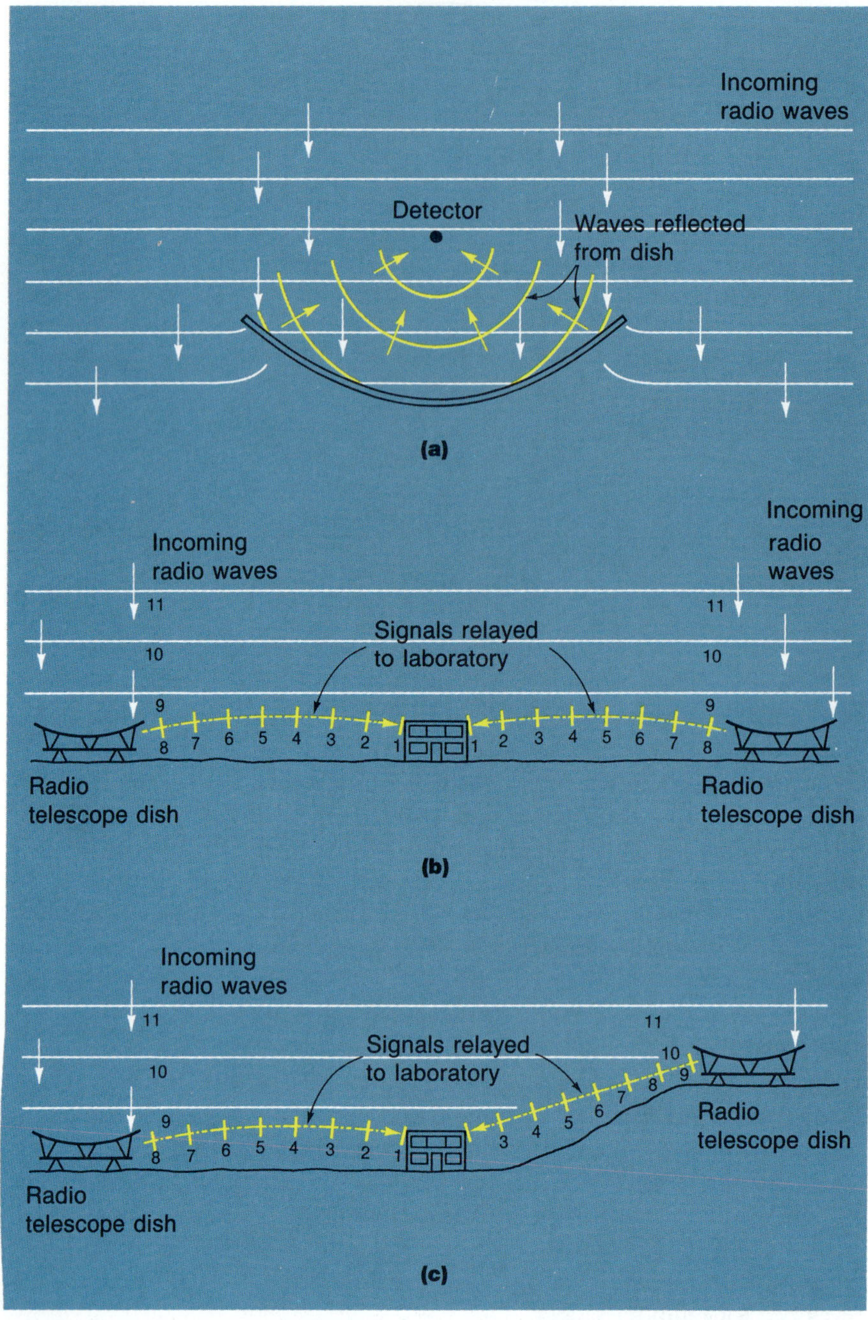

Figure 13–39. *(a) All portions of an incoming wave reach the detector of a radio telescope at the same time. (b) If two radio telescope dishes are to function as a single telescope, the waves must likewise reach the detector at the same time or at least the time difference must be corrected for. (c) This is an obvious (and oversimplified) case of waves whose difference in reception times must be accounted for in the laboratory.*

Figure 13–40. *The Very Large Array is spread over 15 miles of the New Mexico desert.*

experience when you feel the heat of the Sun. Most of the infrared radiation from the Sun is blocked by the atmosphere, however; primarily by the water vapor there. In order to get above much of this water vapor and observe celestial sources of infrared radiation from Earth's surface, we locate infrared observatories on the highest mountains. A major research center for infrared astronomy is the group of observatories on 14,000-foot Mauna Kea in Hawaii (Figure 13–41).

Another way to get above the atmosphere's water vapor is to fly a telescope in a plane. The Kuiper Airborne Observatory (Figure 13–42) is an airplane modified to carry a one-meter infrared telescope at altitudes up to 40,000 feet (12 kilometers). This gets the telescope above about 99 percent of the atmosphere's water vapor.

To completely eliminate the absorption of radiation by the atmosphere, we place a telescope in orbit. The Infrared Astronomical Satellite (IRAS), which was launched by NASA in 1983, was a cooperative effort of Holland, England, and the U.S. Its 57-centimeter telescope has provided information on what the sky looks like in the very long wavelengths of infrared. These long wavelengths hardly penetrate our atmosphere at all. The IRAS ceased operation in 1984 but three orbiting infrared telescopes are now in the planning process. In addition, there are plans to put three radio telescopes into space. These include the interferometry telescope mentioned above as well as the Cosmic Background Explorer (COBE) now scheduled for an 1989 launch by NASA. We will discuss cosmic background radiation in Chapter 18.

In spite of recent setbacks in NASA's space program, in mid-1988, 18 U.S. (or joint U.S. and other nations) scientific spacecraft were still operating productively.

When we turn to wavelengths shorter than visible light, we also find that little of this radiation penetrates our atmosphere. And again it is the space program that allows us a look at the sky in these regions of the spectrum. The International Ultraviolet Explorer was launched in 1978 in a cooperative effort

CLOSE UP

ET Life Part V—CETI

If our search for radio signals from a race of intelligent extraterrestrials is successful it might succeed by detecting their stray, wasted radio signals (discussed in a Close Up in Chapter 7). On the other hand, those beings may already be transmitting messages into space with the purpose of announcing their presence and telling others something about themselves. The same radio telescopes that are used to receive signals from space can be used to transmit radio signals. Considering the probable differences between the beings in different parts of the galaxy, what could one race of these beings communicate to another? The study of this question is called CETI, for Communication with Extraterrestrial Intelligence.

First, we must point out that if another intelligent race were found, the tremendous distances between stars would prohibit a dialog. The very nearest star is nearly five light years away from us. It would require five years for our radio signals to reach a planet circling that star and another five years for the signal to return from beings on that planet. And the likelihood that life is so common that it exists near the very nearest star is extremely remote. If extraterrestrial life exists, the closest life sites are likely to be much farther away; making it impossible to get a reply to our signal during the span of generation on earth.

At first glance, the language problems might seem insurmountable. But we do have something in common with every other race of beings that might exist: the physical universe and its laws, which are the same everywhere. Every intelligent race knows that hydrogen is the most common element and that an atom of hydrogen is made up of one proton and one electron. The prime numbers, those numbers that cannot be divided evenly by any other numbers but one and themselves, are the same in every language. Those studying the problem have concluded that an understandable message could indeed be sent if enough time were devoted to its transmission.

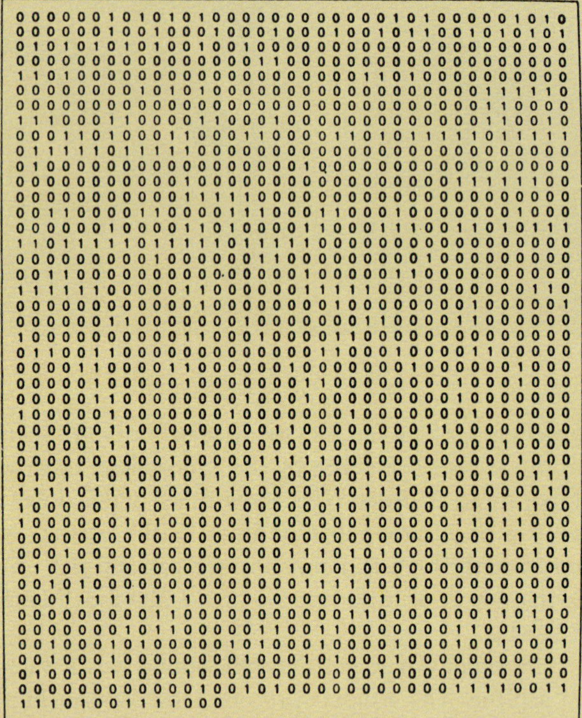

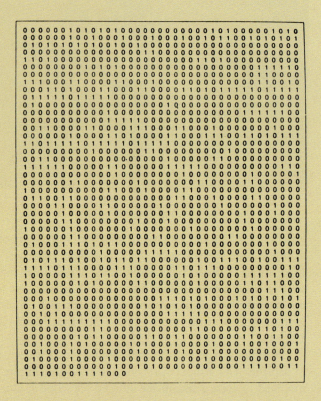

Figure T13–3. *The message consisted of 1679 "on" and "off" signals, here represented as zeros and ones.*

We Earthlings have already sent a very short message. In 1974 the reconditioned reflecting surface of the Arecibo telescope was rededicated and at the ceremony, the telescope was used to transmit a message toward a cluster of 300,000 stars in the constellation Hercules. This transmission lasted only about ten minutes, and so the information that could be sent was very limited. The signal consisted of a series of Morse Code-type pluses containing 1679 data points. Figure

Figure T13–4. *If the 1679 signals are arranged into 23 columns of 73 signals each, the pattern in (a) emerges. Part (b) explains the drawing.*

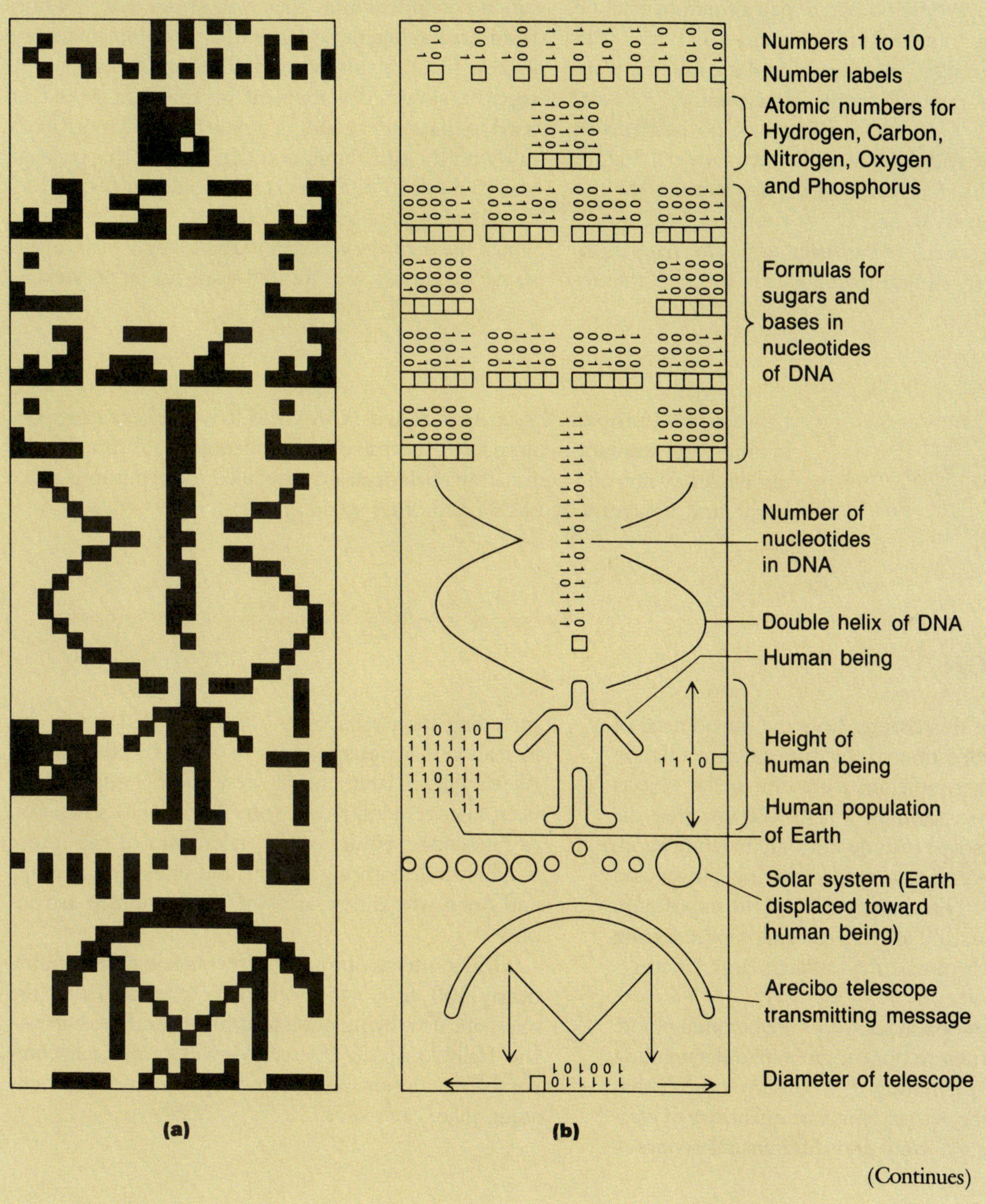

(Continues)

CLOSE UP

T13–3 shows the message as a series of zeroes and ones. The number 1679 was chosen because, except for 1 and 1679 only one other pair of number can be multiplied to obtain 1679: 23 and 73. If the data points are arranged into a rectangular array, there are only two ways to do it: either 23 across and 73 down, or vice versa. One way will produce no pattern but the other way will result in the pattern shown in Figure T13–4. A brief description of the meaning of the pattern is shown in part (b) of the figure.

If extraterrestrial beings near some star happen to detect our ten-minute message, will they be able to decipher it? Who knows? Although we cannot know how much of it they will be able to understand, we can be confident that they will know that it comes from an intelligent source rather than an inanimate object. If more time were available to transmit messages, a slower development of language would be used so that understanding would be much more likely but with the time limitation that existed, this message was thought to be the best that could be done.

When might we receive a reply? The cluster toward which the message was sent is 26,000 light years away, so we need only wait 52,000 years for an answer!

between the European Space Agency and NASA and it is still operating, obtaining high-resolution ultraviolet spectra of celestial objects. A number of ultraviolet, X ray, and gamma ray telescopes are scheduled to be put into space in the next few years by NASA and other space agencies.

CONCLUSION

We have seen in this chapter how the phenomena of refraction and reflection of electromagnetic radiation allow us to gather radiation from dim stellar objects and to focus it to form an image. We saw that the powers of a telescope include not only magnification, but—more importantly—light-gathering power and resolving power. This analysis showed the importance of large telescopes and led to a discussion of reflecting telescopes, which can be made much larger than refractors.

The other-than-optical telescopes are becoming more and more important to our progress in understanding our universe by permitting us to observe objects that are invisible to the eye yet emit vast quantities of electromagnetic energy. New and different telescopes—including space telescopes—have provided us with information that was impossible to obtain by other means. As we will see later, entirely new celestial objects have been discovered in recent years by the new generation of telescopes. Undoubtedly, telescopes of the future will continue to bring us new and unexpected results and open whole new areas of exploration in astronomy.

Galileo's telescope began a revolution in astronomy nearly 400 years ago. Some have predicted that the telescope now being put into orbit around the Earth—the Hubble Space Telescope—will produce a comparable revolution. Astronomers, professional and amateur alike, can't wait.

Figure 13–41. *Mauna Kea in Hawaii contains a number of important telescopes on its summit, including both visible and infrared telescopes. The 10-meter Keck Telescope will be located here.*

Figure 13–42. *NASA's Kuiper Airborne Observatory carries a 1-meter reflecting telescope above most of the atmosphere so that it can receive infrared radiation that does not penetrate to Earth. The telescope is located below the portal seen just in front of the wings.*

RECALL QUESTIONS

1. Who was the first person to use a telescope to systematically study the heavens?
2. Define refraction and chromatic aberration.
3. How is an achromatic lens made? Describe the defect it corrects.
4. Define magnifying power. Why do we not always use the highest magnification available?
5. What is meant by light gathering power and what is it about a telescope which determines its light gathering power?
6. Define diffraction. Why do we not normally notice it?
7. What two things determine the amount of diffraction that occurs when light passes through a hole?
8. Explain what is meant when we say that a certain telescope can "resolve" a particular pair of stars.
9. Aside from the quality of the optical components of a small telescope, what determines its resolving power?
10. Sketch a Newtonian reflecting telescope, showing the relative positions of the primary mirror, the secondary mirror, the eyepiece, and the image produced by the objective.
11. Distinguish between a front-surface mirror and a rear-surface mirror.
12. How large is the largest refractor? The largest single-mirror reflector? Explain why one type can be made larger than the other.
13. Why do we not have to correct the objective of reflectors for chromatic aberration?
14. What is an equatorial mount for a telescope and why is it used?
15. Why must radio telescopes be made so large? About how large is the largest?
16. Why are most research telescopes located on mountains?
17. What is interferometry, and what is its advantage?
18. What is the primary advantage of locating a telescope in space?
19. What is spectroscopy?

QUESTIONS TO PONDER

1. Most backyard telescopes have a finderscope mounted on them—a small telescope that allows one to more easily find a celestial object. How would you expect the magnification and field of view of a finderscope to compare to that of the main telescope?
2. In the drawings showing light rays that come from very distant objects, the rays are represented as being parallel. If two rays come from a single point, how can they ever be parallel?
3. Which light bends more in going through a prism, light of long wavelength or short wavelength? How does the focal length of a lens depend upon the wavelength of light?
4. When you are search in for a stellar object in a backyard telescope, should you choose an eyepiece with a short or a long focal length? Why?
5. How is the field of view of a telescope changed when one changes to an eyepiece of longer focal length? How is the magnification changed?
6. Telescope A has a resolving power of 1.5 seconds, and telescope B has a resolving power of 2.0 seconds. Considering only their resolving power, which is the better telescope?
7. When we build a telescope on the Earth's surface larger than today's largest, it will not have higher resolving power. Why is this?
8. The Hubble Space Telescope will be smaller than many earthbound telescopes, yet it will have better resolution. Why is this?
9. There are people who argue that we should not beam radio messages into space because this will announce our location to any hostile beings who may detect the signal and they may then come and destroy or enslave us. What do you think?

CALCULATIONS

1. Suppose you have lenses o the following focal lengths: 30 cm, 10 cm, and 3 cm. If you wish to construct a telescope of maximum magnification, which two lenses would you use? Which would be the objective and which the eyepiece? What magnification would this telescope produce?
2. The pupil of your eye is the opening through which light enters. The maximum diameter of the pupil of a human eye is about 0.5 centimeter. How does the light-gathering power of two eyes compare to that of a telescope with a 10-centimeter objective?
3. Pizzas are a lot like telescope objectives in that both are described by their diameter while it is their area which is important. How many times as much pizza is in a 14-inch pizza than in a 10-inch pizza?
4. The telescope at Mount Pastukhov in the USSR has a six-meter objective. Compare its light gathering power to that of the five-meter Hale Telescope on Palomar Mountain.
5. In order to have four similar mirrors with the same light-gathering power as one 16-meter circular mirror, what must be the diameter of each of the four?

CHAPTER 14

Mauna Kea Observatory, Mauna Kea, Hawaii.

Astronomer Sidney Wolff and her husband were involved in the building of the Observatory on Mauna Kea, a 14,000-foot high mountain in the Hawaiian Islands, in the late 1960s. In an essay in The Scientist, *she writes:*

*"We had a wonderful time in those early days, developing a site without even such basic amenities as a source of water; a site where all power had to be generated locally because there was no power line, a site where blizzards raged in winter and where even in summer temperatures dipped to freezing every night. But we learned. We learned about altitude sickness and the best strategies for forcing our bodies to acclimate. We learned first aid so we could cope with accidents, since professional help was hours away. We learned how to handle heavy machinery and how to maintain generators. We learned more about telescope gears and worm drives and how to repair scored gears than we ever wanted to know. Nearly every one of us who was involved in those early days can tell—loves to tell—stories of being nearly trapped on the mountain during a blizzard, of hiking to the summit because the road was blocked by snow, of climbing to the top of the dome to remove snow so that not a moment of observing was lost. It was a great adventure, an adventure that surely I had not envisioned when I planned a life of research alone in my office."**

*Sidney C. Wolff, National Optical Astronomy Observatories

Measuring the Properties of Stars

O NLY RECENTLY IN HUMAN history have we even become aware of the astonishing fact that each of the thousands of stars we see is another sun similar to the one that rules our sky. This realization makes obvious the immense distances to those stars; just imagine how far away our Sun would have to be in order to be as dim as a star. Can we hope to learn much about such faraway objects? In Chapter 11 we saw that from spectral analysis we can determine both the temperatures of stars and the elements of which their atmospheres are composed. Galileo and Newton would have been amazed that we can learn such things.

Temperatures and chemical compositions are only the beginning, however. In this chapter we will discuss how parallax allows us to calculate the distances to many stars and how once we know their distances we can determine other quantities, including their luminosities, their motions, their sizes, and their masses. We find relationships among the various properties and these relationships give us clues as to why one star differs from another, how stars are formed, how their lives progress, and how they die. We will delay discussion of the life cycle of stars until the next chapter, turning our attention now to how we measure those properties of stars that will divulge information about their life cycles.

STELLAR LUMINOSITY

When we speak of the brightness of a star, we must be careful to distinguish between its apparent brightness (Figure 14–1) and its luminosity. In Chapter 12 we discussed how the luminosity of the Sun can be calculated if we know the Sun's distance and the amount of solar radiation striking a given area of Earth in a certain amount of time. Suppose two stars differ in apparent brightness so that different amounts of light reach the Earth from the two stars. The cause for this might be any combination of three things: (1) One star may be inherently brighter than the other (its luminosity may be greater); (2) one star may be closer, making it appear brighter (look back at Figure 1–6, which indicates distances to stars of the Big Dipper); or (3) there may be more interstellar material absorbing light from one star than from the other.

In our discussion of the two quantities, *brightness* refers to the apparent brightness seen from Earth and *luminosity* refers to the total amount of energy emitted by a star. You might occasionally find the term *absolute luminosity;* this simply emphasizes that we are speaking of the radiation actually being emitted and not the radiation reaching Earth.

Apparent Magnitude

The quantity described here is apparent magnitude rather than absolute magnitude, which refers to the luminosity of a star and will be discussed later.

One of the greatest astronomers of the pre-Christian era was Hipparchus, a Greek thinker who lived in the second century B.C. Hipparchus made a catalog of some 850 stars, listing for each its location in the sky along with a number that designated its brightness. To indicate stars' brightnesses, he divided all visible stars into six groups, calling the brightest stars in the heavens *magnitude*

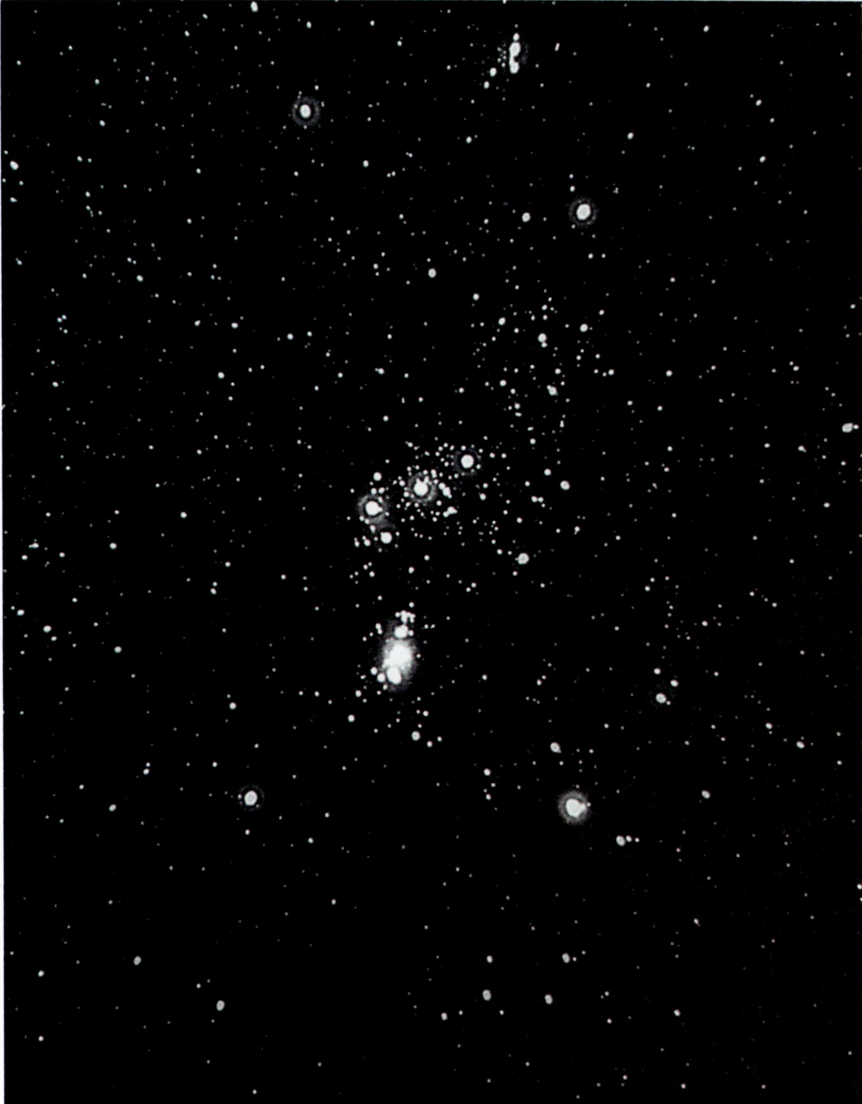

Figure 14–1. *Some of the stars of the constellation Orion appear bright because of their proximity, while others are inherently so luminous that they look bright from Earth even though they are very far away.*

one stars and the dimmest he could see ***magnitude six*** stars. Other stars fell in between, with differences between magnitudes representing equal differences in brightness. Thus the brightness difference between a third-magnitude star and a fourth-magnitude star on his scale was visually the same as that between a fifth- and a sixth-magnitude star. Although it may seem odd to assign the larger number to the dimmer star, we might appreciate his reasoning by thinking of the brighter star as a first-class star and the dimmest as a sixth-class star.

Today we use a slightly revised version of Hipparchus's ***apparent magnitude*** scale, for we measure the brightnesses of stars by photographic and electronic methods. Photographic techniques began to be used for such measurements in the mid-1800s. When these measurements were made, astronomers found that when two stars differ by one magnitude, we receive 2.5 times as much light

Apparent magnitude: A measure of the amount of light received from a celestial object.

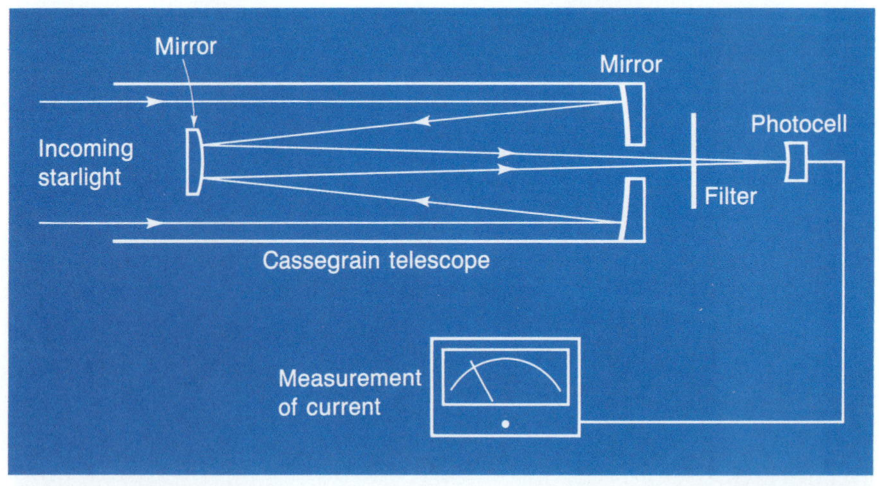

Figure 14–2. *This illustrates the basic idea of photometry, in which the light from a star is focused onto a photocell that measures the amount of light. Often a filter is used to allow entry of light only of a certain wavelength range.*

from the brighter one than from the dimmer. A fifth-magnitude star is about 2.5 times brighter than a sixth-magnitude star. Moving up from fifth to fourth magnitude means that the brightness increases another 2.5 times, making a fourth-magnitude star 6.25 times brighter than a sixth-magnitude star ($2.5 \times 2.5 = 6.25$). A magnitude change of five would mean a change in brightness of 2.5 raised to the fifth power, 2.5^5, or about 100. They recreated the scale accordingly, defining a difference of five magnitudes to correspond to a factor of 100 in the light reaching us. This means that stars differing in apparent magnitude by one have a brightness ratio equal to the fifth root of 100, or 2.512.

Photometer: An electronic device to measure the intensity of light received.

Today we detect the amount of light coming from a star by the response of a ***photometer*** (Figure 14–2), which works on the same principle as the light meter in a camera. Although the eye is able to distinguish only a few different classes of stars (Hipparchus distinguished six), photometric methods allow us to discern the difference in brightness between two stars that may appear identical to the eye. The ability to measure magnitudes accurately has changed Hipparchus's unit-step system into a continuous one so that the magnitudes of stars are now measured to fractional values; to an accuracy of 0.001 magnitudes or better.

Later in the chapter we'll study stars that vary in brightness.

When the magnitude scale is defined as described, we find that some stars in Hipparchus's first-magnitude group are much brighter than others in that group. If a star were 2.5 times brighter than first magnitude, it would have to be assigned a magnitude of zero. A star 2.5 times brighter than this would have a magnitude of -1, a negative number. Sirius, the brightest star in the night sky, is about ten times brighter than the average first-magnitude star and has an apparent magnitude of -1.47. Two other stars (and the Sun) have negative magnitudes. Figure 14–3 shows the approximate magnitude of a number of objects and Appendices I and J contain tables of the brightest and nearest stars, including magnitude values.

Table 14–1 shows the ratio of light received for a given difference in apparent magnitude. The example below shows how to use the chart.

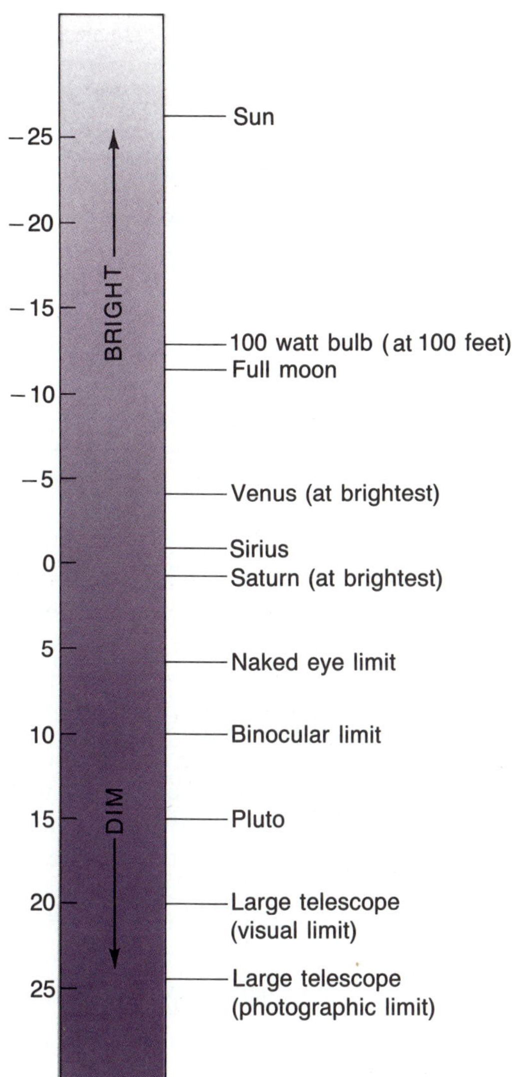

Table 14–1 Magnitude Difference Versus Ratio of Brightness

Magnitude Difference*	Ratio of Light Received	
1.0	2.5	
2.0	6.3	(2.5×2.5)
3.0	16	$(2.5 \times 2.5 \times 2.5)$
4.0	40	(2.5^4)
5.0	100	(2.5^5)
10.0	10,000	(2.5^{10})

*If the difference in magnitude between two stars is 3, we receive 16 times more light from one star than from the other.

Figure 14–3. *Apparent magnitudes of a number of objects.*

EXAMPLE

The star W Pegasi is within the square of the constellation Pegasus (Figure 14–4). This star varies in brightness, getting as bright as eighth magnitude. Fomalhaut (the bright star just south of Aquarius) is a first-magnitude star. Which star appears brighter, and how many times more light do we receive from that star than from the other?

Solution

To answer the first question, remember that the star with the lesser magnitude is the one that appears brighter. Thus Fomalhaut is the brighter star. The difference in magnitude provides the information we need to answer the second question. The difference is 7. Table 14–1 does not list a difference of 7, but

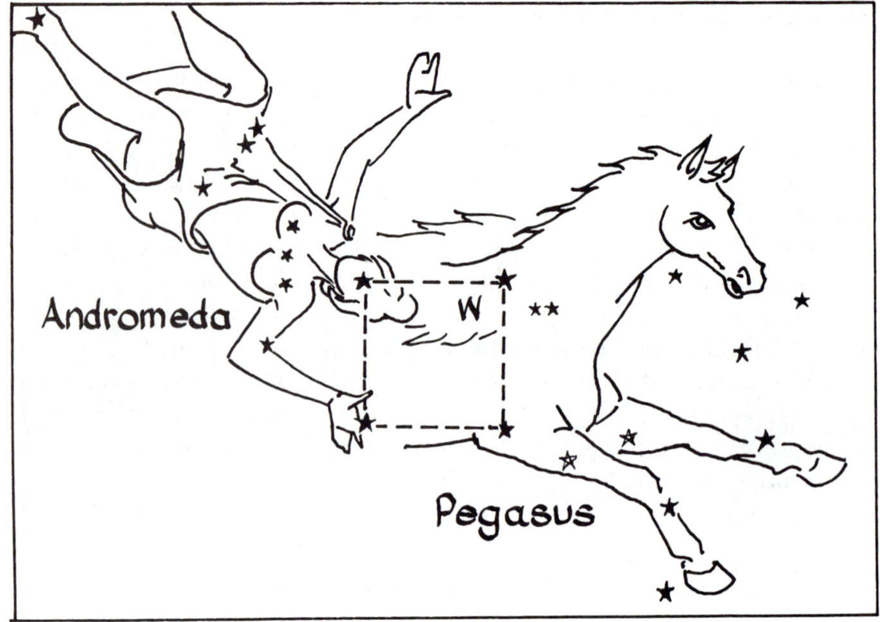

Figure 14–4. *The great square of Pegasus is an obvious feature of the fall sky. The constellations Pegasus and Andromeda share one bright star, Alpheratz, shown as an eye of Andromeda. The star W Pegasi is located approximately at the* W.

Fomalhaut actually has a magnitude of 1.19 and W Pegasi gets as bright as magnitude 7.9. We have rounded the magnitude values for this example.

we can use the values in the table to determine what the light ratio is. To do this we note that a difference of magnitude of 5 corresponds to a light ratio of 100 and a difference of 2 corresponds to 6.3 Thus a difference of 7 means that the ratio of light received is 100 × 6.3, or 630. We receive 630 times as much light from Fomalhaut than from W Pegasi (when the latter is at its brightest).

Try One Yourself. Barnard's star has an apparent magnitude of about 10. Suppose that on some particular night Mars is measured to have an apparent magnitude of 2. How many times more light do we receive from Mars than from Barnard's star?

As indicated in the example, telescopes allow us to see stars that are much dimmer than sixth magnitude. A telescope with a diameter of 12 centimeters (about five inches) might permit one to see, under perfect conditions, stars of about thirteenth magnitude. With a five-meter telescope one can photograph stars as dim as twenty-fifth magnitude.

It is valuable to know the brightnesses—the apparent magnitudes—of stars, but this is not a quantity that is inherent in the stars. Rather it depends partially upon their distances from Earth and therefore upon the positions of the Sun and the Earth in the galaxy. We would like to know the *actual luminosity* of each star, for this would tell us something about the star itself independent of the Earth's location. To do this, however, we must know the distance to the star.

DISTANCES TO STARS—PARALLAX

Recall (from Chapter 2) that in our discussion of the heliocentric/geocentric debate, the fact that stellar parallax had not been observed was an argument against the heliocentric system. Galileo, however, held that it was not observed simply because the stars are too far away. Not until the mid-1800s was stellar parallax first observed, for the maximum parallax angle of the nearest star is only about 1.5 seconds of arc. Although the angle shown between lines from opposite sides of the Earth's orbit toward that star is 1.52 seconds, astronomers define the parallax angle as half of that, 0.76 seconds (so that there is a straightforward application of the properties of right triangles—see Figure 14–5).

It is interesting that at the time of Galileo, the fact that parallax could not be observed was used as an argument against the heliocentric theory. Now its observation is an argument for the theory.

The accompanying Close Up shows that the formula used to measure the distance to a star by parallax is the same basic formula we used in Chapter 6 to calculate the size of the Moon. The formula can be written so as to express the distance to a star in light years as follows:

$$\text{distance to star (light years)} = \frac{3.26 \text{ light years}}{\text{parallax angle in seconds of arc}}$$

Astronomers usually prefer, however, to express it in a different distance unit, the ***parsec.*** With distance in parsecs, the equation becomes

Parsec: The distance of an object that has a parallax angle of one arcsecond.

$$\text{distance to star (parsecs)} = \frac{1}{\text{parallax angle in seconds of arc}}$$

In fact, this equation defines the parsec: One parsec (abbreviated pc) is the distance to a star that has a parallax angle of one second. The parsec is the unit astronomers normally use to express stellar distances. One parsec is equal to about 3.26 light years, so a light year is about one-third of a parsec.

Only stars within about 400 light years have parallax angles large enough to allow accurate calculations of their distances. Since it is often important to know the distance to an astronomical object to determine other quantities about it, developing techniques to measure greater and greater distances is a major

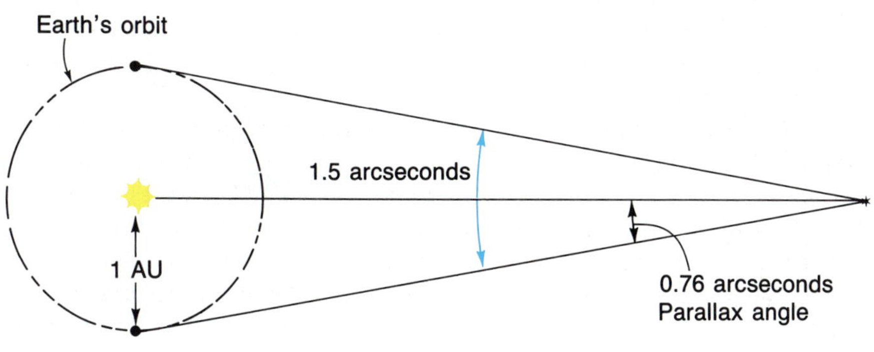

Figure 14–5. *The nearest star has a maximum parallax angle of about 1.5 arcseconds. The parallax angle is defined as half of this, using one astronomical unit as the baseline. The drawing is far out of scale.*

CLOSE UP

The Small Angle Formula

THE SAME BASIC FORMULA, involving the use of small angles, is used at least four ways in astronomy. The origin of the formula is found in geometry and the formula can be changed into its various forms using simple algebra.

Figure T14–1 shows a small angle, θ (theta), as part of a complete circle. Recall that the circumference of a circle is calculated by $2\pi r$, where r is the radius of the circle. The angle subtends arc distance A (colored in the drawing), which is the same fraction of the entire circle as angle θ is of 360 degrees:

$$\#1 \qquad \frac{A}{2\pi r} = \frac{\theta}{360^\circ}$$

As long as only small angles are involved, the distance A along the arc is very close to the same as the distance W straight across the ends of the arc, as shown in Figure T14–2.

In using the equation to determine the size of the Moon, the Moon's width is simply the distance across the arc and the distance to the Moon is the radius of the circle. Putting in values for π and using W for the width of the Moon and d for the distance to the Moon, we get

$$\#2 \qquad \frac{W}{2(3.14)d} = \frac{\theta}{360^\circ}$$

Solving this for the width of the moon

$$\#3 \qquad W = \frac{d\theta}{57.3}$$

$$= .0175\, d\theta \qquad (\theta \text{ in degrees})$$

W and d can be expressed in any distance units, but they must be the same. Thus, if we express the distance to the Moon in kilometers, the width of the Moon must also be in kilometers. The first part of the expression is the same equation we used in Chapter 6 to calculate the Moon's size.

In using this equation for planet sizes (or for the tangential distance between binary stars), the angle is usually expressed in seconds of arc instead of degrees. Using θ in seconds in the first equation above, 360 degrees becomes 1,296,000 seconds (which is 360 degrees × 60 min/degree × 60 seconds/min). W now stands for the diameter of the planet. We get

$$\#4 \qquad W = \frac{d\theta}{206{,}000}$$

$$= .00000485\, d\theta \ (\theta \text{ in arcseconds})$$

This is the equation shown in Chapter 7 for the size of planets. (The number 206,265 is a more accurate value to use in the denominator when more precision is justified.)

The equation above also is the form that might be used to calculate the tangential velocity of a star. In this case, W is the velocity in km/sec and θ is the star's proper motion in seconds of arc per second of time. Normally, though, proper motion is expressed in seconds of arc per year. Thus we must include a factor for the number of seconds in a year (3.15×10^7) and use v for velocity. We obtain

$$v = \frac{d\theta}{3.15 \times 10^7 \times 206{,}000} = \frac{d\theta}{6.50 \times 10^{12}}$$

$$\#5 \qquad v = 1.54 \times 10^{-13}\, d\theta$$

(v in km/s, θ in s/yr, and d in km)

When parallax is used to measure the distance to a star, the terms in the equation take on different meaning as illustrated in Figure T14–3. Here, W is the

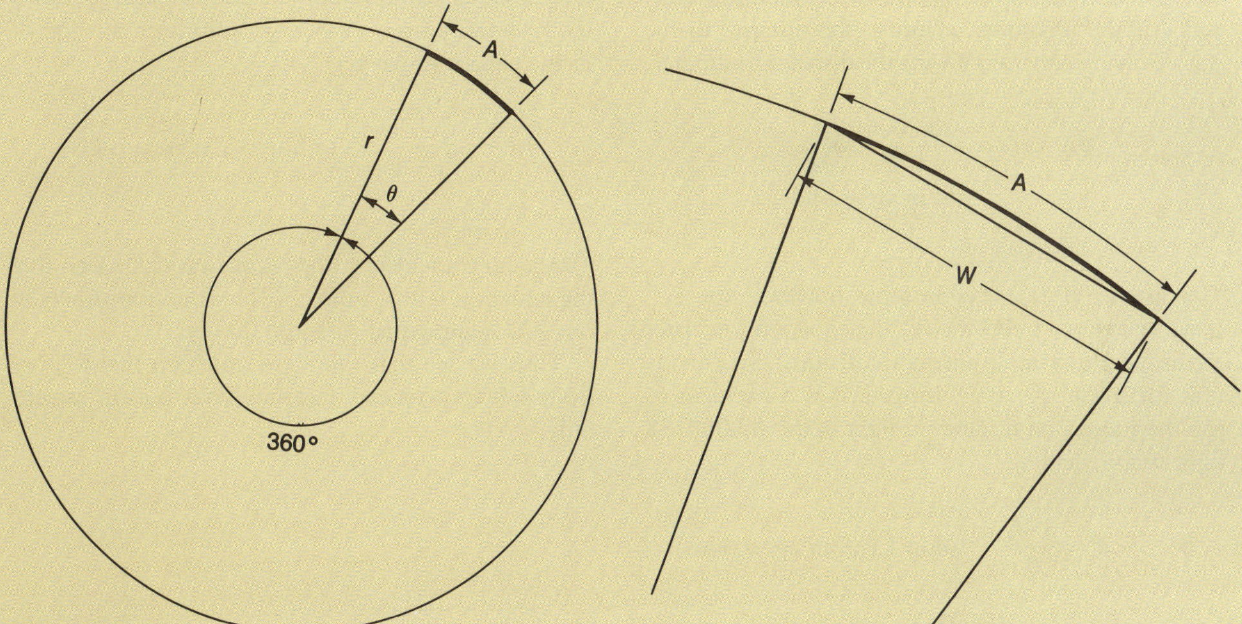

Figure T14–1. *The arc subtended by the angle θ is the same fraction of the entire circumference as angle θ is of 360°.*

Figure T14–2. *For small angles, the arc distance* A *is very nearly equal to the straight line distance between its ends* (W).

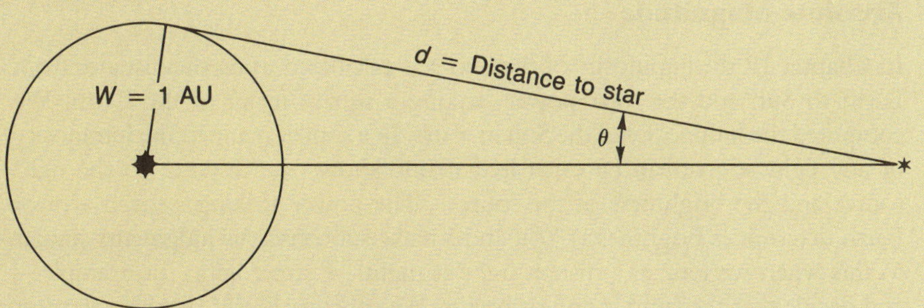

Figure T14–3. *Angle θ is very small since distance d is much greater than one astronomical unit. Therefore the equation worked out for small angles applies here.*

(continues)

CLOSE UP

distance from Earth to Sun (one astronomical unit) and d is the unknown quantity (the distance to the star). Solving equation #4 for the distance to the star,

$$\#6 \qquad d = \frac{206{,}000\ W}{\theta} \qquad (\theta \text{ in arcseconds})$$

The distance W is always the same, however: one AU. If we substitute 1 AU for W, the equation calculates the distance to a star in astronomical units; not a handy unit for stellar distances. Instead, it is convenient to use the Earth-Sun distance in light years (0.0000158 light years) yielding

$$\#7 \qquad d = \frac{3.26}{\theta} \qquad (d \text{ in LY}, \theta \text{ in arcseconds})$$

On the other hand, the distance from Earth to Sun is 0.00000485 parsecs. When we substitute this for W in the equation, we get

$$\#8 \qquad d = \frac{1}{\theta} \qquad (d \text{ in pc}, \theta \text{ in arcseconds})$$

This is no coincidence; the parsec was defined so that the equation is this simple. (The definition results in one AU being equal to 1/206,000 pc.)

Thus we see that one basic equation that follows from simple geometry has numerous uses in astronomy.

endeavor in astronomy. Later we will explore methods of measuring distances to more remote objects; right now we will see how knowing the distance to a star allows us to calculate its luminosity.

Absolute Magnitude

In Chapter 12 the luminosity of the Sun was calculated using the distance from Earth to Sun and the solar power striking a square meter of the Earth. We computed the luminosity of the Sun in watts. In a similar manner, the luminosity of any light source can be calculated if one knows the distance to the light source and the brightness of the source. (The power striking a given area of Earth determines brightness.) You and I make subconscious judgments similar to this when we look at a distant light at night—a street light, for example—and decide that the lamp is dim or bright. We subconsciously take into account how bright the light appears as well as its distance from us in order to decide (qualitatively) the wattage of the bulb. If we had the tools to measure brightness and distance quantitatively, we could calculate the lamp's power in watts.

Power: The amount of energy exchanged per unit time. *In Chapter 12 we spoke of the amount of energy that strikes the Earth in one second. The power of the radiation striking the Earth means the same thing.*

Figure 14–6 emphasizes that if we know any two of the following three factors we can calculate the third: apparent brightness (or, in astronomy, apparent magnitude), luminosity, and distance. The connection between these three will be used a number of times in the chapters to come. Brightness is most easily measured and it is always one of the two known quantities. When the other known quantity is distance we can calculate the luminosity; when the luminosity is known, we can calculate the distance.

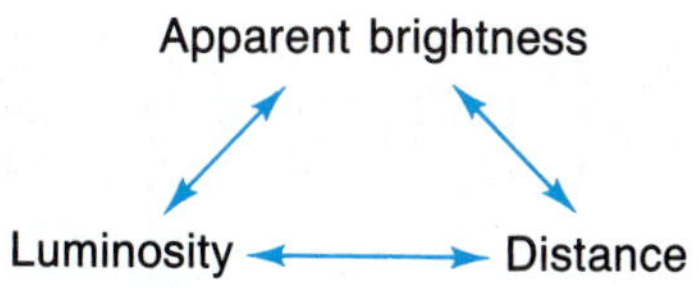

Figure 14–6. *If any two of the three quantities are known, the other can be calculated.*

The triple connection between brightness, distance, and luminosity allows astronomers to calculate the luminosity of some 1000 nearby stars, for that many stars are close enough to determine their distance fairly accurately by parallax. There are times when we want to know a star's luminosity in watts, but very frequently astronomers use another method to state the intrinsic luminosity of a star. The ***absolute magnitude*** of a star is defined as the apparent magnitude that the star would have if it were located ten parsecs from the Earth. We will not calculate values of absolute magnitude from apparent magnitude and distance, but will show one example to illustrate the relationship.

Absolute magnitude: The apparent magnitude a star would have if it were at a distance of 10 parsecs.

Sirius has an apparent magnitude of −1.47 and is 2.7 parsecs from Earth. To determine its absolute magnitude, imagine that we were to move it to a distance of ten parsecs. We would then be increasing its distance from us by a factor of about four ($^{10}/_{2.7}$). The inverse square law tells us that if we move something four times farther away, it will appear one-sixteenth (which is $1/4^2$) times as bright. Now refer to Table 14–1 and notice that a brightness ratio of 16 corresponds to a magnitude change of 3. We therefore add 3 to Sirius's apparent magnitude of −1.47 and obtain a value of +1.53 for its absolute magnitude. Appendix I lists Sirius's absolute magnitude as +1.45; very close to our result.

The error in our value occurred because we rounded $^{10}/_{2.7}$ to 4.

MOTIONS OF STARS

You sometimes hear reference to the "fixed stars." In fact, the stars are not fixed but are moving relative to the Sun. This motion is obviously not visible to the naked eye, for the constellations have retained their shape fairly well over the centuries. In 1718, Edmund Halley discovered, however, that stars do move with respect to one another and therefore constellations do gradually change their shapes.

Figures 14–7 and 14–8 are two photos, taken 22 years apart, of a greatly magnified portion of the sky. The arrows point to a particular star, Barnard's star, that has moved noticeably during that time.

Barnard's star is the second closest star to the Sun and shows the greatest motion as observed from Earth. It moves at the rate of 10.25 seconds of arc per year. Motion expressed as the angle through which a star moves each year is called the ***proper motion*** of the star. (You might speculate as to how a star could have *improper* motion, but the name comes from an old use of the word "proper" to mean "belonging to," for this is the motion that actually belongs to the star as opposed to observed motion that is due to Earth's movement.) Only relatively nearby stars show proper motion, so that the background stars might still be called *fixed* stars although they only appear to be fixed because

Proper motion: The angular velocity of a star as measured from the Sun.

Figure 14–7. *The arrow indicates a star named Barnard's Star.*

Figure 14–8. *This photo, taken 22 years later, shows that Barnard's star has moved noticeably during that time.*

UFO reports often state the speed of the unidentified object, although the proper motion of the object is what is actually observed. This is one example of how a poor understanding of astronomy can lead to inaccurate conclusions concerning celestial objects.

their proper motion is too small to detect. Figure 14–9 shows how proper motion causes the Big Dipper to change its shape over thousands of years.

Proper motion does not tell us the actual velocity of a star in normal units of velocity. We can calculate the speed (or velocity—the distinction is not important here) of a star by using the small angle formula we used to determine the Moon's size and to measure distances to stars by parallax. The first Close Up in this chapter shows the form of the equation that will yield units of kilometers per second. Of course, we must know the distance to the star to make such a calculation. In doing the calculation, we are computing only the speed of the star *across* our line of sight; its *tangential velocity.*

The velocity of a star toward or away from the Earth—its *radial velocity*—is easier to detect and measure than its tangential velocity. In Chapter 11 we discussed how this measurement is made and emphasized the fact that the Doppler Effect measures only the star's radial velocity.

Space velocity: The velocity of a star relative to the sun.

A star's actual motion relative to the Earth—its ***space velocity***—is a combination of its radial and tangential velocities. Since these two are at right angles to one another, they must be added as shown in Figure 14–10 to calculate the star's space velocity.

Naturally, the Earth's movement in its orbit affects the observed motions of stars and must be taken into account in calculating the velocity of a star. Once this is done, we have the star's velocity relative to the Sun. We will see later that there are methods for determining the Sun's movement relative to the distant, "fixed" stars. If this motion is taken into account, we can determine a star's velocity relative to the distant stars.

50,000 years ago

Today

50,000 years from now

Figure 14–9. *The center drawing shows the Big Dipper as it is today; illustrating the proper motions of its stars. From its motion, we can conclude that it once had the shape shown in the top photo and will someday have the shape shown in the bottom one.*

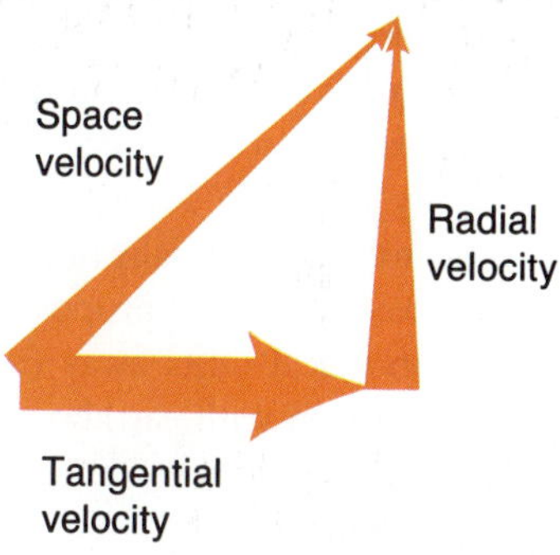

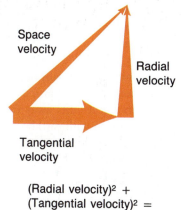

(Radial velocity)^2 +
(Tangential velocity)^2 =
(Space velocity)^2

Figure 14–10. *Since radial velocity and tangential velocity are at right angles, the Pythagorean theorem is used to add them.*

STELLAR COLOR AND TEMPERATURE

As we have seen, the temperature of a star can be determined from the star's color. More specifically, the wavelength at which maximum energy is emitted from the star provides us with an accurate temperature measure. Figure 14–11 is the intensity/wavelength graph of the Sun, showing that the line representing its energy output peaks near the center of the visible region. The graph indicates a temperature of about 6000° C for the surface of the Sun.

Another method of analyzing the spectrum provides an independent measure of temperature. This method depends upon the absorption of radiation at various wavelengths; the absorption spectrum. As we have seen, the absorption

CLOSE UP

A Long-Range Proposal

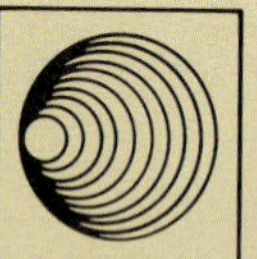

WE SEE IN THIS chapter the importance of knowing the distances to stellar objects and that parallax provides the most direct method of measuring those distances. As telescopes and techniques improve, astronomers are able to observe stellar parallax at greater and greater distances. The procedure is limited, however, by the baseline of the parallax triangle: the diameter of the Earth's orbit. Scientists and engineers at the Jet Propulsion Laboratory (JPL) in Pasadena, California have proposed that we use a specially designed space probe to extend the baseline. They propose sending a telescope 1000 astronomical units away from Earth and using it in conjunction with earthbound telescopes to allow parallax measurements of stars much more distant than is possible with our present short baseline. The project is called TAU for Thousand Astronomical Unit.

The JPL group hopes for a launch of the telescope in 2005, but 50 years would pass before the telescope reaches its destination. Thus it would not be in operation during its designers' lifetimes. Such a long-term project is exciting to some but others argue that it is not politically realistic to ask for funding for something that will not bear fruit for so long.

TAU probably has a potential payoff date farther into the future than any other project seriously proposed for NASA thus far but it is only one of a number of space endeavors which, by their nature, cannot be completed during the terms of the elected officials who determine their fates. Somehow our political systems must provide for such long-range planning.

of certain wavelengths is what allows us to determine the chemical elements in a star's atmosphere but the absorption by an atom depends not only upon what element it is, but also upon the state of its electrons (that is, whether some have been moved to higher energy levels or stripped from the atom). In a gas at higher temperature, more atoms will be at higher energy levels and the transitions that occur from these atoms will be different from the transitions that take place in atoms of a cool gas (whose atoms are at lower energy levels). This provides a method of determining temperature other than by the intensity/wavelength graph.

Annie Jump Cannon (Figure 14–12), an astronomer at Harvard College Observatory, devised a classification scheme for spectra and she separated several hundred thousand stars according to their spectral type. The classes were first labeled alphabetically but later their order was rearranged by temperature and some classes were dropped. The spectral types used today are designated, from hottest to coolest, as O B A F G K M (Figure 14–13).

The hottest stars, O stars, range in temperature from 30,000° C to 60,000° C. The coolest stars, on the other hand, are the M stars, with temperatures less than 3500° C.

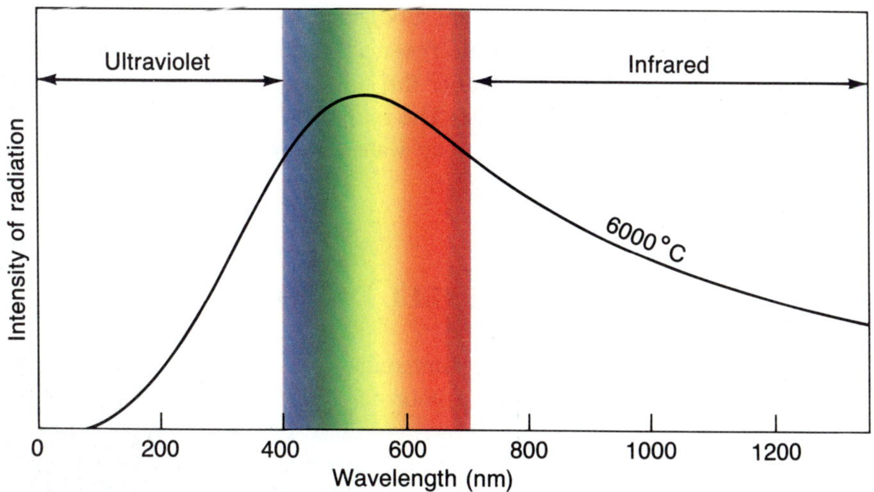

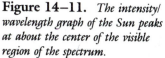

Figure 14–11. *The intensity/wavelength graph of the Sun peaks at about the center of the visible region of the spectrum.*

Figure 14–12. *Annie J. Cannon was a member of the Harvard College Observatory for almost 50 years and was the founder of the spectral classification scheme in use today.*

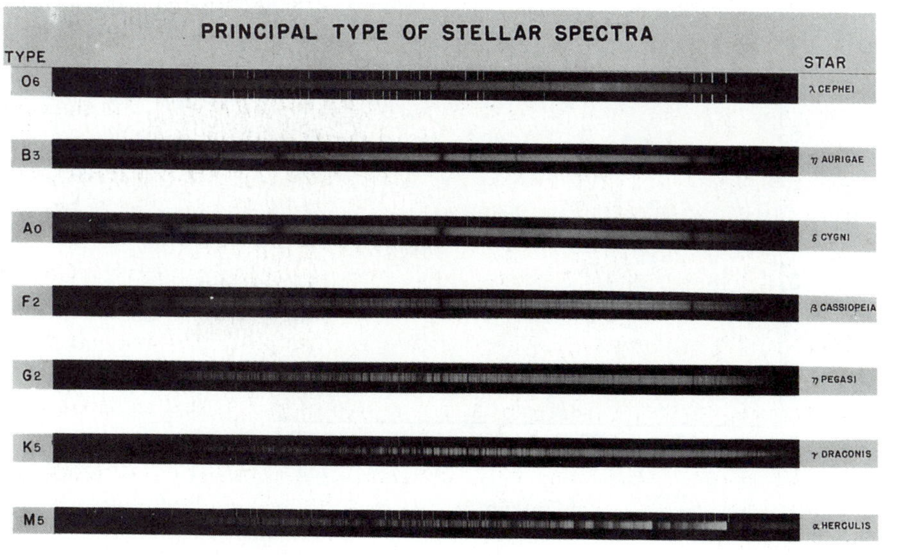

Figure 14–13. *The spectra of stars of various classes. The classes are shown at left and the stars are identified at right..*

Within each spectral class, stars are subdivided into ten categories by number, so our Sun is listed as a G2 star. You need not be concerned about these subdivisions but you should remember the classification scheme from hottest to coolest as being O B A F G K M. The traditional mnemonic is "Oh, Be A Fine Girl, Kiss Me" although you may want to devise your own less sexist memory aid (or substitute *guy* for *girl*).

THE HERTZSPRUNG-RUSSELL DIAGRAM

Suppose that an extraterrestrial being, unfamiliar with humans, somehow learns the age and height of each person in your neighborhood and then uses these values to plot a graph such as that shown in Figure 14–14. The first thing that he would notice is that a pattern exists; that there is some relationship between height and age for humans. Then if our imaginary alien is very analytical, an important fact about the life cycle of humans could be learned even though no one person had been analyzed through his or her entire life. The alien could logically hypothesize that an individual spends most of his or her life at about the same height; that height being the tallest achieved by the person. In our study of stars, we have a similar chart; the ***Hertzsprung-Russell Diagram.***

Hertzsprung-Russell Diagram: A plot of absolute magnitude versus temperature (or spectral class) for stars.

Early in this century, Ejnar Hertzsprung, a Danish astronomer, and Henry Norris Russell, an American astronomer at Princeton University, independently developed the diagram now named for them and often called simply the H-R Diagram. The diagram shows that a pattern exists when stars are plotted by two properties: their temperature (or spectral class) and their absolute magnitude. Figure 14–15 is similar to the diagram Russell plotted in 1913. The stars are not evenly distributed over the entire chart, but seem to group along a diagonal line. If we include many more stars than those on Russell's first diagram, we obtain a plot as shown in Figure 14–16. (Notice the Sun's position on the diagram.)

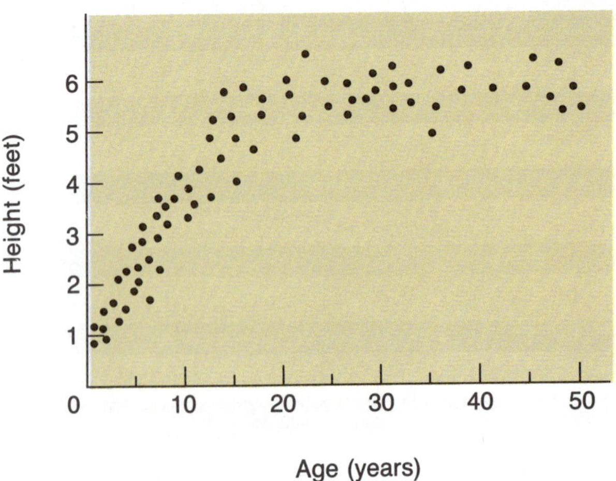

Figure 14–14. *A plot of the age versus height of people in a neighborhood might look like this. Each dot represents one person.*

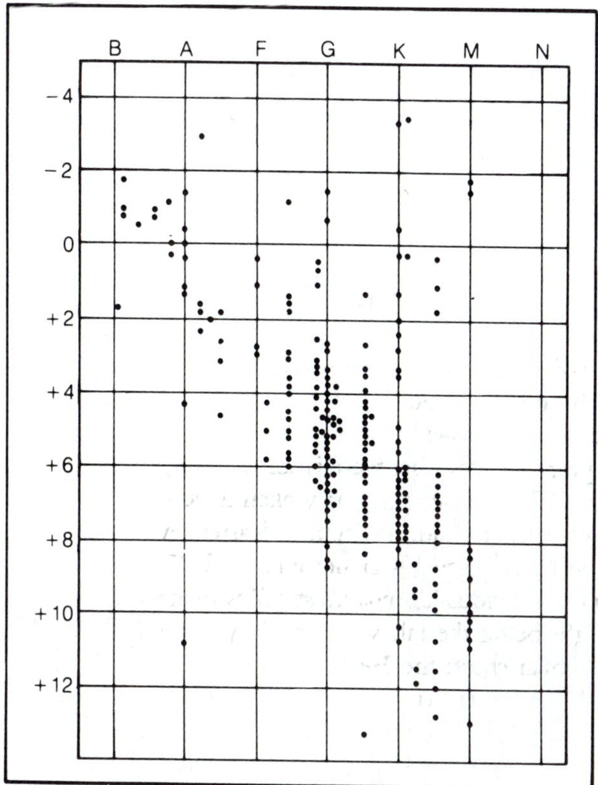

Figure 14–15. *This is similar to the first H-R diagram, plotted by Henry Norris Russell in 1913. He plotted absolute magnitude versus spectral class and saw an obvious pattern in the distribution of stars. Notice that his category labeled N contained no stars.*

About 90 percent of all stars fall into a group running diagonally across the diagram and are called ***main sequence*** stars. Other groups are named as shown in the diagram. We will discuss the significance of the groups later in this chapter and in the next few chapters, when we describe the life cycle of stars.

Main sequence: The part of the H-R diagram containing the largest majority of stars, forming a diagonal line across the diagram.

Spectroscopic Parallax

The distances to the nearest stars can be measured using stellar parallax. The H-R diagram provides us with another method of measuring such distances; a method not confined to neighboring stars. Although this method does not involve parallax, it is called ***spectroscopic parallax.***

Spectroscopic parallax: The method of measuring the distance to a star by comparing its absolute magnitude to its apparent magnitude.

It is possible to determine from the spectrum of a star whether that star is a main sequence star, a giant (or a supergiant; the term *giant* is often used to cover both), or a white dwarf. Suppose that we observe a particular star and determine that it is a main sequence star. As we have seen, it is a fairly routine procedure to ascertain the temperature of a star. Let's suppose that our star has a temperature of 10,000° C. Refer to the H-R diagram of Figure 14–17. The region of the main sequence where stars have a temperature of 10,000° C is marked on the diagram. It is now a simple matter to determine that our chosen star has an absolute magnitude between about +1 and +2.

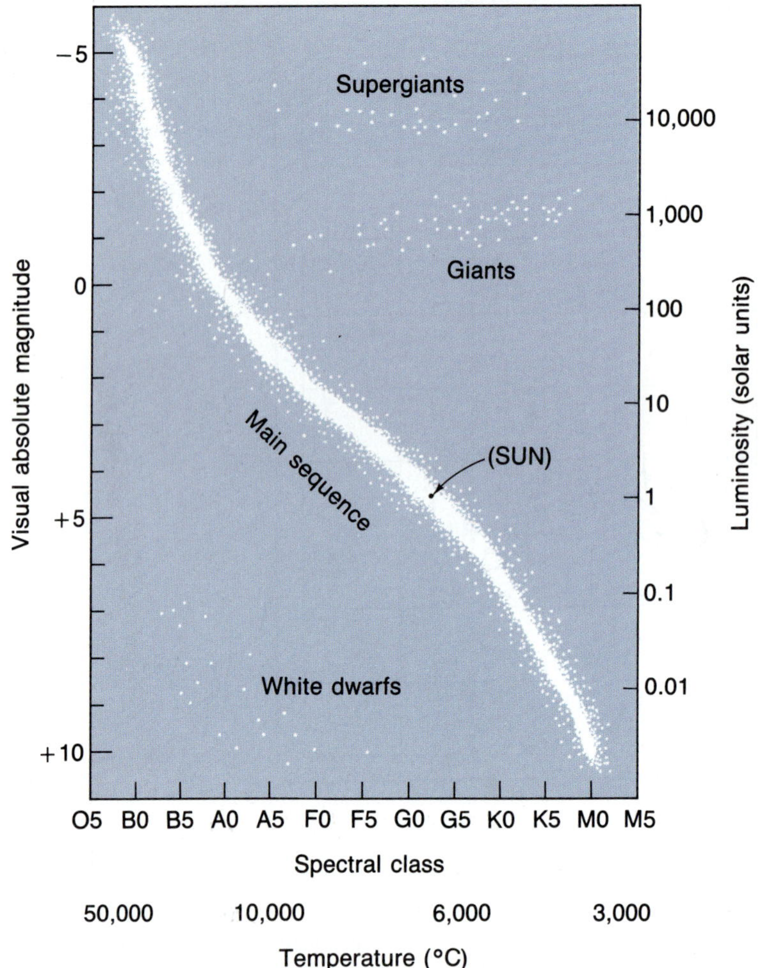

Figure 14–16. *A modern H-R diagram shows that stars fall into various categories, including main sequence stars, white dwarfs, red giants, and supergiants.*

Recall the connection between absolute magnitude, apparent magnitude, and distance. If we know two of these, we can calculate the third. Use of the H-R diagram has allowed us to determine, within a small range, the absolute magnitude of the star. Since apparent magnitude is directly measurable, we can calculate the star's distance.

As indicated, we know the absolute magnitude only within a certain range and therefore our precision in determining the distance is limited. Keep in mind, however, that this is *always* the case with a measurement. In using spectroscopic parallax, the source of the error is obvious but we have seen that the determination of distance by trigonometric (stellar) parallax is also limited in accuracy; both because of the difficulty in measuring the small angles involved and in the fact that other motions must be taken into account, such as the proper motion of the star being measured. All measurements contain error. The important thing in science is to be cognizant of how great the likely error is.

We will do a simplified example to illustrate the method of spectroscopic parallax.

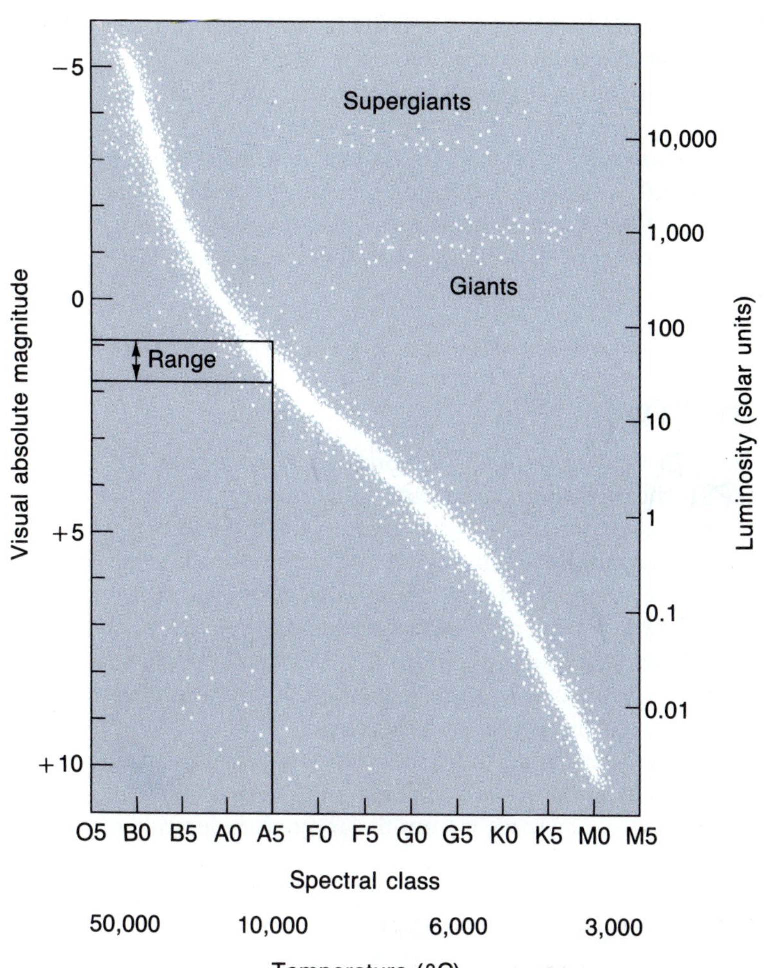

Figure 14–17. *The absolute magnitude of a main sequence star with a temperature of 10,000° C can be determined from the H-R diagram.*

EXAMPLE

Spica, the brightest star in the constellation Virgo, is a B1 type main sequence star with a temperature of about 20,000° C. Its apparent magnitude is 0.91. From this data, determine whether its distance from Earth is less than ten pc, approximately ten pc, greater than ten pc, or much greater than ten pc.

Solution

Referring to Figure 14–17 and using the fact that Spica is a B1-type star, we determine that its absolute magnitude is between about −3 and −4.5 (although a safer range might be from −2.5 and −5.0).

Next, let's compare our absolute magnitude values with Spica's observed apparent magnitude. Spica's apparent magnitude is 0.91, and we can reason that since its apparent magnitude has a value greater than its absolute magnitude, it is farther than ten parsecs away. (Remember, greater magnitude means a dimmer star and so its apparent brightness is less than its brightness would be at ten pc.) In fact, from the difference between +0.91 and −3 or −4, we can conclude that we would have to move it fairly far to bring it to a distance of ten pc. So it is much farther away than ten pc.

Actually, Spica is about 75 parsecs, or 240 light years, from us.

Try One Yourself. The star 40 Eridani is a main sequence star of spectral type K0, which means that its temperature is 5100° C. Its apparent magnitude is +4.4. (Actually, this is the magnitude of the brightest one of a triple star system. We will discuss multiple stars later.) What can you conclude about the distance to 40 Eradani?

Luminosity Classes

Luminosity class: A classification of a star by the amount of electromagnetic energy it radiates.

In the examples in the last section, we worked only with stars on the main sequence. In fact, the procedure can be used with any stars, but the ***luminosity class*** of the star must be determined first. Figure 14–18 shows the division of stars on the H-R diagram by luminosity class. This is normally indicated by a Roman numeral after the spectral class. Sirius is an A1V star, showing that it is a fairly hot star (A1) on the main sequence (luminosity class V); but Rigel (Figure 14–19 shows its location in Orion) is a B8Ia star, a bright supergiant. As we stated before in the case of main sequence stars, we can determine the luminosity class of a star by analysis of its spectrum.

The method for determining the distance to a non-main sequence star is essentially the same, then; but instead of reading the star's absolute magnitude from the main sequence, one reads it from the appropriate location on the H-R diagram.

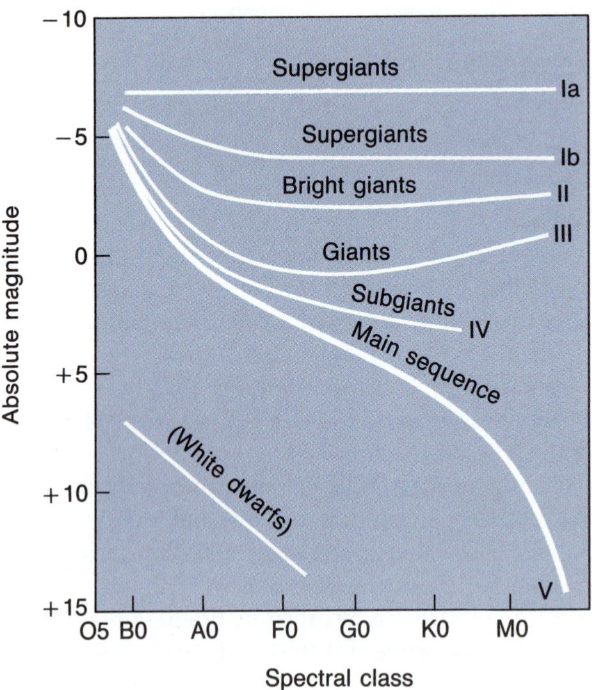

Figure 14–18. *Stars can be classified as to their luminosity class, which is determined by analysis of their spectrum. Luminosity classes are denoted by Roman numerals.*

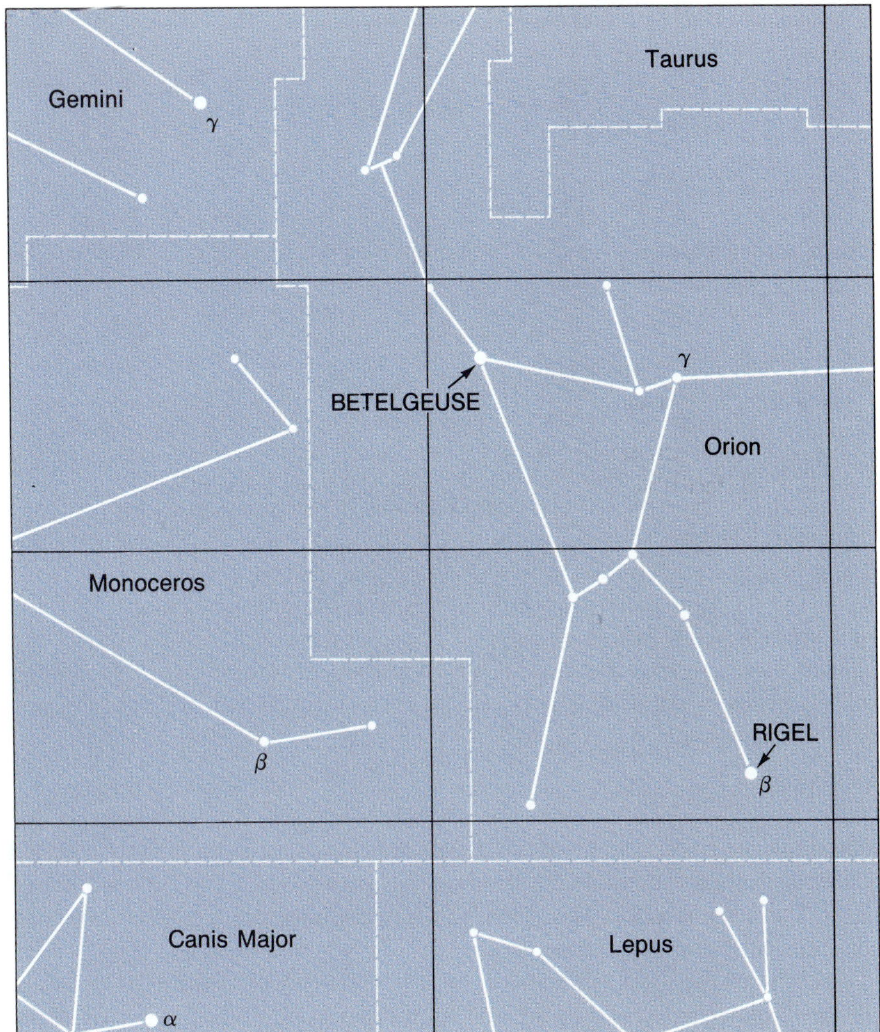

Figure 14–19. *Compare this drawing of Orion with Figure 14–1, a photo of the constellation. The brightest star (Betelgeuse) and the second brightest (Rigel) lie at opposite corners of the main figure of Orion.*

Analysis of the Procedure

Observe that in using the method of spectroscopic parallax to find a star's distance from us, the following two triple connections are used (Figure 14–20): By knowing the temperature of a star and the star's type, its absolute magnitude is determined; then, by knowing its absolute magnitude and its apparent magnitude, its distance is calculated. This last triple connection is used in a different way than it was before. For nearby stars, distance and apparent magnitude were used to determine absolute magnitude.

It is interesting that absolute magnitudes calculated for nearby stars enabled astronomers to draw the H-R diagram. Once the patterns of that diagram were known, they could be applied to stars that are too far away to permit distance measurement by parallax. It is reasonably assumed that these stars fit the same H-R diagram pattern as do nearby stars, so the diagram is then used to determine their absolute magnitude.

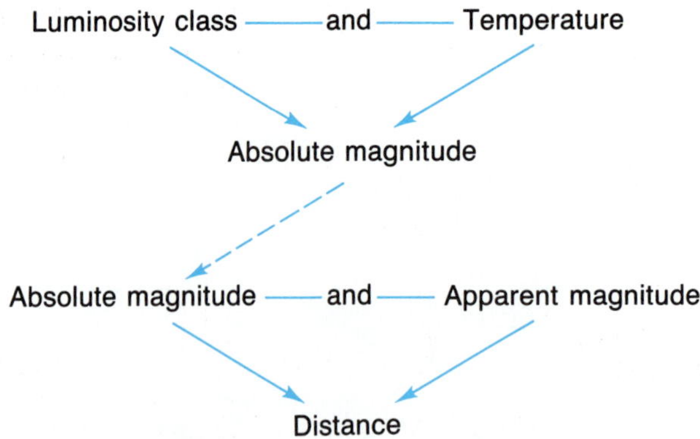

Figure 14–20. *From the luminosity class and temperature of a star we can determine its absolute magnitude. From this and its apparent magnitude, we determine its distance.*

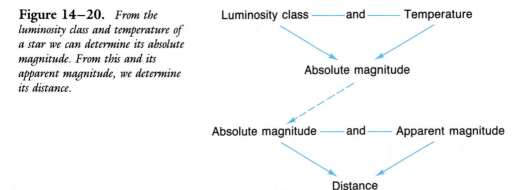

The basic procedure here is one used often in astronomy (and in other sciences). By observing familiar objects (nearby stars in this case), we see patterns and formulate laws (or statements of relationships). We then assume that these patterns and laws hold for more distant objects of the same type. This allows us to use the same patterns to learn more about those distant objects. At times it appears that the whole system is a house of cards ready to fall down, but usually there are cross-checks available; measurements can be taken in a number of ways, thereby permitting us to verify theories by independent measurements.

The Size of Stars

Although the sizes of a few of the largest stars have now been directly detected by interferometry methods, in general a star is observed to be only a point of light. It shows no size. How, then, can we determine the size of an object so distant that it appears to have no size?

White dwarf: A very small, hot star.

Consider the group of stars in the lower left corner of the H-R diagram, the ***white dwarfs.*** These stars are hot but their location in the lower part of the diagram indicates that they are intrinsically dim. It would seem that a hot star would be bright, for this is not only the pattern seen in the main sequence but it also follows from what we said about the intensity/wavelength graph: As an object becomes hotter, the wavelength of the maximum radiation emitted becomes shorter and the *total amount of radiation from the star increases.* How, then, can these stars be dim? The answer is that they are small. Being hot, each square meter of their surface emits more energy than a cooler star but they simply have small total surface areas. The name "white dwarf" indicates their temperature (white-hot) and their size.

Giant star: A star of great luminosity and large size.

Supergiant: A star of very great luminosity and size.

On the other hand, consider the ***giants*** and ***supergiants.*** How do we know that these are large stars? Simply because they are very bright in spite of their low temperatures. The giants are often called *red giants,* paralleling the name given the white dwarfs. (A single stove burner set to "warm" does not yield much light but if enough of these relatively cool burners were lit, they would provide quite a lot of red light.)

In the examples given of white dwarfs, giants, and supergiants, we reasoned qualitatively to learn something about their sizes. Determination of stellar sizes

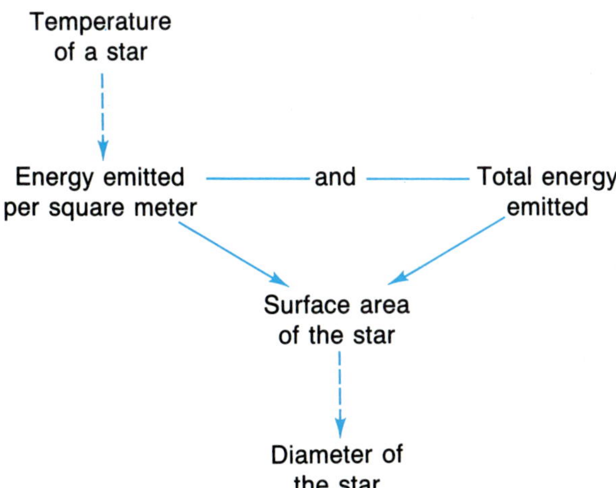

Figure 14–21. *If the temperature of a star and the total energy emitted by it are known, one can use the triple connection shown here to calculate its diameter.*

is not limited to these classes of stars, however; nor is it limited to qualitative methods. Knowing the temperature of an object, we know the amount of energy emitted per square meter of its surface. Then if we know the total energy emitted by the object (by knowing the absolute magnitude), it is easy to calculate the area of its surface and therefore its diameter. Figure 14–21 illustrates the triple connection in the chain of reasoning. We find that stars on the main sequence range from about 0.1 times the Sun's diameter up to as much as 20 times the Sun's diameter. Non-main sequence stars, however, differ in size even more; for white dwarfs may be as small as 0.01 of the Sun's size and supergiants may be 100 times larger in diameter than the Sun.

It is always desirable to have a second, separate method of measuring a quantity. This allows us not only to check our theories regarding the object being measured but also to check our measurement techniques. Fortunately, there is a second method for measuring the sizes of stars, although only a few stars can be measured by this technique. To see how this works, we must discuss multiple star systems.

MULTIPLE STAR SYSTEMS

A few decades ago astronomy books reported that about one-fourth of the objects that appear to be single stars really contain two or more stars in a close grouping. Books published a decade ago reported this as about one-third. A few years ago, it was considered to be about half. Now we can safely say that *more than half* of what appear as single stars are in fact multiple star systems. We are not speaking here of pairs of stars that appear close together as a result of being nearly in the same line of sight from Earth (Figure 14–22). These ***optical doubles,*** as they are called, are merely chance alignments of stars. The stars in which we are interested are gravitationally bound so that they revolve around one another. When two stars are gravitationally linked they are said to be a ***binary star system.*** Figure 14–23 illustrates how the two stars of a binary

Optical double: Two stars that have small angular separation as seen from Earth but that are not gravitationally linked.

Binary star system: A system of two stars that are gravitationally linked so that they orbit one another.

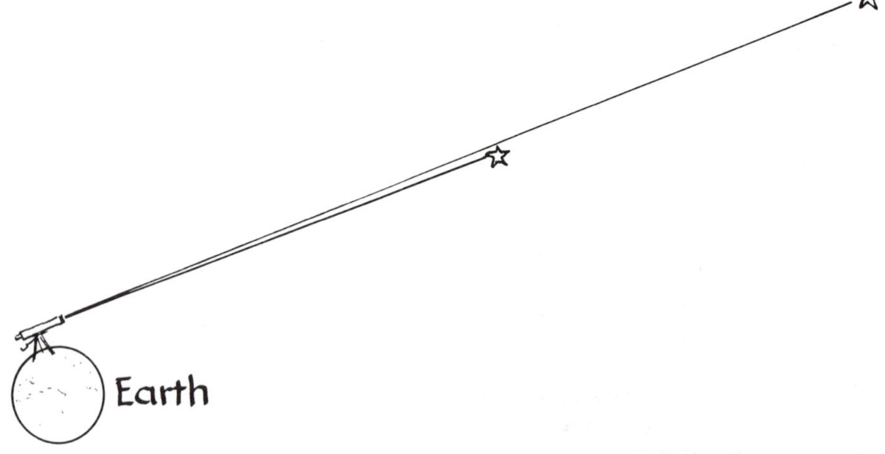

Figure 14–22. *Two stars are said to form an optical double if they appear close together but have no actual relationship to one another.*

pair revolve about their common center of mass, as discussed in a previous chapter for a star and a planet. Although groups of more than two stars are common, we will concentrate on binary systems since multiple star systems are simply combinations of a binary system and a single star or of two binary systems (Figure 14–24).

In 1802 William Herschel obtained the first observational evidence that some double stars orbit one another. (This was also the first direct evidence of gravitational force at work outside our solar system.) Herschel had taken an interest in double stars in hopes of using them for parallax measurements but his discovery of binary star systems turned out to be much more important to astronomy, for it is by such systems that we are able to determine the mass of stars.

Binary systems are classified into a number of categories according to how they are detected. We will discuss each in turn and then explore what can be learned from binary stars.

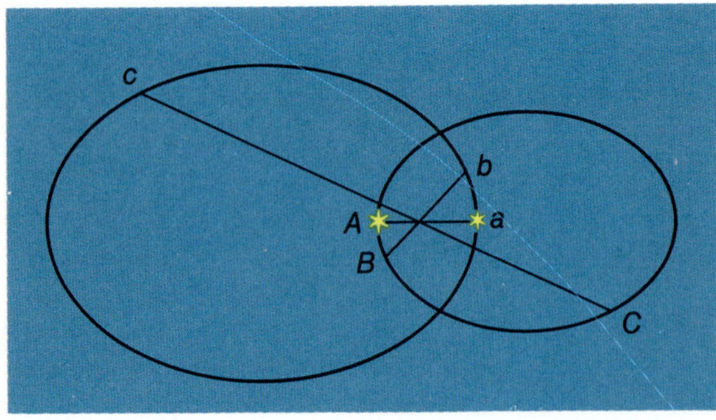

Figure 14–23. *Binary stars orbit their common center of mass. When one star is at* A, *the other is at* a, *then* B *and* b, *etc. Can you tell which is the more massive star?*

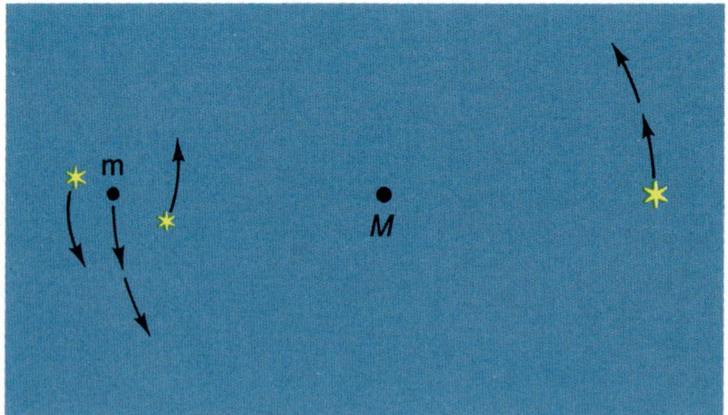

Figure 14–24. *A triple-star system. The two stars at the left orbit their center of mass (point* m). *This pair and the star at right orbit the overall center of mass at* M.

(a)

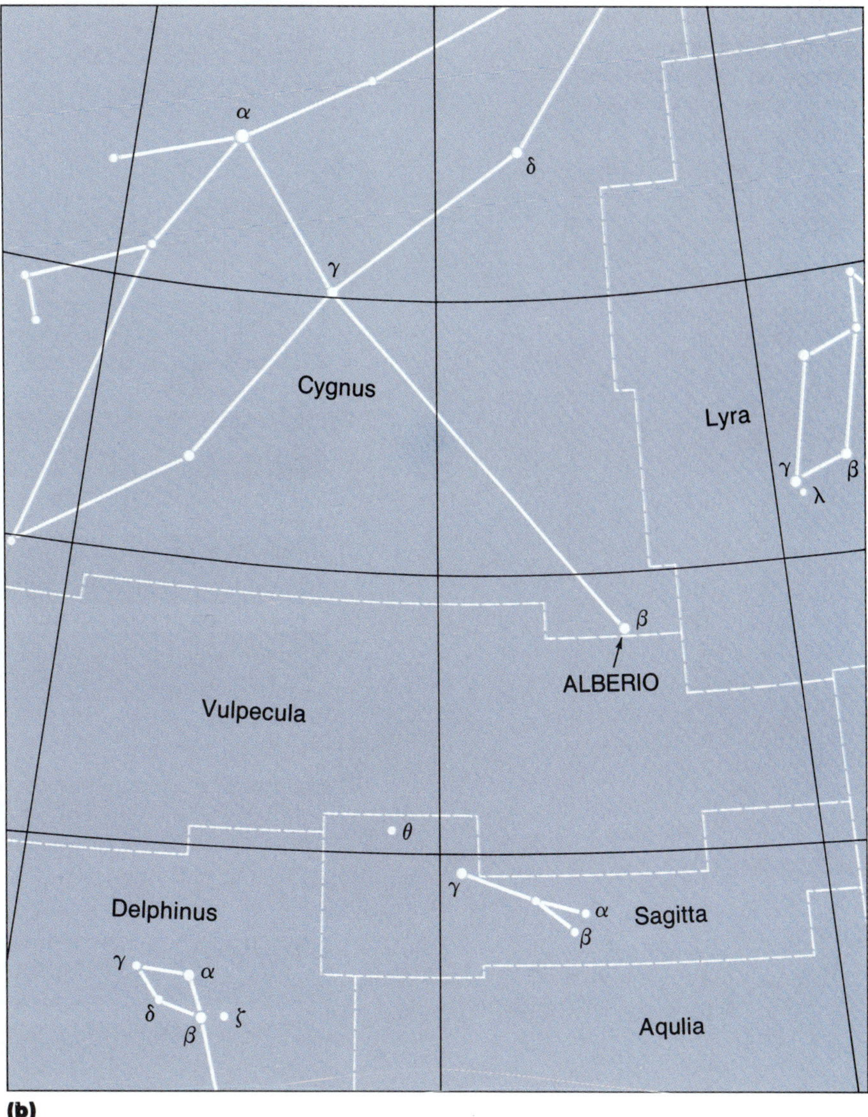

(b)

Figure 14–25. *(a) Albireo is a binary pair that shows an obvious color difference between the two stars. (b) Albireo is the bottom star of the cross of Cygnus. (Normally the designation β would mean that the star is the second brightest in the constellation (after α), but labeling in Cygnus does not follow the rule strictly and β Cygni and is the fourth brightest star in Cygnus.)*

Visual Binaries

A ***visual binary*** is one in which the pair of stars can be resolved as two in a telescope. Using the largest telescopes, perhaps 10 percent of the stars in the sky are visual binaries. In a small telescope, only a small fraction of these can be resolved but some are beautiful sights, particularly when the two stars are very different in color. Figure 14–25(a) is a photo of Albireo, a visual binary in Cygnus, taken by an amateur astronomer. If you have access to a telescope, find Albireo some clear summer night.

Visual binary: An orbiting pair of stars that can be resolved (normally with a telescope) as two stars.

If two stars appear close together in the sky, how can astronomers tell if they are a binary system? The surest way is simply to observe the pair over a period of time and look for signs of revolution. Figure 14–26 shows a binary (at the right in the photo) that reveals obvious orbital motion over a number

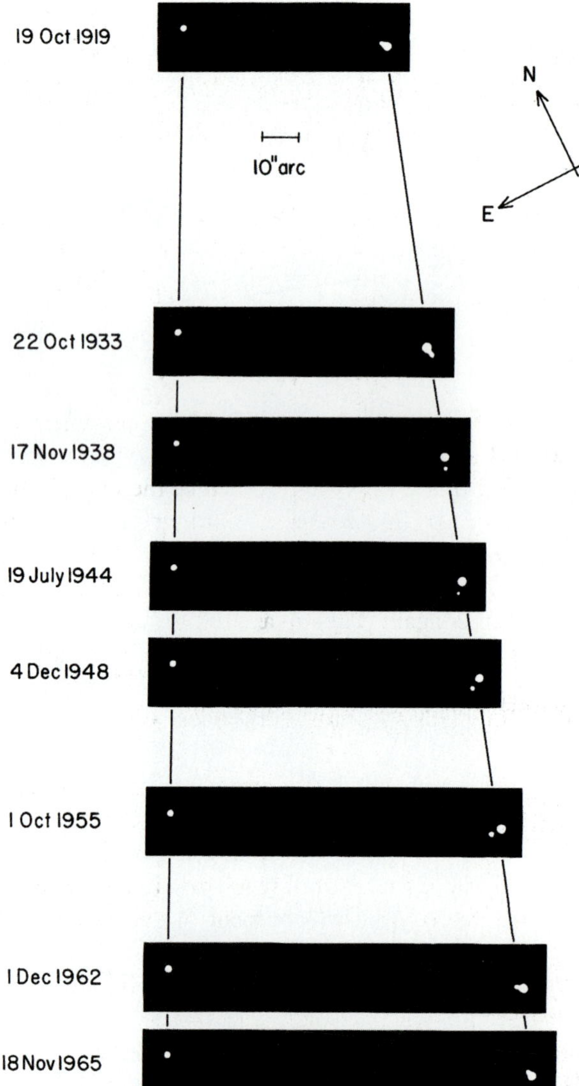

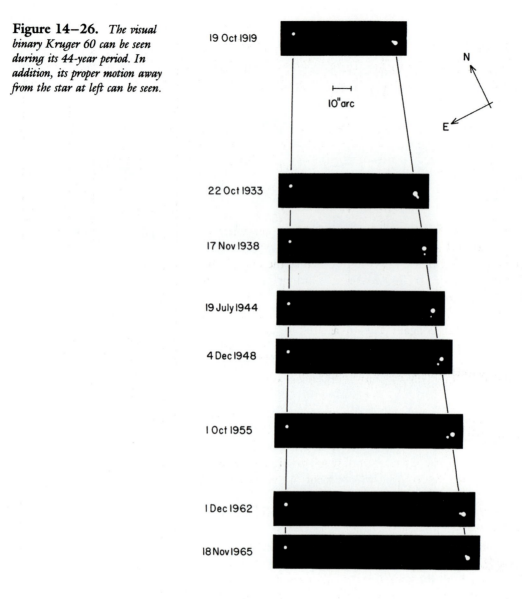

Figure 14–26. *The visual binary Kruger 60 can be seen during its 44-year period. In addition, its proper motion away from the star at left can be seen.*

of years. Things are not usually this easy, however; for in order for us to be able to resolve a binary pair, the stars must either be very close to Earth or be very far apart. The Albireo pair, for example, is separated by some 4500 AU. As indicated by Kepler's laws, a great distance of separation indicates a long period of revolution and the Albireo binary is thought to have a period of many tens of thousands of years. No detectable orbital motion has occurred in the relatively short time over which we have photographic records and it is therefore difficult to confirm that such a system is indeed a binary pair. We do know, however, that the two stars are about the same distance from us and it is therefore thought that they are gravitationally linked.

A second method of determining whether a pair is indeed a binary system employs that powerful astronomical tool, the Doppler effect, and leads us to the next type of binary system.

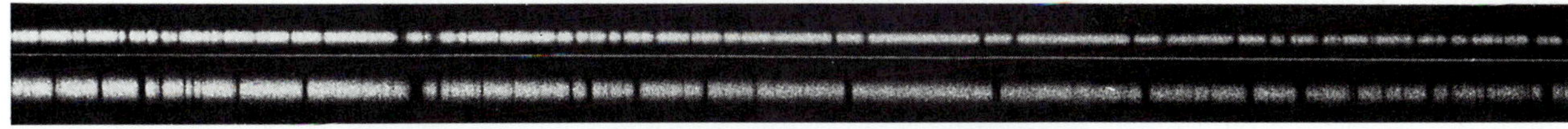

Figure 14–27. *Two spectra of the spectroscopic binary k Arietis are shown here. Notice that lines that are single in the bottom spectrum are split in the upper one. This is particularly evident for the lines to the right of center. When single lines appear, the two stars are moving at right angles to the line of sight and where they are double, one star is moving toward us and the other away.*

Spectroscopic Binaries

Figure 14–27 shows two spectra of the star κ (Greek symbol for Kappa) Arietis, taken at different times. The upper spectrum contains more lines than the lower. Look closely and you will see that this is the result of each of the lines in the lower spectrum having broken into two in the upper. Continuing observation of this star reveals that the spectrum repeatedly goes through a cycle in which each line gradually breaks into two, which spread until they reach a maximum separation, and then coming together again. The explanation for this is that we are seeing not one, but two stars and that they are revolving around one another so that the Doppler effect causes us to see separate spectral lines at the point where one star is moving toward us and the other away. Such a binary system is called a ***spectroscopic binary.***

Spectroscopic binary: An orbiting pair of stars that can be distinguished as two due to the changing Doppler shifts in their spectrum.

The Doppler effect shows only the radial component of the stars' motion; the motion toward or away from us. If a binary pair is oriented so that its plane of revolution is perpendicular to a line from Earth, no Doppler shift is observed in its spectral lines. On the other hand, if the plane of revolution is tilted directly toward the Earth, the Doppler effect allows measurement of the stars' actual velocities during the part of their orbits when one is moving directly toward us and the other directly away. At any other orientation of the plane of revolution, we detect only the radial component of the stars' motions.

In the handle of the Big Dipper there is a particularly interesting example of binary stars (Figure 14–28). As pointed out in the last chapter, good eyesight reveals two stars at that location, the brighter one named Mizar and the dimmer Alcor. Although it had been thought that Alcor and Mizar form an optical double rather than a gravitationally linked binary, more recent observations of the radial velocities of the two make it seem that they are orbiting one another. In any case, if you view Mizar and Alcor through a telescope, you will see that Mizar is itself two stars. Following the standard practice for naming double stars, we call the brighter of these two Mizar A and the other Mizar B. Mizar A and B are a widely separated visual binary and have a period of at least 3000 years.

In 1889 a Harvard Astronomer, Edward Pickering, found that Mizar A is a spectroscopic binary with a period of only 104 days. When it was discovered that Mizar A was binary, Mizar B was scrutinized. In this case, the spectral lines did not separate into two parts but instead they moved back and forth; first redshifted and then blueshifted. Mizar B is indeed a binary star but the spectrum of its companion is not bright enough to be observable. The shifting spectrum

Mizar B is called a single line spectroscopic binary system *rather than a double line system such as Alcor.*

Figure 14–28. *Even a small telescope reveals that Alcor and Mizar are actually three stars.*

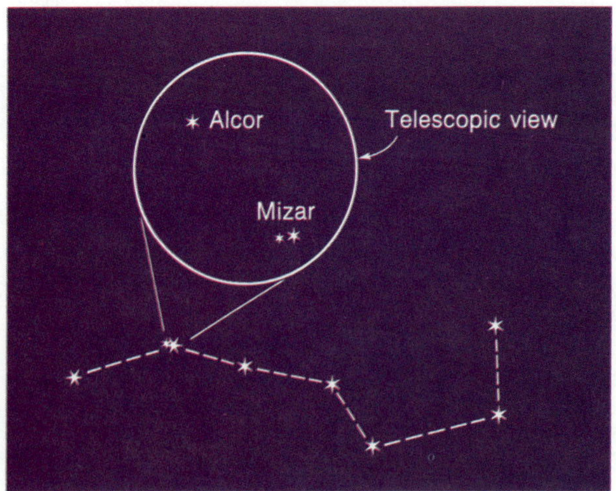

that is observed is the spectrum of the brighter star of the pair. We deduce the existence of the companion from that motion.

Finally, it has been found that Alcor is a spectroscopic binary. Thus the dot in the handle of the Big Dipper is in fact six stars (Figure 14–29).

Eclipsing Binaries

A special case of binary stars occurs when their plane of revolution is along a line from Earth so that one star moves in front of the other as they orbit. The star Algol, in the constellation Perseus, is a good example of an eclipsing binary. Ancient Islamic astronomers called Algol the *demon star,* apparently having observed that it periodically decreases its brightness. In 1783, John Goodricke, an amateur astronomer, explained the brightness change as being due to a dimmer companion passing in front of the brighter. Figure 14–30(a) shows a

Figure 14–29. *Spectroscopic evidence reveals that Alcor and Mizar actually form a six-star system.*

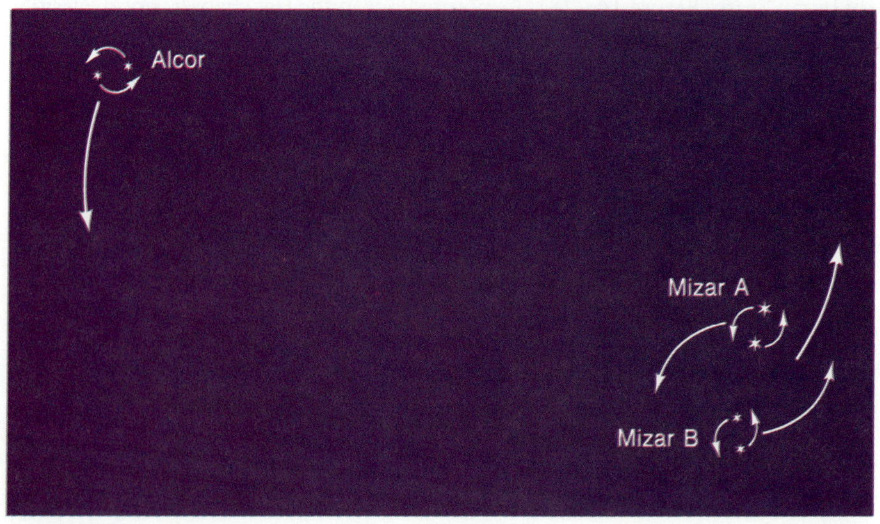

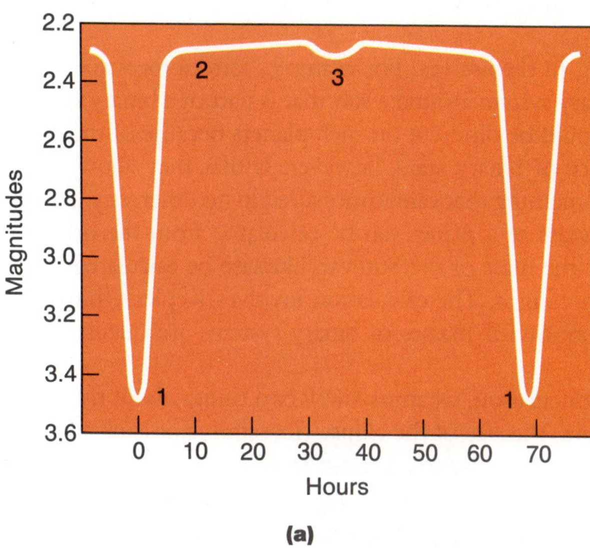

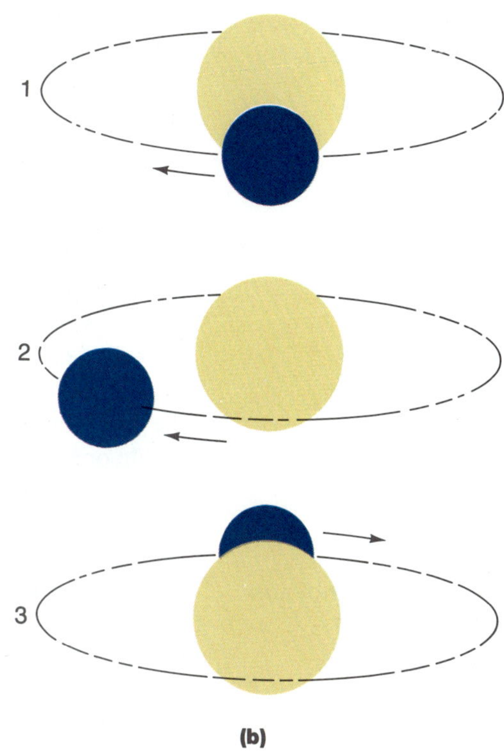

Figure 14–30. *The light curve of Algol (part a) is explained by the eclipsing of its components, one of which is much darker than the other. Positions 1, 2, and 3 correspond in the two drawings.*

graph of the light received from Algol as time passes, the system's ***light curve.*** Notice that every 69 hours the apparent magnitude of Algol changes from 2.3 to 3.5. This happens when Algol A is partially eclipsed—see Figure 14–30(b). Midway between these dips in the light curve smaller dips are seen, caused by the dimmer companion being eclipsed.

Light curve: A graph of the light received from a star versus time.

Since Goodricke's discovery that Algol is a binary star, its nature has been confirmed from spectroscopic evidence. Like Mizar B, its companion is too dim to show its own spectrum but the spectral lines of Algol A move back and forth in rhythm with its cycle of brightness changes. It can now be classed as both an eclipsing binary and a spectroscopic binary.

Other Binary Classifications

There are at least two other ways to detect binary stars. Sometimes a star is seen to shift back and forth in its position among the other stars, indicating that it is revolving around an unseen companion. A number of such systems have been found.

Astrometric binary: An orbiting pair of stars in which the motion of one of the stars reveals the presence of the other.

If one star of a binary system is much hotter than the other, their spectra differ enough from one another that it is possible to ascertain that the spectrum we see is not from a single star but is a composite of two spectra. This provides another method of detecting binary stars but nothing can be learned of the motions of the stars in such a system.

Composite spectrum binary: A binary star system with stars having spectra different enough to distinguish them from one another.

STELLAR MASSES AND SIZES FROM BINARY STAR DATA

Binary stars are interesting in themselves. For example, astronomers speculate on the stability of a planetary system around a star that is part of a binary system and on how conditions would be different on such planets because of the extra sun. The major importance of binary stars, however, is that they allow us to measure masses of stars; something that can be measured in no other way. Recall from Chapter 7 that the mass of a planet can be calculated from the orbit of one of its moons and that the mass of the Sun can likewise be calculated from the orbits of the objects circling it. The calculation involves Kepler's third law as revised by Isaac Newton. Stellar masses of binary systems are calculated in the same way.

In order to do such a calculation, we must know two things about the orbit of one (or both) of the stars: the size of the ellipse (its semimajor axis) and the period of revolution. The latter is easy to ascertain in all cases except when the period is extremely long. The former is a little more complicated.

Barycenter: The center of mass of a binary pair of stars.

Figure 14–31 shows the orbit of a visual binary as it might be observed. The center of mass of the two stars can be determined from the fact that it must always lie along a line connecting them (as was illustrated in Figure 14–23). Each of the orbits is an ellipse but notice that the center of mass of the two stars does not lie symmetrically in either ellipse—it is not at the foci of the ellipses. The explanation for this lies in the fact that we are not seeing the ellipses straight-on. A circle, when viewed at an angle, yields an elliptical shape (Figure 14–32) and an ellipse viewed at an angle also yields an ellipse but one of a different shape than the original. The fact that the center of mass is not at the ellipses' foci can be used to determine how much the ellipses are tilted with respect to our line of sight (Figure 14–33).

Once the true shape of the ellipse is determined, its size can be calculated using the small angle formula. (The distance to the pair must be known to do this, of course.) Knowledge of the size of one of the stars' ellipses, along with knowledge of the period of its motion, allows us to calculate the *total mass* of

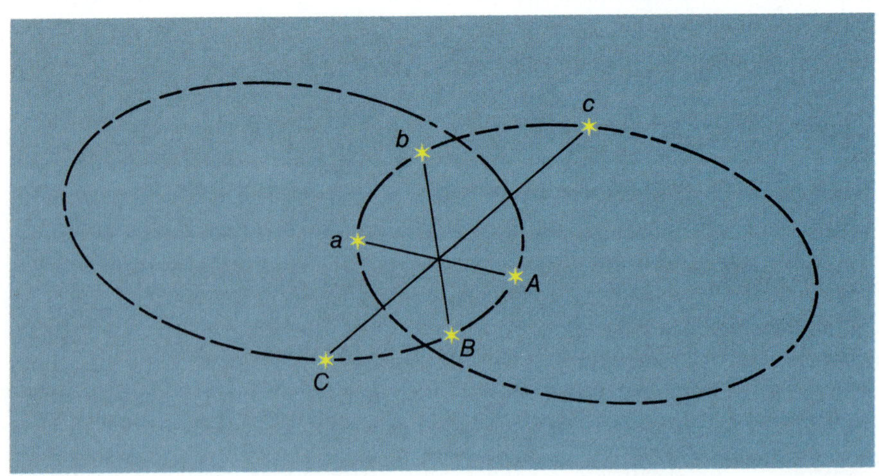

Figure 14–31. *This shows the observed orbits of a binary pair along with the locations of the stars at three times. The center of mass must be located where lines cross that connect the stars. Notice, however, that the center of mass is not at the focus of either ellipse. (Compare this to Figure 14–23.)*

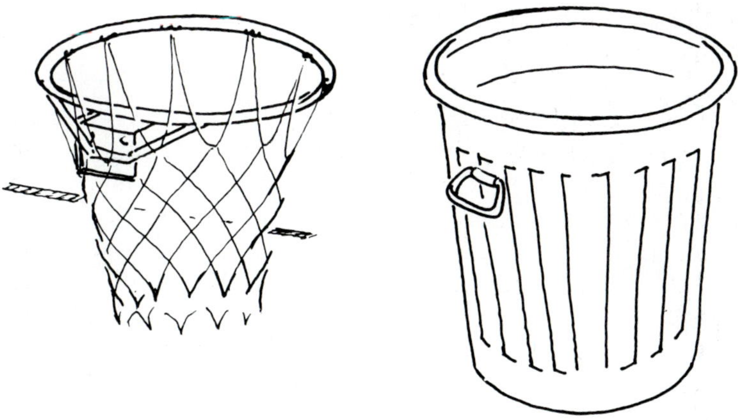

Figure 14–32. *A circular shape, such as the top of a trash can or a basketball rim, appears elliptical when viewed at an angle.*

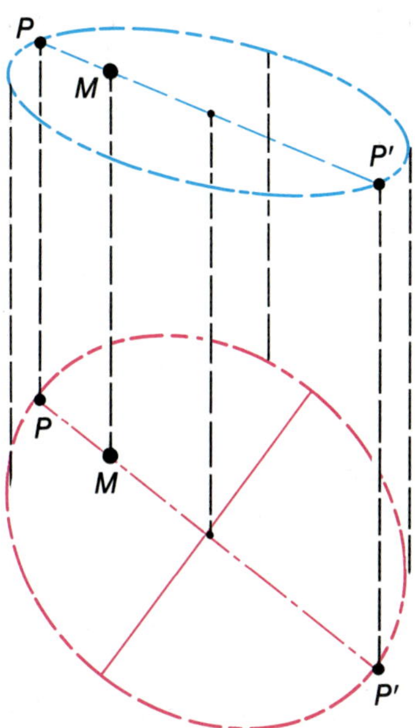

Figure 14–33. *This figure shows the apparent path (above) and the true path (below) of a binary star. Although both shapes are ellipses, if you turn your book, you can see that point M is at the focus of the lower ellipse.*

the two stars. To determine how the mass is distributed between the two, we need only consider the ratio of the two stars' distances to the center of mass. This is analogous to the way that we can calculate the weight of each of the people on a see-saw if we know the total weight of two people and know how far each is sitting from the center if they are balanced. (See Figure 14–34.) This is exactly the same method we used in calculating the mass of Saturn in a Close Up in Chapter 7.

In the case of a spectroscopic binary, if we can be confident that the plane of the stars' revolution lies very close to a line from Earth, we can do a similar calculation but instead of calculating the size of the orbit with the small angle formula, we calculate it from a knowledge of the maximum speed of the star in orbit and the period of the orbit. We are usually not able to know the inclination of the orbit, however; so a mass calculation cannot be done. Valuable information about *average* masses of stars in a great number of spectroscopic systems can be obtained, however, by assuming an average inclination of the

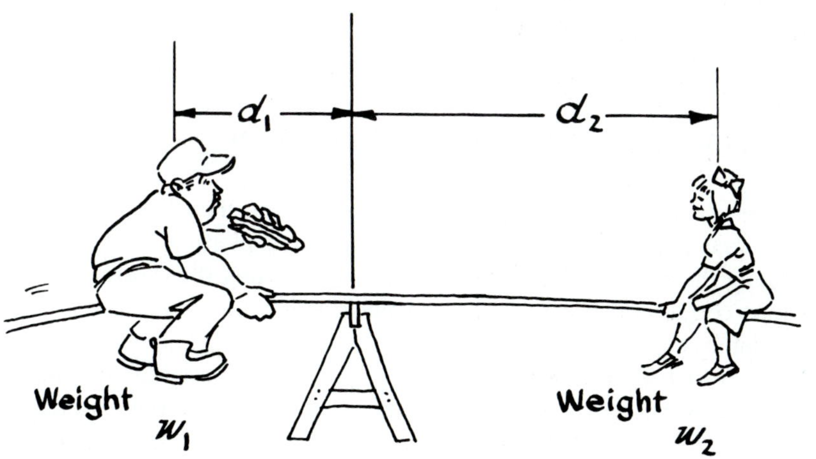

Figure 14–34. *Starting from the simple relationship* $w_1d_1 = w_2d_2$, *it can be shown that, if* W *is the total weight of the two people,*

$$w_1 = \frac{W \times d_2}{d_1 + d_2}$$

This same method allows us to determine the mass of each star in a binary system if we know the total mass and the distances of each star from the center of mass.

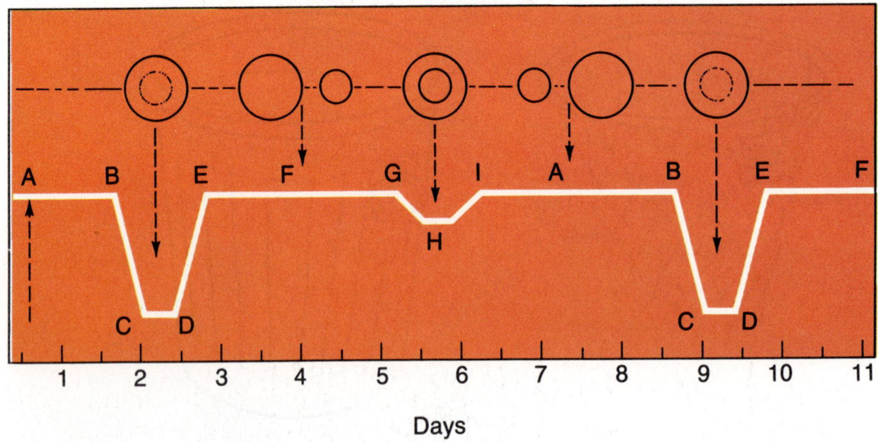

Figure 14–35. *Suppose Doppler effect data tells us that the relative velocity of these two stars as they pass one another is 8.0 × 10^2 km/s (which is 6.9 × 10^7 km/day). Assume that in the leftmost drawing of the stars, the small one is moving to the right. At point B on the light curve, it began to be hidden by the large star. At point D it starts to emerge. From B to D is about 0.8 days, so it took the small star this long to cross the large one. The diameter of the larger star must then be (0.8 days) × (6.9 × 10^7 km/day), or 5.8 × 10^7 km. In a similar manner you can calculate the diameter of the smaller star.*

orbits. (Since any orbit inclination from 0 degrees to 90 degrees is equally likely, if we include enough systems in our analysis the average inclination will be 45 degrees).

Eclipsing binaries that are also spectroscopic binaries provide us with a way of measuring not only the masses of the two stars but also their sizes. Since the fact that they eclipse one another means that the inclination of their orbit with our line of view is zero (or very nearly so), their Doppler shift tells us their velocities. Knowing their velocities and the time it takes to complete an eclipse, we can calculate the size of each star as well as the luminosity of each. Figure 14–35 shows a simplified case that illustrates this calculation. Recall that we can also determine the size of a star if we know its luminosity, distance, and apparent magnitude. Eclipsing binary systems give us an independent method of measuring star sizes.

THE MASS-LUMINOSITY RELATIONSHIP

Suppose we compare the masses of a number of stars to their luminosities by plotting a graph. Figure 14–36 shows such a graph, plotting mass in solar masses along the *x* axis and luminosity in solar units along the *y* axis. There is an obvious correspondence for main sequence stars. More massive stars are more luminous.

Mass-luminosity diagram: A plot of the mass versus the luminosity of a number of stars.

This ***mass-luminosity diagram*** was produced from knowledge of binary stars that are close enough to Earth to yield the necessary data but it is reasonable to assume that more distant stars also follow this same pattern. (Otherwise, we

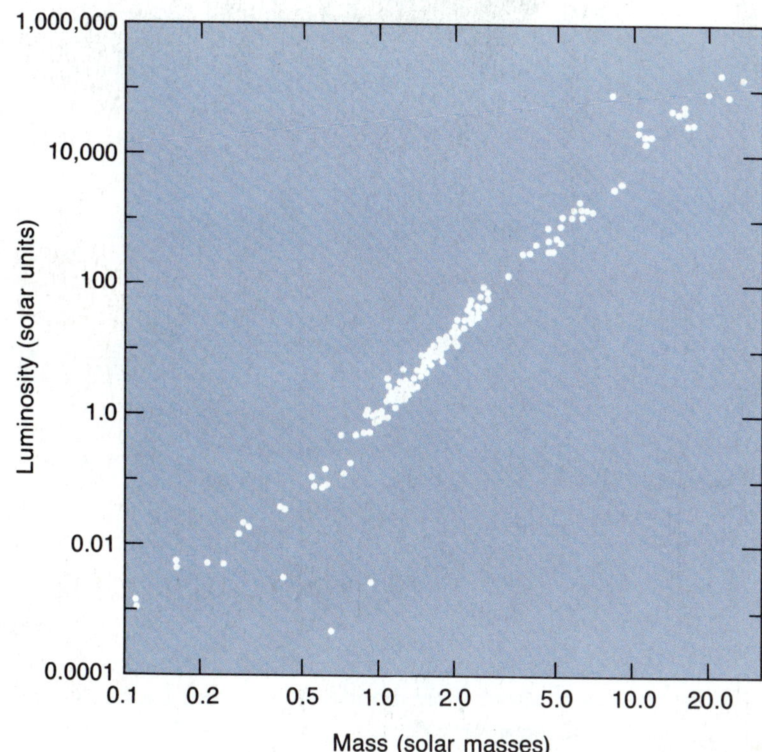

Figure 14–36. *When the luminosities of stars are plotted against their masses, an obvious relationship is seen for most stars—main sequence stars. The three at the bottom that do not fit here are white dwarfs. (Notice that the scale at bottom is a logarithmic scale.)*

would be claiming that we live in some special place in the universe where stars behave differently.) The mass-luminosity relationship is valuable to astronomers in investigating less accessible stars and in constructing and evaluating hypotheses concerning the life cycle of stars. This will be discussed in the next chapter.

CEPHEID VARIABLES AS DISTANCE INDICATORS

Eclipsing binary stars are sometimes called eclipsing *variables,* since their light intensity varies over time. Other types of variable stars are found in the heavens and one particular type is of importance to us here because it provides a method of measuring distances.

John Goodricke, the same astronomer who explained Algol's variations, discovered in 1784 (when he was 19 years old and just two years before his death) that the star δ (Greek symbol for Delta) Cephei varies in luminosity in a regular way but that its variations cannot be explained by an eclipse of a binary companion. This star changes its apparent magnitude between 4.4 and 3.7 every 5.4 days, with a light curve as shown in Figure 14–37. Soon thereafter, other stars exhibiting this characteristic light curve—brightening quickly and then dimming more slowly—were seen. Doppler effect data show that such stars are actually pulsating—changing size—in rhythm with their changes in luminosity and it is fairly easy to identify this class of stars, now called ***Cepheid variables,***

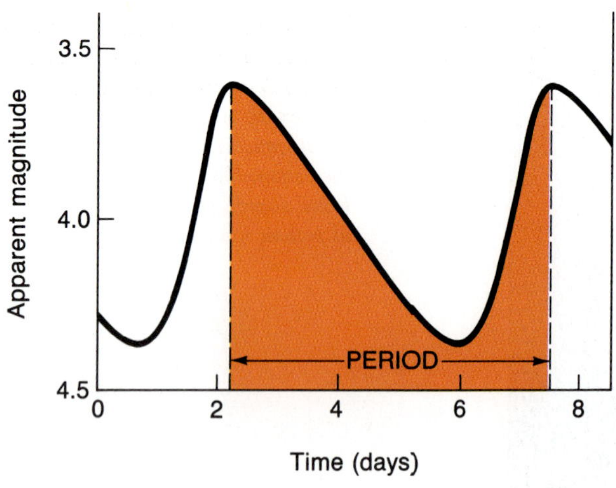

Figure 14–37. *This is the light curve of Delta Cephei, the prototype of a class of stars that have light curves with this shape and are therefore called Cepheid variables.*

Figure 14–38. *Henrietta Leavitt.*

Cepheid (pronounced SEF-e-id) variable: One of a particular class of pulsating stars.

or ***Cepheids.*** Each Cepheid has a very constant period of variation; the periods range from about one day to about three months for different Cepheids.

In 1908 a Harvard College astronomer, Henrietta S. Leavitt (Figure 14–38), published data on variable stars in the Small and Large Magellanic Clouds (Figure 14–39). The Magellanic Clouds appear to the naked eye as hazy cloud-like patches in the sky of the southern hemisphere and were first reported to Europeans by Magellan after his voyage around the world. Telescopes, however, reveal the Clouds to be gigantic groups of stars now known to be separate galaxies. Leavitt discovered that in the case of Cepheid variables, the brighter variables have the longer periods. Figure 14–40 shows a more recent graph of Cepheids in the Magellanic Clouds.

It may seem strange that there is a relationship between period and *apparent* magnitude, for apparent magnitude is not a quantity intrinsic to the star; rather, it depends on our distance from it. The reason the relationship exists is that all of the stars of the Magellanic Clouds are about the same distance from us; at least compared to the distance to the Clouds. To see this, suppose that you are on a hill outside your town at night, observing the street lights of the town (each of which, we will assume, has the same absolute luminosity). The lights will *appear* to be many different brightnesses, depending upon each light's distance from you. Now suppose that the hill is high enough so that you can use a small telescope to observe the lights of a town 200 miles away. You will find that the lights of this town all appear about equally bright, since each of them is 200 miles away, give or take only a few miles.

The stars of the Magellanic Clouds do not all have the same intrinsic luminosity but since they are all at about the same distance, their apparent magnitudes are related directly to their absolute magnitudes. For example, if the absolute magnitude of a certain star in the Large Magellanic Cloud is five magnitudes less than its apparent magnitude, the absolute magnitude of each star in the cloud will be five magnitudes less than its apparent magnitude.

Figure 14–39. *The Large Magellanic Cloud is on the left of this photo and the Small Magellanic Cloud is on the right. (The bright star Achernar is at the top right.)*

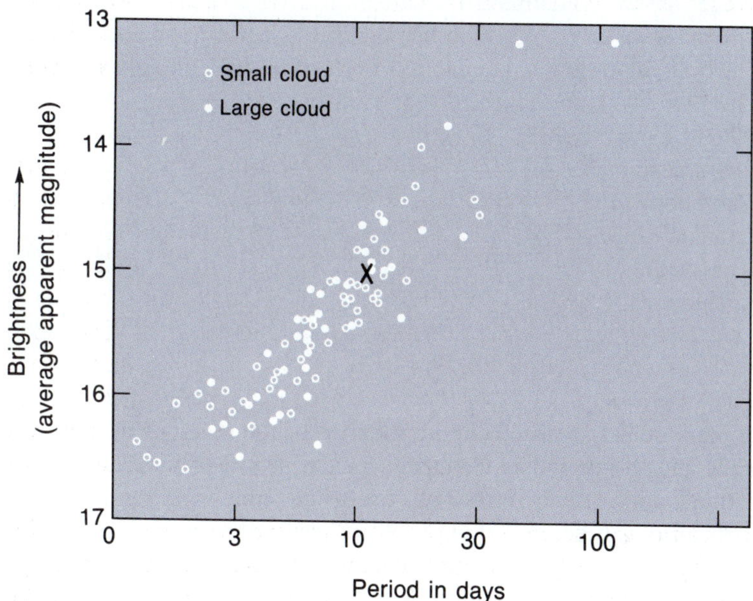

Figure 14–40. *A period-brightness diagram of Cepheids in the Large and Small Magellanic Clouds. Notice that when this early diagram was plotted, only the apparent magnitudes of the Cepheids was known.*

HISTORICAL NOTE

Henrietta Leavitt

HENRIETTA SWAN LEAVITT WAS born on July 4, 1868; one of seven children. She graduated from Radcliff College (then known as the Society for the Collegiate Instruction of Women) in 1892. In her senior year she took a course in astronomy and after teaching and travelling for a few years, she joined the Harvard College Observatory in 1895 as a student and volunteer research assistant. Her work caught the attention of those in charge and in 1902 she was given a permanent position on the staff. Soon she was named chief of the photographic photometry department.

Miss Leavitt's primary work was with variable stars and she discovered large numbers of them. In 1908 she published a list of 1777 variable stars in the Magellanic Clouds; the list included a table of 16 Cepheid variables on which she had very precise data. Concerning these she inserted a comment: "It is worthy of notice that in Table VI the brighter variables have the longer periods."[1] This brief comment went unnoticed by other astronomers. Later that year, because of ill health, she was forced to return to her family home in Wisconsin.

During her years of illness in Wisconsin, Leavitt continued her work (she was mailed stellar photographs from the observatory). When she returned to Cambridge in 1912, she published a report on 25 Cepheid variables. In it she stated, "A remarkable relation between the brightness of these variables and the length of their periods will be noticed."[2] She included a graph showing the relationship between period and brightness and although she recognized the importance of the discovery, her duties at the observatory (and the nature of the work being done there) prohibited her from following up on the discovery.

Henrietta Leavitt was one of a group of some 40 women hired by Edward Pickering, starting in the 1880s, when it was unusual for women to work outside the home. Although women had made contributions to astronomy prior to this time, science was considered by many to be an inappropriate field for them. Pickering was not really a progressive thinker in hiring the women; he did so because they would work for less money than men. Nonetheless, he was rewarded by the significant advances of many members of the group, particularly Annie Jump Cannon and Henrietta Leavitt.

Although her health was never good and her hearing was impaired from childhood, Leavitt worked at the Harvard Observatory until her death of cancer in 1921.

1. Henrietta Leavitt, "1777 Variables in the Magellanic Clouds," *Annals of the Harvard College Observatory* 60 (1908), p 107.

2. Henrietta Leavitt, *Periods of 25 Variables in the Small Magellanic Cloud,* Harvard College Observatory Circular no. 173 (March 3, 1912).

Astronomers quickly realized the importance of the relationship between the magnitude and the period of Cepheids, for the period of a variable star is easy to determine and if the period allows one to determine the absolute magnitude, this quantity can be used along with apparent magnitude to determine the distance to the variable. The problem was that only the *apparent* magnitudes of the Cepheids in the Magellanic Clouds were known. The absolute magnitudes of these stars could not be determined because the clouds are too far away for parallax to be observed. As indicated by Figure 14–40, the apparent magnitudes of Cepheids in the clouds is around 15. They appear dim. It was obvious from Cepheid variables nearer the Earth, however, that Cepheids, as a class, are very luminous stars. This provided qualitative evidence that the Magellanic Clouds are at a great distance from Earth.

All that was needed was to find one Cepheid variable near enough to the Earth that its distance could be measured by parallax. But there is none. Beginning in 1917, Harlow Shapley (whom we will discuss more fully when we study galaxies) worked out a complex statistical method to determine distances to Cepheids within our galaxy. Using such methods, astronomers by the late 1930s were confident that they had determined correct distances to a number of Cepheids. This allowed them to plot a graph of period against *absolute* magnitude, shown in Figure 14–41.

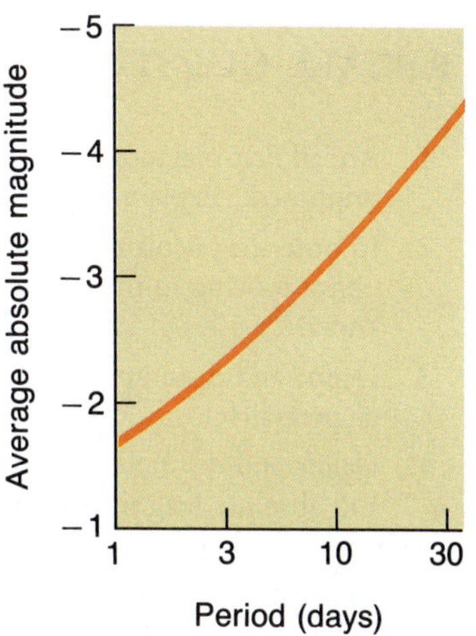

Figure 14–41. *The period-luminosity diagram for Cepheid variables. This figure is made by considering many more Cepheids than in the previous one and it plots* absolute *magnitude versus period.*

To show how the period/luminosity chart can be used to determine distances, suppose that we consider the Cepheid variable marked with an x in Figure 14–40. This Cepheid has a period of ten days and an apparent magnitude of 15. Figure 14–41 shows that a Cepheid with a period of ten days has an absolute magnitude of about −3. Now that both apparent and absolute magnitude are known, distance can be calculated. (By our nonmathematical method, you can see that the star must be extremely far away, for its apparent magnitude is greater than its absolute magnitude by 18.) Use of Cepheid variables tells us that the Large and Small Magellanic Clouds are, in fact, 48,000 and 56,000 parsecs away, respectively.

One of the reasons that it took from 1912 to the late 1930s for astronomers to determine accurately the correct relationship between Cepheids' periods and absolute magnitudes is that there are actually two different types of Cepheid variables; differing from each other in luminosity by about two magnitudes (for the same period). We can now detect the difference between the two, however, and so the problem no longer exists.

Although astronomers have detected less than a thousand Cepheid variables, their importance greatly outweighs their number. Because they are among the very brightest of stars, we can see them not only in distant parts of our own galaxy but in other galaxies, giving us a method of measuring distances to these faraway island universes.

Corresponding to normal metric usage, 1000 parsecs is one kiloparsec, abbreviated kpc, so the Large Magellanic Cloud is 48 kpc away.

CONCLUSION

In this chapter we have seen that once we know the distance to stars, many other avenues open up to us. Knowing distance and proper motion, we calculate a star's tangential velocity. The Doppler effect allows us to measure radial velocity and from these two velocities we can determine a star's actual motion relative to the Earth and to other stars. From distance and apparent magnitude, we can calculate a star's absolute magnitude. This knowledge can be combined with knowledge of the star's temperature to calculate its size.

On the other hand, the patterns seen in the H-R diagram allow us to determine a star's absolute magnitude without first knowing its distance, so that we can use the distance versus absolute magnitude versus apparent magnitude relationship in reverse to determine stellar distances.

The discovery and analysis of binary stars has provided a second method of learning a star's size and also a method of measuring a very fundamental property of a star, its mass.

Finally, a particularly useful type star, the Cepheid variable, provides a tool to measure even greater distances; opening up the entire field of galactic astronomy.

RECALL QUESTIONS

1. Are all first-magnitude stars the same apparent brightness? Explain.
2. In order to calculate (or judge) the velocity of an object moving in the sky, what quantities must we know?
3. Define and distinguish between radial velocity and tangential velocity.
4. Define and distinguish between apparent magnitude and absolute magnitude. To which is luminosity most closely related? Brightness?
5. Describe the most direct method of measuring the distances to stars.
6. How is the absolute magnitude of a Cepheid variable determined?
7. How is the radial velocity of a star measured?
8. What two quantities are plotted on an H-R diagram?
9. Sketch an H-R diagram, labelling the axes and showing the general location of the following: main sequence, red giants, supergiants, and white dwarfs.
10. At what part of the main sequence are the most massive stars located?
11. Explain the method of spectroscopic parallax. Why is it called "parallax?"
12. It would seem that a hot star should be a very luminous one. How can white dwarfs be dim stars?
13. Distinguish between optical doubles and binary stars. Why is one of these much more important to astronomers than the other?
14. Name and describe at least three methods of detecting binary stars.
15. Sometimes we can tell by the spectrum of a star that it is part of a binary system. Explain how this is done.
16. John Goodricke made two important discoveries during his short life. What were they?
17. Describe the problem presented by the fact that the plane of revolution of a binary is unlikely to be parallel to a line from Earth.
18. What type binary yields the most information on the sizes of stars?
19. Name two quantities that must be known in order to determine the absolute magnitude of a star (other than a Cepheid variable).
20. Explain how Cepheid variables are used to determine distances.

QUESTIONS TO PONDER

1. Why do only nearby stars show measurable proper motion? How would nearby and distant stars compare as to tangential velocity?
2. In Figure 14–26 (Kruger 60), the binary pair is getting farther from the single star. From this series of photos it would be impossible to tell which has the proper motion. Still, the binary pair would be taken as the most likely candidate. Why?
3. It is possible for a binary star system to fall into two or more classifications (such as visual binary and eclipsing binary). Describe some situations in which this might be the case.
4. Explain why there is such a simple relationship between the apparent and absolute magnitudes of stars in the Magellanic Clouds.
5. Suppose that two given stars differ by three in apparent magnitude. Is their difference in absolute magnitude less than three, equal to three, greater than three, or can the difference even be determined from the information given? Explain your answer.
6. Do you think that it is reasonable to use the H-R diagram, which is plotted from data on nearby stars, for stars much more distant from us? Why or why not?
7. Explain how an extraterrestrial could logically make conclusions about a typical person's lifetime from a chart such as Figure 14–14.
8. Would a mass-luminosity diagram (Figure 14–36) show the same relationship if the scales were independent of the Sun instead of being multiples of the Sun's values?
9. Why was it important to find the distance to a Cepheid variable after Leavitt had plotted the data for Cepheids in the Magellanic clouds?

CALCULATIONS

1. What is the ratio of light received from stars that differ in magnitude by 15?
2. About how many times as much light reaches us from Antares, the brightest star in Scorpius (apparent magnitude −1.0) than from τ (Greek symbol for Tau) Ceti (apparent magnitude 3.5)?
3. How many times as much light reaches us from Sirius (apparent magnitude −1.47) than from Barnard's star (apparent magnitude 9.5)? You may round the magnitude of Sirius to −1.5 to do the calculation.
4. The star Ross 128 has a parallax angle of 0.30 seconds. How far away is it in parsecs? In light years?
5. Alpha Centauri is the nearest naked-eye star to the Sun. Its apparent magnitude is −0.01 and its distance is 1.35 parsecs. Will its absolute magnitude be greater or less than 0?
6. Use data from the last question to calculate the parallax angle of Alpha Centauri.
7. Altair, one of the summer triangle stars, is on the main sequence, is spectral type A7 (8000° C), and has an apparent magnitude of 0.77. Is Altair closer than ten parsecs, about ten pc away, somewhat farther than ten pc, or much farther than ten pc?
8. Rigel, a bright supergiant, is a B8Ia star with a temperature of about 15,000° C. Its apparent magnitude is +0.14. What can you conclude about its distance from us?

CHAPTER 15

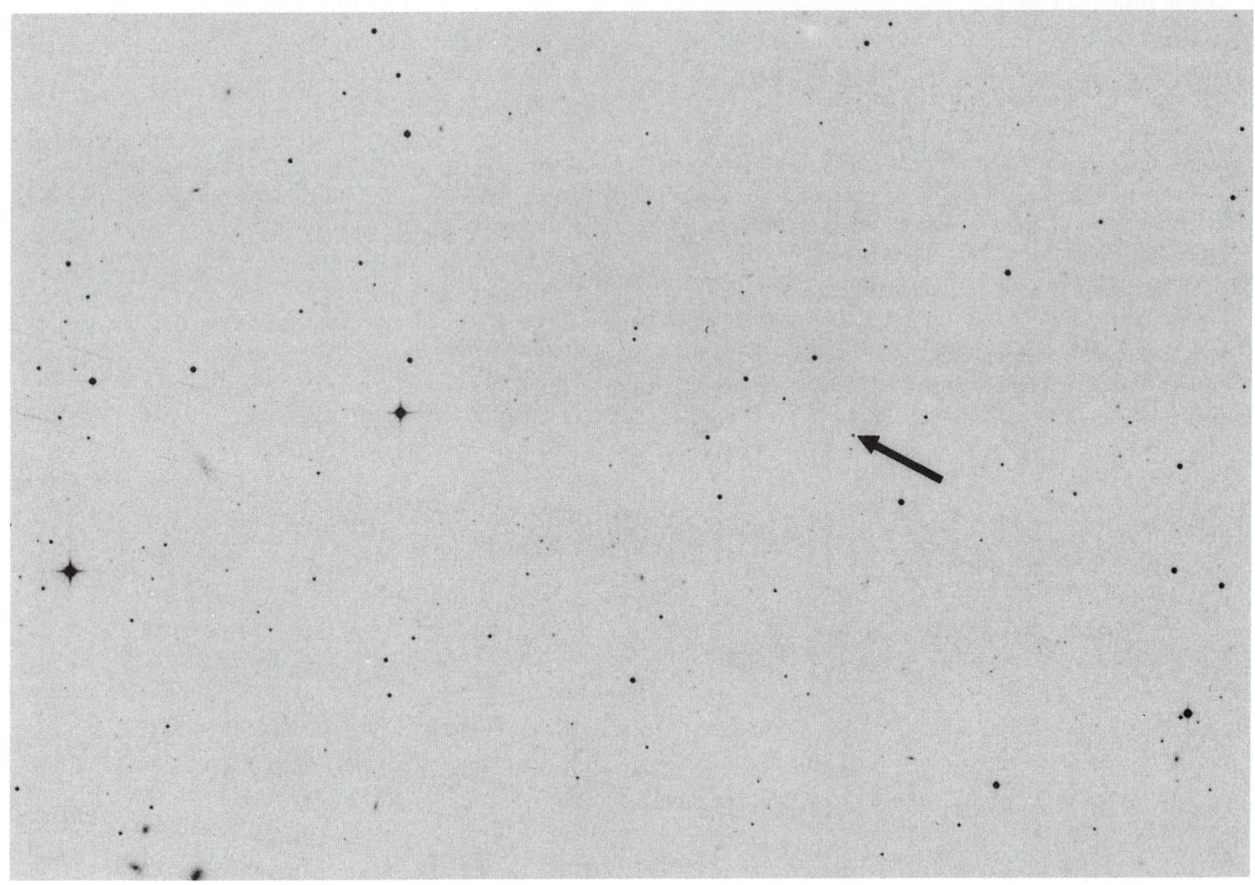

The smallest star within 80 light years of Earth does not have a glamorous name. It is designated RG 0050–2722, named for its celestial coordinates. Its mass is only 2.3 percent that of the Sun. RG 0050–2722 is located in the constellation Sculptor, southeast of Aquarius and not visible to much of the northern hemisphere. The star is invisible to the naked eye everywhere, of course, for its magnitude is only about 21, or 200 million times dimmer than Sirius.

RG 0050–2722 is about 80 light years from us. Its distance is not the major factor in making it appear dim, however, for its absolute luminosity is only 1/470,000 that of the Sun. And while the Sun's surface has a temperature of about 6000 degrees C, this smallest star is a cool 2350 degrees C. Its low luminosity and temperature earn it a unique place on the H-R diagram—at the very bottom of the main sequence.

From David W. Hughes, "The Lowest Mass Star We Know," *Mercury,* Vol XVII, No. 3, May/June 1988, pp. 93–95.

The Lives of Stars and the Solar System

IN RECENT CHAPTERS, PARTICULARLY the last one, we discussed the tools and techniques by which astronomers measure the properties of stars and discern relationships among those properties. In this chapter and the next, we see how these tools and techniques have enabled them to learn how stars form and live out their lives. In addition, we will consider the formation of a star that is particularly important to us—the Sun—and we will examine how its formation is related to the birth of the rest of the solar system.

During the discussion of stellar lives, we will not be able to spend much time on explanations of how present theories developed but in studying the origin of the solar system we will see a good example of competing theories. Finally, we will ask the question of whether ours is one of only a few planet systems in the galaxy or whether there are many such systems around other stars.

A BRIEF WOODLAND VISIT

Imagine that you are from another star system and that you land in a forest on Earth with your visit limited to two days. What could you learn about the life cycles of the woodlands during your short visit? You would see small and large trees but would you be able to determine that trees progress from small to large? In two days you have no chance to see any growth. You would see decayed material on the forest floor but it may not be obvious that this is what remains of once-standing trees. Perhaps you may be lucky enough to see a change take place. You might see a limb fall from a tree. Would you conclude that trees get smaller by dropping their branches? Or perhaps you may see an entire tree fall. What would that tell you?

Like an extraterrestrial forest visitor, we humans are brief tourists in the universe. During our short lifetimes, we see very little change in the heavens. As we measure time, the stars change slowly. We do have an advantage over the forest visitor in that previous generations have handed down to us their observations so that the our learning cycle is longer than one lifetime. The earliest reports of astronomical observations are only a few thousand years old, however; this is almost nothing when compared to the total life of the heavens.

It is possible, though, to learn about the life cycle of stars. We are fortunate that there are a few stellar events that happen quickly enough for us to observe them over human lifetimes or (in some cases) over much shorter times. Although astronomers cannot experiment directly with their subject matter, they can observe tremendous numbers of stars in various stages of development. This, along with earthbound experiments on the material of which stars are made, allows scientists to develop theories of stellar life cycles in which we can have reasonable confidence.

STARBIRTH

Like other scientific theories, today's theories of starbirth and stardeath developed slowly. Successful theories don't spring fully developed into a scientist's

Figure 15–1. *What could an extraterrestrial learn about trees during a quick visit to an Earthly forest?*

mind. Theories in astronomy result from long struggles with data from the heavens, often accompanied with experimentation here on Earth.

The story of today's understanding of the life cycles of stars starts early in this century with astrophysicist Henry Norris Russell playing the major role. The H-R diagram is the key to understanding the lives of stars but it is not simple to read. Russell's early theories held that stars begin their lives as red giants, become 0 and B type main sequence stars, and then move down the main sequence; gradually dimming as they live out their lives. When new evidence arrived, particularly from H-R diagrams of clusters and from new knowledge of nuclear energy processes, Russell realized that stars live most of their lives on the main sequence with very little change in their positions on it and that the red giant stage is not the beginning of a star's life but is near the end. We will start at the beginning.

Clouds of Gas and Dust

Stars have their beginnings in cool dark clouds of dust and gas. There are many such clouds visible in the sky, blocking out the light from stars behind them. Figure 15–2(a) shows such a ***dark nebula*** and Figure 15–2(b) shows a more famous one, the Horsehead Nebula, named for its distinctive shape. Dark nebulae appear as they do because they block light from background stars (as the nebula in part (a) of the figure) or from bright nebulae behind them (as occurs in case of the Horsehead Nebula). We will discuss the origin of bright nebulae later.

Dark nebula: A cloud of interstellar dust that blocks light from stars on the other side of it.

A dark nebula may be many hundreds of light years across and have a total mass of up to 1000 solar masses, yet the particle density in these clouds is extremely low; as low as 10^7 to 10^9 particles per cubic meter. They are called *clouds* only by comparison to normal interstellar space, which has a density as low as 10^5 particles/m^3. On Earth, material with a density as low as these clouds would be considered a vacuum, for the atmosphere at sea level has a particle density of 2×10^{25} particles/m^3. The particles in dark nebula are primarily atoms; about 75 percent hydrogen and 23 percent helium with the remaining 2 percent being heavier elements. A tiny fraction of the cloud consists of dust

Perhaps you recall from Chapter 12 that the composition of the sun is about 75 percent hydrogen and 23 percent helium. (These are percentages by mass; not by number of particles.)

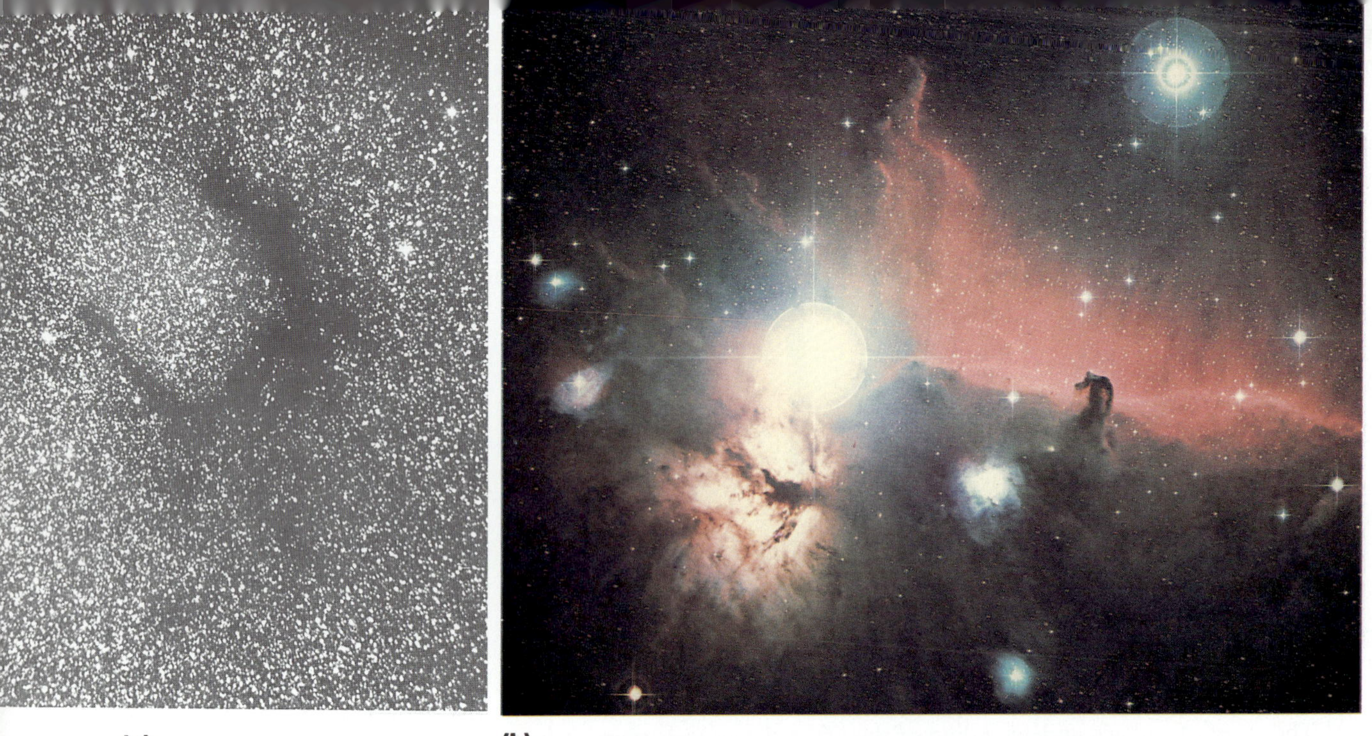

(a) (b)

Figure 15–2. *A dark nebula is an interstellar cloud that obscures light from stars behind. (a) This C-shaped one is in the constellation Aquila. (b) The Horsehead Nebula is a dark nebula in Orion.*

particles the size of cigarette smoke particles and these are responsible for blocking the light from behind.

What Triggers the Collapse?

Although there is mutual gravitational attraction between each of the particles of an interstellar dust cloud, the particles have enough random motion to prevent them from coalescing. Yet at some point, they do contract to form stars. The question of what triggers a cloud to begin its collapse is still an unsolved mystery, although there is evidence that it results from shock waves in the interstellar medium.

An earthly example of a shock wave is the pressure wave produced by an explosion or the sonic boom produced by an object travelling faster than the speed of sound. Such pressure waves on Earth can shatter windows if they are waves in the air (Figure 15–3) or crack foundations if they are in the ground.

The material flowing from stars is commonly called the stellar wind. *The* solar wind *that we detect (and that results in auroras on Earth) is far weaker than the bursts of stellar wind described here.*

There are at least three possible sources of interstellar shock waves. As we will see in the next section, during the formation of very massive stars the stars undergo a period during which they release enormous quantities of material at great speeds. These supersonic bursts cause pressure waves in the surrounding interstellar cloud and may trigger the birth of new stars.

Another source of interstellar shock waves is a supernova: the explosion of a star as it nears the end of its life. We will discuss supernovae when we discuss the manner in which stars die. A third trigger for the collapse of material to form stars, considered to be the predominant one, is the tremendous shock

Figure 15–3. *The pressure wave that travels through the air from an explosion can crack windows.*

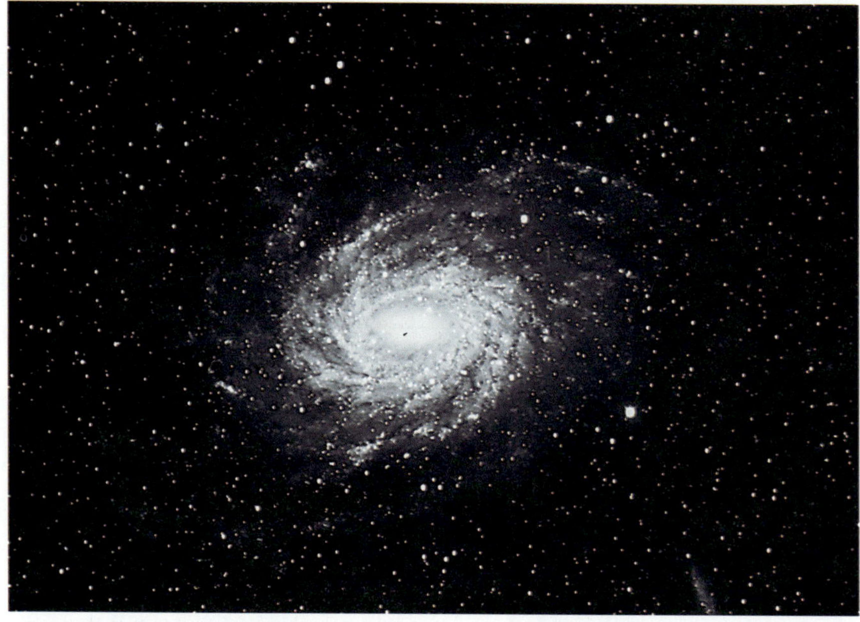

Figure 15–4. *A pressure wave exists on the leading edge of each arm of a spiral galaxy such as this one (M-101) in the constellation Ursa Major.*

waves that move around the entire galaxy, forming its galactic arms (Figure 15–4 and Chapter 17).

Protostars

Computer models of the effect of a shock wave on a gas cloud show that in compressing the cloud, the wave causes it to fragment into various segments (Figure 15–5). If a particular segment becomes compressed sufficiently, gravitational forces between particles within it can be strong enough to cause it to

ACTIVITY

Deep Sky Objects with a Small Telescope

MANY BEGINNING TELESCOPE USERS limit themselves to viewing the Moon and planets but a number of interesting objects outside the solar system are accessible with a small telescope or even with binoculars. The star maps in this book will help you find the objects listed below. At the end of each listing is the time of year when the object is highest in the sky in the evening but each may be seen in roughly the same place later at night earlier in the year.

Orion Nebula—A cloud of dust and gas 1500 LY away. Its diameter is some 15 LY. Find the Trapezium, four stars in a small dark area of the nebula (sometimes called the Fish Mouth). (January through March)

Alcor and Mizar—A naked-eye double star in Ursa Major. Your telescope will reveal that Mizar is itself a double star. (March through July)

Sagittarius—Such a rich area of the sky for interesting objects that if you slowly scan it with your telescope, you will find a number of clusters and nebulae. (July and August)

Albireo—A binary pair with obviously different colors. (Figure 14–25.) It is the "beak" star of the swan Cygnus, or the bottom star of the Northern Cross. (August through October)

Andromeda Galaxy—A spiral galaxy similar to the Milky Way. Can you detect an oval shape? (October through December)

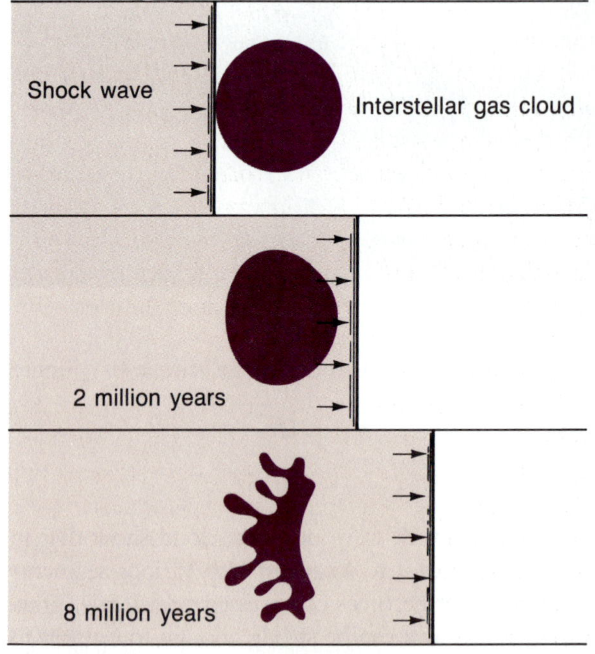

Figure 15–5. *This drawing is made from computer models (by Paul R. Woodward) and shows how a shock wave striking a gas cloud will result in the collapse of sections of the cloud. A real cloud would not be this symmetrical and thus would not form a regular pattern of protostars.*

Pleiades—An open cluster visible to the naked eye. It contains a few hundred stars (although only six or seven are visible with the naked eye) and is about 400 Ly away. See if you can detect the nebula around the stars. (November through February)

h and χ (Greek symbol for chi) Persei—A pair of open clusters visible to the naked eye (though not as easily as the Pleiades). In the sword handle of Perseus, they are an excellent view in binoculars. (November through February)

To fully enjoy your telescope (or binoculars), a good star atlas is almost a necessity. The following are recommended:

The Night Sky. Ian Ridpath and Wil Tirion. London: Collins Gem Guides, 1985. (about $5)

Mag 5 Star Atlas. Barrington, NJ: Edmund Scientific Co. ($6.95)

Whitney's Star Finder (4th edition). C. Whitney. New York: Alfred A. Knopf, 1985. ($6.95)

A Field Guide to the Stars and Planets (2nd edition). D. Menzel and J. Pasachoff. Boston: Houghton Mifflin, 1983. (about $10)

Star Gazing Through Binoculars: A Complete Guide to Binocular Astronomy. S. Mensing. Blue Ridge Summit, PA: TAB Books, 1986. ($14.95).

continue its collapse. As the particles get closer and closer together, the gravitational forces between them become stronger and they fall inward even faster.

Part of the gravitational energy of the falling particles becomes heat energy so that the inner portion of the object—now called a ***protostar***—becomes hot. This occurs simply because the falling object (atom, molecule, or dust grain) speeds up as it falls. As the particles get closer together, collisions become more frequent; causing dust particles to be broken apart. Because of the collisions, the random motion of the atoms and molecules increases, thereby increasing the temperature of the material. (This is similar to the way a falling object here on Earth increases its temperature when it collides with the floor. In both cases gravitational energy is converted to thermal energy.)

Protostar: A star in the process of formation, before reaching the main sequence.

As the cloud contracts and heats, about half of its gravitational energy is radiated away from the heating center and as it gets hotter and hotter, it radiates more and more energy. In only a few thousand years the center is as bright as the Sun. We cannot see the center directly, however, because the outer portion, which continues to fall relatively slowly toward the center, blocks most of the radiation. This ***cocoon,*** or ***cocoon nebula,*** absorbs the radiation from the center and as it does so the cocoon becomes warmer.

Cocoon nebula: The dust and gas that surrounds a protostar and blocks much of its radiation.

A warm object emits infrared (heat) radiation and it is the infrared radiation emitted from the cocoon that we see from Earth and that gives evidence for

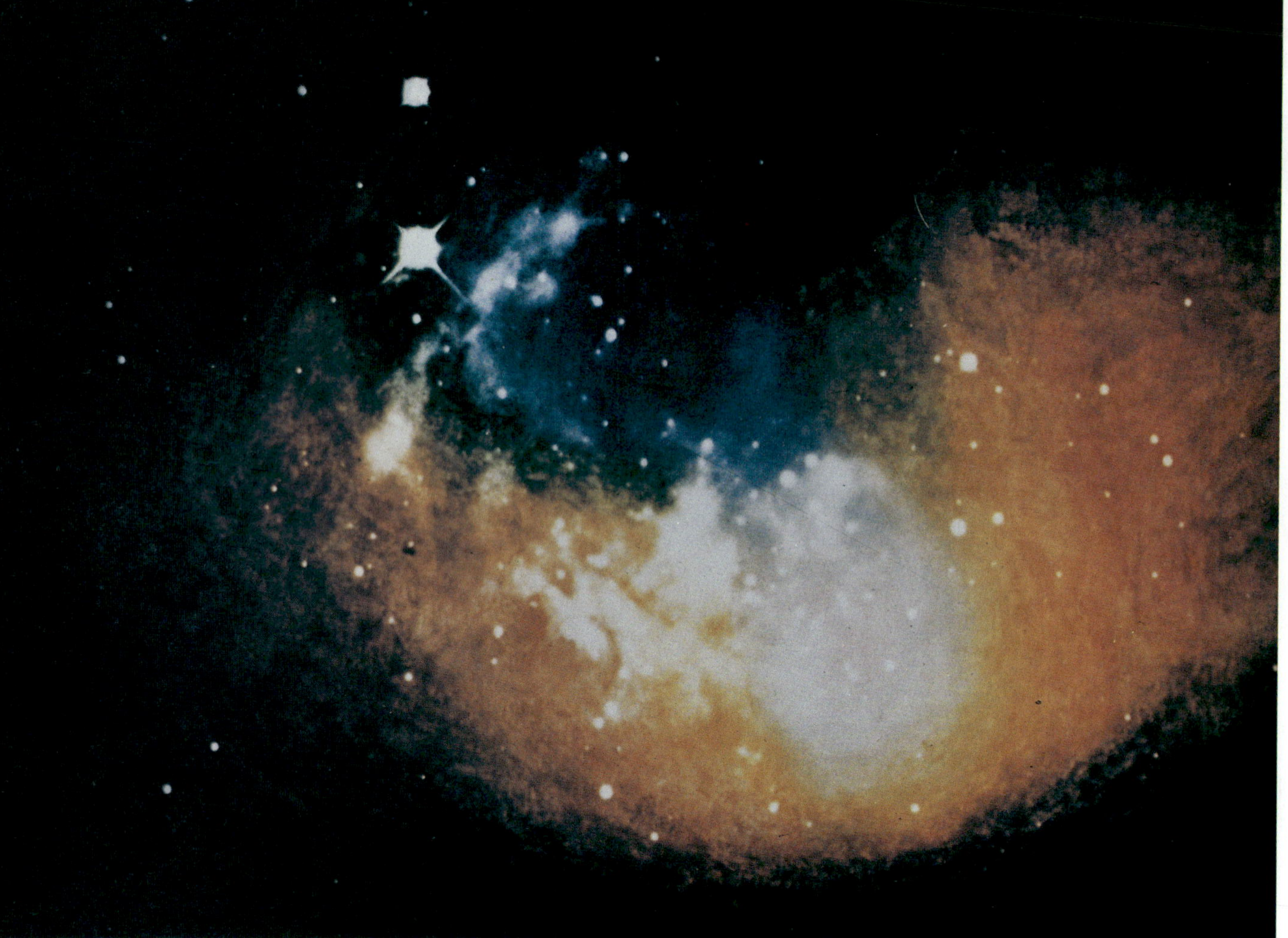

Figure 15–6. *This is an infrared photo of a small region of the Orion Nebula where new stars are forming.*

the existence of protostars. The Infrared Astronomy Satellite (IRAS) has observed many such small infrared sources. Figure 15–6 is an infrared photo and is therefore shown in false color, with each color representing a different intensity of infrared radiation. Such false color photos of nonvisible sources are very useful tools in astronomy but we must not be fooled into thinking that the objects have colors like the photos.

Evolutionary track: The path on the H-R diagram taken by a star as its luminosity and color change.

Figure 15–7 shows the progress of a protostar of one solar mass on an expanded H-R diagram. (Notice that the main sequence is now compressed into the left side of the figure.) The expansion of the diagram beyond the coolest stars on the main sequence was necessary because a protostar begins its life much cooler than any main sequence star. The details of this part of the star's ***evolutionary track*** are not known, as indicated by the question mark, but the protostar begins as a cool, dim object, warms up by gravitational contraction, and moves toward the main sequence. Finally, as the cocoon continues to contract, its smaller size results in it appearing dimmer so the star begins to move downward on the diagram as it nears the main sequence.

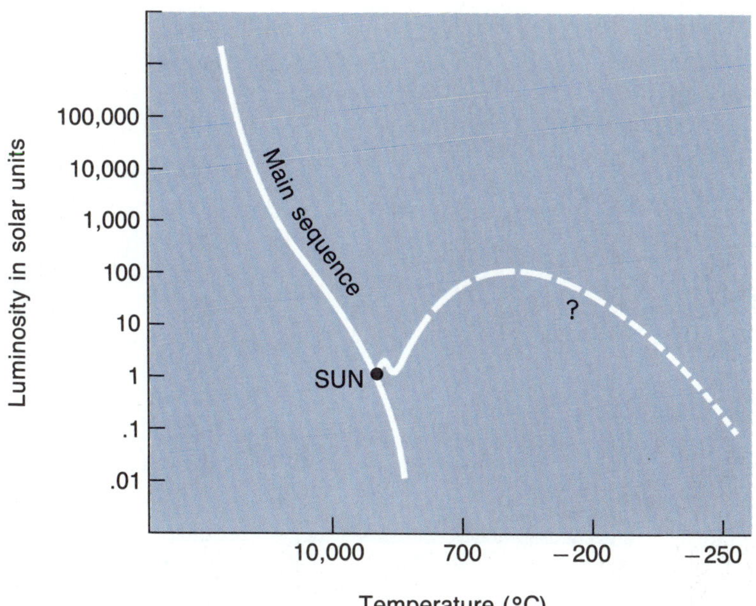

Figure 15–7. *This H-R diagram shows temperatures much lower than on previous diagrams in order to include the early, cool stages of a star of one solar mass.*

Motion on the H-R diagram does not mean actual motion of the star in space, of course. Recall from the last chapter the diagram of people's height versus age. As a person ages, his or her dot moves on this diagram. This is analogous to the motion we are referring to on the H-R diagram.

Astronomers, in speaking of evolutionary tracks and stellar evolution, are using the term evolution *in a very different way than biologists do. One star is said to "evolve" as it lives its life. We are not speaking of an entire species evolving through generations.*

As time passes, the center of the protostar continues to shrink and become hotter, emitting more radiation all the time. This radiation gradually blows away the outer portions of the cocoon. Some of the cocoon may not be blown away but may condense to form planets, as we will discuss later. It appears that in some stars the cocoon is blown away by sudden bursts of solar winds. Certain stars, named *T Tauri* stars after the variable star T in the constellation Taurus, appear to be young stars undergoing some sort of instability that causes enormous flares. This is thought to play a part in blowing away the cocoon of newly forming stars, particularly K and M type stars, which—being at the bottom of the main sequence—are of low mass.

T Tauri stars: A certain class of stars that show rapid and eratic changes in brightness.

The most massive stars, the O and B type stars, follow a track on the H-R diagram toward the top of the main sequence (Figure 15–8). These stars are much fewer in number than stars of lesser mass, but they are so energetic that they signal their presence by emitting ultraviolet light that stimulates the hydrogen of their cocoon (and the nebula beyond the cocoon) to emit its own light. The emission of light here is by a fluorescence process. That is, ultraviolet light is absorbed by the atoms of the gas, causing the atoms to become energized. The atoms then lose this extra energy by emitting light—*fluorescing*. (Ultraviolet light is likewise emitted inside our fluorescent lamps here on Earth. A coating on the inside of the glass of the lamp absorbs this radiation and fluoresces, emitting visible light.) Since the light from such nebulae (Figure 15–9) are emission spectra, they are called an ***emission nebulae***. New stars may be forming

Emission nebula: Interstellar gas that fluoresces due to ultraviolet light from a star near or within the nebula.

Figure 15–8. *A more massive star ends up at a higher location on the main sequence—hotter and brighter.*

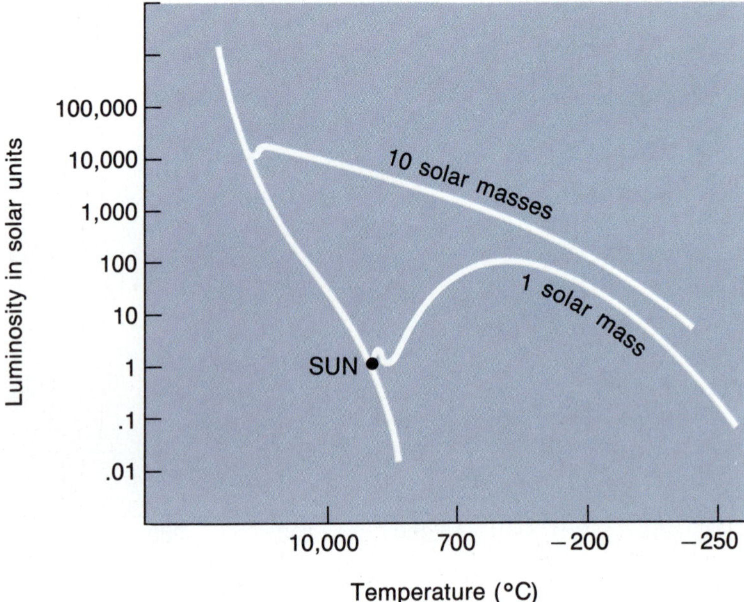

in the dark Horsehead while the bright emission nebula behind it is being lit up from new stars within.

The infalling particles of massive stars experience much greater gravitational force than do particles of low-mass stars and they reach the main sequence much faster. M-type stars remain protostars for hundreds of millions, perhaps billions, of years. The Sun, with more mass, spent about 30 million years in this phase but the most massive stars remain protostars for only tens of thousands of years. Recall that stars of low mass may undergo a period of instability just before joining the main sequence. Massive stars do likewise, but their instability is more violent and O and B stars blow material out at supersonic speeds during this time; creating a shock wave in the surrounding material. This shock wave may be one of the triggers that starts the collapse of other portions of the interstellar cloud to form more stars.

The great amount of radiation from massive stars is what puts a limit on how massive a star can be. Astronomers calculate that a star with a mass greater than 100 to 200 stellar masses will emit radiation so intense that this radiation will prevent more material from falling into the star and thereby limit its size.

On the other hand, protostars with masses of less than about 0.08 solar masses do not have enough internal pressure to ignite hydrogen fusion. They heat up due to gravitational contraction but never become main sequence stars. Eventually they contract as far as they can and then begin to cool, becoming planetlike objects; cold cinders in space. Recall that Jupiter and Saturn emit more energy than they receive from the Sun and that this comes from gravitational energy that remains from their formation.

Star Clusters

The Pleiades, a beautiful cluster visible to the naked eye, is shown in Figure 15–10. This is a group of stars that formed from a single interstellar cloud.

Figure 15–9. *(a) The Lagoon Nebula in Sagittarius is an emission nebula. (This is a 40-minute exposure using an eight-inch telescope.) (b) The Orion Nebula is an emission nebula containing a dark nebula where new stars are forming. It is easily visible in small telescopes or even binoculars.*

The dust that remains around the stars reflects their light in all directions and so the Pleiades are seen in a glowing nebula. In reflecting this light, the small particles of the cloud scatter light of short wavelengths more than they do light of long wavelength. This process is what causes the Pleiades's bright blue ***reflection nebula.***

Reflection nebula: Interstellar dust that is visible due to reflected light from a nearby star.

Galactic (or open) clusters: A group of stars that share a common origin and are located relatively close to one another.

The Pleiades are typical in many ways of a group of young stars. The stars exist in clusters because they are formed at the same time from a single interstellar cloud. Figure 15–11 shows some other ***galactic clusters,*** or ***open clusters.*** Galactic clusters generally contain hundreds of stars but their mutual gravitational attraction is not enough to restrain their random motions from causing them to separate and go their own ways. In a sense, the individual stars "wander out of their nursery."

Much of what we know about star formation has come from examination of clusters. The fact that all the stars in a cluster began forming at about the same time allows us to learn how stars of different mass have progressed at different rates. Figure 15–12 is a representative H-R diagram of a very young cluster. The diagram indicates that stars of low mass have not yet reached the main sequence. All of the stars of the cluster began their formation at essentially the same time, when their interstellar cloud began its collapse. H-R diagrams of clusters provide evidence that stars of low mass spend much more time in the protostar stage than do more massive stars. As we will see a little later,

Figure 15–10. *The Pleiades is an open cluster 410 light years from Earth. Photographed with a telescope, the reflection nebula surrounding the Pleiades is visible. The dimmer stars are part of the cluster and are the same distance from us as the bright 0 and B type stars. (The spikes from each star are caused by diffraction around secondary mirror supports in the telescope.)*

H-R diagrams of older galactic clusters can be used in a similar manner to reveal when stars end the main sequence part of their lives.

STELLAR MATURITY

When the center of a collapsing star becomes hot and dense enough, hydrogen fusion begins. Gravitation serves as the energy source for protostars, heating them and causing the emission of radiation, but main sequence stars—including our Sun—have as their energy source the fusion of hydrogen into helium.

The Stellar Thermostat

Thermostats on the walls in our homes have two features: a thermometer to read the temperature and a switch to turn on and off the furnace or air con-

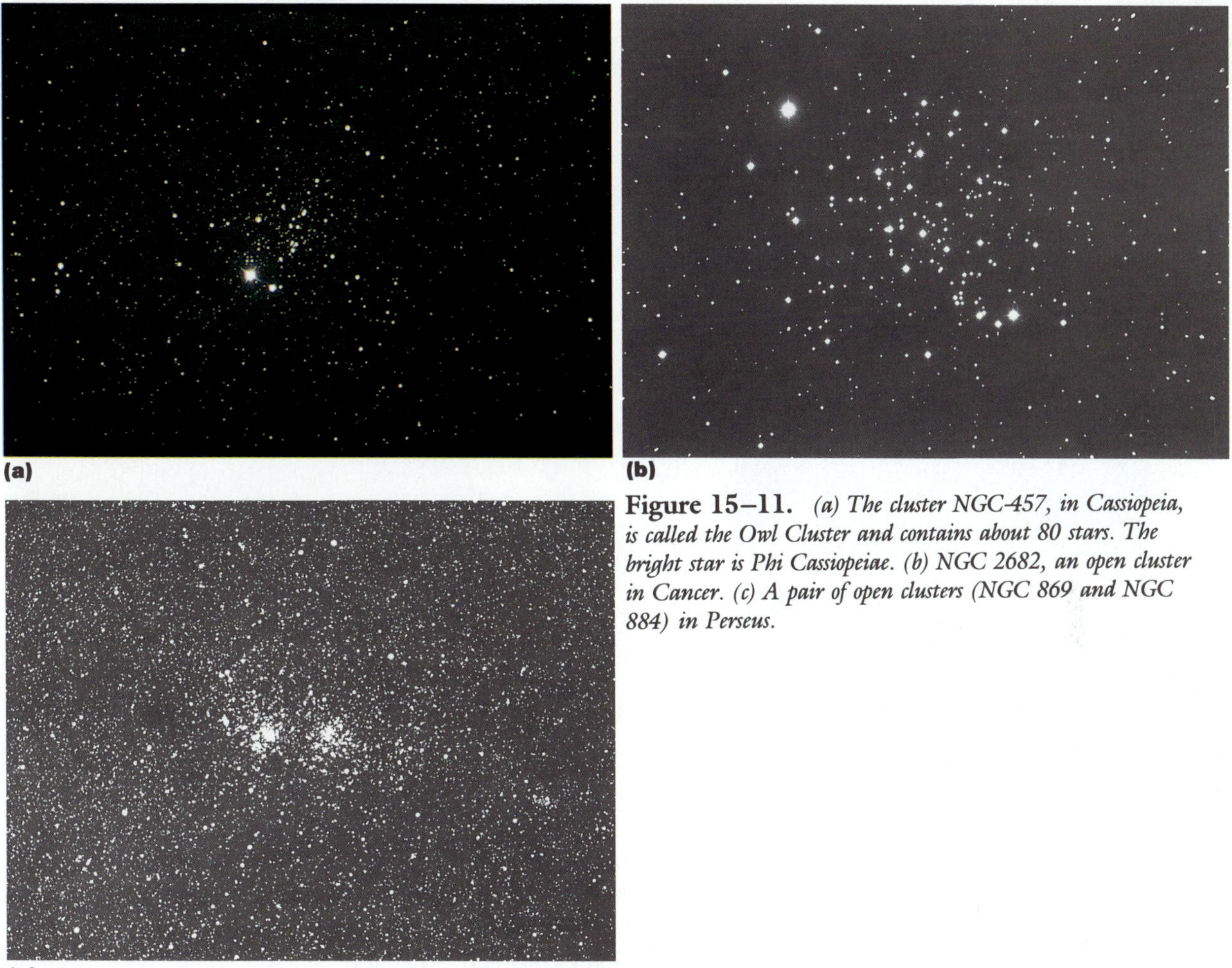

Figure 15–11. *(a) The cluster NGC-457, in Cassiopeia, is called the Owl Cluster and contains about 80 stars. The bright star is Phi Cassiopeiae. (b) NGC 2682, an open cluster in Cancer. (c) A pair of open clusters (NGC 869 and NGC 884) in Perseus.*

ditioner. If the thermometer indicates that the temperature is below some preset level, the switch turns on the furnace until the thermometer indicates that the temperature is high enough, whereupon the switch turns off the furnace. The core of a main sequence star has an analogous regulating mechanism and it is this mechanism that controls the rate of consumption of the hydrogen fuel.

The stellar thermostat is an alternative description of hydrostatic equilibrium, discussed in Chapter 12.

Suppose that somehow the rate of hydrogen fusion in a star begins to increase. This causes the temperature of the core to increase. When the temperature of a gas increases, however, the gas expands and the expansion serves as a switch to decrease the rate of fusion. This occurs because the expanded gas has a greater average distance between its hydrogen nuclei and the increase in distance means fewer collisions and therefore fewer fusions. The result is that the expansion causes the rate of energy production to decrease, which causes the expansion to stop. In this way equilibrium is reached and fusion reactions occur at a uniform rate.

On the other hand, if the rate of fusion in a star's core somehow decreases, the core will contract; decreasing the distance between nuclei. This causes the

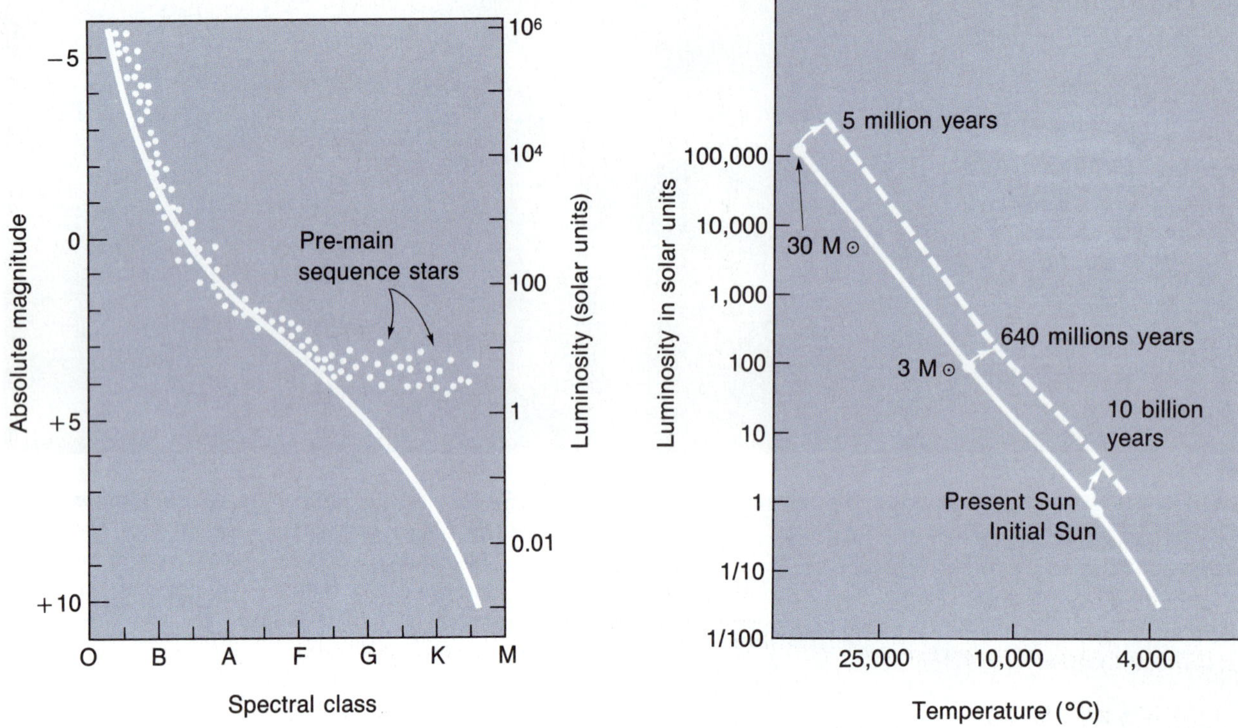

Figure 15–12. *An H-R diagram of a very young cluster shows that stars of low mass have not yet reached the main sequence.*

Figure 15–13. *As stars age on the main sequence, they change slightly in temperature and luminoisity. ($M_{\odot}$ is the Sun's mass.)*

fusion rate to increase. Another effect that occurs when the core contracts is the conversion of gravitational energy into heat, just as occurred when the original interstellar cloud collapsed. Therefore, energy comes both from an increase in fusion and from gravitational energy and both of these energies contribute to the thermostat that brings the star once again to equilibrium.

Main Sequence Life

The stellar thermostat is important during the main sequence life of a star. In a main sequence star, hydrogen in the core of the star is continually being converted to helium as the nuclear reaction occurs. This causes the number of particles in the core to gradually decrease (since four hydrogen nuclei are converted to one helium nucleus) and this results in the core shrinking slightly. As this occurs, both the increase in the rate of fusion and the conversion of gravitational energy into heat come into play, raising the core's temperature and making the star more luminous. Although the luminosity change is slight, the star's position changes slightly on the H-R diagram (Figure 15–13). This is what causes the main sequence to have a perceptible width. Stars start their main sequence lives on the left side of the strip, then move up and to the right as they age.

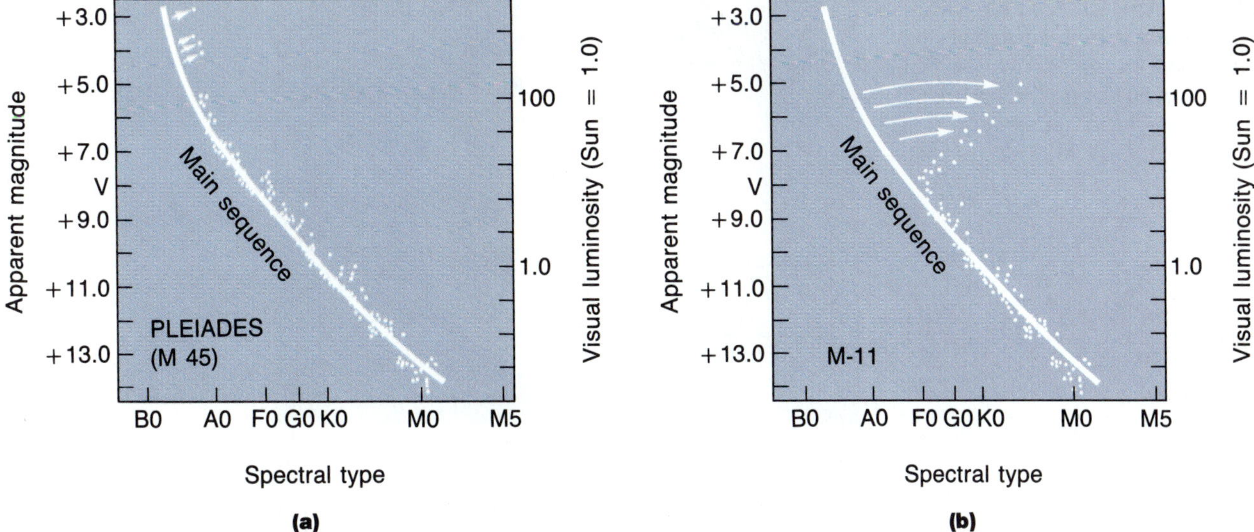

Figure 15–14. *(a) An H-R diagram of the Pleiades reveals that its most massive stars have started leaving the main sequence. The arrows indicate the evolutionary paths they must have taken. (b) The cluster M-11 is older than the Pleiades and stars are turning off the main sequence at a lower point.*

The Sun will continue hydrogen burning on the main sequence for a total of some 10 billion years. In more massive stars the fusion rate is much greater because of the greater pressures and core densities; the thermostat is set higher in massive stars. This means they are more luminous. (Recall the mass-luminosity relationship for main sequence stars: the more massive, the more luminous.) The greater fusion rate also causes them to use up their core hydrogen in a much shorter time. The most massive stars fuse hydrogen so quickly that their cores run out of hydrogen in only a few million years. Calculations for the least massive stars, on the other hand, show that they will continue hydrogen fusion on the main sequence for hundreds of billions of years.

Observational evidence for the shorter lifetimes of massive stars comes from galactic clusters. Figure 15–14(a) is the H-R diagram of the Pleiades, which, although it is considered a young cluster, already has its most massive stars leaving the main sequence. Arrows on the figure indicate their evolutionary paths after they end their main sequence lives. Figure 15–14(b) is a similar diagram for the cluster M-11; a cluster older than the Pleiades. Notice that the most massive stars have moved even farther to the right and that stars of less mass are now leaving the main sequence.

The name M-11 *means that this cluster is number eleven in the Messier Catalog of Nebulae and Star Clusters, originally developed in 1781 by the French astronomer Charles Messier (1730–1817).*

The Red Giant Stage

We saw in the last chapter that about 90 percent of the stars in the sky are on the main sequence. This tells us that a typical star spends 90 percent of its luminous lifetime there. In all cases except for stars of very low mass (less than

Figure 15–15. *As the core of a star shrinks it heats up, causing additional hydrogen fusion in a shell surrounding it. The core gets hot enough to fuse helium to carbon.*

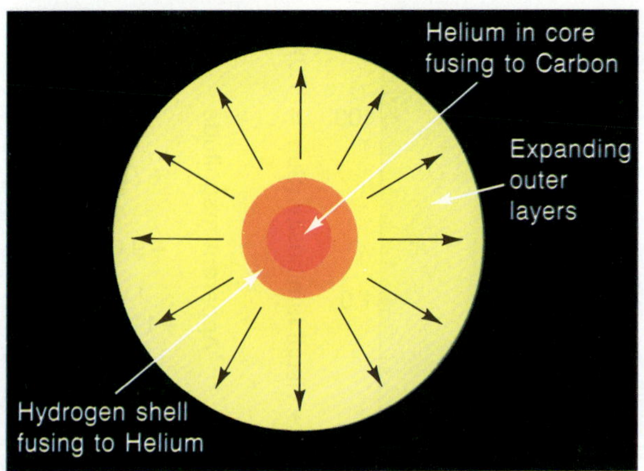

about 0.4 solar mass) the next step for stars after leaving the main sequence is essentially the same. They become giants. Stars with masses less than about 0.4 solar mass die without becoming red giants and we will put off further discussion of them until the next chapter when we discuss stardeath.

The process of becoming a red giant starts when the core begins to run low on hydrogen fuel. Until then, the nearly constant production of fusion energy kept the core from collapsing. Now, lacking a source of energy to fight gravitational collapse, the core starts to shrink dramatically. Just as in the case of the protostar, however, the contraction converts gravitational energy into heat and radiation. In fact, the energy produced in the core actually *increases* over what it was when the energy source was fusion. The resulting increase of radiation from the core causes the shell of material around the core to heat up enough that hydrogen fusion begins there (Figure 15–15).

When our sun becomes a red giant, it will expand until it envelopes Mercury, Venus, and the Earth (Figure 15–16).

We now have gravitational energy in the core providing heat for fusion in a shell surrounding the core. These two sources of energy in the star's center result in a great increase in radiation from the center and cause the outer part of the star to expand. When a gas expands, it cools, and the outer portion of the star does just that. Thus the star moves to the right on the H-R diagram; toward the red giant region. At the same time it moves upward because its total luminosity is increasing.

Table 15–1 Helium Fusion Reactions that Occur in Red Giants

Reaction	Explanation
${}^{4}_{2}He + {}^{4}_{2}He \rightarrow {}^{8}_{4}Be + \gamma$	Two helium nuclei combine to produce a beryllium-8 nucleus and a gamma ray (γ).
${}^{4}_{2}He + {}^{8}_{4}Be \rightarrow {}^{12}_{6}C + \gamma$	Another helium nucleus combines with the beryllium nucleus to form a carbon-12 nucleus and another gamma ray.
${}^{4}_{2}He + {}^{12}_{4}C \rightarrow {}^{16}_{8}O + \gamma$	Carbon in the core fuses with helium and produces oxygen and another gamma ray (and more energy).

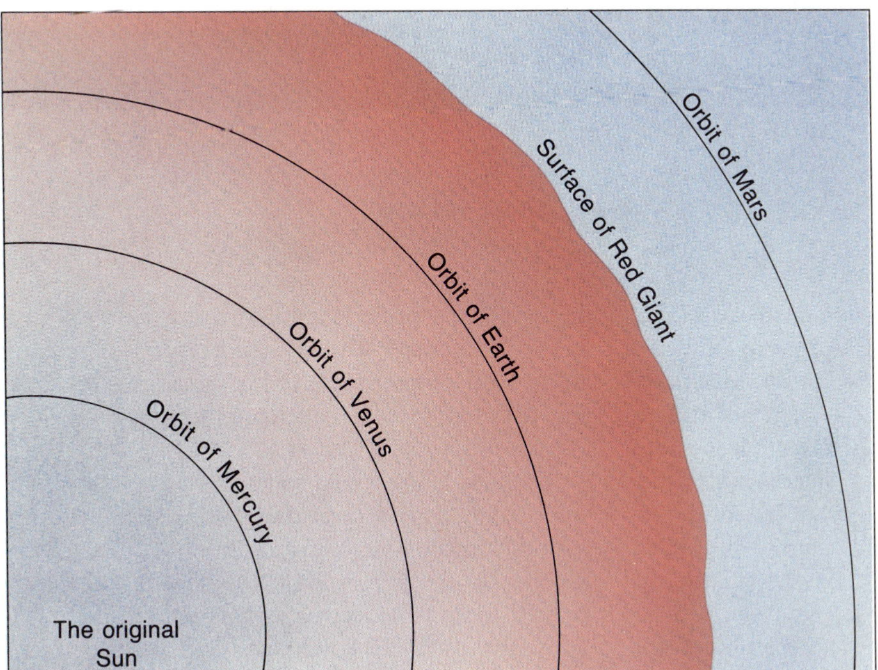

Figure 15–16. *When the Sun becomes a red giant, it will expand until its surface is somewhere between Earth and Mars. Compare this to its present size.*

The core of a red giant continues to contract and heat until it becomes hot enough for helium to fuse into carbon. This reaction, shown in more detail in Table 15–1, has the effect of converting three helium nuclei into one carbon nucleus; releasing gamma rays in the process. The core of the star once more has a source of energy other than gravitational contraction. The new and relatively constant source of energy resets the stellar thermostat and brings the star into a somewhat stable state.

In the more massive red giants, the temperatures and pressures are high enough to cause carbon to fuse to heavier elements such as oxygen, silicon, and magnesium. (A fusion that produces oxygen is shown in Table 15–1.) At still higher temperatures, these elements may fuse to form even heavier ones.

When the source of a star's energy changes from one reaction to another, the star experiences instability. This occurs as one reaction slows, the core contracts and heats, and then the next reaction begins. Thus a red giant goes through cycles of cooling as it expands and then heating as it shrinks. On the H-R diagram, the star traces a track back and forth in approximate horizontal lines. The most massive giants undergo a period of instability in which they become Cepheid variables—important to astronomers as distance indicators.

Notice that as the universe ages, hydrogen is gradually changed into heavier elements. We will see in the next chapter that at the end of the life of a red giant, processes occur that produce the very heavy elements—as heavy as uranium (with 92 protons). It is from these elements—made in stars—that the Earth and its various life forms are created.

We will return to the story of the stellar life cycle in the next chapter. We now turn our attention to another process which took place as the Sun formed. It is a process of immense importance to us: the origin of our solar system.

CLOSE UP

Velikovsky's Pseudo-Theory

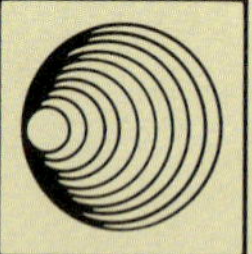

In 1950, a Russian immigrant to the U.S. named Immanuel Velikovsky published a book titled *Worlds in Collision,* which proposed a radically different catastrophic theory for portions of the formation of the solar system. It included explanations for a wide variety of phenomena, including the unusual direction of Venus's rotation, the existence of petroleum on Earth, and craters on the Moon. In addition, it explained many events reported in the Old Testament and other ancient writings.

Velikovsky's theory held that at some time in the past a part of Jupiter was ejected from that planet, becoming a comet as it moved through the solar system. The comet passed very close to the Earth on two different occasions. It also passed close to Mars, temporarily changing Mars's orbit so that the red planet had a near-collision with Earth. One of these close encounters caused the Earth to stop rotating for awhile; explaining the Biblical report that the Sun stood still when Joshua commanded it to. Another near-miss caused the Red Sea to part so that Moses and his people could escape the Egyptian army. Eventually the comet settled into a near-circular orbit around the Sun; becoming the planet Venus. A major tenet of Velikovsky's argument was that the primary force acting in the solar system was electrical in nature rather than gravitational. (He had written a previous book titled *Cosmos Without Gravitation.*)

Worlds in Collision sold millions of copies and caused quite a stir among both the general public and the scientific community. His theories were not well-accepted by scientists. In fact, the scientific community reacted against his ideas in an extreme manner. Scientists ridiculed Velikovsky and even went so far as to threaten to boycott the publisher of *Worlds in Collision*.

Why were these new ideas not accepted by scientists? It would seem that a theory that united so many phenomena would be welcomed with open arms. Velikovsky's followers claimed that the reason he was not accepted by scientists was that he was a threat to the scientific establishment and that scientists rejected him because his training was in psychiatry rather than in the physical sciences. While there may be an element of truth in this claim, the reasons for rejecting the new theory were many.

THE FORMATION OF THE SOLAR SYSTEM

A study of the beginnings of the solar system is interesting as an example of the way science in general (and astronomy in particular) progresses, because the theory is still in its early development and many gaps remain. The search for answers here resembles a mystery story where there are many clues; new ones appear all the time and some of the clues seem to contradict others.

There are two main categories of competing theories to explain the origin of the solar system: evolutionary theories and catastrophic theories. We'll look at the evidence for each and see why one is gaining in favor among astronomers. First, however, we will review the clues; most of which were discussed in earlier chapters.

First, Velikovsky's theories rejected great portions of today's science. But so did Copernicus's model of the solar system and Einstein's theories of relativity. What is the difference? For one, Velikovsky's writings showed that he did not understand the theories that his own were proposed to replace. When he attempted to show how his theory worked better than older theories, he misstated or misapplied the old theories. Copernicus, on the other hand, certainly understood the Ptolemaic system and his writings are clear on that point. No one can question Einstein's knowledge of the science of his day. (Dr. Henry Bauer, in his 1984 book *Beyond Velikovsky—The History of a Public Controversy,* says that in *Cosmos Without Gravitation,* the author appears to be "an ignoramus masquerading as a sage.")

An important criteria of a model is that it produce testable predictions. In fact, some of Velikovsky's predictions have since been found to be true. He predicted, for example, that Venus is a hot planet. He predicted that Jupiter has a magnetic field and emits radio waves. Both have since been found to be true. On the other hand, there are so many predictions in *Worlds in Collision* that some of them are bound to be true. In addition, all of Velikovsky's predictions are very qualitative in nature so their accuracy cannot really be checked. In the case of the temperature of Venus, he did not indicate how hot it should be; he simply said that it was "hot." This gave him at least a ⅓ chance of being correct; a planet is either hot, cold, or Earth-like in temperature.

The Velikovsky controversy is interesting not only as a study in the nature of science and the acceptance of scientific theories but as a study of human behavior, for it is certainly true that scientists overreacted to Velikovsky. Their reaction was such that it appeared to some of the general public that scientists have something to hide; that they are jealous of anyone who challenges their ideas. While there may be an element of truth in this accusation—after all, scientists are people first and scientists second—the real reason for rejecting Velikovsky's ideas is simply that they are ridiculous.

Evidential Clues from the Data

There are a number of patterns among the members of the solar system, and these patterns must be explained by any successful theory of the system's origin. In addition, a theory should be able to account for exceptions to the patterns. Here is a list of significant data that must be explained.

1. All of the planets revolve around the Sun in the same direction (which is the direction the Sun rotates) and all planetary orbits are nearly circular except that of Pluto (and, to a lesser degree, Mercury).

2. All of the planets lie in nearly the same plane of revolution and this plane is close to the equatorial plane of the Sun.

3. Most of the planets rotate in the same direction as they orbit the Sun; the exceptions being Venus, Uranus, and Pluto.

4. The majority of planetary satellites revolve around their parent planet in the same direction as the planets revolve around the Sun (and as the planets rotate). In addition, most satellites' orbits are in the equatorial plane of their planet.

5. There is a pattern in the spacing of the planets as one moves out from the Sun, with each planet being about twice as far from the Sun as the previous planet.

6. There are similarities in the chemical compositions of the planets but a pattern of differences also exists, in that the outer planets contain more ***volatile elements*** and are less dense than the inner planets.

Volatile element: A chemical element that exists in a gaseous state at a relatively low temperature.

Nonvolatile element: An element that is gaseous only at a high temperature and condenses to liquid or solid when the temperature decreases.

7. At least three of the four Jovian planets have rings (and perhaps Neptune does also).

8. Asteroids, comets, and meteroids populate the system along with the planets and each category of objects has its own pattern of motion and location in the system.

9. The planets have more total angular momentum (to be described later in this chapter) than does the Sun, even though the Sun has most of the mass.

10. Recent evidence indicates that planetary systems in various stages of development may exist around other stars.

As astronomers try to solve the mystery of the origin of the Sun and its companions, they must be sure their theories fit these clues. In addition, any successful theory of the origin of the solar system should coincide with the theory of the formation of stars.

Evolutionary Theories

There is no single evolutionary theory for the solar system's origin, but there are a number of theories that have in common the idea that the solar system came about as part of a natural sequence of events. The theories have their beginning with one proposed by René Descartes in 1644. He proposed that the solar system formed out of a gigantic whirlpool, or vortex, in some type of universal fluid and that the planets formed out of small eddies in the fluid. This theory was rather elementary and contained no specifics as to the nature of the universal fluid. It did, however, explain the observation that the planets all revolve in the same plane; that plane being the plane of Decartes's vortex.

After Isaac Newton showed that Decartes's theory would not obey the rules of Newtonian mechanics, Immanuel Kant (in 1755) used Newtonian mechanics to show that a rotating gas cloud would form into a disk as it contracts under gravitational forces (Figure 15–17). Thus the philosophical "universal fluid" of Descartes became a real gas working under natural laws to explain the disk aspect of the solar system (our clue #2). In 1796, Pierre Simon de Laplace, a French mathematician, introduced the idea that such a rotating disk would break up into rings similar to the rings of Saturn and that perhaps these rings

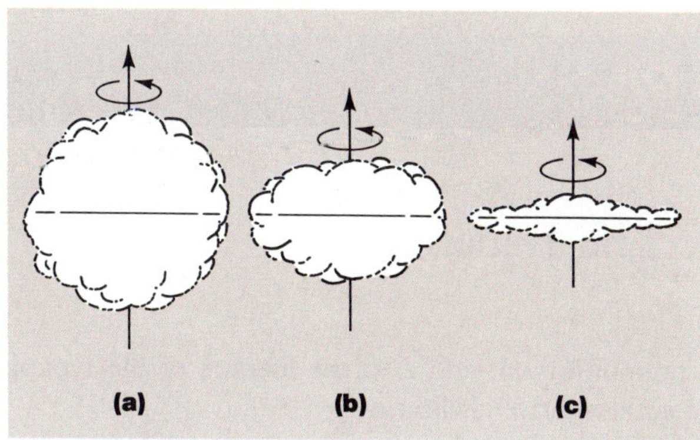

Figure 15–17. *As a cloud contracts, its rotational motion will cause it to form a disk.*

could form into the individual planets while the Sun was being formed from material in the center.

Application of Newtonian mechanics to such a contracting gas cloud, however, caused another problem. To understand this problem, consider what happens when a spinning ice skater pulls in his arms as illustrated in Figure 15–18. His rotation speed increases greatly. This increase in speed is predicted by Newton's laws and is a result of the law of ***conservation of angular momentum.*** We will not define angular momentum mathematically, but will simply state that the angular momentum of a rotating (or revolving) object is greater if the object is rotating (or revolving) faster or if the object is farther from the axis of its rotation/revolution. As the skater pulls in his arms, he decreases their distance from the axis of rotation; in the process decreasing their angular momentum. To make up for this, his entire body increases its rotation speed, keeping the total angular momentum approximately constant ("conserved"). (The total angular momentum would remain perfectly constant if it were not for small friction forces with the ice.)

The law of conservation of angular momentum must also apply to a contracting, rotating cloud of gas. Like the ice skater, the cloud speeds up its rotation as its parts come closer to the center of rotation. When calculations are made for a cloud contracting to form the Sun and planets, we find that the Sun should rotate much faster than it does; spinning around in a few hours instead of its observed period of about a month. In fact, the total angular momentum of the planets (because of their greater distance from center) is observed to be much greater than the angular momentum of the Sun. This should not occur, according to Newton's laws.

The contradiction of this well-established principle caused the evolutionary theories to lose favor early in this century. The opposing theory was a catastrophic theory.

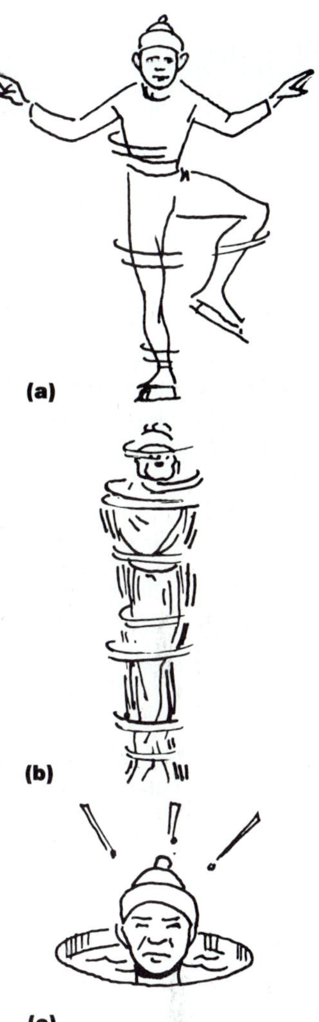

Figure 15–18. *As a spinning ice skater pulls in his arms, his rotational speed increases.*

Conservation of angular momentum: A law that states that the angular momentum of a system will not change unless an outside force is exerted on the system.

Catastrophic Theories

Contrary to what the name may imply, a ***catastrophic theory*** does not refer to a disaster, but rather to an unusual event; in this case the formation of the solar

Catastrophic theory: A theory of the formation of the solar system that involves an unusual incident such as the collision of the Sun with another star.

CLOSE UP

ET Life VI—The Life Equation

ASTRONOMER FRANK DRAKE OF Cornell proposed an equation for calculating the number of communicative civilizations in the galaxy. The equation has been written in a number of forms, including this one:

$$N = R_* f_p n_p f_l f_i f_c L$$

where R_* is the rate at which stars form in our galaxy, f_p is the fraction of these stars that have planet systems, n_p is the average number of planets per system that are Earth-like enough to support life, f_l is the fraction of those on which life develops, f_i is the fraction of life forms that evolve to intelligence, f_c is the fraction of the intelligent races who are interested in interstellar communication, and L is the lifetime of the typical communicative civilization.

The factors in the equation, as you might surmise, are far from well known. Let's take a quick look at each. Based on the average lifetime of stars and on the number of stars in the galaxy, we have a pretty good idea of the rate of star formation; the first term on the right is known to be in the range of one to ten stars per year of the Sun's type. The more we learn about how planetary systems are formed, the better we can estimate what fraction of star have planets circling them. Better knowledge of the origin of our own system will help in this regard and you will see in this chapter that we are beginning to understand the Solar System's formation. In addition, evidence is accu-

system by an unusual incident. In 1745 Georges Louis de Buffon proposed such an event: the passage of a comet close to the Sun. Buffon proposed that the comet pulled material out of the Sun to form the planets (Figure 15–19). In Buffon's time, comets were thought to be much more massive but in this century we learned that a comet's mass is not great enough to cause this breakup of the Sun. However, his basic idea—that a massive object exerted gravitational forces on the Sun, pulling material out and causing it to sweep around the Sun until it eventually coalesced to form the planets—still seemed a reasonable hypothesis. Such an event as the passage of a massive object so near the Sun would be very unusual but not impossible.

More recently it was suggested that the Sun was once part of a triple star system, with the three stars revolving around one another. As we have seen, such star systems are common, so this in itself is not a far-fetched idea. This particular catastrophic theory holds that the configuration was unstable and that one of the stars came close enough to cause a tidal disruption of the Sun that produced the planets. The close approach of this star also caused the Sun to be flung away from the other two stars.

Starting around the 1930s, astronomers began to find major problems with catastrophic theories. First, calculations showed that material pulled from the Sun would be so hot that it would dissipate rather than condense to form planets. A second problem involved deuterium, a form of hydrogen. Even the mulating that planetary systems are common. We might suspect that if this is so, many systems probably have life-supporting planets. (After all, our system *almost* has three planets capable of supporting Earth-type life.) Thus we can make reasonably confident estimates of the values for the first three terms in the equation. Beyond this, things get fuzzier.

As we will see in the next ET Life Close Up, an experiment has been performed that indicates that the most basic molecules of which life is made are formed easily and naturally under conditions expected to be prevalent on some young planets. Many biologists believe that, given the right conditions and enough time, life will develop from these components but they have no way of testing this hypothesis. The last three terms are even less well known. Does life evolve easily to intelligence, and if so, is it common for intelligent races to be interested in communication with aliens? (We seem to enjoy fiction on this topic, but we hesitate to spend money for a search.) Finally, how long does the typical civilization last? How long will ours last?

Various astronomers have inserted their best guesses into the equation, and—depending upon whether they are optimists or pessimists—have come up with answers anywhere from "only a few" to "millions." The equation's value, however, lies less in its mathematical answer than in showing us what we must learn in order to calculate an answer. It shows that we have a lot to learn.

outer portions of the Sun are too hot for deuterium to be stable and so not much deuterium exists in the Sun. Much more deuterium is found on the planets than in the Sun, indicating that the material of the planets could not have been part of the Sun.

Finally, as we will see later, evidence is accumulating that other nearby stars have planetary systems around them. A catastrophic theory would predict that such systems are rare since they are produced by unusual events. If we find planetary systems elsewhere, there must be some common process that forms them.

At the same time as these problems were being realized, a solution appeared for the angular momentum problem of the evolutionary theories and therefore catastrophic theories have been nearly abandoned in favor of modern evolutionary theories.

PRESENT EVOLUTIONARY THEORIES

In the 1940s the German physicist Carl von Weizsäcker showed that a gas rotating in a disk around the Sun would rotate differentially (the inner portion moving faster than the outer) and that this would result in the formation of

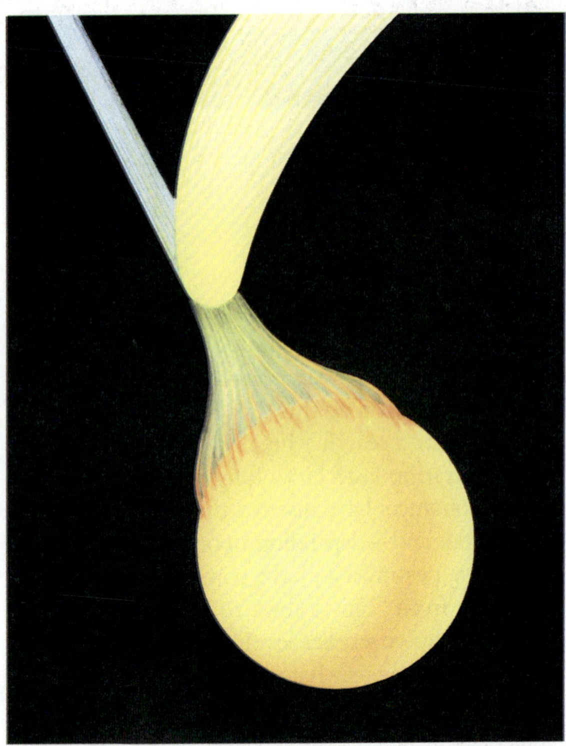

Figure 15–19. *Georges de Buffon hypothesized that a comet might have passed close enough to the Sun to pull material from it. We now know that comets don't have nearly enough mass to cause such disruption of the Sun. (In fact, comets can be destroyed by the Sun.)*

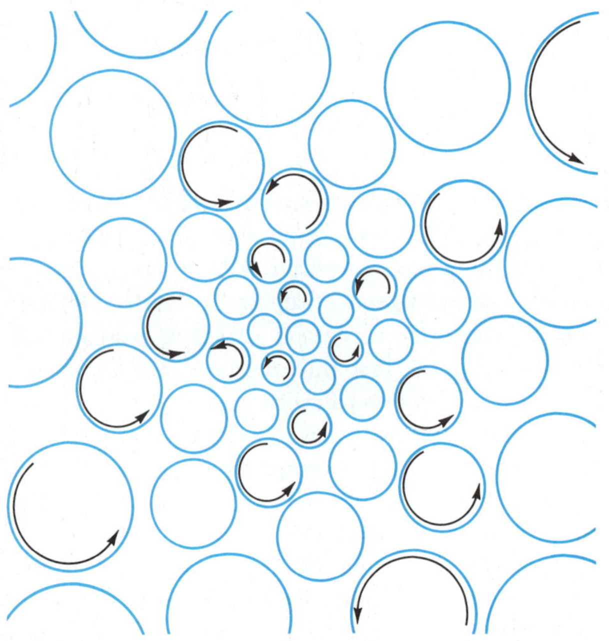

Figure 15–20. *Carl von Weizsäcker showed that eddies would form in a rotating gas cloud and that eddies nearer the center would be smaller.*

eddies as shown in Figure 15–20. As the figure shows, the eddies would be larger at greater distances from the Sun. According to his view, these eddies are the beginnings of planet formation and the eddies therefore explain the pattern of distances between the planets.

A real breakthrough occurred when it was realized that a mechanism exists to account for why the Sun does not rotate faster than it does. Before we explain this, let's start with the beginning of the scenario that is envisioned in today's theory.

As we have seen, protostars form from enormous interstellar clouds of gas and dust. Figure 15–21 shows the Orion Nebula, a stellar nursery where newly formed stars can be seen.

Recall that when a cloud collapses, any slight rotation it had at the beginning will result in a greatly increased speed of the central portion and the formation of a disk as the cloud condenses (Figure 15–17). At the center of the disk is the protostar, or the *protosun* in our system.

As the gases in the disk cool, they begin to condense to liquids and solids, just as water vapor condenses on the cool side of your iced drink glass. Nonvolatile elements such as iron and silicon condense first to form small chunks

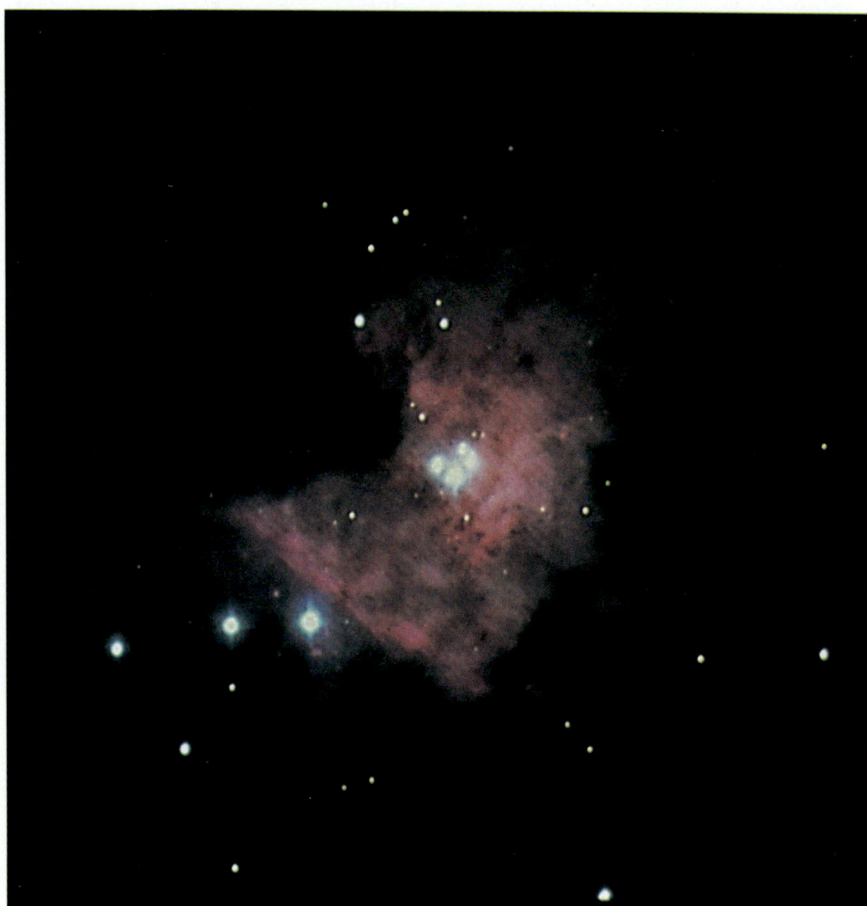

Figure 15–21. *A birthplace of stars, the Orion nebula contains four very bright stars that provide the light for the nebula and that must be very young, since such stars live short lives. (This is the same nebula shown in Figure 15–6 in infrared light.)*

of matter, or dust grains. Each of these grains has its own elliptical orbit about the center and as time passes, more matter condenses onto its surface. The orbits of these tiny objects are elliptical, so that they intersect one another. The resulting collisions between particles have two effects: (1) Particles involved in gentle collisions (as if they were rubbing shoulders with one another as they orbit) occasionally stick together and form larger particles, and (2) particles are forced into orbits that are more nearly circular.

As the matter sticks together, small chunks grow into larger chunks. Their increased mass causes nearby particles and molecules of gas to feel a greater gravitational force toward them. Since this force is still very small, the coalescing is a very slow process but gradually, larger particles—now called ***planetesimals***—sweep up smaller ones. Some planetesimals, resembling miniature solar systems, have dust and gas orbiting them; material that eventually condenses to become the moons we know today.

Planetesimal: One of the small objects that formed from the original material of the solar system and from which a planet developed.

In the inner regions of the new system, the high temperatures near the Sun will not permit condensation of the more volatile elements and so the planets that form there are made primarily of nonvolatile, dense material.

Farther out from the Sun, matter orbiting the new star is moving at a more

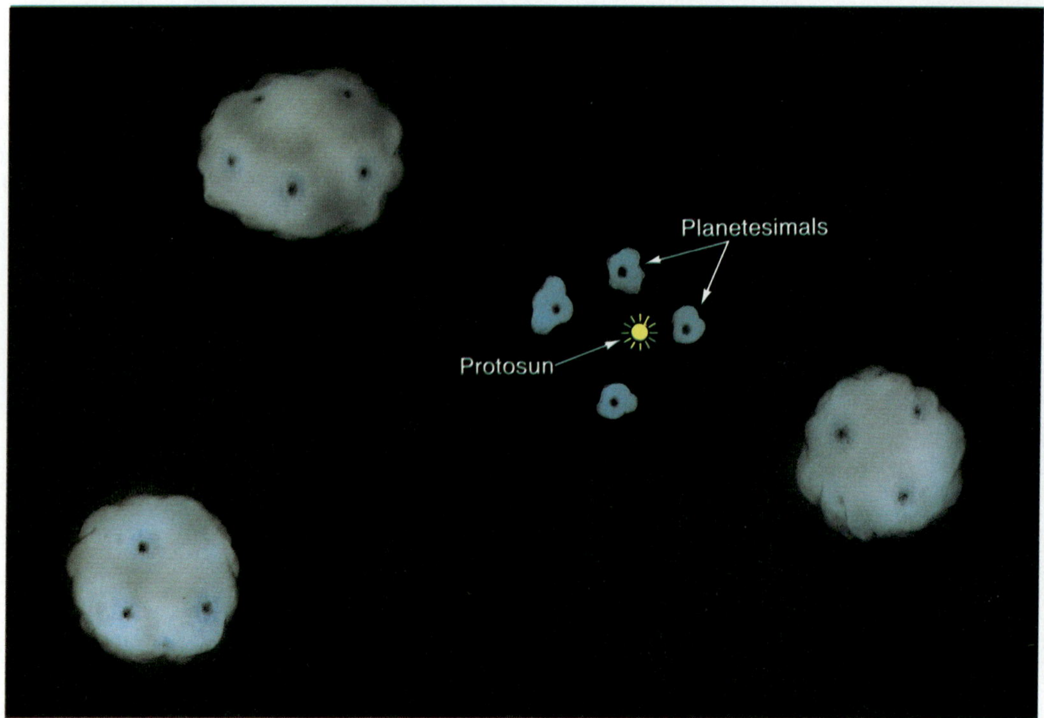

Figure 15–22. *At the stage of development shown here, planetesimals have formed in the inner solar system and large eddies of gas and dust remain at greater distances from the protosun.*

leisurely pace and the swirling eddies around protoplanets are more prominent. The situation at this time is illustrated in Figure 15–22.

A particularly large outer planet, Jupiter, gravitationally stirs the nearby planetesimals of the inner system so that the weak gravitational forces between them cannot pull them together. Today's asteroids are the remaining planetesimals.

Ion: An electrically charged atom or molecule.

While planet formation is taking place, the Sun continues to heat up. It heats the gas in the inner solar system and causes electrons there to be separated from their atoms, forming charged atoms (***ions***) and electrons. Recall that a magnetic field does not exert a force on an uncharged object but that if a magnetic field line sweeps by a charged object, a force will be exerted on that object. This is what must have slowed the Sun's rotation; the magnetic field of the rapidly rotating Sun exerted a force on the ions in the inner solar system, tending to sweep them around with it. Newton's third law tells us, however, that if the Sun's magnetic field exerts a force to increase the rotational speed of these particles, they must exert a force back on the Sun to decrease its rotational speed. So it is the magnetic field of the Sun, discovered rather recently, that provides the explanation for the fact that the Sun rotates so slowly; a fact that was once a stumbling block for evolutionary theories.

The solar system of our story is getting close to what is seen today. The system still contains greater amounts of gas and dust between the planets, however, than does today's solar system. It contains much more hydrogen and other volatile gases in the inner solar system than exists today. To help answer the question of how these gases were moved to the outer solar system and how in general the system was "cleaned up," we can again look into space at interstellar dust clouds.

In these clouds we see stars at various stages of formation. There is recent evidence that many stars, as they reach the main sequence, go through a period of instability like T Tauri stars and that during that time they experience tremendous solar flares; increasing the strength of the stellar wind. If our Sun went through such a period, the pulses of solar wind would sweep the volatile gases from the inner solar system. Even without this increased activity, it is expected that the solar wind would gradually move this material outward but if the Sun did go through this period, there is certainly no difficulty explaining why hydrogen and helium exist on the outer planets but not the inner. Once in the outer system, this material would gradually be swept up by the giant planets there.

Explaining Other Clues

As millions of years passed, remaining planetesimals fell to the planets and moons, resulting in the craters we see on these objects today.

Comets are thought to be material that coalesced in the outer solar system; remnants of small eddies. These objects would feel the gravitational forces of Jupiter and Saturn and many would fall into those planets. Small objects that formed beyond the giant planets' orbits, however, would be accelerated by Jupiter and Saturn as those planets passed nearby and would be pushed outward. These objects are the comets in today's Oort cloud.

Notice that the theory explains that nonvolatile elements would condense in the inner solar system but that volatiles would be swept outward by the solar wind. This accounts for the planets' differences in chemical composition. In fact, astronomers find that when compression forces are taken into account in calculating density, planets closest to the Sun contain the most dense and least volatile material, as would be expected from the theory.

Further confirmation of evolutionary theory is found in Jupiter's Galilean satellites. When we discussed these in Chapter 9 we saw that they also decrease in density and increase in volatile elements as we move outward from Jupiter. The formation of Jupiter and its moons must have resembled the formation of the solar system and we see in Jupiter's system the same pattern.

Some exceptions to the patterns that exist in the solar system can be explained by evolutionary theories. For example, in discussing Venus in a previous chapter, we pointed out that its rotation rate is gravitationally coupled with the Earth's motion. Thus the fact that it rotates in a direction different from the other planets is explained by the small gravitational tugs exerted by the Earth over long periods of time.

Other anomalies are not explained by evolutionary theories. Evolutionary theories predict that the equator of a planet will be in roughly the same plane as the planet's orbit around the Sun, but Uranus's equatorial plane is tilted 98 degrees from its orbital plane. It could well be that exceptions such as this were

The prediction concerning the equator's plane results from predictions about the motion of the swirling material which condensed to form the planet.

indeed catastrophic in origin. Perhaps a collision with another object resulted in an increase in the tilt of Uranus's axis of rotation. We have seen that one of its moons, Miranda, appears to have undergone a major collision in its past. Perhaps the two collisions are related.

Catastrophies may well have played a part in the formation of the solar system, but the evidence is that it was a fairly minor part, involving a relatively few objects, and that the overall formation of the system in which we live was evolutionary in nature. Nonetheless, the origin of the solar system is poorly understood. Pieces continue to fall into place but we still have much to learn.

PLANETARY SYSTEMS AROUND OTHER STARS?

In the last chapter we discussed a method of determining that a star is part of a binary system even when its companion is too dim to be seen. If we see a star that appears to wiggle in its position or that exhibits an elliptical motion, we can conclude that it is in orbit with another object. This provides us with a possible method of detecting the presence of a large planet in orbit around another star. If we hope to find a planet by such means, we must look at nearby stars.

A star named Barnard's Star is the second closest star to the Sun. In the first half of this century a back-and-forth motion was reported for Barnard's star. The motion is very slight, however, and detecting it involves comparing photographs taken over long periods of time under changing conditions. Most astronomers have considered the data very suspect, for the photos were taken under different conditions with instruments that had been changed over the course of the observations. Recent measurements of changes in the *radial* velocity of Barnard's star, however, agree with the original conclusions and indicate that the star does indeed have large planets revolving around it. (Large planets, of course, would cause the star to move more than small ones would.)

Recall that radial velocities are measured by the Doppler effect and that radial motion can thereby be measured with greater sensitivity than can tangential motion.

Other evidence exists. As stated above, astronomers continue to find indications of newly forming stars. The Infrared Astronomical Satellite has been especially useful in this search, since stars begin their lives glowing in the infrared region of the spectrum. This satellite instrument has recently found young stars with disks of gas and dust surrounding them. The most prominent of these is the star Vega, a bright star in the late summer sky. This finding fits perfectly with our picture of the young solar system. (Figure 15–23 is an artist's conception of the disk around Vega.)

The star T Tauri has a companion that emits significant radiation only in the infrared region. The evidence indicates that this companion has too little mass to have the pressure necessary to start fusion reactions but yet it emits significant infrared radiation. A possible explanation for its high temperature is that it is a giant planet in the process of formation, with dust and gas still falling into it. If so, this scenario lends support for evolutionary theories of planet formation; at least large Jupiter-size planets. (Recall that Jupiter still emits more energy than it absorbs).

Dust clouds of solar-system proportions have been detected around a number of other stars. Figure 15–24 shows one, β (Greek symbol for beta) Pictoris.

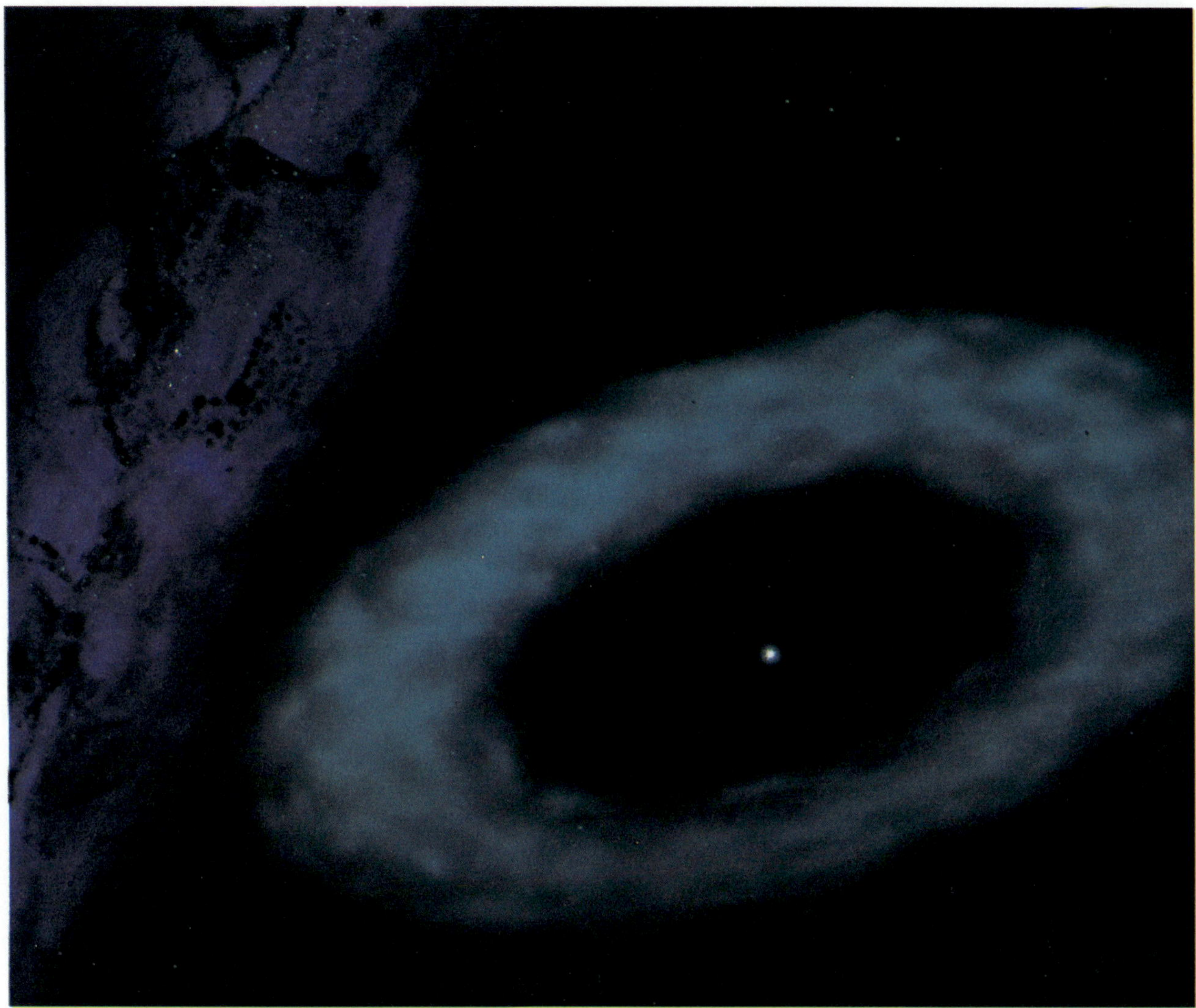

Figure 15–23. *A ring of dust appears to circle Vega, as shown in this painting.*

The central star appears dark in that photo because it was taken with a stellar coronagraph. The spectrum indicates larger-than-dust-sized objects surrounding the star in a sort of Oort cloud.

It is hoped that the Space Telescope can more definitely answer the question about the existence and nature of other solar systems, for with that telescope we may be able to see planets the size of Jupiter circling nearby stars.

If someday we find that planetary systems are common around other stars, we will be even more confident that our solar system formed by an evolutionary rather than a catastrophic process. The reasoning is as follows. If catastrophic processes are necessary for the formation of planetary systems, we expect there to be relatively few such systems since the special circumstances needed are, by their nature, rare. On the other hand, if stars form by the process described by evolutionary theories, we would expect planetary systems to be common. Notice

CLOSE UP

ET Life VII—The Origin of Life

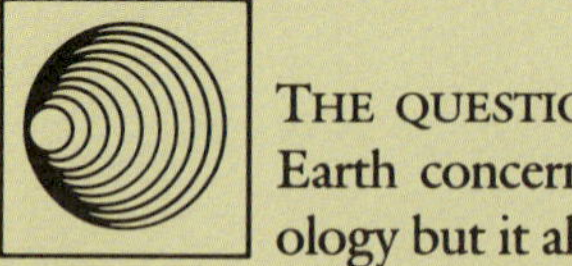

THE QUESTION OF THE origin of life on Earth concerns not only the study of biology but it also touches astronomy, for its answer will help us determine the probability of the existence of extraterrestrial life. Before the mid-1800s it was thought that life is formed spontaneously; in a short period of time. The evidence for this came from the common experience that worms, maggots, and various other creatures often appear in grain that has been stored for some time, even if the grain is stored in a closed container. It was thought that the various forms of animal life were generated from the non-animal life of the grain. This hypothesis was disproven when eggs and larvae were discovered in grain. (Some people even claimed that the mice that sometimes appeared in crumpled bedclothes and old rags had been spontaneously created there. When rags are stored in closed containers, however, no mice showed up.)

In the latter half of the nineteenth century, the theory of evolution was developed, explaining how higher forms of life evolve from more primitive forms; but this still did not answer the question of the beginnings of life. Then some 50 years ago, J. B. S. Haldane, a Scottish biochemist, and A. P. Oparin, a Russian biochemist, proposed that soon after the Earth's formation, the necessary chemical elements were present for complex molecules to form; molecules that are needed for life. The seas of the early Earth were thought to be made up primarily of water, methane, ammonia, and hydrogen and the two men hypothesized that these molecules would spontaneously collect into more complex, organic (carbon-based) molecules.

The Haldane/Oparin hypothesis remained an interesting conjecture until 1950, when Harold Urey (the Nobel Prize winner who discovered deuterium) suggested that his graduate student, Stanley Miller, test the hypothesis by simulating the conditions of the early Earth. Miller put a mixture of water, hydrogen, methane, and ammonia into a sealed container. He heated the liquid—the young Earth was hot—and he used an electrical source to produce sparks in the gas above the liquid. The sparks simulated lightning, which must have been common in the Earth's young atmosphere.

After Miller's apparatus had heated and sparked for a week, the mixture had turned dark brown. Analysis showed that it now contained large amounts of four different amino acids (complex organic molecules that form the basis for proteins). He also found fatty acids and urea (a molecule that is necessary in many life processes).

The Miller-Urey experiment showed that given the right chemicals and a source of energy, chemical reactions will occur that produce the building blocks of life. As you might recall from our discussion of Venus and Earth, it has since been learned that the early Earth's atmosphere was made up primarily of carbon dioxide, nitrogen, and water vapor rather than the four compounds used in Miller's experiment. When these compounds are used in the sealed apparatus, however, even more complex organic molecules are produced. Furthermore, if instead of using electrical sparks for energy, one uses ultraviolet light (which strikes Earth from the Sun), the same results are achieved.

It must be emphasized that the Miller experiment did not produce life, only organic molecules. The experiment confirmed the Haldane/Oparin hypothesis, however, and showed that such molecules will form easily in a short time. The results of the experiment would indicate that the early Earth must have contained seas made up of an organic goo; perhaps like chicken bouillon.

More recently, traces of amino acids have been found in some meteorites. In addition, there is spectroscopic evidence that organic molecules exist in intersteller clouds. Thus the molecular building blocks of life seem to be common in the universe. How does life form from these blocks? We don't know, but it seems that nature's deck may be stacked in the direction of life.

Figure 15–24. *This photo of Beta Pictoris shows a disk of sold particles around it.*

that although the formation of planetary systems by catastrophic events is unlikely, this in itself is not an argument against such a scenario in the case of the solar system. If there is any possibility at all that a catastrophic event can cause a planetary system, it could well have happened here. On the other hand, if we find that evolutionary development of planetary systems is common, the discovery will lend support for hypothesis that this is what occurred in our own system.

CONCLUSION

Stars are formed from giant interstellar clouds of gas and dust. After one of a number of possible mechanisms triggers a cloud to fragment and collapse, gravity takes over within some of the segments, causing the matter to fall inward and increase in temperature as it does so. Finally, the inner portion of the collapsing protostar becomes hot and dense enough that nuclear fusion starts. The cloud is now beginning its life on the main sequence as a normal star.

During its approach to the main sequence, at least one star—the Sun—experienced another process. The material that was left in orbit around the central star began to cool and coalesce into planetesimals. These objects then became planets circling the star.

The length of a star's life depends upon its mass, with more massive stars living shorter lives. Eventually, however, every star runs out of the hydrogen that fuels its nuclear furnace. This results in further collapse of the core and the resulting increase in temperature of the core causes the outer portions of the star to expand so that the star becomes a red giant; many times larger than its previous self. This is the end of the star's peaceful life on the main sequence.

In the next chapter we will investigate how stars die. We will see that they become some of the most unusual objects imaginable.

RECALL QUESTIONS

1. What is a protostar?
2. List three possible events that may be responsible for beginning the collapse of interstellar clouds.
3. Which stars take longest to complete their protostar stage? Why?
4. What is the source of the energy that powers protostars?
5. What energy source provides the high temperature needed to start the fusion of hydrogen?
6. What causes there to be a lower limit to the possible mass of a star? An upper limit?
7. Explain how galactic clusters provide observational evidence for theories of stellar evolution.
8. Explain what is meant by a *stellar thermostat* and explain how it works.
9. Which stars live the longest lives on the main sequence? The shortest? Why?
10. What causes a star to stop its gravitational collapse once it has reached the main sequence?
11. What causes the main sequence to have a width rather than being a simple line?
12. What causes a star to get larger as it becomes a red giant? What causes its surface to cool?
13. About how large will the Sun become when it reaches the red giant stage?
14. In what way did the discovery that the Sun has a magnetic field impact the question of the origin of the solar system?
15. How do evolutionary theories explain the pattern of spacings between planets?
16. How do evolutionary theories explain the pattern of chemical abundance among the planets?
17. If definite evidence were found that planetary systems exist around other stars, would this lend support to either the catastrophic or evolutionary theory? If so, which?
18. Why have catastrophic theories been (nearly) abandoned today?
19. How do evolutionary theories account for the asteroids? The Oort cloud?
20. Describe the evidence that planetary systems exist around other stars.

QUESTIONS TO PONDER

1. Imagine yourself as an extraterrestrial two-day visitor to an Earthly forest. What might you be able to learn about the life cycle of trees? Be careful to base all your conclusions only on the two days of observation.
2. Explain how an examination of the H-R diagram for a large group of stars can tell us the relative length of the various stages of a star's life.
3. Why do astronomers feel that there must be a triggering mechanism for the beginning of an interstellar cloud's collapse? Why couldn't it happen on its own?
4. Why do we think that most stars would be formed from a *rotating* disk of gas? Couldn't it just as well be nonrotating?
5. Massive stars have much more hydrogen in their cores than less massive stars. Why do they run out of hydrogen faster than stars of low mass?
6. The compositions of the Sun and of the Earth are in some ways similar and in some ways different. How are these similarities and differences explained by the evolutionary theory of the solar system's formation?
7. How do evolutionary theories explain observations 1, 2, and 4 (as given in the section "Evidential Clues from the Data")? How do catastrophic theories explain these observations?
8. Hypothesize as to how Venus acquired a retrograde rotation. (Hints: One possible explanation was given in a previous chapter in discussing synchronous rotation. Another involves planetesimals.)
9. Explain the problem presented to evolutionary theories by the law of conservation of momentum and describe how present theory accounts for the observations.
10. Describe similarities in properties between the Jupiter system and the entire solar system. How does the evolutionary theory explain these similarities?
11. Consult other books and report on the catastrophic theory of M. Woolfson, proposed in 1960.
12. Which factors in the Drake equation (Close Up: The Life Equation) are known with most certainty and which are least known?

CALCULATION

1. Insert your own best guesses into the Drake equation and calculate the corresponding number of communicative civilizations.

CHAPTER 16

The great supernova of 1987 appears to the lower right of the Tarantula nebula.

The following quote from Stephen Maran indicates the excitement generated among astronomers by the great supernova of 1987.

"February 25, 1987, was a quiet day at the Goddard Space Flight Center outside Washington. I was walking down a hallway toward the Coke® machine when I overheard the excited comment, 'It's the worst one in 300 years!' A group of scientists were milling around, handing a telegram back and forth, and talking excitedly as new arrivals joined them. 'My God,' I thought, 'there must have been a horrible volcanic eruption.' But I had misheard. It was not an Earthly disaster but a cosmic cataclysm, spotted the day before from South America, that was arousing normally unflappable astronomers to a level of excitement that, for some, approached ecstasy. It was not the 'worst' but the 'first' and not a volcanic eruption but the explosion of a star in our own backyard."

*Stephen P. Maran, "A Blue Supergiant Dissects Itself in a Cosmic Explosion," *Smithsonian,* April 1988, pp. 46–47.

The Death of Stars

ON THE NIGHT OF February 24, 1987, every major telescope in the southern hemisphere was pointing at the same object. A NASA satellite, the International Ultraviolet Explorer (IUE), had been turned toward the object. Soviet cosmonauts aboard the Mir space stations used instruments to detect X rays from it, as did the Japanese satellite Ginga. Astronomers who had reserved telescope time planning to study other objects—if they were fortunate enough not to have been bumped from the telescopes by their seniors—studied this object instead. The focus of all this attention had been discovered just the night before, although the light they were seeing had left the object 170,000 years before. It was a supernova, the brightest one to have appeared in our skies in 383 years. This cataclysmic death of a massive star will be the subject of intense study for decades to come and is providing astronomers the best opportunity yet to test theories of stellar evolution and, particularly, of the death of stars.

We have seen how stars form and how they join the main sequence. We wonder if it is common during a star's formation for planets to form around it and if while a star is on the main sequence it is common for life to evolve on one of those planets. Overall, though, stars on the main sequence are a very tranquil lot, especially compared to what follows after they leave the main sequence.

When stars end their lives on the main sequence, most of them swell up into red giants; enveloping any nearby planets or stellar companions they may have. Some puff their outer layers away, but some blow them away in cataclysmic supernova explosions such as the one that is still under intense study in the skies of the southern hemisphere. Some end up as supercompacted neutron stars; perhaps sending out powerful lighthouse beams across space. Some end their lives by swallowing themselves; becoming black holes.

As we study the unusual events accompanying the ends of stars' lives, we will see yet another example of the formation and confirmation of a theory. In addition, we will see that the major supernova of 1987 is presenting new questions to astronomers as quickly as it is answering old questions.

STAR DEATH

Up until the red giant stage, the primary difference between the evolution of stars of various masses is in the amount of time they spend as protostars and as main sequence stars. From this point on, however, the mass of a star determines which of a number of very different paths its life will take. We will discuss each of these in turn, beginning with those stars of very low mass. Since mass classifications are fairly arbitrary and astronomers have not agreed on official names for the different classes, we will group them like boxers; calling them—in order from least massive to most massive—flyweights, lightweights, middleweights, and heavyweights. Remember, though, that we are really referring to their mass rather than their weight.

Flyweight Stars

Recall from Chapter 12 that in the discussion of energy transport within the Sun it was stated that except for the outer layers of the Sun, little convection takes place in its interior. This is why, once hydrogen is used up in the core, the core is not replenished with fresh hydrogen from outside. In the least massive stars (those with a mass of less than about 0.4 solar masses), however, convection occurs throughout most or all of the volume of the star (Figure 16–1). Such low-mass stars use their fuel extremely slowly and since the universe is thought to be not more than 20 billion years old, none of these stars has had time to end its main sequence life.

When, finally, such a star does run low on its hydrogen fuel, it will contract and heat up just as do the cores of more massive stars. In this case, however, the added heat from the core will be transmitted outward by convection rather than by radiation. Thus the entire star will contract; a red giant will not form. In addition, in the flyweight star there will not be enough gravitational force to compress the star sufficiently to cause helium fusion. Instead, the star will heat up due to contraction and in so doing it will move toward the lower left portion of the H-R diagram and become a white dwarf. It will be white hot because of its energy of contraction but it will still be a dwarf; even smaller than it was before.

Once a white dwarf shrinks to its minimum size, perhaps about the size of the Earth, it will no longer have an energy source but it will continue to radiate its energy, cooling gradually and becoming a burned-out cinder; a ***black dwarf***. No low-mass stars have yet become white dwarfs (much less black ones) but as we will see, there is another source of white dwarfs and these are the ones we observe and plot on H-R diagrams.

Black dwarf: The theorized final state of a star, in which all of its energy sources have been depleted so that it emits no radiation.

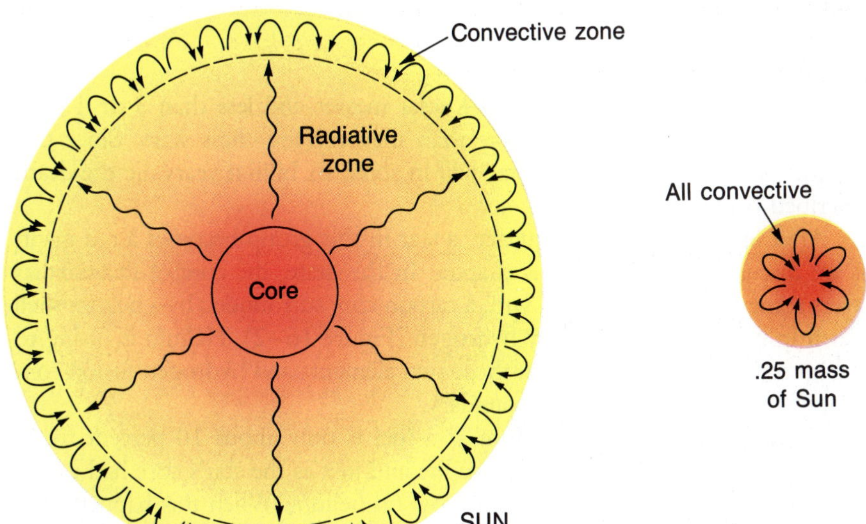

Figure 16–1. *The Sun (at left) contains a large radiative zone. A star of very low mass, on the other hand, consists of one convective zone, meaning that all of its material mixes.*

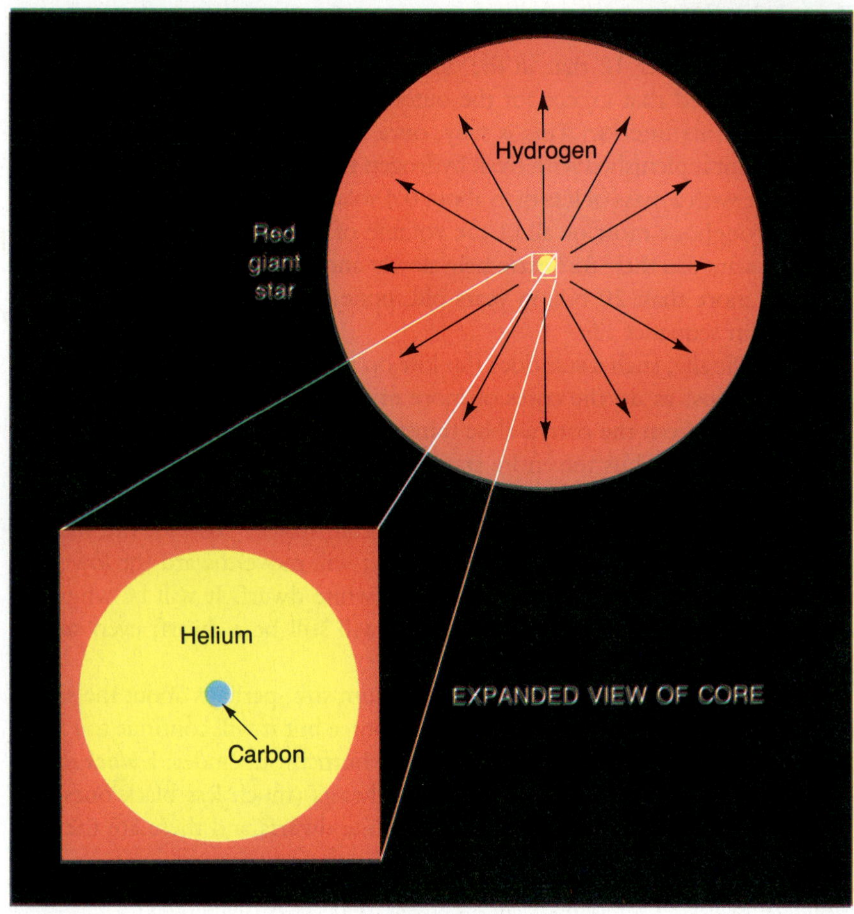

Figure 16–2. *When a red giant reaches the stage where it has a carbon core, the heat from that shrinking core ignites helium fusion in a shell around it. At the same time hydrogen is fusing in a shell beyond that. This activity occupies very little of the volume of the star.*

Lightweight Stars

Stars greater in mass than about 0.4 solar masses and less than 3 or 4 solar masses die in a more exciting way than the very low-mass stars. Since this category includes the Sun, our star will in some six billion years die the death described below.

After leaving the main sequence, a star in this group becomes a red giant, as we discussed in the previous chapter. In this state, the core of the star has layers somewhat like an onion, with a carbon core surrounded by a helium shell which in turn is surrounded by hydrogen (Figure 16–2). Helium is fusing to carbon at the boundary between those two elements and hydrogen fusion continues at the helium-hydrogen boundary.

A red giant's life using helium as a fuel is only about 10 percent to 20 percent of its previous main sequence life and just as the star's core had earlier run out of hydrogen, it likewise runs low on helium. When this happens, the core again shrinks and heats up due to gravitational energy. The star becomes even brighter—a supergiant.

Recall that when the core earlier ran low on hydrogen and collapsed, it was

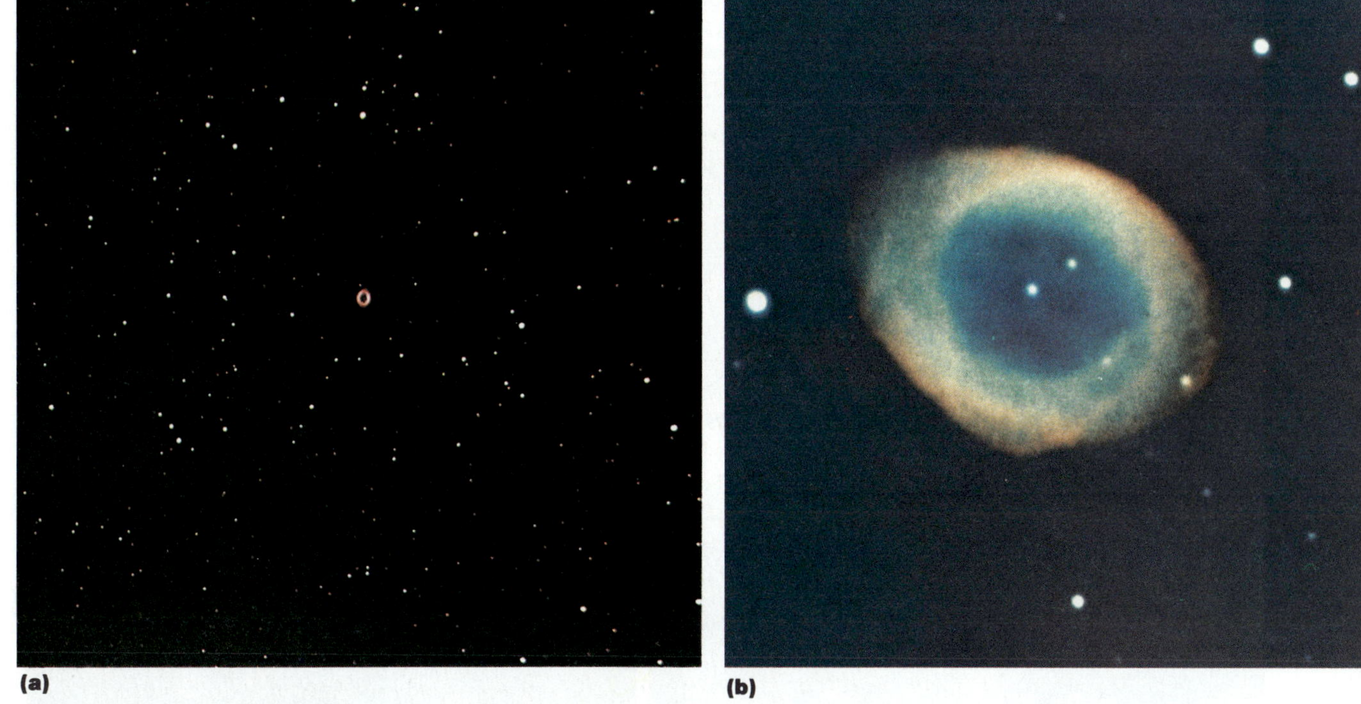

(a) (b)

Figure 16–3. *(a) The Ring Nebula as seen in a small telescope. (b) The five-meter Hale telescope reveals much more detail.*

unstable until it settled down into fairly constant helium fusion. Now as it runs out of helium, the star again becomes unstable. Helium fusion results in carbon building up in the core but the core does not become hot enough to allow the fusion of carbon into more massive elements. The core therefore begins to shrink again and heat up quickly. The energy released by this heating causes the outer portions of the star to be blown outward at a speed of 20 to 30 kilometers/second. Although this material cools down, the core left behind is extremely hot; perhaps 100,000 degrees C.

We have seen that an object at a very high temperature emits large amounts of ultraviolet radiation. The ultraviolet radiation emitted by the hot core of the star causes the escaping part of the former red giant to glow as it blows away. Figure 16–3 shows the Ring Nebula in Lyra, the most famous of a group of objects that were named ***planetary nebulae*** when they were discovered in the nineteenth century. They were given this name because they seemed to resemble Uranus and Neptune when viewed in the relatively small telescopes available at the time. In fact, they have nothing to do with planets. Doppler effect analysis reveals them to be material moving outward from a very hot central star. Although many appear doughnut-shaped in a photo, this simply may be the result of our looking through a shell of material as shown in Figure 16–4. Figure 16–5 shows other planetary nebulae and illustrates that not all of them appear as rings.

Planetary nebula: The shell of gas that is expelled by a red giant near the end of its life.

Astronomers are now convinced that planetary nebulae are the outer portions of a star like the Sun after it has exhausted the helium in its core. The expelled material will continue to disperse, leaving behind a hot core of carbon and oxygen. On the H-R diagram this star moves as shown in Figure 16–6;

Figure 16–4. *Three lines of sight through the spherical shell of the Ring Nebula. Lines X and Z pass through much more of the glowing nebula than does line Y and thus the nebula will appear brighter here.*

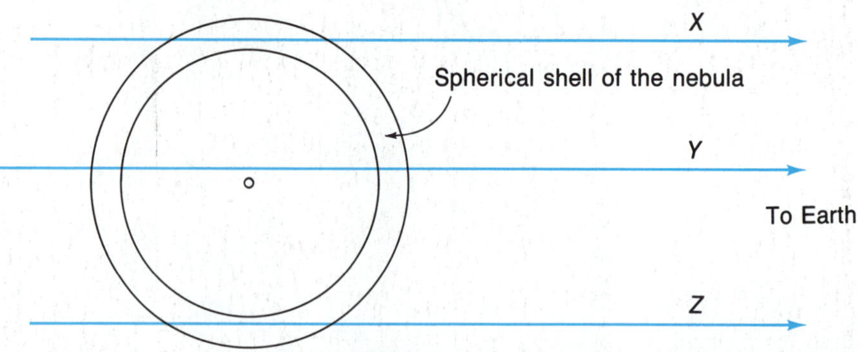

Figure 16–5. *The planetary nebula NGC 6781 (above) looks somewhat like a ring but the Dumbbell Nebula (M27) is more irregular. The bright star in the center of NGC 6781 is the remaining core of the star.*

the change in position on the diagram is due to the emission of the star's outer portion. When the core is revealed, it is extremely bright and hot and emits large amounts of ultraviolet radiation. The core remains at its very luminous stage for a relatively short time; quickly moving down the H-R diagram to become a white dwarf.

White dwarfs are approximately the size of the Earth yet they contain much of the material of the original star. Although the part that is blown away in the planetary nebula makes up almost all of the *volume* of the star, the outer layers of a star have very low density and so the planetary nebula leaves behind the major portion of the star's original mass. The density of the material of a white dwarf is unimaginably great; about a million grams per cubic centimeter. If

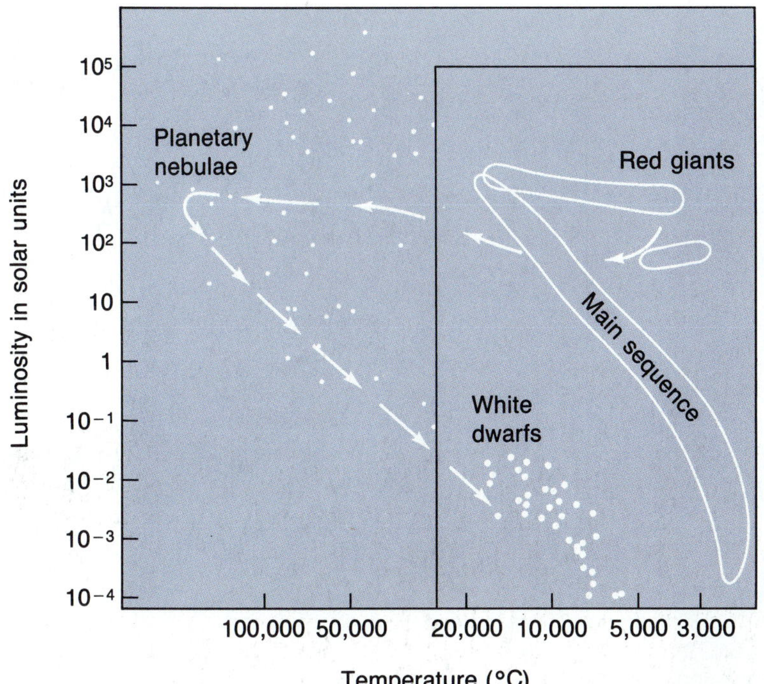

Figure 16–6. *The H-R diagram we have seen before is shown at the right. In order to show the path taken by stars as they pass through the planetary nebula stage and become white dwarfs, we must extend it to higher temperatures. Fewer than 2000 planetary nebulae are known.*

Table 16–1 A Typical White Dwarf

Absolute Luminosity	+11
Mass	0.8 solar masses
Diameter	0.01 solar radii (1 Earth radius)
Density	10^6 gm/cm^3
Surface Temperature	15,000°C

such material existed on Earth, a cubic centimeter of it would weigh a ton! Table 16–1 contains data for a typical white dwarf.

Our galaxy contains billions of white dwarfs. Because their low luminosities allow us to see only those nearby, we have identified fewer than 500. The first white dwarf to be discovered was a binary companion of the brightest star in our sky, Sirius (Figure 16–7). Sirius B was discovered in 1844 by observing the motion of Sirius (now called Sirius A) around the center of mass of the system. It was not until 1862 that Sirius B actually was seen. Sirius B is 1/10,000 as bright as Sirius A and its orbital motion reveals its mass to be about one solar mass. Analysis of its color and luminosity shows that it is slightly smaller than the Earth and so it is thought to be a fairly typical white dwarf.

Alvan Clark, an American telescope maker, discovered Sirius B by accident while testing a telescope.

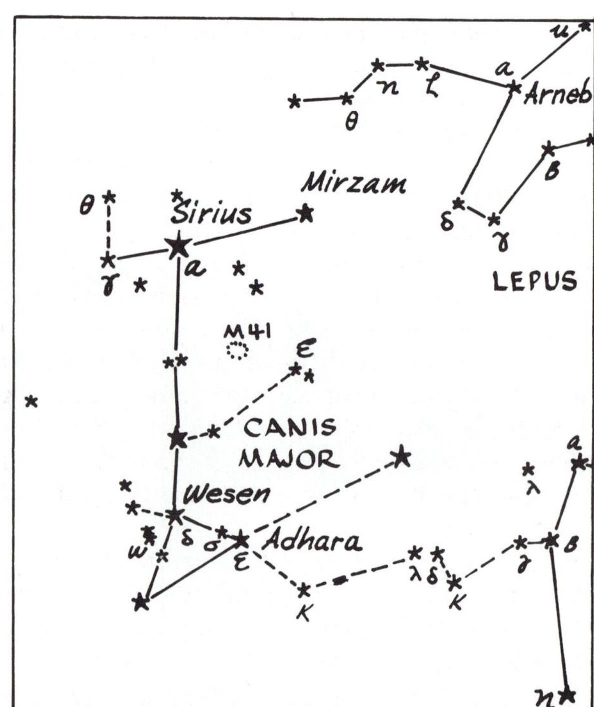

(a)

Figure 16–7. *(a) Sirius is the brightest star in Canis Major (and in our sky, for that matter). (b) Sirius is revealed as two stars in this telescopic view. The spikes radiating from Sirius A are caused by diffraction around mirror supports within the telescope.*

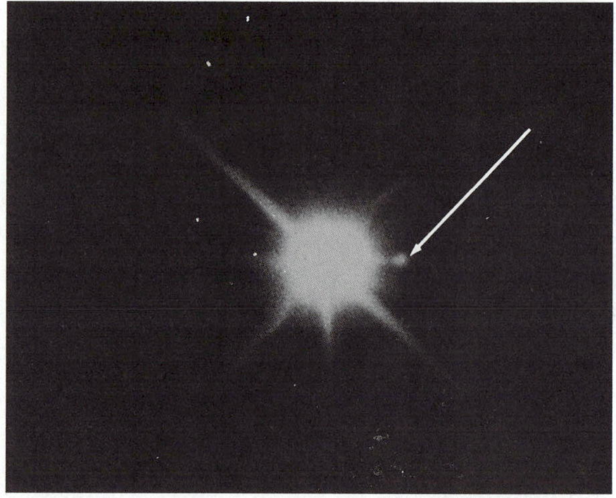

(b)

Electron degeneracy: The state of a gas in which its electrons are packed as densely as nature permits. The temperature of such a high-density gas is not dependent on its pressure as it is in a "normal" gas.

Contrary to rumor, the degenerate core of a star does not undergo improper motion.

Figure 16–8 is a diagram showing the steps by which lightweight stars become white dwarfs.

Degeneracy and the Chandrasekhar Limit

Before proceeding to discuss middleweight and heavyweight stars, we must pause to look at a major difference between the core of a main sequence star and that of a giant or a white dwarf. Recall that the core of a main sequence star is a gas and therefore has a built-in thermostat that controls the rate of its reactions.

If enough pressure is exerted on the core of a star, its matter becomes so tightly compressed that electrons are no longer associated with individual nuclei but roam freely among the nuclei. The gas consists—instead of atoms and molecules—of nuclei and electrons. A gas in this state acts differently from a normal gas and is said to be ***degenerate,*** or ***electron degenerate***. A degenerate gas no longer acts as if it has a thermostat controlling its temperature. The core of a star made up of electron-degenerate gas can continue to heat up without the expansion that accompanies a normal gas when it is heated.

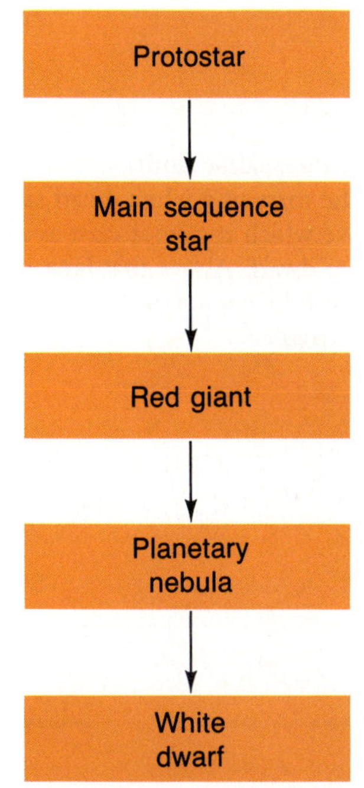

Figure 16–8. *These are the steps a lightweight star takes in going from protostar to white dwarf.*

The core of a star grows dense enough to become degenerate at the onset of the helium burning stage. This explains two things we have already noted: Helium fusion proceeds more quickly than did hydrogen fusion and great instabilities occur in the star's core during this change. The stellar thermostat is no longer in control of the size of the core. Thus—contrary to the way a normal gas works—the core can continue to contract as it heats up. The core left behind by the planetary nebula is made up of degenerate gas. White dwarfs are degenerate.

The size of a white dwarf is therefore determined by the nature of electron degeneracy rather than by the stellar thermostat. Figure 16–9(a) is a graph that shows how the radius of a main sequence star changes with its mass. Notice that stars of greater mass have larger radii (which probably fits your intuition as to how it should be). Part (b) of the figure shows the same relationship for white dwarfs. Here we see just the opposite—the more massive the star, the smaller it is! This is very different from our experience with nondegenerate matter but it is the case for white dwarfs.

In 1930, astrophysicist Subrahmanya Chandrasekhar of the University of Chicago calculated that there is a limit to the mass a white dwarf can have. If a white dwarf were more massive than 1.4 solar masses, its gravitational force

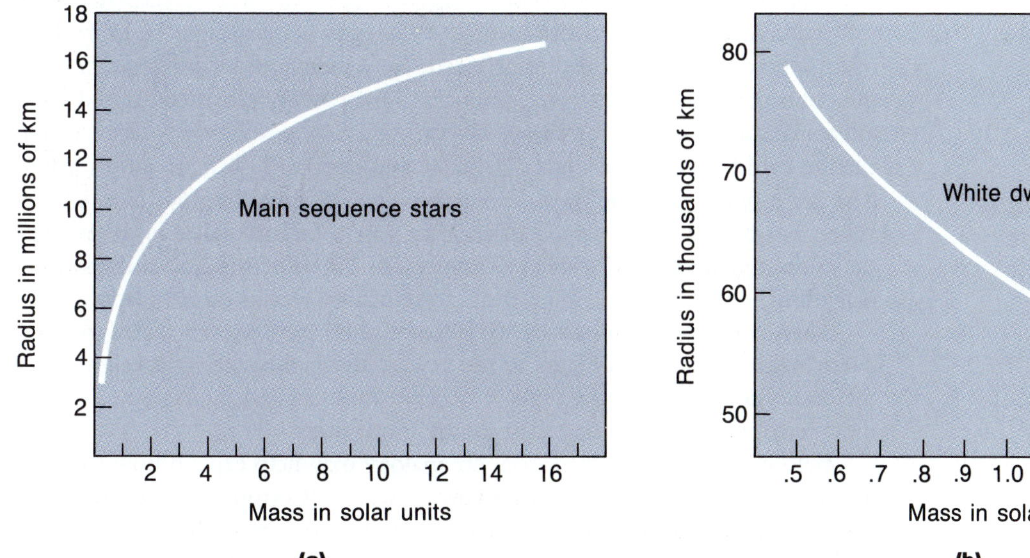

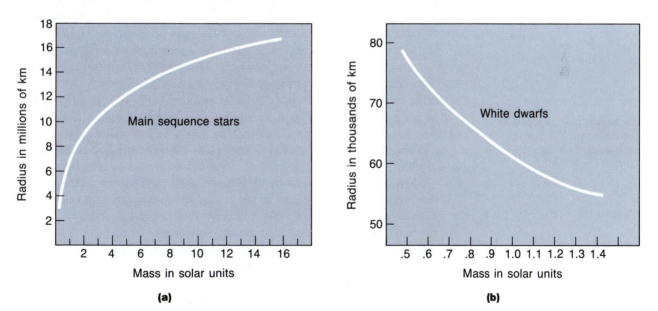

Figure 16–9. *Graph (a) shows that more massive main sequence stars have larger radii. In the case of white dwarfs, however, the more massive the star is, the less its radius.*

would be too great for it to be supported by the pressure of its degenerate gas. This means that all white dwarfs must have masses less than 1.4 solar masses. Stars with masses of up to perhaps three or four times the Sun's mass can end up as white dwarfs only because they lose mass during the red giant stage; particularly at the end, when the planetary nebula is blown off. Stars more massive than this retain a core of more than 1.4 solar masses, the so-called ***Chandrasekhar limit.*** These stars cannot form white dwarfs and such middleweight and heavyweight stars end their lives in a dramatically different way than do less massive stars.

Chandrasekhar limit: The limit to the mass of a white dwarf star above which it cannot exist as a white dwarf. Above that limit it will not be supported by electron degeneracy.

MIDDLEWEIGHT AND HEAVYWEIGHT STARS—SUPERNOVAE

Although the final destinies of middleweights (perhaps four to eight solar masses) and heavyweights (greater than about eight solar masses) are far different, their next step beyond the red giant stage is similar, so we will consider them together for now.

Recall that the core of a lightweight red giant star is made up of carbon and oxygen. If this core has a mass of less than 1.4 solar masses, the reaction will shut down when the helium fuel is exhausted and the star will die a long, slow death as a white dwarf supported by electron degeneracy. In more massive stars, temperatures in the core are greater and further fusion produces heavier elements, including neon, silicon, and even iron. As each of these elements is produced, energy is released, the core heats, and the star expands and shifts on the H-R diagram.

Finally, a cataclysmic event occurs that we are only now beginning to understand. A tremendous explosion sends the rest of the star blasting away from the core. A supernova has occurred. A supernova differs greatly from the way material leaves a red giant during the planetary nebula stage, for in the case of a supernova the initial explosion occurs in a matter of seconds, rather than the outer parts of the star being gradually blown away. When contrasted with a massive star's lifetime of millions of years, it is astonishing that such a major change could occur in seconds. The star continues to brighten, and in a few days (or months, in some cases) it may become 10^8 times brighter than it was. When a supernova is observed in another galaxy, it may shine brighter than the entire remainder of the galaxy (Figure 16–10)! The supernova then gradually dims; typically taking a few years to return to near its original brightness.

An increase of brightness of 10^8 times corresponds to a decrease in magnitude of 20.

There are three reasons why we observe only a few supernovae. First, the overwhelming majority of stars in the sky are flyweights and lightweights and these stars do not experience supernova explosions. Second, supernovae are very short-lived, so that not many exist at any given time.

Before listing the third reason, an analogy may help explain how the two factors above affect how many supernovae we see. Consider a supermarket in your town: Supernova Grocery. What determines the average number of people in the market at one time? First, if only a few of the townspeople shop at that store, it will be less crowded than if everyone in the town goes there. (Only a small fraction of stars become supernovae.) Second, if the average shopper

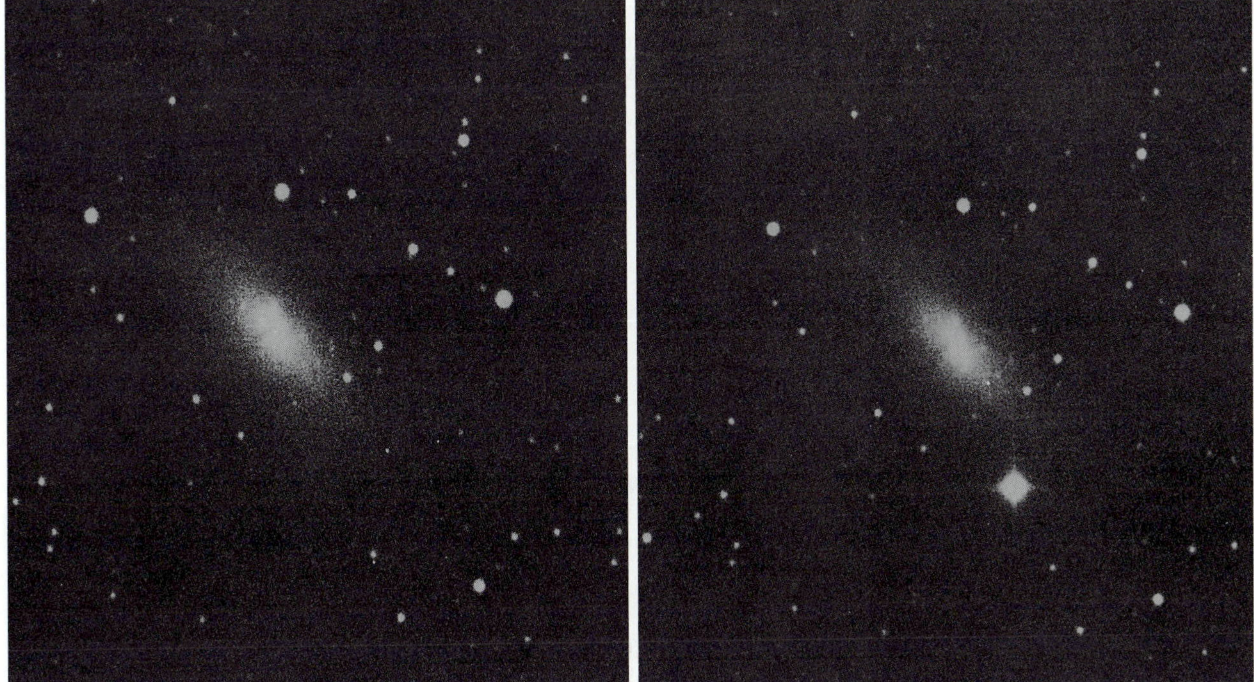

Figure 16–10. *The photo at left was taken in 1959 and shows the central part of NGC 5253 (a magnitude 11 galaxy in Centaurus). In the photo at right (taken in 1972), we see a supernova in the outer part of the galaxy. The supernova is almost as bright as the rest of the galaxy.*

spends only a few minutes in Supernova Grocery, it will be less crowded than if "Mr. Average" wanders around the store for hours. (For stars, the supernova stage is short-lived.)

We know, however, that the galaxy contains so many stars that a supernova occurs in it every few years. Why don't we see more in spite of the two reasons given above? Enter the third consideration: Interstellar dust clouds prevent us from seeing the entire galaxy. Thus, even though supernovae are not really uncommon events in the galaxy, it is seldom that we see one.

We know of three supernovae that have been seen in our galaxy, in the years 1054, 1572, and 1604. The most spectacular on record occurred in the constellation Taurus on July 4, 1054, and was described by Chinese astronomers who reported that it was bright enough to be seen in daylight and to read by at night. It remained bright for a few weeks, then gradually faded until it disappeared from view after about two years. Invention of the telescope was centuries away, of course, so the Chinese of the eleventh century had no way to continue observing their "guest star" (as they called the object). In 1731 an amateur astronomer reported a small nebula in Taurus, near the bull's horns (Figure 16–11), and two hundred years later Edwin Hubble discovered that the nebula consists of material expanding at a rate such that it must have begun its expansion at the time reported by the Chinese for the guest star. Figure 16–12 is a photograph of the supernova remnant, called the Crab Nebula because of its shape. Telescopic observations taken over the last half-century show its growth; revealing that its outer portions are still expanding outward at about 1400 km/s and that it is now about 4.4 light years in diameter.

Since the Crab Nebula is 6500 light years from us, the Supernova actually occurred 7400 years ago (6500 + the 900 years since 1054), but we speak of it as if it happened about 900 years ago.

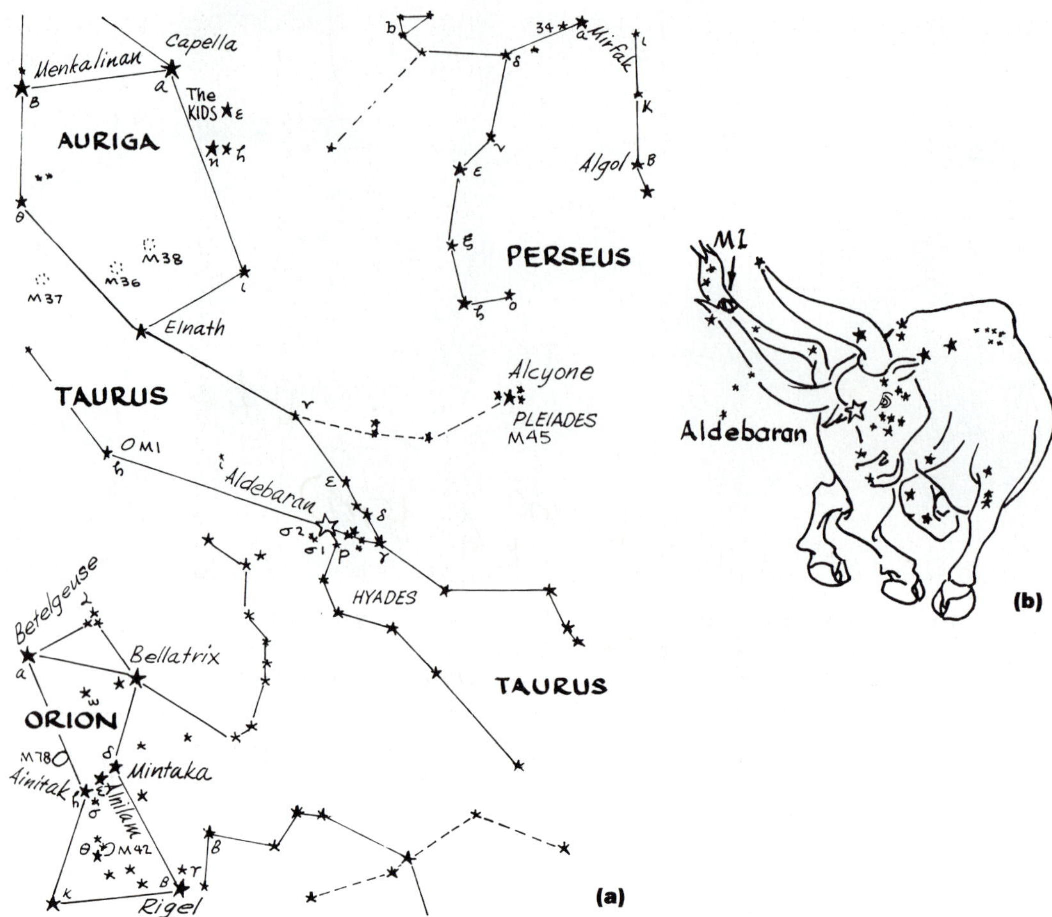

Figure 16–11. *The Crab Nebula is M1 (found between the horns of Taurus the bull) at left center in part (a).*

We have also been able to find remnants of the supernova seen by Tycho Brahe in 1572 (which provided evidence for him that the heavens were not unchanging) and the one seen by Johannes Kepler in 1604. Kepler's supernova occurred just before the invention of the telescope and since that time there have been no supernovae visible in our galaxy.

Table 16–2 lists some major supernova remants. Figure 16–13(a) and (b) are photos of two of these.

Until recently, most supernovae have been discovered by accident; perhaps one per month as astronomers who were doing other investigations happen to observe them in distant galaxies. In most cases, supernovae are observed days after light from the initial explosion has reached us and only in a few cases have they been observed before reaching peak intensity (which normally takes a day or two). We would like to observe them as soon after the explosion as possible and presently two astronomers, Carl Pennypacker of the University of California at Berkeley and Sterling Colgate of New Mexico Institute of Technology, are leading systematic searches for supernovae in other galaxies. Their telescopes

Figure 16–12. *The Crab Nebula is the remnant of a supernova which was seen in the year 1054.*

Table 16–2 Some Supernova Remnants

Remnant	Distance (Light Years)	Diameter (Light Years)	Age (Years)
Crab Nebula	21,000	4.4	930
Cygnus Loop	2500	100	20,000
Gum Nebula	300	2300	11,000
Tycho's Supernova	9800		400
Kepler's Supernova	20,000		370

are driven by computers and are programmed to scan selected portions of the sky rich in galaxies. The computer compares each field of view to what was seen the night before, looking for points of increased brightness. At the time of writing, each of the systems was just being set into operation. Astronomers hope to find a couple of new supernovae per month by these systematic searches.

Any new supernovae found by these searches will be unlikely to match in excitement or importance the discovery of the major supernova of 1987, however. We will examine this supernova after discussing types of supernovae and some predictions made by supernova theory.

(a)

Figure 16–13. *(a) The Veil Nebula, part of which is shown here, is part of the Cygnus Loop, a supernova remnant 2500 light years away. (b) The Gum Nebula.*

(b)

TYPES OF SUPERNOVAE

There are at least two ways that supernovae occur. One is that which we have discussed: the explosion of a massive star near the end of its life. Our limited understanding of the processes that cause the explosion was increased by the discovery that there is another type of supernova; one that occurs in binary star systems. Such systems are also responsible for a more common, but less spectacular, event—the nova.

Novae and Supernovae in Binary Systems

In a binary star system in which one star is more massive than another, the more massive star will end its lifetime sooner. Suppose that this star is in our lightweight star category; it will then end up as a white dwarf. Later, its companion ends its main sequence life and swells up toward its red giant stage. At some point, the material of the outer portion of the red giant is attracted to the white dwarf with more force than toward the center of the bigger star. This material will be pulled from the giant. Although some might fall directly into the white dwarf, analysis shows that most of the material pulled from the red giant will go into orbit around the dwarf (Figure 16–14). Eventually, some of it will fall to the surface. Since it came from the outer portion of the red giant, it is mostly hydrogen and when it strikes the extremely hot surface of the dwarf it flares up in a rapid fusion reaction; causing the dwarf to suddenly become brighter. (The dwarf, remember, has exhausted its supply of hydrogen but is hotter than necessary to cause hydrogen fusion.)

The amount of material falling into the white dwarf depends upon a number of factors, including the rotational speed of the giant, the distance between the two stars, and how fast the giant is growing. In some cases the material falls onto the hot surface in a fairly regular cycle; causing a flareup each time. An occurrence such as this is called a recurrent ***nova*** (the Latin word for *new*) because they were thought by people of Galileo's time to represent new stars. This was a natural conclusion because the increase in brightness that occurs in a nova results in a star appearing where previously none was visible to the naked

Nova: An event in which matter from the giant component of a binary system falls onto the white dwarf component, causing a sudden brightening of hundreds to thousands of times.

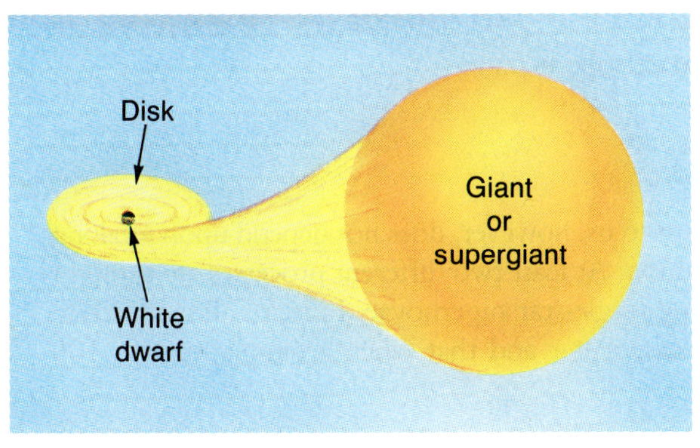

Figure 16–14. *In a binary system composed of a white dwarf and a red giant, material from the growing giant will fall toward the dwarf; going into orbit around it.*

Figure 16–15. *These photos of Nova Cygni 1975 show its dimming from magnitude 2 at maximum light to magnitude 15.*

Table 16–3 A Typical Nova

Luminosity Increase	10,000 times
Absolute Magnitude	-8, or 10^5 Suns
Time to Brighten	Few days
Time to Dim	6 months to one year

eye. A given nova may reoccur after a few weeks, months, or even years and such an event does not result in the destruction of the star as in the case of a supernova. Figure 16–15 shows one of the brightest of recent novae, and Table 16–3 shows data for a typical nova.

There are various types of novae, including classical novae, dwarf novae, *and* recurrent novae. *The differences are not important to us here.*

In some cases, material from the red giant falls slowly and regularly onto the white dwarf; increasing the mass of the dwarf. Recall, however, that as the electron-degenerate white dwarf gains mass, it *loses* size and therefore becomes even more compacted. When its mass reaches the Chandrasekhar limit of 1.4 solar masses, the degenerate electrons can no longer withstand the force of gravity and the star collapses suddenly. If the rapidly collapsing star contains significant amounts of carbon, the fusion of carbon will start quickly and proceed vigorously when the star's temperature reaches 100 million degrees Celsius. The sudden release of energy by the fusion of carbon into heavier elements is thought to be responsible for this type of supernova.

Single-Star Supernovae

The supernovae of most interest to us, however, does not depend upon infalling material from a binary companion. At least two different processes are thought to be responsible for producing single-star supernovae. First, recall that carbon builds up in the core of a massive star and that when a temperature of 100

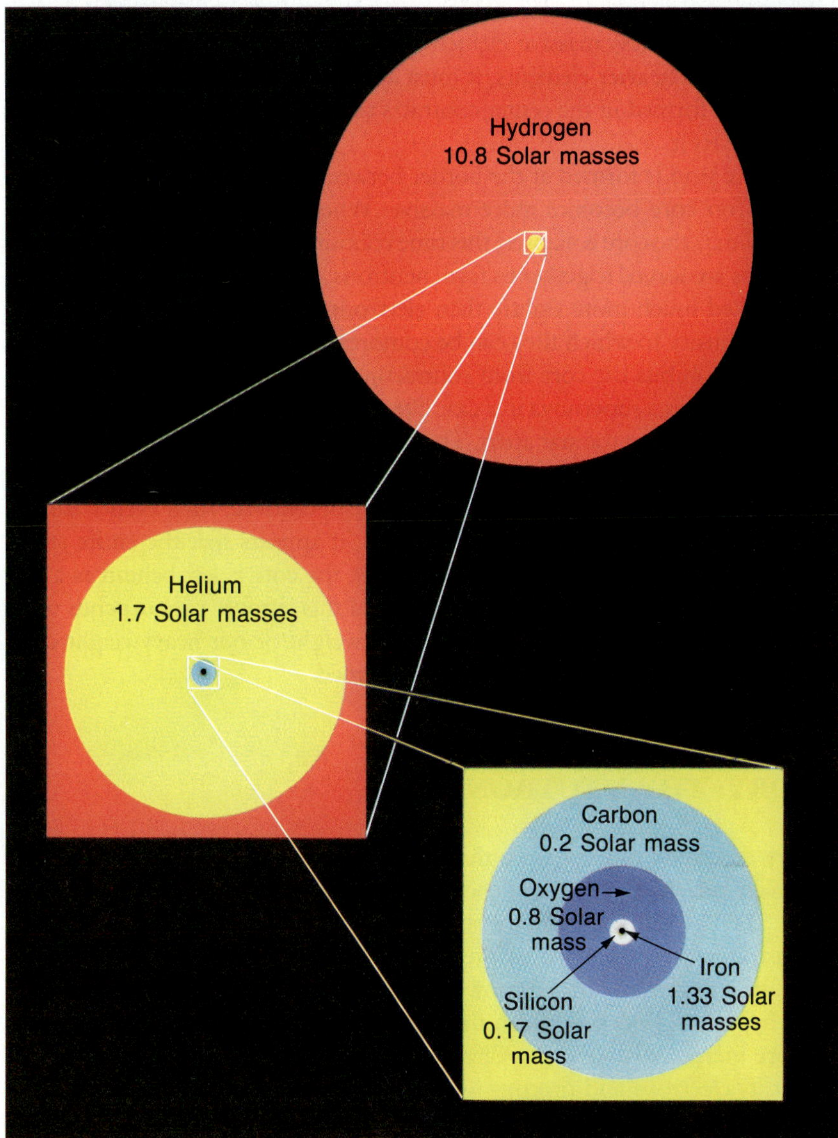

Figure 16–16. *Most of the volume of a 15-solar mass red giant is hydrogen but the central part is multilayered. The hypothesized mass of each layer is shown.*

million degrees Celsius is reached, the carbon begins suddenly to fuse to heavier elements. The beginning of carbon fusion may be violent enough to cause a supernova explosion (as described above for a binary system). ***Carbon detonation*** is thought to be the trigger of the less massive of middleweight stars and therefore these supernovae are similar to those in binary star systems.

Carbon detonation: The sudden onset of the fusion of carbon nuclei to form heavier elements.

In more massive middleweights and in heavyweights, carbon detonation is not explosive enough to cause a supernova. In fact, in these stars the core continues to fuse, in steps, to heavier and heavier elements. Carbon fuses to oxygen, oxygen to silicon, and silicon to iron. Each new process begins with a flourish, causing the center of the star to become hotter, and each fusion continues in layers around the center, as illustrated in Figure 16–16. Now the core resembles a multilayered onion. This does not last long, however, for iron is different from the less massive elements that formed previously. In order for iron to fuse into heavier elements, it must *absorb* energy (rather than release it). Thus the iron formed in the center contracts and heats but no more fusion steps are available to it.

As more iron is formed at the border between the iron core and the silicon shell, the iron core becomes more massive. When its mass exceeds the Chandrasekhar limit, the core collapses; this time continuing to collapse until electrons and protons are forced together to form neutrons. Physicists know that neutrons can be packed much more tightly than electrons but that neutrons also have a maximum density to which they can be compressed. When this occurs, ***neutron degeneracy*** prohibits the core from contracting further. A rebound in the rapid contraction causes a supernova in a case like this. The rebound generates massive shock waves; blowing the star apart. The energy source here is gravitation rather than nuclear energy.

Neutron degeneracy: The state of a gas in which its neutrons are packed as densely as nature permits.

Whatever the details of the explosions of supernovae, such events are singular ones in the lives of stars. In some cases, it appears that the entire star is blown apart, including the core; but in others, the core is left behind as a tiny remnant of the once-mighty star. The nature of this leftover core depends upon whether the star was originally in our middleweight or our heavyweight class. In either case, a unique, peculiar object is formed.

THEORY: THE NEUTRON STAR

We have described much of the theory concerning stars' lives but have only occasionally provided bits of evidence for it. In addition, we have not been able to show the derivation of the theory. There is an interesting story, however, that illustrates a portion of this theory, how it was confirmed, and how the confirmation provided further knowledge.

An hypothesis worked out in the 1930s predicted that after the mass of a star's core increases beyond the Chandrasekhar limit, the star collapses further and that its electrons and protons will combine to form neutrons, resulting in a ***neutron star.*** Just as electron degeneracy prohibits a white dwarf from collapsing under gravity, neutron degeneracy does the same for a neutron star. The hypothesis predicted that the remains of a middleweight star's collapsed core would become a neutron star; a tremendously compressed star with a mass between 1.4 and about 3 solar masses. Table 16–4 shows the properties of

Neutron star: A star that has collapsed to the point at which it is supported by neutron degeneracy.

Table 16–4 A Typical Neutron Star

Mass	1.5 solar masses
Diameter	20 km (width of a small city)
Density	10^{15} gm/cm^3
Temperature	10,000,000°C

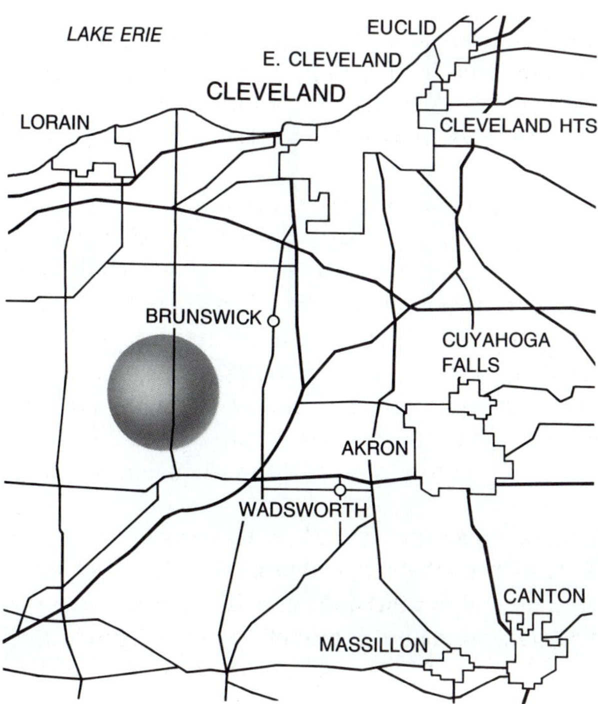

Figure 16–17. *The mass of a typical neutron star is about twice that of the sun but it is not large enough to cover Cleveland.*

such a star. To try to imagine it, picture the entire mass of a star larger than the Sun compressed into a ball the size of a small city (Figure 16–17).

Astronomers had little hope of finding such a star because its extremely small size results in it being very dim, even with its high temperature. Thus the idea was put on astronomy's back burner as an interesting hypothesis but one which seemed beyond our ability to confirm or deny. In 1967 an accidental discovery changed this.

Observation—The Discovery of Pulsars

In 1967 Jocelyn Bell (now Jocelyn Burnell) was a graduate student in astronomy at Cambridge University in England and was working with Antony Hewish and a group of researchers who were searching for quasars, energetic stellar sources we will discuss in a later chapter. The radio telescope she was using for her research did not look at all like the giant radio telescope dishes normally associated with radio astronomy. Instead, it looked more like a field of clothesline and it covered a total of 4½ acres (Figure 16–18). It was designed to be able to detect faint radio sources and to see quick changes in their energy. In the course of research for her dissertation, Bell found a new, unexpected, and unexplained source of radio waves. The signal from it pulsed rapidly; about once every 1.3 seconds. This was a much more rapid pulsation than ever had been observed from a stellar energy source.

In 1974 Antony Hewish was awarded a Nobel prize for discovering pulsars. Bell had made the actual discovery (reporting it in an appendix of her dissertation), but perhaps because she was a graduate student, she was not awarded a share of the prize.

The first thought was that the waves she had received were of terrestrial origin rather than celestial. A check of local radio transmitters, however, failed to indicate a source. In addition, the signal was detected about four minutes

Figure 16–18. *Part of the radio telescope that first detected a pulsar.*

earlier each night than the previous night. Recall that a given star sets four minutes earlier each night as a result of the Earth's moving around the Sun. The researchers concluded that the source was in the sky and was not of human origin.

Their next thought was that a signal from an extraterrestrial race had been detected. In fact, the source was referred to for a short while as a LGM (the initials for "Little Green Men," a reference to science-fiction-type extraterrestrials). This speculation was abandoned for a couple of reasons. First, the pulsations continued in a very regular fashion instead of changing as they would if they contained a message. More convincing, however, was the discovery of three more such sources in other directions in the sky, each with its own characteristic rate of pulsation. It was highly unlikely that a number of different civilizations were sending such signals toward us at the same time, so the sources had to be natural ones. They were renamed ***pulsars.***

Pulsar: A celestial object of small angular size that emits pulses of radio waves with a regular period between about 0.03 and 5 seconds.

The first pulsar detected had a period of 1.3373011 seconds. Such great precision is possible in measuring the rate of a regularly repeating cycle because one can measure over a great number of the cycles and divide by that number to obtain the time for a single one. Figure 16–19 shows a record of pulses from this pulsar and indicates that although they are extremely regular, they do vary in intensity.

In the case of each of the pulsars that were found, the duration of the pulse was about 0.001 second. (Figure 16–19 also illustrates the difference between pulse duration and period.) This immediately revealed to the astronomers an upper limit to the size of the object emitting the signals. The objects could be

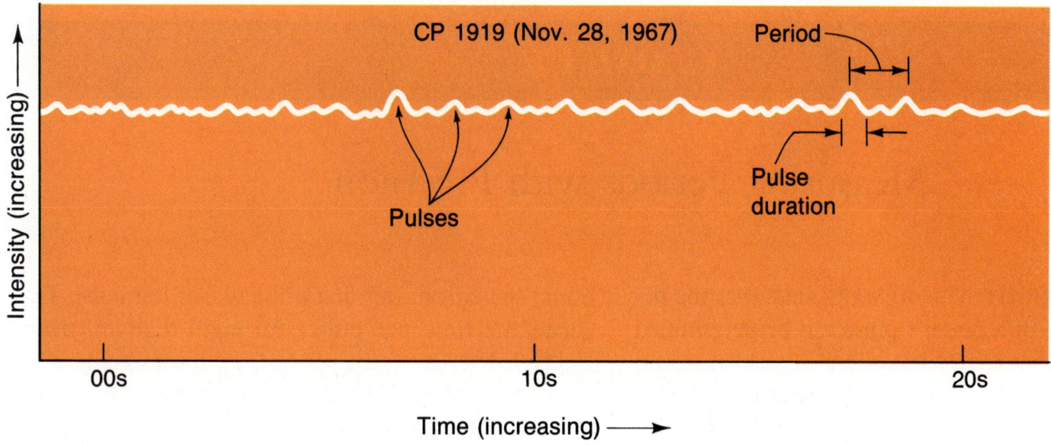

Figure 16–19. *The chart of pulses from the first pulsar indicates their regularity. The difference between pulse duration and pulse period is illustrated.*

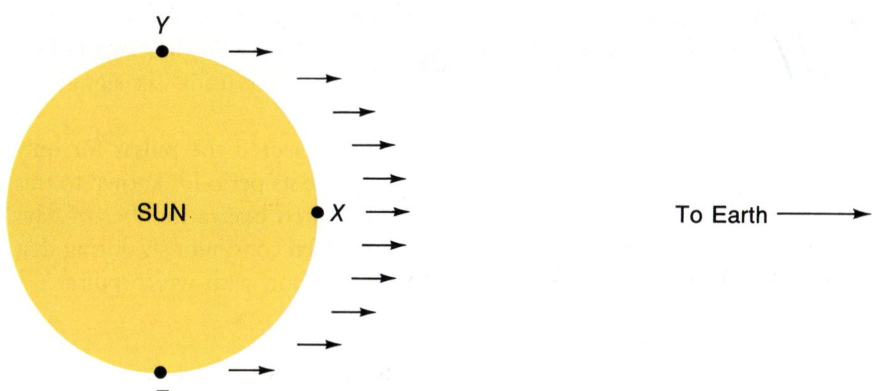

Figure 16–20. *The arrows indicate light that left the Sun at the same time. Since point X is about two light seconds closer to Earth than points Y and Z, its light reaches us two seconds sooner. Thus if the Sun were to increase in brightness instantaneously, we would see the light increase gradually over two seconds.*

no greater than about 0.001 light seconds in diameter—a few hundred kilometers. To see how such a conclusion could be reached even before the nature of the pulsation was known, consider what we would observe if the Sun were to brighten instantaneously. We would not see this instantaneous brightening as being instantaneous at all, for light from the part of the Sun closest to us would reach us about two seconds before light from its edge. Figure 16–20 illustrates this. Since the Sun is about two light-seconds in radius, we would see the intensity of light build up over two seconds. Likewise, if the Sun suddenly shut off, the dimming would appear to take two seconds, rather than appearing to happen all at once.

The observed stretching out of any sudden change in the Sun would not be due to our distance from it but rather from its size. Be sure to understand this idea; it is a useful tool in astronomy that tells us the maximum size of some objects.

The smallest star known in 1967 was the white dwarf, but these are Earth-size objects, not small enough to emit pulsations which lasted only 0.001 second; at least if the pulsation is caused by a change in the light emitted from the entire object. Thus the pulsar must be even smaller than a white dwarf. How could a star be so small? Enter the hypothesized neutron star.

CLOSE UP

Measuring Periods with Precision

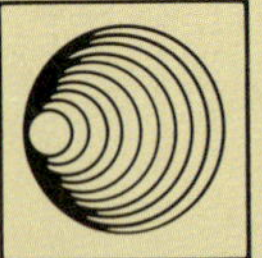

IT MIGHT SEEM REMARKABLE that the period of a pulsar's pulse can be determined to the accuracies given by astronomers; down to 0.0000001 second or less. To show how this is done, we will work an example that will illustrate that great precision can be achieved in determining the period of any regularly repeating cycle.

Observe the chart of pulses in Figure 16–19. Note that they occur very regularly and can be counted with absolute precision. That is, if you pick a pulse on the left side of the chart, you can count how many pulses occur between that one and another you select at the right. You can determine the exact number with no uncertainty. (Ignore minor pulses. Count only those that are part of the regular cycle.)

Now suppose that a pulsar is being observed with a radio telescope and the pulses recorded and counted. Suppose that we count until a total of 2450 pulses are observed and that 3637.58 seconds—a little over an hour—pass from our first pulse to our last pulse. Calculate the time per pulse. An eight-digit calculator yields 1.4847265 seconds but we must decide what accuracy is allowable here. The simple rule for significant figures is that one may carry only as many significant figures in the result of a multiplication or division as in the number with fewer significant figures. The value of time has six significant figures and although it might appear that the number of pulses is given to only three or four figures, this number is exact and has an unlimited number of significant figures. Thus we may state the answer using six significant figures, or 1.48473 seconds.

In the example, we measured the pulsar for only one hour. In practice, once its period is known to this accuracy, it can be measured over a number of days even if it cannot be observed continuously during that time. This yields the precision with which pulsar periods are stated.

THEORY: THE LIGHTHOUSE MODEL OF NEUTRON STARS/PULSARS

Let us consider how an object might emit pulses of radiation. One way would be for its surface to vibrate up and down. (A number of variable stellar objects are known to do this, including Cepheid variables.) Not only did the short duration of flashes from pulsars seem to indicate that they were not white dwarfs, but when astronomers considered the nature of the material of a white dwarf and the gravitational force on its surface, calculations showed that the surface of a white dwarf could not vibrate as quickly as once per second. Neutron stars are much more dense than white dwarfs and have a much greater gravitational force on their surface, so their surface should beat up and down more quickly. In fact, calculations showed that they should be unable to oscillate as *slowly* as once a second.

A second way for an object to emit radiation in repeated bursts is by an eclipsing binary process but for two objects to be in orbit with a period as low

Figure 16–21. *A sailor sees the lighthouse beam as being a series of blinks even though the beam actually shines continually as it rotates.*

as one second would require objects smaller and more dense, even, than neutron stars. The eclipsing binary was ruled out.

A final mechanism for producing pulses is by radiation that comes from a small part of the surface of a rotating object. This is the mechanism that causes a sailor at sea to observe pulses of light from a lighthouse. On a foggy night, he might see the lighthouse beam sweeping through the fog but on a clear night he sees the light only when it shines directly at him. This makes it appear to blink on and off (Figure 16–21). Could a star rotate with a period as short as that observed for pulsars? A star the size of the Sun would be torn apart by such fast rotation but a white dwarf or a neutron star would have two advantages in this regard: Their smaller size would mean that less force would be needed to retain their surfaces under fast rotation and their small size and great mass would result in a much greater gravitational force on their surfaces than is experienced on the surface of the Sun. Calculations showed that a white dwarf might be able to withstand the forces involved in rotating with a period of one second, and perhaps with a period of one-fourth second, but certainly no quicker than that. A neutron star, on the other hand, would have no difficulty in rotating with a period of a fraction of a second.

It is easy to see what would cause either a white dwarf or a neutron star to rotate so fast. Recall that as a spinning object decreases in size, its rotation rate increases. White dwarfs, and especially neutron stars, are so small that they would be expected to be spinning very fast.

We saw that the lack of a shorter rotation period for the Sun presented astronomers with problems in understanding of the formation of the solar system from a nebula.

Logic seemed to be pointing more and more to the neutron star as the explanation for pulsars. In analogy with a sailor's lighthouse, the model developed to explain how neutron stars create pulses is called the ***lighthouse model.***

Lighthouse model: The theory that explains pulsar behavior as being due to a spinning neutron star whose beam of radiation we see as it sweeps by.

Recall that the Sun has a magnetic field. When the theory of the existence of neutron stars was developed back in the 1930s, it was suggested that such a star might have an extremely strong magnetic field since the star is the compacted core of a main sequence star that would be presumed to have had a magnetic field. The neutron star's strong magnetic field is a necessary part of the lighthouse model.

Figure 16–22. *The Earth's magnetic poles are not located at its poles of rotation. The location of the north magnetic pole is shown here; illustrating also that its location changes with time.*

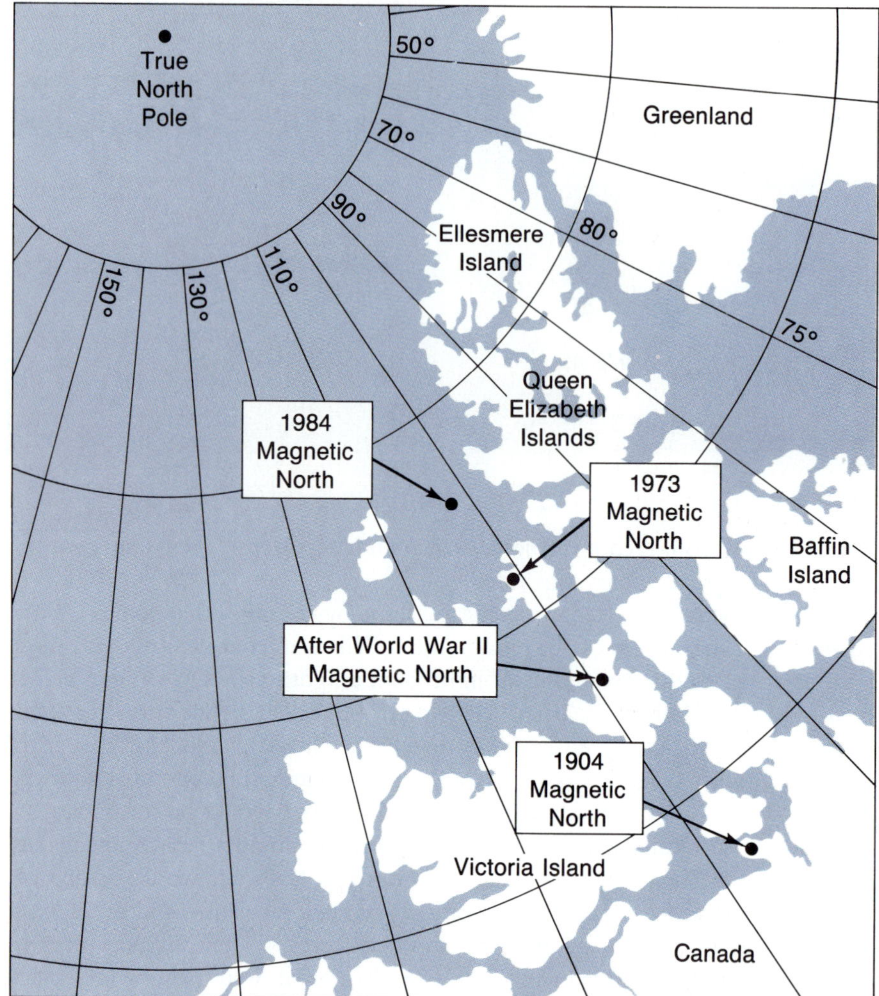

As we discussed when describing the solar system, it is common for the magnetic poles of a planet to be out of alignment with the axis of the planet's rotation. Figure 16–22 shows the location of the Earth's magnetic pole in the northern hemisphere; some 1400 kilometers from the pole of rotation. In addition, we saw that the magnetic field of the Earth traps charged particles and that these particles result in the auroras seen near the magnetic poles of the Earth. In the case of the theorized neutron star's extreme magnetic field, the energy associated with trapped charged particles would be much more intense and a beam of radio waves and other radiation would be emitted near each magnetic pole. If the star's magnetic poles were located off the rotation axis, this beam would sweep through space as the star spins (Figure 16–23). Then if the Earth were located in the path of the beam, we would see a pulse of radio waves each time the beam sweeps by us.

Notice that the lighthouse model would also predict that there are many pulsars we could not observe on Earth, for we would only see those whose lighthouse beam happens to sweep by us. Since the length of every flash from

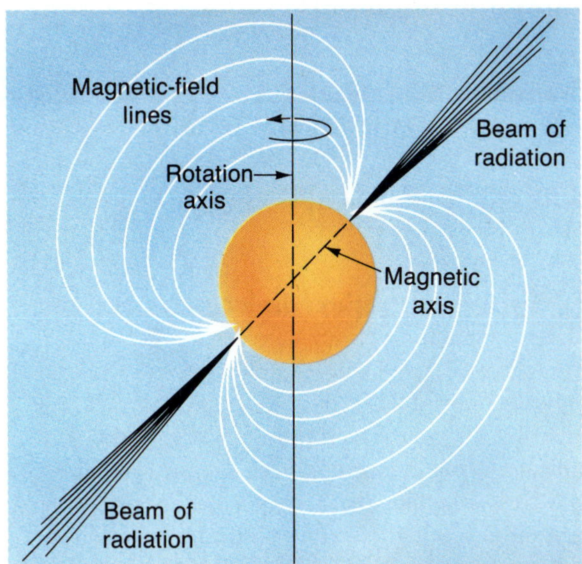

Figure 16–23. *A beam of radiation is emitted near each magnetic pole of the pulsar. As the star rotates, the beam sweeps through space.*

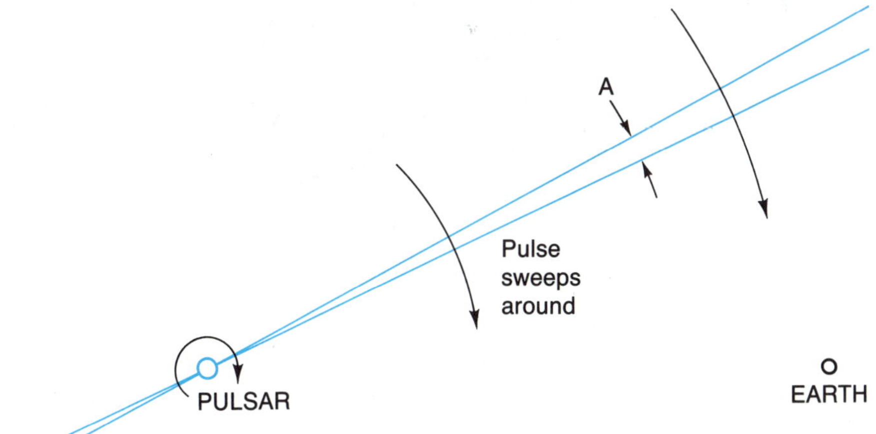

Figure 16–24. *The angular width of the beam from a pulsar determines how long we see its pulse. No pulsar has a long-duration signal, indicating that their angular beam widths are narrow and therefore that we are seeing only a small fraction of the pulsars that exist.*

a pulsar is very short compared to its period, we can conclude that the angular size of the beam is small (Figure 16–24). This means that from Earth we would see only a small percentage of the pulsars that exist.

In 1967, the lighthouse model seemed to be a logical explanation for the pulsar observations, but in order to test it, more pulsars would have to be found, perhaps including one which is related to the remnants of a supernova, where neutron stars are theorized to be.

Observation—The Crab Pulsar and Others

It was only a matter of months after the discovery of the first pulsar that one was found in the Crab Nebula. It might have been found sooner but its rate

CLOSE UP

The Distance/Dispersion Relationship

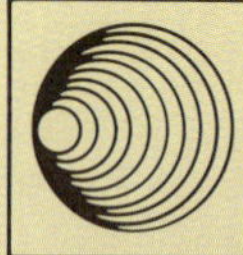

In Chapter 14, various methods were described to measure distances to stellar objects. Pulsars give us another method. It relies on two phenomena: Each burst of radiation from a pulsar contains an entire spectrum of wavelengths and different wavelengths of electromagnetic radiation travel through space at slightly different speeds. For many purposes, it can be assumed that all wavelengths travel at exactly the same speed in space—the so-called *speed of light*—but in practice, since space is not a perfect vacuum, longer wavelengths travel slightly slower than shorter wavelengths. Recall from Chapter 11 that this difference in speed—called *dispersion*—is the same property that causes chromatic aberration in lenses. Its effect here is that as each pulse of radiation from a pulsar travels through space, the longer wavelengths get slightly behind.

When it was stated that the pulses from a pulsar are typically only 0.001 seconds in length, this referred to pulses of a single wavelength. If we consider, for example, a pulsar that emits visible light as well as radio waves, we will find that the visible-light portion of each pulse reaches us before the radio portion, although for any one wavelength of visible light or any one wavelength of radio the pulse length is the same; perhaps 0.001 second. The relative amount of dispersion might be stated as the time that elapses between the detection of a given wavelength of visible light and the detection of a given wavelength of radio energy in the same pulse.

of pulsation is much faster than expected, with a period of 0.033 second so that it blinks 30 times per second. This short period finally ruled out completely the possibility of a spinning white dwarf as being the source of pulses; a white dwarf would tear apart if it rotated 30 times per second.

Not only was the Crab pulsar flashing more frequently than others yet discovered but it was emitting great amounts of energy in the radio region of the spectrum; 100 times more than the total energy emitted by the Sun. Astronomers at the University of Arizona then observed that it also pulses in visible light. Since that time, it has been found that the Crab pulsar emits radiation in all regions of the spectrum, from radio waves to X rays (which were observed by the orbiting High-Energy Astronomy Observatory—HEAO-2, the Einstein Observatory). Adding up the energy emitted by the Crab in all the various regions of the spectrum, it was found that the total energy from the Crab pulsar is more than 25,000 times the energy from the Sun.

Astronomers wondered what was the source of the Crab pulsar's energy. This question had been asked even before its pulsar had been discovered. Astronomers had long been puzzled by the luminosity of the nebula itself; it emits more energy than was thought to have been possible. Both of these energy problems were solved when it was discovered that the Crab pulsar is slowing down; its pulses were found to be growing slightly less frequent. It was hypothesized that the source of the energy that powers the nebula is the rotational

Two factors determine the amount of dispersion: the distance the pulse travels through the interstellar medium and the dispersion properties of that medium. We have here another triple connection: Distance to the pulsar, dispersion of the pulse, and dispersive nature of the interstellar material (Figure T16–1). The amount of dispersion can be measured directly. Thus, to the degree that we know one of the other quantities, we can calculate the final one. If we know the dispersion properties of the interstellar matter between us and a pulsar, we can determine the distance to the pulsar. On the other hand, if we can determine distances by another method, this provides a means of learning more about the interstellar material.

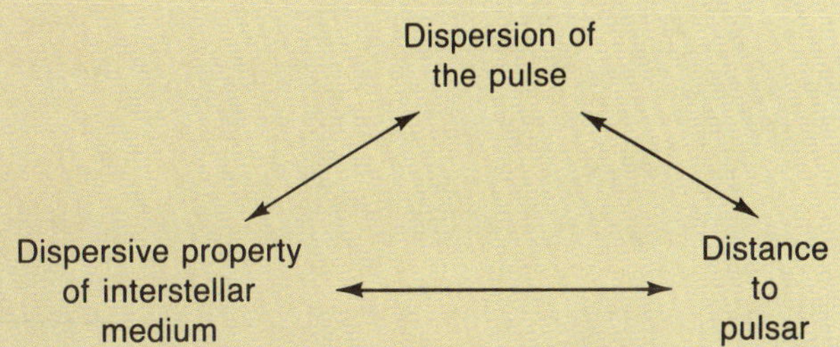

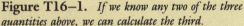

Figure T16–1. *If we know any two of the three quantities above, we can calculate the third.*

energy of the pulsar. As the pulsar spins, its magnetic field propels electrons out into the nebula. These electrons are the cause of the nebula's great luminosity but in being swept from the pulsar, the electrons in turn slow down the pulsar. If the hypothesis was correct, the amount of rotational energy lost as the object slowed its spinning should correspond to the amount of energy emitted by the nebula. Calculations showed that the two amounts of energy did indeed correspond. The theory was confirmed quantitatively.

The pulsar in the Crab Nebula is spinning faster than most others because the supernova in which it had its start was so recent; only 900 years ago. As time passes, this pulsar will gradually slow down. As it loses its rotational energy, the intensity of its pulses will also decrease and it will no longer emit X rays. Finally, in tens of thousands of years, it will be just another radio pulsar; its nebula spread so far that it can no longer be seen.

The pulse rate for a pulsar is sometimes observed to increase suddenly (this is called a glitch*). This occurs because as the neutron star slows its spinning, forces on its surface change, causing "starquakes" and sudden changes in its shape.*

More than 440 pulsars have been discovered; a few with periods less than 0.1 second, but most with periods between 0.1 second and 4 seconds. Normally, no nebula is found surrounding pulsars for the nebula has long since dispersed. To further confirm that pulsars are indeed the neutron stars predicted to be left behind in supernovae, more instances of pulsars associated with expanding nebulae were sought. Since the Crab's pulsar was found, another has been found in the Gum Nebula (Figure 16–13). Astronomers are now confident that they have found the neutron stars that theory predicted 50 years ago.

Figure 16–25. *These are the steps a middleweight star takes from protostar to its final stages.*

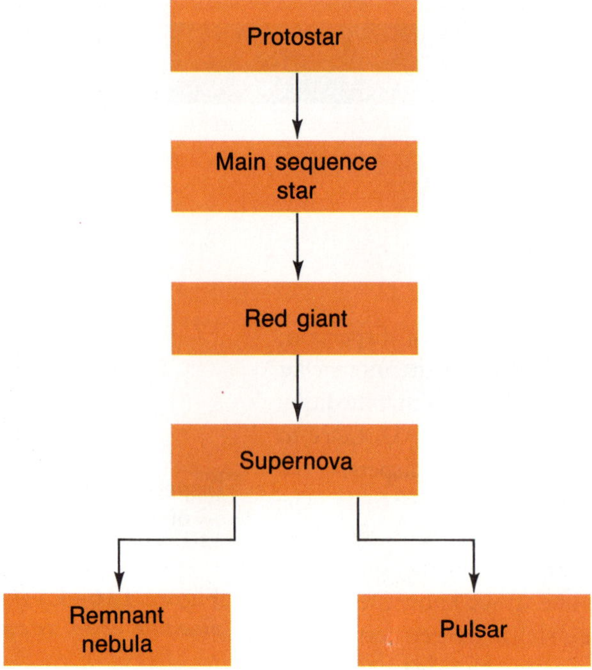

MIDDLEWEIGHT CONCLUSION

Figure 16–25 reviews the steps a middleweight star takes from protostar to pulsar/neutron star. Only a small fraction of all stars are in the middleweight category, however, for two reasons. First, middleweights' lifetimes are short compared to lifetimes of stars with lower mass. Thus even if they are being formed at the same rate, not as many will be in existence at any given time. Secondly, only stars in a small range of mass end up with a core too massive to be supported by electron degeneracy and of low enough mass to be supported by neutron degeneracy. We know that neutron stars, which are the final ends of middleweight stars, are more massive than the Chandrasekhar limit of 1.4 solar masses but we are not sure of the upper limit to the mass of a neutron star. (We certainly do not have any neutron degenerate matter on Earth with which to experiment.) Theory indicates, however, that the limit falls somewhere between two and four solar masses. Thus only stars that end up with masses greater than—but not much greater than—1.4 solar masses can ever become neutron stars and send out their characteristic lighthouse beam of radiation.

THE HEAVYWEIGHTS

A heavyweight star proceeds through its life in basically the same manner as a middleweight, although each stage occurs more quickly. Figure 16–26 is a plot that shows how the density in the core of a 25-solar-mass star increases through the star's life and how long the various types of fusion last.

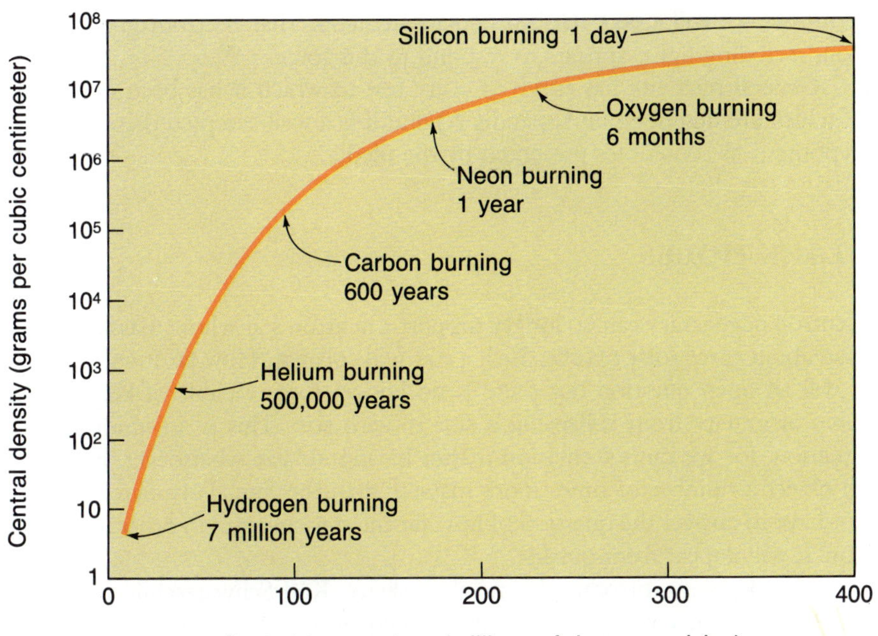

Figure 16–26. *This shows how the density and central temperature of a 25-solar mass star change as the star goes through its stages. Notice the relative time spent in each phase.*

Heavyweights differ from middleweights primarily in what happens to them when their core is compressed to a greater density than electron degeneracy can support. When this occurs in a middleweight, the resulting supernova leaves a neutron star at its center. In a heavyweight, an even more spectacular thing happens: The core swallows itself as a black hole. Before considering these strange objects directly, we must take a quick look at relativity theory.

Light and Gravity

Special theory of relativity: A theory developed by Einstein that predicts the behavior of matter moving at constant speeds relative to the observer.

General theory of relativity: A theory developed by Einstein that expands special relativity to accelerated systems and that presents an alternative way of explaining the phenomenon of gravitation.

Albert Einstein stated that in his youth he wondered whether, if he were moving at the speed of light, he could see himself in a mirror. Such questions led him in 1905 to develop the ***Special Theory of Relativity.*** This theory makes the perception of electromagnetic waves independent of the motion of the observer. It allows us to answer Einstein's mirror question with a "yes," but it reveals a link between the nature of space and time that does not appear in our everyday perception of the universe. Special relativity has some interesting effects (and we will discuss some of them in Appendix A) but it is Einstein's expansion of the theory to the ***general theory of relativity,*** or ***general relativity,*** that must be used to describe the fate of massive stars.

The general theory of relativity replaces the idea of a force of gravity with the idea of space-time curvature. Saying that light is affected by gravity mixes the two theories.

The general theory of relativity presents a different way of looking at the phenomenon we call gravitation; explaining it not as a force but as being the result of a curvature of space. The particular result of interest to us here is that the theory predicts that electromagnetic radiation will respond to this curvature in a way that will make it seem to be responding to the force of gravity. In the present discussion we will continue to speak of the force of gravity in Newtonian terms but we will accept the Einsteinian prediction that electromagnetic radiation, including light, appears to respond to this force.

General relativity has survived every test to which it has been put (some of which are described in Appendix A) and it is a well-accepted theory—not a hypothesis as sometimes presented by the media.

BLACK HOLES

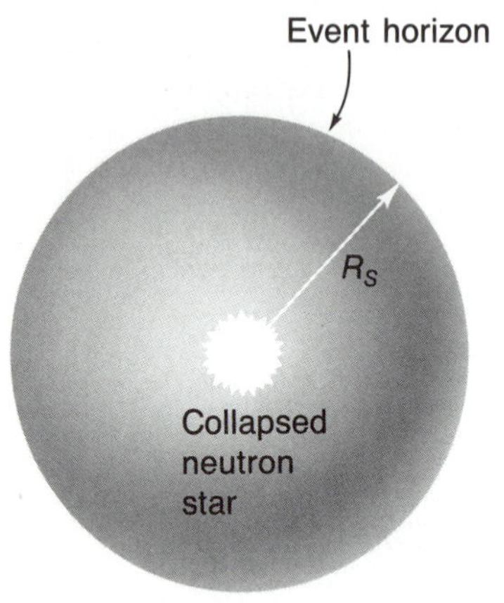

Figure 16–27. *Once a star has shrunk to inside the Schwarzschild radius, light can no longer escape its surface. The Schwarzschild radius forms a sphere called the event horizon.*

Neutron degeneracy can no longer support a neutron star whose mass is greater than about three solar masses. Such a star will collapse. How far it will collapse is still an open question but there is no known force capable of keeping the force of gravity from collapsing a star to zero size. This is an unimaginable situation, for we cannot envision matter having no size whatsoever, especially an object a number of times more massive than the Sun. Fortunately, we do not have to answer the question of how far the star collapses in order to predict how it will appear from outside.

After Einstein introduced general relativity, Karl Schwarzschild calculated that when a star collapses to a dimension equal to or less than what is called the ***Schwarzschild radius,*** light will be unable to escape the object. Recall that in discussing the planets, the escape velocities of the various solar system objects were stated. The escape velocity of an object is the minimum velocity a missile must have in order to escape the gravitational field of the object. (The escape velocity from the Earth's surface is about 11 km/s.) As a star decreases in size, the escape velocity from its surface becomes greater. When its radius becomes so small that its escape velocity is greater than the velocity of light, the star has reached the Schwarzschild radius (Figure 16–27).

The size of the Schwarzschild radius depends upon the mass of the star, for it is this mass that determines the force of gravity. The radius (R_s) is given in kilometers by the formula

$$R_s = 3M \qquad (R_s \text{ in km; } M \text{ in solar masses})$$

Schwarzschild radius: The radius of the sphere around a black hole from within which no light can escape.

Black Hole: An object whose escape velocity exceeds the speed of light.

where M is the mass of the star expressed in multiples (or fractions) of the Sun's mass. Thus a star with a mass five times greater than the Sun's will have a Schwarzschild radius equal to 15 kilometers.

A star whose radius is less than the Schwarzschild radius is called a ***black hole***; "black" because no light escapes from it and "hole" because matter—or light—that falls into it can never be retrieved. A black hole exists if the star's radius is equal to or less than the Schwarzschild radius but the radius of the black hole is considered to be the Schwarzschild radius.

The equation tells us that if matter falls into a black hole, the Schwarzschild radius becomes larger. This is reasonable, for as matter falls into the black hole, more mass is inside exerting more gravitational force and increasing the escape velocity at any given distance.

Event horizon: The sphere around a black hole from within which nothing can escape. Its radius is the Schwarzschild radius.

A name given to the sphere formed at the Schwarzschild radius illustrates an important and interesting feature of black holes. The sphere is said to be the ***event horizon.*** Just as we cannot see beyond the Earth's horizon, we cannot see inside the event horizon. More than that, there is no way we can know about an event inside that sphere. Nothing that happens there is accessible to us and may just as well be in another universe, since we can have no experience of it.

Do Black Holes Exist?

The prediction of the existence of black holes was made in the 1930s when it was realized that a star's mass may cause it to collapse beyond neutron degeneracy. A prediction of something as unusual as a black hole certainly calls for observational verification. But how?

Actually, Pierre Simon Laplace proposed in 1798 that a very massive star might have such a strong gravitational force that light could not escape it. This was 100 years before Einstein's theory linked gravitation with the travel of light!

There is, of course, no hope of seeing a black hole directly, for we cannot see something from which no light escapes. If, however, matter is falling into a black hole, we should expect some of that matter to orbit the black hole in a manner similar to the way matter falling into a white dwarf orbits it (causing a nova as it falls onto the white dwarf). Since the gravitational field near a black hole is so strong, the orbital speed of nearby matter would be extremely great and as collisions between particles turned the regular orbital motion to random thermal motion, the matter would reach temperatures of hundreds of millions of degrees. Figure 16–28 illustrates material being pulled into orbit around a black hole from a companion binary star. Such hot material would radiate great amounts of energy and since it is not yet inside the event horizon, we should be able to detect it. We can predict the characteristics of the radiation and we know that the object should appear as an X ray source.

Numerous X ray sources are found in the heavens; particularly by NASA's HEAOs and by the Japanese and the European orbiting X ray observatories. Are all of these black holes? Probably not. Only if one of these sources is found to be associated with a particularly massive star can we hope that it is a black hole. When we wish to know the mass of a star, we search for a binary system.

Figure 16–28. *If a black hole and red giant or supergiant form a binary system, material will be pulled from the giant (left) and will swirl around the black hole, causing a release of X rays from the heating material in the disk.*

CLOSE UP

Black Holes in Science, Science Fiction, and Nonsense

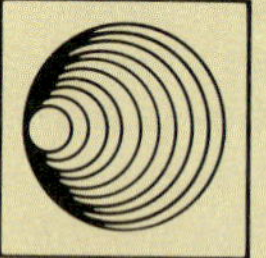

BLACK HOLES ARE FANTASTIC objects and are fruitful subjects for science fiction as well as for a lot of nonsense. We will try to separate the science from the nonsense and the science fiction.

Nonsense

There is a common misconception that a black hole is a giant vacuum cleaner, sweeping up matter across wide portions of space. In fact, the gravitational force of a black hole is unusually great only near the black hole. To make this clear, let's suppose that the Sun could magically become a black hole without losing any mass. If it did so, there would be no change in the gravitational force it exerts on the Earth. The force exerted on the Earth would remain the same and the Earth would continue in its same orbit. The only difference to us on Earth would be that the lights would go out: no radiation would arrive from the Sun. Newton's law of gravity states that the force of gravitational attraction between two objects depends only upon the masses of the two objects and the distance between them. Although predictions from Einstein's theory differ from those of Newton's, these differences show up only where the forces are extremely great, so that we can still use Newton's theory to discuss Earth's orbit. The strength of Newton's gravitational force does not depend upon the Sun's size, only its mass.

Where is the increased gravitational field, then? To explain this, note that with the Sun in its present state, the closest one can get to it (and still be outside) is obviously its surface—some 700,000 kilometers from center. If a person could go down inside the Sun, the force of gravity on him would become *smaller,* for there would then be a gravitational force back toward the matter near the surface (Figure T16–2). In fact, at the center of the Sun, the person would be weightless, for he would be attracted equally in all directions.

The difference in the case of the solar-mass black hole is that now one can get closer to the center of the star while remaining outside its surface (Figure T16–3). For such a black hole (only six kilometers across) one could get within a few kilometers of the center and—as predicted by Newton's law of gravity—the force of gravity would be very great at these small distances. (Actually, Newton's law would not make accurate predictions this close to the black hole, but it correctly predicts that the force would be extremely large.)

Predictions From Science

One thing that makes black holes unique is their simplicity. In order to describe an object like the Earth, many quantities must be specified: size, mass, color, composition, temperature, etc. A black hole, on the other hand, requires only three quantities to describe it completely: mass, rotation, and electric charge. These three quantities determine every other observable quantity—its Schwarzschild radius, for example. In the text we do not mention that black holes can have a rotation or electric charge but in fact these quantities determine the observable properties of black holes.

Stephen Hawking, whose brief biography appears in Chapter 18, has made some interesting predictions concerning black holes. For one, he calculates that the density of the universe at its beginning (Chapter 18 again) was great enough that *mini* black holes would have been produced. A mini black hole might be the size of a pinhead and have a mass like that of an asteroid. Such nonstellar black holes have not been detected and they will be very difficult to detect, but theory predicts the possibility of their existence.

Another finding by Hawking is that it is indeed possible for material to escape from a black hole. He

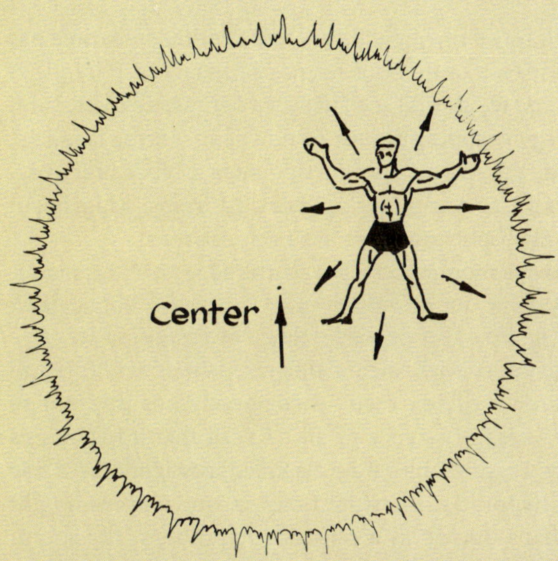

Figure T16–2. *If a person could exist inside the sun, he would weigh less than he did on the surface, for gravitational forces would be exerted on him in all directions by parts of the sun.*

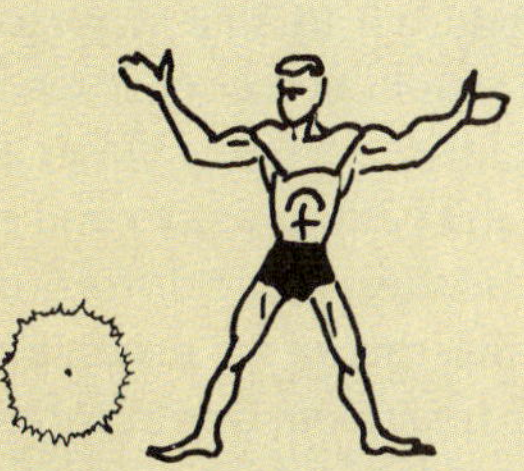

Figure T16–3. *If the entire sun could be shrunk to a ball small enough, a person could be the same distance from its center as shown in the last figure and still be outside its surface. He would then weigh much more than in the previous figure.*

applied the laws of quantum mechanics, which govern the behavior of atoms and nuclei, to black holes and determined that black holes radiate nuclear particles. The amount of radiation depends inversely on the mass of the black hole, however, and for stellar black holes the radiation is negligible. If mini black holes were formed at the beginning, the smallest of these would have already radiated itself to oblivion. Only mini black holes with masses of billions of kilograms—asteroid mass—would still be in existence today.

What about the upper limit of mass for a black hole? There is none. What might be surprising is that as black holes get larger—as their event horizon expands—their average density decreases. While a stellar black hole of ten solar masses has a density equal to that of the nucleus of an atom, if somewhere there is a black hole with a mass of a few million stars, its average density will be that of normal earthly rock.

Unusual phenomena—by earthly standards—occur as an object falls into a black hole. For one thing, as the object nears the black hole, it is pulled apart by tidal forces. Recall that tidal forces result because of the difference between gravitational forces on one side of an object and the other. An object approaching a black hole would feel a much stronger gravitational force pulling on the side of it nearest the hole than on the other side. The force difference would pull the object apart. It would be impossible for a person to fall into a black hole and remain intact.

Even more strange would be the observation of

something falling into a black hole. Forget for now the destruction caused by tidal forces. Einstein's theories of relativity tell us that if we could watch the object fall, we would never see it reach the event horizon. We would see it getting closer and closer to the event horizon and getting redder and redder (Doppler shift-like) but because of the distortion of time that is predicted by the theory of relativity, it would take forever—according to our observation—for the object to reach the event horizon. As time is reckoned on the object, however, it would fall into the hole very quickly. We will have to wait until Chapter 19 to appreciate such unusual effects of time distortion.

Science Fiction

Where does an object go when it falls into a black hole? It has been speculated that it may appear elsewhere or "elsewhen"—at another place or another time. Such travel through "hyperspace" or through time has lent itself to numerous science fiction plots. If it indeed occurs, we should see "white holes" where matter and energy is appearing out of nowhere. (In the language used, the matter and energy enter a black hole, pass through a "worm hole" and emerge from a white hole). No such phenomenon has been observed.

Even more speculative is the idea that the matter may come out in another universe; not another galaxy, but in a parallel universe. Since by definition we have no contact with such a universe, there would be no way of verifying such speculation. Thus it is not in the realm of science at all. (Recall that a hypothesis must be verifiable to be classified as scientific.) While the hypothesis of white holes in our universe might perhaps qualify as a scientific hypothesis, the speculation of a parallel universe must remain pure science fiction.

Then if we find a binary system in which one of the stars is an invisible one with a mass greater than four or five solar masses, we can conclude that the star must be collapsed (or otherwise it would be visible). Finally, if the star emits X rays characteristic of those predicted for a black hole, we would have good evidence for claiming to have found a black hole.

Cygnus X-1 was the first candidate for a black hole. From it we find the predicted black hole X rays and near its location we have found HDE 226868, a supergiant with a mass of about 30 solar masses. The supergiant forms a binary with an unseen companion and the two have a period of 5.6 days (as determined from their Doppler spectrum). The orbit indicates that the companion is a star with a mass as much as perhaps 15 solar masses. This fits what we would expect of a black hole.

Found in 1971, Cygnus X-1 is still considered the most likely black hole. There are other candidates, however, and as we learn more about them, they seem more and more to fit our expectations for black holes. A few are even part of eclipsing binary systems, thus providing us with further information on these corpses of giant stars. Primarily from data gathered with the orbiting X ray

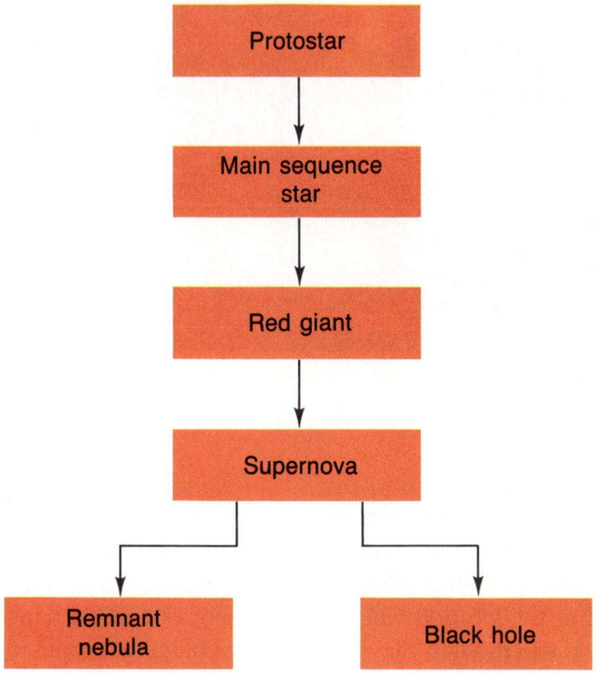

Protostar

Main sequence star

Red giant

Supernova

Remnant nebula

Black hole

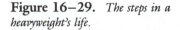

Figure 16–29. *The steps in a heavyweight's life.*

telescopes, astronomers are convinced that black holes have been found; confirming our theory of the death of the most massive stars (and serving as another confirmation of the theory of general relativity).

Figure 16–29 reviews the steps taken by heavyweight stars as they progress from protostars to black holes.

SN1987A

When a new discovery is made in astronomy, a telephone call or telegram is sent to the International Astronomical Union's Central Bureau for Astronomical Telegrams in order to establish priority of discovery. Telegram number 4316, received on February 24, 1987, read:

> W. Kunkel and B. Madore, Las Campanas Observatory, report the discovery by Ian Shelton, University of Toronto, of a mag 5 object, ostensibly a supernova, in the Large Magellanic Cloud. . . .

Astronomers had been waiting for 383 years, since Kepler saw his supernova, to see a naked-eye supernova. Their wait was over. The supernova (Figure 16–30) that has excited astronomers worldwide is officially known as SN1987A (*A* because it was the first discovered that year) but is often called simply "the Magellanic supernova."

The star that formed SN1987A (the "progenitor" star) was Sanduleak—69°202, so named because it is the 202nd star in the 69th degree south of the celestial equator (−69° declination).

One of the first things that astronomers did after the discovery of SN1987A was to examine recent photographs of the region where it occurred in order to determine which star was its source. It is fortunate that the supernova occurred

Figure 16–30. *The photo at left was taken before SN1987A occurred. The supernova is obvious in the other photo.*

in a region of the Large Magellanic Cloud where new stars were known to be forming, for this meant that astronomers already had an interest in the region and many photos of it could be found. Two very luminous blue stars were found very close together at the location where the supernova occurred. This caused confusion, because it was thought that red giants, not blue stars, explode as supernovae. Astronomers are now convinced, however, that it was indeed a blue supergiant that exploded. Although this star had been 100,000 times brighter than the sun before it exploded, it was visible only in a telescope because it is 170,000 light years distant from us.

Reevaluation of Supernova Theories

The explosion of a giant blue star of about 20 solar masses, but only one-tenth the diameter of a red giant, is causing us to reevaluate theories of the deaths of heavyweights. Two hypotheses are being considered. First, recall that a red giant is formed when the outer portion of a star absorbs energy radiated when helium fusion (and then carbon fusion, etc.) starts in the core. Although the outer part of a star is made up almost entirely of hydrogen (and some helium), it does contain small amounts of heavier elements, and heavier elements, in general, absorb more radiation than do hydrogen and helium. It has been proposed that red giants form only in stars that have enough heavier elements to absorb significant radiation from the core's increased output. If this is true, SN1987A simply may have had too little heavy material to ever become a red giant.

A second hypothesis is that SN1987A was once a red giant but that the great stellar wind associated with this massive star expelled some of its outer, cooler material. A tentative confirmation of this hypothesis was made when an increase of light from the region of the supernova was detected by the International Ultraviolet Explorer. This light appears to have been produced when ultraviolet radiation from the explosion reached the material that had previously been expelled.

It may be possible to check this hypothesis further, for as the shock wave from the supernova continues to expand outward, it will overtake the expelled material and when it does so it will cause that material to glow. Perhaps by the

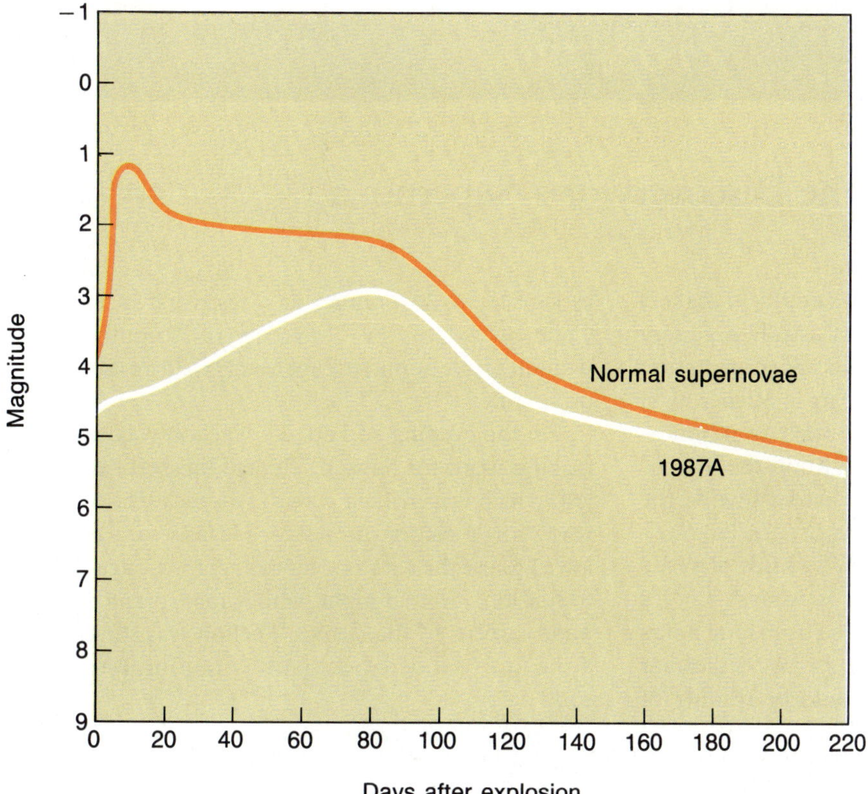

Figure 16–31. *The light curve of Supernova SN1987A differed from what was expected for the first three months. Once it reached its maximum luminosity, it started behaving like other supernovae of its type.*

time you read this the glow will have been observed and the hypothesis confirmed.

Further data from SN1987A will undoubtedly tell us more about the most massive stars and the supernova process. Astronomers are now looking forward to the time when the debris from the supernova will have dispersed enough that they can detect the star left at the center. At the time of writing, X rays from it had already been detected, both by a Japanese satellite and by the Russian space station. It will be years before radio waves are transmitted through the nebula but then we hope to detect the predicted radio waves.

SN1987A has been an unusual supernova in many regards. Four hours before Ian Shelton's discovery photo, a photo had been taken of the Large Magellanic Cloud by another astronomer. (The photographic plate had not been examined right away and so that astronomer did not get credit for the discovery.) From this photo, and from the fact that a very reliable witness in New Zealand had not seen the supernova while looking at the Large Magellanic Cloud just before the photo, we conclude that the supernova brightened very quickly during the first three hours. It then leveled off for a while, before brightening again over three months. It then began to dim. Its maximum brightness was not as great as was expected, however; probably because SN1987A resulted form the explosion of a blue supergiant rather than a (larger) red star. Figure 16–31 shows the light curve of the supernova over the months following its explosion and compares it to a typical curve of a red giant supernova.

We must use our largest telescopes to study the dim supernovae found in other galaxies. SN1987A was too bright for these telescopes and adjustments were required. Using the large telescopes was like trying to drink from a firehose.

CLOSE UP

The Discovery of a Supernova

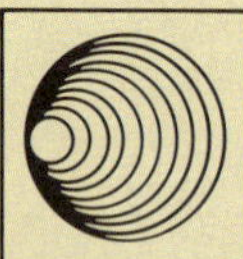

WE HAVE SEEN MANY examples in the text of the hard work that is usually necessary if one is to make a significant contribution in science. Although luck plays a part in science as in most activities, its importance is often overemphasized. In the words of the late Adolph Rupp, the "Baron" of college basketball coaches, "The harder I work, the luckier I get." We present here a report* of Ian Shelton's discovery of Supernova 1987A. Luck played a part in Shelton's discovery but if it weren't for his familiarity with the sky which resulted from long hours with a telescope and if it weren't for his dedication the night of the discovery, we would be reading of someone else's discovery.

It was 1572 when the sight of a supernova, an expiring star, startled Tycho Brahe, the great metal-nosed Danish astronomer. Thirty-two years later Brahe's famous pupil, Johannes Kepler, saw another supernova explode. That was it for the next 383 years. No one spotted another supernova near our galaxy until February 1987, when Ian Shelton—in the course of a miserable all-nighter—discovered one while listening to Pink Floyd on his Walkman.

Shelton isn't exactly in Tycho and Kepler's league. He's a 29-year-old Canadian who dropped out of two graduate schools. He makes less than $15,000 a year as a resident astronomer at a telescope in northern Chile owned by the University of Toronto, which wouldn't have admitted him to its graduate astronomy program.

On the evening of Feb. 23, Shelton was, as usual, making work for himself. He had finished his official job, which was making sure the 24-inch telescope was ready for a visiting professor. He was sometimes allowed to use the telescope himself, but not often enough to suit him. So that night, while others peered at computer screens in the domes. Shelton was shivering in a one-story cinderblock shack with a broken sliding roof.

The shack housed a telescope called an astrograph that had been abandoned years earlier. It was a mere ten inches in diameter. By the time Shelton got to Las Campanas, the astrograph was broken and the manual for it lost, and no one knew how to work it. Shelton fixed it and used it to photograph Halley's comet.

The roof, however, remained a problem on Feb. 23. The winch that was supposed to slide it out of the way was stuck again. Shelton had to climb up and push the corrugated sheet open. He pointed the telescope south toward the Milky Way's closest companion, the Large Magellanic Cloud. There was no particular reason to photograph it. Shelton just wanted

Various hypotheses have been proposed as to the source of the energy that caused the supernova to increase in brightness for a full three months. One hypothesis is that radioactive decay of nuclei produced in the explosion provided the energy. Another is that the energy came from radiation from a rapidly rotating neutron star. Different hypotheses make different predictions as to the future behavior of the supernova and as this is being written, the radioactive decay hypothesis seems to better fit recent observations. Further data will decide which hypothesis is best; or perhaps none of those yet proposed is sufficient to explain this aspect of the supernova.

to see how well his equipment worked during a three-hour exposure.

He stood alone on a ladder in the dark, squinting into a viewfinder and delicately moving the telescope to keep pace with the earth's rotation. The work got especially onerous as the winds across the peak picked up to 40 miles per hour and roared against the shack. Shelton began to wish he hadn't left his coat at the house.

When the three-hour exposure was finished, Shelton went to get his coat and returned to the shack to begin taking another picture. After he returned, at about three in the morning, the roof was blown shut and the telescope knocked horizontal. Stumbling around in the dark—a flashlight would have ruined the photograph—he found that the telescope had survived the roof's blow. At this point Shelton was tired and disgusted enough that he did something uncharacteristic. He decided to go to bed before dawn.

First, though, he had to develop the plate he had exposed. In the darkroom of his little stone house he got very confused when a large round blotch appeared in the photo. His first thought was that it must be a flaw in the plate—there weren't any stars that bright in that part of the sky. "My next reaction was—that's one *hell* of a flaw," he says. "I knew there couldn't have been a flaw that big."

For some reason he didn't rush outside to check the sky. He spent twenty minutes telling himself that it couldn't be an exploding star, that everything else had gone wrong this night, that he was too tired to know what he was doing. "I was trying to think of every other possible explanation so I wouldn't look too stupid when someone else suggested it," he says. That wasn't unlike Brahe's reaction. 'I began to doubt the faith of my own eyes,' wrote Brahe, who started flagging down carriages and asking the occupants if they too saw the bright star. They did—and so did Shelton when he finally put on his coat and went outside.

To anyone used to looking at the Large Magellanic Cloud, the faint glittering point was a glaring anomaly. By another remarkable stroke of luck, Shelton had taken a picture of this area the night before, and there was no bright blotch then. So he knew that in one 24-hour period 170,000 years ago this point had suddenly lit up.

---

*Reprinted with permission from John Tierney, "Exploding Star Contains Atoms of Elvis Presley's Brain," *Discover,* July 1987, pp. 46–61. © 1987, Discover Publications, Inc.

Theoretical astronomers were especially encouraged by SN1987A for many of their predictions were right on target, especially the prediction of large numbers of neutrinos from the explosion. Neutrinos from the supernova were recorded by neutrino detectors in Japan and in Cleveland, Ohio; confirming astronomers' basic supernova theory.

The observations of SN1987A are still coming from a number of sources. Within a few days of the original observations, all rocket and balloon experimenters were sent a letter asking them to turn their attention to the supernova and to gather what data they could. If the explosion of the Challenger space shuttle had not so delayed the shuttle program, the Hubble Space Telescope would have been available to examine the rare supernova. As it is, the supernova gives astronomers one more reason to look forward to the launch of this instrument.

SISTERS OF THE STARS

Astronomers divide the stars into two classes, called *population I* stars and *population II* stars. The difference in the two populations is the amount of heavy elements they contain. Population II stars contain very little material other than hydrogen and helium. We have seen that heavier elements are produced in the cores of stars as fusion takes place and so heavy elements do exist in their cores, but we see little or none in their atmosphere.

The spectra of population I stars, on the other hand, reveal that their atmospheres contain heavier elements. The separation of stars into these two groups is somewhat arbitrary since a continuum actually exists: from stars that have no elements beyond hydrogen and helium in their atmospheres to those that contain the greatest amounts; but nonetheless, the distinction is convenient.

There is an easy explanation for there being different amounts of the heavy elements in different stars. Recall that stars are formed from interstellar clouds of gas and dust and that most stars end their lives by blowing (or exploding) much of their mass back out into space. The material they expel into space therefore contains some of the heavier elements produced within the star. Thus population II stars are those old stars that were formed from interstellar material long ago in the history of the universe, before the interstellar material became enriched in heavy elements. Population I stars, on the other hand, are young stars formed from material that contained the remains of previous generations of stars.

Which population is our Sun? We can answer this question without knowing anything about the spectrum of the Sun. Recall that the solar system was formed from material that did not fall all the way into the Sun as the interstellar cloud collapsed; it was formed from the same material that made the Sun. The fact that the Earth (and the other planets) contain heavy elements means that the cloud from which they formed contained those materials. The Sun is a population I star (although it contains less heavy material than some other stars).

Recall that fusion in the core of stars continues to release energy as heavier and heavier elements are formed until iron is produced. In order for iron nuclei to fuse with other nuclei to form heavier elements, there must be an *input* of energy. The reaction that fuses iron to heavier elements does not release energy; it absorbs it. Although most of the matter of which the Earth is made is less massive than iron, we find many elements more massive than iron. Such matter could not have been formed by fusion within the core of a star. It was instead formed during a supernova explosion, when tremendous amounts of energy were available. Most of the energy of a supernova is used up in the release of radiation (especially neutrinos) and in blasting away the outer parts of the star

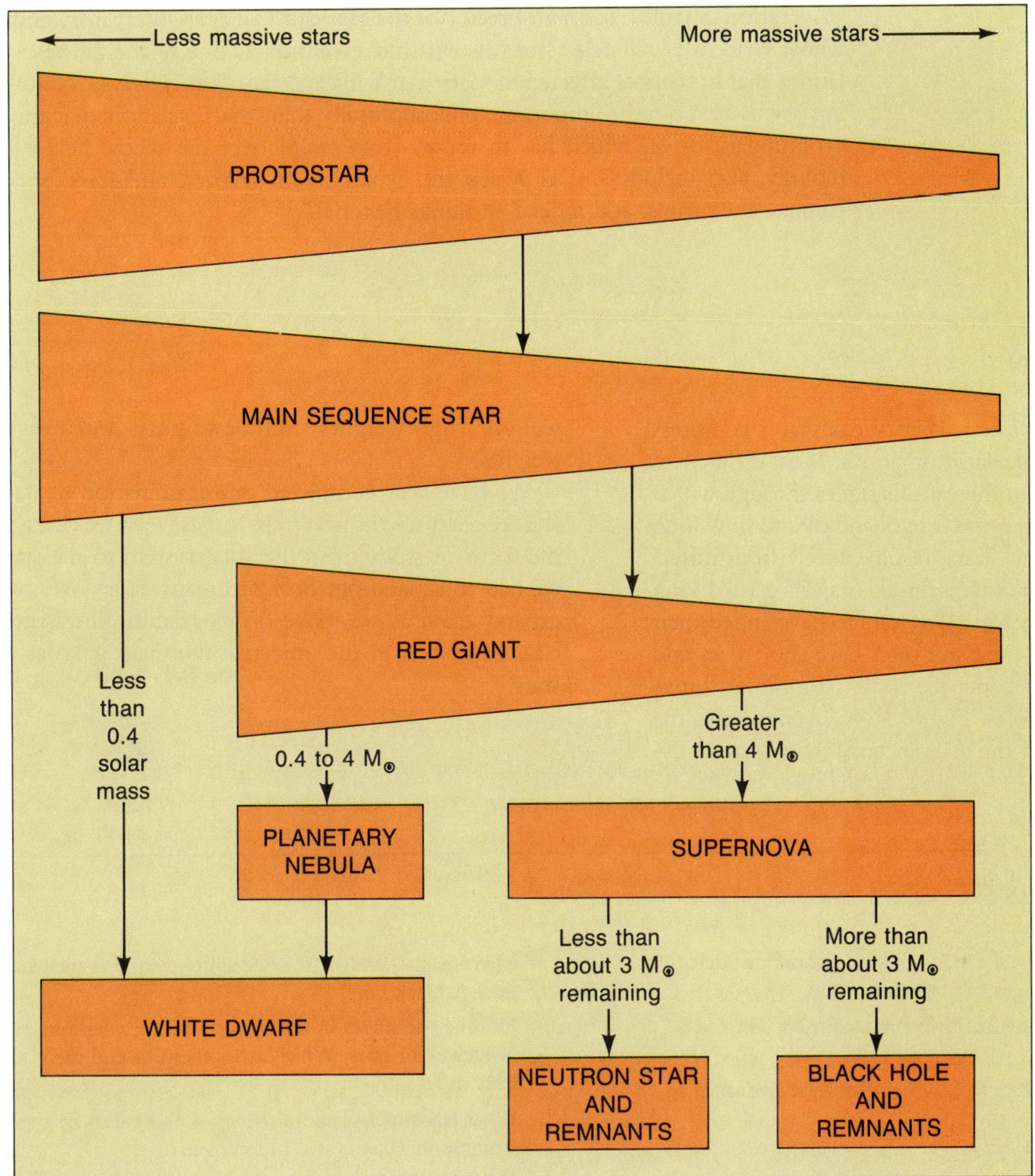

Figure 16–32. *This summarizes the steps in the life cycles of stars of various masses. Some boxes are wider at the left, representing the longer times low-mass stars spend in those phases.*

but a small fraction is absorbed by the material of the star; fusing it into heavy elements.

The above discussion has implications for us humans. The material that makes up our bodies is from the Earth. Where was it before it was part of the Earth? In an interstellar cloud. And before that? In a star!

Harlow Shapley, former Director of the Harvard College Observatory and whose work we will study later, lists cosmic evolution as one of the ten revelations that have most affected modern man's life and thought. He says, "Nothing seems to be more important philosophically than the revelation that the evolutionary drive, which has in recent years swept over the whole field of biology, also includes in its sweep the evolution of galaxies, and stars, and comets, and atoms, and indeed all things material." *

CONCLUSION

All stars get their starts when shock waves compress parts of cold interstellar dust clouds. Most differences between stars, from the various stages through which each star will progress as it lives and dies to how long each of those stages lasts, result entirely from difference in mass. The universe itself has not existed long enough for the least massive stars to have ended their lives, but the most massive ones have short lives and at the end they become the most astonishing things in nature: black holes. Figure 16–32 summarizes the life cycles of all of the stars in our boxing stable: flyweights, lightweights, middleweights, and heavyweights.

We have now completed our examination of stars and are ready for the next step. Just as when we changed the focus of study from the solar system to the stars we had to expand our horizons many times over, we expand again as we move on to examine the largest class of objects in the universe: the giant galaxies of stars.

RECALL QUESTIONS

1. What property of a star is most important in determining the stages of its evolution?
2. Why do stars of the lowest mass not become red giants?
3. What causes stars to heat up after their hydrogen fusion ceases?
4. What is a black dwarf?
5. What is a planetary nebula and what causes it? Describe the appearance of planetary nebulae.
6. Describe a white dwarf with regard to size, mass, luminosity, and temperature.
7. Sketch the stages in the life cycle of lightweight stars.
8. What supports a white dwarf from further collapse? What supports a neutron star?
9. What happens to the size of a main sequence star when mass is added to it? A white dwarf? A neutron star? A black hole?
10. Outline the stages in the lives of middleweight and heavyweight stars. What causes them to end their lives differently?
11. What leads us to believe that the object seen by the Chinese in 1054 was a supernova?
12. What is a nova, and how does it differ from a supernova? Why were novae originally given that name?
13. Describe how pulsars were discovered. What led astronomers to conclude that they were neutron stars?
14. When astronomers were searching for the nature of pulsars, why, if one assumes that the pulses from a pulsar are caused by vibrations in their surfaces, did the observed rates of pulsation seem to rule out both white dwarfs and neutron stars?
15. Describe the lighthouse model of pulsars. What ob-

*Harlow Shapley, *Beyond the Observatory*, New York: Scribner, 1967, pp. 15–16.

servation(s) led astronomers to select this model from among other suggested models?

16. What is the origin of the tremendous energy released by pulsars and their nebulae?
17. What is a black hole? How can we expect to observe one?
18. Define *Schwarzschild radius* and *event horizon.*
19. What general statement can be made about the escape velocity from a black hole?
20. If the Sun (magically) became a black hole, what would be the effect on the Earth? Explain.
21. What is the observational evidence that black holes exist?
22. What is SN1987A?

QUESTIONS TO PONDER

1. We see relatively few white dwarfs compared to main sequence stars, but we are confident that they are very common. Explain this discrepancy.
2. Explain why main sequence stars, white dwarfs, neutron stars, and black holes respond differently with regard to changes in their sizes when matter is added to them.
3. Consult an astronomy magazine such as *Astronomy, Sky and Telescope,* or *Mercury* to find some of the latest data concerning the Magellanic supernova of 1987 (SN1987A).
4. Explain why the observation that radiation from the first pulsar was appearing four minutes later each night led astronomers to conclude that they were not of Earthly origin (or, if they were, that astronomers had produced them).
5. How can it be legitimate for the period of a pulsar to be stated to an accuracy of eight significant figures?
6. Compare the mechanism that has slowed the rotation of the Sun (Chapter 15) to the mechanism that slows pulsars.
7. When astronomers were looking for pulsars in supernova remnants they automatically searched the Crab Nebula, for it is from a recent supernova. If the lighthouse model is correct, why was it unlikely, and therefore lucky, that we were able to find a pulsar there?
8. In order to calculate how fast a white dwarf (or neutron star) would pulse by vibrating its surface, we must know the strength of the gravitational field at its surface. What other property of the object must be known?
9. It is predicted that supernovae explode at the rate of one per second over the entire universe. Why don't we see more?

CALCULATIONS

1. What is the value of the Schwarzschild radius of a star of seven solar masses?
2. If the mass of a black hole is doubled, how does the size of its event horizon change?
3. Betelgeuse is a red giant 650 LY away. The Crab Nebula is 21,000 LY away. If Betelgeuse were to become a supernova, how many times brighter than the Crab would it appear?

CHAPTER 17

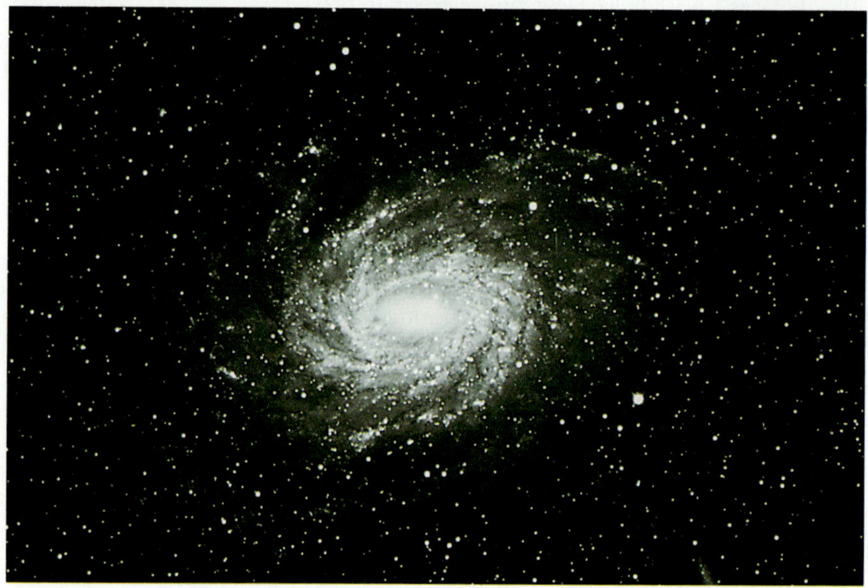

This was the first spiral nebulae investigated by von Maanen.

Spiral nebulae were discovered in the late 1700s but their nature remained a mystery for nearly 150 years. Early in this century, Adriaan van Maanen, an astronomer at Mount Wilson Observatory, was convinced that they were clouds of gas rather than galaxies similar to the Milky Way. He determined to settle the question based on evidence and between 1916 and 1923 he made careful measurements of pairs of photographs taken up to twelve years apart. He found that the nebulae rotated with periods of tens of thousands of years. Such a great rotation rate for a galaxy would mean that stars on its outer edge would have to be moving faster than the speed of light. Thus his data proved that the nebulae were not galaxies.

Other astronomers obtained conflicting data but von Maanen was using the largest telescope in existence and he found a great rotation speed for each of the seven nebulae for which he had the necessary pairs of photos.

In 1924 Edwin Hubble (also working at Mount Wilson Observatory) proved conclusively, by the use of Cepheid variables, that the nebulae were separate galaxies.

Why did von Maanen obtain incorrect results? In Science and Objectivity *(Iowa State University Press, 1988), Norriss Hetherington argues that although von Maanen was attempting to be impartial, his preconceived conclusion distorted his findings. Hetherington cites other examples of unintentional nonobjectivity in astronomy (including Lowell's Martian canals) and he warns scientists that they must be constantly on guard against allowing their anticipated results to cloud their data.*

Galaxies

T HOUGHOUT THIS TEXT WE have seen how our experience of the universe has grown from one that was centered on our immediate environment to one whose center was thought to be located farther and farther away. In the next few chapters we will see that today it appears that the center simply does not exist: the universe has no center. Hand-in-hand with this receding center has come a tremendous expansion of our realization of the size of the universe. Opposition to Galileo occurred not only because he placed Earth off-center but because in his universe the Earth played such a small part.

Galileo's contemporary, Giordano Bruno, did propose that the stars were suns and he paid for his belief with his life; being burned at the stake for heresy in 1601.

Yet Galileo did not realize that the Sun that rules our sky is just one of countless suns; that each of the stars he saw in the sky is another sun. When we changed our study from the solar system to the stars, we had to expand our entire scale of thinking. An indication of this is that the unit of measure used in the solar system, the astronomical unit, was found to be very inconvenient for stellar distances; we changed to the light year and parsec to describe such distances.

Now we turn our attention away from considering stars as individuals to considering them in galaxies. We will again have to make a mental leap in our thinking about distances and sizes. The leap is so great that it may be impossible for us to truly understand the distances involved. It is fun to try, however.

As we study objects at greater and greater distances from us, our knowledge of their nature and properties is more recent and less certain. In this chapter we will discuss the galaxy of stars of which the Sun is such an insignificant part and then we will consider other galaxies. We will compare two conflicting theories astronomers are now debating to explain the structure of galaxies like ours. In addition, we will look again at the question of accuracy in scientific measurements, for we will find that among the galaxies, measurements are much less precise than we might be comfortable with.

OUR GALAXY

Milky Way Galaxy: The galaxy of which the Sun is a part. From Earth, it appears as a band of light around the sky.

Until fairly recently, humans were not even aware of the existence of the galaxy within which we live. We now know that our Sun is one of about 200 billion stars making up what we call the ***Milky Way Galaxy,*** the Milky Way, or simply the Galaxy. Most stars in the Galaxy are arranged in a wheel-shaped disk that circles around a bulging center (Figure 17–1). The diameter of the galaxy is about 100,000 light years and the Sun is about two-thirds of the way from the center.

Discovery of the Galaxy and the Sun's Position

Figure 17–2 is a photo of the region of the sky that includes the constellation Auriga. The hazy area stretching though Auriga was called the *via lactae* (the Milky Way) by the Romans. You are encouraged to find a good clear sky away from city lights and look at the beautiful Milky Way, a sight many of us in the modern world never get a chance to appreciate. The Milky Way completely encircles the Earth, passing through Sagittarius, Aquila, Cygnus, Cassiopeia,

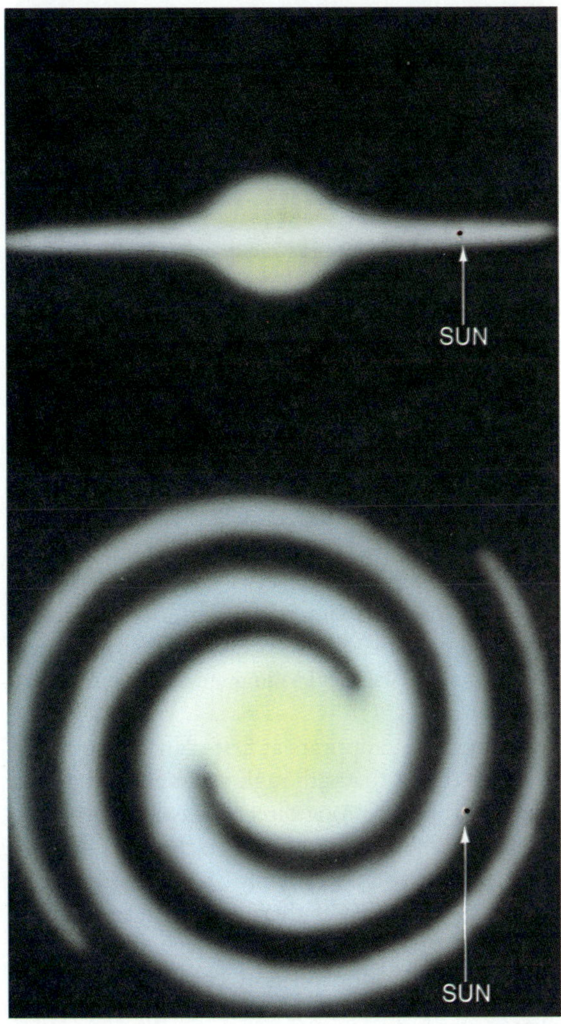

Figure 17–1. *The Sun is located about two-thirds of the way out along the disk of the galaxy, shown here both edge-on (a) and face-on (b).*

Figure 17–2. *The constellation Auriga is located in the Milky Way.*

Auriga, and between Gemini and Orion on the northern hemisphere of the celestial sphere. Then it passes through Monoceros, near Canis Major, and through Vela and Crux on the southern hemisphere.

Galileo turned his telescope to the Milky Way and discovered that it was made up not of haze (or even milk!), but of stars far too numerous to count. The stars cause the misty appearance of the Milky Way because so many are so far away that although the naked eye cannot distinguish the individual stars, it can see the overall illumination from them. In Galileo's telescopic view, he saw more haze behind the many stars his telescope revealed and he concluded that it was caused by even more stars too faint to see individually.

Notice that the term Milky Way *is being used here to refer to the effect seen in the sky. Whether the term refers to this phenomenon or to the galaxy of stars itself should be clear from the context.*

William Herschel, discoverer of Uranus, wrote in 1849 that through a telescope

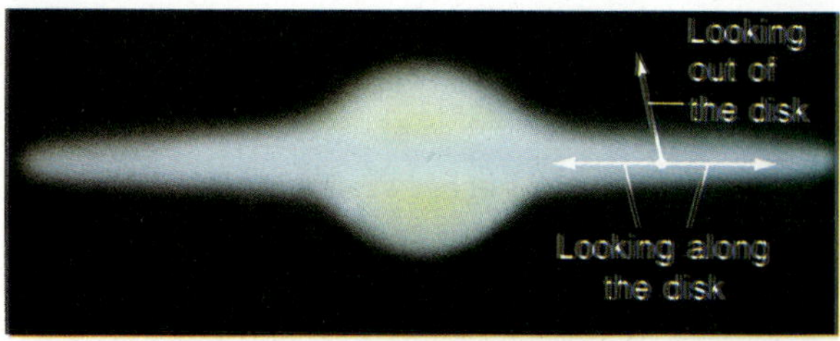

Figure 17–3. *When we see the Milky Way in the sky, we are looking along the disk of the galaxy. Otherwise we are looking out of the disk.*

Figure 17–4. *The Herschels' counting of stars led them to conclude that the Galaxy is shaped like this. The Sun is located at the bright spot within the Galaxy.*

> We find [the stars] crowded beyond imagination along the extent of [the Milky Way]; . . . so that, in fact, its whole light is composed of nothing but stars of every magnitude from such as are visible to the naked eye down to the smallest points of light perceptible with the best telescope.

The telescopic view of the Milky Way led astronomers to conclude that we live in a disk of stars. When we view the Milky Way in the sky, we are looking along that disk (Figure 17–3) and when we are looking in other directions, we are looking out of the disk.

As we view the Milky Way from Earth it appears at first glance that we are at the center, for to the naked eye it seems that although there are some local variations, the Milky Way is about as bright in one direction as in another. In the 1780s, in order to determine where the Sun lies relative to the disk, William Herschel and his sister Caroline made star counts in nearly 700 selected regions distributed around the sky. They reasoned that if there is a pattern of more stars in one direction than in another, that direction could be assumed to be toward the center of the disk. Their conclusions not only confirmed the disk-like shape of the Galaxy but indicated that the Sun is indeed at the center; for they saw about the same number of stars in all directions in the Milky Way. Figure 17–4 shows the shape they concluded for the Galaxy. Notice that the Sun is nearly at its center.

In the early part of this century, Jacobus C. Kapteyn sought to find the Sun's location by analyzing the density of stars in various directions from the Sun. He did this by measuring not only the number of stars in each direction but their distances from us. He found that the density of stars decreases in every direction from the Sun and it was logical for him to conclude, like William and Caroline Herschel, that the Sun is at the center of the disk.

The conclusion that the Galaxy centers on the Sun was viewed with skepticism, as you might expect. We once thought that the Earth is the center, only to find that it circles the Sun. Are we now discovering that the Sun is the center? The evidence from the Herschels and from Kapteyn pointed to an affirmative answer but the finding seemed to be contrary to the trend that began before written history and continued through Copernicus and Galileo. Is our own Sun truly at the center?

Figure 17–5. *This wide-angle photo shows the Milky Way in Sagittarius (left) and Scorpius (right).*

Figure 17–6. *The Pleiades, a galactic (or open) cluster.*

Today we know that the Sun is not at the center. To understand why the Herschels and Kapteyn obtained their erroneous results, imagine yourself standing in a large forest. Suppose you try to decide whether or not you are at the center of the forest by counting trees in all directions. Unless you are very near an edge, you will see the same number of trees in all directions even though you are nowhere near the center. The reason is that you are prohibited by the trees themselves from seeing beyond a certain distance. If you cannot see to the edge of the forest, this method of determining your location is not valid.

The situation in the case of the stars is somewhat different, for the stars do not fill our view as do the trees in a forest. When we look out among the stars, though, interstellar dust and gas places a limit on how far we can see. The Herschels were unaware of the existence of this material. They assumed that they could see all the way to the edge of the group of stars within which our Sun lies. Since they could only see a limited distance and since the Sun is not near an edge, they concluded that it was at the center.

Another analogy: If you are surrounded by dense fog, the fact that you can see the same distance in every direction is not evidence that you are in the center of anything.

Likewise, interstellar dust kept Kapteyn from counting the stars correctly. The density of stars at greater and greater distances from the Sun appeared to decrease because when Kapteyn counted stars at great distances, the interstellar dust prohibited him from seeing them all. Figure 17–5 is a photo of part of the Milky Way as seen from Earth. The dark areas stretching across it are dust and gas clouds.

These investigators reached erroneous conclusions simply because one of their assumptions—that there is nothing in space to block the view of distant stars—was wrong. Incorrect assumptions cause trouble not only in science but in everyday life. Often we are not even aware of what our assumptions are and this prevents us from even accepting the possibility that our conclusion may be in error.

Figure 17–7. *Two globular clusters. The one shown at left is M13 (in Hercules) and one at right is M3, in Canes Venatici. Such clusters contain hundreds of thousands of stars.*

Globular Clusters

Globular cluster: A spherical group of up to hundreds of thousands of stars, found primarily in the halo of the Galaxy.

In a previous chapter we saw that some stars begin their lives in galactic clusters. These clusters are called "galactic" because they are found within the disk of the Galaxy. The Pleiades (Figure 17–6) is the most prominent example of this type of cluster, that typically may have hundreds of stars. Figure 17–7 shows examples of a much larger type of cluster, the ***globular cluster.*** These beautiful, symmetrical clusters may look like they have solid centers but they are actually groups of hundreds of thousands of stars. The stars are so densely packed in the center of the clusters that we do not see the individual stars but simply see a white area. The average separation of stars near the center of a globular cluster is about 0.5 light year. (In the Sun's region of space, stars are separated by an average distance of about 4 or 5 light years.) Globular clusters are not confined to the disk of the galaxy but are seen outside the disk.

While Kapteyn was seeking to determine the Sun's location in the Milky Way by studying star locations, Harlow Shapley was trying to do the same using globular clusters. He had a problem determining the distances to globular clusters, however, for they are much farther away than the stars Kapteyn was analyzing. Just a few years earlier Henrietta Leavitt had discovered the relationship between the periods and the apparent magnitudes of Cepheid variables in the Magellanic Clouds. Shapley observed Cepheid variables in globular clusters but could not use them as distance indicators until he had determined the relationship between their periods and absolute magnitudes. He used a statistical method with Cepheid variables within the galactic disk to do so and then he turned his attention to Cepheids in globular clusters.

In 1917 Shapley published results of his survey of distances and directions to globulars. He showed that they are not distributed evenly around the sky but tend to be located more on one side; centered about the constellation Sagittarius. In fact, they seemed to be distributed in a sphere centered on a point thousands of light years away from the Sun. Figure 17–9 shows the approximate distribution of globular clusters compared to Herschel's model of the Galaxy. Shapley assumed that the clusters revolve around the center of the Galaxy and concluded that the Galaxy's center therefore lies at the middle of the group of globular clusters. This meant that the Galaxy is much larger than

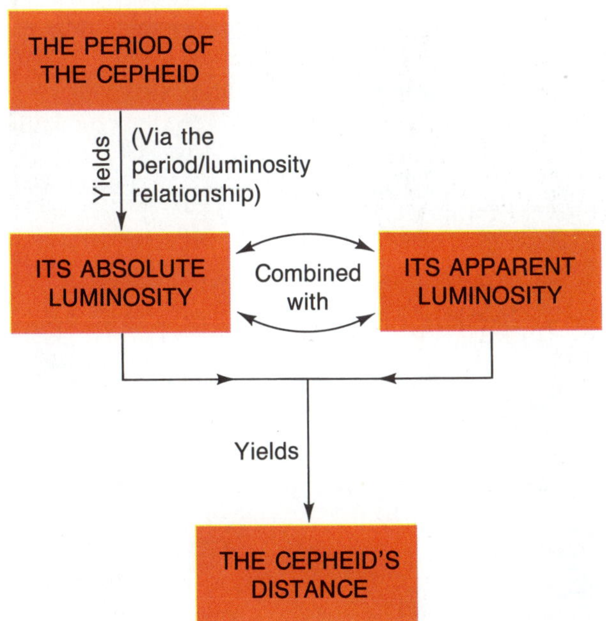

Figure 17–8. *Once Shapley had determined the relationship between Cepheids' periods and absolute luminosities, he could use a chain like that above to determine the distance to any Cepheid variable.*

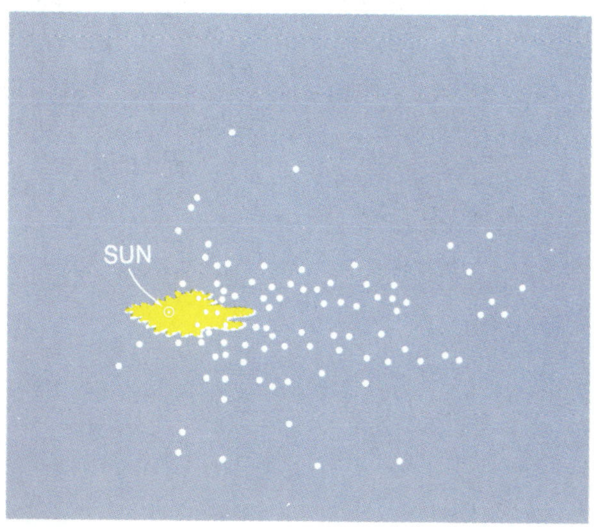

Figure 17–9. *The distribution of the globular clusters is shown relative to the Sun and to the Herschels' model of the Galaxy.*

indicated by Herschel's model. It also meant that the Galaxy is not centered about the solar system; that our Sun is not at the center of the galaxy.

In the 1920s, further evidence indicated that the Sun is not at a unique position in the Galaxy. Jan Oort (who proposed the comet cloud that bears his name) and Bertil Lindblad studied the motions of great numbers of stars near the Sun. They found that there is a pattern in the velocities of stars, depending upon their directions from the Sun. Kepler's third law, when applied to stars revolving around the center of the Galaxy, predicts that stars closer to center should move faster and those farther from center should move slower. This is the reason for the pattern of velocities shown in the Oort-Lindblad analysis. They concluded, as had Shapley, that the center of the Galaxy was thousands of light years away in the direction of Sagittarius.

It should be pointed out that Oort and Lindblad saw a pattern only after analyzing very great numbers of stars, for stars have random motions along with their pattern of motion around the galactic center. This is similar to the way that cars on a multilane freeway have a pattern of motion in one direction but at any given time certain cars may be changing lanes or otherwise deviating from the pattern. Oort and Lindblad ignored individual stellar motions and concentrated on patterns of average motions.

Finally, in 1930 the interstellar dust was discovered. This resolved the conflict between the conclusions of Herschel and Kapteyn on the one hand and Shapley, Oort, and Lindblad on the other. Although Shapley's original values

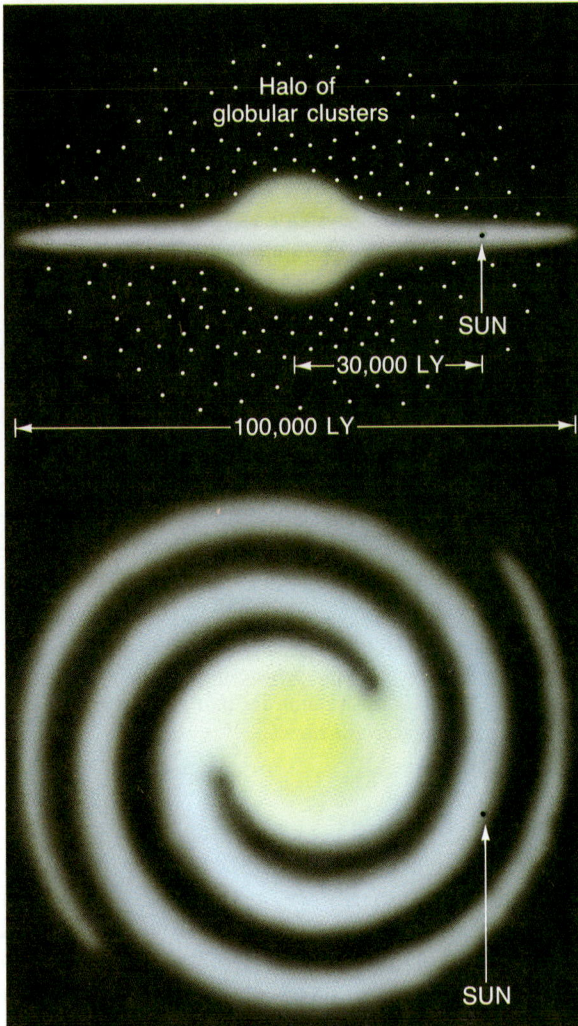

Figure 17–10. *The Galaxy, showing the halo and the Sun's approximate position.*

Table 17–1 Galactic Data

Radius of disk	50,000 LY
Radius of nuclear bulge	15,000 LY
Sun's distance from center	30,000 LY
Sun's orbital period	250,000,000 years
Thickness of disk	3000 LY
Number of stars	200 to 400 billion

for the size of the Galaxy had to be revised when it was discovered that there are two types of Cepheid variables, his basic deductions were correct. He had shown that the Galaxy was much larger than previously thought. The boundaries of the universe were again pushed back.

Halo (around the Galaxy): The outermost part of the Galaxy, fairly spherical in shape, beyond the spiral component.

Figure 17–10 is a drawing of the Galaxy that shows the Sun's position and the location of globular clusters in a ***halo*** around the central bulge and nucleus. The figure also indicates the spiral nature of the Galaxy, which we will discuss shortly. Table 17–1 shows today's values for various Milky Way data. Keep in mind though, that the numbers here are approximate; not only because of uncertainties inherent in any measurement but because there are no specific boundaries for the various parts of the Galaxy. The disk has no sudden end, for example, and so any statement of its thickness is necessarily approximate.

CLOSE UP

Calculating The Mass of the Inner Galaxy

WE WILL USE THE equation for Kepler's third law, along with the Sun's motion, to calculate the mass of the inner part of the Galaxy. First, we must express the Sun's 30,000 LY distance from the Galaxy's center in astronomical units. From Appendix B we find that one light year is equivalent to 63,000 astronomical units. Thus,

$$30{,}000 \text{ LY} \times \frac{63{,}000 \text{ AU}}{1 \text{ LY}} = 1.9 \times 10^9 \text{ AU}$$

We have carried the answer to two significant figures, although this accuracy may not be justifiable since the possible error in the distance is considered to be about 5000 light years.

Solving the Kepler equation for mass, and substituting the values,

$$d^3 = P^2 (M + m)$$
$$M + m = d^3/P^2$$
$$= \frac{(1.9 \times 10^9 \text{ AU})^3}{(2.5 \times 10^8 \text{ yr})^2}$$
$$= 1.1 \times 10^{11} \text{ solar masses}$$

The value obtained is the total of the mass of the Sun and the mass of the inner galaxy. The mass of the Sun, of course, is negligible compared to the value obtained, so that our answer is simply the mass of the part of the Galaxy that lies within the Sun's orbit.

The Mass of the Galaxy

Back in Chapter 7 we saw that Kepler's third law was revised by Isaac Newton to read

$d^3 = P^2 (M + m)$ d in AU, P in years, and M and m in solar masses

where d is the distance between the two objects, P is the period of the motion, and M and m are the individual masses of the two objects. We saw there that the relationship holds not only for objects in orbit about the Sun but for satellites in orbit around planets. Later we saw application of the law to binary star systems.

The discovery by Oort and Lindblad that the Galaxy in the Sun's neighborhood undergoes differential rotation meant that Kepler's third law might be applied to it to calculate the masses involved. In the case of a star revolving around the Galaxy's center, one of the masses in the equation is the mass of the star and the other is the mass of the entire inner galaxy, including all objects in the Galaxy that are closer to the center than that star is. This may seem strange since the inner portion of the Galaxy is made up a number of objects rather than one as it is in the case of the Earth's orbit around the Sun. It can be shown, however, that this is the correct application of the equation.

Figure 17–11. *Globular clusters orbit the galactic nucleus, passing through the disk twice during each orbit.*

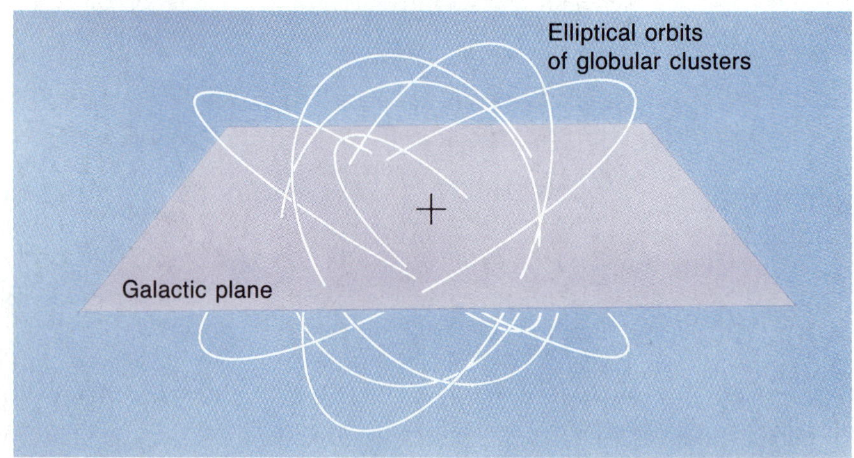

In the accompanying Close Up, Kepler's law is used to calculate the mass of the inner galaxy. The value obtained, 110 billion solar masses, must be taken as approximate. Keep in mind also that this is not the mass of the entire galaxy but simply the mass of the inner portion. Recent analysis of the pattern of rotation in the outer parts of the Galaxy indicate that the total mass of the Galaxy is about 10^{12} (a thousand billion) solar masses; about ten times more mass than is calculated for the inner galaxy assuming the applicability of Kepler's third law. At present, the nature of this additional mass is unknown for there are not enough stars to account for that much mass. We will see later that a similar problem exists in the case of other galaxies.

Galactic Motions

The motion of the Sun around the galactic center can be determined by methods such as Oort and Lindblad used to measure motions of stars near the Sun. We find that the Sun is travelling in a nearly circular path around the galactic center at a speed of about 250 kilometers per second. It is now moving toward the constellation Cygnus. Knowing that the radius of the Sun's path is 30,000 light years, we can calculate that the circumference of its orbit is about 200,000 light years and that it takes about 250,000,000 years to complete one revolution. Although this seems a tremendously long time, the Sun has completed some 20 orbits during its 5-billion-year lifetime.

Saying that the Sun is moving toward Cygnus does not mean that it is getting closer to Cygnus; it means that the Sun is moving in that direction through space. Besides, each star of the constellation Cygnus has its own motion; some toward us and some away from us.

We have shown that the mass of the entire inner portion of the galaxy can be calculated assuming that the Sun moves according to Kepler's third law. Doppler shift data of stars in the Galaxy show that in the inner part of the Galaxy, stars do not show differential rotation and do not obey Kepler's third law in their motion. Near the Galaxy's center, stars are very close together and the gravitational force between neighboring stars is more important than in regions where stars are widely separated. The result is that the center portion of the Galaxy rotates almost as a solid object; a phonograph record, perhaps.

As we might expect, however, the orbits of stars are not perfectly circular. Along with the general orbiting motion of each portion of the Galaxy, each

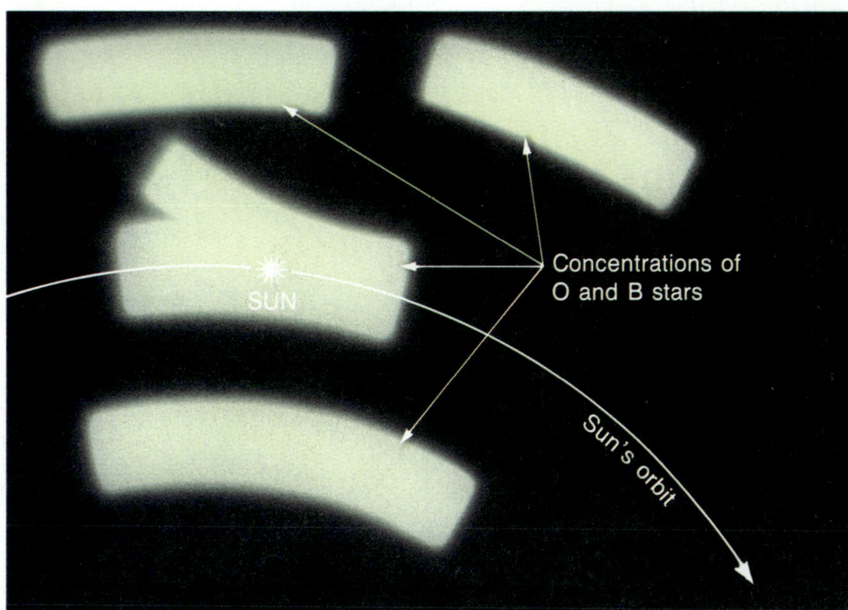

Figure 17–12. *O and B type stars are found at greatest concentrations in the areas shown as shaded.*

star has its own peculiar motion; perhaps having a component of motion from one side of the disk to the other or perhaps moving closer to the galactic center at one time and moving farther away at another time as it follows an elliptical path. In general, however, the paths of stars seem to be nearly circular.

The 100 or so globular clusters do not, of course, follow the rotation of the disk because their planes of orbit about the galactic nucleus do not lie within the disk. This means that during each orbit, a cluster passes twice through the disk (Figure 17–11).

THE SPIRAL ARMS

The spiral nature of the Galaxy is not obvious from observations in visible light because we can see only limited distances along the plane of the Galaxy. In 1951, however, astronomers at Yerkes Observatory detected that as one looks either toward or away from the galactic center, the distribution of 0 and B type stars is not uniform. They seem to be clustered at certain distances. Figure 17–12 illustrates the concentrations of these type stars. This was the first hint of the spiral nature of our galaxy. More evidence came with the discovery of the 21-centimeter line of hydrogen during the same year.

Spiral galaxy: A disk-shaped galaxy with two arms in a spiral pattern.

Cool hydrogen gas emits radiation of a particular wavelength; 21.1 centimeters. This is part of the radio portion of the spectrum and is not as readily absorbed by the interstellar medium as is light. Radio telescopes can therefore use ***21-centimeter radiation*** to detect high concentrations of cool hydrogen such as exists in interstellar clouds. Since hydrogen is the main component of

Twenty-one centimeter radiation: Radiation from atomic hydrogen, with a wavelength of 21.11 centimeters.

Figure 17–13. *If hydrogen gas were distributed uniformly in the galactic disk, 21-centimeter radiation from the Sun's forward direction would show a fairly regular blueshift (c).*

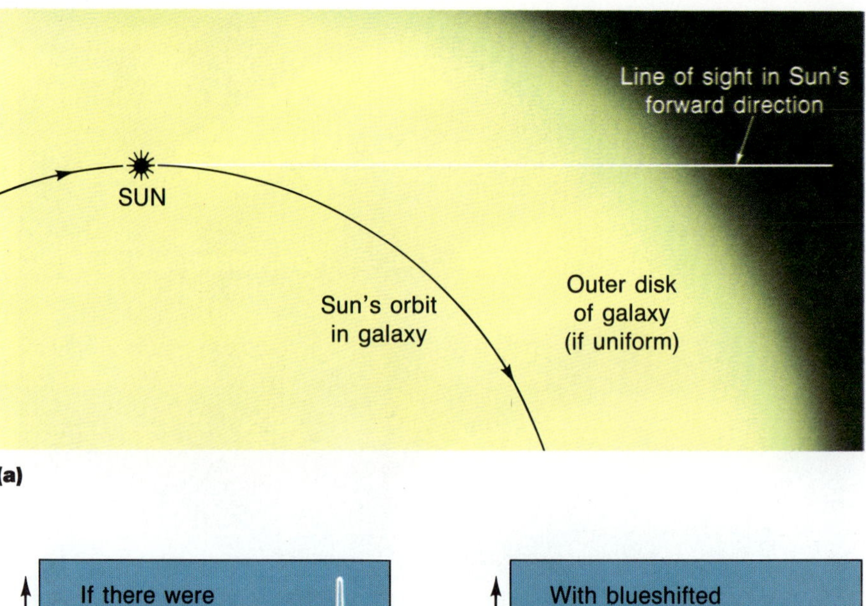

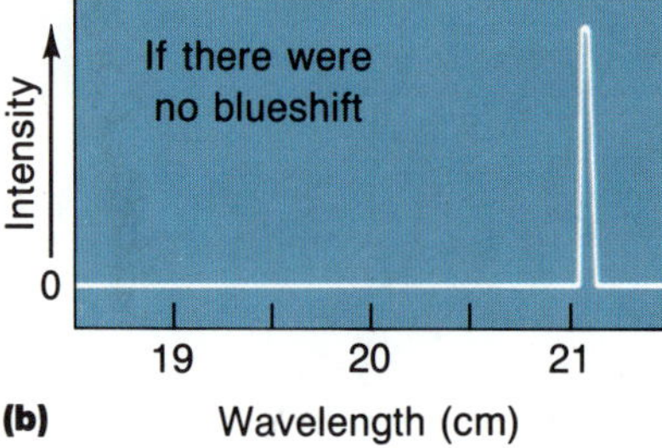

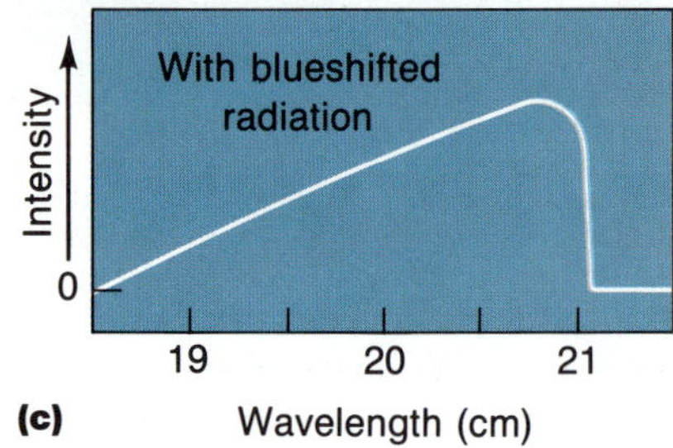

interstellar material—and the entire universe—this is an ideal method of detecting cool hydrogen clouds at great distances.

As we have seen, it is gas and dust from which new stars arise and to which some of the material of a massive star returns at the end of the star's lifetime. Therefore hydrogen gas clouds detected by 21-centimeter radiation are located at the same place as newly forming stars. Recall that 0 and B stars are very massive and are therefore short-lived. This means that they will be found only where stars have recently formed and that we might expect to find hydrogen clouds at locations identified by the astronomers at Yerkes Observatory as having high concentrations of O and B type stars.

It may seem that it would be impossible to determine the distance of a source of 21-centimeter radiation, for a radio telescope would seem to reveal only the direction to the source of the waves. In a sense this is true. The Doppler effect, however, allows us to determine the radial motion of the hydrogen with respect to us. Suppose, for example, that a radio telescope is pointed in a direction across the Galaxy as shown in Figure 17–13(a). Because of Keplerian motion, we would be moving toward the hydrogen along this direction, so a buleshift of the 21-centimeter radiation would be seen. If the hydrogen were distributed uniformly, the radiation would be Doppler shifted so that its wavelength would range from just less than 21.1 cm to quite a bit less. A graph of

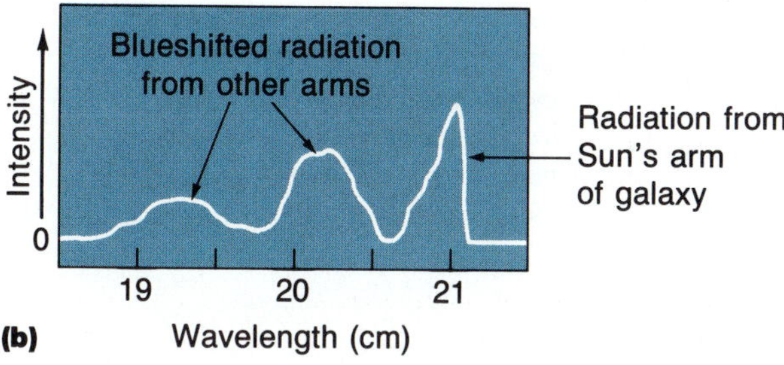

Figure 17–14. *Twenty-one centimeter radiation observed in the Sun's forward direction actually shows peaks of blueshifted radiation (b). This occurs because the hydrogen that emits it is concentrated in spiral arms (a).*

radio intensity versus wavelength might then appear as shown in part (c) of this figure.

On the other hand, consider how the radiation would appear if the hydrogen is located in spiral arms as shown in Figure 17–14(a). In this case, each spiral arm would have a fixed value for its Doppler shift and the graph would appear as shown in Figure 17–14(b). This is how it appears in an actual case and so 21-centimeter radiation gives us further evidence for the spiral nature of the Galaxy.

Figure 17–15 shows a map of the Milky Way Galaxy made by analysis of 21-centimeter radiation. Notice the blank areas on the far side of the nuclear bulge and in the opposite direction from us. This occurs because galactic matter in these locations is moving nearly perpendicular to our line of sight and therefore does not show a Doppler shift. Without the Doppler shift data, we have no way of untangling the various signals we receive from those directions.

Mapping the Galaxy in 21-centimeter radiation contains plenty of room for error, so the map must be considered approximate. There is certainly no doubt that our galaxy is spiral, though. As we will see later, such a spiral appearance is a common one for other galaxies. Figure 17–16 shows a spiral galaxy that is similar to the Milky Way.

Figure 17–15. *A map of the galactic spiral arms based on 21-centimeter radiation. The Sun is located at the small circle and the galactic center at the large one.*

Figure 17–16. *If we could get outside our galaxy and view it face-on, it might look somewhat like this one (NGC 2997) in the southern constellation Antlia. The bright stars in the photo are in our galaxy; much closer to us than is NGC 2997, like flies on a car windshield.*

SPIRAL ARM THEORIES

It may seem at first glance that the spiral arms could be explained by the differential rotation that results from Kepler's third law. Suppose that we consider a group of stars that at a given time are in a straight line across the center of the Galaxy as shown in Figure 17–17(a). As indicated by the arrows in the figure, stars closer to center would orbit faster. This would wind up the line of stars, forming them into a spiral. The problem is that the line would wind up too much. Recall that the Sun has made some 20 revolutions around the Galaxy; Figures 17–17(b), (c), and (d) show the line of stars winding up so much that it would no longer be distinguishable; the Galaxy would appear as a fairly uniform disk. Yet we do perceive a spiral nature to our own galaxy and we see it in many other galaxies. This simple differential rotation hypothesis cannot be correct.

There are presently two theories competing to explain the spiral nature of galaxies: the ***density wave model*** and the ***chain reaction model.***

Density wave model: A model for spiral galaxies that proposes the arms are the result of density waves sweeping around the galaxy.

Chain reaction model: A model for spiral galaxies that explains the arms as resulting from a series of supernovae, each triggering the formation of new stars.

The Density Wave Theory

This theory was first proposed in 1960 by Bertil Lindblad of Sweden and holds that it is not a simple fixed line of stars we see in a spiral arm of a distant galaxy but a line formed by the brightest stars and the glowing nebulae surrounding them. The theory holds that stars revolve around the Galaxy independent of the spiral arms and that the arms are simply areas where the density of gas is greater than at other places. According to this theory, there are almost as many

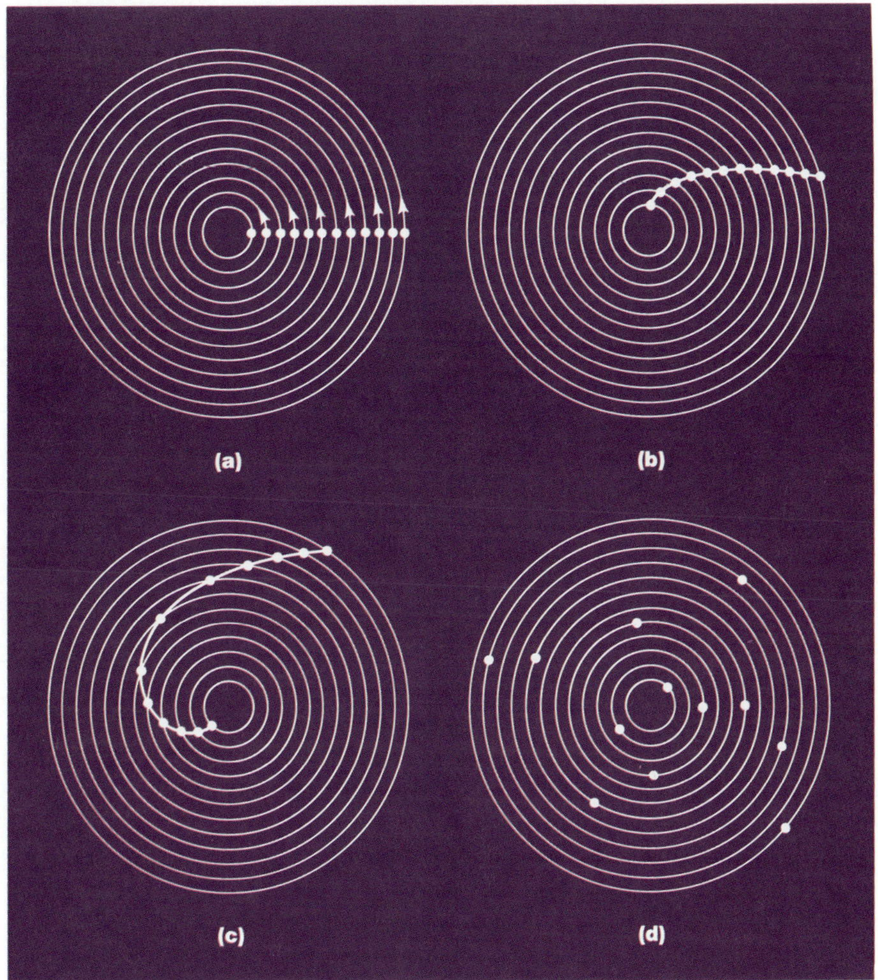

Figure 17–17. *Stars that started out in a straight line from the galaxy's center (a) and orbit according to Kepler's laws would begin to show a spiral pattern (b) and (c), but after a few revolutions, the pattern will be lost (d).*

stars per volume between the arms as in the arms but the arms contain more of the brightest stars and a higher density of gas and dust. The areas of denser gas move around the Galaxy in ***density waves,*** causing the formation of new stars and glowing emission nebulae. This is best explained by an analogy.

Density wave: A wave in which areas of high and low pressure move through the medium.

Suppose cars are travelling on a long superhighway at a speed exceeding the speed limit; perhaps leaving campuses for spring break. Also travelling along the road, at less than the speed limit, is a traffic patrol car with its radar on. We are observers in a helicopter high in the sky and we see the cars fairly evenly distributed along the highway except around the police car. For a short distance behind and in front of the police vehicle the cars are bunched up. As a given car approaches the police cruiser from behind, the car slows down, slowly passing the feared patrol. Then when the driver feels that his or her car is safely in front of the police cruiser, it again speeds up. This causes there to be a high density of cars around the police car even though the cars making up that group change all the time. Figure 17–18 illustrates the situation. The police car moves along,

Figure 17–18. *This represents an overhead view of a highway seen as time passes. Car X is a police car, moving just slower than the speed limit. Bunched up around that car are other slow-moving cars that only gradually pass the police car. Thus as time goes by the cars near X are different but there remains a high density of cars there.*

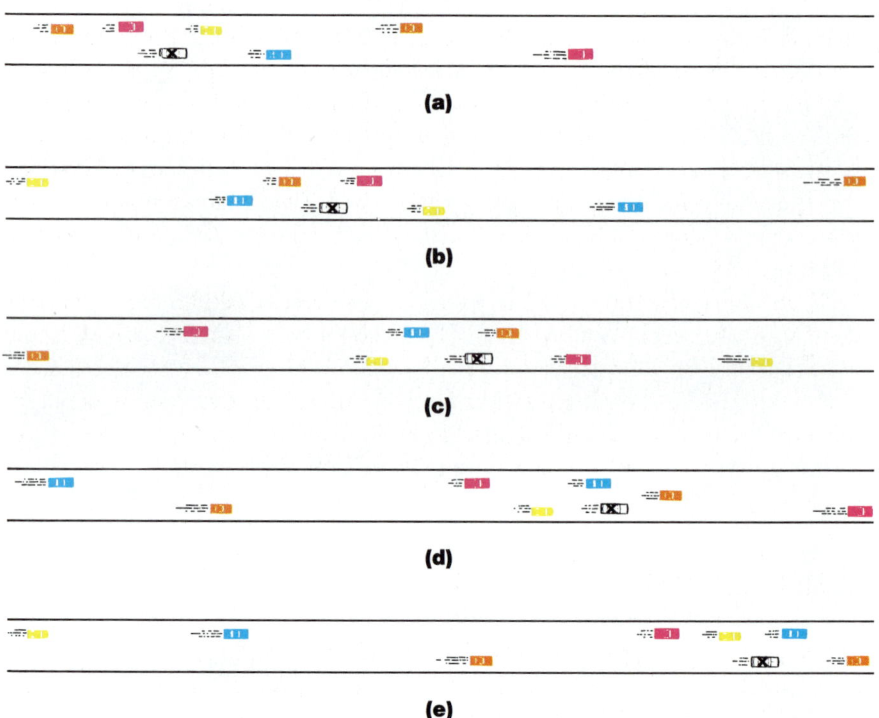

seeming to carry its high-density group along with it in what we see from the helicopter to be a density wave.

In a galaxy, the density wave consists of a region of gas and dust that is more dense than normal. Density waves are common here on Earth, for every sound wave is a density wave with regions of high and low density making up the wave. The major difference between sound waves and the density waves of the spiral galaxy is that in the atmosphere of Earth, a sound wave travels faster than the particles of the gas itself but in the near-vacuum of a galaxy, the wave travels more slowly than the particles. The gas and dust particles—as well as stars—catch the wave from behind and pass through it in much the same way that cars in the analogy passed through the density pulse around the police car.

Now recall that new stars are formed when an interstellar cloud of gas collapses. The density wave theory holds that the trigger for this collapse is the wave. As interstellar clouds approach the density wave from behind, they are compressed and stars are formed. Although stars of all masses are formed along the edge of the density wave, the brightest, most massive stars have ended their lives before they pass far from this edge. This means that when we look at a spiral galaxy, the spiral arms are obvious to us because they are the areas containing the bright stars.

The main problem with the density wave model is the question of what starts the wave in the first place. This has led some astronomers to propose another model: the chain reaction model.

The Chain Reaction Theory

According to the chain reaction model of galactic spiral arms, the triggers that start the collapse of most interstellar clouds are nearby supernova explosions. Then as the more massive of the stars finish their lives and become supernovae, they trigger more star formation, and so on. Thus the formation of new stars is confined to areas where this process is taking place. Now differential rotation enters the picture. Computer analysis shows that at the rate at which massive stars would be formed, Keplerian motion would cause spiral arms to be formed and sustained.

The chain reaction model is a much more recent development than the density wave model. Time will tell which model best fits further analysis and new observations. (Or perhaps parts of each will have to be combined into a new model.) At present both models are able to explain the spiral structure of this type galaxy but the chain reaction model has the advantage of being able to explain how a spiral arm would begin since the chain reaction starts with a single supernova.

THE GALACTIC NUCLEUS

Observations by Shapley, Oort, and Lindblad in the early part of this century revealed that the center of the Galaxy lies in the direction of Sagittarius. Because visible light from the galactic nucleus does not reach us, astronomers had to await the development of nonvisible light telescopes to learn more about that nucleus.

In 1960, observations of the 21-centimeter radiation pinpointed the galactic nucleus and revealed that hydrogen clouds there are in a very turbulent state; moving at very high speeds. In addition, the radiation indicated that there are two great expanding arms of hydrogen moving away from the center: one on this side moving toward us and another on the other side moving away. The Doppler effect cannot tell us about material moving across our line of sight but the assumption is that a ring or a shell of hydrogen has been blown out of the center.

The orbiting infrared observatories have revealed that the galactic nucleus is a very bright infrared source; equivalent to hundreds of Suns. The size of the source (judged from its angular size and its distance) seems to be only a few light years across; very small for the amount of infrared radiation emitted.

Finally, X rays detected by satellite receivers reveal that an even smaller source in the nucleus emits tremendous amounts of energy in this region of the spectrum.

What is the source of energy in the nucleus of the Galaxy? The only source that seems feasible is gravity and current theory holds that in the center of the Galaxy there is a gigantic black hole, orbited nearby by a great number of stars. As matter falls into the black hole, losing its orbital motion due to collisions, energy is released. The release of energy here is similar to the energy released as matter falls into stellar black holes, discussed in the previous chapter. In the case of the galactic nucleus, however, a *super*massive black hole is at the center and it is surrounded by a great amount of matter, causing enormous quantities of energy to be released.

Recent discoveries of fast-moving stars and gas near the center of the Galaxy lend support for the black hole hypothesis, for the Doppler shift indicates that the gas is orbiting the center with a speed that would require a central mass of perhaps 5 million solar masses in order to hold the gas in orbit. It seems that only a black hole could be this massive.

Some other spiral galaxies also have tremendously powerful energy sources at their center and we see signs of violent activity near those centers. The supermassive black hole hypothesis is rapidly gaining support among astronomers and seems to be the best explanation for the energy source in our galaxy and in similar galaxies. We will see in the next chapter that there are galaxies that produce even more energy in their nuclei. We are far from understanding the nucleus of our galaxy and further knowledge must come from observations of other galaxies as well as our own.

BEYOND THE MILKY WAY

Look up into the sky in the part of the Andromeda constellation indicated in Figure 17–19. If you are under a good dark sky on a clear night, you will see a fuzzy little patch of light. Through binoculars, the fuzzy patch looks larger but is still a fuzzy patch. Back through the ages this spot must have been a great source of wonder to curious people. It was called the Andromeda ***Nebula*** because of its nebulous, or fuzzy appearance.

After the invention of the telescope, many more such nebulous objects were found but their nature remained a source of wonder. In 1924, Edwin Hubble found Cepheid variables in three of what had been called ***spiral nebulae*** and he thereby showed that they in fact were spiral ***galaxies***.

The importance of Hubble's discovery was tremendous, for it greatly expanded our appreciation of the size of the universe. Just as, centuries before, the realization that our planet is just one of many planets was a giant leap in understanding our place in the universe, the realization that our Milky Way Galaxy is just one of a number of galaxies was another giant leap. We know now that the Andromeda Galaxy is a spiral galaxy much like our own and that it is about 2.2 MLY (million light years) away. When you view the Andromeda Galaxy, give some thought to the fact that the light reaching your eye has been travelling for more than 2 million years.

TYPES OF GALAXIES

The serious study of galaxies began with observation and classification of the objects. This is a common starting place in science. When we come upon an assortment of objects about which we know little or nothing, we begin by grouping them into classes according to their observable properties.

It was Edwin Hubble who provided the classification scheme that is the basis for the one still in use today. He divided galaxies into three types: spiral, ***elliptical,*** and ***irregular,*** with subdivisions within each. In more recent times,

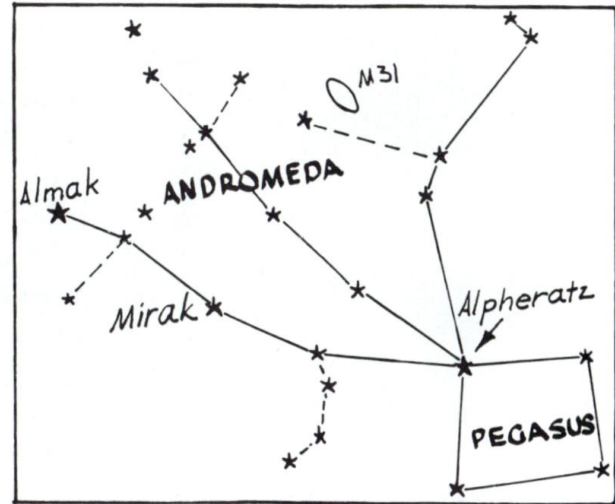

Figure 17–19. *The constellation Andromeda, showing where the Andromeda galaxy (M31) is located.*

astronomers have discovered objects in intergalactic space that fit none of Hubble's categories, not even irregular. We will discuss each category in turn, then look at how properties of galaxies are measured. In the next chapter we will turn our attention to the objects that are too peculiar to be called "irregular."

Elliptical galaxy: One of a class of galaxies that have smooth spheroidal shapes.

Irregular galaxy: A galaxy of irregular shape, that cannot be classified as spiral or elliptical.

Spiral Galaxies

Figure 17–20 shows a number of spiral galaxies. Some have their arms wound very tightly while others are more loosely bound. A few—called ***barred spirals***—reveal "bars" of stars across their nuclei and the spirals start from those bars.

When a spiral galaxy is seen edge-on, it often shows the lane of dust and gas clouds which we see in our own galaxy. Figure 17–20 shows two spiral galaxies (NGC 4565 and 4594) with prominent dark lanes.

Most spiral galaxies are from 50,000 to 2,000,000 light years across and contain from 10^9 to 10^{11} stars. (We will see later how such measurements are made.) For comparison, recall that the Milky Way is about 100,000 light years across and contains about 2×10^{11} stars.

Barred spiral galaxy: A spiral galaxy in which the spiral arms come from the ends of a bar through the nucleus rather than from the nucleus itself.

Spiral galaxies, class "S," are subdivided into groups from Sa to Sc, depending upon how loosely bound are their arms. Barred spirals are likewise classed as SBa through SBc. We need not be concerned with these labels, but you will see them on some photos.

Elliptical Galaxies

Figure 17–21 shows some elliptical galaxies and makes it obvious why they were given that name. Various elliptical galaxies show different eccentricities in their elliptical shape, depending in part on their orientation to Earth. (A football appears round if viewed end-on, for example.) The actual eccentricity of an elliptical galaxy is difficult to determine because its orientation is unknown.

Most of the galaxies in existence are ellipticals but most galaxies listed in catalogs are spirals. The reason for this is that although a few giant elliptical galaxies are larger than any spiral galaxy (having 100 times more stars than the Milky Way), most are small and dim (with one millionth as many stars as the Milky Way). Only relatively nearby dwarf elliptical galaxies are visible from Earth.

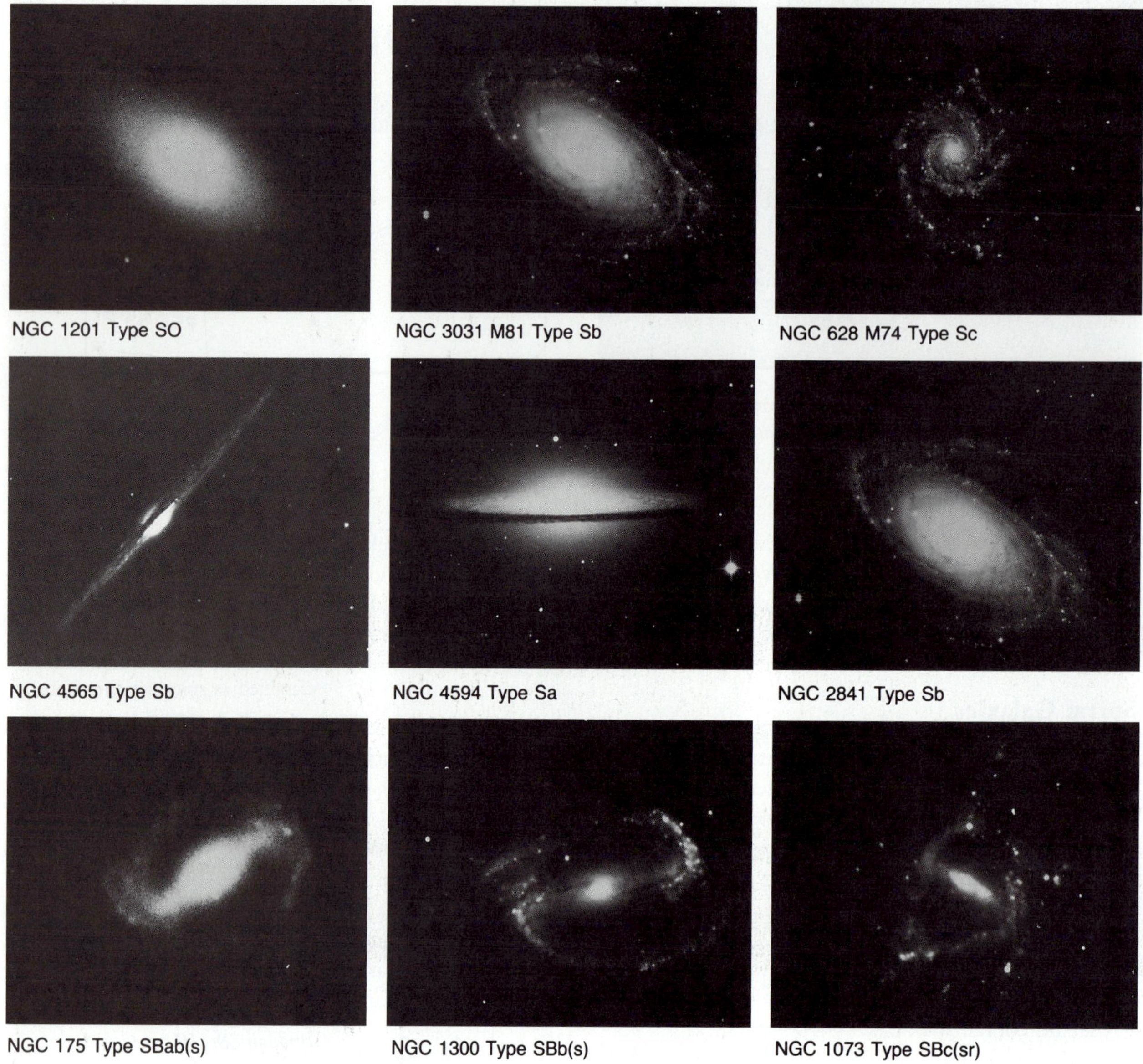

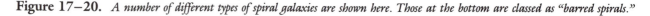

Figure 17–20. *A number of different types of spiral galaxies are shown here. Those at the bottom are classed as "barred spirals."*

Irregular Galaxies

Hubble was unable to classify some galaxies into either of the above categories, nor did they exhibit other common characteristics. Reference to Figure 17–22 indicates why these were called irregular galaxies. Fewer than 20 percent of all galaxies fall in the category of irregulars and they are all small; normally having fewer than 25 percent of the number of stars in the Milky Way.

The Magellanic Clouds are usually classed as irregular galaxies, although some astronomers think that the Large Magellanic Cloud is a barred spiral that has been disrupted by its near proximity to the Milky Way and perhaps by a past collision with the Small Magellanic Cloud. It should not be unusual for

NGC 205 NGC 4486 NGC 147

Figure 17–21. *Elliptical galaxies show no spiral structure.*

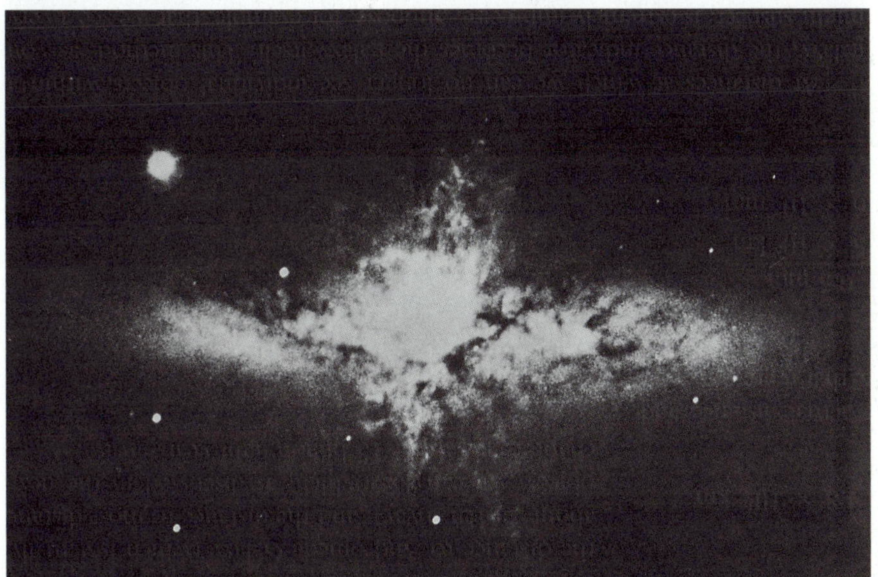

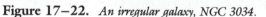

Figure 17–22. *An irregular galaxy, NGC 3034.*

Figure 17–23. *This odd-shaped object is thought to be two galaxies colliding. The system is NGC 4038 and 4039.*

galaxies to collide, because on the average they are separated by distances only about 20 times their diameter. Stars within a galaxy, on the other hand, are separated by millions of times their diameter and therefore collide infrequently. Because of their great distances from Earth, however, galaxies exhibit no proper motion (motion across our line of sight), so we have to deduce past collisions from their present appearance. Figure 17–23 shows two galaxies that appear to be in the process of colliding.

Computer simulations of collisions between galaxies indicate that they pass through one another with few collisions between individual stars, although interstellar dust and gas collisions become important in the collisions and the overall shape of each galaxy is changed drastically by the collision.

Figure 17–24. *The photo at left is the Andromeda Galaxy. The square in the photo is enlarged at right and the arrows point to two different Cepheid variables in that galaxy.*

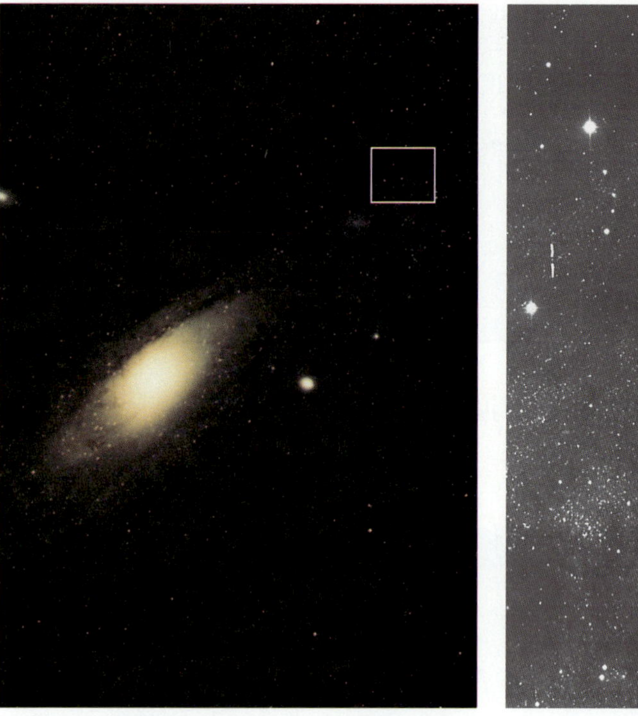

Sometimes galaxies may merge, especially in the case of a collision between a large and a small galaxy. In this case, the larger one swallows up the smaller, cannibalizing it.

MEASURING GALAXIES

The most important properties of a galaxy we can measure are its distance, mass, and motion. As we have seen on a number of occasions, if the distance to an extended object is known, its size and absolute luminosity can be calculated from its angular size and apparent magnitude respectively.

Distances to Other Galaxies

Distances to galaxies are measured using a number of different distance indicators. We have already described one of these, the Cepheid variable. The distances to the Magellanic Clouds was the first such application of Cepheid variables on an extragalactic (outside the Milky Way) object. Figure 17–24 shows two Cepheid variables in the Andromeda Galaxy, the galaxy Hubble used to show that the so-called spiral nebula were indeed galaxies.

Although Cepheid variables are very bright stars, we can distinguish them only in relatively nearby galaxies; out to perhaps 20 MLY. If we wish to measure the distance to a galaxy in which we are unable to see Cepheids, we must find other distance indicators. By analysis of giants, supergiants, and novae in our

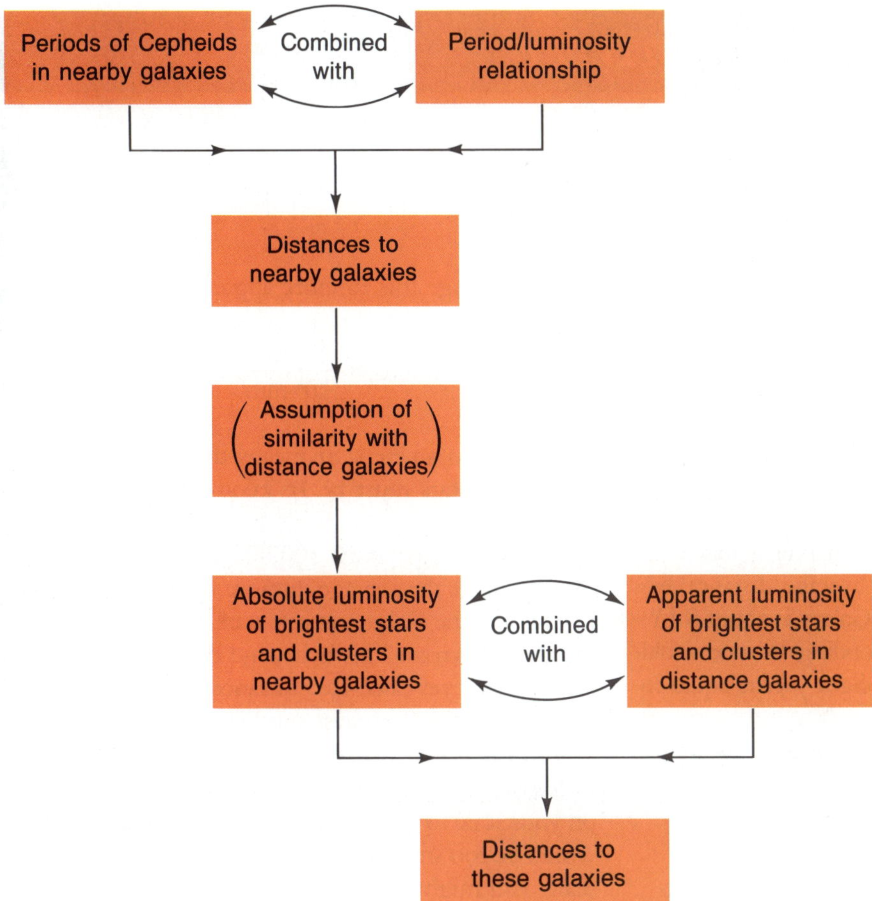

Figure 17–25. *Starting with the period/luminosity relationship of Cepheids, astronomers are able to follow a chain of reasoning and observation that allows them to determine the distances to galaxies too far away for their Cepheids to be visible.*

galaxy and nearby galaxies, we learn that all have approximately the same range of luminosity. In addition, we find that large globular clusters and supernovae are of consistent brightness from one galaxy to another. This allows us to use these bright stars and clusters as distance indicators in more distant galaxies.

The logic here is interesting. By determining the luminosities of nearby Cepheid variables, astronomers determined the period/luminosity relationship for Cepheids. Then they assume that Cepheids in other galaxies are basically the same as the ones in our galaxy and so they use that same period/luminosity relationship in reverse to calculate the absolute luminosity and thereby the distances to Cepheids in the nearer galaxies. Knowing these distances, they learn that the brightest stars and globular clusters in galaxies of each type have the same luminosity. That is, the brightest 50 stars in one spiral galaxy have the same average luminosity as the brightest 50 in every other spiral galaxy. This also is assumed to be true for globular clusters. Now they can use these objects as distance indicators to learn the distances to galaxies in which they can see the individual objects. Figure 17–25 summarizes this chain, which allows the measurement of distances as far as about 100 MLY.

ACTIVITY

The Scale of the Universe

CHAPTER 7 INCLUDES A drawing of the solar system with the Sun and planets drawn to scale. In that figure, the Sun is a speck on the page, and Pluto—although invisible—is at the other side of the page. It is instructive to expand this drawing to include other stars, the Galaxy, and other galaxies.

The dot that represents the Sun (actual diameter: 1,500,000 km) in the figure in Chapter 7 has a diameter of about 0.5 mm. The average distance between stars in our region of the Galaxy is about 5 LY. We will calculate what the corresponding distance would be in the figure, given that one light year is equal to 9.5×10^{12} km.

First we set up the ratio of the distance on the figure to the actual distance:

$$\frac{\text{distance on figure}}{\text{actual distance}} = \frac{0.5 \text{ mm}}{1.5 \times 10^{6} \text{ km}}$$

Next we will calculate the distance between stars in kilometers:

$$5 \text{ LY} \times \frac{9.5 \times 10^{12} \text{ km}}{1 \text{ LY}} = 4.8 \times 10^{13} \text{ km}$$

Now let's use that value as the actual distance and solve for the distance in the figure that represents separation of stars:

$$\frac{\text{distance on figure}}{4.8 \times 10^{13} \text{ km}} = \frac{0.5 \text{ mm}}{1.5 \times 10^{6} \text{ km}}$$

$$\text{distance on figure} = 1.6 \times 10^{7} \text{ mm}$$

Finally, change this to more appropriate units:

$$1.6 \times 10^{7} \text{ mm} \times \frac{1 \text{ m}}{10^{3} \text{ mm}} = 1.6 \times 10^{4} \text{ m}$$

This is 16 kilometers! It means that on the scale of the figure, stars must be 16 kilometers, or 10 miles, apart.

The Galaxy is about 100,000 LY in diameter, or 20,000 times the average distance between stars. This means that the Galaxy on our scale would be 320,000 kilometers across. This is beyond our imagining, and we have not yet calculated distances to other galaxies. We therefore will suggest another scale and leave it to you to calculate the distances using this scale.

Suppose you take the average distance between stars on your scale to be five centimeters. Although the stars would be too small to see on an actual drawing to this scale, you might cheat and make each star a very tiny speck.

On your drawing, five centimeters will correspond to five light years. Now calculate the size of the Galaxy, the distance to Andromeda Galaxy (which is actually about 2 MLY away), and the distance to most distant objects we can see (which, as discussed in the next chapter, are about 15 billion light years away).

At distances at which we can no longer see individual objects within a galaxy, the distance indicator becomes the galaxy itself. This method, as you might suspect, is extremely imprecise. Suppose, for example, that we see a very distant spiral galaxy. Since we know the range of luminosities spiral galaxies have, we can make some judgement as to that galaxy's distance by assuming that it is an average spiral galaxy. If we assume that it is among the dimmest galaxies of its type, we find its nearest probable distance. On the other hand, if we assume that it is among the brightest, we find its farthest probable distance. In this way, we find the range of distances within which we can be fairly certain the galaxy falls.

When applied to an individual galaxy, the whole-galaxy method of assessing distance is very imprecise. Fortunately, the method need not be restricted to individual galaxies. As we will discuss later, galaxies exist in clusters; from a few galaxies to thousands of galaxies per cluster. If we consider a cluster that contains a great number of spiral galaxies, we can logically assume that the brightest spiral galaxy in that cluster has about the same luminosity as the brightest spiral in another cluster. Thus we use the brightest galaxies as a distance indicator to the cluster.

Notice that one measurement builds on another in a series of steps. The result of this is that if there is an error in a beginning step, the error will be transmitted up through the chain of steps. For example, if somehow there is an error in our understanding of the period/luminosity relationship for Cepheids, the stated distances to the farthest galaxies will have to be adjusted.

Fortunately, there are constant checks made as new data arrives. The Hubble Space Telescope will allow us to begin near the bottom step and check some of the results up the line. For example, the Hubble Telescope will be able to detect Cepheid variables out as far as 75 MLY instead of just 20 MLY. This will allow us to use Cepheids to measure distances that previously had been measured using bright stars and globular clusters as distance indicators. The telescope will therefore allow us to refine—or reevaluate—these previous measurements.

The above analysis makes an important assumption: Galaxies in our neighborhood of the universe are basically the same as those farther away. This may seem reasonable, but remember that we are seeing distant galaxies not as they are today but as they were in the past. Light coming from a galaxy 100 million light years away has been traveling 100 million years and we cannot rule out the possibility that galaxies have changed in that time.

THE MASSES OF GALAXIES

A galaxy's mass can be determined in a number of ways, all of them limited in precision primarily because of the tremendous distances involved. One method of measuring the mass of a galaxy is by observing the rotation periods of some parts of it. Naturally, we cannot wait for part of a galaxy to complete a revolution, for this takes millions of years. Instead, we use Doppler shift data to measure the velocity of part of a galaxy. Then knowing the distance of that part from the center, we can determine the period of revolution of the stars located there. As in the case of our own galaxy, Kepler's third law then allows us to calculate the mass of the material within the orbit of the part of the galaxy being studied.

Another method of measuring galactic masses is similar to that used to measure stellar masses. There are many cases of a pair of galaxies revolving around one another. Again, we cannot wait the long times necessary to measure

Figure 17–26. *In order to use the Doppler shift to know the speeds of two galaxies that orbit one another, we would have to know the tilt of their plane of revolution to our line of sight.*

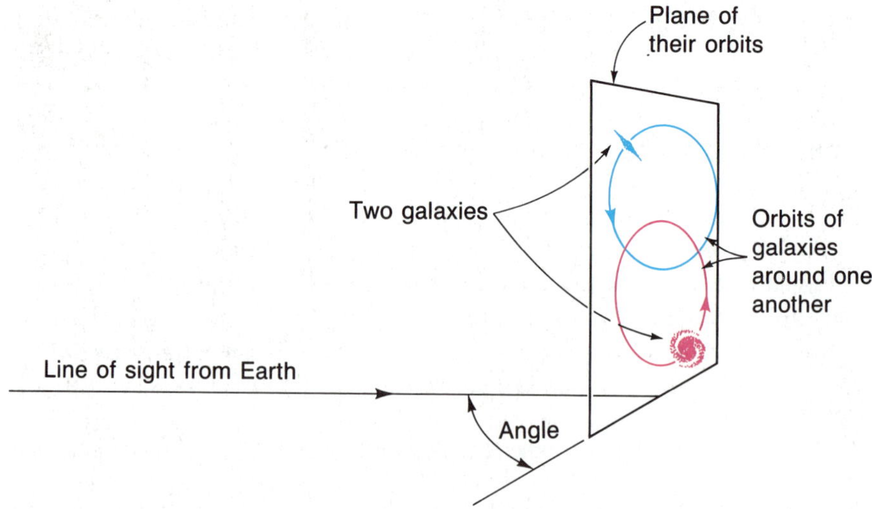

the period of revolution but we use the Doppler shift to gauge the speeds of the galaxies. The problem with this method is that it is difficult to determine the angle of the plane of revolution to our line of sight (Figure 17–26) and knowledge of this is necessary if we are to make an accurate measurement. Making measurements for great numbers of such binary galaxies allows us to determine an average mass for a given type of galaxy, however.

Clusters of Galaxies

Be careful with the terminology here. A galactic cluster *is an open cluster of stars. It is not a* cluster of galaxies.

Local group (of galaxies): The cluster of 20 or so galaxies which includes the Milky Way Galaxy.

Most galaxies are part of clusters rather than drifting companionless though space. Figure 17–27 shows a cluster of galaxies. Thousands of such clusters are known and as many as 10,000 galaxies may be contained in the largest clusters.

Our Milky Way Galaxy is part of a group of about two dozen galaxies, including the Magellanic Clouds and the Andromeda Galaxy, that form a cluster called the ***local group***. The Andromeda Galaxy and the Milky Way are by far the largest members of this cluster, which appears to contain only one other spiral galaxy.

A third method of measuring the masses of galaxies takes advantage of their clustering. There is only one force that could hold a galaxy within its cluster: the gravitational force. If we observe a galaxy near the outside of a large cluster and assume that it is in a relatively circular orbit, we can use the Doppler effect to measure its speed and therefore its period. This allows us to calculate the total mass of the galaxies within its orbit. If we were to calculate the mass of the cluster based on measurements for only one galaxy, there would be too many uncertainties to make the determination meaningful but the same measurement, repeated for a number of galaxies on the outer portion of the cluster, gives a useful value for the mass of the cluster and therefore an average value for the masses of the galaxies within the cluster.

When measurements by the various methods are made, we find that the cluster method reveals that clusters have much more mass than is accounted for by the visible stars within the galaxy. Within our own galaxy, we know that the interstellar gas and dust accounts for at least 10 percent to 20 percent of the total mass but if other galaxies contain a like amount, it is not near enough to account for their ***missing mass.*** We know that intergalactic dust and gas exists between galaxies in a cluster but this is also not nearly enough to account for the calculated mass. There are recent indications that the halo of our own galaxy contains more material than previously thought and perhaps galaxies' halos account for much or all of the missing mass. This is another of the questions about which astronomers hope to obtain data from the Space Telescope when it is finally in operation.

Missing mass: The difference between the mass of clusters of galaxies as calculated from Keplerian motions and the amount of visible mass.

THE PRECISION OF SCIENCE

It is sometimes said that science is an exact study. If those who make such statements mean that the measurements of science are exact, they are wrong. It is obvious that measurements of distances to galaxies are not exact. What is not so obvious is the fact that *every* measurement in science is to some degree an approximation. No measurement is exact. What scientists attempt to do is to be aware of how *inexact* their measurements are. For example, in calculating the distance to a galaxy, parallel calculations are done to determine the probable error in the measurement. Measurements may show that a galaxy is 200 MLY away, with a likely error of 50 MLY. This means that the galaxy is measured to be between 150 and 250 MLY away. Calculating the likely error is a common practice in all natural sciences. Thus scientists attempt to be specific about their inexactness.

Figure 17–27. *The Hercules cluster of galaxies.*

In addition, we have mentioned that scientists must take into account assumptions that are consciously or unconsciously included in the measurements. In the case of determining distances to faraway galaxies, for example, assumptions are made that there is negligible intergalactic material that might diminish the light from those galaxies. Early measurements of the size of the Galaxy were in error because they did not take into account the interstellar dust and gas. A scientist tries to be aware of the assumptions involved in each measurement. There is no claim that a measurement is exact. In fact, there is little room in science for the word "exact."

CONCLUSION

Our Sun is one of some 200,000,000 stars in an enormous spiral galaxy that is itself just one of countless galaxies in the observable universe. The nature of our galaxy was discovered only rather recently because our vision is obscured by gas and dust along the Galaxy's disk. Although measurements of the Galaxy's characteristics are necessarily imprecise, some of its properties can be measured using ideas dating from the time of Isaac Newton, such as the calculation of the Galaxy's mass using Kepler's third law. Detection and measurement of many other features of the Galaxy must rely on more modern methods and equipment, such as radio astronomy, which has opened many doors to understanding both the Milky Way and other galaxies.

There is a plethora of unanswered questions about galaxies. Some questions, such as the source of the tremendous energy at the Galaxy's center, seem to have only one possible explanation but in other cases there are competing theories to explain a phenomenon, such as the theories explaining galaxies' spiral arms. Galactic astronomy is now an extremely active field and researchers anxiously await data from larger Earth-bound telescopes and from new orbiting telescopes. Of particular interest are questions about the unusual and poorly understood galactic objects to be discussed in the next chapter.

RECALL QUESTIONS

1. State two meanings of the term *Milky Way*.
2. How did William and Caroline Herschel seek to determine the Sun's position in the Galaxy? What measurements did Kapteyn make to answer the same question?
3. Explain why 21-centimeter radiation reveals the same parts of the Galaxy in which O and B type stars are located.
4. Use an analogy to explain the nature of a density wave.
5. Sketch the Galaxy both edge-on and face-on, showing its prominent features and the approximate location of the Sun.
6. What incorrect assumptions did early astronomers make that caused them to be in error as to the location of the Sun in the Galaxy?
7. Describe the Milky Way as seen from Earth.
8. Distinguish between galactic clusters, globular clusters, and clusters of galaxies.
9. Describe Shapely's method of determining the Sun's location in the Galaxy.
10. Describe Oort and Lindblad's method of determining the Sun's location in the Galaxy. What was the approximate date of their analysis?
11. What is the galactic halo?
12. Describe the Sun's motion in the Galaxy. How was this determined?
13. What two observations give evidence of spiral arms in the Galaxy?
14. Name the two most accepted models for the spiral arms in galaxies.
15. What is the present theory for the origin of the energy produced in the galactic nucleus?
16. Identify Harlow Shapely and Edwin Hubble, listing at least one contribution made by each to galactic astronomy.
17. What is meant by a distance indicator, and what serves as a distance indicator for nearby galaxies? For more distant galaxies?
18. Describe three methods for measuring the mass of a galaxy other than the Milky Way.
19. What is the local group?
20. What is meant by *missing mass* in galactic astronomy?
21. Are scientific measurements exact? Explain.

QUESTIONS TO PONDER

1. Describe the nighttime sky of a planet orbiting a star in a globular cluster.
2. If stars revolve around the galactic center independent of the spiral arms, it would seem that there should be as many stars per unit volume between the arms as inside them. Yet the text states that there is *almost* the same density of stars between the arms as inside them. Why would there be a difference at all?
3. The orientation of the plane of revolution of binary galaxies is difficult to determine, thus limiting the usefulness of Doppler shift measurements of veloc-

ity. Why does the same problem not exist in using Doppler shift data to determine the speed of part of an individual spiral galaxy?

4. Explain how Kepler's law is used to calculate the mass of the Galaxy. Discuss the limitations of this method.
5. If a radio telescope cannot reveal directly the distance to a source of 21-centimeter radiation, how can this radiation be used to reveal spiral arms in the Galaxy?
6. Consult an elementary physics text as to the nature of sound waves in a gas and contrast this to the density wave theory for spiral galaxies.
7. Explain how we determine the distances to galaxies that are too far away to allow us to see Cepheids in them.
8. Discuss the limitations on the accuracy of measurements of distances to galaxies. Doesn't the lack of precision in these measurements make galactic astronomy less a science than other sciences? Discuss.
9. It was stated that no measurement is exact. Does this mean that we cannot get an exact value for the height of a page in this book? Explain.
10. Compare the use of bright galaxies as distance indicators to grading a class "on the curve."

CALCULATIONS

1. Show how one can calculate the period of the Sun's motion around the galaxy, given that the Sun's distance to the center is 30,000 LY and its speed is 250 km/sec.
2. Suppose that a portion of another galaxy is observed to move with a speed of 200 km/sec and to be at a distance of 40,000 LY from the center of our galaxy. What is the period of revolution of that portion of the Galaxy?
3. Use the data from the last question to calculate the mass of that galaxy that is inside the radius of the portion described.
4. Suppose we wish to construct a physical model of the Galaxy. If the Sun is made a small grain of sand 0.1 millimeter across, what will be the diameter of our model?

CHAPTER 18

Albert Einstein:

"The existence and validity of human rights are not written in the stars."

"The unleased power of the atom has changed everything save our modes of thinking, and we thus drift toward unparalleled catastrophes."

(Asked by the press what would happen in the event of nuclear war): "Alas, we will no longer be able to listen to the music of Mozart."

"If A is success in life, then A equals X plus Y plus Z. Work is X, Y is play and Z is keeping your mouth shut."

"One thing I have learned in a long life—that all our science, measured against reality, is primitive and childlike—and yet it is the most precious thing we have."

"The most beautiful thing to be felt by man is the mysterious side of life. There is the cradle of Art and real Science."

"The most incomprehensible thing about the universe is that it is comprehensible."

Active Galaxies and Cosmology

IN THIS CHAPTER WE study two topics that are closely related: the most distant objects in the universe and the nature of the universe as a whole. We will see that the most distant objects contain messages about the past history of the universe, although the messages are very difficult to read.

The fields of astronomy we study here are changing very rapidly, for new telescopes and new techniques are resulting in frequent discoveries in these fields. Because of this there are probably more unanswered questions about the subjects of this chapter than in any other branch of astronomy. We will encounter a number of these unanswered questions and we will see how important it is to distinguish between scientific data and the conclusions that arise from that data.

THE EXPANDING UNIVERSE

In 1912 Vesto M. Slipher, an astronomer working in Percival Lowell's observatory in Flagstaff, Arizona, was assigned to examine the spectrum of some of the spiral nebulae to learn about their chemical composition. A then-current hypothesis regarding the nature of these objects was that they were planet systems in formation and Lowell was seeking to discover life on other planets. Instead of finding evidence for life, Slipher found something else; something that also had profound implications. He found a redshift in the spectra of most of the nebulae he examined. If the redshift was due to the Doppler effect, it meant that most of the other nebulae were moving away from us; as fast as 1800 kilometers/second. There seemed to be no reasonable explanation for this strange finding.

Lowell's telescope, on which Slipher worked, had only a 24-inch aperture. The 100-inch telescope used by Hubble and Humason therefore had 16 times more light-gathering power.

In 1924, Edwin Hubble presented evidence that Cepheid variables exist in some of these nebulae, showing that the nebulae were in fact galaxies. Hubble, working with Milton Humason, then used the 100-inch telescope on Mount Wilson to photograph the spectra of these objects. Their work not only confirmed the findings of Slipher but showed that there is a pattern in the speeds with which galaxies are receding from us.

Figure 18–1 is a diagram by Hubble and Humason on which they plotted the distances to a number of galaxies against the galaxies' recessional velocities. It indicates that the more distant a galaxy is, the faster it is moving away from us. This data, taken during the 1920s, used distance indicators that relied on incorrect Cepheid variable values and the data had to be adjusted when more was learned about Cepheids. As Figure 18–2 indicates, however, the relationship still holds when more and more galaxies are added using today's distance values.

If other galaxies are moving away from ours, does this not mean that our galaxy must be at the center? Have we finally discovered that we are at a special location after all? Hubble published his findings in 1929, 12 years after Shapley's work with globular clusters had taken our Sun out of the central position within the Galaxy. By this time it had become a working premise—an article of faith—that our location within the cosmos is not central. To see why Hubble's discovery does not in fact conflict with this basic premise, consider the following analogy.

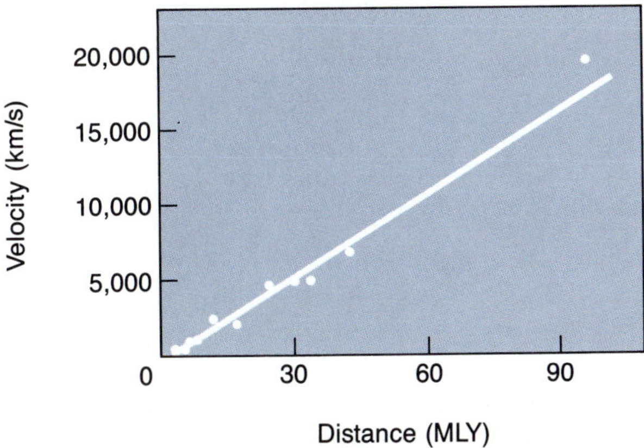

Figure 18–1. *This is similar to a diagram prepared by Hubble and Humason in 1931. It shows a relationship between the recessional velocities of galaxies and their distances.*

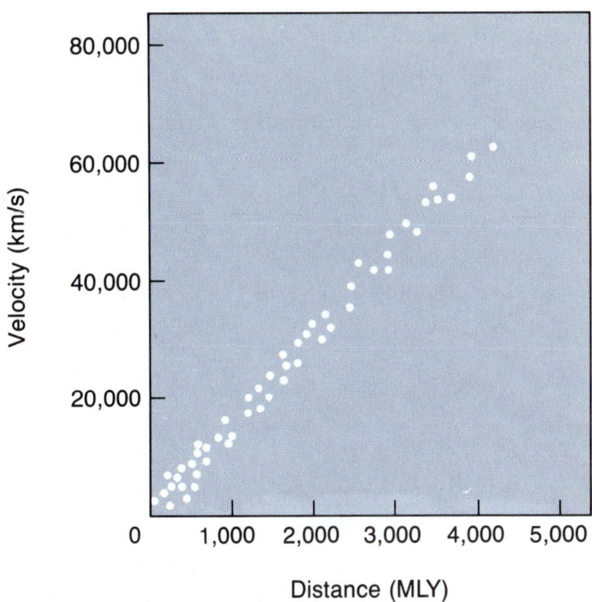

Figure 18–2. *The relationship found by Hubble and Humason is valid using more recent data but reevaluations of distances have caused the slope of the line to change.*

Imagine that you are trainer in a flea circus and that you put a number of educated fleas on a balloon that your assistant is blowing up. You have instructed these fleas to hold their positions on the balloon. Figure 18–3 shows the balloon being blown up with the fleas in place.

Now imagine what a particularly intelligent flea sees when it looks out toward neighboring fleas. (Assume either that the flea can see around the curvature of the balloon, or that the fleas are close enough together that the balloon seems flat to them.) The intelligent flea will see every other flea getting farther and farther away as the balloon is blown up. In addition, the fleas more distant from him would be moving away at a greater speed than the ones nearby. To see this, refer to the magnified view of a portion of the balloon show in Figure 18–4. Note that during the one second which elapsed from part (a) to part (b) of the figure, flea X has moved 0.1 cm, flea Υ has moved 0.2 cm, and flea Z has moved 0.3 cm away. So our smart flea sees X moving at 0.1 cm/s, flea Υ moving twice as fast, and flea Z moving three times as fast. The more distant a given flea is from the observer, the faster it is moving.

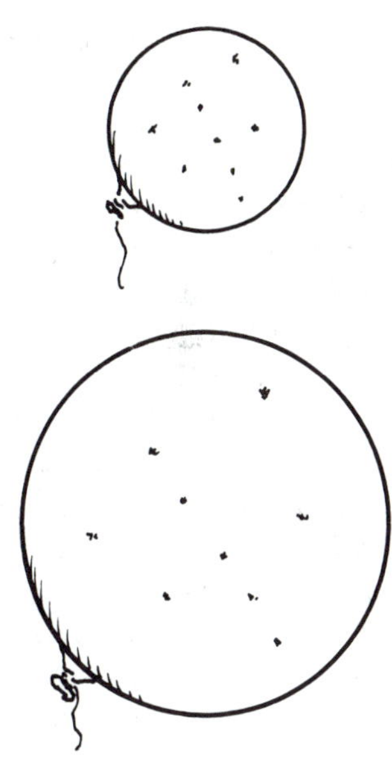

Figure 18–3. *Fleas on a balloon get farther apart when the balloon is blown up.*

The important point to see is that the same result would be obtained no matter which one of the fleas is the intelligent flea. The smart one in the drawing was not at a central location on the surface of the balloon, for there *is* no central location. While the balloon is being blown up, every flea sees every other flea moving away with a velocity that depends upon the flea's distance from the observer.

The analogy shows that if galaxies are part of an expanding universe, we will see exactly what Hubble and Humason observed: every galaxy will be seen

Figure 18–4. *One second passes between (a) and (b). The flea at left sees the more distant fleas receding at a greater rate than those nearby.*

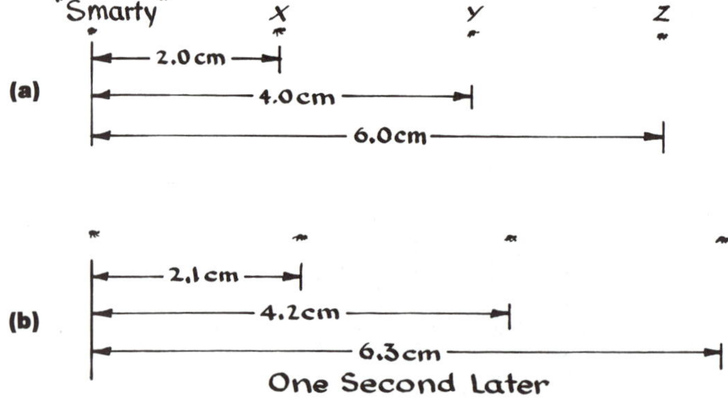

to be moving away from us at a speed that depends upon how far away it is. A number of astronomers since Hubble have surveyed galaxies at greater and greater distances and have come to the same conclusion: The universe is expanding.

Separating Data from Conclusions

It is always important in science to separate one's observations from the conclusions derived from those observations. The present case serves as a good example of this rule. It has been observed that the spectrum of light from distant galaxies is shifted toward the longer wavelengths. In addition, there is a pattern of the dimmer galaxies having greater shifts than the brighter galaxies. This is indisputable data that has been confirmed by a number of researchers. Figure 18–5 shows photos, all taken with the Palomar telescope, of five different galaxies and the spectra of those galaxies. Compare the location of the two prominent absorption lines in each of the spectra; noting that the dimmer the galaxy, the farther to the right (toward longer wavelengths) is the pair of lines.

Notice that in the last paragraph we did not refer to the distance to galaxies but rather to their brightness. Such distances are determined by various methods, as seen in the last chapter, and are not direct observations. They therefore are not data; much less indisputable data.

In particular, a redshift that would result from a phenomenon described by the theory of general relativity was tried. We'll see it later in this chapter and it is discussed more fully in Appendix A.

Once data has been obtained, we draw conclusions from it. In this case we try to decide what causes the redshift. The only possible explanation astronomers can find is the Doppler effect. Other explanations have been tried but none are consistent with the observations.

Here is an example of a simpler explanation that might seem to work (at first glance, anyway): Recall that light is reddened as it passes through our atmosphere. This is what causes the Sun to look red at sunset. Perhaps the light from distant galaxies is reddened by the dust and gas in interstellar space. If this is true, light from more distant galaxies would exhibit more reddening—exactly what is observed. The explanation fails, however, when the redshifted spectrum is examined. To see why, we will look briefly at the two processes.

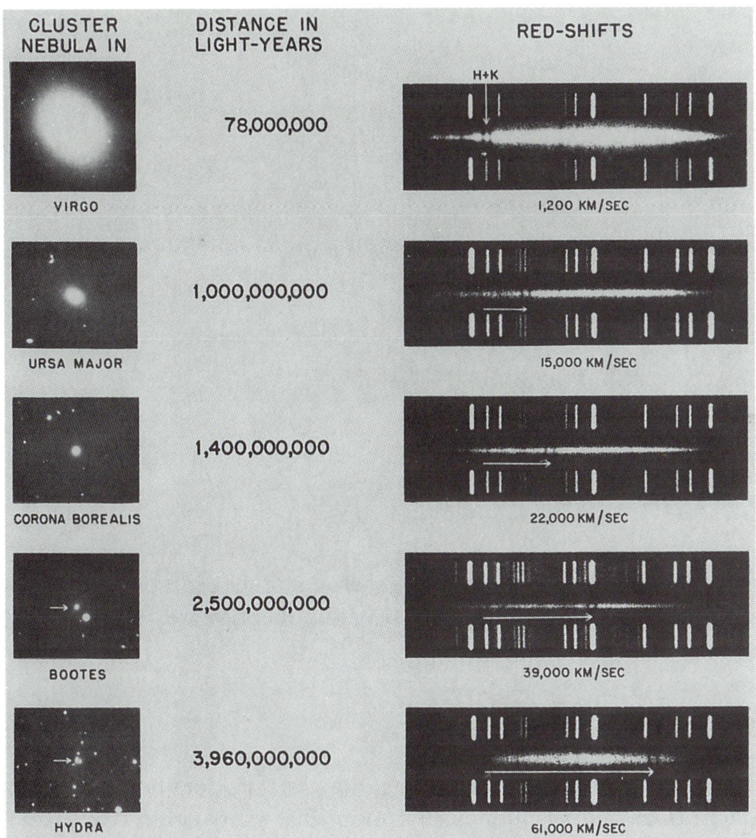

Figure 18–5. *The photos at left are of galaxies at various distances. At right, their spectra are shown (the white streak in the center of the reference lines in each case). Arrows indicate how far the two prominent absorption lines have shifted.*

Reddening by interstellar matter occurs because of the absorption of shorter wavelengths. Longer wavelengths are absorbed less and therefore penetrate to the observer. Any absorption lines or emission lines in the original spectrum remain at the same wavelength as shown in Figure 18–6(b). In the case of the Doppler shift, on the other hand, all wavelengths are lengthened, including those of spectral lines as shown in Figure 18–6(c). Observation definitely indicates that the reddening-by-absorption hypothesis is incorrect, for the entire spectrum is shifted as predicted by the Doppler explanation.

All spectral lines of distant (dim) galaxies show this ***cosmological redshift,*** not just those of the visible spectrum. The observations are 100 percent in agreement with what the Doppler effect predicts for light from a receding object. Thus we *conclude* that the galaxies are moving away from us. We do not, however, *observe* this.

Cosmological redshift: The shift toward longer wavelengths that is due to a galaxy's motion related to the expansion of the universe.

The redshifts observed for distant galaxies are called cosmological redshifts *because they have major implications in the study of cosmology, as we will see later.*

The above discussion certainly does not mean that astronomers are unsure of what causes the cosmological redshift and that the Doppler effect is just one hypotheses of many. In fact, the Doppler effect is the *only* explanation astronomers can find for the observations and there is (almost) no doubt that the Doppler effect is responsible for the observations. The word *almost* is inserted only because we must always remember that this is a reasoned conclusion and is not the original observation. Although such an event is considered extremely

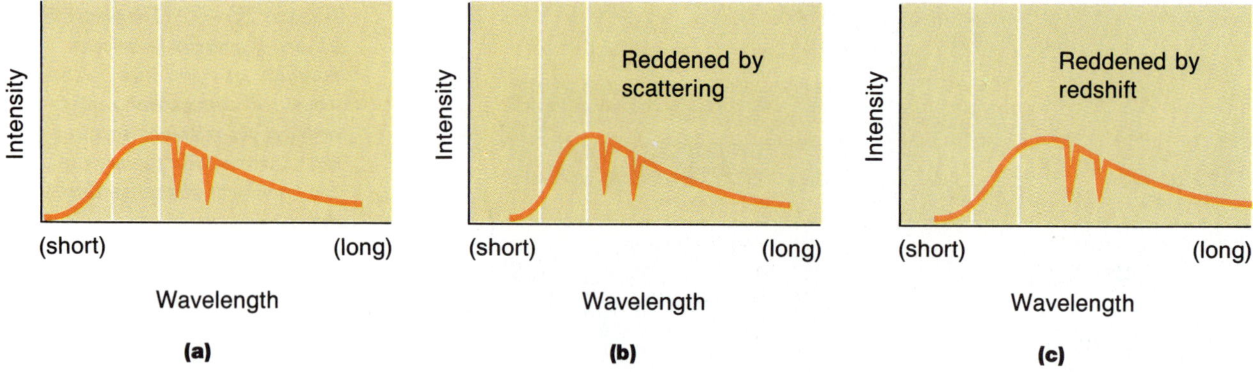

Figure 18–6. *(a) An intensity/wavelength curve showing two (exaggerated) absorption lines. (b) This is the same source with its light reddened by scattering. Notice that the absorption lines remain in the same place. (c) Again, the same source but with its light reddened by Doppler shift. The entire curve is shifted to the right in this case.*

unlikely, we can never rule out the possibility that another explanation will be found. It is important to remember that scientific theories are tentative and subject to constant testing.

What Is Expanding and What Is Not?

When it is said that the universe is expanding, this certainly does not mean that the solar system is expanding. Nor does it mean that stars within our galaxy are getting farther apart. In fact, it does not even mean that all galaxies are getting farther apart.

The Doppler shift leads us to conclude that other *clusters* of galaxies are moving away from ours. Some individual galaxies are actually moving towards us. This occurs in two ways. First, galaxies in the local group are moving randomly within that group; each responding to the overall gravitational force of the others. Thus the Andromeda Galaxy and a half-dozen others in the local group are at this time moving towards the Milky Way Galaxy.

Second, in nearby clusters, the same random motion of individual galaxies results in some of the galaxies moving toward us at the present time, even though the cluster in which each galaxy exists is moving away from the local group. Figure 18–7 illustrates how this can happen. In fact, no individual cluster—as far as we can tell—is expanding.

Discoveries of superclusters of galaxies—clusters of clusters—may mean that it is the superclusters that are moving away from one another.

It is the clusters of galaxies that are moving farther apart. Thus although we often say that other galaxies are moving away from us, we should really say that other clusters of galaxies are moving away from our cluster.

THE HUBBLE LAW

Refer to Figure 18–8, which is a repeat of Figure 18–2 but with a straight line drawn through the data. Whenever data can be represented by a straight line on a graph, there is a direct proportionality between the quantities on the

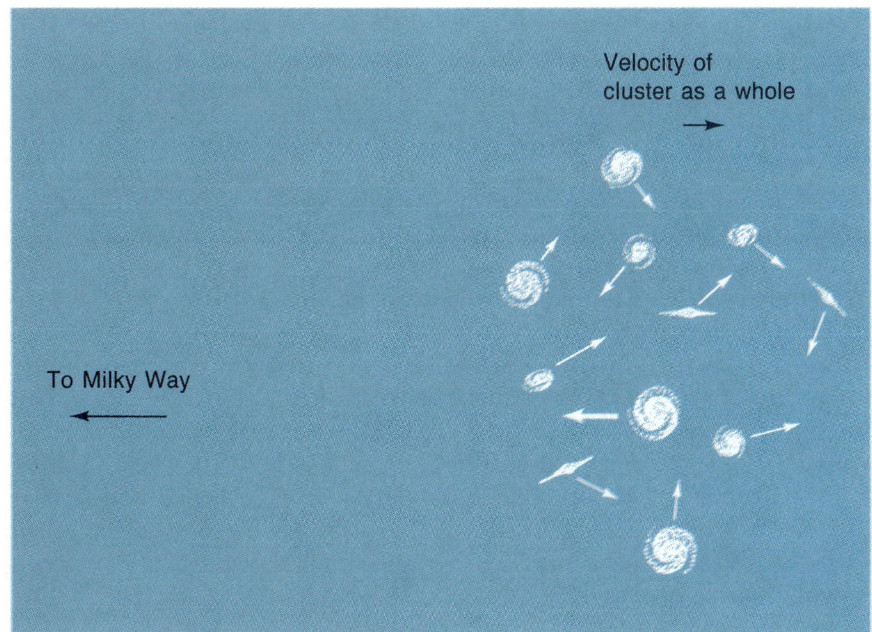

Figure 18–7. *Galaxies within a cluster have random motions so that although the cluster is moving away from us, a particular galaxy within it may be moving toward us at this particular time.*

two axes. Thus we can write the relationship between the two quantities, radial velocity, v, and distance, d, as

$$v = Hd$$

where H is the constant of proportionality, called the ***Hubble constant*** in honor of the man who discovered this ***Hubble law.*** The value of the constant is simply

Hubble constant: The proportionality constant in the Hubble law; the ratio of recessional velocities of galaxies to their distances.

Hubble law: The relationship that states that a galaxy's recessional velocity is directly proportional to its distance.

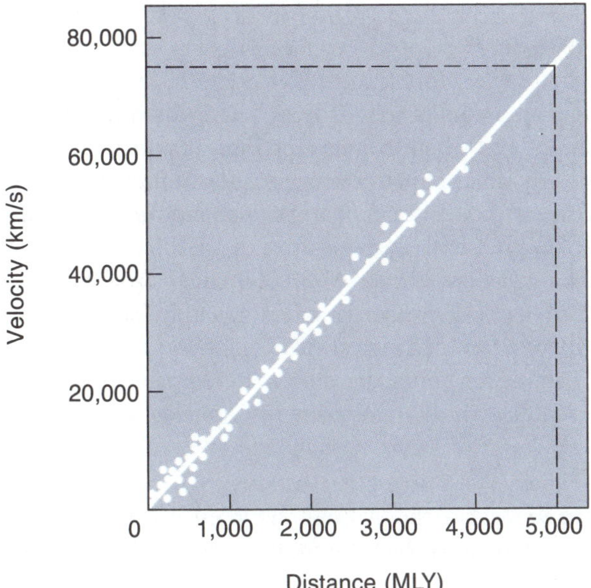

Figure 18–8. *A line is drawn through the data points of the earlier graph. A point near the top of the line is marked in order to calculate the slope. (See the example in the text). For the calculation, one point is taken to be the origin (0,0) and the other is (5000, 75000).*

the slope of the line on the graph. The slope of a line on a graph is defined as the ratio of the change in the quantity plotted on the y axis to the change in the quantity plotted on the x axis. We will calculate the Hubble constant as an example of calculating the slope of a line.

EXAMPLE Calculate the slope of the line in the graph of Figure 18–8, thereby calculating the Hubble constant

Solution In Figure 18–8 a point has been marked near the top of the line. Any point on the line could be chosen. Reading the values of that point, we obtain a velocity of 75,000 km/s and a distance of 5000 MLY. The slope must be calculated using the *changes* in both quantities, so we must choose another point in order to subtract and get the change. On this graph, the line passes through the point where both quantities are zero and we will choose that as our second point. Thus the slope is

$$\text{Slope} = \frac{75{,}000 \text{ km/s} - 0}{5000 \text{ MLY} - 0}$$

$$= 15 \text{ km/(s} \times \text{MLY)}$$

Try One Yourself. Calculate the slope of the line in the 1931 Hubble-Humason graph (Figure 18–1) in order to find the value of the Hubble constant as determined with that data.

Astronomers usually use megaparsecs (million parsecs) as their distance unit. Fifteen km/(s × MLY) is 50 km/(s × Mpc).

Thus according to the graph, the Hubble constant is about 15 km/(s × MLY) (read as "kilometers per second per million light years"). It is important to see what Hubble's constant means and what its units represent. A value of 15 km/(s × MLY) means that for each million light years a galaxy is from us, its speed is greater by 15 kilometers per second. Refer again to the graph to check that a galaxy at a distance of 1000 MLY has a speed in km/s of 15,000 km/s. In like manner, a galaxy 2000 MLY away has a speed of 30,000 km/s.

The slope of the 1931 Hubble-Humason graph is much greater than the more modern graph discussed above. Their calculation for the constant was about 180 km/(s × MLY), about ten times the more modern value. This was because of their misunderstanding of the luminosity of Cepheids, as discussed in the previous chapter.

The data shown in Figure 18–8 and referred to as more modern data corresponds to that of Allan Sandage of the Mt. Wilson Observatory and Gustav Tammann of the University of Basel, who have worked for years on the question. Another group of researchers uses slightly different methods for determining distances and obtains values close to 25 km/(s × MLY). Today, astronomers are fairly confident that the true value lies somewhere between these two findings, between 15 and 25 km/(s × MLY) [or 50 and 80 km/(s × Mpc)].

The Hubble constant is very difficult to determine. We can calculate the distance to galaxies in nearby clusters fairly accurately using established distance indicators but the motion of a galaxy within a nearby cluster is significant compared to the motion of the cluster (so that the galaxy may be even approaching us, as we saw above). Thus the velocity of a nearby galaxy is not primarily cosmological but is due in large part to random motions within its cluster. It therefore does not fit the Hubble law. A distant galaxy, on the other hand, is in a cluster whose motion away from us is much greater than the galaxy's individual random motion, so that its redshift is primarily cosmological. For these galaxies, however, determination of the distance to the cluster is much less precise. So in cases where we can measure the distance accurately, the value obtained for velocity is less meaningful and where we can have faith in our velocity measurement, the distance measurement is imprecise.

It is important to know the value of the Hubble constant because knowledge of the expansion rate of the universe is fundamental to our understanding of the universe as a whole. We will return to this idea later in this chapter.

The Hubble Law Used to Measure Distance

We saw in the last chapter that distances to nearby galaxies can be measured using Cepheid variables as distance indicators but that as we progress to more distant galaxies, Cepheids can no longer be seen; requiring us to use globular clusters and the brightest stars as distance indicators. At even greater distances, these cannot be seen and the brightest galaxies serve as distance indicators to clusters of galaxies. Finally, at the farthest distances, even this method cannot be used and we turn to the Hubble law to indicate a galaxy's distance. This idea is illustrated with an example.

EXAMPLE

Suppose a very faint source is observed to have a redshifted spectrum that indicates a recessional speed of 120,000 km/s. Assume that the Hubble law applies to this object and calculate its distance.

Solution

First, we must decide which value of the Hubble constant to use. Let's calculate the value using both 15 km/(s × MLY) and 25 km/(s × MLY), so that we will have calculated the range of distance within which we can be fairly confident the object lies. Using 15 km/(s × MLY):

$$\begin{aligned} v &= Hd \\ 120{,}000 \text{ km/s} &= 15 \text{ km/(s} \times \text{MLY)}d \\ d &= 8000 \text{ MLY} \end{aligned}$$

Now using 25 km/(s × MLY):

$$120{,}000 \text{ km/s} = 25 \text{ km/(s} \times \text{MLY)} \times \text{d}$$
$$d = 5000 \text{ MLY}$$

Thus we conclude that the object is between 5000 MLY and 8000 MLY away. This is 5 to 8 billion light years!

You may be uncomfortable with such great uncertainty in an answer, but remember that without the Hubble law and this calculation, we have no idea whatsoever of the distance to the object. The Hubble law does provide us with an idea of its distance.

Try One Yourself. Suppose an object is observed with a redshift that indicates a speed of 90,000 km/s. What is the range of distance within which we can feel reasonably sure that the object lies?

Figure 18–9 reviews the various methods used to measure distances in astronomy, from radar within the solar system to the Hubble law for the most distant objects.

Look-Back Time

We are referring now to distant objects *here because, as we will see later, we cannot be sure that they are galaxies.*

As new techniques are developed, we are seeing objects farther and farther away. We have now detected objects which may be as far away as 18 billion light years. If these objects are truly this far, the light we see left them 18 billion

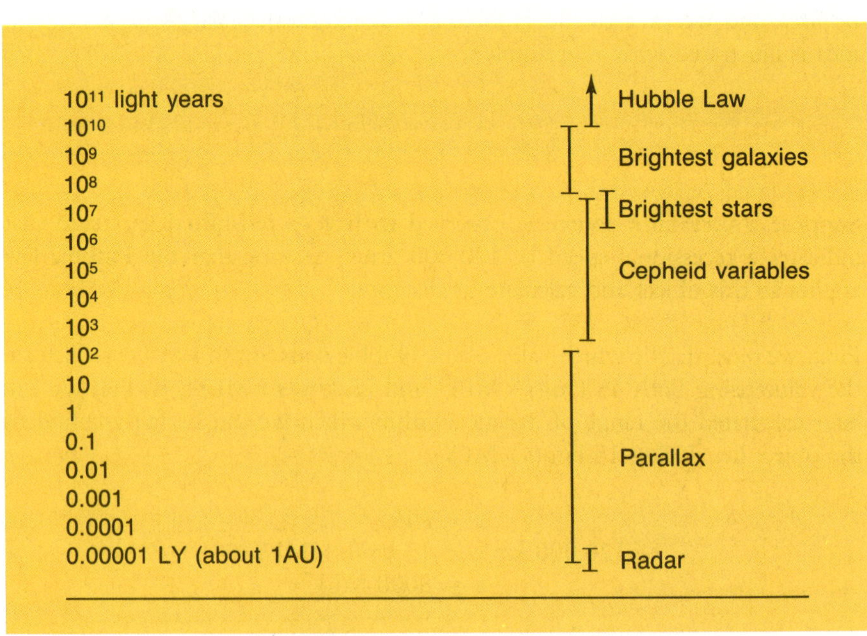

Figure 18–9. *A review of the various methods of measuring distance in astronomy.*

years ago. We are seeing far into the past. Astronomers speak of ***look-back time,*** which emphasizes this idea. An object 18 billion light years away has a look-back time of 18 billion years.

Look-back time: The time light from a distant object has traveled to reach us.

One of the problems with using large galaxies as distance indicators relates to the idea of look-back time. When we see these galaxies, we are not seeing them as they are today but as they were in the past. There is some evidence that within large clusters of galaxies, some galaxies combine to form supergalaxies. That is, some galaxies are "cannibalized." If this occurs, the largest galaxies in nearby clusters are not the same size as the largest galaxies in distant clusters because we are seeing nearby clusters at a later stage of their life than we see distant ones. In the distant ones, not enough of their life has passed for galaxies to have gobbled up one another. This would invalidate the assumption that distant clusters are similar to nearby clusters and would call into question the practice of using galaxies as distance indicators. Although this is a possible problem, the fact is that the Hubble law seems to apply equally well to faraway clusters where galaxies are used as distance indicators as it does with nearby galaxies where other methods are used for determining distance. This indicates that the dilemma introduced by look-back time is not a major obstacle in this case.

Among the objects with the greatest look-back times we find a class of objects not yet discussed: active galaxies. These objects play a major part in helping us to determine the scale of the universe.

ACTIVE GALAXIES

All galaxies emit some radio waves; we have seen that such waves are emitted from the nucleus of our galaxy. The radio waves from a normal galaxy constitute only about 1 percent of that galaxy's total luminosity, however. In the late 1940s, astronomers began observing strong extragalactic radio sources and in 1951 it was discovered that a radio source in the constellation Cygnus is actually a double source associated with a galaxy. Cygnus A, as the source is called, emits about a million times more radio energy than does the Milky Way.

Since the discovery of Cygnus A, many more such ***radio galaxies*** have been discovered, typically emitting millions of times more radio waves than a normal galaxy. Figure 18–10 is Centaurus A. Observe that the visible galaxy in this case is an elliptical galaxy with a prominent lane of interstellar gas and dust. The double-lobed feature seen in Centaurus A is common in radio galaxies, and most of the galaxies associated with double-lobed radio sources are either giant ellipticals or spirals. The size of the radio lobes is enormous; sometimes extending 15 MLY from the visible galaxy.

Radio galaxy: A galaxy having greatest luminosity at radio wavelengths.

Recall that the Milky Way is about 100,000 LY, or 0.1 MLY in diameter. From end to end, large double lobes are 300 times this size!

Radio galaxies often appear unusual when viewed in visible light. Figure 18–11 shows a jet of hot gas being emitted from Virgo A. This feature is seen in a number of cases. More recently, galaxies have been observed that have properties like those of radio galaxies but that have their primary emission at

(a)

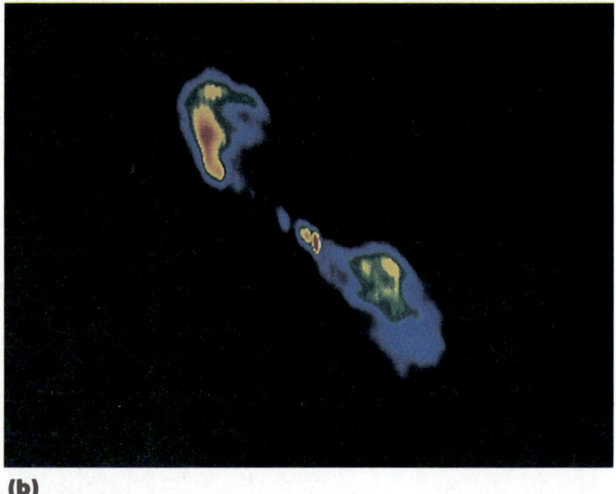

(b)

Figure 18–10. *Centaurus A is an elliptical galaxy, shown in (a) in visible light. In (b) we see its radio image. The radio sources are primarily in two lobes at opposite sides of the galaxy.*

Active galaxy: A galaxy with an unusually luminous nucleus.

wavelengths other than the radio region of the spectrum. The entire class of these unusual galaxies are often called ***active galaxies*** rather than radio galaxies.

Astronomers hypothesize that the source of the tremendous energy of active galaxies is an immense black hole at the nucleus of the galaxy. The great amounts of energy involved cannot be explained in any other manner. It is thought that the black hole is surrounded by a dense accretion disk and that the jets result from explosive events near the black hole that cause material to be ejected at great speeds in both directions along the axis of the disk.

Active galaxies are a subject of large amounts of interest and research in the astronomical community and many fundamental questions remain to be solved.

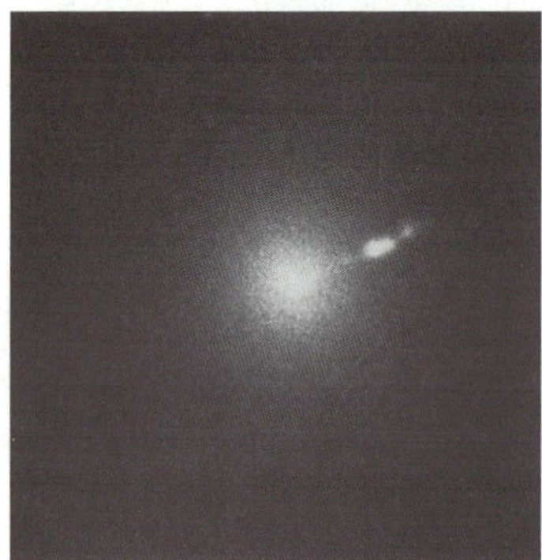

Figure 18–11. *Virgo A (M87) is ejecting a jet of hot gas.*

QUASARS

Individual stars are very weak radio sources. Prior to 1960 the Sun was the only star from which radio waves had been detected. In that year, however, two radio sources were found in the sky that were so small that they appeared to be stars. Recall that planets' sizes can be measured when they occult a star. The same phenomenon can be used in reverse to measure the size of a distant object. In the case of one of these radio sources, 3C 273 (so named because it is the 273rd object listed in the third Cambridge catalogue), its visual and radio images were observed as it was occulted by the Moon. By observing 3C 273 as its light and radio waves are blocked out by the Moon and again as they reappear on the other side, details of the source could be determined. The object appeared to be very small; like a star rather than like a galaxy. In addition, it had a small jet protruding from it like the jets that have been observed from some radio galaxies.

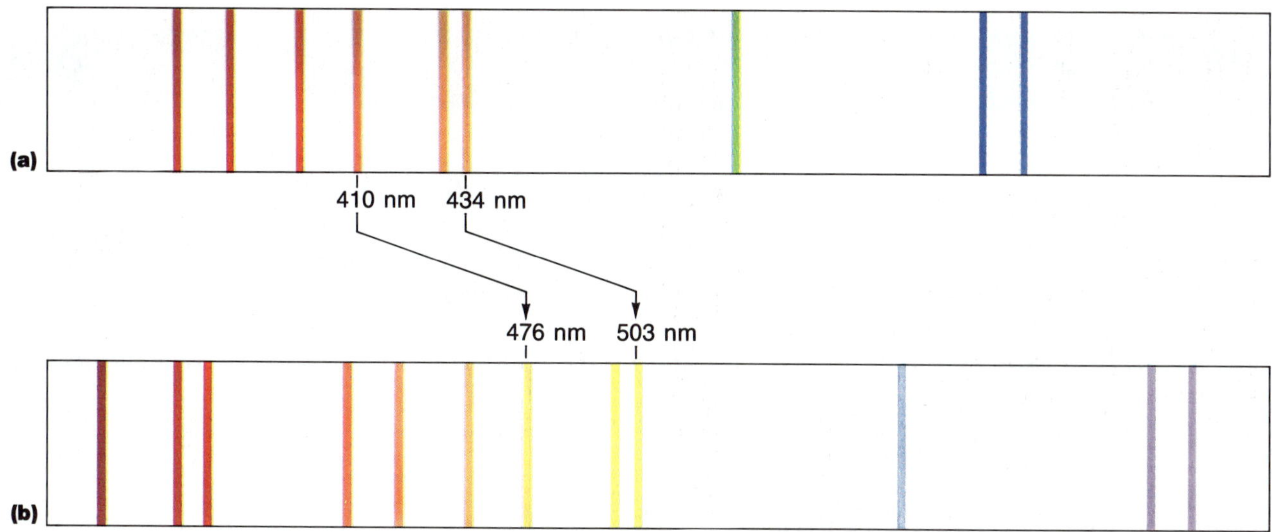

Figure 18–12. *(a) A "normal" spectrum. The wavelengths of two lines are indicated. (b) The same spectrum shifted 16 percent. Observe that the pattern of lines is the same.*

The radio waves from 3C 273 seemed to have two sources; one the jet and one the main body of the object. We have seen that double-lobed radio sources are common for active galaxies.

The spectra of both 3C 273 and 3C 48 (the other source mentioned above) were found to be extremely unusual. Although spectral lines were prominent, the lines could not be identified with any known chemical element. Were we seeing objects that had entirely different chemical elements than are known to us? Because of their unusual nature, the objects were called ***quasistellar radio sources,*** or **quasars.**

Quasar (quasi-stellar radio source): An object with a very large redshift but that is approximately stellar in size.

In 1963 Maarten Schmidt found the solution to the puzzle of the unusual spectrum of 3C 273. He found that the prominent spectral lines that had been seen are simply hydrogen spectral lines that are very greatly redshifted. Each wavelength is redshifted to a value 16 percent greater. (That is, the ratio of the change in wavelength to the normal, nonshifted wavelength is 16 percent. $\Delta\lambda/\lambda = .16$). Figure 18–12 illustrates the situation. Schmidt's colleague Jesse Greenstein soon found that when the spectrum of 3C 48 was examined, its spectral lines were found to be shifted even farther; by about 37 percent.

If the redshifts of the two quasars is caused by the Doppler effect, one quasar is moving at 15 percent the speed of light and the other is moving at 30 percent the speed of light (or 90,000 kilometers per second). These were speeds far greater than any yet encountered for celestial objects (or for terrestrial objects, other than nuclear particles).

If the great redshifts are cosmological redshifts—that is, if they follow the Hubble law—the closest of the two quasars must be as far away as the farthest galaxies. Yet it is brighter than those galaxies, even though in size it is more like a star than a galaxy.

Since the early 1960s, hundreds of quasars have been discovered. Unlike the first two, most are not sources of radio waves. Quasars are bluish-white

CLOSE UP

Relativistic Redshift Velocities

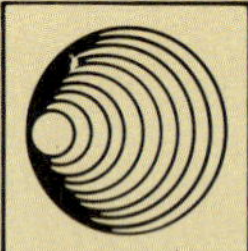

WE SAW IN CHAPTER 11 that the redshift of an object is stated as the ratio of the amount of the redshift of a particular wavelength to the value of the nonshifted wavelength, $\Delta\lambda/\lambda$. Stated this way, it is very easy to calculate the velocity of the object, for the relationship is simply

$$\Delta\lambda/\lambda = v/c$$

The redshift is usually given the symbol z, so that

$$z = \Delta\lambda/\lambda \text{ and}$$
$$v = zc$$

That is, the redshift multiplied by the velocity of light yields the velocity of the object. This equation applies only if the velocity of the object is significantly less than the velocity of light, however. We will see in Appendix A that when velocities approach that of light, many of our normal rules do not apply. Instead, we must use the ideas of Einstein's special theory of relativity. Quasars have recessional velocities close enough to the speed of light that special relativity must be used here. According to the theory of relativity, the equation relating the velocity of an object and its redshift (z) is

$$v = \left(\frac{(z + 1)^2 - 1}{(z + 1)^2 + 1}\right)c$$

We will calculate the velocities of two objects with different redshifts using both the standard and the relativistic equation in each case.

Suppose that an object has a redshift of 2.0. First, we will calculate its velocity according to the nonrelativistic equation:

$$v = zc$$
$$= 2c$$

The result tells us that the velocity is twice the speed of light. In the next chapter we will see that the special theory of relativity tells us that such a velocity is im-

Recall that red giants are typically light-minutes in diameter but that galaxies are thousands, or hundreds of thousands, of light years in diameter.

objects and are X ray emitters. In addition, many vary in intensity in an irregular way; typically changing intensity in weeks or months. This latter observation gives us confirmation of their small size; they cannot be greater than a few light-weeks or light-months in diameter. A few have been found with intensity variations of less than a day.

The first two quasars seen, with redshifts of 16 percent and 37 percent, have smaller redshifts than most; the 16 percent redshift being the least yet observed. To date, the quasar with the greatest redshift ($\Delta\lambda/\lambda = 4.11$) indicates that its speed is about 93 percent the speed of light.

If quasars' redshifts are cosmological, the objects are extremely far away. Hubble's constant is known with limited precision but if we calculate distances to quasars using the smallest and largest likely values for the constant, we obtain distances of 11 and 18 billion light years respectively for the most distant quasar.

possible. Let's calculate the object's velocity according to special relativity.

$$v = \left(\frac{(z + 1)^2 - 1}{(z + 1)^2 + 1}\right)c$$

$$= \left(\frac{(2 + 1)^2 - 1}{(2 + 1)^2 + 1}\right)c$$

$$= \left(\frac{3^2 - 1}{3^2 + 1}\right)c$$

$$= 0.8c$$

Thus the correct result is 80 percent the speed of light and not twice the speed of light.

Now let's consider an object with a redshift of 0.1. By the nonrelativistic equation, we simply obtain a velocity of $0.1c$, or 10 percent the speed of light.

Finally, we calculate the velocity of this object using the relativistic equation.

$$v = \left(\frac{(z + 1)^2 - 1}{(z + 1)^2 + 1}\right)c$$

$$v = \left(\frac{(0.1 + 1)^2 - 1}{(0.1 + 1)^2 + 1}\right)c$$

$$= \left(\frac{1.1^2 - 1}{1.1^2 + 1}\right)c$$

$$= \left(\frac{0.21}{2.21}\right)c$$

$$= 0.095c$$

Notice that the answer (9.5 percent c) is not greatly different from that obtained by nonrelativistic means (10 percent c). When we deal with redshifts below 10 percent, the difference is even less. Since most objects have redshifts much less than 10 percent, usually we can use the nonrelativistic equation with no loss of accuracy.

Now consider what the luminosity of such an object must be for us to be able to detect it at that distance. It must be far more luminous than an entire galaxy. Finally, remember that we are speaking about a small object; far smaller than a galaxy.

Competing Theories for the Quasar Redshift

When astronomers were faced with the prospect of such a luminous object being so small, they were forced to reexamine their assumptions. Could they be sure that quasars' redshifts are due to the Doppler effect? We raised this question earlier with regard to the redshifts from galaxies but the discovery of the tremendous redshifts from quasars made astronomers look at the situation once again. As we will discuss in Appendix A, the general theory of relativity predicts

Figure 18–13 *(a) The local hypothesis holds that quasars are relatively nearby objects that have been ejected from the Galaxy at great speeds. (b) If quasars are emitted by galaxies, however, some should be moving toward us from nearby galaxies and these would show blueshifts. No quasar has a blueshift.*

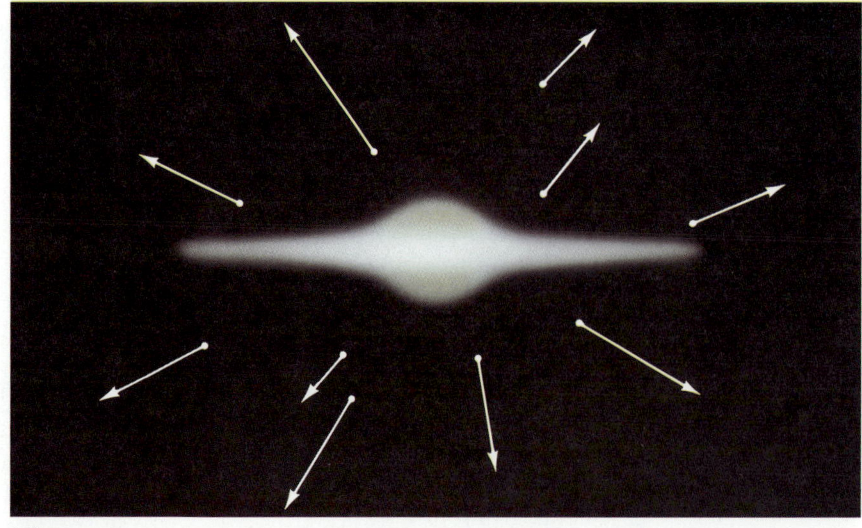

(a)

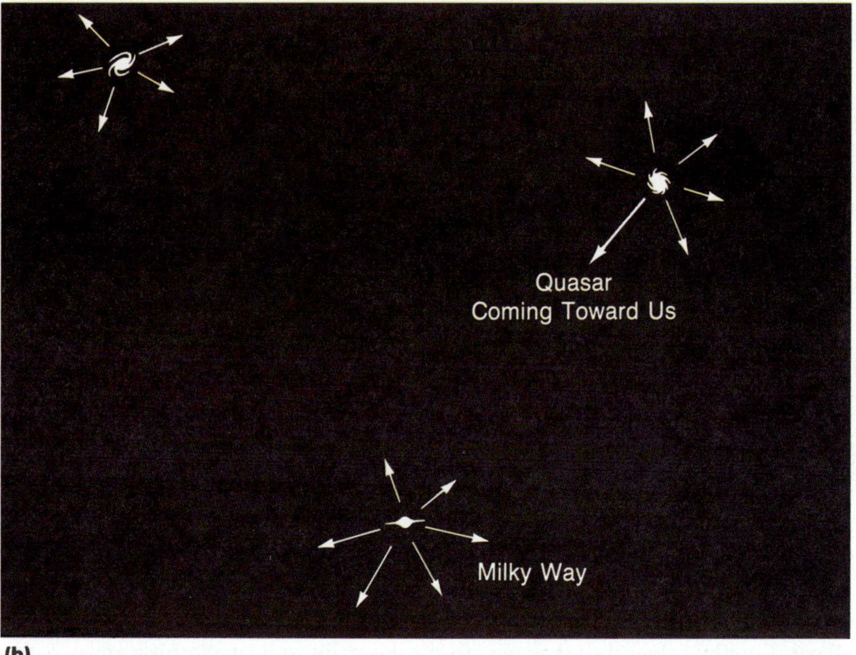

(b)

that light leaving a massive object will be redshifted due to the mass of the object. Calculations show that in order to produce redshifts such as those seen in quasars, the gravitational field near the object must be far greater than that near the most massive neutron star, and well-grounded nuclear theory tells us that there is a limit to the mass of a neutron star. After a certain mass is reached, a neutron star cannot exist. A black hole is formed. The theory of relativity simply cannot be used to explain the redshift.

Perhaps there is some other explanation. This explanation would require drastically different laws of physics than those of today, however. We never can rule out such a possibility but we must progress with what we know until it is definitely ruled out. The Doppler effect *does* explain the redshift, even if it leads us to conclude that quasars have velocities greater than we are accustomed to.

Almost all astronomers agree that the redshift must be due to the Doppler effect. Now for the next step in the logic: Is the redshift cosmological? Perhaps quasars are nearby and do not follow the Hubble law. Figure 18–13(a) shows fast-moving objects ejected from our galaxy; some at nearly the speed of light. If this is what quasars are, then they do not have such tremendous luminosities after all.

The problem with this ***local hypothesis*** is not only that we can imagine no source for such objects but that we must ask why we do not see similar objects from other galaxies. We cannot expect our galaxy to be unique in this regard. Yet if other galaxies emit such fast-moving objects, we would see some of them with tremendous *blue*shifts as they move toward us, as shown in Figure 18–13(b).

The local hypothesis has lost favor among astronomers, not so much because conflicting evidence has been found but because new evidence is leading us in another direction.

Figure 18–14. *The Seyfert galaxy NGC 4151 has a redshift of 0.001.*

Local hypothesis: A proposal stating that quasars are much nearer than a cosmological interpretation of their redshifts would indicate.

GALAXIES AND THE GRAVITATIONAL LENS

Recall that one of the first two quasars seen has a double-lobed radio source and a jet from its center. In this regard, it resembles an active galaxy. As astronomers found more and more quasars, they discovered that most of the nearer ones are associated with clusters of galaxies. The quasars are more luminous than the galaxies but the association causes us to look for more similarities.

One particular type of spiral galaxy, a ***Seyfert galaxy*** (Figure 18–14), has a very luminous nucleus which—although it is not as luminous as a quasar—in some cases varies in intensity in time periods even shorter than those of quasars. Similarities in the spectra of Seyfert galaxies and quasars point further to a link between galaxies and quasars, indicating that perhaps a quasar is a galactic nucleus.

Seyfert galaxy: One of a class of galaxies with active nuclei and whose spectrum contains emission lines.

Recently, high-resolution photos of nearer quasars show a fuzz on the image near the quasar. Again, it appears that quasars may be at the nuclei of some type(s) of galaxies and that the fuzz is caused by the stars of the galaxies.

Finally, twin quasars were discovered in 1979 in Ursa Major. The two quasars are very close together, separated by only 6 seconds of arc. Both have the same luminosity, the same redshift ($\Delta\lambda/\lambda = 1.4$), and identical spectra. The explanation for these identical twins lies with the theory of general relativity. That theory predicts the possibility of a large mass bending light from a more distant object so that two images of the distant object appear. The predicted phenomenon is called a ***gravitational lens*** but until the discovery of the twin quasars, not much attention was paid to the prediction. Since the discovery of the twin quasars, the intervening galaxy has been found and we are now confident

Gravitational lens: The phenomenon in which a massive body between another object and the viewer causes the distant object to be seen as two.

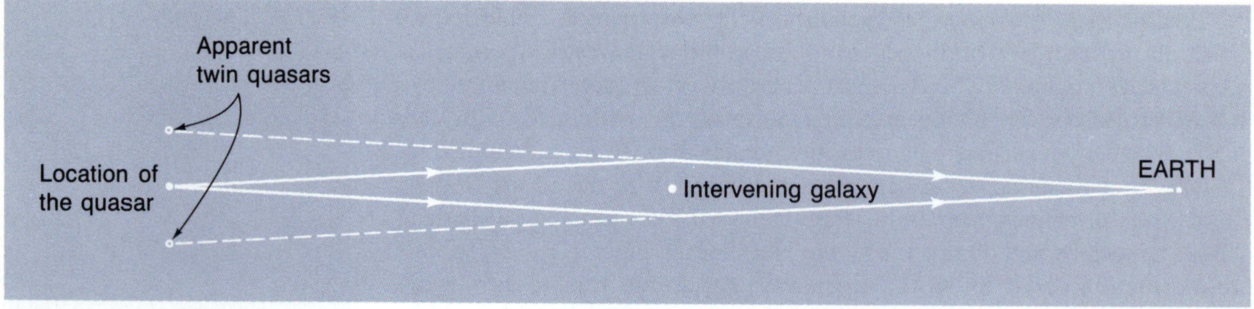

Figure 18–15. *Light from the quasar at left is bent as it passes by the galaxy at center. This causes the light to appear to come from twin quasars, one on either side of the galaxy. Such gravitational lenses are predicted by the theory of general relativity.*

that the twins are actually one quasar that has been made to appear double by the galaxy that lies between it and us (Figure 18–15). At least six more examples of the gravitational lens effect have been found since 1979.

The importance of the discovery of gravitational lenses is not only that they provide another confirmation of the general theory of relativity but that they indicate that quasars are indeed very distant; that their redshift is cosmological in nature. They really are the most distant objects yet observed.

The Nature of Quasars

If quasars are at the distances indicated by their redshifts, we are left with two major questions: What accounts for their enormous luminosity and why do we find them only at great distances from us?

To answer the luminosity question, we point to the same source of energy that is thought to be responsible for the energy in nuclei of active galaxies: black holes. There is essentially no limit to the amount of energy that can be produced by a black hole, for the energy is determined by the rate at which material is falling into the hole. We calculate that if matter falls into a black hole at the rate of a dozen solar masses per year, its luminosity will match that of a quasar.

This ties in with the second question, concerning why all quasars are at great distances. The answer involves the idea of look-back time: We see quasars only at great distances because they existed only in the distant past and are not a part of today's universe. It seems that quasars may be at the nuclei of young galaxies. The idea is that the nucleus of such a galaxy is very crowded with matter and that this matter supplies fuel for its black hole. As the matter is depleted by falling into the black hole, the galaxy becomes what we call a Seyfert galaxy. Then it becomes an active galaxy. Finally, it settles down to be a "normal" galaxy, like the Milky Way.

It must be emphasized that this scenario is very speculative. It does, however, tie together the many observations of the unusual objects we see. As we learn more about each class of object, we see more and more similarities between them; increasing astronomers' confidence that when they see quasars, Seyfert galaxies, active galaxies, and normal galaxies they are seeing examples of successive stages in a galaxy's life and not fundamentally different types of objects.

COSMOLOGY

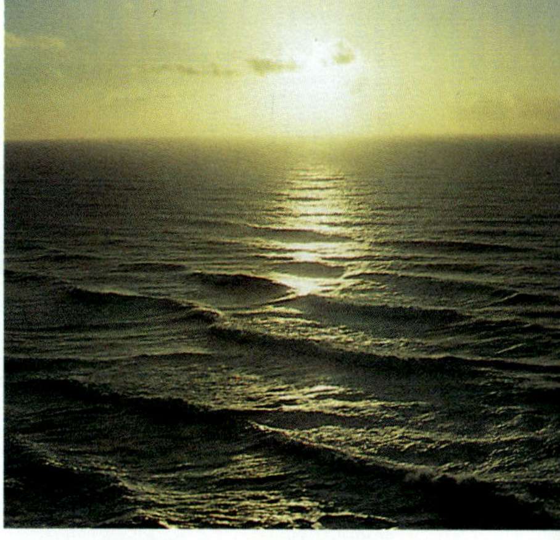

Figure 18–16. *Although the ocean surface is homogeneous, you see something different when you look in different directions because the surface is not isotropic.*

Cosmology is the study of the universe as a whole. This entire book has been a study of the universe but now we consider the universe as a single entity: asking from whence it came and to where it is heading.

Before one can proceed with a study of cosmology, certain assumptions must be agreed upon. Although the assumptions we will list cannot be proven, we have already seen some evidence that they are reasonable. The first assumption is that the universe is ***homogeneous***. This means that no matter where an observer is positioned in the universe, the universe will look essentially the same. This does not mean, of course, that there is another Earth from which an observer can look to see another Mars in the sky. It does not even mean that there is another local group of galaxies similar to our local group. It means, rather, that on the largest scale, the universe has about the same density and composition of matter at one location that it has at another. We have seen throughout this text that humans have come to learn that they are not in a favored place in the universe. In cosmology, this becomes an underlying assumption.

Homogeneous: Having uniform properties throughout.

The second assumption is somewhat similar to the first. The assumption of ***isotropy*** (I-SOT-rah-pee) states that the universe looks the same in all directions. Figure 18–16 shows that this assumption is different from homogeneity. The ocean shown is homogeneous, for no matter where a person is located on the surface of the ocean, the view would be the same. However, someone on the surface sees a different view when she looks across the waves than she sees looking along the waves. This means that the ocean is not isotropic.

Isotropy: The property of being the same in all directions.

Just as in the case of homogeneity, isotropy applies only on the largest scale. Recall that when we look out from Earth, we can see farther when we look out of the disk than if we look along the disk. The universe is not isotropic on this small scale. To apply the idea of isotropy, one must imagine being outside the Galaxy and even outside the local group of galaxies. In this case, the assumption is reasonable.

The final assumption is ***universality.*** This means that the same physical laws apply everywhere in the universe. Although this may seem so obvious that it need not be stated, recall that it was fairly recently in human history that Isaac Newton found the first law of nature that could be shown to apply beyond the Earth. Now we have vastly more evidence for this assumption as we analyze radiation from the most distant objects but we must remember that it is still an *assumption* that this principle holds throughout the universe.

Universality: The property of obeying the same physical laws throughout.

The three assumptions of cosmology are so much at the heart of the subject that they form what is called the ***cosmological principle.*** The principle formalizes the idea that we are not located at a special place in the universe. It brings to final completion the revolution of Copernicus; the revolution that proposed that the Earth is not central. The cosmological principle not only states that the Earth is at no special place in the universe but that there *is* no special place in the universe.

Cosmological principle: The basic assumption of cosmology that holds that on a large scale the universe is the same everywhere.

CLOSE UP

Stephen Hawking, the Ultimate Theoretician

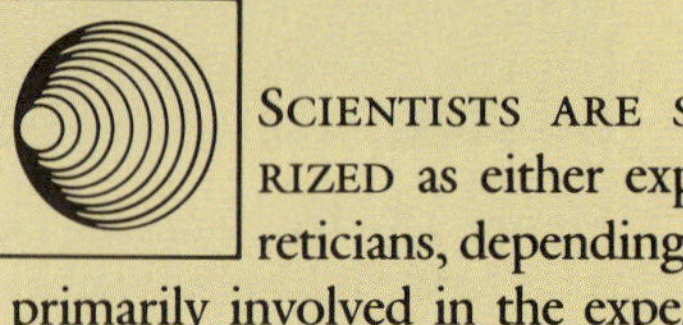

SCIENTISTS ARE SOMETIMES CATEGORIZED as either experimentalists or theoreticians, depending upon whether they are primarily involved in the experimental or theoretical aspects of their science. William Herschel, who we highlighted in a Close Up in Chapter 9, was primarily an experimental astronomer—an observer—while Albert Einstein is commonly agreed to be the premier theoretical physical scientist in the first half of this century. This half-century's answer to Einstein is generally considered to be Stephen Hawking, a professor of mathematics at Cambridge University.

Soon after entering Cambridge he began having physical problems; stumbling and slurring his speech. His condition was diagnosed as amyotrophic lateral sclerosis, commonly called Lou Gehrig's Disease. This disease, usually fatal, disables the voluntary muscles of the body. Hawking became depressed, abandoned his studies, and began drinking heavily. Then in January 1963 his life took another turn when he met Jane Wilde, a student of language who was more attracted to Stephen's mind than she was bothered by his deteriorating physical condition. They were married in 1965 and he became a father.

His marriage gave him the determination Hawking needed to continue with his life and work and his brilliance soon came to the fore. He wrote his dissertation on "The Properties of the Expanding Universe" and was awarded the Ph.D. in 1966.

Hawking continued working on the relationship of general relativity to black holes and to the big bang and he wrote papers on these subjects even as his physical condition deteriorated. By the early 1970s he was confined to a wheelchair. The disease does not affect the mind, however, and the fact it is not particularly painful left Hawking free to use his mind fully. His inability to use a pencil required him to work with long, complicated equations in his mind. He reports that it is impossible to handle the very complicated equations and so he has developed geometrical ways of thinking instead; giving problems a diagrammatic interpretation.

New discoveries continued to pour from his mind, including the preposterous idea that the laws of quantum mechanics require that black holes emit a stream of radiation. This concept, slow to be accepted by other astronomers, has since been verified and the radiation is now known as "Hawking radiation."

THEORIES OF COSMOLOGY

Steady state theory: A cosmological theory that holds that on a large scale, the universe does not change with time.

The cosmological redshift is the central observation of cosmology and the redshift is assumed to be due to the Doppler effect. The Hubble law is therefore of basic importance. There have been a number of theories to explain what Hubble expansion tells us about the past and future of the universe. We will consider two categories of theories: ***steady stage theories*** and ***big bang theories.*** The categories are plural because there have been derivatives of the originals; improving them to fit new observations.

Hawking's disease affects not only his skeletal muscles, but also the muscles used in speech. Until 1986, he often used an interpreter who was able to understand his distorted language. That year, Hawking developed pneumonia with severe complications. In order to save his life, surgeons removed his trachea. Now he has no voice at all. Instead, he uses three fingers (which are all he can control) to call up words on a computer screen. A voice synthesizer delivers a sentence once he has completed it. Using this method of communication, he has recently completed a book intended to explain his ideas to the general public: *A Brief History of Time: From the Big Bang to Black Holes* (Bantam Books, 1988).

After learning of his disabilities, one understands why Stephen Hawking is a theoretician rather than an experimentalist. His brilliance was apparent even in his youth but it may be that his lack of physical abilities has served to make him even more brilliant; a nearly perfect cerebral being, a mind contemplating the universe.

Figure T18–1. *Stephen Hawking in front of a statue of Isaac Newton.*

The Big Bang

The fundamental idea of big bang theories is a fairly simple one. The present motion of galaxies leads us to believe that at one time in the past, they were all together. Suppose a large firecracker is placed at the center of small stack of sand. A moment after the explosion of the firecracker, we analyze the situation. Figure 18–17 shows what might appear. Soon after the explosion, grains of sand are at various distances from their original position and—here is the important point—grains that are farther from their respective starting points are moving faster. That is how they got farther from one another. No matter which grain one considers, all others are moving farther from it.

Big bang theory: A theory of cosmology that holds that the expansion of the universe began with an initial explosion.

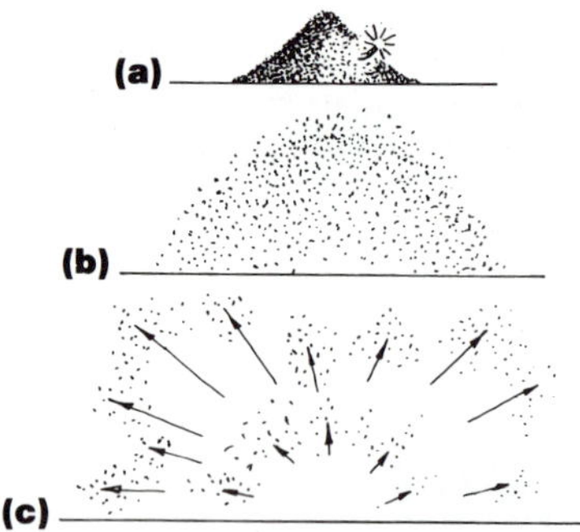

Figure 18–17. *Part (a) shows a firecracker about to explode inside a small pile of sand. Parts (b) and (c) show the sand later, with a few arrows included to illustrate that the sand grains that are farthest from the initial explosion are moving fastest.*

The explosion that started the expansion of the universe is called the ***big bang,*** an irreverently trivial name for the event that began the universe. In its most basic sense, the standard big bang model is simply the idea that every bit of the matter and energy in the universe was once compressed to an unimaginable density. It was all together in what is sometimes called the ***primoridal egg.*** In the big bang, the material exploded apart.

We must be careful not to think of the primordial cosmic egg as existing at a certain place, at one point in the universe. It *was* the universe—the entire universe. It did not exist as part of the universe.

After the initial explosion, the nuclear particles formed into atoms of low mass (hydrogen and helium) then clustered under the force of gravity into stars, clusters of stars, galaxies, and clusters of galaxies. The clusters of galaxies are still moving apart.

We will return later to a discussion of conditions shortly after the big bang and to predictions that arise from those conditions. First, we will take a quick look at an alternative theory.

Big bang: The theorized initial explosion that began the expansion of the universe.

Primordial egg (or primordial atom): The original matter that is theorized to have exploded to form the universe of today.

Perfect cosmological principle: The assumption that on a large scale the universe is the same everywhere in space *and time.*

Reference to a special *time does not imply a* better *time; only a time that is different form past and future times.*

Steady State Theories

The cosmological principle holds that we are at no special place in space. It does, however, allow for us being at a special place in time; for at the present moment, the universe is less dense than it was in the past and more dense than it will be in the future because expansion reduces its density. In 1948, a ***perfect cosmological principle*** was proposed by astronomers Fred Hoyle, Hermann Bondi, and Thomas Gold. It states that our location in time is no more unique than our location in space, and that the universe of the past was the same as now and the same as it will be in the future. Since the overall universe does not change in this cosmology, it is called the *steady state theory.*

Before looking at the implications of the perfect cosmological principle, we pause a moment to consider why such a thing might be proposed. Remember that a theory should be esthetically pleasing. The perfect cosmological principle is neater because it expands the nonspecialness to include time in addition to space. The perfect cosmological principle carries the nonspecialness idea through both space and time.

It seems to be almost undeniable that the universe is expanding. How can this be squared with the steady state theory? It seems that if clusters of galaxies are getting farther apart, the universe will necessarily be less dense in the future than it is now. Hoyle, Bondi, and Gold proposed that matter is created in the space between galaxies and that this new matter is sufficient for the universe to maintain its density (Figure 18–18). Such spontaneous creation of matter would violate a long-established principle of physics: the conservation of mass/energy, which holds that the total amount of mass and energy in the universe cannot change.

The principle of the conservation of mass/energy is a dear one in science but it has changed in the past. Before this century there were two such conservation principles: the conservation of mass and the conservation of energy. It was thought that neither matter nor energy could be created or destroyed. Then in 1905, Einstein's theory proposed that mass could be changed into

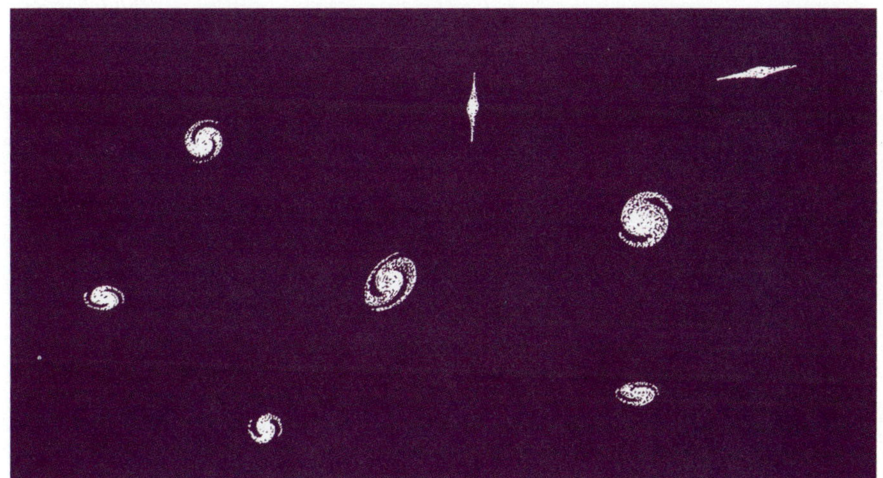
(a) Now

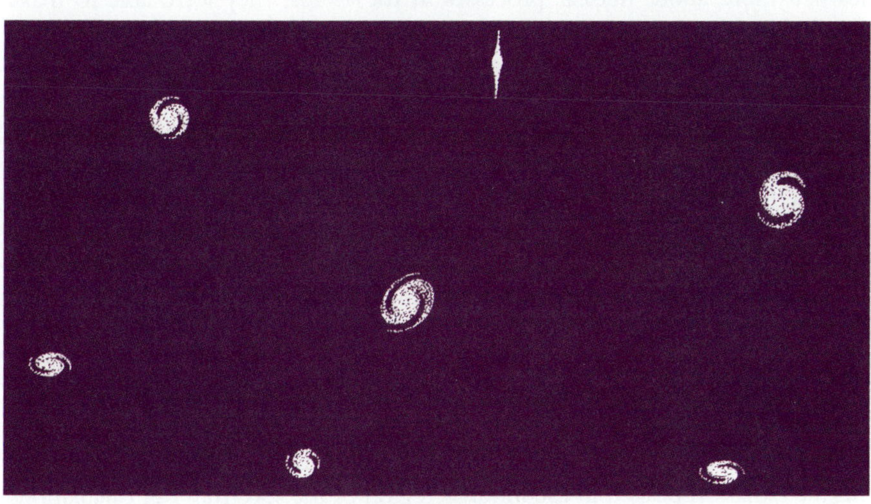
(b) Later

Figure 18–18. *If the universe does not decrease in density as galaxies move apart, matter must be created in the space between galaxies.*

energy; a prediction that was dramatically demonstrated in the atomic bomb. Einstein combined the two theories into one, the conservation of mass/energy, which sees mass and energy as two forms of the same thing and holds that the total mass and energy in the universe remains the same.

The amount of mass that would have to be created in order to fill the gaps between galaxies as they move apart would be very small; only one new hydrogen atom in each cubic meter of space every 10 billion years or so. This means that if one new atom were created every 10,000 years in the volume of the Houston Astrodome, the density of the universe would remain the same. The theory holds that as millions of years pass, newly formed atoms collect to form new stars and galaxies, and that the expanded space is thereby refilled, so that the universe is no different now than it was in the past and than it will be in the future.

The fact that steady state theories violate the principle of conservation of mass/energy should not be considered a strike against the theories. First, according to the predictions, the necessary creation events are so rare that they would be far below any possible detection limit and therefore far below the accuracy to which the principle has been verified. Secondly, big bang theories may also involve a violation of the principle. In this case it is a gigantic one-time violation, as all matter and energy are created in one big burst at the beginning.

Evidence: Background Radiation

In order to decide which theory better fits the observations, we examine predictions made by each theory and see which ones we can test. The spontaneous creation prediction in the steady state theory, for example, cannot be tested because there is no way to detect the creation of one new hydrogen atom in the Astrodome every 10,000 years. On the other hand, a prediction made by the big bang theory was, quite accidentally, found to be correct. As scientists learned more about nuclear processes in the 1940s, they were able to make predictions about the character of the early universe shortly after the big bang. The material of the big bang was originally extremely hot but it cooled as it expanded outward, in much the same way that a gas cools when it expands. The hot matter of the very early universe was opaque to radiation, but when it cooled to about 3000 degrees C, it became transparent. At this point, the gas that made up the universe was emitting radiation that had the characteristics of radiation from a 3000 degree object and this radiation existed over the entire universe.

The universe has remained transparent to radiation since that time, so it was predicted that this radiation should still be in existence. Remember that as we look at objects at greater and greater distance, we are looking back in time. If we were to detect the radiation left over from the big bang, it would be coming from far back in time and therefore from a great distance. According to the Hubble law, then, it would be greatly redshifted. The original 3000 degree radiation peaked in the infrared region of the spectrum but calculations showed that it would by now have been redshifted to the microwave region, with wavelengths a millimeter or so. Such radiation would not be characteristic of radiation from a 3000 degree object but it would appear the same as radiation from a very cold object; one at about −260 degree C to −270 degree C. It was predicted that this ***background radiation*** should be striking Earth from all directions but that it should be very faint.

Background radiation (or cosmic background radiation): Long-wavelength radiation observed from all directions; believed to be the remnants of radiation from the big bang.

On the Kelvin temperature scale, a scale normally used in science, −270 degrees C corresponds to 3K. The radiation is therefore often called 3 degree radiation.

In 1948, when the prediction of background radiation was made, there was no way to detect such weak waves so the prediction was laid aside and forgotten. Then in the mid-1960s a group of physicists at Princeton University were studying the big bang and again predicted the existence of background radiation, not realizing that the prediction had been made some 15 years before. They were confident that it was then possible to build a radio receiver that would be able to detect the radiation and they set about building one on the roof of the Princeton biology building. They were too late.

At the same time as the predictions were being made at Princeton, only a few miles away two employees of Bell Telephone Laboratories, Arno Penzias and Robert Wilson, were doing applied research on microwave transmission in hopes of improving the transmission of messages. They were frustrated, however, by some low-intensity radio waves they observed to be coming from all directions toward their receiver.

As you might suspect, the radiation they had found turned out to be the cosmic background radiation predicted by the physicists at Princeton and by astronomers years before—the prediction that was based on the big bang theory. Penzias and Wilson were awarded the Nobel Prize in 1978 for their discovery of the radiation; radiation for which they had not been searching and the cosmological implications of which they were unaware.

Largely because of the discovery of the background radiation, the steady state theory has been almost abandoned today in favor of the big bang theory.

Recall that we said that at the time of the big bang, the condensed material made up the entire universe. This is why the background radiation now comes to us from all directions. The radiation fills the universe just as it did at the beginning. The matter and energy *is* the entire universe, and there is no space left over outside of it.

THE BIG BANG TODAY

Today the big bang theory, in some form or another, is accepted by almost all astronomers simply because the evidence indicates it. It better explains the observations than any other theory yet devised. Not only that, it makes predictions that are verifiable and many of these have been verified. On the other hand, not only does the steady state theory fail to explain background radiation but it has difficulties explaining some newer observations such as quasars. Recall that quasars are found only at great distances; indicating that they once existed in the universe but no longer do. If this is true, the universe is not unchanging, as is proposed by the steady state theory.

Astronomers are now discussing details of the big bang, such as what was going on shortly—very shortly—after the beginning. They are describing conditions after a very small fraction of a second, in fact. The latest theories, called ***inflationary universe theories,*** involve uneven expansion in those first fractions of a second. You might question the meaning and value of this but the calculations are based on our knowledge of nuclear particles and the calculations do produce predictions about the effects such early conditions would have on today's universe.

Inflationary universe theory: A model of the expanding universe that involves a very rapid expansion near the beginning.

The Age of the Universe

If the big bang theory is correct we can use the Hubble constant to determine how long ago the big bang occurred, for the constant tells us how fast galaxies are spreading apart. Instead of writing v for velocity in the Hubble law, let us substitute the definition of velocity: the ratio of distance to time.

$$v = Hd$$

$$\frac{d}{t} = Hd$$

CLOSE UP

Science, Cosmology, and Faith

WHERE DID THE MATTER and energy of the big bang originate? Science cannot answer that question at present. In fact, it seems possible—perhaps likely—that it will never be able to answer that question. For if the matter existed in atomic form before the big bang, it did not carry information with it through the big bang. Thus the matter itself can tell us nothing about its pre-big bang history.

Does this then mean that science can say that there must be a creator, and that science has, after all, found a need for God? An individual scientist may believe this, but science itself cannot use God as the explanation for the big bang. Science cannot use God for any explanation. It has been said that science avoids God. It does, indeed. The reason that science does not use God for explanations is basically that science has been successful in explaining the material world without reference to a God. Science, by intention, uses natural causes to explain natural effects.

We say that science is successful in its method because scientific explanations of the workings of the material world have led us to further understanding of that world. The fact that the success has come without reference to God indicates that the material universe seems to be describable by completely natural principles.

What about cases where science is unable to find an answer? If science, when it comes to something which it cannot explain at the time were to explain it by reference to God, the search for an explanation would end. If Newton had used God as an explanation for why things fall to Earth, he would never have developed the theory of gravitation. If we use God as an explanation for the big bang, there would be no reason to look further for a natural explanation. Use

Now choose two widely separated galaxies and let d be the distance they are apart. Thus t is the time taken to go that distance, or the age of the universe. Solving the equation for t, we get

$$t = 1/H$$

This tells us that the age of the universe is simply the reciprocal of the Hubble constant. This constant is normally expressed as km/(s × MLY) or km/(s × Mpc). Before we can substitute numbers to calculate the age of the universe, we must express both of the distances in the same units rather than one being expressed in kilometers and the other in millions of light years. One million light years is about 9.5×10^{18} kilometers, so a Hubble constant of 15 km/(s × MLY) corresponds to 1.6×10^{-18} km/(s × km). The age of the universe is the reciprocal of this, or

$$t = \frac{1}{\dfrac{1.6 \times 10^{-18}}{s}} = 6.3 \times 10^{17}\,s = 2 \times 10^{10}\text{ yr.}$$

of supernatural explanations would shut down science. History, however, tells us that it is profitable to look for natural explanations. Supernatural explanations cannot be used in natural science.

Testability

There is another reason that science cannot use God for an explanation and this relates to the reason that traditional science does not accept creationism as a science: A theory of science must be able to be shown to be wrong. A theory must be testable. Every theory must be regarded as tentative; as being only the best theory we have at present. It must contain within itself its own possibility of destruction. The 1948 prediction concerning cosmic background radiation was such a case. If the background radiation had not been found, the big bang theory would have had to have been adjusted, or—if enough such contradictions appeared—the theory would have had to be dropped and replaced with another. This has happened a number of times in science. It happened to the steady state theory and we have seen other examples in this text.

On the other hand, if science relied on a creator to explain the inexplicable, there would be nowhere to go; no way to prove that explanation wrong. The question would have already been settled.

This is not to say that God might not be the explanation. Many people believe that a creator is the ultimate explanation of everything. Perhaps some questions, like the origin of the material for the big bang, simply cannot be answered without reference to a creator. But in that case, science simply cannot answer the question. The question is beyond the realm of science. Science does not deny the existence of God. God is simply outside its realm.

This is 20 billion years. On the other hand, if we take the value of the Hubble constant to be 25 km/(s × MLY), we calculate 12 billion years as the age of the universe.

The calculation above assumes that galaxies have continued to move apart at the same rate back through their history; that is, that the Hubble constant does not change over time. In fact, we would expect the speeds of galaxies to change. Because of the gravitational force they exert on each other, the recessional speeds of galaxies should be decreasing. This would mean that in the past, they were moving apart faster than they are now and that the age of the universe is therefore less than we calculated above. For this reason, and because of the fact that the Hubble constant is not known precisely, 20 billion years is taken as the *upper limit* to the age of the universe. The universe is this age or younger.

Earlier we said that the most distant quasar yet found has a redshift of 4.11. This indicates a look-back time of 93 percent of the age of the universe. Although we know the age of the universe fairly imprecisely (12 to 20 billion years), Doppler shift data is accurate enough that we can be confident that the farthest

Figure 18–19. *The Cosmic Background Explorer (COBE) will orbit the Earth to examine the characteristics of the background radiation. It is scheduled for launch in May 1989.*

quasar thus found has a look-back time of 93 percent of whatever is the correct value for the age.

THE FUTURE: WILL EXPANSION STOP?

What came before the big bang? As discussed in the accompanying Theme Box, that question may not be a scientific one. But it might. To see how science might get a hint at what came before the big bang, we look into the ***future***.

If galaxies are moving apart, they will obviously be farther apart in the future than they are now. Where does this stop? There *is* an agent to stop it: gravity. The force of gravity must be slowing the expansion of the universe. The question is, will it slow it enough to stop it and bring the galaxies back together?

There are two possible answers: Either the expansion stops or it doesn't. First, suppose the expansion does not stop. This will mean, simply, that the clusters continue to get farther apart. Gradually, the stars use up the hydrogen fuel that powers them, the glowing stars get fewer, the glow fades, and the universe fizzles out. Not a very attractive idea, but a possibility.

On the other hand, maybe the expansion stops. Then gravity will begin to pull the galaxies back toward one another. Intelligent beings (our descendants?) living on some other planet, will be able to use the Doppler effect to observe that the galaxies are then getting closer together. The infall will continue until all matter in the universe is condensed into a tremendously dense ball; the "big crunch." What might happen then? Perhaps another big bang; another universe.

Oscillating universe theory: A big bang theory that holds that the universe goes through repeating cycles of explosion, expansion, and contraction.

If this is the future, perhaps it was the past. Perhaps this is what preceded the big bang: a previous universe containing the same matter and energy as ours. Such a scenario is called an ***oscillating universe.*** It is a much more exciting prospect than a fizzling-out universe.

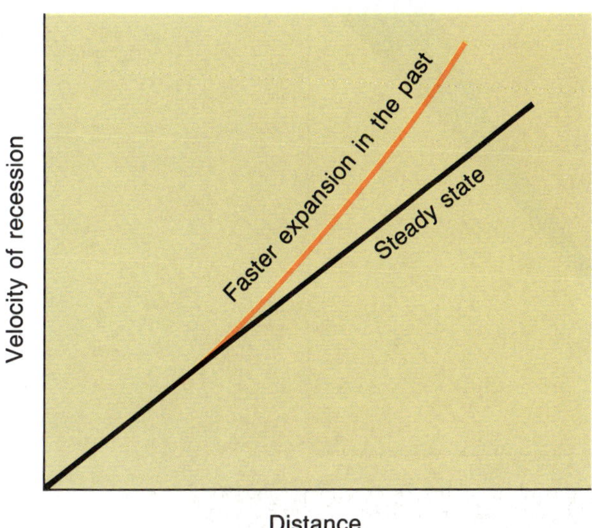

Figure 18–20. *If the Hubble data falls along the straight line, a steady state universe will be indicated. If it falls along a curve such as the one shown, it will indicate that the universe was expanding faster in the past and has slowed its expansion.*

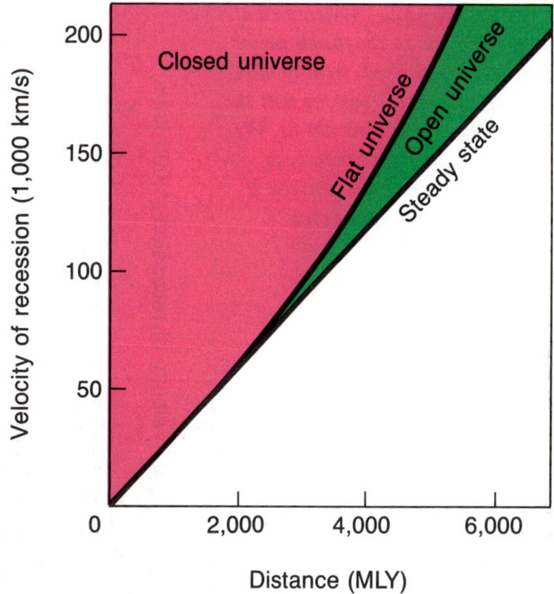

Figure 18–21. *If the data fall in the green area, the universe is expanding too fast to stop and we live in an open universe. If the true case is in the red area, this indicates that the expansion speed changes greatly as time passes and that the expansion will stop.*

Evidence: Distant Galaxies

There are two ways to search for an answer to the question of whether the universe will stop expanding. One way is to see if we can measure how much the expansion is slowing with time. We can't, of course, use the few years of our observations and expect to see a change. However, we are seeing distant galaxies as they were in the past, not the present. When we look at the most distant galaxies—at active galaxies and quasars—we are detecting light that left them billions of years ago and the Doppler shift of that light tells us their speed then, not now. Therefore we should be able to tell how much they're slowing down by comparing their speeds to the speeds of galaxies nearer to us. If we can determine this, we can calculate whether their rate of slowing is enough to bring them to a stop.

In practice, the observation is tough to make. The problem is that in order to look far enough back in time to get a significant change in speed, we are looking at objects so distant that the light is extremely dim. The Doppler effect is easily observable but we cannot get accurate measurements of the distances to these galaxies. And the farther away they are, the less precise is our measurement of their distances.

Figure 18–20 is a graph of velocity versus distance for galaxies. On it one line is drawn to show the relationship as predicted by the steady state theory. This is the straight line, for the theory holds that the expansion is exactly the same now as it was in the past (and will be in the future.) On the other hand, if gravity is slowing the expansion, the actual case will be a line that curves;

Figure 18–22. *When present data is plotted on the graph we see that recessional speeds seem to have been greater in the past, so that the steady state theory does not fit. We do not have enough accuracy to determine whether the universe is open, closed, or flat, however.*

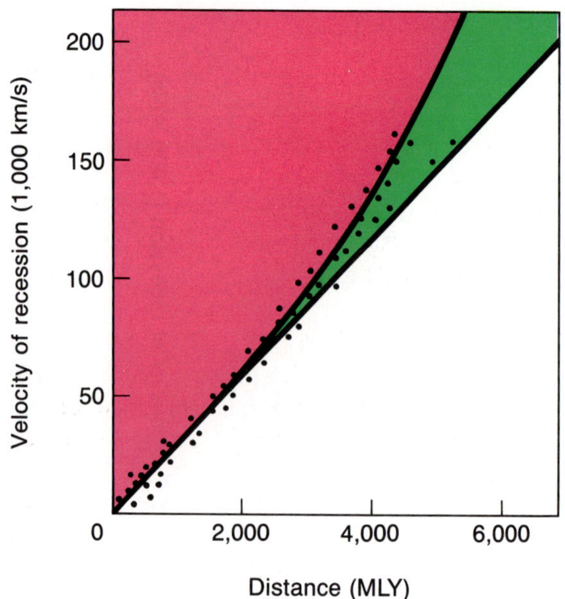

perhaps like the other line. Along this line, the velocities of the most distant galaxies is faster than the steady state theory would predict, for the velocities were greater in the distant past.

Open universe: The state of the universe if it continues expanding without stopping.

Closed universe: The state of the universe if it stops expanding and begins contracting.

Flat universe: The condition of the universe if gravity just balances its expansion so that it stops expansion only in an infinite amount of time.

The curved line in Figure 18–21 represents the borderline case between a universe that will expand forever (an ***open universe***) and one that will stop its expansion and start falling inward on itself (a ***closed universe***). Suppose we plot measured values on the graph. If they fall on the straight line, they will indicate that we live in a steady state universe. If they fall in the green area of the graph, it will indicate an open universe; one which will expand forever. On the other hand, if the data indicate a curvature greater than the curved line (in the red area) the universe is closed.

The borderline case between an open and a closed universe is called a ***flat universe.*** Inflationary universe theories predict that the universe is exactly flat and that the data should fall on the curved line. If the universe is flat, it will gradually slow its expansion but will stop only after an infinite amount of time; therefore it will never fall back inward.

Figure 18–22 shows present data plotted on the graph. Notice that the data indicate that the steady state theory is unsatisfactory because they favor a line curving upward; showing that galaxies are moving more slowly now than in the past. The data are not accurate enough, however, to tell us whether or not the expansion will stop; although one might say that there is a hint that the slowdown is not great enough to stop the expansion. That is, there are more points plotted to the right of the borderline case than to the left. So although the conclusion must be considered very tentative, we would say that indications are that galaxies are not losing speed fast enough to ever stop moving apart. We must emphasize, however, that this is an extremely tentative conclusion.

The Density of Matter in the Universe

The theory of general relativity (Appendix A) predicts that both matter and energy cause what we call gravitational force.

The second method of seeking an answer to the question is to determine the overall density of matter (and energy) in the universe. If there is enough mass/energy in each volume of space, there will be enough gravitational force to stop the expansion. So we look for the mass/energy. The result is that we find only about 10 percent of the mass/energy needed to stop the expansion. However, we know that we are presently unable to detect all of the matter in a given volume of space. Black holes are difficult to detect and we don't know how many are out there. They could account for a great amount of mass. Probably smaller amounts of mass could be hidden as black dwarfs and planets. Other matter could be hidden in neutrinos—nuclear particles that are very difficult to detect—although evidence obtained by analyzing travel times of neutrinos from the Magellanic Supernova agrees with previous theory and indicates that neutrinos are either massless or very close to massless.

We saw in the last chapter that the motion of galaxies within clusters indicate that the clusters have more mass than we can account for. This observation is the basis for references to "missing mass," for we know that there must be more mass than we see. In cosmology we see that if the universe is either flat (as predicted by inflationary theories) or closed, we again have "missing mass."

Both types of data—the results of looking at distant galaxies in order to look back in time and the calculation of the density of the universe—indicate that the expansion will not stop. In this the answer? It is much to early to say, for the data is too tentative. What perhaps is most exciting is that although astronomers have not been able to decide on the fate of the universe, they now know some of the questions that must be answered in order to determine that fate.

CONCLUSION

As astronomers look more and more deeply into space, they are looking farther and farther back into time and seeing objects that are very strange when judged by the standards of today's world. The distant, mysterious quasars have been explained tentatively as being near the beginning of a chain of evolution of galaxies that progresses from quasars through Seyfert galaxies to the familiar galaxies in our neighborhood.

We are becoming more convinced all the time that black holes play an extremely important role in our universe, serving as the energy source for quasars and active galaxies, and that they are at the center of many—or all—normal galaxies.

Investigation of distant objects is helping us answer the cosmological questions about the origin and fate of the universe. We have few answers, but our theories tell us what observations must be made by present and future telescopes in order to get more answers.

One might ask what practical use can be made of the study of cosmology. The answer is simply that not all pursuits need to have practical or useful consequences. Some, like music, art, philosophy (and cosmology), are engaged in purely for the sheer intellectual or artistic pleasure they afford. It is part of the nature of human beings to be curious and questions such as those raised in this chapter are among the most fundamental that can be asked—and perhaps answered—by science.

RECALL QUESTIONS

1. It is said that the universe is expanding. Explain just what this means.
2. Describe the evidence that the universe is expanding.
3. Explain how, if almost all galaxies are moving away from ours, we are not necessarily at or near the center.
4. What is meant by a "cosmological" redshift?
5. State the Hubble law. What is the presently accepted range of the Hubble constant? Include units in your answer.
6. Describe some difficulties encountered in determining the Hubble constant with accuracy.
7. Explain how the Hubble law can be used to determine the distance to a galaxy. Discuss the limitations of this method of determining distance.
8. What is meant by look-back time?
9. In what way are active galaxies "active?"
10. After the spectra of quasars were understood, why were the objects still so difficult to explain?
11. What is a possible relationship between active galaxies, Seyfert galaxies, quasars, and normal galaxies?
12. What do we mean when we say that a quasar has a redshift of 25 percent? Which quasar is moving away faster, one with a redshift of 25 percent of one with a redshift of 30 percent?
13. Which of the quasars of the last question are further away? Upon what law or principle is this answer based?
14. Why has the local hypothesis for quasars been nearly abandoned?
15. What is the present theory of the origin of the energy of quasars?
16. Why are there no nearby quasars?
17. Name and explain the three assumptions of the cosmological principle.
18. Name and describe two major cosmological theories.
19. What is background radiation and how does it enter the discussion of cosmological theories?
20. How does the steady state theory explain that the universe could be expanding but still not decreasing in density?
21. How is the age of the universe calculated?
22. Describe two methods of determining how quickly the expansion of the universe is slowing.
23. What is the oscillating universe?
24. What is meant by "missing mass?"
25. Identify: Cygnus A, Arno Penzias and Robert Wilson, Fred Hoyle, Hermann Bondi and Thomas Gold, Edwin Hubble, inflationary theory, open universe.

QUESTIONS TO PONDER

1. Give an example from material of an earlier chapter that shows the distinction between an observation and a conclusion.
2. Some galaxies in nearby clusters are moving towards us but no galaxy in a distant cluster does this. Why?
3. How can we determine the motion of the Milky Way galaxy within the local group? (Hint: Consider redshifts of galaxies in different directions from us.)
4. If the slope of the Hubble graph were greater (with distance plotted on the x axis), would the Hubble constant be larger or smaller? Explain.
5. What is a gravitational lens and what does the fact that they have been observed tells us about quasars?
6. Quasars do not exist in our part of the universe. Why is this not a violation of the homogeneity principle?
7. Give an example from nature (other than the one given in the chapter) of something that is homogeneous but not isotropic, and then give an example of something that is isotropic but not homogeneous.
8. Discuss the observation of background radiation as an example of the confirmation of a theory, comparing it to the prediction of stellar parallax in the 1600s.
9. The age of the universe calculated form the Hubble constant assumes a constant velocity of recession for galaxies. If in fact galaxies were moving faster at an earlier time, is the universe older or younger than calculated? Show your reasoning.
10. Where in the universe did the big bang occur? Explain.
11. It has been said—perhaps only half in jest—that when astronomers come upon a problem they cannot solve, they use a black hole. Give two examples that may have lead someone to say this.

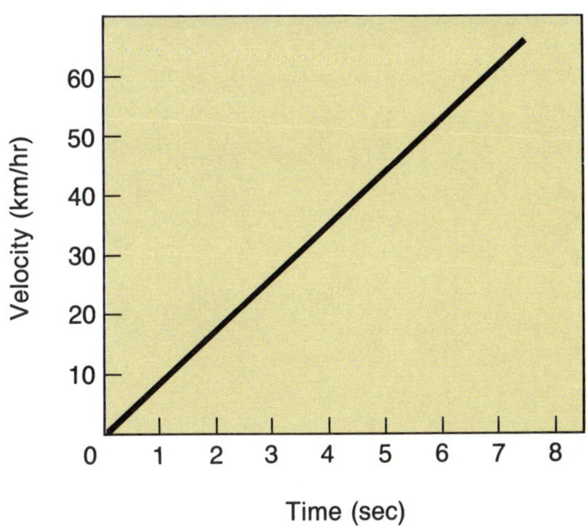

Figure 18–23. *This graph might represent the changing speed of a race car. Determine the slope of the graph. (Calculation #1)*

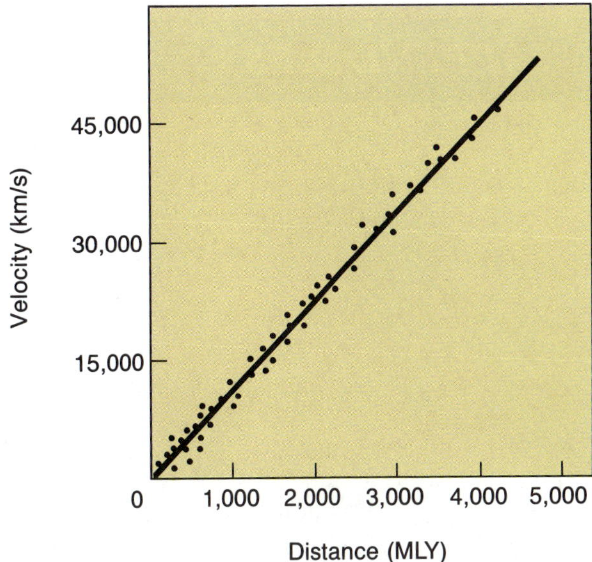

Figure 18–24. *If this were correct data for the Hubble graph, what would be the Hubble constant? (Calculation #2)*

CALCULATIONS

1. Determine the slope of the graph shown in Figure 18–23. Be sure to include units in your answer.
2. If the data for he Hubble expansion were as shown in Figure 18–24, what would be the value of the Hubble constant?
3. Suppose that an object is seen with a redshift that indicates a speed of 100,000 km/s away from us. If this redshift is cosmological, how far away is the object? (Assume some reasonable value for the Hubble constant.)
4. Suppose that someone is moving away from you at a constant speed of 15 mi/hr and is at the moment 60 miles away. If he had maintained this speed since leaving you, how long ago were you together?
5. For the motion in question 4 above, what is the "Hubble constant?"
6. If the Hubble constant is 20 km/s × MLY, what is the maximum age of the universe?
7. Using methods discussed in the Close Up (Relativistic Redshift Velocities), calculate the speed of an object with a redshift of 5 percent using both the relativistic and the nonrelativistic equation.

APPENDIXES

Relativity

IN THIS APPENDIX, WE will study Einstein's theories of relativity and their application to astronomy. In many ways Einstein's theories do not fit our common-sense notions. Einstein defined common sense as "the deposit of prejudice laid down in the mind before the age of 18." We should not automatically reject a theory just because it contradicts common sense.

Discussion of relativity relies heavily on thought experiments, or—from the German—*gedanken* experiments. We will describe a number of these; not expecting that they can actually be performed but using them to investigate the logical conclusions of a theory in circumstances that are impossible to achieve in practice.

We will see that general relativity has replaced Newtonian gravitation as the theory best able to fit the observations. However, just as in our everyday life it is sometimes convenient to think in geocentric terms even when we know the heliocentric theory is a better one, it is still permissible to explain most events using Newton's ideas of gravitation. It is only when we consider unusual cases, such as the very strong gravitational fields near massive stellar objects, that the theories of Newton no longer work and Einsteinian relativity must be used.

Einstein developed two theories of relativity. One, the special theory, applies only in cases where there is no acceleration. The other, the general theory, was developed later and involves accelerated objects.

THE SEARCH FOR A STATIONARY REFERENCE FRAME

In Chapter 18, we saw that clusters of galaxies are moving apart, presumably as they continue the expansion that began with the big bang. It was explained that we must not consider the location of the big bang as a point in the universe from which the expansion is taking place, for at the time of the big bang the material of the explosion made up the entire universe. This idea tends to leave us unsettled. We intuitively feel that there must be something that forms the framework of the universe; something that exists independent of the galaxies and into which the galaxies expand as they get farther apart. We sense that some position must be at rest; perhaps the point where the big bang occurred. As we will see, however, this common-sense idea is incorrect.

To examine the progression of ideas concerning motion and space, we will begin with a search for the framework that is truly at rest. The search begins with the physics of Isaac Newton.

Newtonian Relativity

Although the word *relativity* is normally associated with Einstein, some of the first ideas of relative motion date back to Galileo and Newton. We will take a quick look at ***Newtonian relativity*** before proceeding to study the giant step taken by Einstein.

Newtonian relativity: The principle that states that all positions and velocities are stated with respect to a specific frame of reference.

Suppose you are riding along in a car on a long, straight, smooth highway. The cruise control is set at 60 miles per hour and your speed remains constant.

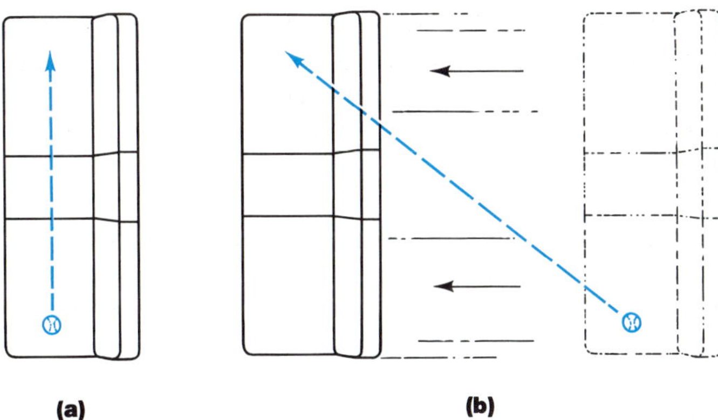

Figure A–1. *In a stationary car, a ball tossed across the back seat simply moves across (a) but in a moving car it is moving forward as it moves across the car (b).*

Children in the back seat entertain themselves by tossing a ball back and forth. Does it matter to them that the car is moving? Must they toss the ball differently than they do when the car is sitting still in a parking lot? I hope that your experience tells you that there is no difference between the way the ball responds in a moving car from the way it acts in a stationary car. Yet, as Figure A–1 shows, when viewed from outside the car in the two situations, the ball has a much different path. Viewed from outside the moving car it not only has a speed across the back seat but is also moving 60 miles per hour along with the car.

To discuss such situations, we talk about the ***reference frame*** of the people involved. The two children throwing the ball in the moving car are in a reference frame that is moving at 60 miles per hour with respect to the Earth. The principle of Newtonian relativity generalizes the experience of the ball-throwers:

Reference frame: The location from which a particular motion or other location is measured.

> Any mechanical experiment performed in a nonaccelerating reference frame will produce the same result in any other nonaccelerating reference frame, even if it is moving with respect to the first.

A mechanical experiment is any activity that involves forces and motions of material objects. The throwing of the ball in the car is a mechanical experiment that produces the same result whether the car is moving at a constant speed or is sitting at a red light. It is important to note that a constant speed is required. If the car accelerates while the ball is in the air, the ball seems to the people in the car to act differently. In the reference frame of an accelerating car it takes a path such as shown in Figure A–2, because while the ball is in the air there is no force on it to accelerate it along with the car. It therefore continues at its previous speed and falls behind.

Newtonian relativity is an everyday part of our experience. If the child behind you in the moving car throws a ball at 5 miles per hour toward the back of your head, with what speed does it hit you? The car is moving 60 miles per hour, so the ball is moving 65 miles per hour relative to the ground. This is not the important speed, however. The speed of importance to you is the speed of the ball relative to you, and this is only 5 miles per hour. The point of this is that speeds add just as we would expect them to and one must always keep in mind the frame of reference of a particular motion.

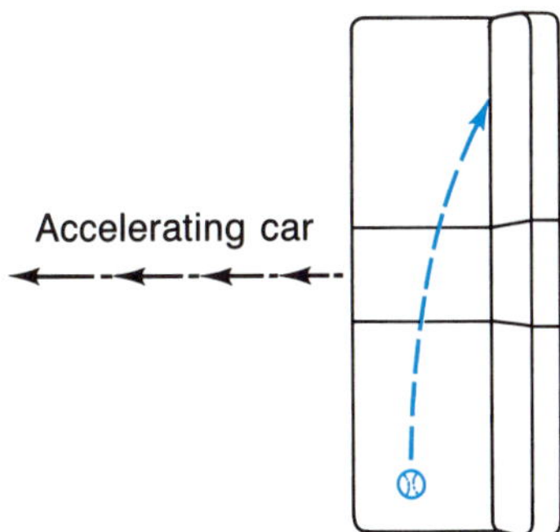

Figure A–2. *If the car accelerates forward as a ball is thrown across the back seat, the ball will seem to move backwards because the car is speeding up while the ball is in the air.*

Newtonian relativity makes it necessary for us to examine what we mean when we say that something is at rest. The terms *at rest* and *in motion* are always given in reference to something else. We can only say that an object is at rest *with respect* to another object, or is in motion *with respect* to another object. We cannot say that an object is truly at rest. When we say that an object is at rest, we usually mean that the object is at rest with respect to the Earth. For example, when we say that a car is not moving, it is understood that the reference frame of our statement is the Earth.

The Earth is not always the natural reference frame, however, for when you tell your little sister in the back seat of the moving car to sit still, you are not telling her to be at rest with respect to the Earth, but rather to the car. The reference frame in this case is the car.

Our intuition may tell us that there must be some object—some reference frame—that is fundamentally at rest in the universe. We know that the Earth is not at rest because it is moving around the Sun. We know that the Sun is not at rest because it is moving around the galactic nucleus. Finally, we know that our galaxy is moving with respect to other galaxies in the local group and that the local group is moving with respect to other clusters of galaxies. The principle of Newtonian relativity leaves us no hope of finding, by the use of a mechanical experiment, a position that is truly at rest. Until early in this century, however, it was thought that electromagnetic radiation can provide us a method of finding the universe's stationary reference frame.

A Swimming Pool Analogy

In order to show how scientists hoped to find the stationary reference frame of the universe, we will analyze a situation involving water waves. Then we will show how the same ideas apply to electromagnetic waves in space.

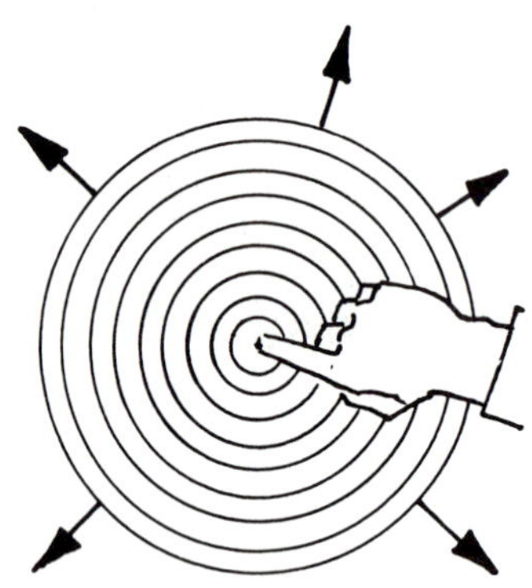

Figure A–3. *If you dip your finger regularly in a pool of liquid, waves will spread evenly in each direction with a speed that depends upon the characteristics of the liquid.*

Think of yourself making small waves in a swimming pool by dipping your hand in and out (Figure A–3). If you do this, the waves will move away from your hand at a speed that depends upon the characteristics of the water. (For example, if the pool were filled with pancake syrup, the waves would move more slowly.) If you move to the other side of the pool and generate the waves there, they will move across the pool in the opposite direction but at the same speed as before.

If, as you dip your hand in and out, you move your hand across the pool, the waves will again travel at the same speed. Your hand does not push the waves forward but merely disturbs the water. The waves are a consequence of the nature of the water surface. (This idea is at the heart of the explanation of the Doppler effect. You might review this in Chapter 11.) In the language of the last section, we would say that the waves have a fixed speed relative to the reference frame of the water.

Now let us consider a gedanken experiment. Suppose that there are tiny extraterrestrials in a spacecraft just above the surface of the water as you make your waves. They know that their craft is not accelerating, but they wish to know whether it is moving across the surface of the water or is stationary with respect to the water. We will allow them to obtain this information, however, only by observing the speed of water waves. That is, they cannot determine

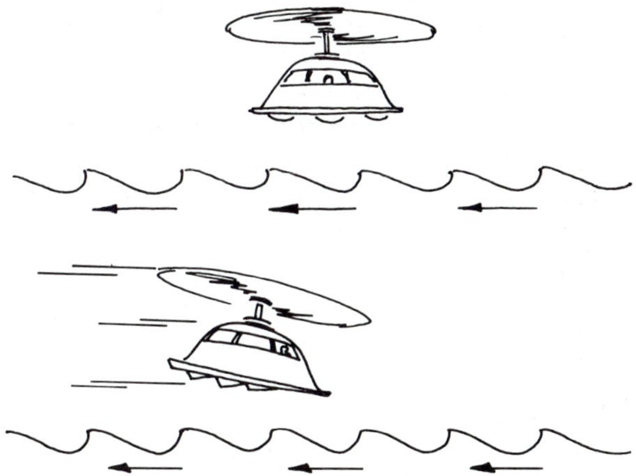

Figure A–4. *The extraterrestrials may be stationary over the water as the waves move under them (a) or they may be moving across the water (b). The only way we allow them to determine their motion relative to the water is by observing the speed of the waves.*

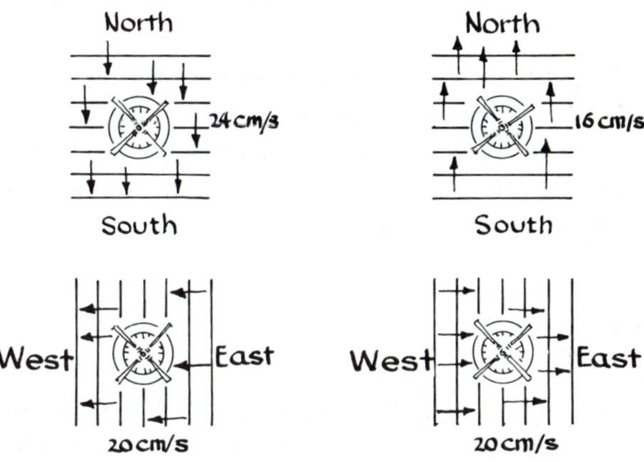

Figure A–5. *These are four overhead views of the alien spacecraft above the waves, showing the speeds of waves coming from various directions.*

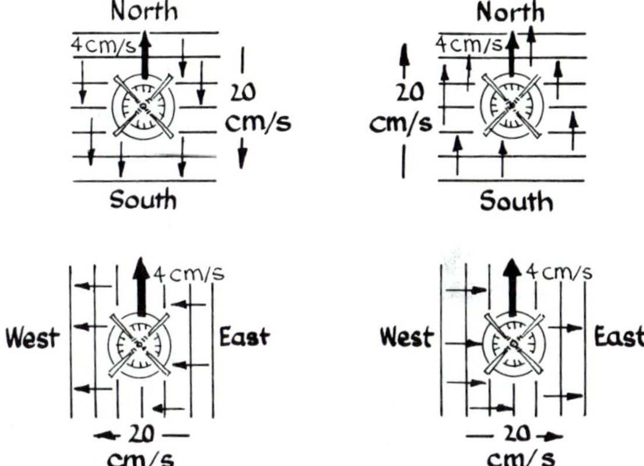

Figure A–6. *As seen from the reference frame of the water, the waves all travel at 20 cm/s and the spacecraft is moving northward at 4 cm/s.*

whether the water itself is moving relative to them but only whether the waves are. Figure A–4 should clarify this distinction.

The extraterrestrials observe waves passing under them, sometimes in one direction and sometimes in another, and they measure the speeds of the various waves. They find that waves from the north pass under them with a speed of 24 cm/s and waves from the south go at 16 cm/s. From both the east and the west, waves pass at the same speed, 20 cm/s. Figure A–5 reviews the situation, showing these speeds. What is the speed and direction of the little spacecraft relative to the water? The only way for the extraterrestrials to explain the observed wave speeds is to conclude that the spacecraft has no motion toward the east or west but is moving northward with a speed of 4 cm/s. The beings in the spaceship conclude that the waves move through the water at 20 cm/s, so that their motion adds to the speed of waves from the north and subtracts from the speed of waves from the south. Figure A–6 shows the situation from the reference frame of the water.

Aether, the Universe's Reference Frame

We on spaceship Earth wish to determine the direction and speed of our motion through the universe. Newtonian relativity tells us that no mechanical experiment can be used for this determination but physicists of the late nineteenth century devised a plan to use light instead. In this way, they hoped to be able to determine the stationary reference frame of the universe in the same way that the imaginary extraterrestrials determined the stationary reference frame of the water. We will follow their logic and examine their results.

One of the first questions that arises concerning the nature of light is what it travels in. Water waves require a water surface for their travel. Sound waves likewise require a medium; normally the air. (Sound also travels in other materials, as is obvious if someone in a distant part of a concrete building hammers on the wall.) Waves in a coiled spring travel along that spring. The question arises as to what is the medium for light waves. Although no other waves in our experience do so, light waves travel in a vacuum. Yet it seems that from the very nature of a wave, there must be something to *wave*; to move back and forth as the wave passes. A vacuum, by its definition, contains no atoms. Thus the thing that carries light waves apparently is not made of atoms. It must be very strange material, indeed, for the planets of the solar system move at great speeds in their orbits and do not seem to be hindered at all by this material. The medium hypothesized in the last century to carry light was given the name ***aether***. The aether was expected to permeate all space, not just the space between objects but also those objects themselves. (Or else, how could light travel through such things as glass and water?)

Aether: The material that was once hypothesized to fill all space and serve as a medium for the propagation of light.

If light travels through the aether and comes to us from the most distant objects in the universe, it was concluded that the aether permeates the entire universe and it seemed logical to conclude that the aether must be the substance that is at rest in the universe. If scientists could detect the aether, they could find that which is truly at rest. This would allow them to determine the *real* speed of an object, rather than being restricted to saying that the object is moving at some speed *relative to* another object.

We are using the older spelling of the world aether *rather than the newer* ether *to avoid confusion with the chemical "laughing gas," which is entirely different. (Aether was the Greek personification of the sky; son of Chaos and Darkness.)*

The aether, of course, was expected to be difficult to detect. Material objects pass right through it, so one could not expect to put his hand out and feel it whizzing by. In our analogy of the extraterrestrials flying over the swimming pool, we did not permit them to detect the water itself but only to detect the motion of waves in the water. At the end of the last century, the situation with regards to the aether was thought to be similar to the extraterrestrials and the water: Scientists would not be able to detect the aether directly, but they should be able to detect waves moving through it. Just as the imaginary extraterrestrials determined their speed relative to the water by examining the speed of water waves, physicists hoped to determine the speed of light waves in different directions and thereby detect Earth's motion through the aether.

Paul Hewitt, in Conceptual Physics, *(Little Brown, 1985) gives a clear, easy-to-read description of the apparatus and the experiment.*

In 1883, Albert Michelson and Edward Morley devised a clever experiment to detect Earth's motion through the aether. Their experiment was sensitive enough to detect speeds less than the speed of the Earth in its orbit, which is about 30 kilometers/second. (It might seem that detecting a speed of 30 km/s should be easy, but remember that the speed of light is 300,000 km/s, and they were seeking to detect a change of 30 km/s in its speed.) They planned to

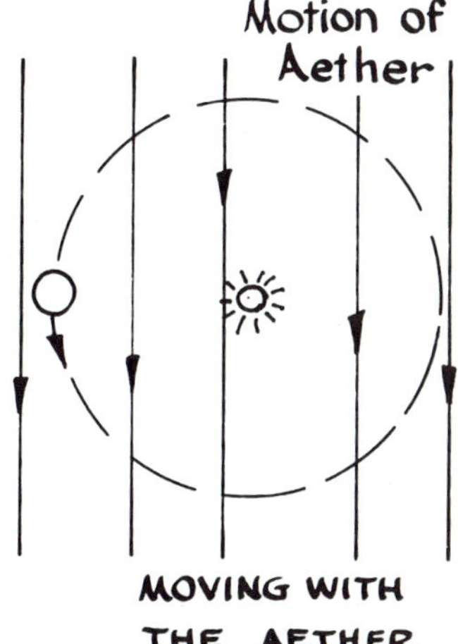

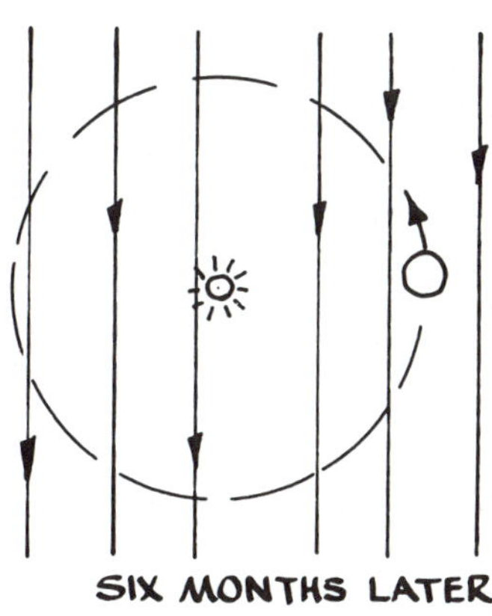

Figure A–7. *If at some time the Earth happens to be moving with the aether, six months later it will be moving against it.*

subtract the Earth's motion from their results in order to obtain the speed of the Sun through the aether.

The results of the ***Michelson-Morley experiment*** were a great surprise, for the researchers could detect absolutely no motion of the Earth through the aether. On the unlikely chance that at the time they were doing the experiment, the Earth happened to be moving just exactly along with the aether (Figure A–7), they tried it six months later when the Earth was moving in the opposite direction through space. Again, the result showed no motion. Later refinements of their apparatus produced the same results. Motion of the Earth through the aether simply could not be detected. The only realistic conclusion was that the aether does not affect the motion of light. If it cannot be detected directly and if it has no effect on the propagation of light, there is no reason to say it exists.

Michelson-Morley experiment: The experiment of Albert Michelson and Edward Morley that sought to measure the motion of light with respect to the aether.

If the aether does not exist, we are left with two major questions: What serves as the medium to carry electromagnetic waves, and does the failure to detect the aether mean that we can find no stationary reference frame for the universe?

SPECIAL RELATIVITY

We saw in Chapter 11 that light is now understood to be a disturbance of the electromagnetic field (which the light itself produces), and that no substance like the aether is required for its propagation. This still leaves us with the question of whether there exists a fundamental, stationary reference frame. If such a reference frame does not exist, we are forced to call into question some very fundamental ideas, including Newtonian relativity. Here's why.

Figure A–8 shows a traveler on a spaceshift approaching Earth. A person on Earth shines a light in the direction of the approaching craft. The Michelson-Morley experiment seems to indicate that the speed of the light detected by the traveler is the same no matter what her speed. Newtonian relativity would have predicted that if the speed of light is measured to be 186,000 mi/s when the

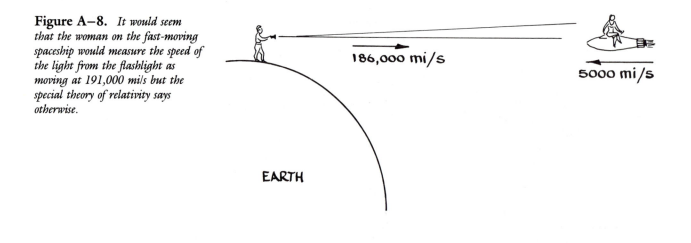

Figure A–8. *It would seem that the woman on the fast-moving spaceship would measure the speed of the light from the flashlight as moving at 191,000 mi/s but the special theory of relativity says otherwise.*

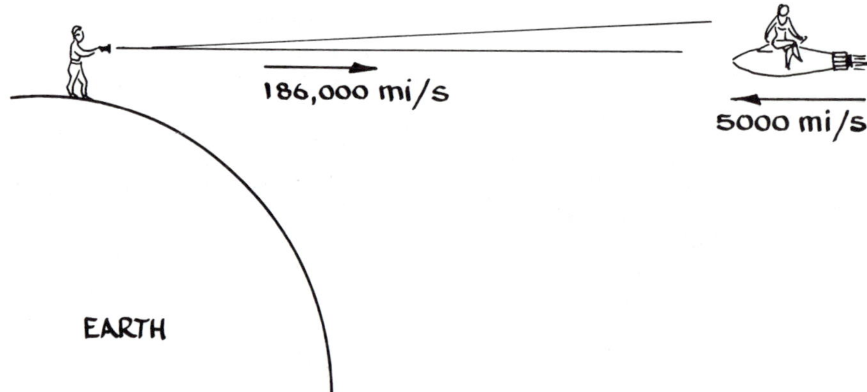

spacecraft is sitting still, then for a spacecraft approaching the light source, the speed of light should be greater by exactly the speed of the craft. That is, if the spaceship is moving at 5,000 mi/s, it should measure the light's speed to be 191,000 mi/s. The Michelson-Morley experiment indicates otherwise, saying that the speed of the observer does not matter and that therefore the person on the spaceship would also measure the speed of light as being 186,000 mi/s. This seems to be nonsense!

In formulating the special theory of relativity, Einstein did not have as his primary motivation the explanation for Michelson and Morley's inability to find evidence for the aether. Nonetheless, his special theory of relativity agreed with the findings of that experiment. We can state the fundamental postulates of his theory in two parts.

1. Any experiment performed in a nonaccelerating reference frame will produce the same result in any other nonaccelerating reference frame.
2. The speed of light is the same for all observers regardless of the motion of the observers and of the motion of the source of the light.

The first of these postulates is an extension of Newtonian relativity to cover *all* experiments, rather than just mechanical experiments. The second is an acceptance of the results of the Michelson-Morley experiment so that it becomes a primary postulate in the theory.

The two postulates of special relativity produce some nonintuitive results when carried to their conclusions. We will show how a couple of strange results arise from the postulates but other results will be described without showing that they must logically follow from the postulates.

The Speed Limit

The theory of relativity accepts the results of the Michelson-Morley experiment at face value. This means that the strange results discussed for the space traveler

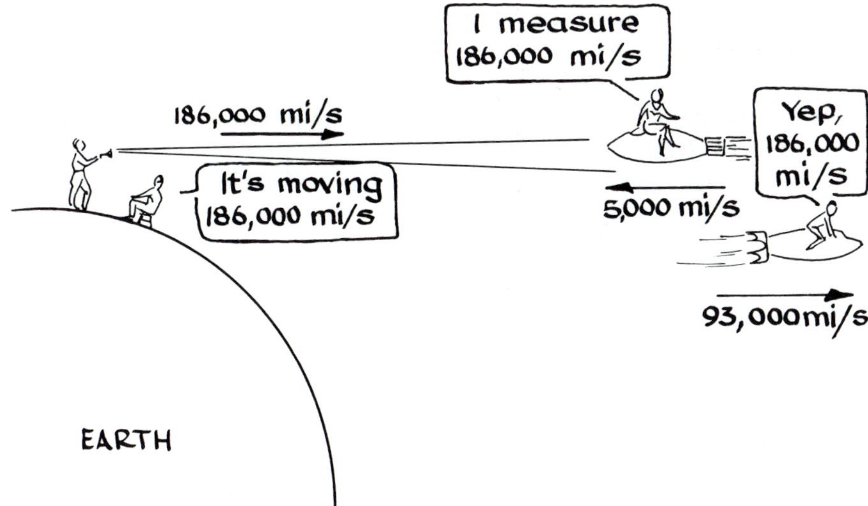

Figure A–9. *All of the people agree on the speed of the light beam despite their speeds relative to one another.*

should indeed occur. Figure A–9 emphasizes that the theory of special relativity tells us that speeds do not add together like common sense tells us they should. In that figure the person on Earth is measuring the speed of the departing light beam as traveling at 186,000 mi/s and the people on the two spacecraft measure exactly the same speed of the beam, despite their motion toward or away from the Earth. The second postulate says that no matter what the motion of the measurer, the speed of light will be the same.

An interesting conclusion from the second postulate is that nature has a speed limit; that *neither matter nor energy can travel faster than the speed of light.*

Perhaps because nature's speed limit seems so contrary to our common-sense ideas, people confuse a statement about the impossibility of going faster than the speed of light with statements concerning technological or physical impossibility. For example, it might be said that it is impossible for a weightlifter to lift 1000 pounds above his head. While this may seem impossible, it is not inconceivable that some giant may someday do so. Or is might be said that it is impossible for a submarine to go below a depth of five miles in the ocean. This may well be a technological impossibility at the present time and in fact it may remain forever a technological impossibility, but it is still not impossible in the same way that it is impossible to exceed the speed of light. It is impossible to exceed the speed of light *by the very nature of things.* While it is true that every scientific law must be considered tentative, the special theory of relativity has passed every experimental test to which it has been put. We will mention some of these tests in the following sections.

THE RELATIVITY OF TIME

One of the most remarkable predictions made as a result of the postulates of special relativity is that the flow of time is not constant; that the rate at which

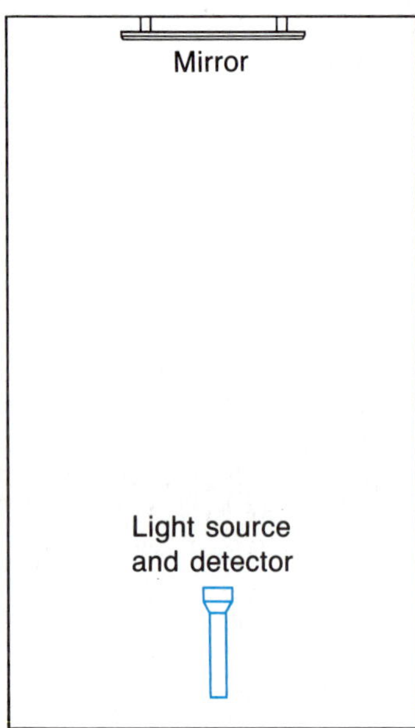

Figure A–10. *The device consists simply of a source of light shining up to a mirror arranged to reflect the light straight back down.*

time passes depends upon the motion of the object that is observed. A thought experiment will show why this occurs.

We will consider a process from two different frames of reference, one the frame of reference of a fast-moving spacecraft and another the Earth's frame. We construct two identical pieces of apparatus for the experiment. Each consists of a source of light and a mirror arranged so that the light leaves its source near the floor, travels up to a mirror on the ceiling, reflects from it, and then travels back down to the source. The "process" is the trip taken by the flash of light. When the light gets back down, the process is finished. Figure A–10 shows the apparatus.

Before one of the devices is put aboard the spacecraft, we check that they are identical by timing how long it takes for light to make its round trip in each. (The time would be fantastically short, but that does not matter in a thought experiment. The important point is that some amount of time would be required and that it would be the same for both devices.)

We put one device on a spacecraft that will travel past the Earth at a tremendous speed; a speed equal to half the speed of light. As the spaceship passes the Earth, the astronaut aboard it flashes the light in her device, and at the same time, we on Earth flash the light in ours. Both we and the observer in the spaceship can see both devices. Do both of the light flashes take the same amount of time to complete their trips? No. Things are not that simple.

Consider the apparatus in the spacecraft as seen by each of the experimenters. The astronaut sees the light go straight up and back down. We on Earth see something else when we look at the device on the spacecraft. As Figure A–11 indicates, we see the beam travel a longer path than the spacefarer sees, for to us, the beam travels not just up and back down, but forward as well.

Now according to the second postulate of special relativity, the speed of light is the same to both observers, the one on Earth and the one in the spaceship. Since we on Earth see the light in the spaceship's device traveling farther than the light in our earthly equipment, we must see the light on the spacecraft taking longer to complete its trip than does the light in the device beside us. That is, the observer on Earth concludes that the same process takes longer in the spaceship.

In the discussion thus far, we have acted as if the Earth is stationary and spacecraft is moving. The first postulate, however, tells us that there is no preferred reference frame, and that we can just as well consider the spacecraft as sitting still and the Earth moving. If we look at the experiment from the point of view of the person on the spacecraft, we realize that she will see things in a parallel-but-opposite manner. She will see the beam of light in the earth-bound device as traveling a diagonal, longer path, and she will conclude that the process takes longer on Earth than in her spaceship.

Time dilation: The slowing of time of an object moving at a relativistic speed (a speed great enough that Einsteinian relativity must be considered).

In each case, the person standing beside his or her device sees nothing unusual about the behavior of that device. It takes the normal amount of time to complete its cycle, but when that person observes the same process in a device which is moving relative to him or her, the cycle is seen as taking longer. A general statement of ***time dilation*** is that a *process is seen as taking longer in a moving reference frame.*

We chose for our "process" a trip made by a flash of light, but if we see light taking longer to complete its cycle, we will see *every* occurrence in that

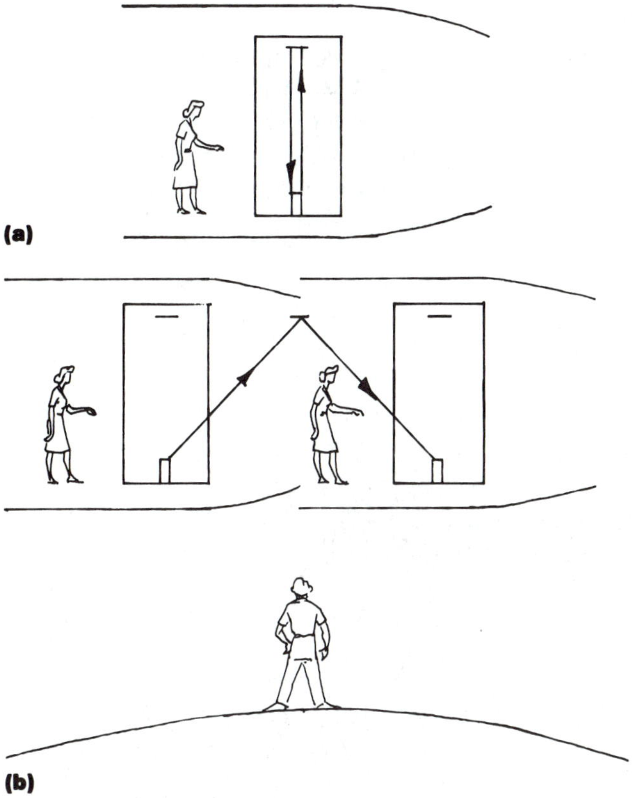

(a)

(b)

Figure A–11. *From the point of view of the spacecraft, the light beam just travels straight up and back down (a) but when seen from Earth, the beam takes a longer path (b).*

reference frame taking longer. This includes the time between ticks of a clock, the growth of hair on a person's head, and the length of a person's life. Not just a light-clock, but *time itself* slows down in a moving reference frame as seen by a stationary observer. (We say *stationary* here only in the sense that the observer can consider himself stationary compared to the moving reference frame.) An alternative statement of time dilation is that *moving clocks run slow.*

Time dilation has been verified experimentally a number of times, including an experiment in which extremely accurate clocks were flown around the world. The principle seems strange to us only because it is not important at everyday speeds and is therefore not observed in everyday life. It is an accepted theory in the scientific world.

The Equation of Time Dilation

Suppose a person (considered stationary) is observing the same process in his own reference frame and in a moving reference frame. The process might be anything: the motion of a light beam to the ceiling and back, the beat of a heart, or the lifetime of a person. The equation which expresses time dilation is as follows:

$$\Delta t = \frac{\Delta t_0}{\sqrt{1 - v^2/c^2}}$$

The symbol Δ simply means that a time interval is involved (rather than a time of day, for example).

where

Δt_0 = the time required for the process as seen by a person in the reference frame of the process

Δt = the time required for the process as seen by a person moving relative to the process

v = the relative speed of the two reference frames

c = the speed of light.

An example will clarify the meaning of the equation.

EXAMPLE An observer on Earth watches the news program "60 Minutes" on the TV of a spaceship that is traveling by at half the speed of light. How long does he see the show lasting on the TV of the spaceship?

Solution The show takes one hour as seen by someone on the spaceship, so "Δt_0" is 1 hour. The velocity v is half the speed of light. When we insert the velocity into the equation, we will substitute $.5c$ instead of putting in a numerical value. This will simplify the calculation.

$$\begin{aligned}\Delta t &= \frac{\Delta t_0}{\sqrt{1 - v^2/c^2}} \\ &= \frac{1 \text{ hr}}{\sqrt{1 - (.5c)^2/c^2}} \\ &= \frac{1 \text{ hr}}{\sqrt{1 - (.5)^2}} \\ &= 1.15 \text{ hr}\end{aligned}$$

Thus, the show takes one hour and nine minutes as seen by the person on Earth. (From the point of view of the astronaut in the spaceship, the show lasts an hour but that person observes the same show on the earthbound TV to last 1.15 hours.)

Try One Yourself. Suppose the spaceship is traveling at 90 percent the speed of light with respect to the Earth. An observer on Earth is able to see the second hand on the watch of a person on the spaceship. How long does it take for the watch hand to tick out one second as seen by the person on Earth? (The tick takes one second when measured by a person holding it.)

Table A–1 Time Dilation

Speed of Clock (km/s)	Fraction of Light Speed	Time Dilation
15,000	.05	1.001
30,000	.10	1.005
75,000	.25	1.033
150,000	.50	1.15
225,000	.75	1.51
270,000	.90	2.29
285,000	.95	3.20
297,000	.99	7.09
298,500	.995	10.01

(Example: A process that takes one second in the observer's reference frame would be seen to take 2.29 seconds in a reference frame traveling 90 percent the speed of light.)

Table A–2 Length Contraction

Fraction of Light Speed	Measured Length
0.00	1.0000
.10	.995
.25	.968
.50	.866
.75	.661
.90	.436
.95	.312
.99	.141
.995	.100

(A one-meter object moving at 90 percent the speed of light is measured by a stationary observer to be only .436 meter long.)

Table A–1 shows the results of a number of such calculations, and illustrates that time dilation is unimportant until high speeds are reached, and that it becomes very important as one nears the speed of light. The answer to the "Try One Yourself" can be found in the table.

Our fastest spacecraft have not begun to approach the speeds discussed here. Pioneer and Voyager achieved perhaps 0.0001 times the velocity of light.

CHANGES IN LENGTH AND MASS

The special theory of relativity also predicts changes in the length and mass of an object as its speed approaches that of light. We will show the results of these predictions without showing that they logically follow from the postulates.

As an object approaches the speed of light, its length as measured by a stationary observer shortens in the direction of motion (Figure A–12). Table A–2 shows the length for various speeds. Relativistic ***length contraction*** requires astronomers to use a more complicated formula for calculating Doppler shifts for quasars and other objects moving at significant fractions of the speed of light. The relativistic Doppler shift was discussed in a Theme Box in the last chapter.

Length contraction: The decrease in length (along the direction of motion) of an object moving at relativistic speed.

If length and time are affected by speeds near that of light, we might not be surprised that mass is also affected. Table A–3 shows that the mass of an object increases as its speed increases and that the mass approaches infinity as its speed approaches the speed of light. Remembering the definition of mass—the resistance to change in velocity—we see why it would be impossible to accelerate a material object to a speed greater than the speed of light. An infinite

(a)

(b)

Figure A–12. *(a) The jogger as seen by someone jogging alongside. (b) The jogger as seen by a person sitting by the jogging path. Length contracts in the direction of motion.*

Table A–3 Mass Increase

Fraction of Light Speed	Mass
0.00	1.0000
.10	1.005
.25	1.03
.50	1.15
.75	1.51
.90	2.29
.95	3.20
.99	7.09
.995	10.01

(An object with a rest mass of 1 kilogram will have a mass of 7.09 kilograms when moving at 99 percent of the speed of light.)

amount of mass would require an infinite amount of force to further increase its speed.

It must be emphasized that it is mass, not size, that increases with speed. Nearing the speed of light, the size of an object actually decreases in the direction of motion. It is the mass—the amount of the object's inertia—that increases.

Like time dilation, relativistic changes of length and mass have been experimentally verified in a number of cases. Most of these involve nuclear particles, which are regularly accelerated to speeds near the speed of light both in nature and in nuclear experimentation.

MASS-ENERGY EQUIVALENCE

Although many of he predictions of the special theory of relativity have helped our understanding of the astronomical world, it was the equivalence of mass and energy that had the earliest, and one of the greatest, effects. Without an understanding of this phenomenon we would not understand the source of the energy of stars.

The electrons moving down the tube of your TV set are moving fast enough that their mass increases a few percent.

The transformation of mass into energy was described in Chapter 12 when we discussed energy production within the Sun. What was ignored in that discussion is that this transformation occurs not only in nuclear reactions but it occurs in every energy change, including regular chemical reactions. When wood burns in a fireplace a small amount of mass goes out of existence, producing the heat and radiation. The mass lost in such a chemical reaction is only about one part in a billion compared to almost one part in a thousand lost in hydrogen fusion. Because so little mass is lost in chemical reactions, the effect is often ignored.

CLOSE UP

The Muon Experiment

ONE CONFIRMATION OF SPECIAL relativity involves a particle that is produced when cosmic rays collide with atoms high in the atmosphere. Measurements of the radioactivity of the particles—muons—confirm the predictions of relativistic time dilation and length contraction and provide us with an example of these features of the theory. I will describe the results of a simple muon experiment.

The half-life of muons is observed to be 1.5 microseconds when they are at rest with respect to an observer. The half-life means that if a person has, say, 1000 muons at one instant of time, then 1.5 microseconds later he will have only 500 left, for the others have changed to other subatomic particles.

Muons can be detected both in the atmosphere and at the Earth's surface. Because their radioactivity has the result that some of them change to other particles, we expect to find fewer at the surface than higher up. Let us consider some muons approaching Earth's surface at 2.9×10^8 meters/second (97 percent of the speed of light). Suppose that 1000 of these muons are observed at a height of 435 meters. Their high speed will get them to the surface in 1.5 microseconds, which corresponds to the half-life of the muons. Thus we would expect to find 500 of the original muons reaching the Earth.

When the experiment is carried out, however, it is found that many more than 500 of them reach the surface (about 840, actually). The reason? The high speed of the muons caused their "clocks" to run slower than ours. While 1.5 microseconds went by on Earth, only 0.38 microsecond passed for the muons. Thus, fewer than half of them disappeared by radioactive decay. The number of muons missing from the original 1000 corresponds exactly to the figure predicted by the theory of special relativity.

Now let's assume that instead of observing the high-speed muons from Earth, we ride along with them. From this point of view, we will not observe any change in the rate of passage of time. Instead, as we look down toward the surface, we see a difference in our *distance* from it. Instead of being 435 meters above the surface, we find this length shortened to about 111 meters. The time required to cover this distance is just 0.38 microsecond (the same time interval that resulted from the analysis above). So, if 1000 muons start out, we are not surprised that more than 500 remain at the Earth's surface.

Notice that a person on Earth attributes the number of remaining muons to time dilation but a person riding with the muons attributes it to length contraction. But even though they differ concerning which relativity effect occurs, the end result is the same whether the observer is on Earth or traveling with the muons and the theory of special relativity is confirmed.

The equivalence of mass and energy was dramatically verified by an experiment done by Enrico Fermi under the football stands at the University of Chicago in 1942 (Figure A–13). The experiment led not only to the atomic bomb but to the peaceful use of nuclear energy. Its importance to us here, however, is that it served as just one more confirmation of Einstein's theory. Every single test of the predictions made by the theory of special relativity has agreed with those predictions. Special relativity is firmly entrenched as an accepted theory.

Figure A–13. *A bronze plaque at Chicago's Stagg Field commemorates the first experimental nuclear chain reaction.*

GENERAL RELATIVITY

The general theory of relativity expands the special theory to include accelerated frames of reference and it changes fundamentally the way we look at the phenomenon of gravity. The theory begins with a statement of the equivalence of gravity and acceleration.

Imagine a girl in an enclosed room in a spaceship far from Earth or any other gravitational influence. If the spaceship is not accelerating, the girl will feel weightless and will float freely about the room. If the spaceship accelerates in a direction toward the ceiling, however, she will feel pushed to the floor. Suppose the acceleration of the spaceship is equal to 9.8 m/s^2. In that case, she will feel the floor pushing up on her feet with a force exactly equal to her weight on Earth, since 9.8 m/s^2 is the acceleration of gravity here on Earth (Figure A–14). Is there any way that the girl can tell whether she is in an accelerating spacecraft far from the Earth or is sitting still on the Earth's surface?

Figure A–14. *The scale will read the same whether it is stationary on Earth or is in deep space accelerating at 9.8 m/s^2.*

Suppose she tries to answer this question by dropping her book and measuring its acceleration toward the floor. She determines the acceleration to be 9.8 m/s². We know that on Earth things fall at this acceleration, but in the accelerating spacecraft the book would act in exactly the same manner; the reason being that the floor now accelerates upward toward the book (Figure A–15). Dropping the book will not let her determine whether she is on Earth or not. In fact, the ***principle of equivalence*** of Einstein's general theory of relativity tells us that there is no experiment whatsoever she can do to distinguish between the two conditions. What we think of as the force of gravity is indistinguishable from an acceleration in the direction we call upward.

Principle of equivalence: The statement that effects of the force of gravity are indistinguishable from those of acceleration.

The principle of equivalence might be just an interesting case of similar observations if it did not lead to a prediction for the behavior of light that is different from previous ideas.

Light and the Equivalence Principle

The girl is now in a spacecraft on the surface of the Earth, not accelerating. The spacecraft has a window on one side and through the window a beam of light enters parallel to the floor. The principle of equivalence leads to a prediction that the beam will bend downward as it crosses the room. To see why this is so, imagine that the room is not on Earth but is in deep space accelerating upward at 9.8 m/s². A quick burst of light enters the room as shown in Figure A–16(a). As the room accelerates upward, gaining speed all the time, the light takes the path shown in parts (b) and (c) of the figure. Notice that from our point of view, in the unaccelerated frame of the page in the book, the light continues in a straight line. From the point of view of the girl in the accelerating craft, however, the light will have bent downwards, as shown in Figure A–17.

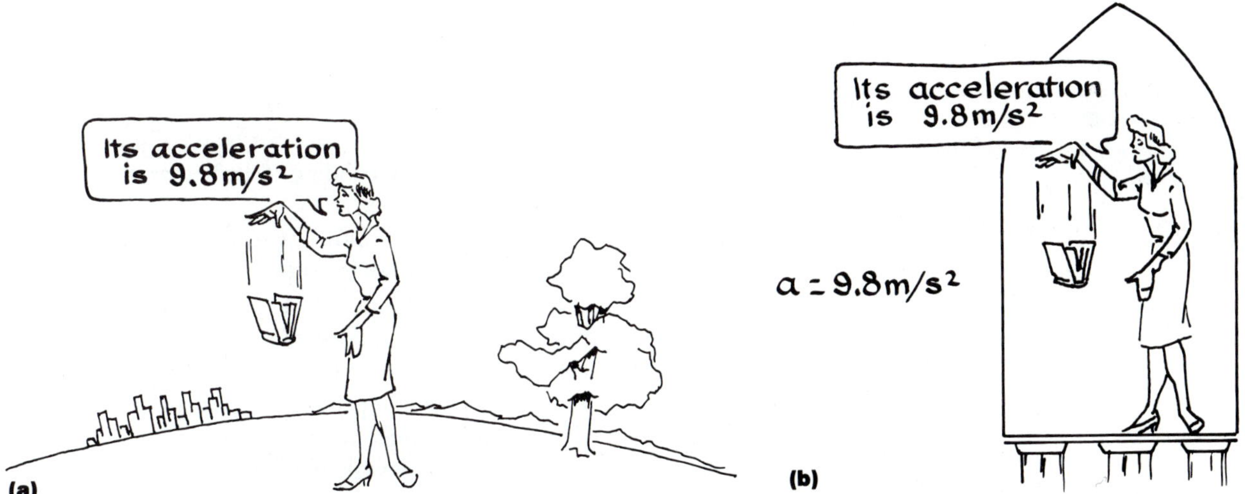

Figure A–15. *Whether on Earth or on a spacecraft far from Earth accelerating at 9.8 m/s², the book will "fall" at 9.8 m/s².*

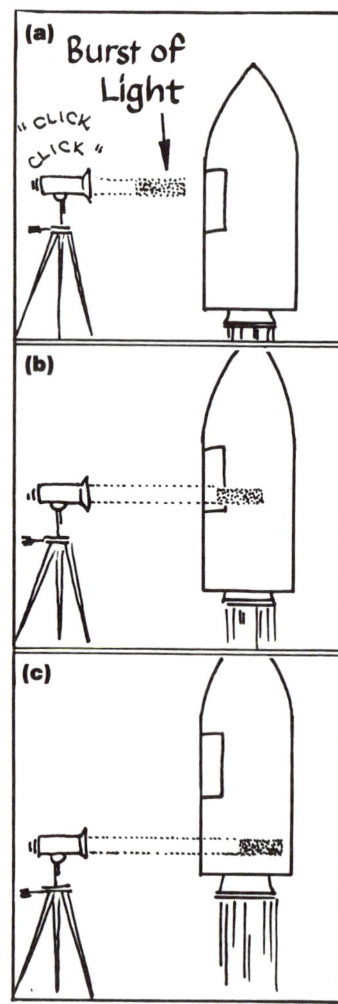

Figure A–16. *A burst of light entering the spaceship as it leaves Earth (a) continues in a straight line but since the spaceship is accelerating upward, the light hits the floor (c).*

Figure A–17. *As seen by a person in the spacecraft, the light beam bends downward as if falling to the floor. The acceleration of the beam will be measured to be exactly equal to the acceleration of the ship.*

The principle of equivalence says that the same bending should happen when the spacecraft is stationary on the surface of the Earth (Figure A–18). In this case, however, the bending will appear to be due to gravity.

The drawings that show the bending of light exaggerate the amount of bending, of course, for in the time required for a beam of light to cross a room, an acceleration of 9.8 m/s^2 would not result in enough curvature of the light's path to make it measurable. If the girl's room were in an extremely strong gravitational field, perhaps on the surface of a neutron star where the acceleration of gravity is 10 billion times greater than on Earth, the bending would be appreciable.

If the principle of equivalence is valid, light should bend in the presence of a massive object. This prediction was made by Einstein's theory in 1907 but the predicted amount of bending near the Earth was very small and no experimental check of the prediction was done until 1919, during a solar eclipse. When the Sun is totally eclipsed by the Moon, stars can be seen in the sky and this provides an opportunity to observe the bending of light that originated at a distant star as the light passes near the Sun.

Suppose a total eclipse occurs while the Sun—as seen from Earth—is between two bright stars. Figure A–19(a) shows the stars as they normally appear. During the eclipse, shown in (b) of the figure, light from the stars must pass near the Sun before reaching Earth. When it does, the theory predicts that the light will be bent and will therefore make the stars appear slightly farther apart as shown in part (c). In practice, the bending was predicted to be very little; only 1.75 seconds of arc. This would produce very little change in the apparent position of a star close to the Sun, much less than indicated in the figure. The eclipse of 1919 did produce results in agreement with the general theory, however, and it provided the first experimental confirmation of the theory. Since then, similar measurements have been taken during other eclipses; always confirming the predictions of Einstein's theory.

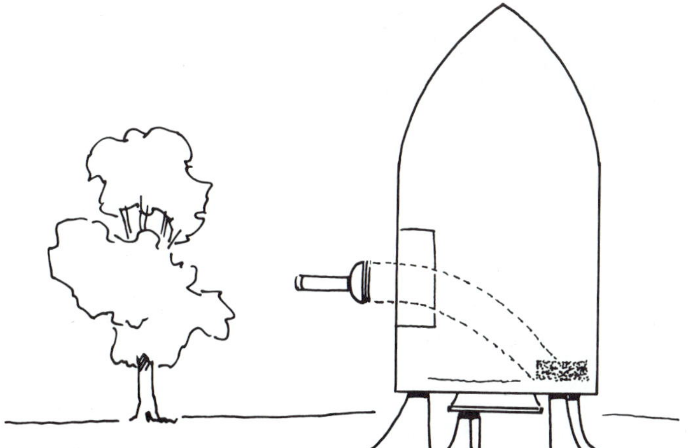

Figure A–18. *Since gravitation is indistinguishable from acceleration, the beam should "fall" due to gravitation.*

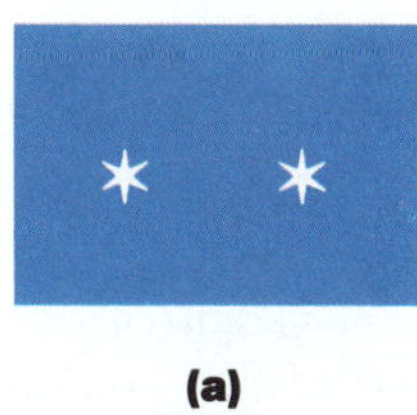

(a)

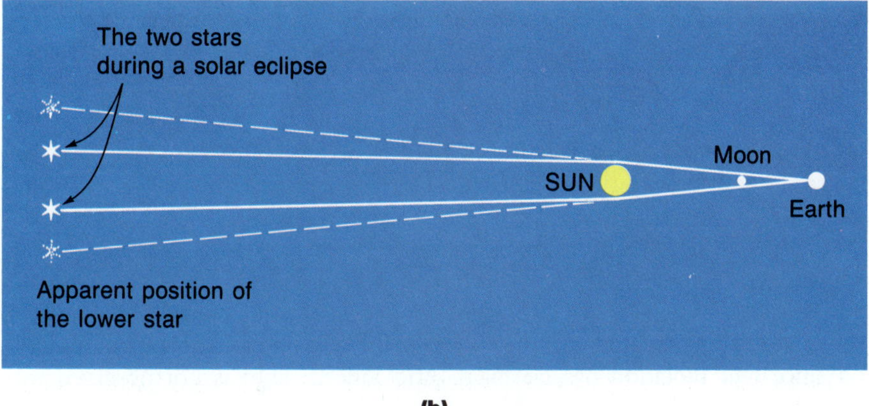

(b)

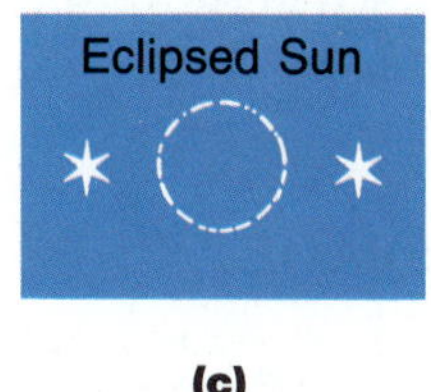

(c)

Figure A–19. *Light from the two stars is bent as it passes near the Sun, causing the stars to appear farther apart.*

HISTORICAL NOTE

Albert Einstein

ONE OF THE GREATEST theoretical physicists of all time was considered a slow learner as a child. The young Albert Einstein found the formal, disciplinary schools of Germany at the end of the last century intimidating and boring and he dropped out before completing high school. He studied at home, however, and learned to play the violin well enough that—although he played it only for his own enjoyment and relaxation—he became an accomplished violinist. His studies of geometry and science led him to conclude, at age 12, that the Bible is not literally true. That shock implanted in him a deep distrust of authority of any kind; a distrust that he carried with him throughout life.

On his second try, Einstein was granted admission to the Swiss Federal Institute of Technology in Zurich. At the university he often failed to attend class, preferring instead to study on his own; reading the classical works of theoretical physics. He was granted a Ph.D. in 1900 but it was two years before he found regular work—as a patent examiner in the Swiss Patent Office.

Einstein was similar to Isaac Newton in his lack of success in formal school and another similarity appears in the timing of their scientific work. Like Newton, Einstein's most revolutionary work was done during a very few years when he was in his early twenties. His work at the patent office was not particularly demanding and in 1905, Einstein published four important papers in the prestigious German physics journal *Annalen der Physik.* Although the one for which he is best known is the one concerning special relativity, he was awarded the Nobel Prize for another, concerning the photoelectric effect of light. Many physicists quickly recognized the importance of his work, but it was not until 1909 that Einstein was given a full-time academic position at the University of Zurich.

In 1903 Einstein married his college sweetheart, Mileva Maritsch, and they had two sons. When World War I began, Einstein had a position in Germany, but his wife and children were vacationing in Switzerland. They were unable to return to Berlin and the forced separation resulted in a divorce in 1919. Later that year, Einstein married his cousin Elsa.

In 1916, Einstein published his general theory of relativity but his relativity theories were slow to gain acceptance due to the lack of experimental verification.

Newton's theory predicts no gravitational effect on electromagnetic radiation since that radiation has no mass. And indeed, light is not observed to respond to gravity in our everyday world. It is only when we consider very strong gravitational fields that the bending of light is observed and since Newton's theory had never been checked in such strong fields, no one realized that it makes incorrect predictions in those cases.

THE CURVATURE OF SPACE

We will return to the special theory of relativity for a moment, because application of that theory to certain circumstances can produce conflicts that seem

In 1919, however, the Royal Society of London announced that its scientific expedition to observe the solar eclipse of that year had verified Einstein's prediction of the amount of bending of starlight near the Sun. The international acclaim that followed changed Einstein's life, for he was suddenly considered a genius. He began to travel more and more, giving lectures over the world.

Although Einstein continued his scientific work until he died in 1955, his fame allowed him to exert influence in world affairs. The fact that he was a Jew, along with his criticism of the political situation in Germany, made it fortunate that he was visiting California when Hitler assumed power in 1933. He renounced his German citizenship and never returned to his home country. The next year he became an American citizen.

Einstein spent the last decades of his life working on grand unified theories. He wrote in 1951 that "the fascinating magic of this work will continue to my last breath." He was unsuccessful in his attempts to find a link between electromagnetic forces and other forces; a link we may be finding today.

It is ironic that Einstein's name is so closely linked to the atomic bomb. It is true that his theories predicted that mass could be converted to energy. Nevertheless, Einstein was an avowed pacifist and he worked untiringly to prevent war; seeing it as the ultimate scourge of humanity. He argued that the establishment of a world government was the only permanent solution.

Einstein disliked fame and the trappings that accompany it. In his travels to the Far East he refused to ride in rickshaws, feeling that to do so would be degrading to the person pulling him along. He preferred to be treated as a common person and he disliked formal attire. He regularly gave important lectures in an open-collar shirt, and it was common for him to be seen near his Princeton home in rumpled clothes, carrying his violin.

Einstein died with two major disappointments: his inability to develop a satisfactory grand unified theory and his lack of success in persuading world leaders to control nuclear armaments.

impossible to solve but that relate to similar problems introduced by the bending of light near a massive object. We will then turn to a problem presented by general relativity before introducing a solution to both problems.

Imagine a large cylindrical room such as shown in Figure A–20. Suppose the boy shown in that room measures the distance around the walls as well as the room's diameter. He then divides the circumference of the room by its diameter. His calculation will result, of course, in 3.14, or π.

Now suppose that the boy is suspended in the center of the room as it rotates around him like a ride in an amusement park, except that this room rotates fast enough that principles of relativity must be applied. If, suspended in the middle of the rotating room, he measures the circumference, he will now obtain a value less than he obtained before, for the walls of the room are moving

Figure A–20. *The boy is in a circular room. He measures the circumference and diameter of the room to determine the ratio between the two numbers. In a nonspinning room, the ratio is 3.14159.*

relative to him and this means that they will undergo length contraction. The diameter of the room, however, will be the same as before, for length contraction occurs only in the direction of motion.

The boy now divides the newly measured circumference by the diameter; the value he obtains is less than π! How can this be? We have a conflict here between a principle of geometry and a prediction of special relativity. Before trying to resolve this conflict, let us look at a similar problem that results from general relativity.

A Triangle Around the Sun

Suppose that we select three points on the orbit of the planet Mercury and draw a triangle around the Sun by connecting these points with straight lines. To do so accurately we must be sure that the lines are straight, so we should first define what we mean by a straight line. One way to define a straight line is to say that light travels in a straight line (unless it is reflected or refracted). We know from general relativity, however, that light bends as it passes near the Sun. Using light to define a straight line would therefore result in a triangle such as that shown in Figure A–21; certainly not a standard triangle. In this triangle, the sum of the internal angles add to more than 180 degrees. We learn in geometry, however, that the internal angles of a triangle must total 180 degrees.

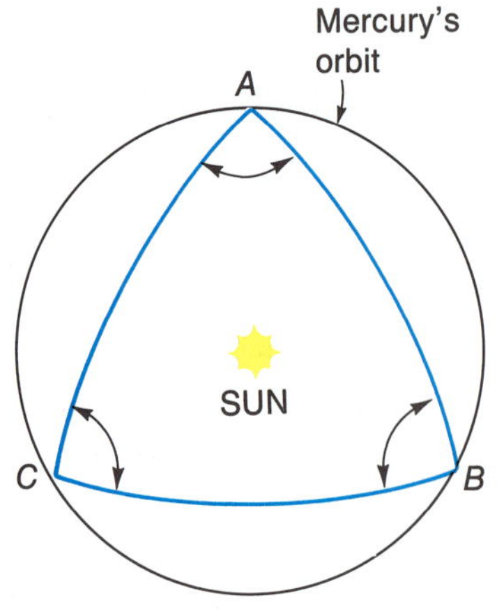

Figure A–21. *If a triangle is drawn in Mercury's orbit using the path of light to determine straight lines, the total of the three angles of the triangle will be greater than 180 degrees.*

Perhaps we should define a straight line as the closest distance between two points. We can determine the closest distance by choosing the path that will get us from one point to the other the quickest. There is an old principle of optics, established long before Einstein's relativity, that light travels from one point to another by the path that takes the least time. Thus this definition of a straight line leaves us with the same distorted triangle. It seems that geometry does not work any better near the Sun than it did in the rotating room.

Space Warp

The solution to both of these conflicts with geometry is to admit to the curvature of space. Space warp is not easy to imagine because the world we experience is a three-dimensional one and we cannot imagine what dimension our world can curve into. We say that our world is one of three dimensions because we use three directions to specify the exact location of something. We can state these directions as north-south, east-west, and up-down. For example, I might state that to get from my office to my bed at home, I must go 4325 feet north, 5843 feet west, and 12 feet up to the second floor. Although other choices might be made for the three coordinates that specify the location of an object relative to me, I must always give three pieces of information in our three-dimensional world.

Imagine a land of two dimensions, "Flatland," on which two-dimensional creatures live. The surface of a desk might represent Flatland. Imagine that these creatures know only the two dimensions of their universe; that they can perceive north-south and east-west but have no conception of up-down. The location of everything in the universe of Flatland can be specified by saying how far it is along an east-west line and how far along a north-south line.

Recall the analogy in the last chapter of the fleas on an expanding balloon. Instead of regular fleas, make the creatures Flatfleas and the balloon Flatland. You might object that the surface of the balloon isn't flat but if the balloon is large enough compared to the size of the Flatfleas, they would not easily perceive its curvature (Figure A–22). If one Flatflea realizes that his universe is curved, how can he explain this idea to his contemporaries? Saying that the universe is curved "downward" would have no meaning, for "up" and "down" are undefined in Flatland. We humans see that the balloon's surface is curved into the third dimension but it would take a great stretch of the Flatfleas's imagination to think of a third dimension for it is not part of their everyday world.

By a similar analogy, we picture the curvature of *our* space. Suppose that the presence of a massive object causes space to be warped. We can picture space near the Sun as being warped analogous to the way the surface of a waterbed is warped by a bowling ball placed in its center. The waterbed's surface would be distorted so that a straight line following the surface would have to

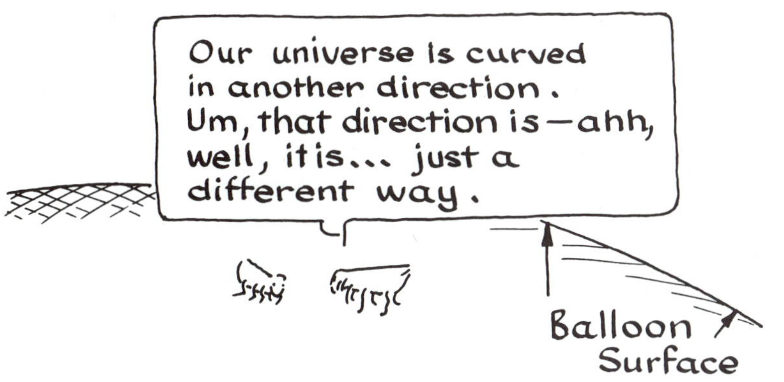

Figure A–22. *If fleas were truly two-dimensional, they would have trouble imagining a curvature of their world. (In fact, fleas are part of our universe and have thickness.)*

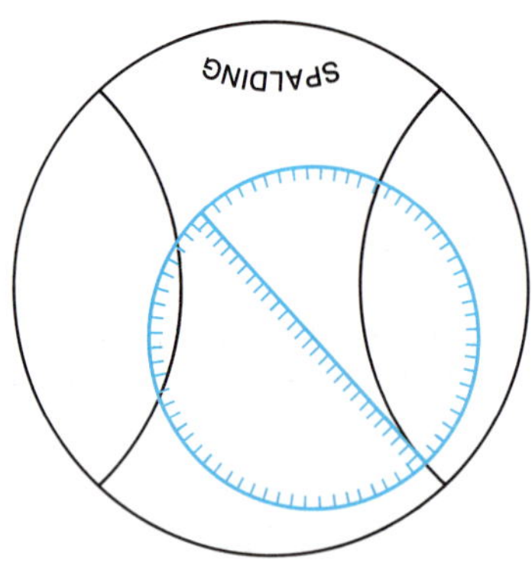

Figure A–23. *If you draw a circle on a basketball, measure its circumference and diameter, and divide the former by the latter you will obtain a value less than 3.14. (To understand this, think of the shape of the diameter line.)*

follow the distortion. A triangle formed from such lines would not obey the rules of Euclidean geometry. In a like manner, as light travels it follows the curvature of space caused by the presence of a massive object.

In this way, we say that space near a black hole has a large amount of curvature. It is warped more than space near a regular star. In fact, we might picture the center of a black hole as distorting the fabric of the waterbed so much that the surface closes back over the object; closing behind it and completely obscuring it. The object is cut off from the rest of the waterbed surface. In this sense, when an object falls into a black hole, it is removed form our universe.

What about the case of the rotating room, then? Imagine someone standing near the wall in this room. That person would feel a great force outward from center. (We discussed such centrifugal forces in a Close Up in Chapter 5; pointing out that although they seem real in the rotating frame of reference, they are imaginary forces.) There are many parallels between the force felt by this person and the Newtonian gravitational force, including the fact that both can be considered a curvature of space. Figure A–23 shows a circle drawn on a basketball. If the circumference of such a circle is divided by its diameter measured along the surface, the result is less than π. Space in the rotating room is warped in this way.

The reason that normal Euclidean geometry does not work in curved space is that it applies only to flat surfaces. The geometry of curved space is different. In this non-Euclidean geometry the internal angles of a triangle need not add to 180 degrees nor does the ratio of a circumference of a circle to its diameter equal π.

THE CORRESPONDENCE PRINCIPLE

Correspondence principle: The idea that predictions of a new theory must agree with the theory it replaces in cases where the previous theory had been found to be correct.

There is a general principle in science concerning the replacement of old theories by new ones. The ***correspondence principle*** states that the predictions of the new theory must agree with those of the previous theory where the old one yields correct results. This is entirely reasonable; in the case we are discussing it simply says that general relativity must not disagree with Newton's gravitational theory where the latter theory provided correct results. For example, Newton's theory predicts that the planets orbit the Sun in elliptical paths but that there are minor deviations from perfect ellipses (caused by gravitational attractions to other planets). We have seen that irregularities in Uranus's orbit led to the discovery of Neptune. In fact, the discovery of Neptune after it was predicted by Joseph Leverrier and John Adams (Close Up, Chapter 9) provided a dramatic confirmation of Newton's theory of gravitation. Einstein's theory must also predict elliptical orbits and it must not contradict the confirmed irregularities predicted by Newton's theory.

The general theory of relativity is in accord with the correspondence principle, for it does predict elliptical orbits with the variations that have been observed. The difference is that it views the orbits as being due to planets following their natural paths in a space that is warped by the presence of the Sun.

ASTRONOMICAL APPLICATIONS: THE ORBIT OF MERCURY

In 1859, fourteen years after predicting the existence of Uranus, Joseph Le-Verrier reported that Mercury's elliptical orbit ***precesses;*** that it does not keep the same orientation in space. Figure A–24 illustrates a greatly exaggerated precession, showing the near point in the orbit (the perihelion) gradually sliding around the Sun. Mercury's orbit precesses very slowly—at the rate of 574 arcseconds per century—which is less than a degree per century. Calculations show that the theory of gravity can account for the basic effect, for gravitational pulls by other planets, particularly Venus and Jupiter, would cause it. The total precession accounted for by these gravitational tugs amounts to 531 arcseconds per century, however, and this is 43 arcseconds short of the observed 574 arcseconds.

Precession (of an elliptical orbit): The change in orientation of the major axis of the elliptical path of an object.

The unaccounted-for 43 arcseconds of precession per century was a mystery to astronomers. One hypothesis held that there is another planet in the inner solar system that is responsible for the extra precession. This planet was given the name Vulcan and extensive searches for it were carried out but it was never found.

One of the first problems to which Einstein applied his new general relativity theory was the precession of Mercury's orbit. He found that the theory accounted precisely for the 574 arcseconds of precession. Einstein later wrote that "for a few days, I was beside myself with joyous excitement." Like Newton's theory, general relativity predicts the precession effect caused by the other planets but unlike Newton's theory, it predicts additional precession due to properties of curved space. Thus the Sun itself caused the extra 43 arcseconds of precession. This was the first direct test of conflicting predictions made by the two theories and the test showed Einstein's theory to be the better one.

The quote is from Was Einstein Right? *(Clifford M. Will, Basic Books, New York, 1986). Recommended.*

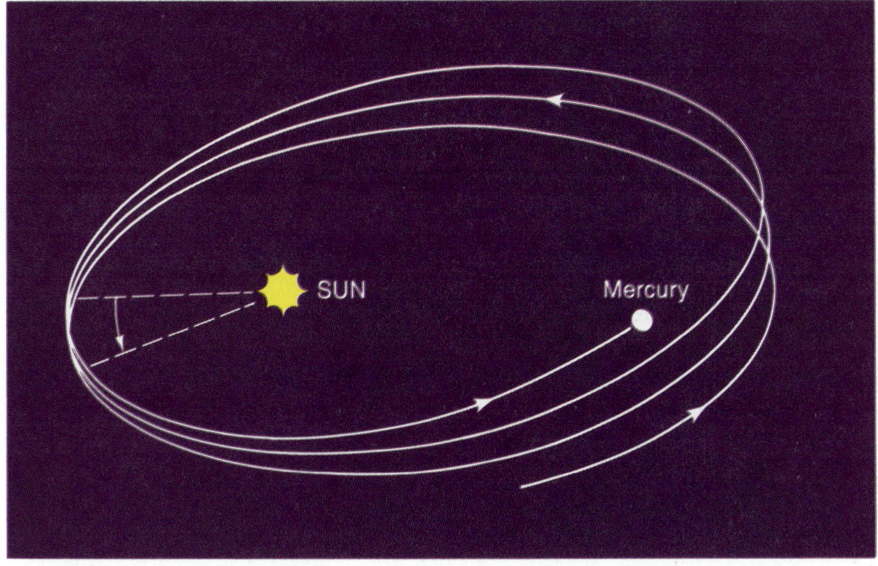

Figure A–24. *This exaggerates the actual precession of the perihelion (the closest point to the Sun of Mercury's orbit.)*

A Binary Pulsar

In 1974, Joe Taylor, a professor at the University of Massachusetts, and Russell Hulse, a graduate student working with him, were using the giant radio telescope at Arecibo to study pulsars. They discovered a very unusual pulsar; one that seemed to be changing its pulsation rate regularly. After nearly a month studying the object they deduced that the changes were due to the Doppler effect as this pulsar revolved around a companion in a binary system. The companion, however, could not be detected. As we saw in previous chapters, binary star systems are useful in determining the masses of stars. Recall, however, that when the spectrum of only one star is observed, we cannot calculate the object's mass. Astronomers were very interested in determining the masses of the objects in this case, however, for it would provide the first measurement of the mass of a pulsar and thereby serve as a check on neutron star/pulsar theory.

The fortunate thing was that the binary system had a short period of revolution (less than eight hours) so that many orbits could be observed over a few weeks. This permitted Hulse and Taylor to determine that the pulsar's orbit was precessing at the rate of four degrees per century; a precession much greater than Mercury's because of the great mass of the pulsar and its nearby companion.

The amount of precession predicted by general relativity depends only on the masses of the two objects in orbit and their distance of separation. If the masses of the objects had been known, the binary pulsar would have served as one more experimental check on general relativity. On the other hand, if the applicability of the theory was assumed, general relativity could be used to calculate the masses of the two objects. Four years of observation and study of the system were required but finally the orbit of the pulsar was determined with sufficient accuracy to allow Hulse and Taylor to use relativity theory to determine both its mass and that of its companion. This was the first instance of the use of general relativity to calculate a stellar property and the first precise determination of the mass of a neutron star.

The Gravitational Redshift and Other Predictions

Another prediction made by general relativity is that electromagnetic radiation should experience a redshift as it leaves a massive object. Looking at the situation from a Newtonian point of view, we can say that light leaving a massive object loses energy because of the force of gravity. When an object is thrown upwards, it loses speed (and therefore kinetic energy) but light cannot lose speed and the loss of energy in this case shows up in a redshift of the light.

Gravitational redshift: The increase in wavelength of radiation leaving an object, as predicted by the theory of general relativity.

The ***gravitational redshift*** was once suspected as the cause of the redshift of the light from quasars but calculations have shown that this cannot be the case. Although the effect is present in the radiation leaving quasars, it is not the primary cause of the great redshift seen in the radiation from these objects.

An even more unusual prediction of general relativity is that the rate at which clocks run will be affected by the presence of a massive object. (This is a different effect than time dilation caused by speed.) In discussing special relativity, we pointed out that one check of that theory involved flying very accurate clocks around the world. In fact, the experiment that checked the special relativity effect did observe the slowing of a clock due to general relativity. The theory of general relativity was verified.

In the last chapter we also mentioned the prediction that galaxies should act as gravitational lenses for objects behind them. Indeed, this effect has been observed for distant quasars.

General relativity, although seeming to contradict our common sense, has been verified in every experimental test to which it has been put. It is a well-established part of today's science.

COSMOLOGICAL SPACE CURVATURE

Recall that in the last chapter we discussed the future of the universe, saying that various possibilities exist. One possibility is that the universe is closed—having enough mass to stop its expansion. If so it will someday begin to collapse; perhaps resulting in an oscillating universe. On the other hand, the universe may be open—with too little mass ever to stop the expansion. A third possibility exists: the flat universe predicted by inflationary theories. The flat universe is the condition of the universe if the amount of mass is just enough to balance the expansion but not enough to cause the universe to fall in on itself. The idea of the curvature of space allows us to see why the three scenarios are called *closed, open,* and *flat.*

If the universe is slowing its expansion fast enough that it will eventually stop and then start collapsing in on itself, the universe can be represented by a spherical two-dimensional surface as shown in Figure A–25(a). Notice that the curvature results in the universe being closed on itself. The amount of curvature determines how quickly the expansion will stop and the collapse will start.

An open universe is represented by a *saddle* surface such as shown in Figure A–25(b). This shape corresponds to a universe that will never stop its expansion. Although a drawing cannot show this, an open universe also has no center and no boundaries.

The third possibility, that of a universe that approaches a stop but will not collapse, is represented by a flat plane as in Figure A–25(c). Although a sphere and a saddle surface might have any number of different curvatures, a plane has no curvature. It is therefore a unique case and it would be unlikely that the universe just happens to have that shape. Inflationary theories, however, predict that a flat universe is the necessary result of conditions shortly after the big bang.

BEYOND RELATIVITY—GUTs

The purpose of science is to unite the myriad of observations of the physical world into some coherence. In this sense, Newton's law of gravity related not only the trajectory of a cannonball and the fall of a piece of chalk but it showed

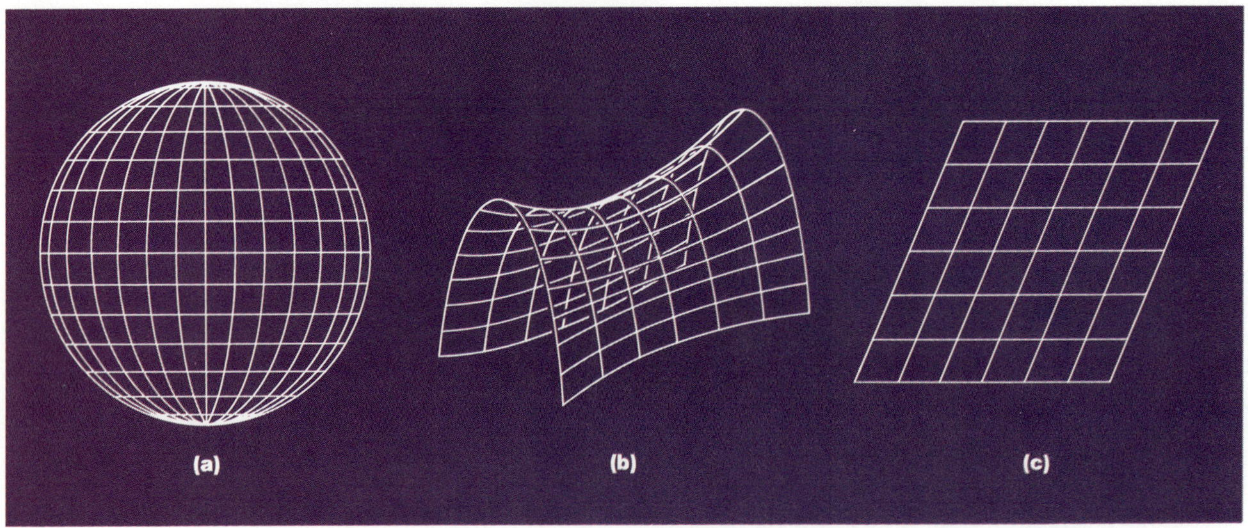

Figure A–25. *(a) A closed universe is represented by a sphere. The surface is limited in area but has no boundary and no center. (b) An open universe is represented by a saddle shape. The shape extends forever in each direction. (c) A regular plane represents the flat universe.*

that the orbiting of the Moon and the planets is also related to the fall of that piece of chalk. Einstein's theories of relativity correct Newtonian gravitation in cases of extreme forces but in addition they show a connection between the phenomenon of electromagnetic radiation and that of graviation, using the vehicle of curved space. They show a relationship that previously was not understood between space, mass, and time.

It has long been known that all material interactions can be explained by four forces. These include gravity and the electromagnetic force as well as two types of nuclear forces, called the strong nuclear force and the weak nuclear force. All interactions between the objects in our world can be explained by these forces. The contact force you experience when you push against a wall, for example, is simply the result of electromagnetic forces between the atoms of your hand and those of the wall.

Scientists are searching for links between the four known forces. There are many similarities in the nature of the force of gravity and the force that exists between electric charges, for example, and the hope is that these two forces could be found to be two aspects of some more universal theory.

Grand Unification Theories (GUTs): Theories that unify the electromagnetic force with nuclear forces.

In recent times, theorists have been able to relate or unify electromagnetic forces to the two types of nuclear force. Such unification schemes are known generally as ***grand unified theories,*** or—less grandiosely—***GUTs***. These theories have basic application in the study of the big bang, for in the big bang matter was so highly compressed that one can no longer study it by consideration of four separate forces. Attempts continue to relate the gravitational force and curved space-time to the other forces. It is hoped that as our knowledge grows about the early universe, we can proceed towards the ultimate unified theory.

CONCLUSION

During the times of Copernicus and then of Kepler and Galileo, people were confused and scandalized by the new theories proposed by these men. People's common sense told them that the Earth was the center of the universe and they were convinced that anything else was nonsense. Today, as children, we are told by confident parents and teachers that the Sun is the center of the solar system and that the Earth is but one planet orbiting it. Later in life we learn of our Galaxy and other galaxies but although these ideas certainly expand our horizons, they probably do not conflict with our common-sense ideas of the universe. The theory of relativity, however, is different. It seems to upset the order of the universe, even though scientists tell us that it adds order. It seems completely nonsensical and does not fit our experience at all.

Will children some day grow up thinking of relativity as the normal, everyday way of viewing things? This may seem far-fetched to us, for we say that relativity does not apply to our everyday lives. That sounds like what people must have said about Galileo's ideas. In our struggle with new ideas, are we really any different from the people of Galileo's time?

RECALL QUESTIONS

1. What is a gedanken experiment?
2. Name Einstein's two theories of relativity and state the different conditions under which each applies.
3. State the principle of Newtonian relativity and give an example of its application.
4. The sign by the road says "Speed limit 65 MPH" but it does not state what the speed is relative to. Explain.
5. What is the aether? Does it exist?
6. What was the Michelson-Morley experiment and what was it designed to show?
7. Suppose a space traveler is moving toward Earth at a speed equal to half the speed of light and shines a light beam toward Earth. With what speed will the light strike the Earth? With what speed does it leave the spaceship?
8. What does the special theory of relativity predict concerning the rate at which time passes as one nears the speed of light.
9. What happens to the size and mass of an object as it increases its speed to relativistic values, as observed by a person not moving with it? How about its size and mass as seen by a person who is moving along with the object?
10. State and explain the principle of equivalence between gravitation and acceleration.
11. Describe a thought experiment that shows that the principle of equivalence leads to the conclusion that the path of light is affected by the presence of a massive object.
12. Describe two experimental checks of the theory of general relativity.
13. Describe a thought experiment that shows that if special relativity is correct, space must be non-Euclidean.
14. What is the correspondence principle? Give an example.
15. Does general relativity explain the precession of Mercury's orbit better than gravitational theory? Explain.
16. Explain how general relativity was used to determine the mass of a binary pulsar.
17. What does a grand unified theory attempt to do?

QUESTIONS TO PONDER

1. If an airplane that can fly at a maximum speed of 200 miles per hour relative to the air is flying into a 40-miles-per-hour headwind, what is its maximum speed relative to the ground?
2. Explain the logic by which Michelson and Morley hoped to determine the Earth's motion with respect to the aether.
3. One postulate of the special theory of relativity simply expanded on Newtonian relativity. Explain.
4. Explain why the aether was thought not to be detectable by such methods of capturing it, measuring its density, or detecting its spectrum.
5. Would a person who spends his life in a spaceship traveling to a distant star at nearly the speed of light live a longer life? Explain.
6. Does the theory of special relativity allow for time travel in the sense of going backwards or forward in time such as we see in science fiction movies? Explain why or why not.
7. If Newton's gravitational theory has been replaced by the general theory of relativity, why do we continue to use it?
8. Discuss the problems involved in drawing a simple triangle around the Sun.
9. Give an example of the correspondence principle in the case of the heliocentric theory replacing the geocentric theory.
10. Venus is the greatest contributor to Mercury's orbital precession because the two are neighbors. The planet causing the next most effect, however, is Jupiter. Hypothesize as to why this occurs.
11. In doing gedanken experiments in relativity, how can it be permissible to consider one person to be stationary?
12. Study Chapter 11 of Lewis Epstein's *Relativity Visualized* (Insight Press, San Francisco, 1983) and summarize his discussion of warped space.
13. One postulate of special relativity states that the speed of light is the same for all observers. Light, however, goes slower in glass than in air. Explain.

CALCULATIONS

1. Suppose I spend five minutes reading the comic page of my newspaper. If I am observed doing this by someone traveling at 75 percent the speed of light relative to me, how long will that person see me reading the comics? Answer this by doing the appropriate calculation and then by referring to Table A–1.
2. In the case described in question 1, suppose the other person turns to the sports pages and spends 20 minutes reading them. How much time will I see her spend reading about sports?
3. Suppose you are moving at 90 percent the speed of light relative to Earth. You hold a pencil so that it is pointed in the direction of your motion and you measure its length to be 20 cm. How long will it appear to a person on Earth? (Use Table A–2).
4. If the pencil in the last question has a mass of 30 grams, what will its mass appear to be to a person on Earth? (Table A–3).

APPENDIX B Units and Constants

Metric Prefixes

nano (n)	$= 10^{-9}$
micro (μ)	$= 10^{-6}$
milli (m)	$= 10^{-3}$
centi (c)	$= 10^{-2}$
kilo (k)	$= 10^{3}$
mega (M)	$= 10^{6}$

Metric-English Conversion

1 kilometer (km)	= 0.6214 miles
1 meter (m)	= 39.37 inches
2.54 centimeter (cm)	= 1 inch
1 kilogram (kg)	= 1000 g (weighs 2.2 pounds)
1 gram (g)	= 0.001 kg (weighs 0.035 oz)

Units of Astronomical Measure

1 astronomical unit (AU)	$= 1.49598 \times 10^{11}$ m
	$= 92.96 \times 10^{6}$ miles
1 light year (LY)	$= 6.324 \times 10^{4}$ AU
	$= 9.461 \times 10^{15}$ m
	$= 5.879 \times 10^{12}$ miles
1 parsec (ps)	$= 2.063 \times 10^{5}$ AU
	$= 3.086 \times 10^{16}$ m
	= 3.262 LY

Temperature

	Kelvin (K)	Celsius (°C)	Fahrenheit (°F)
Absolute zero	0	−273	−459
Freezing point of water	273	0	32
Boiling point of water	373	100	212

$$K = {}^\circ C + 273$$

$$\frac{{}^\circ C}{{}^\circ F - 32} = \frac{5}{9}$$

Constants

Speed of light	$= 2.9979 \times 10^{8}$ m/s
Electron mass	$= 9.1095 \times 10^{-31}$ kg
Proton mass	$= 1.6726 \times 10^{-27}$ kg

APPENDIX C Math Review

Algebra is essentially a way of using symbols to express arithmetical relationships. For example, suppose Keith is 4 inches taller than Chris. Using the first letter of each person's name to express his height in inches, we can write

$$K = C + 4$$

The equation does not tell us how tall either person is; only that whatever Chris's height, Keith is 4 inches taller. If we learn that Chris is 59 inches tall, we can substitute that value for C in the equation:

$$K = 69 + 4$$
$$K = 73$$

Thus we calculate that Keith is 73 inches tall.

Suppose we wish to arrange the original equation so that it is easy to solve for Chris's height. It is logical that if Keith is 4 inches taller than Chris, then Chris is 4 inches shorter than Keith. There is a rigorous algebraic way to derive this, however, based on the following rule:

> **If whatever mathematical operation is performed on one side of an equation is also performed on the other, the equation is still valid.**

In this case, let's start with the original equation and subtract 4 from each side.

$$K = C + 4$$
$$K - 4 = C + 4 - 4$$
$$K - 4 = C$$

Or, writing it another way,

$$C = K - 4$$

This tells us that whatever Keith's height is, Chris is 4 inches shorter.

MULTIPLICATION AND DIVISION

Suppose that Kim has five times as much money as Mandy. We can express this algebraically as

$$K = 5 \times M$$

Normally, though, we omit the multiplication sign in the expression:

$$K = 5M$$

The equation tells us that if we know how much money Mandy has, we multiply by 5 to find out how much Kim has. Now suppose we want to change the expression so that we can solve it easily if we learn how much money Kim has. Using the rule above, we divide both sides of the equation by 5:

$$\frac{K}{5} = \frac{5M}{5}$$

The expression on the right, however, is simply equal to M, because if we multiply something by 5 and then divide by 5 we end up where we started. Thus

$$\frac{K}{5} = M$$

Or

$$M = \frac{K}{5}$$

This tells us that we divide Kim's money by 5 to calculate Mandy's money.

Suppose we take our original expression relating Kim's and Mandy's finances and divide both sides by M.

$$\frac{K}{M} = \frac{5M}{M}$$

This yields

$$\frac{K}{M} = 5$$

which tells us that if we divide the amount of money Kim has by the amount Mandy has, we will obtain the number 5.

COMBINED OPERATIONS

Consider the equation

$$C = \frac{5(F - 32)}{9}$$

(The parentheses tell us that we must multiply their entire contents by 5 and divide the entire contents by 9.) This particular equation shows how to change a Fahrenheit temperature F to a Celsius temperature C. For example, suppose we wish to know the Celsius equivalent of 68 degrees F. We substitute 68 for F.

$$C = \frac{5\ (68 - 32)}{9}$$

Now we do the indicated subtraction:

$$C = \frac{5\ (36)}{9}$$

Next we multiply 36 by 5 and divide by 9 (in any order).

$$C = 20$$

Thus 68 degrees F corresponds to 20 degrees C.

Now let us write the equation in a manner that will allow easy calculation of F when C is known. First, we will multiply each side of the equation by 9:

$$9C = 5\ (F - 32)$$

Then divide each side by 5:

$$\frac{9C}{5} = F - 32$$

(The parentheses have been removed since there is no longer any multiplication or other operation indicated on the right side.) Finally, we add 32 to each side:

$$\frac{9C}{5} + 32 = F$$

Or

$$F = (9/5)C + 32$$

Compare this expression with the original one. Notice that the first one indicated that 32 was to be subtracted from F before multiplying and dividing. In this case, 32 is added only after C has been multiplied by the fraction.

EXPONENTS

The rule stated earlier applies also to raising numbers to powers. For example, suppose we know that one inch is equivalent to 2.54 centimeters and we wish to know how many square centimeters are in a square inch. We will write the relationship between the length units and then square it to get the relationship between the area units:

$$1 \text{ in} = 2.54 \text{ cm}$$
$$(1 \text{ in})^2 = (2.54 \text{ cm})^2$$

Notice here that we have included the units in our expression. Recall that the parentheses tell us that the operation must be performed on everything inside. Thus

$$1^2 \text{ in}^2 = 2.54^2 \text{ cm}^2$$

or

$$1 \text{ in}^2 = 6.45 \text{ cm}^2$$

One square inch, therefore, is equivalent to 6.45 square centimeters.

APPENDIX D Celestial Coordinates

The celestial sphere is an imaginary sphere centered on the Earth. Astronomers specify locations of objects in the sky by a coordinate system on this sphere. The ***equatorial coordinate system*** describes the location of objects by the use of two coordinates, ***declination*** and the ***right ascension***.

The declination of an object on the celestial sphere is its angle north or south of the celestial equator. Angles north of the equator are designated positive and those south are negative. Thus the scale ranges from +90 degrees at the north pole to −90 degrees at the south pole. For example, Sirius (the second brightest star in our sky after the Sun) has a declination of −16°43′.

The right ascension of an object states its angle around the sphere, measuring eastward from the vernal equiniox (the location on the celestial equator where the Sun crosses it moving north.) Instead of expressing the angle in degrees, however, it is stated in hours, minutes, and seconds. These units are similar to units of time, with 24 hours around the entire circle. Sirius has a right ascension of 6h 45m 6s, or—as it is usually written—$6^h\ 45^m.1$, since 6 seconds is 0.1 minute.

APPENDIX E Solar Data

	Value	Ratio to Earth
Diameter	1,392,530 km	109
Mass	1.989×10^{30} kg	330,000
Average Density	1.41 gm/cm^3	0.26
Surface Gravity	270 m/s^2	28
Escape Velocity	617 km/s	55
Surface Temperature	6000°C	
Luminosity	3.9×10^{26} watts	
Tilt of equator to ecliptic	7.25°	
Rotation Period		
Equator	25.38 days	
40° latitude	28.0 days	
80° latitude	36.4 days	

APPENDIX F Physical Properties of the Planets

Planet	Equatorial Diameter		Mass* (Earth = 1)	Density (gm/cm³)	Surface Gravity (Earth = 1)
	km	*Earth = 1*			
Mercury	4,878	0.382	0.055	5.43	0.38
Venus	12,104	0.95	0.82	5.24	0.90
Earth	12,760	1	1	5.52	1
Mars	6,794	0.53	0.107	3.93	0.38
Jupiter	143,000	11.2	318	1.32	2.69
Saturn	120,000	9.4	95	0.70	1.19
Uranus	52,000	4.1	14.5	1.25	0.93
Neptune	48,400	3.8	17.2	1.77	1.22
Pluto	2,260	0.2	0.002	2	0.05

*Mass of Earth = 5.9742×10^{24} kg

Planet	Escape Velocity (km/s)	Equatorial Tilt to Orbital Plane (degrees)	Sidereal Rotation Period (days or hours)
Mercury	4.25	0	58.65 d
Venus	10.36	178	243 d
Earth	11.18	23.44	23.934 h
Mars	5.02	25.20	24.623 h
Jupiter	59.6	3.12	9.842 h
Saturn	35.6	26.73	10.665 h
Uranus	21.1	97.86	16 – 28 h
Neptune	24.6	29.56	18 – 20 h
Pluto	1	118?	6.39 d

APPENDIX G Planetary Orbit Data

Planet	Semimajor Axis		Orbital Period	Orbital Eccentricity	Inclination to Ecliptic (degrees)
	(10^6 km)	(AU)			
Mercury	57.9	0.387	87.97 d	0.2056	7.0
Venus	108.2	0.723	224.7 d	0.0068	3.39
Earth	149.6	1	365.26 d	0.0167	0
Mars	227.9	1.524	1.881 y	0.0934	1.85
Jupiter	778.3	5.203	11.86 y	0.0485	1.3
Saturn	1427	9.539	29.46 y	0.0556	2.49
Uranus	2870	19.19	84.01 y	0.0472	0.77
Neptune	4497	30.06	164.79 y	0.0086	1.77
Pluto	5900	39.44	248.5 y	0.250	17.2

APPENDIX H Satellites

Planet Satellite	Av. Distance (1000 km)	Sidereal Period (days)	Diameter (km)	Mass (10^{21} kg)
Earth				
Moon	384.4	27.322	3476	7,350
Mars				
Phobos	9.38	0.319	28 × 22 × 18	0.0096
Deimos	23.46	1.262	16 × 12 × 10	0.0019
Jupiter				
Metis	127.96	0.295	40?	0.095
Adrastea	128.90	0.298	24 × 20 × 16	0.019
Amalthea	181.3	0.498	270 × 170 × 150	7.2
Thebe	221.9	0.675	110 × 90	0.76
Io	421.6	1.769	3630	89,200
Europa	670.9	3.551	3138	48,700
Ganymede	1,070	7.155	5262	149,000
Callisto	1,880	16.689	4800	108,000
Leda	11,094	238.7	16?	0.0057
Himalia	11,480	250.6	186?	9.5
Lysithea	11,720	259.2	36?	0.076
Elara	11,737	259.7	76?	0.76
Anake	21,200	631	30?	0.038
Carme	22,600	692	40?	0.095
Pasiphae	23,500	735	50?	0.19
Sinope	23,700	758	36?	0.076
Saturn				
Atlas	137.67	0.602	38 × 26	
Prometheus	139.35	0.613	140 × 100 × 80	
Pandora	141.70	0.629	110 × 90 × 80	
Epimetheus	151.42	0.694	140 × 120 × 100	
Janus	151.47	0.695	220 × 200 × 160	
Mimas	185.54	0.942	392	45
Enceladus	238.04	1.370	500	74
Tethys	294.67	1.888	1060	740
Telesto	294.67	1.888	34 × 28 × 26	
Calypso	294.67	1.888	34 × 22 × 22	
(no name)	330?	2.2?	15-20	
Dione	377.42	2.737	1120	1,050
Helene	378.1	2.737	36 × 32 × 30	
(no name)	380?	2.8?	15-20	
(no name)	470?	3.8?	15-20	
Rhea	527.04	4.518	1530	2,500
Titan	1,221.86	15.945	5150	135,000
Hyperion	1,481.1	21.277	410 × 260 × 220	17.1
Iapetus	3,561.3	79.330	1460	1,880
Phoebe	12,954	550.0	220	

Planet Satellite	Av. Distance (1000 km)	Sidereal Period (days)	Diameter (km)	Mass (10^{21} kg)
Uranus				
1986U7	49.7	0.33	40	
1986U8	53.8	0.38	50	
1986U9	59.2	0.43	50	
1986U3	61.8	0.46	60	
1986U6	62.7	0.48	60	
1986U2	64.6	0.49	80	
1986U10	66.1	.051	80	
1986U4	69.9	0.56	60	
1986U5	75.3	0.62	60	
1986U1	86.0	0.76	170	
Miranda	129.4	1.413	480	75
Ariel	191.0	2.520	1330	1,400
Umbriel	266.3	4.144	1110	1,300
Titania	435.9	8.706	1600	3,500
Oberon	583.5	13.463	1630	2,900
Neptune				
Triton	354.3	5.877	3800	130,000
Nereid	5,512	360.2	300	21
Pluto				
Charon	19.7	6.387	1200?	

APPENDIX I The Brightest Stars

Star	Popular Name	Apparent Magnitude	Absolute Magnitude	Distance (LY)	Spectral Type
	Sun	−26.8	4.83	0.000015	G2
α CMa A	Sirius	−1.45	1.4	8.7	A1
α Car	Canopus	−0.72	−3.1	98	F0
α Boo	Arcturus	−0.06	−0.3	36	K2
α Cen	Rigel Kentaurus	0.01	4.4	4.2	G2
α Lyr	Vega	0.04	0.5	26.5	A0
σ Aur	Capella	0.05	−0.6	45	G8
β Ori A	Rigel	0.14	−7.1	900	B8
α CMi A	Procyon	0.37	2.7	11.4	F5
α Ori	Betelgeuse	0.41	−5.6	520	M2
α Eri	Achernar	0.51	−2.3	118	B3
β Cen AB	Hadar	0.63	−5.2	490	B1
α Aql	Altair	0.77	2.2	16.5	A7
α Tau A	Aldebaran	0.86	−0.7	68	K5
α Vir	Spica	0.91	−3.3	220	B1
α Sco A	Antares	0.92	−5.1	520	M1
α PsA	Fomalhaut	1.15	2.0	22.6	A3
β Gem	Pollux	1.16	1.0	35	K0
α Cyg	Deneb	1.26	−7.1	1600	A2
β Cru	(Beta Crucis)	1.28	−4.6	490	B0.5
α Leo A	Regulus	1.36	−0.7	84	B7

APPENDIX J The Nearest Stars

Star	Apparent Magnitude	Absolute Magnitude	Distance (LY)	Spectral Type
Sun	−26.8	4.83	0.000015	G2
Proxima Centauri	11.5	15.5	4.2	M5
α Centauri A	0.01	4.4	4.2	G2
α Centauri B	1.5	5.8	4.2	K5
Barnard's Star	9.5	13.2	5.9	M5
Wolf 359	13.5	16.8	7.6	M6
Lalande 21185	7.5	10.4	8.1	M2
Luyten 726-8A	12.5	15.4	8.2	M5
Sirius A	−1.5	1.4	8.7	A1
Sirius B	7.2	11.5	8.7	white dwarf
Ross 154	10.6	13.3	9.4	M4
Ross 248	12.2	14.8	10.3	M5
ε Eridani	3.7	6.1	10.7	K2
Ross 128	11.1	13.5	10.8	M4
Luyten 789-6	12.2	14.6	11.0	M6
61 Cygni A	5.2	7.6	11.2	K5
61 Cygni B	6.0	8.4	11.2	K7
ε Indi	4.7	7.0	11.2	K5
τ Ceti	3.5	5.7	11.4	G5
Lacaille 9352	7.4	9.6	11.4	M1
Procyon A	0.4	2.7	11.4	F5
Procyon B	10.8	13.1	11.4	white dwarf

APPENDIX K The Constellations

Name	Genitive	Abbreviation	Approximate position		English Meaning
			Right Ascension	*Declination*	
Andromeda	Andromedae	And	01^h	+40	Andromeda*
Antlia	Antliae	Ant	10	−35	Air Pump
Apus	Apodis	Aps	16	−75	Bird of Paradise
Aquarius	Aquarii	Aqr	23	−15	Water Bearer
Aquila	Aquilae	Aql	20	+05	Eagle
Ara	Arae	Ara	17	−55	Altar
Aries	Arietis	Ari	03	+20	Ram
Auriga	Aurigae	Aur	06	+40	Charioteer
Bootes	Bootis	Boo	15	+30	Herdsman
Caelum	Caeli	Cae	05	−40	Chisel
Camelopardus	Camelopardis	Cam	06	−70	Giraffe
Cancer	Cancri	Cnc	09	+20	Crab
Canes Venatici	Canum Venaticorum	CVn	13	+40	Hunting Dogs
Canis Major	Canis Majoris	CMa	07	−20	Big Dog
Canis Minor	Canis Minoris	CMi	08	+05	Little Dog
Capricornus	Capricorni	Cap	21	−20	Sea Goat
Carina	Carinae	Car	09	−60	Keel of Ship
Cassiopeia	Cassiopeiae	Cas	01	+60	Cassiopeia*(Queen of Ethiopia)
Centaurus	Centauri	Cen	13	−50	Centaur*
Cepheus	Cephei	Cep	22	+70	Cepheus*(King of Ethiopia)
Cetus	Ceti	Cet	02	−10	Whale
Chamaeleon	Chamaeleonis	Cha	11	−80	Chameleon
Circinis	Cirini	Cir	15	−60	Compass
Columba	Columbae	Col	06	−35	Dove
Coma Berenices	Comae Berenices	Com	13	+20	Berenice's Hair*
Corona Australis	Coronae Australis	CrA	19	−40	Southern Crown
Corona Borealis	Coronae Borealis	CrB	16	+30	Northern Crown
Corvus	Corvi	Crv	12	−20	Crow
Crater	Crateris	Crt	11	−15	Cup
Crux	Crucis	Cru	12	−60	Southern Cross
Cygnus	Cygni	Cyg	21	+40	Swan (or Northern Cross)

Name	Genitive	Abbreviation	Approximate position		English Meaning
			Right Ascension	*Declination*	
Delphinus	Delphini	Del	21	+10	Dolphin or Porpoise
Dorado	Doradus	Dor	05	−65	Swordfish
Draco	Draconis	Dra	17	+65	Dragon
Equuleus	Equulei	Equ	21	+10	Little Horse
Eridanus	Eridani	Eri	03	−20	River Eridanus*
Fornax	Fornacis	For	03	−30	Furnace
Gemini	Geminorum	Gem	07	+20	Twins
Grus	Gruis	Gru	22	−45	Crane
Hercules	Herculis	Her	17	+30	Hercules*
Horologium	Horologii	Hor	03	−60	Clock
Hydra	Hydrae	Hya	10	−20	Hydra*(water monster)
Hydrus	Hydri	Hyi	02	−75	Sea serpent
Indus	Indi	Ind	21	−55	Indian
Lacerta	Lacertae	Lac	22	+45	Lizard
Leo	Leonis	Leo	11	+15	Lion
Leo Minor	Leonis Minoris	LMi	10	+35	Little Lion
Lepus	Leporis	Lep	06	−20	Hare
Libra	Librae	Lib	15	−15	Scales of Justice
Lupus	Lupi	Lup	15	−45	Wolf
Lynx	Lincis	Lyn	08	+45	Lynx
Lyra	Lyrae	Lyr	19	+40	Harp or Lyre
Mensa	Mensae	Men	05	−80	Table (or mountain)
Microscopium	Microscopii	Mic	21	−35	Microscope
Monoceros	Monocerotis	Mon	07	−05	Unicorn
Musca	Muscae	Mus	12	−70	Fly
Norma	Normae	Nor	16	−50	Carpenter's Level
Octans	Octantis	Oct	22	−85	Octant
Ophiuchus	Ophiuchi	Oph	17	00	Ophiuchus*(serpent bearer)
Orion	Orionis	Ori	05	+05	Orion*(hunter)
Pavo	Pavonis	Pav	20	−65	Peacock
Pegasus	Pegasi	Peg	22	+20	Pegasus*(winged horse)

Name	Genitive	Abbreviation	Approximate position		English Meaning
			Right Ascension	*Declination*	
Perseus	Persei	Per	03	+45	Perseus*
Phoenix	Phoenicis	Phe	01	−50	Phoenix
Pictor	Pictoris	Pic	06	−55	Easel
Pisces	Piscium	Psc	01	+15	Fishes
Piscis Austrinus	Piscis Austrini	PsA	22	−30	Southern Fish
Puppis	Puppis	Pup	08	−40	Stern of Ship
Pyxis	Pyxidis	Pyx	09	−30	Compass of Ship
Reticulum	Reticuli	Ret	04	−60	Net
Sagitta	Sagittae	Sge	20	+10	Arrow
Sagittarius	Sagittarii	Sgr	19	−25	Archer
Scorpius	Scorpii	Sco	17	−40	Scorpion
Sculptor	Sculptoris	Scl	00	−30	Sculptor
Scutum	Scuti	Sct	19	−10	Shield
Serpens	Serpentis	Ser	17	00	Serpent
Sextans	Sextantis	Sex	10	00	Sextant
Taurus	Tauri	Tau	04	+15	Bull
Telescopium	Telescopii	Tel	19	−50	Telescope
Triangulum	Trianguli	Tri	02	+30	Triangle
Triangulum Australe	Trianguli Australi	TrA	16	−65	Southern Triangle
Tucana	Tucanae	Tuc	00	−65	Toucan
Ursa Major	Ursae Majoris	UMa	11	+50	Big Bear
Ursa Minor	Ursae Minoris	UMi	15	+70	Little Bear
Vela	Velorum	Vel	09	−50	Sails of Ship
Virgo	Virginis	Vir	13	00	Virgin
Volans	Volantis	Vol	08	−70	Flying Fish
Volpecula	Vulpeculae	Vul	20	+25	Fox

*These are proper names.

APPENDIX L The Optics of Telescopes

REFRACTION EXPLAINED

The bending of light as it passes into and out of a piece of glass can be explained by the fact that light travels more slowly in glass than it does in air. To see why a change in speed of a light beam crossing boundary at an angle results in its bending, imagine two wheels on an axle rolling down a board as shown in Figure L–1(a). As they roll, they roll onto a carpet, encountering it at an angle as illustrated in Figure L–1(b). As should correspond to your experience, the wheels will roll more slowly on the carpet. Since one wheel strikes the carpet first, that wheel slows down first, and this results in the device changing its direction of motion.

Figure L–2(a) illustrates a light wave encountering an area where it moves more slowly. Like the two wheels, the side of the wave that strikes the boundary first will slow down first and result in the light bending at the boundary as shown in Figures L–2(b) and (c). Likewise, when the wave enters a region

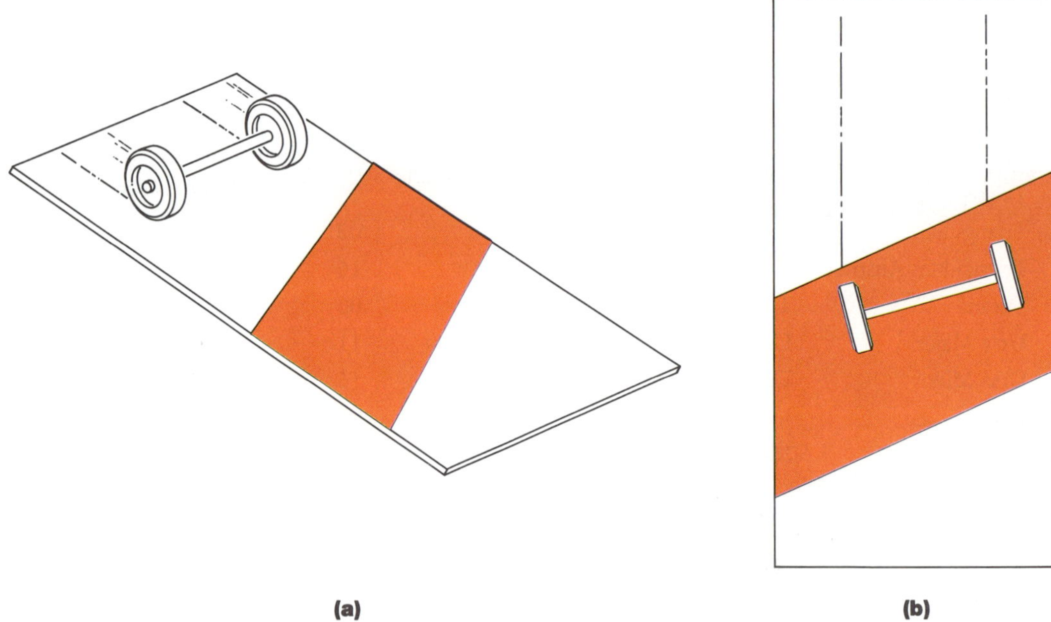

Figure L–1. *(a) Two wheels rolling down a plane. The colored area toward which they are heading is a carpet. (b) An overhead view of the two wheels later. They changed direction as they rolled onto the carpet, because the wheel that hit the carpet first slowed down first.*

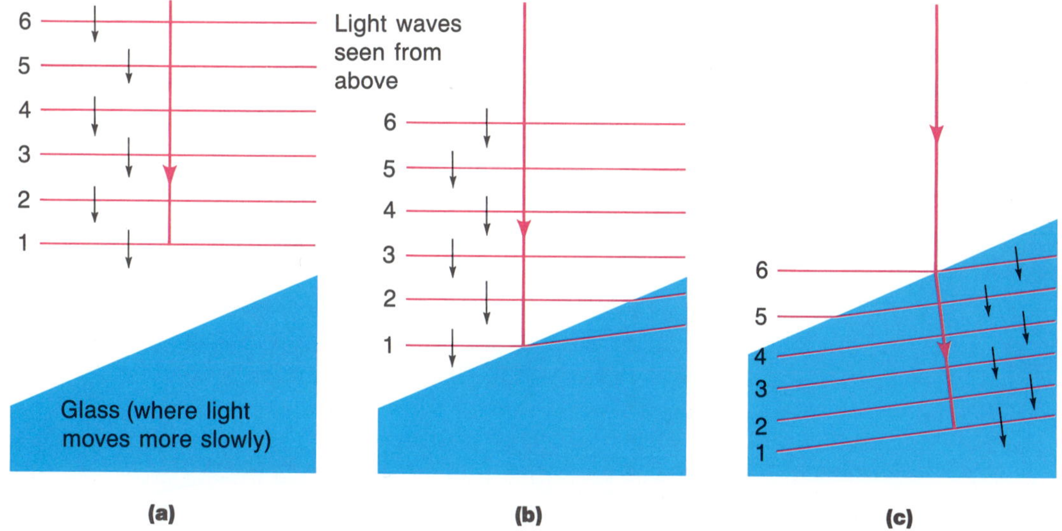

Figure L–2. *(a) This is an overhead view of crests of light waves moving down the page, about to encounter a piece of glass. (b) The edges of the first two waves enter the glass, slowing down as they do so. (c) The slowing of the waves results in a curve in their paths so that they no longer move straight down the page.*

where it moves faster, it bends in the opposite direction. As shown in Figure L–3(a), this results in a light ray experiencing no change in direction due to glass that has parallel sides (such as a window pane).

THE FOCAL LENGTH OF A LENS

Figure L–4(a) is like those of Chapter 13, showing parallel incoming rays of light passing through a lens. The fact that the rays are parallel indicates that

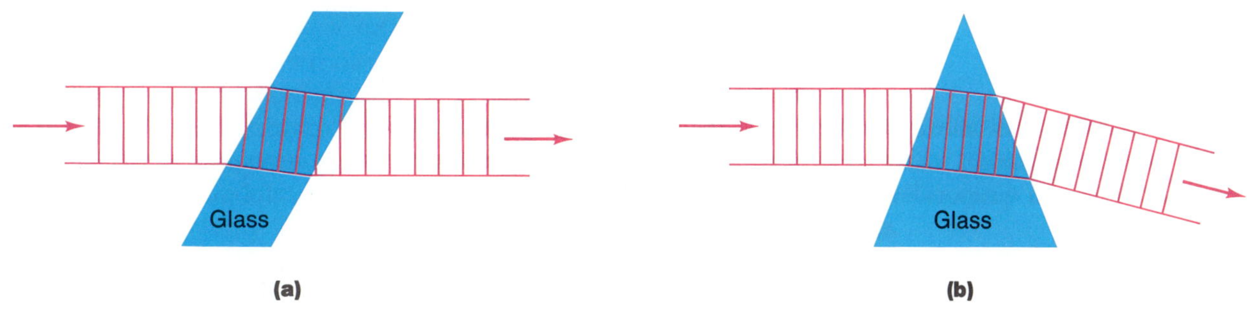

Figure L–3. *(a) Light waves passing through a piece of glass that has parallel sides are deflected from their path upon entering but when they exit and speed up again, they continue in the same direction as before. (b) Light waves going through a prism. We are considering light of a single wavelength here, so that we see no color separation but only a change in direction.*

Figure L–4. *(a) Rays from a very distant object are essentially parallel. The point where they are brought to a focus is defined as the focal point. (b) Rays from a nearby object are spreading out (diverging) when they reach the lens and do not converge at the focal point.*

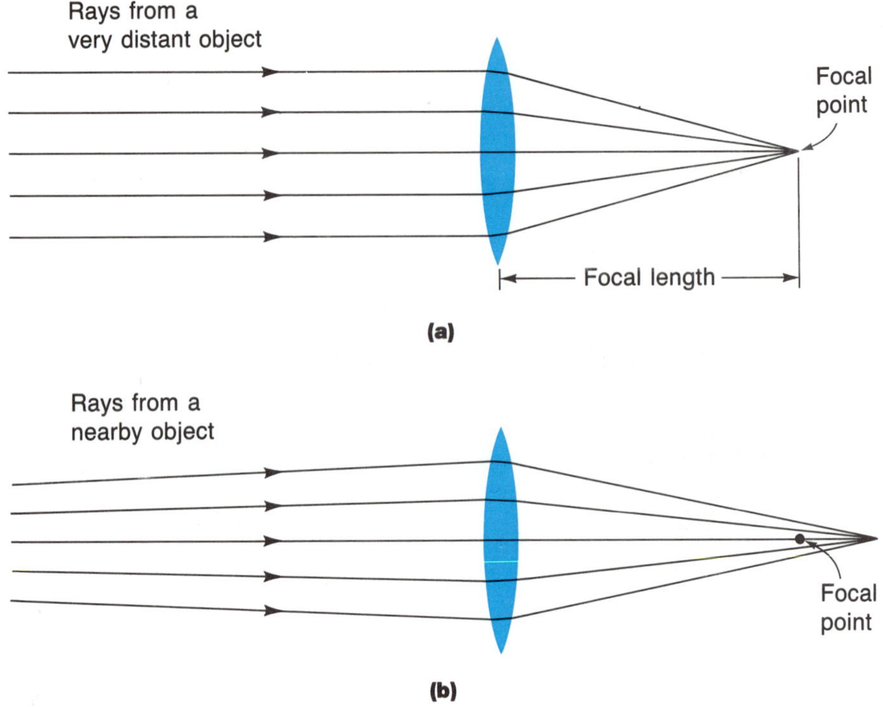

they come from a distant object. If we represent the same lens focusing rays of light from an object that is closer, the drawing would look like Figure L–4(b). Notice the difference in the orientation of the incoming rays of each case. In both cases the lens causes the rays to bend toward one another, but when the object is nearby, the rays come to a focus farther from the lens. (This is why you might have to focus a camera to take pictures at different distances.) The focal point of a lens is that point where light from a very distant point directly in front of the lens comes to a focus, and it is not the same point where light from a nearby point focuses.

When the object is far enough away that we can consider the incoming rays to be parallel to one another, we say that the object is "at infinity." This is why the lens of an adjustable-focus camera is marked with ∞, the symbol for infinity. On the lens of my camera, the longest distance marked is 30 ft. The next mark is ∞. This means that when the object is beyond 30 feet, the rays entering the camera lens are very close to parallel and the object can be considered to be at infinity.

Figure L–5 illustrates how a number of differently shaped lenses bend the light from an infinitely distant object. There are four things worth noticing:

1. Some of the lenses form their images at greater distances than others. Since the object (a house) is at infinity, the distance from the lens to the image in each case is the focal length. Lens B has a shorter focal length than lenses A and C.

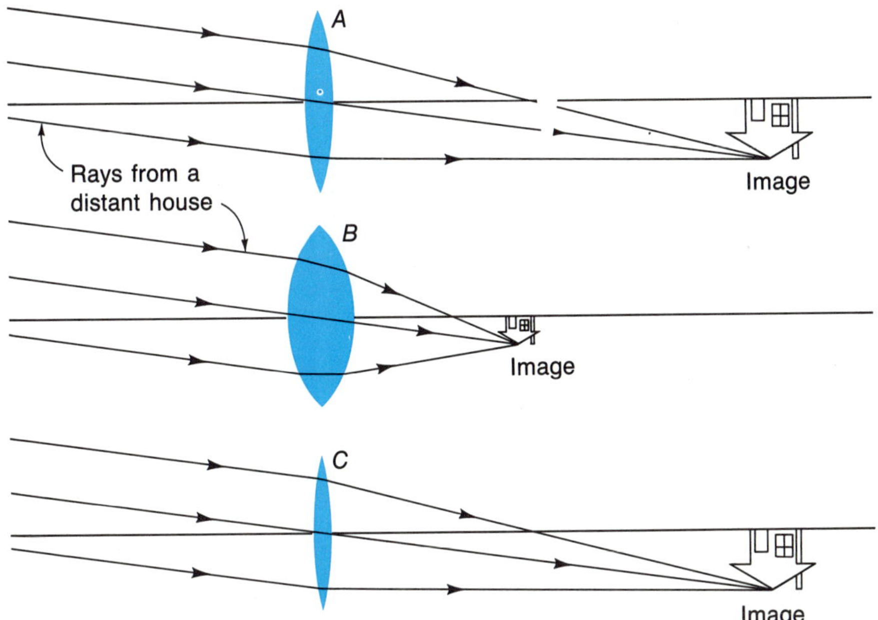

Figure L–5. *Lenses* A *and* B *have the same diameter (height) but the more curved lens has a shorter focal length. Lens* C *has a smaller diameter than* A *and* B *but the same curvature as lens* A *and thus it has the same focal length.*

2. The focal length of a lens is determined by the curvature of the glass. The lens with more curvature has a shorter focal length. (The type of glass is also a factor since different kinds of glass refract light different amounts, but we'll assume that all of the lenses illustrated are made of the same kind of glass.)
3. The diameter of the lens does not determine the focal length. The curvature of the lens is important, not its size.
4. The longer the focal length, the larger the image. This is why a telephoto lens on a camera is so long; in order to form a large image on the film, the lens must have a long focal length.

THE CAMERA

A camera basically is just a lens that brings light to a focus on a sheet of photographic film. It is enclosed in order to keep out unwanted light and has a shutter in front of the film so as to control the amount of time during which starlight (or light from a friend's beaming face) enters. Figure L–6 shows a simple camera.

Notice in the figure that the light from the upper star forms its image below that of the lower star. Indeed, the image on the film in a camera is inverted; upside down.

When the camera is used to take a picture of objects other than stars, the two groups of light rays shown in Figure L–6 can be thought of as coming from two points on the object being photographed. The figure shows light from only two points on the object but for every point on the object, there will

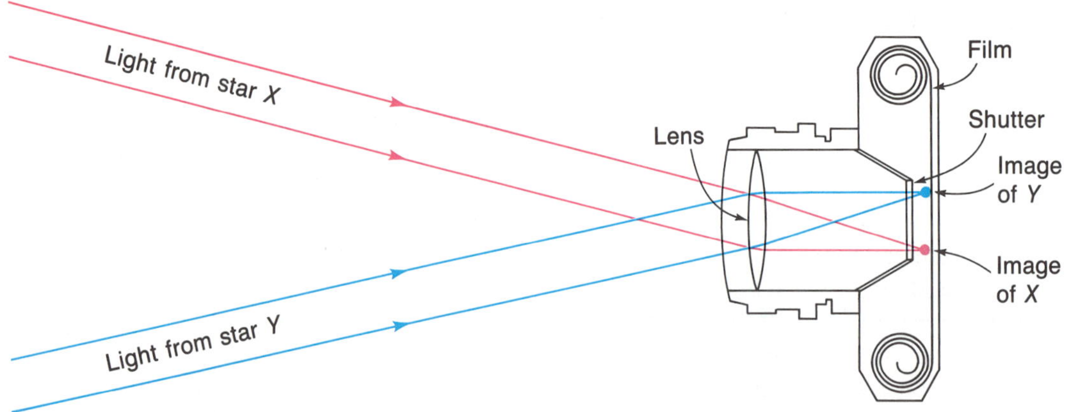

Figure L–6. *Light from the two stars is focused by the lens onto the film at the back of the camera. The shutter must be opened to allow the light to get to the film.*

be a corresponding point on the film and an inverted image of the entire object will be formed.

When we focus a camera, we are adjusting the distance from the lens to the film. This distance must be changed when the distance from the object to the film changes. Figure L–7 shows what occurs when a camera is slightly out of focus. The light from a single point on the object is spread out over more than one point on the image. Some simple cameras have a fixed focus in which the lens-to-film distance cannot be adjusted. These are adjusted for a medium distance and although their small lenses make focusing less critical, they sacrifice clarity in the case of nearby and distant objects.

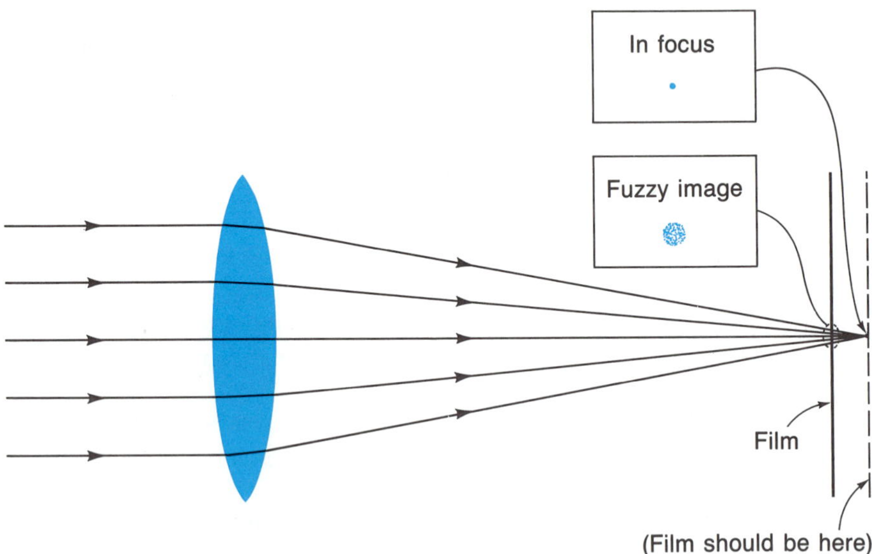

Figure L–7. *If the film is too close to the lens, what should be a single point on the film will be a blurry spot. The same thing happens if the film is too far from the lens.*

LENSES USED AS MAGNIFIERS

If a lens is held very near an object, the lens may not be able to bend the light enough to bring it to a focus. Figure L–8(a) shows this situation. The light from the object does not come to a focus on the other side of the lens. Suppose, though, a person's eye is on that side of the lens, as shown in part (b) of the figure. In this case, the light entering the eye of the person *appears* to have come from a different place than it actually did. The artist has drawn dashed lines backwards along the directions that the rays entered the viewer's eye. Observe that the light that came from the top of the bug appears to have come from higher up; above the bug. And Figure L–8(c) shows that the light that came from the bottom of the bug appears to have come from farther below. The result is that the bug appears larger to the viewer than it actually is. When the object is placed close to the lens so that this situation occurs, we are using the lens as a magnifier; a "magnifying glass."

When a lens is used as a magnifier, the image is not inverted. As should be clear from the drawings, the person sees the bug right side up but larger. Notice that we are using the same type of lens here as we did when we formed an image on a piece of paper or on film in a camera. The difference is that in this case the object has been placed much closer to the lens—closer than the focal point. If the object were farther from the lens, an image would form on the other side of the lens, just as it did in the examples we showed of images of stars and of a house.

The eyepiece of a telescope uses a lens as a magnifier. The image formed by the objective becomes the object for the eyepiece. The objective produces an inverted image and since the eyepiece does not reinvert it, the image seen in an astronomical telescope is inverted. (A telescope designed for terrestrial use has a different optical system.)

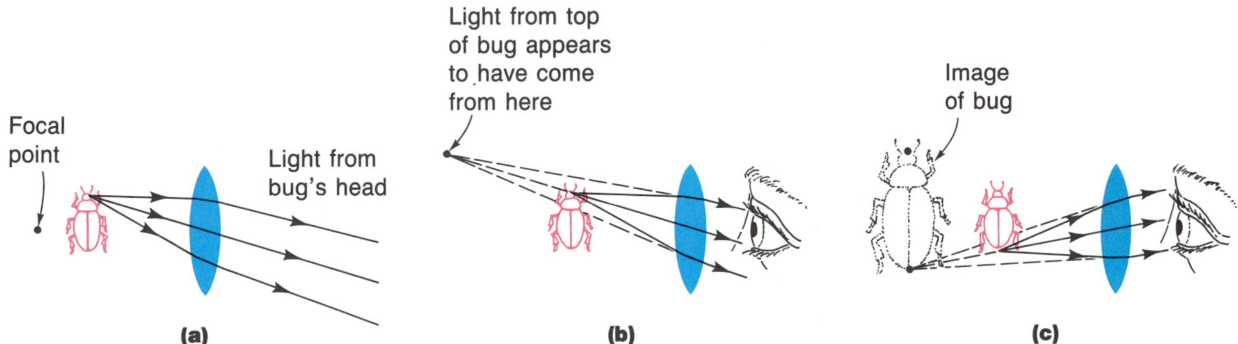

Figure L–8. *In using a lens as a magnifying glass, the object is placed close to the lens. In this case its light is not brought to a focus by the lens. When viewed, however, it appears to have come from an object larger than the actual object.*

Figure L–9. *(a) A spherical mirror does not reflect all rays to the same point. (b) A parabolic shape is the ideal curvature for a mirror but it is more difficult to produce.*

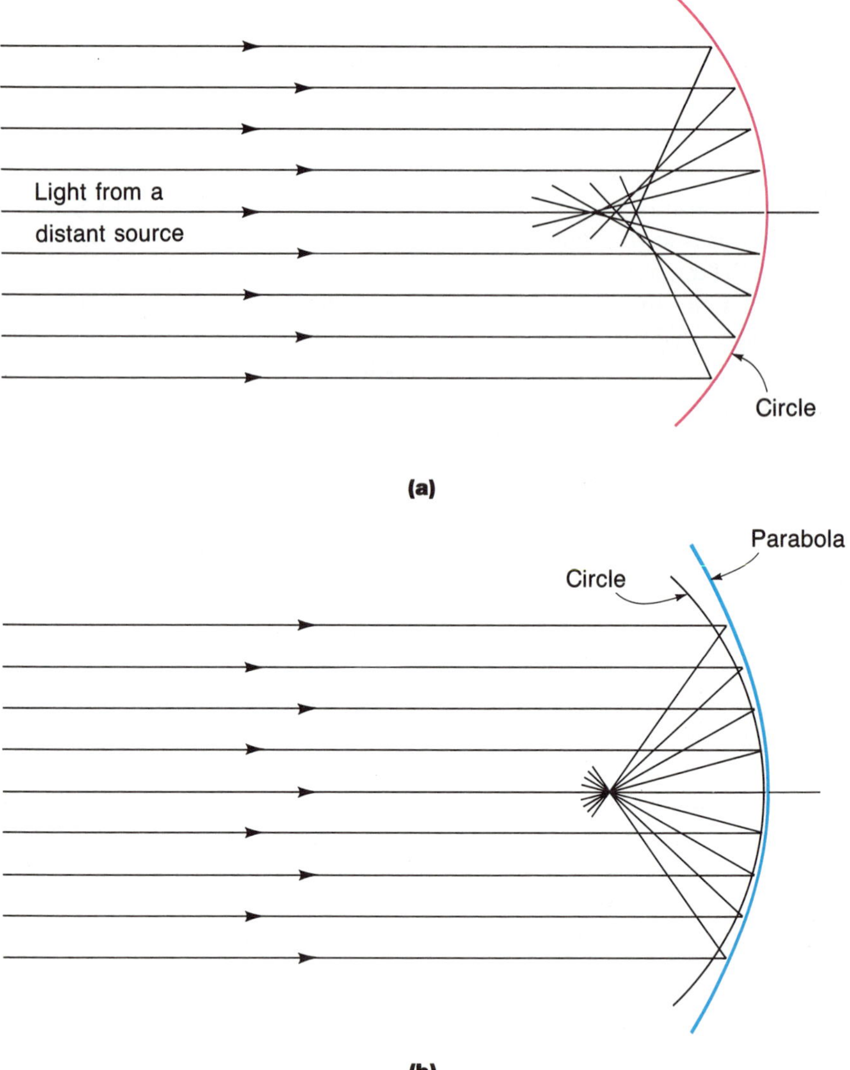

SPHERICAL ABERRATION OF TELESCOPE MIRRORS

The correct shape for the surface of the objective mirror of a reflector is not spherical, although the spherical shape is the one easiest to produce. A spherical surface produces *spherical aberration*, due to the fact that light that hits the outer part of the mirror does not focus at the same place as light hitting the center. The ideal shape for the objective of a reflector is a *paraboloid;* the shape obtained by rotating a parabola around its central axis. Figure L–9 shows spherical aberration as well as how it is eliminated by the use of a parabolic mirror. Large telescopes use parabolic mirrors to eliminate this aberration.

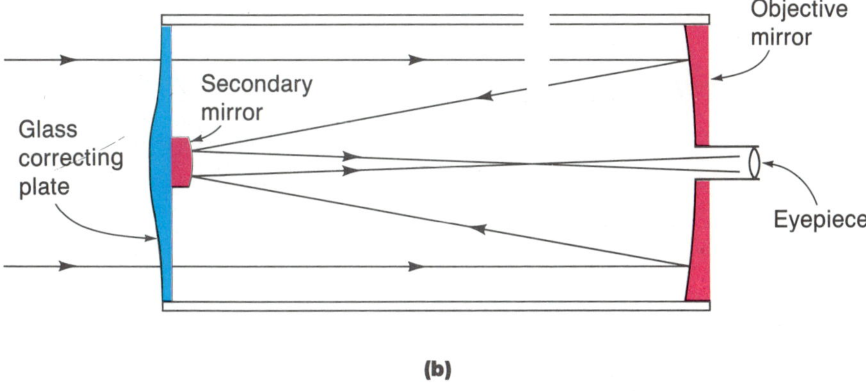

(b)

Figure L–10. *The Schmidt-Cassegrain focal arrangement places a correcting plate on the front of the telescope. This corrects for the spherical objective and also serves as a holder for the secondary mirror. (The curvature of the correcting plate is exaggerated in part (b).)*

THE SCHMIDT-CASSEGRAIN FOCUS

In order to correct for spherical aberration, some small telescopes are constructed with a spherically shaped objective and have a piece of glass across the front of the telescope that corrects for the spherical aberration. This corrector plate has very slight curvature; not enough to cause chromatic aberration. Figures 13–26 and 13–27 show such a Schmidt-Cassegrain telescope; one with a five inch mirror. Figure L–10 shows the optical arrangement. The telescope shown has an effective focal length of 125 centimeters but the length of the main body of the telescope is only about 26 centimeters (about ten inches).

QUESTIONS

1. What is meant when we say that an object is "at infinity?"
2. What is it about a lens that determines its focal length?
3. When a lens forms an image, how is the size of the image related to the focal length of the lens? How is the size of the image related to the diameter of the lens?
4. What determines whether a lens will act as a magnifying glass or form an image on the opposite side of the lens?
5. Why is it not important to an astronomer that the image in a telescope is inverted?
6. How does the eye change in order to focus on objects at different distances? (This is a research question; the answer is not in this text.)
7. Why are refracting telescopes so long?

APPENDIX M

Answers to Selected Questions and Try One Yourself Exercises

Included here are solutions to all Try One Yourself exercises and answers to even-numbered Recall Questions and Calculations. Questions to Ponder are not answered here because they generally do not have definite right or wrong answers.

PROLOGUE

Recall Questions

2. A time exposure of the northern sky has a number of concentric arcs (whose length depends upon the amount of time the film was exposed). A bright, short arc is near the center (caused by Polaris.)
4. The Moon moves westward across the sky. Depending upon its phase, it may be seen to rise in the east and/or set in the west.
6. Isaac Newton developed the law of universal gravitation in the seventeenth century.
8. Composition, size, and mass are mentioned in this chapter. Other discernable attributes of stars include temperature, apparent and absolute luminosity, distance from the Sun, and speed relative to the Sun.
10. Radio telescopes and ultraviolet telescopes are mentioned in this chapter. Other instruments include detectors of radiation from every region of the spectrum.
12. The Galaxy contains about 200 billion stars.

Calculations

2. Answers to the first and third parts are 2.56×10^5 and 5.6×10^{-3}.
4. 5,977,000,000,000,000,000,000,000 kg.

CHAPTER 1

Try One Yourself

$$4.5'' \times \frac{1'}{60''} = 0.075'$$

$$0.075' \times \frac{1^\circ}{60'} = 0.00125^\circ$$

Recall Questions

2. From east to west.
4. A constellation is an arbitrary grouping of stars named for a particular person, place, or thing (often of mythological origin.)
6. They appear to move in a circle around Polaris (or actually, a point very near Polaris.)
8. The east-to-west motion of a planet with respect to the background stars.
10. New, waxing crescent, first quarter, waxing gibbous, full, waning gibbous, third (or last) quarter, and waning crescent.
12. The Earth casts a circular shadow on the Moon during a lunar eclipse.
14. Aries, Pisces, Aquarius, Capricorn, Sagittarius, Scorpius, Libra, Virgo, Leo, Cancer, Gemini, and Taurus. Although the Sun passes through Ophiuchus, it is not considered one of the zodiacal constellations.

16. One of the smaller planetary orbits of the Ptolemaic model, the center of which revolves around the Earth.

18. It is from a Greek word meaning "wanderer." (Thus the Greeks listed the Sun and Moon as planets.)

20. In the east (shortly before sunrise) or the west (shortly after sunset.)

Calculations

2. 240 arcminutes.

CHAPTER 2

Recall Questions

2. Aquinas was a thirteenth-century Christian theologian and philosopher who was later declared a saint by the Roman Catholic Church. He blended the philosophy of Aristotle with Christian theology; accepting the Ptolemaic model as being in accord with the Bible.

4. Because for many centuries the contrary idea had been held and he feared the scorn of others.

6. Retrograde motion of a planet was explained as being due to the relative motion of the Earth and the planet. As the Earth passes an outer planet, that planet appears from Earth to move "backward" among the stars.

8. The aberration of starlight and stellar parallax.

10. Parallax is the apparent motion of an object caused by motion of the observer. To demonstrate it, hold an object in front of you and view it first with one eye and then with another. The object will appear to shift position relative to the background.

12. The apparent change in the direction of a star due to the Earth's orbital motion. It was discovered in 1729.

Calculations

2. 5.20 AU. By the graphical method shown in this chapter you should probably determine it to be between about 4.5 AU and 6 AU.

CHAPTER 3

Try One Yourself

$$\frac{P^2}{d^3} = 1 \text{ year}^2/\text{AU}^3$$

$$\frac{0.615^2}{d^3} = 1$$

$$\frac{0.378}{d^3} = 1$$

$$d^3 = 0.378$$

$$d = 0.723 \text{ AU}$$

Recall Questions

2. The discovery of a "new" star (a supernova).

4. Put two tacks in a board and lay a loop of string loosely around them. Use a pencil or pen to stretch the string taut, and—keeping it taut—trace a curve around the tacks. By using a longer string or putting the tacks closer together, the ellipse will have less eccentricity.

6. An AU is the average distance from Earth to Sun; about 93,000,000 miles or 150,000,000 kilometers.

8. (a) Mountains and valleys on the Moon, (b) more stars than are visible with the naked eye, (c) four moons of Jupiter, (d) sunspots, and (e) Venus's complete cycle of phases.

10. A planet appears as a disk, while a star reveals no size.

12. No. He apparently did not know of it, or at least did not appreciate its value.

14. From slowest to fastest: Saturn, Jupiter, Mars, Earth, Venus, Mercury. In distance from the Sun (closest to farthest): Mercury, Venus, Earth, Mars, Jupiter, Saturn (the opposite order from the first.)

CHAPTER 4

Try One Yourself (Acceleration)

$$Acceleration = \frac{\text{change in velocity}}{\text{time to make the change}}$$
$$= \frac{((7\text{ mi/hr}) - (3\text{ mi/hr}))}{(1.5\text{ sec})}$$
$$= \frac{4\text{ mi/hr}}{1.5\text{ sec}}$$
$$= \frac{2.67\text{ mi/hr}}{\text{sec}}$$

Try One Yourself (Falling Objects Again)

$$\text{Change in velocity} = \text{acceleration} \times \text{time}$$
$$(\text{final velocity}) - (\text{initial velocity}) = \text{acceleration} \times \text{time}$$
$$(\text{final velocity}) - 0 = 9.8\text{ m/s}^2 \times 5\text{ s}$$
$$\text{final velocity} = 49\text{ m/s}$$

(Or use 32 ft/s^2 for acceleration and obtain 160 ft/s)

Recall Questions

2. Earth, water, fire, air, and aether (or quintessence)

4. The distance traveled divided by the time of travel.

6. The speed of a freely falling object increases by 9.8 m/s (or 32 ft/s) during each second of fall.

8. The acceleration of gravity is 9.8 m/s^2, or 32 ft/s^2.

10. Inertia is the tendency of an object to retain its velocity. Countless examples are possible: A car continues to move when the engine is turned off.

12. They stated laws and defined terms, in quantitative fashion so that numerical measurements could be made.

14. Newton lived in the last half of the seventeenth century and the first quarter of the eighteenth.

16. Mass is the more fundamental quantity. As an object changes position with respect to the Earth's center, its weight changes but its mass doesn't.

18. When a hammer exerts a force on a nail (driving it into the wall), the nail exerts a force on the hammer, stopping it.

Calculations

2. 105 km/hr. This corresponds to about 65 miles/hr and therefore is possible.

4. 39.2 m/s or 128 ft/s.

6. After the first second, the object has attained a speed of 9.8 m/s (or 32 ft/s). Its average speed during the second is therefore 4.9 m/s (or 16 ft/s). Falling for one second at this speed yields a distance of 4.9 meters or 16 feet.

CHAPTER 5

Recall Questions

2. The rock continues in a straight line in the direction in which it is going when the string breaks. Gravity, of course, will change its direction.

4. The direction of centripetal force on an object is toward the center of the circle around which the object is moving.

6. He used his law to predict the acceleration of the Moon in its orbit. He also calculated the Moon's acceleration based on the curvature of its motion. The two accelerations were the same, confirming his law.

8. Newton developed calculus in order to solve the problem of the magnitude of the gravitational force that would result from the forces caused by the infinite number of attracting points within the Earth.

10. The centripetal force here is the gravitational attraction of the Earth toward the Sun.

12. During most of the trip to the Moon, a spacecraft is coasting and is using no fuel for propulsion.

14. July 20, 1969.

Calculation

2. (Using the method of the final Close Up):
The difference in radius will increase the person's weight to 4 times its value on Earth. The difference in mass will reduce it to 0.1 of its Earth value. Thus: (110 lb) × (4) × (0.1) = 44 lb.

CHAPTER 6

Try One Yourself

$$W = \frac{\theta\, d}{57.3}$$
$$= \frac{0.52 \times 384{,}000 \text{ km}}{57.3}$$
$$= 3{,}480 \text{ km}$$

Recall Questions

2. Unless other clues are available, we must know its distance (along with its perceived angular size.)
4. On the first day of summer, the Sun was observed to be straight overhead at Syene. That same day, it was seven degrees from the zenith at Alexandria. This meant that a line from each of the cities formed an angle of 7 degrees at Earth-center, and since an entire circle contains 360 degrees, the distance around the Earth is ($^{360}/_{7}$) times the distance between the cities.
6. The Moon's diameter is about ¼ of Earth's.
8. The umbra is the portion of a shadow that receives no light directly from the light source. The penumbra is that portion of a shadow that is blocked from direct light from part of the light source but receives light from other parts.
10. A totally eclipsed Moon has a dark red color. The color is caused by light refracted through the Earth's atmosphere. When light passes through the atmosphere, the blue end of the spectrum is scattered from the beam, leaving the red end.
12. A partial solar eclipse might not be noticed because during it we still receive more light from the Sun than on a cloudy day. One cannot look directly at the Sun and perceive its shape.
14. An annular eclipse occurs when the Moon is far enough from the Earth that its apparent size is less than the apparent size of the Sun.
16. By observing the orbital motions of the Earth and Moon it was determined that the center of mass of the two objects is 81 times farther from Earth than from the Moon, and that therefore the Moon's mass is $^{1}/_{81}$ of Earth's.
18. The capture theory holds that the Moon was once not linked gravitationally to Earth and that it was then captured by the Earth. The double planet theory holds that the Moon formed at the same time as the Earth. The fission theory holds that the material of the Moon was once part of the Earth and that it was torn from the Earth by tidal forces. The large impact theory holds that an object collided with the Earth and blasted the material of the Moon from it.

    The large impact theory seems to best explain the data presently available.

Calculations

2. No, the Moon would have been more than 7 degrees from the zenith at Alexandria because it is close enough to the Earth that parallax is observed in its case.

CHAPTER 7

Try One Yourself (Measuring Distances in the Solar System)

The one-way travel time to Venus is half of 4.8 minutes, or 2.4 minutes. This is 144 seconds. Using the velocity of light:

$$\text{distance} = \text{velocity} \times \text{time}$$
$$= (3.0 \times 10^5 \text{ km/s}) \times (144 \text{ s})$$
$$= 7.32 \times 10^7 \text{ km}$$

Since one AU is 1.5×10^8 km,

$$(7.32 \times 10^7 \text{ km}) \times \frac{1 \text{ AU}}{1.5 \times 10^8 \text{ km}} = 0.288 \text{ AU}.$$

Venus is 0.72 AU from the Sun. Since $(1 - 0.72)$ is 0.28, the above answer is reasonable.

Try One Yourself (Calculating Planetary Sizes)

$$\text{Width} = \frac{\text{angular size} \times \text{distance}}{206{,}000 \text{ arcseconds}}$$
$$= \frac{47 \text{ arcseconds} \times (6.3 \times 10^8 \text{ km})}{206{,}000 \text{ arcseconds}}$$
$$= 1.45 \times 10^5 \text{ km}$$

Try One Yourself (Saturn's Mass by the Revised Kepler's Third Law)

First, change the units to AUs and years:

$$23{,}500 \text{ km} \times \frac{1 \text{ AU}}{1.5 \times 10^8 \text{ km}} = 1.57 \times 10^{-4} \text{ AU}$$
$$1.26 \text{ days} \times \frac{1 \text{ yr}}{365 \text{ days}} = 3.45 \times 10^{-3}$$

Now taking the mass of Mars *(M)* to be approximately equal to the total mass of Mars and Deimos and using Kepler's third law:

$$\frac{d^3}{P^2} = M$$
$$\frac{(1.57 \times 10^{-4})^3}{(3.45 \times 10^{-3})^2} = M$$
$$M = 3.25 \times 10^{-7}$$

This is in terms of the mass of the Sun, which is 2.0×10^{30} kg, so the mass of Mars is

$$(3.25 \times 10^{-7}) \times (2.0 \times 10^{30} \text{ kg}) = 6.5 \times 10^{23} \text{ kg}$$

This compares well with the 6.39×10^{23} kg given in the table.

Recall Questions

2. The Sun's mass is much greater than the total mass of the rest of the solar system, for the Sun makes up more than 99.8 percent of the system's total mass.
4. Jupiter is the largest planet. It's diameter is 11 times Earth's and its mass is 318 times Earth's.
6. Ceres is the largest asteroid. It was discovered by Guiseppe Piazzi, a Sicilian monk, (on January 1, 1801).
8. The density of an object is the ratio of its mass to its volume.
10. All of the planets revolve around the Sun in the same direction. Most (except Venus and Uranus) rotate on their axes in the same direction as they revolve.
12. Compared to the Jovians, the terrestrials are nearer the Sun, smaller, less massive, more solid, slower rotating, more dense, have a thinner atmosphere, fewer moons, and no rings.
14. The Earth is the most dense of the terrestrial planets (by a small margin), and is far more dense than any of the Jovians.
16. The fact that the Earth has an escape velocity of 7 km/s means that an object must be fired upward from above the Earth's atmosphere with at least that speed in order not to fall back.
18. For a given temperature, gases of lower mass have a greater average speed. Hydrogen has a mass much lower than carbon dioxide, and at temperatures prevalent in the Earth's upper atmosphere, a significant fraction of any existing hydrogen atoms have a speed greater than Earth's escape velocity.
20. (Refer to Figure 7–8 for this sketch.)
22. The Van Allen belts are areas above the Earth where its magnetic field have trapped nuclear particles and electrons.

Calculations

2. The predicted planet would be at 77.2 AU from the Sun. (See the Close Up, The Titius-Bode Law.)
4. 7.38×10^{22} kg
6. At closest approach, Mars is 0.52 AU, or 7.8×10^7 km from Earth (assuming circular orbits.)
   The angular diameter of Mars at closest approach is 18 arcseconds.

CHAPTER 8

Try One Yourself

Using Appendix F and multiplying the number found there (0.107) times Earth's mass, we obtain 6.39×10^{23} kg for Mars's mass. The table lists its equatorial diameter as 6742 km. First, express Mars's radius in meters and calculate its volume:

$$\begin{aligned} \text{volume} &= (4/3)\ \pi\ (\text{radius})^3 \\ &= (4/3)\ \pi\ (3{,}371{,}000 \text{ m})^3 \\ &= 1.60 \times 10^{20} \text{ m}^3 \end{aligned}$$

Now calculate density:

$$\begin{aligned} \text{density} &= \text{mass/volume} \\ &= \frac{6.39 \times 10^{23} \text{ kg}}{1.60 \times 10^{20} \text{ m}^3} \\ &= 3.99 \times 10^3 \text{ kg/m}^3 \end{aligned}$$

This corresponds well to 3.93 gm/cm³, as indicated in the table. The slight difference is due to rounding errors.

Recall Questions

2. Venus is closest to Earth in size; Mars in rotation period.
4. Mercury's rotation period is exactly ⅔ of its period of revolution. This has occurred because the mass of the planet is not evenly distributed and since the planet has an eccentric orbit, gravitational forces toward the Sun have changed its rotation rate to produce the resonance.
6. There are two reasons for the extremes in temperature. First, daylight and night periods are very long (88 Earth days each.) Second, there is no atmosphere to block sunlight and to retain heat on the surface.
8. U.S. spacecraft sent to Venus include both Mariners and Pioneer Venus craft. In 1989, the U.S. is scheduled to launch a Magellan Venus spacecraft to the planet. The USSR has sent a number of Venera craft to Venus. The USSR has soft-landed a number of probes, while the U.S. has accomplished some hard landings. The U.S. has soft-landed craft on Mars.
10. Mercury is cratered, much like the Moon, but without large maria. Temperatures on the surface vary greatly and there is essentially no atmosphere. Venus's surface is rock-strewn, with rolling hills, a few craters, and some very large mountains. It is very dry, with an extremely high temperature and a high-pressure atmosphere. Mars has a rocky surface containing some extinct volcanos and what appear to be dried-up riverbeds. The atmosphere is thin and causes frequent dust storms.
12. Opposition occurs when a planet is 180 degrees from the Sun in our sky.
14. The atmospheric pressure on Mars is too low for liquid water to exist. Mars appears to have dried-up riverbeds, indicating that water once flowed there.
16. Large objects have a greater gravitational force on their surfaces, pulling any large bulges down to the surface. The gravitational force on the surface of a small object is not nearly so great, so irregularities can be retained more easily.
20. Irons are made up of 80 to 90 percent iron, with most of the rest being nickel. Stones are made of rock, sometimes with flakes of iron. Stony irons are about half iron and half stone. About 90 percent of meteorites are stones. Most meteorites that are found are irons, however, because they are easier to differentiate from regular earthly rocks.

Calculations

2. Based on Kepler's third law, the semimajor axis is 18 AU.

CHAPTER 9

Recall Questions

2. Jupiter's equator has this rotation period. (Nearer the poles the period is greater.)
4. Jupiter's gaseous atmosphere, made mostly of hydrogen, extends a few thousand miles downward, finally becoming so dense that it must be considered a liquid (but without a definite gas/liquid boundary.) About 10,000 miles below the cloudtops, the liquid becomes liquid metallic hydrogen. At the center is probably a solid core.
6. The plane of its rings is tilted with respect to its orbital plane. Thus we see the rings at different angles at different parts of Saturn's orbit.
8. Calculations indicated that Uranus's orbit was nearly circular, showing that it is not a comet.
10. The rings of Uranus are too sparse to be seen from Earth. They were discovered (in 1977) when they occulted a star.
12. The Jovian planets have much lower densities than do the terrestrial planets.

Calculations

2. The object would have $3/5$ the density of Earth.
4. 6.0×10^4 km.

CHAPTER 10

Recall Questions

2. Pluto's orbit is more eccentric than any other planet's, and it is tilted more with respect to the plane of Earth's orbit.
4. Charon allows us to calculate the mass of Pluto.
6. The masses of the largest asteroids are calculated by observing the effect they have on smaller asteroids. The masses of other asteroids are estimated based on their sizes and expected densities.
8. There was no need to guide the spacecraft through the asteroid belt, for the belt is mostly empty space.
10. It is thought unlikely that a planet could explode. In addition, all of the asteroids together would make up an object smaller than any of the other planets.
12. A comet is made up of solid material along with various ices. On the surface is a solid crust with holes through which the vaporizing ice is forced out when the comet comes close to the Sun.
14. The Oort cloud is a theorized halo of billions of comet nuclei between 10,000 and 100,000 AU from the Sun.
16. First, the vast majority of long-period comets are in elongated elliptical orbits around the Sun. They must therefore spend most of their time far from the Sun. Second, comets are regularly being captured in the inner solar system. The existence of the cloud explains their origin.

CHAPTER 11

Try One Yourself (Wave Motion in General)

$$\text{wave speed} = \text{wavelength} \times \text{frequency}$$

$$335 \text{ m/s} = \text{wavelength} \times 4000 \text{ cycles/s}$$

$$\text{wavelength} = \frac{335 \text{ m/s}}{4000 \text{ cycles/s}}$$

$$\text{wavelength} = 0.08375 \text{ meters}$$

(Notice that the units actually turn out to be *meters/cycle*. That is appropriate, however, for one wave is one cycle.)

Try One Yourself

The difference in the two wavelengths is 0.048 nm. Substituting the values in the Doppler equation:

$$\begin{aligned} v &= c(\Delta\lambda/\lambda) \\ &= (3.0 \times 10^8 \text{ m/s})(0.048/656.285) \\ &= 2.2 \times 10^4 \text{ m/s} \end{aligned}$$

Since the shifted wavelength is shorter than the unshifted one, the star is moving toward the Sun. (Shorter wavelength: blue shift.)

Recall Questions

2. The wave with the lower frequency, 2 cycles/second, has the longer wavelength. (The velocities are the same, and velocity = wavelength × frequency, so the lesser frequency is the greater wavelength.)
4. The least wavelength is about 400 nm and the greatest is 700 nm. (These are 4×10^{-7} m and 7×10^{-7} m respectively.)
6. Sound is not an electromagnetic wave.
8. Electron: The negatively charged object that—in the Bohr model of the atom—orbits the nucleus.

   Nucleus: The small, dense central portion of the atom containing the positive charge and most of the mass of the atom.

   Photon: The smallest amount of electromagnetic energy of a particular wavelength.
10. The energy of a photon is directly proportional to the frequency of the light. (Light of double the frequency has photons of double the energy.)
12. A continuous spectrum is produced by a hot solid (or very dense hot gas.) An emission spectrum is produced by a hot gas of a density less than that which produces a continuous spectrum. An absorption spectrum is produced by light having a continuous spectrum passing through a gas that is too cool to emit light.
14. The Doppler effect is the change of wavelength (and therefore frequency) caused by relative motion of the source and observer. The Doppler effect has nothing to do with the intensity of the wave.
16. To determine the temperature of a star, the overall pattern of the spectrum is observed, with particular attention to the wavelength at which radiation is most intense. The composition of a star is determined by analyzing the pattern of absorption lines. Radial motion is determined by the amount of shift in the absorption lines compared to those from a nonmoving source.
18. The Doppler effect can measure the rotation of an object, the radial motion of an object, or the revolution of two stars around one another.

Calculations

2. 1.3 meters.
4. The star is receding with a speed of 9.1×10^3 m/s.

CHAPTER 12

Try One Yourself

A sphere at Mars's distance has an area of

$$\begin{aligned} \text{Area} &= 4\,\pi\,r^2 \\ &= 4\,\pi\,(2.3 \times 10^{11}\text{ m})^2 \\ &= 6.6 \times 10^{23}\text{ m}^2 \end{aligned}$$

Thus

$$\begin{aligned} \text{Power per square meter} &= \frac{3.9 \times 10^{26}\text{ watts}}{6.6 \times 10^{23}\text{ m}^2} \\ &= 590\text{ watts/m}^2 \end{aligned}$$

Another way to calculate this would be to use the corresponding value for Earth and apply the inverse square law. Since Mars is 1.52 times farther from the Sun, it receives $(1/1.5)^2$ as much radiation per square meter. Thus, $(1380\text{ W/m}^2) \times (1/1.52)^2 = 600\text{ W/m}^2$; in close agreement.

Recall Questions

2. Solar rotation can be detected by observing the motion of sunspots or by the Doppler effect.
4. A chemical reaction involves combining or separating atoms, such as carbon and oxygen combining to form carbon monoxide or carbon dioxide. No change in the nuclei of the atoms is involved. A nu-

clear reaction involves a change in the nucleus of an atom, such as hydrogen nuclei fusing to form a helium nucleus. (Other examples are possible for either reaction, of course.)

6. Hydrogen is consumed and helium is produced in the Sun. The accompanying decrease of mass produces the energy.
8. Hydrostatic equilibrium is the balance between inward and outward pressures in a planet or a star. The pressure at any level within an atmosphere (or within a star) must be enough to support the gas above it.
10. Conduction (example: heat being transferred along a piece of metal), convection (example: hot air rising to the top of a room), and radiation (example: heat transferred from the Sun). Convection occurs in the core of the Sun and radiation is the most important method of transfer through most of the remainder, with convection occurring near the surface.
12. The photosphere is about 500 km thick. Its temperature is about 8000 degrees C at its innermost edge and about 2000 degrees C at its outermost.
14. Sunspots are caused by the magnetic field of the Sun and occur where intense magnetic fields penetrate the photosphere.

Calculations

2. 2×10^{20} kg. (This is about 200,000,000,000,000,000 tons.) It would burn out in 10^{10} seconds, or about 7600 years.

CHAPTER 13

Try One Yourself (Magnifying Power)

1.5 meters is 1500 millimeters.

$$\text{magnification} = \frac{(F_{obj})}{(F_{eye})} = \frac{1500 \text{ mm}}{12 \text{ mm}} = 125$$

Try One Yourself (Light-Gathering Power)

Light-gathering power depends upon the square of the diameter of the objective lens:

$$\frac{5^2}{3^2} = \frac{25}{9} = 2.8$$

Thus, the light-gathering power is increased by 2.8 times.

Recall Questions

2. Refraction is the bending of a wave upon changing its medium of travel.

   Chromatic aberration is the defect of lens systems that results in light of different color focussing at different distances from the lens.
4. The magnifying power of an instrument is the ratio of the angular size of an object when seen through the instrument to the angular size seen by the naked eye.
6. Diffraction is the spreading of waves when they pass an edge of an object.
8. Two stars are said to be resolved if they can be distinguished as two stars.
10. (Figure 13–20 is such a sketch.)
12. The largest reflector at the time of writing is a 6 meter telescope in the USSR. The largest refractor is the 40 inch telescope at Yerkes Observatory in Wisconsin.

    It is economical to make reflectors larger than refractors for a number of reasons, including the following: (1) Four surfaces of the objective must be ground in a refractor; only one for a reflector. (2) Refractors must be corrected for chromatic aberration; reflectors don't need to be. (3) Quality of glass is more important for a refractor than for a reflector. (4) The glass of a reflector can be supported in more places than can that of a refractor.

14. An equatorial mount has its axes tilted so that the plane of one is parallel to the equatorial plane of the Earth. In tracking a star, rotation around only that axis is necessary.
16. Telescopes on mountains are above most of the atmosphere.
18. Blurring effects of the atmosphere are eliminated.

Calculations

2. The eyes gather only 1/200 as much light as the telescope.
4. The telescope on Mount Pastukhov gathers 1.44 times as much light as the Hale Telescope.

CHAPTER 14

Try One Yourself (Apparent Magnitude)

The difference in apparent magnitude between the two objects is 8. Referring to Table 14–1, we see that a difference of 5 corresponds to a ratio of 100 and a difference of 3 corresponds to a ratio of 16. Therefore, we receive 100 × 16, or 1600 times more light from Mars than from Barnard's Star.

Try One Yourself (Spectroscopic Parallax)

Refer to Figure 14–17. It shows that a KO type star has an absolute magnitude between about +5.5 and +6.5. Since its apparent magnitude is +4.4, we can conclude that 40 Eridani is closer to us than 10 pc. (In fact, it is 4.8 pc from the Sun.)

Recall Questions

2. We must know the object's distance as well as its angular velocity.
4. Apparent magnitude is a measure of the amount of light received from an object, while absolute magnitude is a measure of the amount of light emitted by an object.

   Luminosity is closely related to absolute magnitude and brightness relates more closely to apparent magnitude.
6. The period of a Cepheid variable is related to its absolute magnitude. Thus if the period is known, the absolute magnitude can be determined.
8. Absolute magnitude (or luminosity) and temperature (or color or spectral type).
10. The most massive stars are found at the top of the main sequence. They are the hottest and most luminous.
12. Luminosity depends upon both temperature and size. White dwarfs are very small and therefore their total luminosity is low.
14. Visual binaries can be distinguished as two stars in a telescope. Spectroscopic binaries are detected by use of the Doppler Effect.

    Eclipsing binaries change their total brightness regularly when one star eclipses the other.

    Astrometric binaries are detected when only one star of the pair can be seen but that star is observed to move in such a way as to indicate that it has a companion revolving around it.

    Composite spectrum binaries are pairs of stars in which the spectra of the two stars are different enough that examination of the combined spectrum indicates that two stars are present.
16. Goodricke discovered the explanation of Algol's varying in brightness and he discovered the first Cepheid variable.
18. Eclipsing binaries furnish the best information on stellar sizes.
20. From the period of a Cepheid's variation, its absolute magnitude is determined. Then from knowledge of its absolute magnitude and its apparent magnitude, its distance can be determined (by spectroscopic parallax).

Calculations

2. They differ in magnitude by 4.5, so we receive between 40 (which corresponds to a difference of 4) and 100 (which corresponds to 5) times as much light from Antares than from Tau Ceti. (The actual ratio is about 63.)

4. $1/.3 = 3.3$, so the star is 3.3 parsecs away.

6. $1/1.35 = 0.74$, so Alpha Centauri has a parallax angle of 0.74 arcsecond.

8. Figure 14–18 indicates that the absolute magnitude of stars of luminosity class Ia all have absolute magnitudes of about -7. Since Rigel's apparent magnitude is only $+0.14$, it must be very much farther away than 10 pc. (In fact, it is nearly 300 tpc away.)

CHAPTER 15

Recall Questions

2. Collapse is initiated by shock waves caused either by an explosion during the formation of a nearby O or B type star or by a supernova, or a shock wave that is part of the galactic structure.

4. Gravitation is the energy source for protostars.

6. If a star has too little mass, there will not be enough internal pressure to sustain fusion. If a star is too massive, intense radiation emitted by it will limit the amount of material that falls into it.

8. *Stellar thermostat* is the term applied to the process that determines the temperature of a star. If the rate of fusion were to increase, tending to make the core hotter, the core would expand; reducing the rate of fusion. On the other hand, if the rate of fusion decreased, the core would contract; thereby increasing the fusion rate again.

10. Fusion reactions begin, producing heat and tending to cause the star to expand.

12. The increased heat produced by the contracting core causes the outer portion to expand. As the outer part expands, it cools, in accordance with the normal behavior of a gas.

14. Evolutionary theories predicted that the Sun should rotate faster than it does. The discovery of the Sun's magnetic field provided an explanation for how angular momentum was transferred from the Sun to the planets.

16. As the Sun formed, its heat prevented volatile elements from condensing in the inner solar system. Instead, solar wind swept them outward. In the cooler outer regions of the system they were able to condense.

18. (a) Material pulled from the Sun would dissipate rather than condense into planets, (b) Deuterium abundances on the planets are higher than on the Sun and (c) Evidence is accumulating that planetary systems are common, and therefore part of the evolutionary process of stars.

20. Motions of Barnard's Star indicates that large planets exist around it. In addition, disks of gas and dust have been discovered around some young stars.

CHAPTER 16

Recall Questions

2. The material of stars of very low mass completely mixes by convection during their main sequence life. Therefore, there is little or no material outside the core to expand at the end of that life.

4. A black dwarf is the hypothesized cooled remains of a white dwarf.

6. A white dwarf is about the size of the Earth, with a mass of about 0.8 solar masses, an absolute luminosity of $+11$, and a temperature of 15,000 degrees C.

8. Electron degeneracy supports a white dwarf from collapse and neutron degeneracy supports a neutron star.

10. (Figure 16–25 outlines the steps in a middleweight star's life and Figure 16–29 does the same for heavyweights.) The primary difference between the two is that middleweight stars end up as neutron stars and heavyweights end up as black holes.

12. Various types of novae exist. The type discussed in this chapter is the recurrent nova, caused by a red giant and a white dwarf revolving around one an-

other. Periodically, material from the expanding giant spills onto the hot surface of the dwarf, causing it to flare up. A supernova, on the other hand, is the destructive explosion of a star. *Nova* is a Latin word for *new,* for the stars were once thought to entirely new stars.

14. It was calculated that white dwarfs would not be able to vibrate as quickly as pulsars were observed to blink but the same type calculations showed that neutron stars should vibrate faster than some observed pulsars.

16. The energy of pulsars is hypothesized to come from the slowing down of their rate of rotation.

18. The Schwarzschild radius is the minimum distance from the center of a black hole from which light can escape. The event horizon is the sphere around a black hole from within which light cannot escape.

20. If the Sun became a black hole, the Earth would continue in its orbit as it does now. Effects on Earth that result from solar radiation would be changed drastically, of course.

22. SN1987A is the supernova discovered in the Large Magellanic Cloud in February 1987.

Calculations

2. If the mass of a black hole is doubled, the radius of its event horizon (the Schwarzschild radius) also doubles.

CHAPTER 17

Recall Questions

2. The Herschels counted the number of stars in various directions in order to determine whether the Sun is closer to the edge of the Galaxy in some direction. Kapteyn added to this a determination of the density of stars at different distances in various directions.

4. Traffic will bunch up around a slow-moving police car on a freeway. The congested area will follow the police car, but it will be made up of different cars at different times as cars catch up with it, pass it, and slowly pull away from it.

6. They assumed that we were able to see to the end of the disk of the Galaxy.

8. Galactic clusters are loose clusters of relatively few stars found within the galactic plane. Globular clusters are tightly bunched groups of greater numbers of stars—up to hundreds of thousands—usually located outside the disk of the Galaxy. Clusters of galaxies are groups of galaxies, each galaxy perhaps containing hundreds or thousands of galactic clusters and globular clusters.

10. In the 1920s, Oort and Lindblad determined the Sun's position in the Galaxy by analyzing the motions of stars near the Sun. They determined that the center of the Galaxy was in the direction of Sagittarius.

12. The Sun moves in an approximate circle around the center of the Galaxy, with a radius of about 30,000 LY and a period of about 250,000,000 years. This is determined by measurements of stellar motions as well as observations of interstellar dust.

14. The density wave theory and the chain reaction theory.

16. Harlow Shapley, in 1917, showed that the Sun is not in the center of the Galaxy by publishing measured locations of globular clusters. He was involved in the discussion of the nature of the nebulae, arguing that they are part of the Galaxy.

Edwin Hubble discovered Cepheid variables in the Andromeda Galaxy, thereby showing that it was outside the Galaxy. (In the next chapter we will study a more profound contribution by Hubble.)

18. (a) Measurements of rotation rates of various parts of a galaxy and application of Kepler's third law, (b) use of Kepler's third law applied to binary pairs of galaxies, and (c) calculations of the mass necessary to prevent a galaxy from leaving the cluster of galaxies in which it is observed.

20. Not enough mass is directly observed in clusters of galaxies to hold the clusters together. Since they obviously hold together, more mass must exist than we can account for.

Calculations

2. 250 km/s is 8.3×10^{-4} LY/yr. The circumference of the path is 2.5×10^{5} LY. The time for one revolution is therefore 3×10^{8} years.

4. If 0.1 mm corresponds to 1.4×10^6 km (the diameter of the Sun) and the Galaxy's diameter is 10^{18} km, then the galaxy on this scale is about 70,000 km across. We can't build it. (The Earth's radius is only 13,000 km!)

CHAPTER 18

Try One Yourself (The Hubble Law)

On the figure, a distance of 90 MLY corresponds to a speed of about 17,000 km/s. Thus

$$\text{Slope} = \frac{17{,}000 \text{ km/s} - 0}{90 \text{ MLY}}$$
$$= 190 \text{ km/(s}\cdot\text{MLY)}$$

This is the Hubble constant according to that graph.

Try One Yourself (The Hubble Law Used to Measure Distance)

Using a Hubble constant of 15 km/(s·MLY):

$$v = Hd$$
$$90{,}000 \text{ km/s} = 15 \text{ km/(s}\cdot\text{MLY)}$$
$$d = 6000 \text{ MLY}$$

Using 25 km/(s·MLY) in the same manner, we obtain 3600 MLY; thus the distance lies between 3600 and 6000 MLY.

Recall Questions

2. Except for a few nearby ones, all galaxies are moving away from ours. The farther ones are moving fastest and this is consistent with expansion.
4. Cosmological redshift is that redshift that can be attributed to the motion of galaxies resulting from expansion and following the Hubble law.
6. Although distances to nearer galaxies can be measured with reasonable accuracy, random (noncosmological) motion of these galaxies is significant. For more distant galaxies, random motion is not important but their greater distance makes measurement of their distance less precise.
8. When we observe distant object, we are seeing them as they were, not as they are. The amount of time required for light to travel from an object to us is the object's look-back time.
10. If quasars are as distant as their redshift indicates, they must have tremendous luminosity and the source of this energy poses a problem.
12. For a quasar with a redshift of 25 percent, the ratio of the shift in wavelength of its spectral lines to the unshifted wavelength is 0.25.

    A quasar with a redshift of 30 percent is moving away faster than one with a redshift of 25 percent.
14. If our galaxy emits quasars, we would expect other galaxies to do so and if this is the case some of those emitted by other galaxies would be moving toward us, showing a blueshift. No objects with such a blueshift are observed.
16. If they were nearby and moving at the speeds indicated by their redshifts, they would have had to have been produced recently. It is thought, instead, that quasars do not exist in the universe of today, and we see them only because they are very distant and therefore have a great look-back time.
18. Big bang theories hold that the matter and energy of the universe were once tightly compressed, and that they exploded apart to produce the expanding universe of today. Stead state theories hold that matter is being created in the space between galaxies and that this matter gradually coalesces to form new galaxies, so that the universe of the past, present, and future is essentially the same.
20. The steady state theory holds that matter is created in the space between expanding galaxies. This matter gradually forms into galaxies, keeping the overall density constant over time.
22. One method is to determine the speed of the very distant galaxies and compare it to nearer galaxies. Since we are seeing the distant ones in the past, we can determine how much their speed must have changed. The other method is to determine whether there is enough mass in the universe to stop the expansion by gravitational attraction.
24. In order to hold clusters of galaxies together, more mass must exist than we observe. The difference in what is needed and what is observed is the *missing mass*.

Calculations

2. The slope of the line in Figure 18–24 is 11 km/(s·MLY), so that is the value of the Hubble constant from that graph.

4. 4 hours.

6. 4.8×10^{17} seconds, or 1.5×10^{10} years.

APPENDIX A

Try One Yourself

$$t = \frac{t_0}{\sqrt{1 - v^2/c^2}}$$

$$= \frac{1 \text{ s}}{\sqrt{1 - (.9c)^2/c^2}}$$

$$= 2.29 \text{ s}$$

(This is confirmed in Table A–1.)

Recall Questions

2. The special theory of relativity applies when no acceleration is involved. The general theory of relativity applies in all cases.
4. It is assumed that the speed is relative to the surface of the Earth.
6. The Michelson-Morley experiment compared the speed of light in perpendicular directions. It was designed to detect a difference in light's speed depending upon whether the light was travelling perpendicular to the aether's motion or along the direction of the aether's motion.
8. As one nears the speed of light, an observer sees time moving more slowly for the fast-moving person.
10. The effect of gravitation cannot be distinguished from the effect caused by acceleration of the reference frame. The two states are equivalent as far as any measurement that can be made within the reference frame.
12. (a) Light is predicted to bend near the Sun and this has been observed during eclipses, (b) the precession of Mercury's orbit is in accordance with the theory, (c) gravitational lenses, predicted by the theory, have been observed, and (d) the predicted increase in clock speed in a weaker gravitational field has been confirmed.
14. The correspondence principle states that predictions made by a new theory must agree with predictions made by previous theories to the extent that those previous theories proved reliable.
16. General relativity predicts a certain amount of precession of a binary system, depending upon the mass of the two objects. The rate of precession of a particular binary pulsar was measured and the theory was used to calculate the masses of the objects.

Calculations

2. 30.2 minutes
4. 68.7 grams

GLOSSARY

Numbers in parentheses refer to the chapters in which the terms were first introduced.

Aberration of starlight The apparent shift in the direction of a star due to the Earth's orbital motion. (2)

Absolute magnitude The apparent magnitude a star would have if it were at a distance of 10 parsecs. (14)

Absorption spectrum A spectrum that is continuous except for certain discrete wavelengths. (11)

Acceleration The result obtained by dividing the change in velocity of an object by the time required to make that change. (4)

Acceleration of gravity The acceleration of an object due only to gravitational force. Usually the term refers to an object near Earth's surface, where the acceleration is 9.8 m/s^2, or 32 ft/s^2. (4)

Achromatic lens (or achromat) An optical element that has been corrected so that it is free of chromatic aberration. (13)

Active galaxy A galaxy with an unusually luminous nucleus. (18)

Aether (or quintessence) In ancient philosophy, the essence of which the heavens are made; higher in value than the four elements of earth, air, fire, and water. In more modern times, it was hypothesized to serve as a medium for the propagation of light (4 and Appendix A)

Albedo The fraction of incident sunlight that an object reflects. (8)

Altazimuth mount The axis arrangement of the telescope in which one axis changes the telescope altitude and the other its azimuth. (13)

Angular separation Measured from the observer, the angle between lines toward two objects. (1)

Angular size of an object The angle at the viewer between two lines drawn from the viewer to opposite sides of the object. (6)

Annular eclipse An eclipse in which the Moon is too far from Earth for its disk to completely cover that of the Sun, so that the outer edge of the Sun is seen as a ring. (6)

Aphelion The point in its orbit when a planet (or other object) is farthest from the Sun. (8)

Apparent magnitude A measure of the amount of light received from a celestial object. (14)

Apogee The point in the orbit of an Earth satellite where it is farthest from Earth. (6)

Apollo asteroids Asteroids that cross the Earth's orbit. (10)

Asteroid belt The region between Mars and Jupiter where most asteroids orbit. (10)

Asteroid Any of the thousands of minor planets that orbit the Sun. (7)

Astrometric binary An orbiting pair of stars in which the motion of one of the stars reveals the presence of the other. (14)

Astronomical unit A unit of distance equal to the average distance between the Earth and the Sun. (2)

Aurora Light radiated in the upper atmosphere due to impact from charged particles. (7)

Autumnal equinox The point on the celestial sphere where the Sun crosses the celestial equator while moving south. (1)

Average speed The distance traveled by an object divided by the time elapsed during the travel. (4)

Azimuth The angle to an object measured around the horizon from north. (13)

Barred spiral galaxy A spiral galaxy in which the spiral arms come from the ends of a bar through the nucleus rather than from the nucleus itself. (17)

Barycenter The center of mass of two astronomical objects revolving around one another. (6 and 14)

Big bang The theorized initial explosion that began the expansion of the universe. (18)

Big bang theory A theory of cosmology that holds that the expansion of the universe began with an initial explosion. (18)

Binary star system A pair of stars gravitationally bound so that they orbit one another. (11 and 14)

Black dwarf The theorized final state of a star, in which all of its energy sources have been depleted so that it emits no radiation. (16)

Black hole An object whose escape velocity exceeds the speed of light. (16)

Blueshift A change in wavelength toward shorter wavelengths. (11)

Bohr atom The model of the atom proposed by Niels Bohr, containing electrons in orbit around a central nucleus and explaining the emission of light. (11)

Carbon detonation The sudden onset of the fusion of carbon nuclei to form heavier elements. (16)

Cassegrain focus The optical arrangement of a reflecting telescope in which a mirror is mounted so that it intercepts the light from the primary and reflects it back through a hole in the center of the primary. (13)

Catastrophic theory (of solar system formation) A theory of the formation of the solar system that involves an unusual incident such as the collision of the Sun with another star. (15)

Celestial equator A line on the celestial sphere directly above the Earth's equator. (1)

Celestial pole The point on the celestial sphere directly above the poles of the Earth. (1)

Celestial sphere The sphere of heavenly objects that seems to center on the observer. (1)

Center of mass The average location of the various masses in a system, weighted according to how far each is from that point. (6)

Centripetal acceleration The center-directed acceleration of an object moving in a circle. (5)

Centripetal force The force directed toward the center of the curve along which the object is moving. (5)

Cepheid variable A star of a particular class of pulsating stars. (14)

Chain reaction model A model for spiral galaxies that explains the arms as resulting from a series of supernovae, each triggering the formation of new stars. (17)

Chandrasekhar limit The limit to the mass of a white dwarf star above which it cannot exist as a white dwarf. Above that limit it will not be supported by electron degeneracy. (16)

Chromatic aberration The defect of optical systems that results in light of different colors being focused at different places. (13)

Closed universe The state of the universe if it stops expanding and begins contracting. (18)

Cocoon nebula The dust and gas that surrounds a protostar and blocks much of its radiation. (15)

Coma (of comet) The part of a comet's head made up of diffuse gas and dust. (10)

Composite spectrum binary A binary star system with stars having spectra different enough to distinguish them from one another. (14)

Conduction (of heat) The transfer of heat in a solid by collisions between atoms and/or molecules. (12)

Conservation of angular momentum A law that states that the angular momentum of a system will not change unless an outside force is exerted on the system. (15)

Constellation An area of the sky containing a pattern of stars named for a particular object, animal, or person. (1)

Continuous spectrum A spectrum containing an entire range of wavelengths rather than separate, discrete wavelengths. (11)

Convection The transfer of heat in a gas or liquid by means of the motion of the material. (12)

Corona The outer atmosphere of the Sun. (6)

Coronagraph A telescope designed to photogragh the atmosphere of the Sun (when there is no eclipse). (12)

Correspondence principle The idea that the predictions of a new theory must agree with the theory it replaces in cases where the previous theory has been found to be correct. (Appendix A)

Cosmological principle The basic assumption of cosmology that holds that on a large scale the universe is the same everywhere. (18)

Cosmological redshift The shift toward longer wavelengths that is due to a galaxy's motion related to the expansion of the universe. (18)

Crust (of the Earth) The thin, outermost layer of the Earth. (7)

Dark nebula A cloud of interstellar dust that blocks light from stars on the other side of it. (15)

Declination The angle from the celestial equator to an object, stated as positive when north and negative when south. (Appendix D)

Density The ratio of mass to volume. (6)

Density wave mode A model for spiral galaxies that proposes that the arms are the result of density waves sweeping around the galaxy. (17)

Deuterium The hydrogen nucleus containing a neutron along with the proton. (12)

Diamond ring The bright ring seen just before and after totality in a solar eclipse. (6)

Differential rotation Rotation of an object in which different parts have different periods of rotation. (9)

Differentiation The sinking of denser materials toward the center of planets or other objects. (7)

Diffraction The spreading of light upon passing the edge of an object. (13)

Diffraction grating A device that uses the wave properties of electromagnetic radiation to separate the radiation into its various wavelengths. (13)

Dispersion The separation of light into its various wavelengths upon refraction. (13)

Doppler effect The observed change in wavelength from a source moving toward or away from the observer. (11)

Eccentricity (of an ellipse) The ratio of the distance between foci to the length of the major axis. (3 and 7)

Eclipse season A time of the year during which a solar or lunar eclipse is possible. (6)

Ecliptic The apparent path of the Sun on the celestial sphere. (1)

Elliptical galaxy One of the class of galaxies that have smooth spheroidal shapes. (17)

Electromagnetic spectrum The entire array of electromagnetic waves. (11)

Electron degeneracy The state of a gas in which its electrons are packed as densely as nature permits. The temperature of such a high-density gas is not dependent on its pressure as it is in a "normal" gas. (16)

Ellipse A geometrical shape, every point of which is the same total distance from two fixed points (the foci). (3)

Elongation The angle of the Moon (or a planet) from the Sun in the sky. (1)

Emission nebula Insterstellar gas that fluoresces due to ultraviolet light from a star near or within the nebula. (15)

Emission spectrum A spectrum made up of discrete wavelengths rather than a continuous band of wavelengths. (11)

Epicycle The circular orbit of a planet in the Ptolemaic model, the center of which revolves around the Earth in another circle. (1)

Equatorial coordinate system A system that expresses the position of a celestial object in terms of right ascension and declination. (Appendix D)

Equatorial mount The axis arrangement of a telescope in which one of the axes is parallel to the Earth's axis of rotation and the other is aligned with the equatorial plane. (13)

Escape velocity The minimum velocity something must have in order to escape the gravitational attraction of an object such as a planet. (7)

Event horizon The sphere around a black hole from within which nothing can escape. Its radius is the Schwarzschild radius. (16)

Evolutionary track The path on the H-R diagram taken by a star as its luminosity and color change. (15)

Eyepiece The magnifying lens (or combination of lenses) used to view the image formed by the objective of a telescope. (13)

Fiber optics The transmission of light through small filaments of glass (or plastic). (13)

Field of view The actual angular width of the scene viewed by an optical instrument. (13)

Fireball An extremely bright meteor. (10)

First quarter The phase of the Moon which occurs when the Moon's elongation is 90 degrees east. (1)

Flat universe The condition of the universe if gravity just balances its expansion so that it stops expanding only in an infinite amount of time. (18)

Focal length The distance from the focal point of a lens or mirror to its center. (13)

Focal point (of a converging lens or mirror) The point at which light from a very distant object converges after being reflected or refracted. (13)

Focus of an ellipse One of the two fixed points that define an ellipse. (3)

Free fall The condition of an object when there is no force exerted on it except for gravitational force. (4)

Frequency The number of repetitions per unit of time. (11)

Full Moon The phase of the Moon when the Moon's elongation is 180 degrees. (1)

Fusion (nuclear) The combining of two nuclei to form one. (12)

Galactic (or open) cluster A group of stars that share a common origin and are located relatively close to one another. (15)

Gamma rays The portion of the electromagnetic spectrum containing frequencies greater than about 10^{18} Hz. (11)

General theory of relativity A theory developed by Ein stein that expands special relativity to accelerated systems and that presents an alternative way of explaining the phenomenon of gravitation. (16 and Appendix A)

Geocentric model A model of the universe with the Earth at center. (1)

Geostationary orbit An orbit in which a satellite stays at a fixed location above the Earth's equator. (13)

Giant (star) A star of great luminosity and large size. (14)

Gibbous Moon The phases between first quarter and full Moon and between full Moon and third quarter. (1)

Glitch The sudden increase in the pulse rate for a pulsar, thought to be the result of a "starquake." (16)

Globular cluster A spherical group of up to hundreds of thousands of stars, found primarily in the halo of the Galaxy. (17)

Grand Unification Theories (GUTS) Theories that unify the electromagnetic force with nuclear forces. (Appendix A)

Granulation Division of the Sun's surface into small convection cells. (12)

Gravitational lens The phenomenon in which a massive body between another object and the viewer causes the distant object to be seen as two. (18)

Gravitational redshift The increase in wavelength of ra diation leaving an object, as predicted by the theory of general relativity. (Appendix A)

Greenhouse effect The effect by which infrared radiation is trapped within a planet's atmosphere by reflection from particles (for example, carbon dioxide molecules) within that atmosphere. (8)

Halo (galactic) The outermost part of the Galaxy, fairly spherical in shape, beyond the spiral component. (17)

Heliocentric Centered on the Sun. (2)

Hertz (abbreviated Hz) The unit of frequency equal to one cycle per second. (11)

Hertzsprung-Russell Diagram (H-R Diagram) A plot of absolute magnitude versus temperature (or spectral class) for stars. (14)

Homogeneous Having uniform properties throughout. (18)

Hubble constant The proportionality constant in the Hubble law; the ratio of recessional velocities of galaxies to their distances. (18)

Hubble law The relationship that states that a galaxy's recessional velocity is directly proportional to its distance. (18)

Hydrostatic equilibrium A balance between the pressures in a planet's atmosphere or within a star. (12)

Hypothesis A tentative theory. (1)

Image The visual counterpart of an object, formed by refraction or reflection of light from the object. (13)

Inertia The property of an object whereby it tends to maintain whatever velocity it has. (4)

Inferior conjunction The configuration of a planet when it is as nearly between the Earth and the Sun as it can get during a given orbit. (8)

Inflationary universe theory A model of the expanding universe that involves a very rapid expansion near the beginning. (18)

Interferometry A procedure that allows a number of telescopes to be used as one by taking into account the time at which individual waves from an object strike each telescope. (13)

Infrared The portion of the electromagnetic spectrum containing frequencies from about 10^{12} to about 10^{14} Hz. (11)

Ion A charged atom resulting from the atom's loss or gain of an electron. (10)

Irregular galaxy A galaxy of irregular shape that cannot be classified as spiral or eliptical. (17)

Isotropy The property of being the same in all directions. (18)

Kinetic energy The energy an object possesses due to its motion. (10)

Large impact theory A theory that holds that the Moon formed as the result of an impact of a large object upon the Earth. (6)

Length contraction The decrease in length (along the direction of motion) of an object moving at relativistic speed. (Appendix A)

Light curve A graph of the light received from a star versus time. (14)

Light-gathering power A measure of the amount of light collected by an optical instrument. (13)

Lighthouse model (of pulsars) The theory that explains pulsar behavior as being due to a spinning neutron star whose beam of radiation we see as it sweeps by. (16)

Limb (of the Sun or Moon) The apparent edge of the object as seen in the sky. (12)

Local group (of galaxies) The cluster of 20 or so galaxies that includes the Milky Way Galaxy. (17)

Local hypotheses A proposal stating that quasars are much nearer than a cosmological interpretation of their redshifts would indicate. (18)

Look-back time The time during which light from a distant object has traveled to reach us. (18)

Luminosity The rate at which electromagnetic energy is being emitted. (12)

Luminosity class A classification of a star by the amount of electromagnetic energy it radiates. (14)

Lunar eclipse An eclipse in which the Moon passes into the shadow of the Earth. (6)

Magnetic field A region of space where magnetic forces can be detected. (7)

Magnetosphere The volume of space in which the motion of charged particles is controlled by the magnetic field of the planet rather than by the solar wind. (9)

Magnifying power (or magnification) The ratio of the angular size of an object when it is seen through the instrument to its angular size when seen with the naked eye. (13)

Main sequence The part of the H-R diagram containing the majority of stars, forming a diagonal line across the diagram. (14)

Major axis (of ellipse) The longest distance across an ellipse. (3)

Mantle (of the Earth) The thick, solid layer between the crust and the core of the Earth. (7)

Mare (MAH-ray, plural maria) Any one of the lowlands on the Moon or Mars that resemble a sea when viewed from Earth. (3)

Mass The quantity of inertia possessed by an object. (4)

Mass-luminosity diagram A plot of the mass versus the luminosity of a number of stars. (14)

Meridian An imaginary line that runs from north to south, passing through the observer's zenith. (2 and 7)

Meteor The phenomenon of a streak in the sky caused by the burning of a rock or dust particle as it falls. (10)

Meteor shower The phenomenon of a large group of meteors seeming to come from a particular area of the celestial sphere. (10)

Meteorite An interplanetary chunk of matter after it has hit a planet or moon. (10)

Meteoroid An interplanetary chunk of matter smaller than an asteroid. (10)

Michelson-Morley experiment The experiment of Albert Michelson and Edward Morley that sought to measure the motion of light with respect to the aether. (Appendix A)

Milky Way Galaxy (or "the Galaxy") The group of a few hundred billion stars, of which our Sun is one. From Earth, it appears as a band of light around the sky. (Proloque and 17)

Minute of arc One-sixtieth of a degree of arc. (1)

Missing mass The difference between the mass of clusters of galaxies as calculated from Keperian motions and the amount of visible mass. (17)

Nanometer (abbreviated nm) A unit of distance equal to 10^{-9} meter. (11)

Neap tide The time of least difference between high and low tide, occurring when the solar tide partly cancels the lunar tide. (6)

Neutrino An elementary particle that has little or no rest mass and no charge but carries energy from a nuclear reaction. (12)

Neutron The massive nuclear particle with no electric charge. (12)

Neutron degeneracy The state of a gas in which its neutrons are packed as densely as nature permits. (16)

Neutron star A star that has collapsed to the point at which it is supported by neutron degeneracy. (16)

New Moon The phase of the Moon when its elongation is zero. (1)

Newtonian focus The optical arrangement of a reflecting telescope in which a plane mirror is mounted along the axis of the telescope so that it intercepts the light from the primary and reflects it to the side. (13)

Newtonian relativity The principle that states that all positions and velocities are stated with respect to a specific frame of reference. (Appendix A)

Nonvolatile element An element that is gaseous only at a high temperature and condenses to liquid or solid when the temperature decreases. (15)

Nova An event in which matter from the giant component of a binary system falls into the white dwarf component, causing a sudden brightening of hundreds of thousands of times. (16)

Nucleus (of atom) The central, massive part of an atom. (11)

Nucleus (of comet) The solid chunk of a comet, located in the head. (10)

Objective lens (or objective) The primary light-gathering element—lens or mirror—of a telescope. (13)

Oblateness A measure of the "flatness" of a planet, calculated by dividing the difference between the largest and smallest diameter by the largest diameter. (9)

Occultation The passing of one astronomical object in front of another. (9)

Oort cloud The theorized sphere lying between 10,000 and 100,000 AU from the sun containing billions of comet nuclei. (10)

Open universe The state of the universe if it continues expansion without stopping. (18)

Opposition The configuration of a planet when it is opposite to the Sun in our sky. That is, the objects are aligned as follows: Sun-Earth-planet. (2 and 8)

Optical double Two stars that have small angular separation as seen from Earth but that are not gravitationally linked. (14)

Oscillating universe theory A big bang theory that holds that the universe goes through repeating cycles of explosion, expansion, and contraction. (18)

Paraboloid The surface that results from rotating a parabola around its axis. (Appendix L)

Parallax The apparent shifting of nearby objects with respect to distant ones as the position of the observer changes. (2)

Parsec The distance of an object that has a parallax angle of one arcsecond. (14)

Partial lunar eclipse An eclipse of the Moon in which only part of the Moon passes through the umbra of the Earth's shadow. (6)

Partial solar eclipse An eclipse in which only part of the Sun's disk is covered by the Moon. (6)

Particle density The number of separate atomic and/or nuclear particles per unit of volume. (12)

Penumbra The portion of a shadow that receives direct light from only part of the light source. (6)

Penumbral lunar eclipse An eclipse of the Moon in which the Moon passes through the Earth's penumbra but not through its umbra. (6)

Perfect cosmological principle The assumption that on a large scale the universe is the same everywhere and through all time. (18)

Perigee The point in its orbit when an Earth satellite is closest to Earth. (6)

Perihelion The point in its orbit when a planet (or other object) is closest to the Sun. (8)

Period (of a regularly repeating cycle) The time that elapses between successive repetitions. (3)

Phases (of the Moon) The changing appearance of the Moon during its cycle caused by the relative positions of Earth, Moon, and Sun. (1)

Photometer An electronic device used to measure the intensity of light received. (14)

Photon The smallest possible amount of electromagnetic energy of a particular wavelength. (11)

Photosphere The region of the Sun from which visible radiation is emitted. (11)

Planetary nebula The shell of gas that is expelled by a red giant near the end of its life. (16)

Planetesimal One of the small objects that formed from the original material of the solar system and from which a planet developed. (15)

Positron A positively charged electron that is emitted from the nucleus in some nuclear reactions. (12)

Power The amount of energy exchanged per unit of time. (14)

Precession (related to rotation) The conical shifting of the axis of a rotating object. (6)

Precession (of an elliptical orbit) The change in orientation of the major axis of the elliptical path of an object. (Appendix A)

Pressure The force per unit area. (12)

Prime focus The point in a telescope where the light from the objective is focused. (13)

Primordial egg (or primordial atom) The original matter that is theorized to have exploded to form the universe of today. (18)

Principle of equivalence The statement that effects of the force of gravity are indistinguishable from those of acceleration. (Appendix A)

Prominence The projection of solar material beyond the disk of the Sun. (12)

Proper motion The angular velocity of a star as measured from the Sun. (14)

Proton The massive, positively charged particle in the nucleus of an atom. (12)

Proton-proton chain The series of nuclear reactions that begins with four protons and ends with a helium nucleus. (12)

Protostar A star in the process of formation, before reaching the main sequence. (15)

Pulsar A celestial object of small angular size that emits pulses of radio waves with a regular period between about 0.03 and 5 seconds. (16)

Ptolemaic model The theory of the heavens devised by Claudius Ptolemy. (1)

Quadrature The configuration of a planet when its elongation is 90 degrees. (2)

Quasar (quasi-stellar radio source) An object with a very large redshift but approximately stellar in size. (18)

Quintessence (or aether) In ancient philosophy, the essence of which the heavens are made; higher in value than the four elements of earth, air, fire, and water. (4)

Radial velocity Velocity along the line of sight, toward or away from the observer. (11)

Radiant (of a meteor shower) The point in the sky from which the meteors of a shower appear to radiate. (10)

Radiation (of heat) The transfer of heat by infrared waves. (12)

Radio galaxy A galaxy having greatest luminosity at radio wavelengths. (18)

Radio waves The portion of the electrtomagnetic spectrum containing frequencies up to about 10^9 Hz. (11)

Ray (lunar) A bright streak on the Moon caused by material ejected from a crater. (3)

Real image An optical image formed by the actual convergence of light on a screen. (13)

Redshift A change in wavelength toward longer wavelengths. (11)

Reference frame The location from which a particular motion or other location is measured. (Appendix A)

Reflection nebula Interstellar dust that is visible due to reflection of light from a nearby star. (15)

Refraction The bending of light as it crosses the boundary between two materials. (13)

Resolving power (or resolution) The measure of the ability of a telescope to see fine details in an object. (13)

Retrograde motion The east-to-west motion of a planet against the background of stars. (1)

Revolution The orbiting of one object around another. (1)

Right ascension The angle measured eastward from the vernal equinox to the position of an object, stated in hours, minutes, and seconds. (Appendix D)

Rille A trench-like depression in the Moon's surface. (3)

Roche limit The minimum radius at which a satellite (held together by gravitational forces) may orbit without being broken apart by tidal forces. (9)

Rotation The spinning of an object about an axis passing through it. (1)

Scarps Cliffs in a line. They are found on Mercury, Earth, Mars, and the Moon. (8)

Schwarzschild radius The radius of the sphere around a black hole from within which no light can escape. (16)

Scientific model A theory that accounts for a set of observations in nature. (1)

Second of arc One-sixtieth of a minute of arc. (1)

SETI Acronym for Search for Extraterrestrial Intelligence. (13)

Seyfert galaxy One of a class of galaxies with active nuclei and whose spectrum contains emission lines. (18)

Sidereal day The amount of time that passes between successive passages of a given star across the meridian. (7)

Sidereal period The time of revolution of one object around another with respect to the stars. (2)

Solar day The amount of time that passes between successive passages of the Sun across the meridian. (7)

Solar flare An explosion near or at the Sun's surface, seen as an increase in activity such as prominences. (12)

Solar system The Sun and all objects revolving around it, as well as gas and dust between those objects. (Prologue)

Solar wind The flow of nuclear particles from the Sun. (12)

Space velocity The velocity of a star relative to the Sun. (14)

Special theory of relativity A theory developed by Einstein that predicts the behavior of matter moving at constant speeds relative to the observer. (16 and Appendix A)

Spectrograph A spectrometer that produces a photograph of the spectrum. (13)

Spectrometer An instrument that separates electromagnetic radiation according to wavelength. (13)

Spectroscopic binary An orbiting pair of stars that can be distinguished as two due to the changing Doppler shifts in their spectrum. (14)

Spectroscopic parallax The method of measuring the distance to a star by comparing its absolute magnitude to its apparent magnitude. (14)

Spectrum The order of colors or wavelengths produced when light is dispersed. (11)

Spherical aberration A defect of an optical system whereby rays at different distances from the axis focus at different distances from the objective. (Appendix L)

Spiral galaxy A disk-shaped galaxy with two arms in a spiral pattern. (17)

Spring tide The greatest difference between high and low tide; occurring about twice a month, when the lunar and solar tides correspond. (6)

Stadium An ancient Greek unit of length, with a length of perhaps 0.15 to 0.2 kilometers. Various stadia were in use. (6)

Steady state theory A cosmological theory that holds that on a large scale, the universe does not change with time. (18)

Stellar parallax The apparent annual shifting of nearby stars with respect to background stars. Also, the angle of parallax, using 1 AU as baseline. (2 and 14)

Stellar wind The nuclear particles that flow from stars. (15)

Summer solstice The point on the celestial sphere where the Sun reaches its northernmost position. (1)

Sunspot A region of the photosphere that is temporarily cool and dark compared to surrounding regions. (12)

Supergiant A star of very great luminosity and size. (14)

Supernova (plural: supernovae) The explosive destruction of a star. (Prologue and 16)

Synodic period The time interval between successive similar configurations of a planet-Sun-Earth (for example, successive inferior conjunctions). (8)

Syzygy A straight line arrangement of three celestial objects. (8)

T Tauri stars A certain class of stars that show rapid and erratic changes in brightness. (15)

Tail (of comet) The gas and/or dust swept away from a comet's head. (10)

Tangential velocity Velocity perpendicular to the line of sight. (11)

Terminator The sunrise or sunset line on a moon or planet. (3)

Third quarter The phase of the Moon which occurs when the Moon's elongation is 90 degrees west. (1)

Tidal friction Friction forces that result from tides on a rotating object. (6)

Time dilation The slowing of time of an object moving at a relativistic speed (a speed great enough that Einsteinian relativity must be considered). (Appendix A)

Total lunar eclipse An eclipse of the Moon in which the Moon is completely in the umbra of the Earth's shadow. (6)

Total solar eclipse An eclipse in which light from the normally visible portion of the Sun (the photosphere) is completely blocked by the Moon. (6)

Transit The passage of a celestial object across the meridian. Also, the passage of a small celestial object across the disk of a large one. Also, a device to measure the time of occurrence of a meridian transit. (2)

Twenty-one centimeter radiation Radiation from atomic hydrogen, with a wavelength of 21.11 centimeters. (17)

Ultraviolet The portion of the electromagnetic spectrum containing frequencies from about 10^{14} to about 10^{16} Hz. (11)

Umbra The portion of a shadow that receives no direct light from the light source. (6)

Universality The property of obeying the same physical laws throughout. (18)

Van Allen Belts Region around the Earth where charged particles are trapped by the Earth's magnetic field. (5)

Velocity Speed, along with a statement of the direction in which the object is moving. (4)

Vernal equinox The point on the celestial sphere where the Sun crosses the celestial equator while moving north. (1)

Visual binary An orbiting pair of stars that can be resolved (normally with a telescope) as two stars. (14)

VLA Acronym for the Very Large Array (of radio telescopes) located in New Mexico. (15)

Volatile Capable of being vaporized at a relatively low temperature. (6)

Volatile element A chemical element that exists in a gaseous state at a relatively low temperature. (15)

Waning (Moon phases) The phases between full Moon and new Moon. (1)

Wavelength The distance from a point on a wave to the next corresponding point. (11)

Waxing (Moon phases) The phases between new Moon and full Moon. (1)

Weight The gravitational force between an object and the planet, moon, etc., on—or near—which it is located. (5)

White dwarf A very small, hot star that has ended its main sequence life. (14)

Winter solstice The point on the celestial sphere where the Sun reaches its southernmost position. (1)

X rays The portion of the electromagnetic spectrum containing frequencies from about 10^{16} to about 10^{20} Hz. (11)

Zeeman effect The splitting of spectral lines caused by the presence of a strong magnetic field. (12)

Zenith The point in the sky located directly overhead. (6)

Zodiac The band that lies 9 degrees either side of the ecliptic on the celestial sphere. (1)

INDEX

Photo Credits continued
Fig. 1–2 © 1980, Anglo-Australian Telescope Board © 1980; **Fig. 1–5** Harvard College Observatory; **Fig. 1–17a** Lick Observatory; **Fig. 1–17b** Lick Observatory; **Fig. 1–17c** Lick Observatory; **Fig. 1–17d** Lick Observatory; **Fig. 1–17e** Lick Observatory; **Fig. 1–17f** Lick Observatory; **Fig. 1–17g** Lick Observatory; **Fig. 1–21a** The Bettmann Archive; **Fig. 1–21b** The Bettmann Archive; **Fig. 1–23** Tersch Enterprises—Tiara Observatory; **Fig. 1–24** Tersch Enterprises—Tiara Observatory; **Fig. 1–25** Tersch Enterprises—Tiara Observatory; **Fig. T1–1** Kitt Peak National Observatory; **Fig. T1–2** Kitt Peak National Observatory.

Chapter Two
Chapter Opening Photograph by Sally P. Ragep. Courtesy of The Collection of Historical Scientific Instruments, Harvard University; **Fig. 2–1** The Bettmann Archive; **Fig. 2–2** Harvard College Observatory; **Fig. 2–18** Tersch Enterprises—Tiara Observatory; **Fig. 2–21** NASA.

Chapter Three
Chapter Opening The Mansell Collection; **Fig. 3–1** The Bettmann Archive; **Fig. 3–2** The Bettmann Archive; **Fig. 3–9** Drew University; **Fig. 3–10** Yerkes Observatory Photograph; **Fig. 3–11** Lick Observatory; **Fig. 3–12** Yerkes Observatory; **Fig. 3–14b** Hale Observatories; **Fig. 3–15** Lowell Observatory Photograph; **Fig. 3–18** Photo provided courtesy of Celestron International; **Fig. T3–3** By permission of the Houghton Library, Harvard University.

Chapter Four
Chapter Opening © Fred Espenak/Science Photo Library, Photo Researchers, Inc.; **Fig. 4–2** Education Development Center (EDC); **Fig. 4–13** The Granger Collection; **Fig. 4–16** National Institute of Standards and Technology (NIST), Washington, D.C.; **Fig. 4–17** © Margaret Durrance/Photo Researchers, Inc.; **Fig. T4–1** Culver Pictures.

Chapter Five
Chapter Opening NASA; **Fig. 5–10** NASA; **Fig. 5–11** NASA; **Fig. 5–13** NASA; **Fig. 5–15** NASA; **Fig. 5–16** NASA; **Fig. 5–18** NASA; **Fig. 5–19** NASA; **Fig. 5–21** NASA; **Fig. T5–4** NASA.

Chapter Six
Chapter Opening Workman Publishing; **Fig. 6–1** Johnny Hart, Creator's Syndicate; **Fig. 6–2a** Author photo; **Fig. 6–2b** Author photo; **Fig. 6–8** Author photo; **Fig. 6–10a** Dennis Milon/Photo by Dennis Trail; **Fig. 6–10b** Dennis Milon/Photo by Dennis Trail; **Fig. 6–15a** Celestron International; **Fig. 6–15b** Celestron International; **Fig. 6–18a** Author photo; **Fig. 6–18b** NOAO; **Fig. 6–19** NASA; **Fig. 6–23** © Hans Vehrenberg, Hansen Planetarium; **Fig. 6–28a** Author photo; **Fig. 6–28b** Author photo; **Fig. T6–2a** Author photo; **Fig. T6–2b** Author photo; **Fig. T6–3** Author photo; **Fig. T6–5a** Author photo; **Fig. T6–5b** Joe McTyre, *Atlanta Constitution*.

Chapter Seven
Chapter Opening NASA; **Fig. 7–2** Author photo; **Fig. 7–12** NASA; **Fig. 7–19b** Nancy Rodger/The Exploratorium; **Fig. 7–24** ;© Jack Finch/Science Photo Library, Photo Researchers, Inc; **Fig. 7–25** © Richard J. Quataert/Taurus Photos; **Fig. 7–26a** NASA; **Fig. 7–26b** NASA; **Fig. 7–26c** NASA; **Fig. 7–27** Kitt Peak National Observatory; **Fig. T7–1** Yerkes Observatory.

Chapter Eight
Chapter Opening Painting by Michael W. Carroll; **Fig. 8–4a** NASA; **Fig. 8–4b** NASA; **Fig. 8–6** NASA; **Fig. 8–7** NASA; **Fig. 8–8** NASA; **Fig. 8–9** Planetary Data Facility of U.S. Geological Survey; **Fig. 8–15a** Lowell Observatory; **Fig. 8–15b** Author photo; **Fig. 8–16** *TASS* from Sovfoto; **Fig. 8–18** © Richard J. Quataert/Taurus Photos; **Fig. 8–19** Worlds in Comparison, Astronomical Society of the Pacific (ASP); **Fig. 8–20** Worlds in Comparison, ASP; **Fig. 8–23a** Lick Observatory; **Fig. 8–23b** Lick Observatory; **Fig. 8–24** Worlds in Comparison, ASP; **Fig. 8–26** Official U.S. Navy Photo; **Fig. 8–27** Author photo; **Fig. 8–28** Author photo; **Fig. 8–29** Lowell Observatory; **Fig. 8–31** Worlds in Comparison, ASP; **Fig. 8–32a** Worlds in Comparison, ASP; **Fig. 8–33** NASA/Jet Propulsion Laboratory (JPL); **Fig. 8–34** NASA; **Fig. 8–35** NASA; **Fig. T8–1** NASA/JPL; **Fig. T8–2** NASA/JPL.

Chapter Nine
Chapter Opening Painting by Michael W. Carroll; **Fig. 9–1** Worlds in Comparison, ASP; **Fig. 9–2** California Institute of Technology (CIT) and Carnegie Institute of Washington; **Fig. 9–4** Author photo; **Fig. 9–5a** Author photo; **Fig. 9–5b** Author photo; **Fig. 9–6** NASA; **Fig. 9–7** Worlds in Comparison, ASP; **Fig. 9–8** NASA; **Fig. 9–11** Worlds in Comparison, ASP; **Fig. 9–12** NASA/JPL; **Fig. 9–13** NASA; **Fig. 9–14** NASA; **Fig. 9–15** NASA; **Fig. 9–16** NASA; **Fig. 9–17** NASA; **Fig. 9–19a** Hale Observatories; **Fig. 9–19b** NASA; **Fig. 9–20** Lowell Ob-

servatory; **Fig. 9–23** Worlds in Comparison, ASP; **Fig. 9–25** NASA; **Fig. 9–27** NASA; **Fig. 9–28** Lick Observatory; **Fig. 9–31b** NASA/JPL; **Fig. 9–35a** NASA/JPL; **Fig. 9–35b** NASA/JPL; **Fig. 9–36** Worlds in Comparison, ASP; **Fig. T9–1** NASA; **Fig. T9–2a** © Wayne Miller/Magnum Photos; **Fig. T9–2b** Thomas Nebbia, © National Geographic Society; **Fig. T9–3** NASA/JPL; **Fig. T9–4** NASA; **Fig. T9–5** NASA/JPL; **Fig. T9–6** NASA/JPL.

Chapter Ten
Chapter Opening *Sky and Telescope;* **Fig. 10–1a** Lowell Observatory; **Fig. 10–1b** Lowell Observatory; **Fig. 10–2** U.S. Naval Observatory Photograph; **Fig. 10–4** Dennis Milon/Photo by George East; **Fig. 10–6** Author photo; **Fig. 10–8** Dennis Milon/Photo by George East; **Fig. 10–14** © 1986, Max-Planck-Institute für Aeronome Courtesy of Dr. H. U. Keller; **Fig. 10–15** U.S. Naval Observatory; **Fig. 10–21** Dennis Milon; **Fig. 10–23** NOAO (Kitt Peak Observatory); **Fig. 10–27a** Field Museum of Natural History: #GEO 81638; **Fig. 10–27b** Field Museum of Natural History: #17251; **Fig. 10–27c** Field Museum of Natural History: #39663; **Fig. 10–28** Meteor Crater Enterprises, Inc., Flagstaff, Arizona; **Fig. 10–29** Griffith Observatory; **Fig. 10–30** Dennis Milon/Photo by James M. Baker; **Fig. 10–32** *TASS* from Sovfoto; **Fig. T10–3** Walter Alvarez, University of California, Berkeley.

Chapter Eleven
Chapter Opening Charles H. Phillips/Smithsonian Books; **Fig. 11–1** David Parker/Photo Researchers, Inc; **Fig. 11–4** Courtesy of Bausch & Lomb; **Fig. 11–7a** Author photo; **Fig. 11–8a** Roger Kramer; **Fig. 11–13a** Sargent-Welch Scientific Company; **Fig. 11–13b** Sargent-Welch Scientific Company; **Fig. 11–13c** Sargent-Welch Scientific Company; **Fig. 11–19** Deutsches Museum, Munich; **Fig. 11–23b** Education Development Center; **Fig. 11–29a,b** Lick Observatory; **Fig. T11–1** American Institute of Physics (AIP)/Niels Bohr Library.

Chapter Twelve
Chapter Opening NOAO; **Fig. 12–1** © Larry Fajkus/Taurus Photos; **Fig. 12–3a** Palomar Observatory; **Fig. 12–3b** Palomar Observatory; **Fig. 12–18** Project Stratoscope of Princeton University; **Fig. 12–21** Dennis Milon/Photo by George East; **Fig. 12–22** NASA; **Fig. 12–23** Photo by High Altitude Observatory, National Center for Atmospheric Research. The National Center for Atmospheric Research is sponsored by The National Science Foundation; **Fig. 12–24** NASA; **Fig. 12–25** Hale Observatories; **Fig. 12–26** Science and Engineering Council, J. A. Eddy; **Fig. 12–27** Royal Greenwich Observatory and the Science and Engineering Council; **Fig. T12–2** Author photo.

Chapter Thirteen
Chapter Opening © Doug Johnson/SPL/Science Source, Photo Researchers, Inc.; **Fig. 13–4a** Author photo; **Fig. 13–5a** Author photo; **Fig. 13–10** Author photo; **Fig. 13–11** Illustration by Bruce Bond; **Fig. 13–13** Author photo; **Fig. 13–15** Education Development Center; **Fig. 13–16** Illustration by Bruce Bond; **Fig. 13–17a** Dave Duszynski; **Fig. 13–18** George C. Atamian; **Fig. 13–21** The Granger Collection; **Fig. 13–22a** Author photo; **Fig. 13–22b** Author photo; **Fig. 13–23** Courtesy of Perkin-Elmer Corporation; **Fig. 13–24a** Yerkes Observatory; **Fig. 13–24b** Yerkes Observatory; **Fig. 13–26** Author photo; **Fig. 13–27a** Author photo; **Fig. 13–27b** Author photo; **Fig. 13–27c** Author photo; **Fig. 13–28** © Copyright 1979 by CIT and Carnegie Institution of Washington; **Fig. 13–29** © CIT; **Fig. 13–30** © European Southern Observatory; **Fig. 13–31** MMT Observatory; **Fig. 13–32** © European Southern Observatory; **Fig. 13–33** McDonald Observatory, The University of Texas at Austin; **Fig. 13–34a** Courtesy NRAO/AUI; **Fig. 13–34b** NRAO; **Fig. 13–35** Author photo; **Fig. 13–36** The National Astronomy and Ionosphere Center is operated in Puerto Rico by Cornell University of New York under a cooperative agreement with the National Science Foundation; **Fig. 13–37b** Courtesy NRAO/AUI. Acknowledgment: P. E. Angerhofer, R. Braun, S. F. Gull, R. A. Perley, R. J. Tuffs; **Fig. 13–40** Courtesy NRAO/AUI; **Fig. 13–41** Institute for Astronomy/University of Hawaii; **Fig. 13–42** NASA; **Fig. T13–1** NASA; **Fig. T13–2** NASA/Ames Research Center.

Chapter Fourteen
Chapter Opening Institute for Astronomy/University of Hawaii; **Fig. 14–1** Tersch Enterprises—Tiara Observatory; **Fig. 14–7** Yerkes Observatory; **Fig. 14–8** Yerkes Observatory; **Fig. 14–12** Harvard College Observatory; **Fig. 14–13** Mt. Wilson and Las Campanas Observatories, CIT; **Fig. 14–15** Estate of Henry Norris Russell; **Fig. 14–26** Swarthmore College Observatory; **Fig. 14–27** Lick Observatory; **Fig. 14–38** Harvard College Observatory; **Fig. 14–39** Harvard College Observatory, Bloemfontein, South Africa.

Chapter Fifteen
Chapter Opening © 1981, Royal Observatory, Edinburgh; **Fig. 15–2a** Martin C. Germano; **Fig. 15–2b** © 1980, Royal Observatory, Edinburgh; **Fig. 15–4** NOAO; **Fig. 15–6** NASA; **Fig.**

15–9a Dan Gordon; **Fig. 15–9b** Jack Newton, Victoria, B.C.; **Fig. 15–10** © 1985, Royal Observatory, Edinburgh; **Fig. 15–11a** Jack Newton; **Fig. 15–11b** Palomar Observatory; **Fig. 15–11c** Yerkes Observatory; **Fig. 15–21** U.S. Naval Observatory; **Fig. 15–23** NASA; **Fig. 15–24** NASA.

Chapter Sixteen
Chapter Opening Chris Floyd; **Fig. 16–3a** Celestron International; **Fig. 16–3b** NASA; **Fig. 16–5a** © 1965 California Institute of Technology; **Fig. 16–5b** Lick Observatory; **Fig. 16–7a** Tersch Enterprises—Tiara Observatory; **Fig. 16–7b** Lick Observatory; **Fig. 16–10a** Palomar Observatory; **Fig. 16–10b** Palomar Observatory; **Fig. 16–12** © 1959, California Institute of Technology; **Fig. 16–13a** Jack Newton; **Fig. 16–13b** © 1980, Royal Observatory, Edinburgh; **Fig. 16–15** Lick Observatory; **Fig. 16–18** University of Cambridge, Mullard Radio Astronomy Observatory. With the compliments of Professor Antony Hewish; **Fig. 16–28** Griffith Observatory. Painting by Lois Cohen; **Fig. 16–30a** Chris Floyd; **Fig. 16–30b** Chris Floyd.

Chapter Seventeen
Chapter Opening NOAO; **Fig. 17–2** Tersch Enterprises—Tiara Observatory; **Fig. 17–4** Yerkes Observatory; **Fig. 17–5** Dennis Milon/Photo by Allan E. Morton; **Fig. 17–6** © 1985, Royal Observatory, Edinburgh; **Fig. 17–7a** Palomar Observatory; **Fig. 17–7b** Lick Observatory; **Fig. 17–15** Gart Westerhout, U.S. Naval Observatory; **Fig. 17–16** © 1980, Anglo Australian Telescope Board; **Fig. 17–19a** Tersch Enterprises; **Fig. 17–20 #1,2,3,6,7,8,9** Mt. Wilson and Las Campanas Observatories, Carnegie Institution of Washington; **#4** NOAO; **#5** Palomar Observatory; **Fig. 17–21 #1–3** Palomar Observatory; **Fig. 17–22a** Palomar Observatory; **Fig. 17–22b** Palomar Observatory; **Fig. 17–23** Palomar Observatory; **Fig. 17–24a** © 1959, California Institute of Technology; **Fig. 17–24b** Hale Observatories; **Fig. 17–27a** Palomar Observatory; **Fig. 17–27b** Palomar Observatory.

Chapter Eighteen
Chapter Opening American Institute of Physics, Niels Bohr Library; **Fig. 18–5** Palomar Observatory; **Fig. 18–10a** © 1980, Anglo-Australian Telescope Board; **Fig. 18–10b** NRAO/AUI. Acknowledgment: J. O. Burns, E. J. Schreier, E. D. Feigelson; **Fig. 18–11** Lick Observatory; **Fig. 18–16** Photri; **Fig. 18–19** NASA; **Fig. T18–1** © Ian Berry/Magnum.

Appendix A
Fig. A–13 Argonne National Laboratory.

Star Charts Star charts from *Griffith Observer,* Griffith Observatory, Los Angeles.

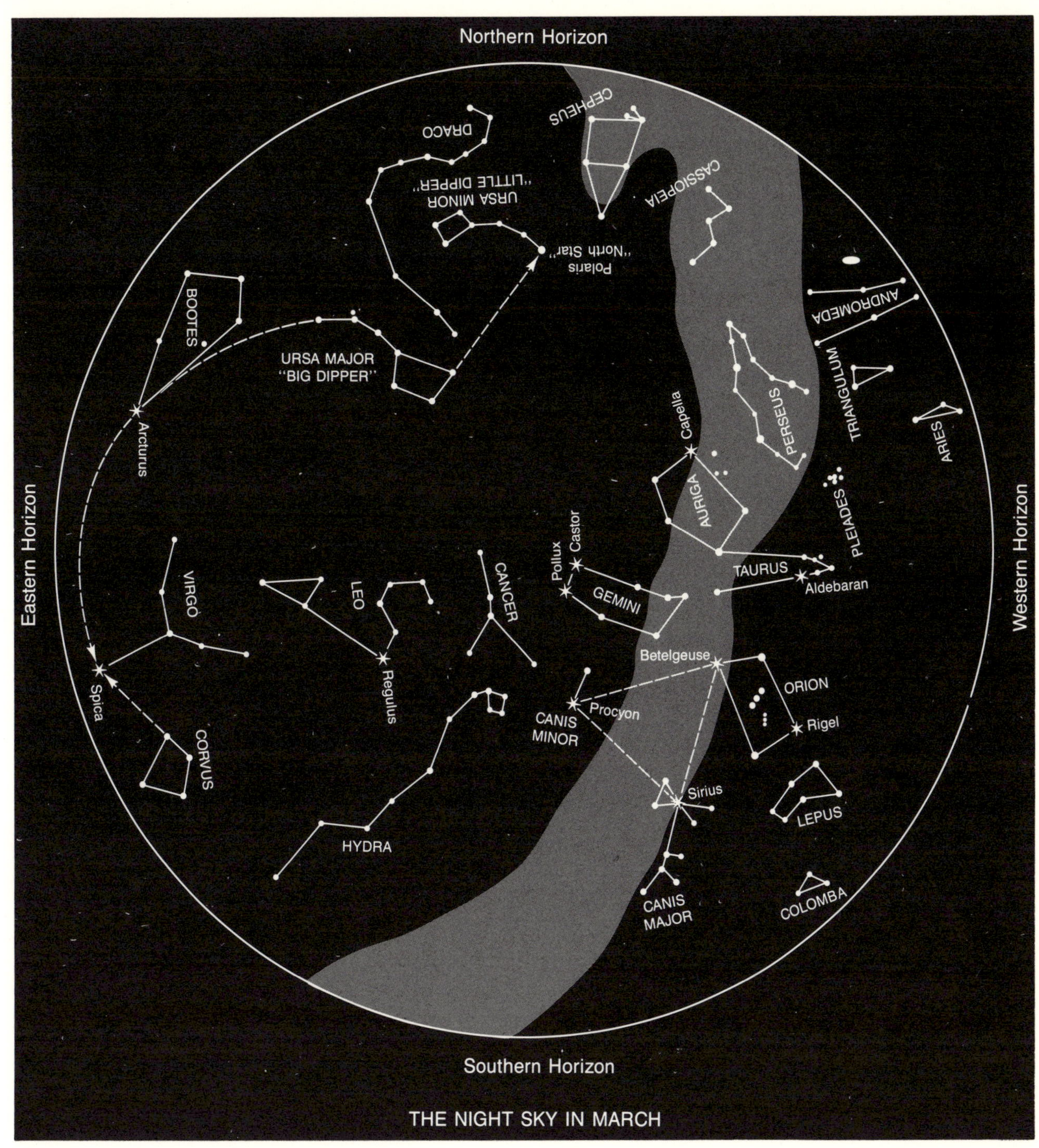

To use: Hold chart vertically and turn it so the direction you are facing shows at the bottom.

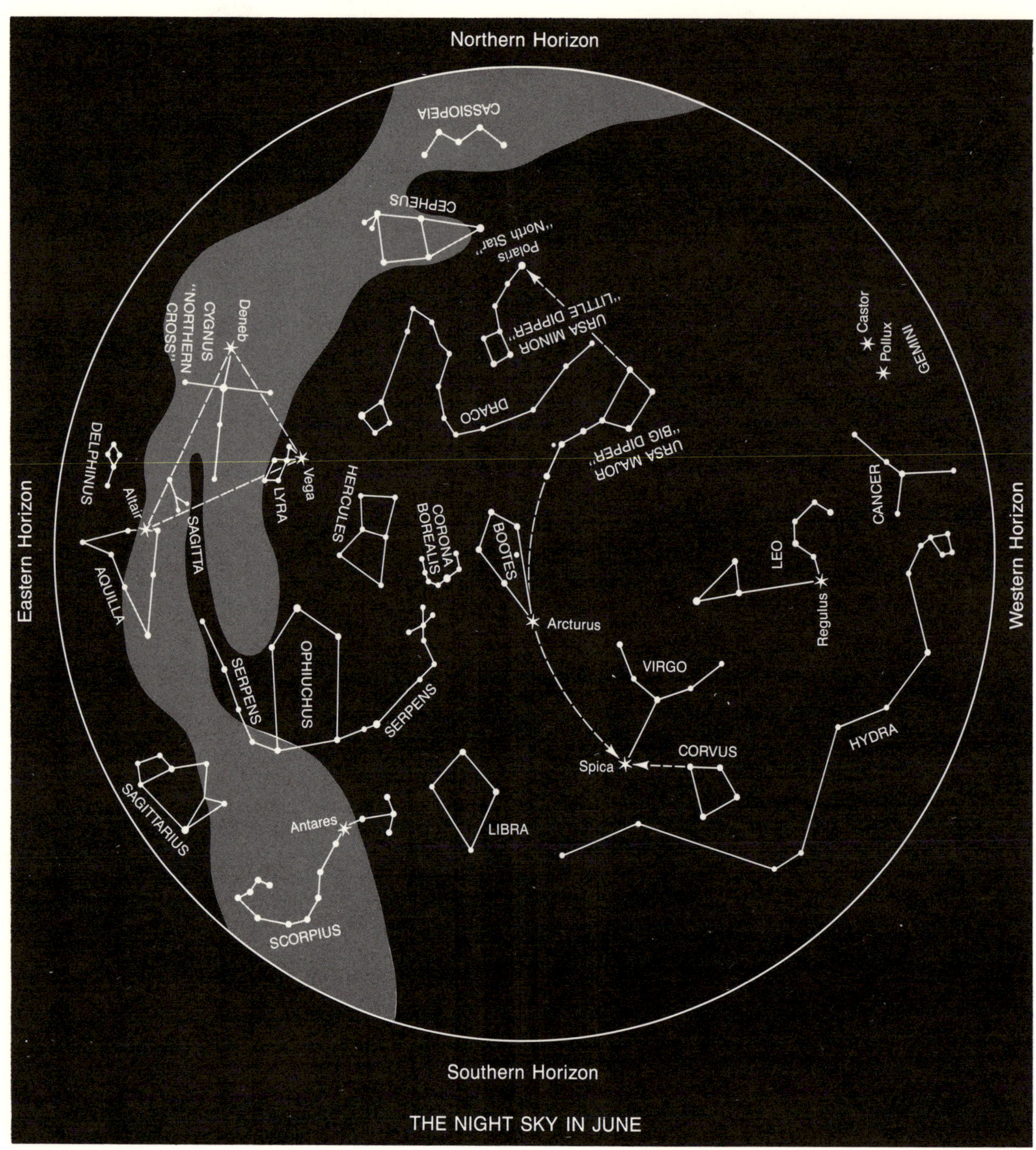

To use: Hold chart vertically and turn it so the direction you are facing shows at the bottom.

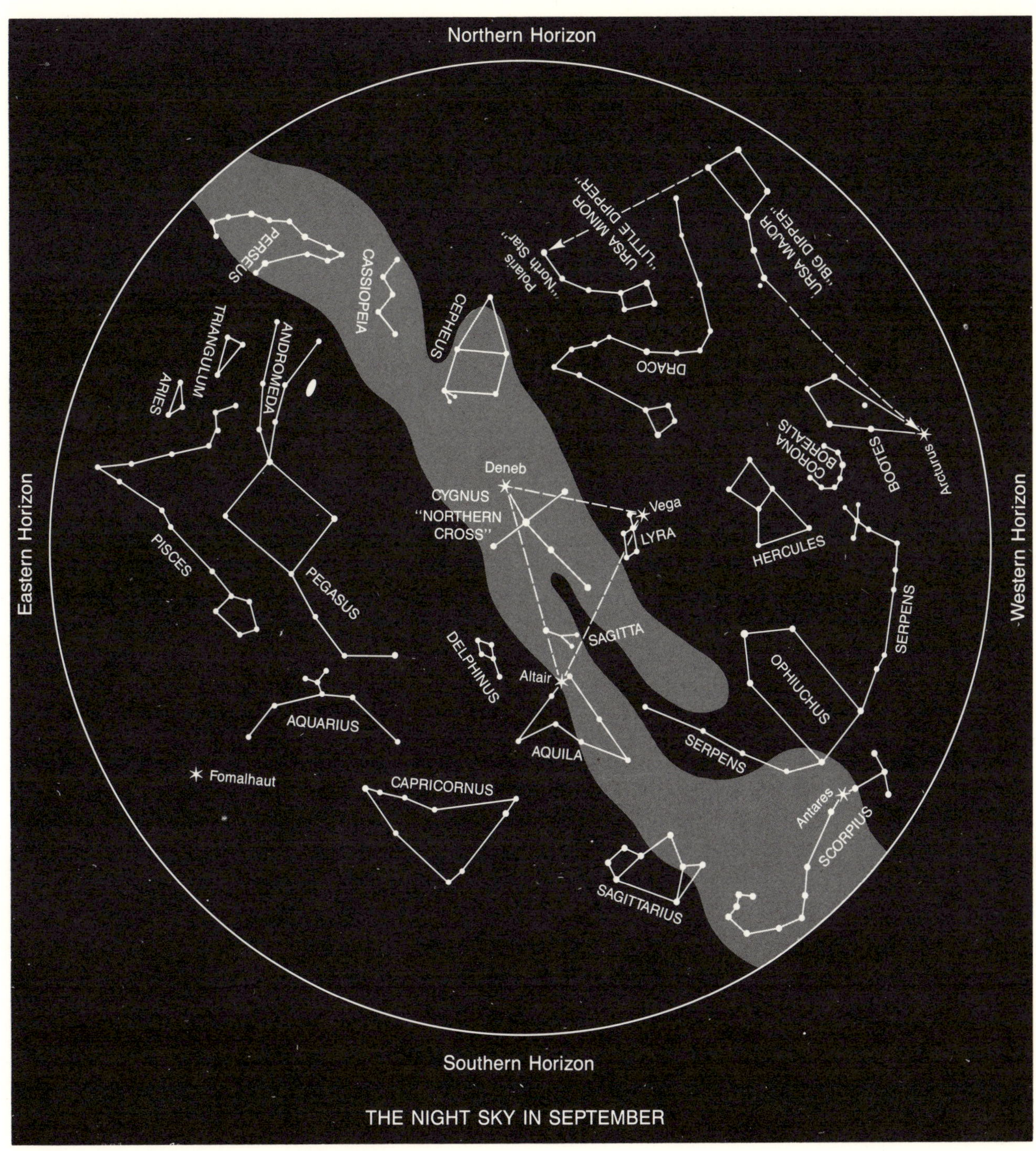

To use: Hold chart vertically and turn it so the direction you are facing shows at the bottom.

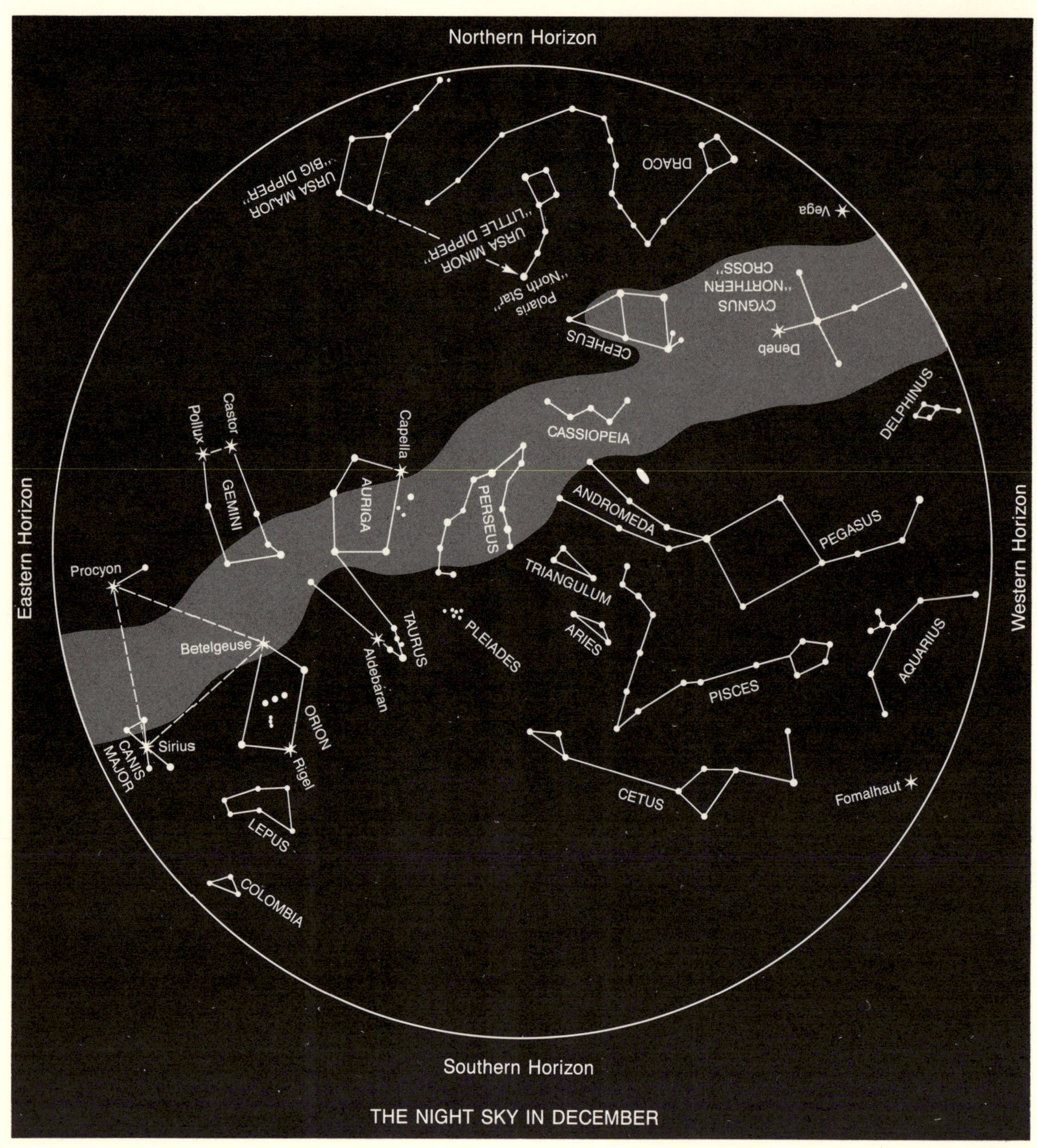

To use: Hold chart vertically and turn it so the direction you are facing shows at the bottom.